USEFUL RULES AND FORMULAS

Integer Exponents   *(see section 5.1)*

$$a^m a^n = a^{m+n}$$

$$\frac{a^m}{a^n} = \begin{cases} a^{m-n} & \text{if } m > n \\ \dfrac{1}{a^{n-m}} & \text{if } m < n \end{cases}$$

$$\left(\frac{a^m b^n}{c^p}\right)^k = \frac{a^{mk} b^{nk}}{c^{pk}}$$

$$a^{-n} = \frac{1}{a^n} \qquad \frac{1}{a^{-n}} = a^n$$

$$\frac{a^{-n}}{b^{-m}} = \frac{b^m}{a^n} \qquad a^0 = 1$$

Rationalizing Denominators
(see section 5.5)

$$\frac{a}{\sqrt{b}} = \frac{a \cdot \sqrt{b}}{\sqrt{b} \cdot \sqrt{b}} = \frac{a\sqrt{b}}{b}$$

$$\frac{a}{\sqrt{b} + \sqrt{c}} = \frac{a \cdot (\sqrt{b} - \sqrt{c})}{(\sqrt{b} + \sqrt{c}) \cdot (\sqrt{b} - \sqrt{c})}$$

$$\underset{\text{conjugates}}{\underbrace{\qquad\qquad}}$$

Rational Exponents   *(see section 5.6)*

$$a^{1/n} = \sqrt[n]{a}$$

$$a^{m/n} = \sqrt[n]{a})^m = \sqrt[n]{a^m}$$

Imaginary Numbers   *(see section 6.1)*

$$i = \sqrt{-1} \quad \text{with} \quad i^2 = -1 \qquad \sqrt{-a} = i\sqrt{a}$$

Extracting Roots   *(see section 6.2)*

If $x^2 = a$, then $x = \pm\sqrt{a}$.

Completing the Square   *(see section 6.3)*

If $x^2 + bx + c = 0$,

then $x^2 + bx + \dfrac{b^2}{4} = -c + \dfrac{b^2}{4}$.

Quadratic Formula   *(see section 6.4)*

If $ax^2 + bx + c = 0$,

then $x = \dfrac{-b \pm \sqrt{b^2 - 4ac}}{2a}$.

Slope   *(see section 7.2)*

$$m = \frac{y_2 - y_1}{x_2 - x_1}$$

Slope-intercept form of a line: $y = mx + b$
Parallel lines: $m_1 = m_2$
Perpendicular lines: $m_1 m_2 = -1$

Deriving Equations of Lines
(see section 7.3)

Point-slope form of a line:

$$y - y_1 = m(x - x_1)$$

Determinants   *(see section 7.7)*

$$\begin{vmatrix} a_1 & b_1 \\ a_2 & b_2 \end{vmatrix} = a_1 b_2 - a_2 b_1$$

$$\begin{vmatrix} a_1 & b_1 & c_1 \\ a_2 & b_2 & c_2 \\ a_3 & b_3 & c_3 \end{vmatrix} = a_1 \begin{vmatrix} b_2 & c_2 \\ b_3 & c_3 \end{vmatrix} - a_2 \begin{vmatrix} b_1 & c_1 \\ b_3 & c_3 \end{vmatrix} + a_3 \begin{vmatrix} b_1 & c_1 \\ b_2 & c_2 \end{vmatrix}$$

INTERMEDIATE ALGEBRA

Norman L. Siever

Los Angeles Valley College

HOUGHTON MIFFLIN COMPANY ■ **BOSTON**

Dallas   Geneva, Illinois   Palo Alto   Princeton, New Jersey

Printed in the U.S.A.
Library of Congress Catalog Card Number: 89-80964

ISBN:
Text: 0-395-45313-5
Exam copy: 0-395-52696-5

ABCDEFGHIJ-D-9543210-89

CONTENTS

Preface   xi

Tips to Success in Algebra   xv

1 REAL NUMBERS   2

1.1 Integers and Rational Numbers   2
1.2 Order of Operations   9
1.3 Formulas and Expressions   14
1.4 Real Numbers   22

Chapter Summary   32

Review Exercises   33

Test   39

2 LINEAR EQUATIONS AND INEQUALITIES   40

2.1 First-Degree Equations   40
2.2 Coin and Ticket Problems   46
2.3 Investment Problems   49
2.4 Distance-Rate-Time Problems   55
2.5 Linear Inequalities   60
2.6 Absolute Value Equations and Inequalities   68

Chapter Summary   75

Review Exercises   77

Test   81

3 POLYNOMIALS 82

3.1 Multiplying Polynomials   82
3.2 Common Factors and Grouping   89
3.3 Factoring Trinomials   93
3.4 Factoring Squares and Cubes   100
3.5 Factoring Second-Degree Equations   105
3.6 Geometry Problems   112

Chapter Summary   120

Review Exercises   121

Test   124

Chapter 1-3   Cumulative Review Exercises   125

4 RATIONAL EXPRESSIONS 127

4.1 Reducing, Multiplying, and Dividing Rational Expressions   127
4.2 Adding and Subtracting Rational Expressions   135
4.3 Complex Fractions   142
4.4 Dividing Polynomials   150
4.5 Fractional Equations and Inequalities   154
4.6 Word Problems   160
4.7 Literal Equations and Formulas   165

Chapter Summary   169

Review Exercises   171

Test   175

5 EXPONENTS, ROOTS, AND RADICALS 176

5.1 Integer Exponents   176
5.2 Roots   186
5.3 Simplifying and Combining Radicals   191
5.4 Multiplying Radicals   198

5.5 Dividing Radicals 202

5.6 Rational Exponents 210

Chapter Summary 216

Review Exercises 218

Test 221

6 QUADRATIC EQUATIONS AND INEQUALITIES   222

6.1 Imaginary and Complex Numbers 222

6.2 Extracting Roots 228

6.3 Completing the Square 236

6.4 The Quadratic Formula 242

6.5 Equations Containing Radicals 249

6.6 Combined Methods of Solving Equations 254

6.7 Quadratic and Rational Inequalities 259

Chapter Summary 266

Review Exercises 268

Test 273

Chapters 1-6   Cumulative Review Exercises 274

7 LINES AND LINEAR SYSTEMS   278

7.1 Graphing Points and Lines 278

7.2 Slope 284

7.3 Deriving Equations of Lines 294

7.4 Linear Systems in Two Variables 298

7.5 Word Problems in Two Variables 307

7.6 Linear Systems in Three Variables 316

7.7 Determinants and Cramer's Rule 321

Chapter Summary 330

Review Exercises 332

Test 338

8 THE CONIC SECTIONS   339

8.1 Parabolas   339
8.2 Distance and Circles   351
8.3 Ellipses and Hyperbolas   359
8.4 Non-Linear Systems   371
8.5 Word Problems   374

Chapter Summary   379

Review Exercises   382

Test   387

9 FUNCTIONS   389

9.1 Relations and Functions   389
9.2 Special Functions   398
9.3 Variations   403
9.4 Composite and Inverse Functions   412

Chapter Summary   420

Review Exercises   422

Test   426

Chapters 1-9   Cumulative Review Exercises   427

10 EXPONENTIAL AND LOGARITHMIC FUNCTIONS   430

10.1 Exponential Functions   430
10.2 Logarithms   443
10.3 Properties of Logarithms   454
10.4 Exponential Equations   462

Chapter Summary   470

Review Exercises   472

Test   476

11 SEQUENCES AND SUMS   478

11.1 Sequences and the Sigma Notation   478
11.2 Arithmetic Sequences   483
11.3 Geometric Sequences   490
11.4 Infinite Geometric Sequences   499
11.5 The Binomial Expansion   507

Chapter Summary   513

Review Exercises   515

Test   520

Sample Final Examinations   522

Selected Topics   A1

A. Sets   A1
B. Synthetic Division   A5
C. Changing the Index of Radical   A7
D. Inequalities in Two Variables   A9
E. Matrix Methods   A15
F. Common Logarithms   A19
G. Calculating with Logarithms   A24

Tables

Common Logarithms   A28
Powers of e A30
Roots and Powers   A30
Algebraic Phrases   A31

Answers   A32

Index   A111

PREFACE

This book is written for students fresh out of Beginning Algebra, for those who have not seen an x in a few years, and for everyone in between. It covers all the material required for more advanced mathematics courses, as well as the skills necessary for courses in science, business, and other related disciplines.

FEATURES OF THE TEXT

Examples

Each topic is illustrated with many examples in which carefully selected problems are worked out clearly and thoroughly. Comments accompany new steps in the series of calculations and enable students to approach problem-solving confidently. For immediate reinforcement, designated exercises following each section are directly keyed to the examples.

Flexibility

The text has been structured with flexibility in mind. The core material is contained in 61 sections, which can be covered comfortably in a semester. If time is limited, Chapter 1 may be omitted entirely. Instructors with additional time may choose from seven selected topics following Chapter 11.

The exercises themselves are organized to accommodate all levels of students. Exercises in an A set are keyed to specific examples and allow students to reinforce their understanding of a newly introduced topic. Exercises in a B set are of greater difficulty and require students to hone problem-solving strategies. For increased challenges, exercises in a C set offer more difficult problems and occasional brain-teasers.

Word Problems

Word problems are frequently drawn from a wide range of topics and disciplines. The diversity of the topics enables students to apply their problem-solving skills to relevant situations. These applications include chemistry, finance, sports, electronics, real estate, ecology, and statistics to name a few.

Reinforcement of Topics

Key topics are consistently reinforced to enhance student learning. Once introduced, key topics are used throughout the text. For example, radicals are introduced in Chapter 5 and then used extensively in the quadratic equations of Chapter 6. Not to be abandoned, they appear unexpectedly in later problems such as systems of equations (Chapter 7) and arithmetic and geometric sequences (Chapter 11). Students who class-tested the manuscript appreciated this "built-in review," which keeps all the material current.

Unique Features

Included in this book are the following learning aids not usually found in algebra textbooks:

1. Tips to Success in Algebra (following the Preface)
2. Tips to factoring *difficult* trinomials (Section 3.3)
3. Factoring flow-chart (Section 3.4)
4. Rational expressions versus rational equations (Exercise 4.5)
5. Tips to solving literal equations (Section 4.7)
6. Tips for peak performance on final examinations (following Chapter 11)
7. Table of Algebraic Phrases (following Selected Topics)

Other Learning Aids

Key features of the book include end-of-chapter summaries with *specific examples* of various formulas, chapter review exercises, and chapter tests. In addition, cumulative review exercises at the end of Chapters 3, 6, and 9 help students retain mathematics skills learned in earlier chapters. For further practice and review, the book offers a selection of sample final examinations at the end of the book.

INSTRUCTIONAL PACKAGE

Student's Solutions Manual

This manual contains complete, worked-out solutions to all *odd-numbered* problems in the exercises and to *all* problems in the chapter review exercises, chapter tests, cumulative review exercises, and sample final examinations.

Instructor's Manual with Testing Program

The Instructor's Manual includes four printed tests that are provided for each chapter, for every two chapters, for a cumulative review of Chapters 1 through 6, and for a final examination. Each set of four tests consists of Form A, which comes in both free-response and multiple-choice formats, and Form B, which also comes in both formats. Answers to these tests are provided at the end of the manual.

Computerized Test Generator

The database contains about 1700 test items from which instructors can select to form examinations. Instructors who do not have access to a computer can use the Printed Test Bank to choose items to be included on a test being prepared by hand. The Test Generator is available for the IBM-PC or compatible computers, for Macintosh computers, and for the Apple II family of computers.

Answer Key Booklet

Also available to the instructor, this booklet contains answers to all the even-numbered problems not provided in the answer section of the text. The booklet can be inserted conveniently at the back of the book for quick reference.

Expert Tutor

The Expert Tutor is an interactive instructional microcomputer program for student use. Each concept in the text is supported by a lesson on the Expert Tutor. Lessons on the tutor provide additional instructions, hints, examples, and worked-out solutions. This instruction and practice can be used in several ways: (1) to repeat instruction on the skill or concept not yet mastered, (2) to review material in preparation for examinations, or (3) to cover material a student missed because of absence from class.

No knowledge of computers is necessary on the part of the student or instructor. The Expert Tutor is available for the IBM-PC (or compatible) with DOS 2.0 or higher and 256K of memory.

Acknowledgments

Many thanks are in order—first, the staff at Houghton Mifflin, who had faith in me and yet hounded me until every last detail was ironed out; next, Patti Confort, my colleagues Phil Clarke and Glen Paget, as well as many of the others listed below who worked on the supplementary material and did the accuracy checks; and especially, my many students who helped to class-test the four preliminary manuscripts and who offered important suggestions and criticisms.

In addition, I would like to acknowledge the following reviewers for their invaluable comments and suggestions:

C. J. Alexander, Des Moines Area Community College; Barbara Buhr, Fresno City College; Philip S. Clarke, Los Angeles Valley College; Patricia F. Confort, Roger Williams College; Douglas B. Crawford, College of San Mateo; Charles C. Edgar, Onondaga Community College; Michael Farrell, Carl Sandburg College; Marie Ferraguto, Northern Essex Community College; Frank Garcia, North Seattle Community College; Frank Gunnip, Macomb Community College; B. J. Harmon, Monroe County Community College; Sarah Kennedy, Texas Tech University; Anne F. Landry, Dutchess Community College; Jerry Matlock, Richmond College; Michael J. Mears, Manatee Community College; Linda J. Murphy, Northern Essex Community College; Nancy Nickerson, Northern Essex Community College; Carol O'Loughlin, Northern Essex Community College; Glen Paget, Los Angeles Valley College; Ross Rueger, College of the Sequoias; Elizabeth M. Wayt, Tennessee State University; Kelly P. Wyatt, Umpqua Community College.

Finally, I would like to thank my family and friends, who never quit asking, "When is your book coming out?"

TIPS TO SUCCESS IN ALGEBRA

Tip 1 Keep Up on a Daily Basis

Don't get behind in class, and then cram the night or weekend before a test. Algebra needs to be learned in small sections and practiced continually in order to be mastered. Cramming will not help you to retain problem-solving skills needed in new topics throughout the course.

Tip 2 Practice New Skills

Exercises in an A set are directly keyed to examples presented in every new section. Doing these problems and repeating those that you have missed will help you understand algebra, solidify your knowledge, and build up your confidence.

Tip 3 Prevent Careless Mistakes

You know what these are: miscopying exercise problems, factoring incorrectly, supplying the wrong positive or negative signs, and so forth. You can avoid unnecessary errors on homework by concentrating on the problem and mentally reviewing your work *before* looking up the answer at the back of the book.

Tip 4 Attempt More Challenging Exercises

Exercises in a B or a C set are more challenging than those in an A set and therefore enable you to test your mastery of various skills. Refresh your skills by doing the chapter review and cumulative review exercises. Prepare efficiently for actual examinations by taking the tests at the end of every chapter and the sample final examinations at the end of the book.

Tip 5 Simulate Test Conditions

You probably do homework problems at home or in the library. Simulate test conditions by going into your classroom when it is vacant and doing your homework in your cramped seat. This is not always possible, but such a "dress rehearsal" can lessen the jolt of an actual test.

Tip 6 Go Slow

Unlike homework, examinations administered by your instructor have time limits. Therefore, pressuring yourself to work fast can lead to needless errors, memory lapses, and poor scores. Simply take the time to work carefully. In the 1960's the U.S. Olympic track coach timed his sprinters in an all-out 100 meters. After they recovered, he told them to run the same distance slower at 90 percent of their original speed. Surprisingly, the sprinters ran even faster than before! Likewise, a relaxed mind works more efficiently during algebra tests. By going seemingly slower, you will go actually faster.

INTERMEDIATE ALGEBRA

1 REAL NUMBERS

1.1 Integers and Rational Numbers

We begin with the numbers used most often in algebra, the **integers:**

$$\ldots, -5, -4, -3, -2, -1, 0, 1, 2, 3, 4, 5, \ldots$$

They are used to represent temperatures ($-26°C$, $42°F$), monetary gains and losses ($-\$20$ is a \$20 loss), elevations (Death Valley, California, has elevation -279 feet), and golf scores (-4 is 4 below par, $+3$ is 3 above par). Certainly you can supplement this list.

A **rational number** is any number that can be put into the form $\frac{a}{b}$, where a and b are integers and $b \neq 0$. Examples of rational numbers, or "ratios of integers," include

$$\text{Fractions: } \frac{1}{2}, \frac{-3}{17}, -\frac{7}{4}, \ldots$$

$$\text{Integers: } 6 = \frac{6}{1}, -3 = \frac{-3}{1}, 0 = \frac{0}{1}, \ldots$$

$$\text{Mixed numbers: } 2\frac{5}{8} = \frac{21}{8}, -1\frac{3}{4} = \frac{-7}{4}, \ldots$$

By convention, the negative rational number $\frac{7}{-4}$ is written as either $\frac{-7}{4}$ or $-\frac{7}{4}$. Note that $\frac{6}{0}$ is undefined (see Problem 67), which necessitates the stipulation that $b \neq 0$ in the definition of a rational number.

These numbers should be familiar to you from beginning algebra. So should a lot of studying. Not that algebra is particularly difficult, but like any discipline, it takes a lot of time to master. "How much time?" you ask. As much time as is necessary to *keep up on a daily basis.* Don't get behind and expect to catch up by cramming on weekends or the night before an exam. This might work in courses requiring lots of reading, but a problem-solving course like mathematics must be learned in small doses over a long period of time. If you haven't done so already, you should read the Preface, "Tips to Success in Algebra," now.

Adding positive and negative numbers has a simple interpretation. A poker player who wins $2 and then loses $5 ends up losing $3. We write

win $2 + lose $5 = lose $3   as $2 + (-5) = -3$.

Likewise, we write the result

lose $2 + win $7 + lose $3 = win $2   as $-2 + 7 + (-3) = 2$.

Each of these can be represented on a number line:

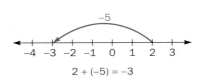   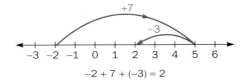

$$2 + (-5) = -3$$ $$-2 + 7 + (-3) = 2$$

No matter which interpretation (gambling or geometry) is used, addition is a simple task. No formal rules will be given for addition.

EXAMPLE 1

a. $8 + (-5) + (-6) = -3$.

b. $\dfrac{-5}{12} + \dfrac{1}{12} + \left(\dfrac{-11}{12}\right) = \dfrac{-15}{12}$

$\qquad\qquad = \dfrac{-5}{4}$ or $-\dfrac{5}{4}$.   Either form is acceptable.

c. $\dfrac{1}{4} + \left(-\dfrac{3}{2}\right) + \dfrac{5}{8} = \dfrac{1 \cdot 2}{4 \cdot 2} + \left(-\dfrac{3 \cdot 4}{2 \cdot 4}\right) + \dfrac{5}{8}$ The least common denominator (L.C.D.) is 8.

$\qquad\qquad = \dfrac{2}{8} + \left(-\dfrac{12}{8}\right) + \dfrac{5}{8}$

$\qquad\qquad = \dfrac{-5}{8}$.

As an example of subtracting one number from another, we note that $8 - 2 = 6$. This can be checked by adding: $6 + 2 \overset{\checkmark}{=} 8$. We get the same result by adding $8 + (-2) = 6$. Thus,

$$8 - 2 = 8 + (-2),$$

which means subtraction can be converted to addition by changing the sign of the second number.

To **subtract,** change the sign of the second number and then add:

$$a - b = a + (-b)$$
$$a - (-b) = a + b$$

EXAMPLE 2

a. $3 - 9 = 3 + (-9)$ Change 9 to -9, then add.
     $\qquad\; = -6.$ **Check:** $-6 + 9 \overset{\checkmark}{=} 3$

b. $\dfrac{5}{8} - \left(-\dfrac{1}{8}\right) = \dfrac{5}{8} + \dfrac{1}{8}$ Change $-\dfrac{1}{8}$ to $\dfrac{1}{8}$, then add.

$\qquad\qquad\quad = \dfrac{6}{8}$ **Check:** $\dfrac{6}{8} + \left(-\dfrac{1}{8}\right) \overset{\checkmark}{=} \dfrac{5}{8}$

$\qquad\qquad\quad = \dfrac{3}{4}.$

c. $-13 - 19 = -13 + (-19)$
$\qquad\qquad\;\; = -32.$ **Check:** $-32 + 19 \overset{\checkmark}{=} -13$

d. $-\dfrac{5}{6} - (-4) = -\dfrac{5}{6} + 4$

$\qquad\qquad\;\; = -\dfrac{5}{6} + \dfrac{4 \cdot 6}{1 \cdot 6}$ L.C.D. = 6

$\qquad\qquad\;\; = -\dfrac{5}{6} + \dfrac{24}{6}$

$\qquad\qquad\;\; = \dfrac{19}{6}.$

In expressions containing both operations, convert all subtractions to additions, as above, but leave the existing additions unchanged.

EXAMPLE 3

a. $9 - 18 - (-6) + 7 = 9 + (-18) + 6 + 7$
$\qquad\qquad\qquad\quad\;\; = 4.$

b. $\dfrac{-4}{5} - \left(\dfrac{-9}{10}\right) + \left(-\dfrac{2}{3}\right) - \dfrac{1}{2} = \dfrac{-4}{5} + \dfrac{9}{10} + \left(-\dfrac{2}{3}\right) + \left(\dfrac{-1}{2}\right)$

$\qquad\qquad\qquad\qquad\qquad\quad = \dfrac{-4 \cdot 6}{5 \cdot 6} + \dfrac{9 \cdot 3}{10 \cdot 3} + \left(\dfrac{-2 \cdot 10}{3 \cdot 10}\right) + \left(\dfrac{-1 \cdot 15}{2 \cdot 15}\right)$

$\qquad\qquad\qquad\qquad\qquad\quad = \dfrac{-24}{30} + \dfrac{27}{30} + \left(\dfrac{-20}{30}\right) + \left(\dfrac{-15}{30}\right)$ L.C.D. = 30

$$= \frac{-32}{30}$$

$$= \frac{-16}{15}.$$

Multiplying positive and negative numbers should be familiar to you from beginning algebra. Certainly $3 \cdot 5 = 15$. Also, $3(-5) = -15$, because $3(-5) = (-5) + (-5) + (-5) = -15$. Finally, $(-3)(-5) = 15$; the proof of this fact will be given in Exercise 1.4, Problem 56. These last two products have a simple interpretation. Suppose you are on a diet, losing 5 pounds per month (represented by -5). Three months from now (represented by $+3$) you will have lost 15 pounds; in symbols, $(+3)(-5) = -15$. On the other hand, three months ago (-3) you were 15 pounds heavier than you are now; in symbols, $(-3)(-5) = +15$.

The rules for division are the same as for multiplication. Thus, $\frac{15}{3} = 5$ because $3 \cdot 5 = 15$; $\frac{15}{-3} = -5$ because $(-3)(-5) = 15$; $\frac{-15}{3} = -5$ because $3(-5) = -15$; and $\frac{-15}{-3} = 5$ because $(-3) \cdot 5 = -15$. We can summarize all of this in a nutshell:

> The product or quotient of two numbers with **like** signs is **positive**.
> The product or quotient of two numbers with **unlike** signs is **negative**.

Closely related to products are powers. The symbol 4^3 means $4 \cdot 4 \cdot 4 = 64$. Here, 4 is called the **base** and 3 the **exponent**. The exponent indicates the number of times the base appears as a factor. The answer 64 is called the **third power** of 4, or "4 cubed." (The exponent 3 is sometimes called the power, but this error in nomenclature causes no harm.) Likewise, $9^2 = 9 \cdot 9 = 81$ and is read "9 squared."

EXAMPLE 4

a. $(-6)(-4) = 24.$

b. $(-2)(-3)(-5) = 6(-5)$
$= -30.$

c. $(-2)^3 = (-2)(-2)(-2)$
$= -8.$

d. $(-1)^{10} = 1.$ Why?   *(continued)*

e. $(-6)^2 = (-6)(-6)$ ⟵——— (-6) appears twice as a factor.
$ = 36.$

f. $-6^2 = -(6 \cdot 6)$ ⟵——— Only 6 appears twice, not -6.
$ = -36.$

g. $\dfrac{2}{3} \cdot \dfrac{-5}{11} = \dfrac{-10}{33}.$

h. $\dfrac{2}{3} \div \dfrac{-4}{15} = \dfrac{2}{3} \cdot \dfrac{15}{-4}$ Invert and multiply.

$\phantom{\dfrac{2}{3} \div \dfrac{-4}{15}} = \dfrac{\overset{1}{\cancel{2}}}{\underset{1}{\cancel{3}}} \cdot \dfrac{\overset{5}{\cancel{15}}}{\underset{-2}{\cancel{-4}}}$

$\phantom{\dfrac{2}{3} \div \dfrac{-4}{15}} = \dfrac{5}{-2}$

$\phantom{\dfrac{2}{3} \div \dfrac{-4}{15}} = \dfrac{-5}{2} \text{ or } -\dfrac{5}{2}.$ Either form is acceptable.

i. $\left(\dfrac{-3}{4}\right)^3 = \dfrac{-3}{4} \cdot \dfrac{-3}{4} \cdot \dfrac{-3}{4}$

$\phantom{\left(\dfrac{-3}{4}\right)^3} = \dfrac{-27}{64}.$

We conclude this section with a final topic. A **prime number** is an integer greater than 1 that is divisible only by 1 and itself. For example,

$$2 = 2 \cdot 1, \quad 3 = 3 \cdot 1, \quad \text{and} \quad 19 = 19 \cdot 1$$

are prime numbers because they have no other factors. On the other hand,

$$4 = 2 \cdot 2, \quad 6 = 2 \cdot 3, \quad 21 = 3 \cdot 7, \quad \text{and} \quad 35 = 5 \cdot 7$$

are *not* prime numbers because they have factors other than 1 and themselves. The first fifteen prime numbers are

$$2, 3, 5, 7, 11, 13, 17, 19, 23, 29, 31, 37, 41, 43, 47, \ldots$$

You should verify that these are prime numbers.

If an integer is not a prime number, it is called a **composite number** and can always be factored into a product of primes. Moreover, the result of factoring a composite number will always be the same, no matter how it is done.

EXAMPLE 5 | Factor these composite numbers into prime numbers.

a. $24 = 8 \cdot 3 = 2 \cdot 2 \cdot 2 \cdot 3 = 2^3 \cdot 3$ ⟵

                                   ⟶ Note the same result.

b. $24 = 6 \cdot 4 = 2 \cdot 3 \cdot 2 \cdot 2 = 2^3 \cdot 3$ ⟵

c. $100 = 10 \cdot 10 = 2 \cdot 5 \cdot 2 \cdot 5 = 2^2 \cdot 5^2$

d. $360 = 36 \cdot 10 = 6 \cdot 6 \cdot 2 \cdot 5$
$= 2 \cdot 3 \cdot 2 \cdot 3 \cdot 2 \cdot 5 = 2^3 \cdot 3^2 \cdot 5$

EXERCISE 1.1 *Answers, page A32*

A

Compute as in EXAMPLES 1, 2, 3, and 4:

1. $9 + (-17) + (-20)$

2. $-18 + 11 + (-13)$

3. $-3 - (-12)$

4. $-5 - 13$

5. $-2 - 9 - (-18)$

6. $5 - 11 - (-6)$

7. $-2 - 5 + (-1) - (-12)$

8. $4 + (-9) - 13 - 21$

9. $\dfrac{-5}{12} + \dfrac{13}{12} - \left(\dfrac{-7}{12}\right)$

10. $\dfrac{1}{10} - \dfrac{7}{10} - \left(-\dfrac{11}{10}\right)$

11. $\dfrac{3}{8} - \left(\dfrac{-5}{6}\right) - \dfrac{1}{4}$

12. $-\dfrac{4}{9} - \dfrac{2}{3} + \dfrac{1}{6}$

13. $\dfrac{-7}{12} - \dfrac{3}{8} + \left(\dfrac{-7}{6}\right) + \dfrac{11}{16} - 2$

14. $\dfrac{1}{9} - \dfrac{3}{4} - \dfrac{7}{8} - \left(\dfrac{-5}{24}\right) + 3$

15. $(-4)(-6)(-12)$

16. $5(-7)(-11)$

17. $(-2)^4$

18. $(-3)^4$

19. $(-5)^3$

20. $(-7)^3$

21. -2^6

22. -3^4

23. $-(-5)^2$

24. $-(-1)^7$

25. $\dfrac{-7}{12} \cdot \dfrac{-6}{35}$

26. $\dfrac{11}{15} \cdot \dfrac{-10}{33}$

27. $\dfrac{8}{15} \div \dfrac{-6}{25}$

28. $\dfrac{-4}{9} \div \dfrac{-8}{36}$

29. $\dfrac{-3}{14} \div 9$

30. $22 \div \dfrac{-33}{36}$

31. $\dfrac{-2}{3} \cdot \dfrac{-9}{14} \cdot \dfrac{21}{4}$

32. $\dfrac{5}{8} \cdot \dfrac{-7}{22} \cdot \dfrac{33}{77}$

33. $\dfrac{-7}{16} \cdot \dfrac{12}{52} \cdot (-39)$

34. $\dfrac{-9}{15} \cdot 12 \cdot \dfrac{-10}{36}$

35. $(-2)^3(-3)^2$

36. $(-4)^3(-2)^3$

37. $\left(\dfrac{-2}{5}\right)^4$

38. $\left(-\dfrac{1}{2}\right)^5$

39. $-\left(-\dfrac{3}{4}\right)^3$

40. $-\left(\dfrac{-5}{4}\right)^2$

41. $\left(\dfrac{-1}{2}\right)^3\left(\dfrac{2}{3}\right)^2$

42. $\left(\dfrac{3}{4}\right)^2\left(-\dfrac{2}{3}\right)^3$

Factor into prime numbers, as in EXAMPLE 5:

43. 36   **44.** 48   **45.** 120   **46.** 270   **47.** 144

48. 108   **49.** 1800   **50.** 450   **51.** 2520   **52.** 1960

53. The prime numbers from 2 to 47 were listed in the text. List the remaining prime numbers less than 100.

54. The **Goldbach Conjecture** states that "every even integer greater than 2 can be expressed as the sum of two prime numbers." For example,

$$4 = 2 + 2, \quad 6 = 3 + 3, \quad 8 = 3 + 5, \quad \text{and} \quad 10 = 3 + 7 = 5 + 5.$$

Express 12, 14, 16, 18, 20, 22, 24, 26, 28, and 30 as the sum of two primes. Be sure to include all possible ways, as was done with the number 10 above.

55. This problem has a surprise answer. Volume I and volume II of a book are standing on a bookshelf (volume I on the left and volume II on the right as you face the shelf). Each hard cover flap is $\frac{1}{16}$ inch thick; the pages of volume I comprise $\frac{1}{2}$ inch, and the pages of volume II comprise $\frac{3}{4}$ inch. A bookworm eats its way from the first page of volume I to the last page of volume II. How far does it travel horizontally? Neglect any blank pages at the beginning and end of each volume.

56. More little critters! A bug is at the bottom of a 40-foot vertical hole in the ground and wants out. It climbs 3 feet each day but slips down 2 feet every night while asleep. When does it reach the top? (The answer is *not* 40 days.)

57. Compute $1 - 2 + 3 - 4 + \cdots + 97 - 98 + 99 - 100$. (The dots mean that the pattern continues; that is, $5 - 6 + 7 - 8$, and so on are included.)

58. Compute $(-1)^1 + (-1)^2 + (-1)^3 + \cdots + (-1)^{99} + (-1)^{100}$.

59. Add up all the numbers from 1 through 100, *without using a calculator*. Even with a calculator, this is not as easy as it seems. *Hint:* Regroup the numbers in a manner that makes your work easier.

60. Three men walk into a hotel and the desk clerk tells them that a room costs $60 a night. Each of them pays him $20. Soon afterwards, the clerk realizes he overcharged them $5, which he promptly gives to the bellboy to return to them. Not wanting to divide this three ways and hassle with change, he gives each man back $1 and keeps $2 for himself. This means each guest ends up paying $19, or a total of $57. This, added to the bellboy's $2, totals $59. What happened to the other dollar?

In the computation of $\dfrac{3}{16} - \dfrac{5}{96} + \dfrac{11}{36},$

the least common denominator cannot be seen "by inspection." To find the L.C.D., we first factor each denominator into prime factors, as in Example 5:

$$16 = 2^4$$
$$96 = 2^5 \cdot 3$$
$$36 = 2^2 \cdot 3^2$$

Take highest powers of 2 and 3.

$$\overline{\text{L.C.D.} = 2^5 \cdot 3^2 = 32 \cdot 9 = 288}$$

Then we obtain the L.C.D. by taking as factors all of the prime numbers that appear, 2 and 3, each raised to the *highest power* of it that appears, 2^5 and 3^2. Our L.C.D. = $2^5 \cdot 3^2 = 288$. We then proceed as follows:

$$\frac{3}{16} - \frac{5}{96} + \frac{11}{36} = \frac{3 \cdot 18}{16 \cdot 18} - \frac{5 \cdot 3}{96 \cdot 3} + \frac{11 \cdot 8}{36 \cdot 8} \longleftarrow \text{L.C.D.} = 288$$

$$= \frac{54}{288} - \frac{15}{288} + \frac{88}{288}$$

$$= \frac{127}{288}.$$

The required building factors were obtained by division: $288 \div 16 = 18$, $288 \div 96 = 3$, and $288 \div 36 = 8$. Using this method, combine the following fractions:

61. $\dfrac{7}{16} + \dfrac{5}{12} - \dfrac{11}{60}$

62. $\dfrac{3}{50} + \dfrac{11}{12} + \dfrac{7}{30}$

63. $\dfrac{13}{18} - \dfrac{2}{45} - \dfrac{7}{30} + \dfrac{1}{12}$

64. $\dfrac{5}{9} + \dfrac{1}{30} - \dfrac{7}{54} + \dfrac{11}{24}$

65. $\dfrac{7}{50} + \dfrac{2}{15} - \dfrac{1}{75} + \dfrac{4}{45}$

66. $\dfrac{9}{64} + \dfrac{11}{24} - \dfrac{7}{16} + \dfrac{1}{96}$

67. Division by zero, such as $\dfrac{6}{0}$, is undefined. To prove this fact, assume that it is defined—say $\dfrac{6}{0} = x$, where x is some number. Show that this does *not* check when you multiply it out.

1.2 Order of Operations

The computations of the previous section involved five operations: addition, subtraction, multiplication, division, and powers. They are simple enough when performed alone, but mix them together in the same problem and you often have another problem: which to do first. For example,

does $\quad 2 + 6 \cdot 7 = 8 \cdot 7 = 56?$ (obtained by first adding and then multiplying)

or does $\quad 2 + 6 \cdot 7 = 2 + 42 = 44?$ (obtained by first multiplying and then adding)

Likewise,

does $5 \cdot 2^3 = 5 \cdot 8 = 40$?   (obtained by first cubing and then multiplying)

or does $5 \cdot 2^3 = 10^3 = 1000$?   (obtained by first multiplying and then cubing)

To avoid these ambiguities, we obviously need a priority system for the operations. The order in which they are to be performed will now be stated:

Order of Operations

1. Perform all operations within **parentheses** or brackets, starting first with the *innermost* of these grouping symbols. A fraction bar is considered a grouping symbol.
2. Perform all **powers**.
3. **Multiply** and/or **divide**, working from left to right.
4. **Add** and/or **subtract**, working from left to right.

We can now answer the questions posed above.

EXAMPLE 1

a. $2 + 6 \cdot 7 = 2 + 42$ Multiply.
$= 44.$ Add.

b. $5 \cdot 2^3 = 5 \cdot 8$ Power.
$= 40.$ Multiply.

c. $5(2 - 6)^3 = 5(-4)^3$ Parentheses.
$= 5(-64)$ Power.
$= -320.$ Multiply.

d. $88 \div 11 \cdot 2 = 8 \cdot 2$ Divide, then multiply, left to
$= 16.$ right.

e. $7 - (-2)^4 = 7 - (16)$ Power, then subtract.
$= -9.$ WARNING! $7 - (-2)^4 \neq 7 + (+2)^4$

f. $\left(\dfrac{1}{2} - \dfrac{2}{3}\right) \div \left(\dfrac{5}{8} + \dfrac{1}{4}\right) = \left(\dfrac{3}{6} - \dfrac{4}{6}\right) \div \left(\dfrac{5}{8} + \dfrac{2}{8}\right)$

$\left.\right\}$ Parentheses.

$= \dfrac{-1}{6} \div \dfrac{7}{8}$

$$= \frac{-1}{\underset{3}{\cancel{6}}} \cdot \frac{\overset{4}{\cancel{8}}}{7}$$ Invert and multiply.

$$= \frac{-4}{21}.$$

g. $12 - 2[5 - 7(1 - 3)] = 12 - 2[5 - 7(-2)]$ $1 - 3 = -2$ (innermost
$ = 12 - 2[5 - (-14)]$ $7(-2) = -14$ parentheses)
$ = 12 - 2[19]$ $5 - (-14) = 5 + 14 = 19$
$ = 12 - 38$ Multiply.
$ = -26.$ Subtract.

h. $8 + \dfrac{12}{6 - \dfrac{16}{1 - \dfrac{14}{5 - 7}}} = 8 + \dfrac{12}{6 - \dfrac{16}{1 - \dfrac{14}{-2}}}$ Start with smallest fraction bar and work upwards:

$\qquad\qquad\qquad\qquad\qquad\qquad\qquad\quad \longleftarrow\ 5 - 7 = -2$

$$= 8 + \dfrac{12}{6 - \dfrac{16}{1 - (-7)}} \qquad \dfrac{14}{-2} = -7$$

$$= 8 + \dfrac{12}{6 - \dfrac{16}{8}} \qquad\longleftarrow\qquad 1 - (-7) = 8$$

$$= 8 + \dfrac{12}{6 - 2} \longleftarrow \qquad \dfrac{16}{8} = 2$$

$$= 8 + \dfrac{12}{4}$$

$$= 8 + 3$$

$$= 11.$$

The results of parts a, b, and c in Example 1 are consistent with those of most scientific calculators.

a. $2 + 6 \cdot 7 = 44$ $\quad$ 2 $\boxed{+}$ 6 $\boxed{\times}$ 7 $\boxed{=}$ 44

b. $5 \cdot 2^3 = 40$ $\quad$ 5 $\boxed{\times}$ 2 $\boxed{y^x}$ 3 $\boxed{=}$ 40

c. $5(2 - 6)^3 = -320$ $\quad$ 5 $\boxed{(}$ 2 $\boxed{-}$ 6 $\boxed{)}$ $\boxed{y^x}$ 3 $\boxed{=}$ −320

Check to determine whether your calculator similarly respects the order of operations.

EXERCISE 1.2   *Answers, pages A32–A33*

A

Compute as in EXAMPLE 1:

1. $5 + 2 \cdot 9$

2. $3 - 7 \cdot 6$

3. $(5 + 2) \cdot 9$

4. $(3 - 7) \cdot 6$

5. $9 - 4(-7)$

6. $-6 + 7(-3)$

7. $(2 - 7)(12 - 3)$

8. $(21 + 3)(1 - 9)$

9. $4(-2)^3$

10. $-8 \cdot 4^2$

11. $-4(1 - 5)^3$

12. $-3(-7 + 2)^2$

13. $-36 \div 6 \cdot 2$

14. $48 \div 3(-6)$

15. $9 - (1 - 4)^2$

16. $-3 + (2 - 7)^3$

17. $\left(\dfrac{1}{8} - \dfrac{3}{4}\right)\left(\dfrac{2}{5} - \dfrac{1}{2}\right)$

18. $\left(\dfrac{5}{6} - \dfrac{7}{3}\right)\left(\dfrac{3}{4} - \dfrac{1}{3}\right)$

19. $\left(2 - \dfrac{3}{4}\right)^3$

20. $\left(\dfrac{2}{3} + \dfrac{3}{2}\right)^2$

21. $\left(\dfrac{1}{2} + \dfrac{2}{3}\right) \div \left(\dfrac{5}{6} - \dfrac{5}{2}\right)$

22. $\left(8 + \dfrac{1}{3}\right) \div \left(\dfrac{1}{6} - 6\right)$

23. $5 - 3(4 - 9)^2$

24. $-2 + 7(1 - 5)^3$

25. $7 - [5 - (6 - 9)]$

26. $-9 + [4 - (1 - 7)]$

27. $6 + 7[5 - 2(1 - 4)]$

28. $4 - 2[8 + 3(2 - 6)]$

29. $18 \div (2 - 11)(-4 + 2)$

30. $-14 \div (1 - 8)(2 - 7)$

31. $\dfrac{5 - 7 \cdot 3}{(1 - 5)^2}$

32. $\dfrac{(7 + 2)^3 - 7^3}{2 \cdot 11 - 4 \cdot 5}$

33. $5 + \dfrac{15}{6 - \dfrac{9}{8 - \dfrac{1}{4 - 5}}}$

34. $7 - \dfrac{14}{5 + \dfrac{10}{3 - \dfrac{8}{1 - 5}}}$

B

35. $4 - 7[6 - 2(3 + 2(1 - 4))]$

36. $2 + 5[9 - 4(5 + 3(2 - 9))]$

37. $9 + 3[11 - 3(2 - 3(1 - 2(2 + 5)))]$

38. $4 - 2[6 - 2(5 - 2(3 - 4(3 + 5)))]$

39. $22 - \dfrac{24}{11 - \dfrac{15}{7 - \dfrac{8}{3 + \dfrac{5}{4 + 1}}}}$

40. $17 + \dfrac{36}{1 - \dfrac{60}{1 - \dfrac{42}{1 - \dfrac{6}{1 - 7}}}}$

41. $\cfrac{\cfrac{\dfrac{16}{8}-2}{\dfrac{9}{8}-1}+6}{\dfrac{8}{1-3}+1}$ (For you left-handers)

42. $\cfrac{\cfrac{\dfrac{10}{12}+7}{\dfrac{8}{10}+5}-2}{\dfrac{10}{2-7}-2}$

C

43. $\dfrac{-(1-7)^2+9[4-(6+1)]}{5+\cfrac{12}{5-\cfrac{6}{2+1}}}$

44. $\dfrac{-(1-3)^2+3\left(\dfrac{8+2\cdot5}{4-1}\right)+7(8-2)}{10-\cfrac{16}{6-\cfrac{4}{1-3}}}$

Numerical expressions can be altered in value by the use of parentheses. For example,

$$4-5\cdot3+8=4-15+8=-3,$$
$$\textbf{but}\quad 4-5\cdot(3+8)=4-5\cdot(11)=4-55=-51,$$
$$\textbf{and}\quad (4-5)\cdot(3+8)=(-1)\cdot(11)=-11.$$

In the following problems, insert parentheses and/or brackets, if necessary, to make the calculations correct:

45. $2+5\cdot7=49$

46. $5-6-9=8$

47. $8-2\cdot3-4=10$

48. $8-2\cdot3-4=-2$

49. $8-2\cdot3-4=14$

50. $8-2\cdot3-4=-6$

51. $8-2\cdot3-4\cdot8=-8$

52. $8-2\cdot3-4\cdot8=-48$

53. $6\cdot8\div4-2=24$

54. $6\cdot8\div4-2=10$

55. $5-7^2=4$

56. $5-7^2=-44$

57. $2-3\cdot4+1^2=-25$

58. $2-3\cdot4+1^2=-73$

59. The number 4, used exactly four times in conjunction with arithmetic operations, can generate many other numbers. For example,

$$4-4+4-4=0,\quad \frac{4+4}{4+4}=1,\quad \text{and}\quad \frac{4}{4}+\frac{4}{4}=2.$$

Using expressions like these, obtain 0, 1, and 2 with other combinations of four 4's. Then obtain 3, 4, 5, 6, and so on, as far as you can go. You may also use parentheses, square roots ($\sqrt{}$), powers (but *not* 4^2, because it contains the number 2), and any other symbols you can conjure up. Remember, you must use exactly **four 4's,** but no other digits.

60. Repeat Problem 59, using **six 6's.**

| 1.3 | **Formulas and Expressions** |

The word *algebra* comes from the Arabic *al-jabr*, which means "the reduction and reunion of parts." It originally applied to solutions of a problem in which the whole was reduced to its constituent parts and then put back together. You do this all the time in mathematics! Algebra was introduced to the Arabs by the scholar Mohammed ibn Mûsa al-Khowârizimî (c. 825 A.D.)*, who reputedly derived it from the Hindus in his travels to India. The Moslems, in turn, spread it throughout Europe, where it was further developed over the centuries by the likes of Fibonacci, Euler, Napier, Pascal, Descartes, Gauss, Laplace, Cardan, Tartaglia, and Hardy, to name but a few. (You will encounter many of these mathematicians in later chapters.) Algebra eventually comes to you, the college student, as an extension of arithmetic, wherein negative numbers appear and letters are used to represent numbers.

Any combination of numbers, letters, and arithmetic operations is called an **algebraic expression.** Examples of such expressions are

$$2x^2 - 7x + 3, \quad u^2v - \frac{v^2}{2u - v}, \quad \text{and} \quad st^2 + \sqrt{3t} - 32.$$

Algebraic expressions have no numerical value until we assign values to the letters. We now show how to evaluate expressions by substituting numbers for the letters.

EXAMPLE 1

Evaluate $2x^2 - 7x + 3$ for $x = 0, 3, -5,$ and $\frac{1}{2}$.

Solution
We substitute each of these values of x as shown.

$$x = 0: \quad 2(0)^2 - 7 \cdot 0 + 3 = 0 - 0 + 3$$
$$= 3.$$

$$x = 3: \quad 2(3)^2 - 7 \cdot 3 + 3 = 18 - 21 + 3$$
$$= 0.$$

$$x = -5: \quad 2(-5)^2 - 7(-5) + 3 = 50 + 35 + 3$$
$$= 88.$$

$$x = \frac{1}{2}: \quad 2\left(\frac{1}{2}\right)^2 - 7\left(\frac{1}{2}\right) + 3 = \frac{2}{1} \cdot \frac{1}{4} - \frac{7}{1} \cdot \frac{1}{2} + 3$$

---

*The name means "Mohammed, son of Musa (or Moses), from Khwarezmi," which is located south of the Black Sea.

$$= \frac{1}{2} - \frac{7}{2} + \frac{6}{2} \qquad\qquad 3 = \frac{3 \cdot 2}{1 \cdot 2} = \frac{6}{2}$$

$$= 0.$$

In this example, we substituted various values of x into the expression. For this reason, letters in algebra are called **variables.** We will see in Chapter 2, when we solve equations, that letters can represent unknown quantities. Letters, variables, unknowns—call them what you will, but they're all the same and they bridge the gap between arithmetic and algebra. Whereas letters can vary, the number 6 can equal only 6, never 5 or −38. Thus numbers do not vary but rather remain **constant,** and they are so named.

Expressions have practical applications when used in **formulas.** For example, the formula $P = 2L + 2W$ expresses the perimeter P of a rectangle in terms of the rectangle's length L and its width W. The following examples contain a few of the many formulas used throughout this course.

EXAMPLE 2 | The perimeter and area of a rectangle are given by the formulas:

Perimeter: $P = 2L + 2W$
Area: $A = LW$

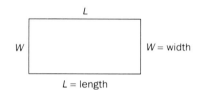

$L = \text{length}$

Compute them for the given lengths and widths. All measurements are in inches.

a. $L = 7$, $W = 4$

$$P = 2 \cdot 7 + 2 \cdot 4$$
$$= 22 \text{ inches.}$$
$$A = 7 \cdot 4$$
$$= 28 \text{ square inches.}$$

b. $W = \frac{3}{4}$, $L = 1\frac{5}{6}$

$$P = 2 \cdot \frac{11}{6} + 2 \cdot \frac{3}{4} \qquad\qquad \text{Write } 1\frac{5}{6} = \frac{11}{6}.$$

$$= \frac{11}{3} + \frac{3}{2} \qquad\qquad \frac{2}{1} \cdot \frac{11}{6} = \frac{11}{3}, \quad \frac{2}{1} \cdot \frac{3}{4} = \frac{3}{2}$$

$$= \frac{22}{6} + \frac{9}{6} \qquad\qquad \frac{11 \cdot 2}{3 \cdot 2} = \frac{22}{6}, \quad \frac{3 \cdot 3}{2 \cdot 3} = \frac{9}{6}$$

$$= \frac{31}{6}, \quad \text{or} \quad 5\frac{1}{6} \text{ inches.}$$

(continued)

$$A = \frac{11}{\cancel{6}_2} \cdot \frac{\cancel{3}^1}{4}$$

$$= \frac{11}{8}, \quad \text{or} \quad 1\frac{3}{8} \text{ inches}^2.$$ "Inches^2" means "square inches."

EXAMPLE 3   The circumference and area of a circle are given by the formulas:

Circumference: $C = 2\pi r = \pi d$

Area: $A = \pi r^2$

r = radius
d = diameter
d = 2r
$\pi \approx 3.141592654...$
$\pi \approx 3.14$ approximately

Compute them, for the given radius and diameter. All measurements are in meters. Express your answers both in terms of π and rounded to 2 decimal places.

a. $r = 6$

$$C = 2\pi(6)$$
$$= 12\pi$$
$$= 37.70 \text{ meters}$$

12 $\boxed{\text{x}}$ π $\boxed{=}$
and then round to 2 decimal places.

$$A = \pi(6)^2$$
$$= 36\pi$$
$$= 113.10 \text{ square meters}$$

b. $d = 9$

Because the diameter is twice as long as the radius, the radius $r = \frac{9}{2}$.

$$C = 2\pi\left(\frac{9}{2}\right)$$

$$= 9\pi$$
$$= 28.27 \text{ meters}$$

$$A = \pi\left(\frac{9}{2}\right)^2$$

$$= \frac{81\pi}{4}$$

$$= 63.62 \text{ meters}^2$$

A **right triangle** is a triangle that contains a 90° angle (a "right angle"). The longest side, which is opposite the 90° angle, is called the **hypotenuse,** and the two shorter sides are the **legs.**

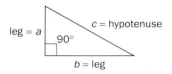

The **Pythagorean Theorem** (and its converse) from geometry state that

"A triangle is a **right triangle** if and only if the sum of the squares of the legs equals the square of the hypotenuse." That is,

$$a^2 + b^2 = c^2$$

For example, the triangle

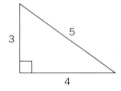

is a right triangle, because $3^2 + 4^2 = 5^2$,

or $9 + 16 \overset{\checkmark}{=} 25$.

EXAMPLE 4   The three numbers below form the sides of a triangle. In each case, decide whether or not it is a right triangle.

a. 7, 9, 11

Testing these numbers in the Pythagorean Theorem gives

$$7^2 + 9^2 \overset{?}{=} 11^2, \quad \text{or} \quad 49 + 81 \neq 121.$$

This is *not* a right triangle.

b. 13, 85, 84

The *largest* number must be tested as the hypotenuse.

$$13^2 + 84^2 \overset{?}{=} 85^2, \quad \text{or} \quad 169 + 7056 \overset{\checkmark}{=} 7225.$$

This *is* a right triangle.

EXERCISE 1.3   *Answers, page A33*

Evaluate these expressions for the given values, as in EXAMPLE 1:

1. $x^2 + 5x - 6$ for $x = 0, 1, -\dfrac{1}{2}, 3$, and -6

2. $3x^2 - 19x - 14$ for $x = -2, 7, 8$, and $-\dfrac{2}{3}$

3. $y^3 - 3y^2 - 4y + 12$ for $y = 3, -2, 2, -5$, and $\dfrac{1}{2}$

4. $2z^3 + z^2 - 18z - 9$ for $z = 3, -1, -3, -\dfrac{1}{2}$, and $\dfrac{1}{4}$

Given $x = 2$, $y = -3$, evaluate these expressions:

5. $3x^2 - 5xy + 4y^2$

6. $2(x - y)^2 - 3(x + y)^2$

7. $x^3 - y^3 + 2x^2y - 7$

8. $(2x + y)(3xy + 18)$

9. $\dfrac{x}{y} + \dfrac{y}{x} + \dfrac{x + y}{x - y}$

10. $[3x + y]^{(x - y)}$

Compute the perimeter and area of the rectangles with the given length and width. Assume all measurements are in feet. See EXAMPLE 2.

11. $L = 8$, $W = 3$

12. $L = \dfrac{7}{8}$, $W = \dfrac{1}{2}$

13. $W = 2\dfrac{2}{3}$, $L = 3\dfrac{3}{4}$

14. $W = 2.7$, $L = 3.8$

15. $L = W = 6.25$

16. $W = L = \dfrac{5}{6}$

Compute the *circumference* and *area* of the circles with the given radius or diameter. Assume all measurements are in centimeters. Express your answers both in terms of π and rounded to 2 decimal places. See EXAMPLE 3.

17. $r = 5$

18. $r = 2.3$

19. $d = 8$

20. $d = 11$

21. $r = \dfrac{3}{4}$

22. $r = \dfrac{1}{3}$

23. $d = \dfrac{4}{5}$

24. $d = 2\dfrac{2}{3}$

Use the Pythagorean relation $a^2 + b^2 = c^2$ to decide which of the following groups of three numbers form the sides of a right triangle. See EXAMPLE 4.

25. 5, 12, 13

26. 8, 6, 10

27. 7, 25, 24

28. 41, 9, 40

29. 4, 6, 8

30. 9, 12, 16

31. 2.1, 7.2, 7.5

32. 1.2, 1.5, 0.9

33. $\dfrac{1}{3}, \dfrac{1}{4}, \dfrac{1}{5}$ Hint: L.C.D.

34. $\dfrac{3}{10}, \dfrac{2}{5}, \dfrac{1}{2}$

35. $\dfrac{3}{4}, 1, 1\dfrac{1}{4}$

36. $2\dfrac{1}{4}, 3, 1\dfrac{1}{2}$

37. When a ball is thrown upward from the ground, its height h above the ground (in feet) after t seconds is given by the formula

$$h = -16t^2 + 64t.$$

a. Find the height of the ball when $t = 0$, 1, 2, 3, and 4.
b. What is the greatest height reached by the ball, and at what time does this occur?
c. When does the ball strike the ground?

38. When \$$P$ is deposited in a bank account paying simple interest at the rate r (expressed as a decimal) and remains on deposit for t years, the interest earned is given by the formula

$$I = Prt.$$

Find the interest earned when

a. $P = \$12{,}000$, $r = 7\% = 0.07$, $t = 3$
b. $P = \$1850$, $r = 8.5\%$, $t = 4$

39. Fahrenheit (F) and Celsius (C) temperatures are related by the formula

$$F = \frac{9C}{5} + 32.$$

Find F when C $= 0°$, $25°$, $62°$, $100°$, and $-40°$.

40. The frictional force F (in kilograms) needed to keep a car from skidding off a curved road is given by

$$F = \frac{kmv^2}{r^2},$$

where k = the coefficient of friction between the tires and the road, m = the mass of the car (in kilograms), v = the velocity of the car (in kilometers per hour), and r = the radius of the curve (in meters).

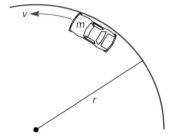

a. Find the frictional force F when $k = 0.62$, $m = 4500$, $v = 65$, and $r = 750$.
b. If the velocity were doubled and all other quantities remained constant, what would happen to the frictional force required?

41. The sum of the terms of an arithmetic sequence (Section 11.2) is given by the formula

$$S = \frac{n(a_1 + a_n)}{2},$$

where n = the number of terms, a_1 = the first term, and a_n = the last term. Find the sum S when

a. $n = 150$, $a_1 = 4$, and $a_n = 1047$

b. $n = 32$, $a_1 = \frac{1}{2}$, and $a_n = 16$

42. The sum of the terms of a geometric sequence (Section 11.3) is given by the formula

$$S = \frac{a_1(1 - r^n)}{1 - r},$$

where $n =$ the number of terms, $a_1 =$ the first term, and $r =$ the common ratio. Find the sum S when

a. $n = 5$, $a_1 = 3$,   and $r = -2$

b. $n = 6$, $a_1 = 4$,   and $r = \dfrac{1}{2}$

43. Right triangles whose sides are natural numbers can be obtained from the following picture:

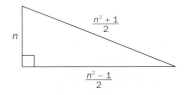

with $n = 3, 5, 7, 9, 11, \ldots$

For example, if $n = 3$, then the sides of the right triangle are

$$3, \quad \frac{3^2 - 1}{2}, \quad \text{and} \quad \frac{3^2 + 1}{2}, \quad \text{or} \quad 3, 4, \text{ and } 5. \qquad \textbf{Check: } 3^2 + 4^2 = 5^2$$
$$9 + 16 \overset{\checkmark}{=} 25$$

You obtain the three sides of a right triangle for each of the following numbers: $n = 5, 7, 9, 11, 13, 15,$ and 17. Check your answers in each case.

Find the shaded area in each of the following figures. The units of measurement are unspecified.

44.

Hint:
outer square – inner square

45.

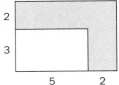

46.

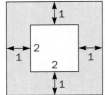

47.

48.

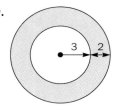

49.

c

50.

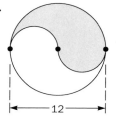

51.

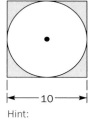

Hint:
square – circle

52. The region shown is to be made into a play area for kids. The plans call for grass to be installed, costing $15 per square foot, and for the entire region to be enclosed by fencing, costing $65 per foot. Find the total cost to make the play area.

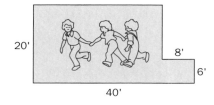

53. A wire 20 inches long is cut into two pieces, each of which is bent into the shape of a square. If one of the pieces is 8 inches long, find the sum of the areas of the two squares.

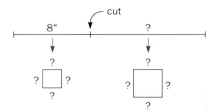

1.4   Real Numbers

Using long division or a calculator, we can express all rational numbers as decimals, either terminating or repeating. For example,

$$\frac{3}{10} = 0.3$$

$$\frac{1}{4} = 0.25 \qquad \left.\right\} \text{Terminating decimals}$$

$$\frac{-13}{8} = -1.625$$

$$\frac{2}{3} = 0.6666\ldots = 0.\overline{6}$$

$$\frac{3}{11} = 0.272727\ldots = 0.\overline{27} \qquad \left.\right\} \begin{array}{l}\text{Repeating decimals:} \\ \text{the bar is over the} \\ \text{repeating digits}\end{array}$$

$$\frac{4163}{3330} = 1.2501501501\ldots = 1.2\overline{501}$$

What about $\sqrt{2} = 1.414213562\ldots$, which is neither a terminating nor a repeating decimal? (We will discuss roots in detail in Section 5.2.) This is called an irrational number. An **irrational number** is a number whose decimal representation is neither terminating nor repeating. Thus an irrational number is a number that is not rational and so cannot be written as the ratio of integers, $\frac{a}{b}$. Examples of irrational numbers include

$$\sqrt{2} = 1.414213562\ldots$$
$$\sqrt{3} = 1.732050808\ldots$$
$$\sqrt[3]{11} = 2.2239801\ldots$$
$$\pi = 3.141592654\ldots$$
$$0.101001000100001\ldots$$

Non-terminating, non-repeating decimals *cannot* be put into the form $\frac{a}{b}$

Taken together, the rational and irrational numbers form the real-number system. Thus a **real number** is a number that is either rational or irrational. We can represent real numbers as points on a number line:

The number systems used in algebra are displayed on the next page. In Section 6.1, we will discuss numbers that are not real—namely, imaginary and complex numbers.

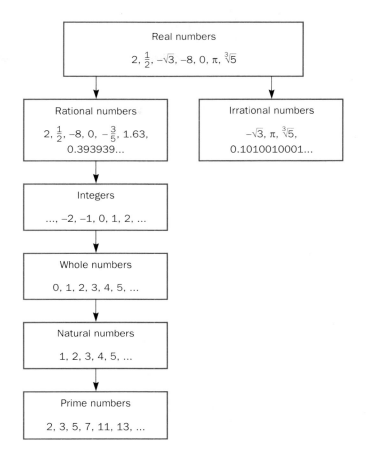

In the previous section we introduced algebraic expressions. The **terms** of an algebraic expression are those quantities separated by addition. Thus the expression

$$5x^2 + 3xy - \frac{17}{9x} + \sqrt{7y} \quad \text{has four terms} \quad 5x^2, \ 3xy, \ \frac{-17}{9x}, \text{ and } \sqrt{7y}.$$

A special type of term that is either a constant or the product of a constant and natural-number powers of variables, with no variable in the denominator, is called a **monomial.** Thus

$$5x^2, \quad 3xy, \quad -9u^2v^3w, \quad \frac{6st^3}{7}, \quad -3, \quad \text{and} \quad z \quad \text{are monomials,}$$

$$\text{but} \quad \frac{-17}{9x}, \quad \sqrt{7y}, \quad \frac{8v^2}{d}, \quad 3x^{-2}, \quad \text{and} \quad -3y^{1/2} \quad \text{are not monomials.}$$

The **degree of a non-constant monomial** is the sum of the exponents of the variables. The degree of a non-zero constant is defined to be zero, whereas the degree of the constant 0 is undefined. Thus the degrees of the monomials listed above are as follows:

monomial	$5x^2$	$3xy$	$-9u^2v^3w$	$\dfrac{6st^3}{7}$	-3	z
degree	2	2	6	4	0	1

The sum of two monomials is called a **binomial,** and the sum of three monomials is a **trinomial.** A **polynomial** is the sum of any number of monomials; thus monomials, binomials, and trinomials are polynomials. The **degree of a polynomial** is the highest degree of its monomial terms. Let us illustrate:

Expression	Degree	Type
$7x^2y^6$	8	monomial
$5x^4 + 3y$	4	binomial
$9x^2 + y^2 - 24$	2	trinomial
$3z + 8w - zw^2 + zw$	3	polynomial
$\dfrac{x}{y} + 2\sqrt{x}$	—	not a polynomial

In a monomial, any factor or product of factors is the **coefficient** of the remaining factors. In the term $3xyz,$ for example, 3 is the coefficient of $xyz,$ $3x$ is the coefficient of $yz,$ and xz is the coefficient of $3y.$ The **numerical coefficient** of a term is the real-number factor. Thus the numerical coefficients of $3xyz,$ $-7a^3b,$ $\dfrac{5x^2y}{8},$ $uv,$ and $-z^3$ are, respectively, 3, -7, $\dfrac{5}{8}$, 1, and -1. Unless otherwise stated, *coefficient* will always mean *numerical coefficient.*

In Section 1.3, Example 1, we evaluated the polynominal

$$2x^2 - 7x + 3$$

for rational numbers $x = 0$, 3, -5, and $\dfrac{1}{2}$. Nothing prevents us from evaluating this expression for irrational numbers, such as $x = \sqrt{2}$:

$$2(\sqrt{2})^2 - 7(\sqrt{2}) + 3$$

Unfortunately, its value will have to wait until Chapter 5. In the meantime, we will be concerned with the expression $2x^2 - 7x + 3$ itself, and we will assume that it represents a real number for any real number x. (See "Closure Property" in the table that follows.)

In order to work with this and other expressions, we need to list some properties exhibited by real numbers. In what follows, *all variables a, b, c, x, y, . . . represent real numbers*.

Properties of Real Numbers

Name	Property	Examples
Closure Properties	$a + b$ is a real number ab is a real number	$\sqrt{2} + \sqrt{5}, \frac{1}{4}\pi$, and $2x^2 - 7x + 3$ are real numbers
Commutative Properties	$a + b = b + a$ $ab = ba$	$y + \sqrt{3} = \sqrt{3} + y$ $5 \cdot \sqrt{7} = \sqrt{7} \cdot 5$
Associative Properties	$a + (b + c) = (a + b) + c$ $a(bc) = (ab)c$	$2 + (3 + 7x) = (2 + 3) + 7x$ $-4(9y) = (-4 \cdot 9)y$
Distributive Property	$a(b + c) = ab + ac$	$2(3x + 7) = 2 \cdot 3x + 2 \cdot 7$
Identity Properties	The unique real numbers 0 and 1 satisfy: $a + 0 = a = 0 + a$ $a \cdot 1 = a = 1 \cdot a$	$\pi + 0 = \pi = 0 + \pi$ $\frac{x}{2} \cdot 1 = \frac{x}{2} = 1 \cdot \frac{x}{2}$
Inverse Properties	Each a has a unique real number $-a$ (its **opposite**) such that $a + (-a) = 0 = -a + a$ Each $a \neq 0$ has a unique real number $\frac{1}{a}$ (its **reciprocal**) such that $a \cdot \frac{1}{a} = 1 = \frac{1}{a} \cdot a$	$2x + (-2x) = 0$ $= -2x + 2x$ $5x \cdot \frac{1}{5x} = 1 = \frac{1}{5x} \cdot 5x$

In beginning algebra, you learned to simplify expressions such as

$$4(2x + 1) - (5x - 9)$$

by first removing the parentheses and then combining like terms. Recall that **like terms** or **similar terms** are terms that have the same variables raised to the same powers; they may differ only in their coefficients. Thus

$$
\begin{array}{ll}
2x \text{ and } 4x & \text{are like terms,} \\
5a^3b^2, 7a^3b^2, \text{ and } -b^2a^3 & \text{are like terms,} \quad (\text{Why does } -b^2a^3 = -a^3b^2?) \\
x^3 \text{ and } x^4 & \text{are unlike terms,} \\
4x^2y \text{ and } 13xy^2 & \text{are unlike terms.}
\end{array}
$$

Using a variation of the Distributive Property, $(a + b)c = ac + bc$, we can combine like terms:

$$
\begin{aligned}
2x + 4x &= (2 + 4)x \\
&= 6x.
\end{aligned}
$$

$$
\begin{aligned}
5a^3b^2 + 7a^3b^2 - b^2a^3 &= (5 + 7 - 1)a^3b^2 \\
&= 11a^3b^2.
\end{aligned}
$$

$$x^3 + x^4 \quad \text{cannot be further combined.}$$

$$4x^2y + 13xy^2 \quad \text{cannot be further combined.}$$

The opposite of the polynomial $5x - 9$ is written $-(5x - 9)$, and by the Inverse Property for addition,

$$(5x - 9) + [-(5x - 9)] = 0.$$

But also,

$$
\begin{aligned}
(5x - 9) + (-5x + 9) &= 5x - 9 + (-5x) + 9 & \\
&= 5x + (-9) + (-5x) + 9 & \text{Write } 5x - 9 \text{ as } 5x + (-9). \\
&= 5x + (-5x) + (-9) + 9 & \text{Commutative Property.} \\
&= 0. & \text{Inverse Property.}
\end{aligned}
$$

Comparing these equations leads to the conclusion that

$$-(5x - 9) = -5x + 9,$$

because the opposite of $5x - 9$ is unique. In general, *a negative sign preceding parentheses changes the sign of each term inside the parentheses*. Let us now simplify the expression mentioned earlier:

$$
\begin{aligned}
4(2x + 1) - (5x - 9) &= 8x + 4 - (5x - 9) & \text{Distributive Property.} \\
&= 8x + 4 - 5x + 9 & \text{Change signs.} \\
&= 8x + 4 + (-5x) + 9 & \text{Write } 4 - 5x \text{ as } 4 + (-5x). \\
&= 8x + (-5x) + 4 + 9 & \text{Commutative Property.} \\
&= 3x + 13. & \text{Combine like terms.}
\end{aligned}
$$

The foregoing discussion illustrates that the real-number properties are the basis of each step needed to simplify polynomial expressions. In our first example below, we cite the property used in each step. Later this becomes too cumbersome, so instead we will rely on your knowledge from beginning algebra. Keep in mind that each step has a reason behind it.

EXAMPLE 1

Simplify each polynomial expression:

a. $7(3x - 2) + 6(2x + 5) = 21x - 14 + 12x + 30$ Distributive Property.
$= 21x + 12x - 14 + 30$ Commutative Property.
$= 33x + 16.$ Combine like terms.

b. $5(3x^2 - 7x + 2) - (2x^2 - 3x + 5) = 15x^2 - 35x + 10 - 2x^2 + 3x - 5$
$= 13x^2 - 32x + 5$

c. $4a - [3a - 2(3b - a)] = 4a - [3a - 6b + 2a]$ Remove inner parentheses.
$= 4a - [5a - 6b]$ Combine like terms.
$= 4a - 5a + 6b$ Remove brackets.
$= -a + 6b.$

Expressions arising from word phrases are of particular importance to you, especially when you solve word problems in Chapter 2. A table of such commonly used phrases is given below. More lists can be found in Sections 3.5 and 4.6. In these phrases, x denotes "a number."

Addition phrases	Algebraic expressions
the sum of x and 6 six more than a number a number increased by six	$x + 6$
Subtraction phrases	
the difference between x and 4 four less than a number a number decreased by four 4 subtracted from a number	$x - 4$
the difference between 4 and x four decreased by a number	$4 - x$
Multiplication and mixed phrases	
the product of 3 and x	$3x$
the sum of twice a number and 7 seven more than twice a number	$2x + 7$
twice the sum of a number and 7	$2(x + 7)$
5 less than four times a number 5 subtracted from four times a number	$4x - 5$

Compare!

Compare!

(continued)

five decreased by four times a number	$5 - 4x$
twice the square of a number	$2x^2$
the square of twice a number	$(2x)^2$
the sum of the squares of x and y	$x^2 + y^2$
the square of the sum of x and y	$(x + y)^2$
eight more than twice the difference between a number and five	$2(x - 5) + 8$
nine less than three times the sum of one and a number	$3(1 + x) - 9$
two numbers whose sum is 32	x and $32 - x$

Tongue-twisters!

In the next example, we make use of this table of phrases.

EXAMPLE 2 Translate into algebraic expressions, and then simplify:

a. Find the sum of $2a$ and seven, increased by the difference between a and -4.

$$(2a + 7) + [a - (-4)] = 2a + 7 + a + 4$$
$$= 3a + 11.$$

b. From the sum of $-3x$ and -8, subtract twice the difference between $-11x$ and nine.

$$[-3x + (-8)] - 2(-11x - 9) = -3x - 8 + 22x + 18$$
$$= 19x + 10.$$

c. Increase twice the square of a number by three times the number, then subtract four times the difference between the number and its square.

$$2x^2 + 3x - 4(x - x^2) = 2x^2 + 3x - 4x + 4x^2$$
$$= 6x^2 - x.$$

EXERCISE 1.4   *Answers, pages A33–A34*

A

**Classify each number below as one or more of the following types:
real, rational, irrational, integer, whole, natural, prime, or none of these.**

1. -15

2. $\dfrac{2}{3}$

3. $\sqrt{7}$

4. $-\sqrt[3]{6}$

5. 1.56

6. -0.48

7. $2 + \sqrt{3}$

8. $7 - \pi$

9. $\sqrt{25}$

10. $\sqrt{\dfrac{4}{9}}$

11. $\sqrt[3]{8}$ **12.** $\sqrt{121}$ **13.** $\sqrt{2} \cdot \dfrac{1}{\sqrt{2}}$ **14.** $-\pi + \pi$ **15.** $0.212121 \ldots$

16. $0.\overline{612}$ **17.** $0.2121121112 \ldots$ **18.** $0.12345678910111213 \ldots$

19. 137 **20.** $\dfrac{\sqrt{3}}{0}$

Simplify each polynomial expression, as in EXAMPLE 1:

21. $5(2a - 3) - 2(a - 5)$ **22.** $-(-2x + 6y) + 7(2x - 4y)$

23. $(5x^2 + 2x - 7) - (x^2 + 9x - 3)$ **24.** $(3a^2 - 2ab - 6b^2) - (-9a^2 - 7ba + 14b^2)$

25. $6(x^2 - 2xy - 3y^2) - 4(2x^2 - 5xy + 7y^2) + 3(2y^2 - xy - x^2)$

26. $-4(a^2 - 2a - 1) + 7(3a - 5 + a^2) - 3(2a^2 - 9a)$

27. $2a - [5a - (a - 4b)]$ **28.** $x - [9 - (2x + 8)]$

29. $3x + 5[2x - 4(x - 2y) + (x - y)]$ **30.** $3a - 4[2b + 5(a - 7b) - 2(a + 3b)]$

31. $2x - 5[9 + 2(4x - 7(1 - 2x))]$ **32.** $y + 6[9y - 4(5y + 2(3y - 7)) - 2y]$

Translate into algebraic expressions, and then simplify, as in EXAMPLE 2:

33. Find the sum of $-7x$ and two, increased by twice the difference between eight and $3x$.

34. Find three times the difference between $5a$ and three, decreased by twice the sum of 5 and $-3a$.

35. Add the difference between $4a + b$ and $a - 7b$ to the sum of $2a + b$ and $b - 2a$.

36. Add the sum of $2x$ and $3y$ to the difference between $-4x$ and $-y$, then subtract the difference between $5y$ and $7x$.

37. Add twice the square of a number to the product of three and the number, then subtract the difference between the number and three times its square.

38. Add the difference between 12 and $3x$ to the result of decreasing the product of -6 and x by seven.

39. Find seven more than twice the difference between three times a number and five.

40. Find eight less then eight times the result of increasing eight by eight times a number.

B

Simplify:

41. $a - 2[a - 2(a - 2(a - 2)) - 2]$ **42.** $5b - 3[5b + 3(5b - 3(5b - 3)) - 3b]$

43. $3x + 2[2x - (y - (x - 2(x - y)))]$ **44.** $8a - 2[a - 3(5 + 4(4a - 3(7 - a))) - a]$

45. According to the text, x^3 and x^4 are unlike terms and, therefore, cannot be added further. Or can they? Show by numerical example that $x^3 + x^4 \neq x^7$. *Hint:* Substitute $x = 2$.

46. In Section 3.1, we will show that unlike terms can be multiplied, yielding such a result as $x^3 \cdot x^4 = x^7$. Verify that this is true for $x = 2$.

 Using the Distributive Property, we can perform certain "quick products" *without using a calculator.* **For example,**

$$14(102) = 14(100 + 2) = 1400 + 28 = 1428$$
$$32(999) = 32(1000 - 1) = 32{,}000 - 32 = 31{,}968$$

 In like manner, obtain the following "quick products" *without using a calculator:*

47. 17(201)   **48.** 15(302)   **49.** 16(99)   **50.** 23(999)

51. 42(998)   **52.** 27(9998)   **53.** 520(1002)   **54.** 810(9999)

55. The Distributive Property has a geometric interpretation based on the formula for the area of a rectangle: area = width × length. Suppose the large rectangle shown below is cut along the dotted line such that two smaller rectangles are formed. By expressing the areas of all three figures in terms of a, b, and c, give the correct geometric meaning of the Distributive Property.

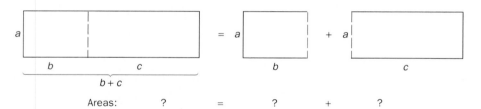

56. In Section 1.1 it was stated that $(-3)(-5) = 15$. In the following proof of this fact, answer each question "Why?" by filling in the correct property of real numbers.

$$5 + (-5) = 0 \qquad \text{Why?}$$
$$(-3)[5 + (-5)] = 0 \qquad \text{Any number times 0 equals 0}$$
$$\text{(see Review Exercises,}$$
$$\text{Problem 122).}$$
$$(-3)5 + (-3)(-5) = 0 \qquad \text{Why?}$$
$$-15 + (-3)(-5) = 0 \qquad (-3)5 = -15 \text{ from Section 1.1.}$$
$$\text{We know that} \quad -15 + \quad 15 \quad = 0 \qquad \text{Why?}$$
$$\text{Therefore} \qquad (-3)(-5) = 15 \qquad \text{Why?}$$

The Commutative, Associative, and Distributive Properties given in the text involved addition and multiplication. If these properties are extended to other operations, they become valid in some cases but not in others. For example, the

Associative Property of subtraction: $a - (b - c) = (a - b) - c$ is *false.*

To illustrate this, we can say that

$$9 - (5 - 2) \neq (9 - 5) - 2, \quad \text{because} \quad 9 - 3 \neq 4 - 2;$$
$$\text{or} \quad a - (b - c) \neq (a - b) - c, \quad \text{because} \quad a - b + c \neq a - b - c.$$

Using specific numbers, or algebra, illustrate the following:

57. The Associative Property of division: $a \div (b \div c) = (a \div b) \div c$ is false.

58. The Commutative Property of subtraction: $a - b = b - a$ is false.

59. The Commutative Property of powers: $a^b = b^a$ is false. (But can you find any numbers for which it *is* true?)

60. The "left-handed" Distributive Property of division:

$$(a + b) \div c = a \div c + b \div c \quad \text{is true.}$$

61. The "right-handed" Distributive Property of division:

$$a \div (b + c) = a \div b + a \div c \quad \text{is false.}$$

62. The Distributive Property of powers: $a^{(b+c)} = a^b + a^c$ is false.

The Closure Properties given in the text say that the sum and the product of two real numbers are always real numbers. This is not always true for other number systems and operations, however. For example,

The sum of two prime numbers is *not always* a prime number.

To exhibit this, note that 3 + 7 = 10, which is *not* prime. In like manner, do the following problems:

63. Exhibit two prime numbers whose sum *is* prime.

64. Exhibit two natural numbers whose difference and quotient are not natural numbers.

65. Exhibit two odd numbers whose sum is not odd. (The *odd numbers* are ± 1, ± 3, ± 5, . . .)

66. Exhibit two irrational numbers whose sum is rational.

True or False? Illustrate your answers with specific numbers.

67. The sum and the product of two rational numbers are always rational.

68. The product of two prime numbers can be a prime number.

69. The Associative Property of powers: $a^{(b^c)} = (a^b)^c$ is true.

70. The Commutative Property of powers: $(a^b)^c = (a^c)^b$ is true.

71. The Commutative Property of division: $a \div b = b \div a$ is false.

72. The Distributive Property of products: $a(b \cdot c) = ab \cdot ac$ is true.

73. A number can be its own reciprocal.

74. A number cannot be its own opposite.

75. The quotient of two real numbers is always a real number.

76. The product of two irrational numbers is always irrational.

CHAPTER 1 ▪ SUMMARY

1.1 Integers and Rational Numbers

Integers: $0, 1, -1, 2, -2, 3, -3, \ldots$

Example:
$$-5 - 7 - (-4) = -5 - 7 + 4$$
$$= -8.$$

Rational numbers: $\dfrac{2}{3}, 0, \dfrac{-5}{4}, 8, -12, 1\dfrac{1}{2}, \ldots$

Example:
$$(-3)\left(\dfrac{-5}{12}\right) = \dfrac{5}{4}.$$

1.2 Order of Operations

1. Parentheses
2. Powers
3. Multiplication/division
4. Addition/subtraction

Example:
$$12 - 2(1 - 4)^3 = 12 - 2(-3)^3$$
$$= 12 - 2(-27)$$
$$= 12 + 54$$
$$= 66.$$

1.3 Formulas and Expressions

Rectangle formulas:
$$P = 2L + 2W$$
$$A = LW$$

Circle formulas:
$$C = 2\pi r = \pi d$$
$$A = \pi r^2$$

Pythagorean Theorem:
$$a^2 + b^2 = c^2$$

1.4 Real Numbers

A real number is either

rational: $-2, 0, 5, \dfrac{1}{4}, \ldots$

or   irrational: $\sqrt{2}, \sqrt{5}, \sqrt[3]{7}, \pi, \ldots$

Properties of Real Numbers:
Closure, Commutative, Associative, Distributive, Identity, and Inverse.

Example:
$$3(2x^2 - x) - 2(x^2 - 5x)$$
$$= 6x^2 - 3x - 2x^2 + 10x$$
$$= 4x^2 + 7x.$$

These properties justify the simplifying of polynomial expressions.

■
CHAPTER 1 · REVIEW EXERCISES   *Answers, pages A34–A35*

1.1, 1.2

Compute:

1. $-4 - (-13) - 15$

2. $-\dfrac{1}{8} - \left(\dfrac{-3}{4}\right) + \dfrac{1}{2}$

3. $(-8)(-2)(-7)$

4. $\dfrac{3(-4)(-6)}{2(-16)}$

5. $\dfrac{7}{24} + \dfrac{9}{16} - \dfrac{5}{72}$

6. $\dfrac{5}{6} \cdot \dfrac{-3}{10}$

7. $\dfrac{-5}{8} \div \dfrac{-15}{56}$

8. $\dfrac{-7}{12} \cdot \dfrac{-16}{12} \div \dfrac{-35}{8}$

9. $(-6)^3$

10. $\left(-\dfrac{3}{4}\right)^4$

11. $\left(\dfrac{-1}{2}\right)^5 \left(-\dfrac{4}{5}\right)^2$

12. $4 - (-3)^3$

13. $\dfrac{2^4 - 4^2}{3^2 - 2^2}$

14. $7 - 5 \cdot 3$

15. $(5 - 9)(2 - 7)$

16. $-48 \div 6 \cdot (-2)$

17. $-9(2 - 7)^3$

18. $-5 - (2 - 8)^2$

19. $5 - 7[6 - (1 - 5)]$

20. $\left(7 + \dfrac{1}{3}\right)\left(5 - \dfrac{1}{8}\right)$

21. $\left(\dfrac{1}{4} - \dfrac{3}{8}\right) \div \left(\dfrac{3}{5} + \dfrac{1}{4}\right)$

22. $12 \cdot \dfrac{7}{8} \div (-14)$

23. $12 + \dfrac{20}{11 - \dfrac{8}{6 - \dfrac{14}{1 - 8}}}$

24. $2 + 3[6 - 3(4 + 9(1 - 6))]$

Factor into prime numbers:

25. 72   **26.** 540   **27.** 1540   **28.** 3528   **29.** 14,175

Insert parentheses, if necessary, to make these calculations correct:

30. $5 - 2 \cdot 3 + 7 = 16$

31. $5 - 2 \cdot 3 + 7 = 6$

32. $5 - 2 \cdot 3 + 7 = -15$

33. $5 - 2 \cdot 3 + 7 = 30$

34. List the prime numbers between 100 and 125.

35. Obtain the sum $\dfrac{5}{144} + \dfrac{7}{480} + \dfrac{11}{360}$. *Hint:* Factor into primes to find the L.C.D.

36. Obtain the product and the sum of $1, \dfrac{1}{2}, \dfrac{2}{3}, \dfrac{3}{4}, \dfrac{4}{5},$ and $\dfrac{5}{6}$.

1.3

Evaluate these expressions for the given values:

37. $x^2 - 7x - 18$ for $x = 0, 9, -3, -2$, and $\dfrac{1}{2}$

38. $2y^2 - 13y + 15$ for $y = 1, -2, 5, -\dfrac{1}{2}$, and $\dfrac{3}{2}$

39. $z^3 - 5z^2 + 2z + 8$ for $z = 2, 0, -1, 4$, and $\dfrac{1}{4}$

40. $3a^2 - 5ab + 2b^2$ for $a = 3, b = -2$

41. $(x - y)^2 - (x + z)^3 + (y - z)^4$ for $x = 2, y = -1, z = -3$

Compute the perimeter and the area of the rectangles with the given length and width:

42. $L = 8$ in., $W = 5$ in.

43. $W = 9$ in., $L = 1\dfrac{1}{2}$ ft

44. $L = W = 3.6$ cm

45. $L = 2\dfrac{1}{6}$ ft, $W = 1\dfrac{2}{3}$ ft

46. $W = \dfrac{3}{4}$ in., $L = 1\dfrac{5}{8}$ in.

Compute the circumference and area of the circles with the given radius or diameter. Assume all measurements are in meters.

47. $r = 7$

48. $r = 5.6$

49. $d = 12$

50. $d = 7$

51. $r = \dfrac{2}{3}$

52. $d = 2\dfrac{3}{8}$

53. When a ball is thrown upward from the top of a building, its height h above the ground (in feet) after t seconds is given by

$$h = -16t^2 + 64t + 80.$$

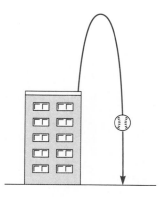

    a. Find the height of the ball when $t = 0, 1, 2, 3, 4$, and 5.
    b. What is the greatest height attained by the ball, and at what time does this occur?
    c. When does the ball hit the ground?
    d. What is the height of the building?

54. Celsius and Fahrenheit temperatures are related by the formula

$$C = \frac{5(F - 32)}{9}.$$

    Find C when $F = 32°, 212°, 98.6°, -40°$, and $68°$.

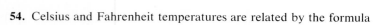

55. Statistics show that marriages are least likely to end in divorce when the bride's age (B) is seven more than half the groom's age (G):

$$B = \frac{G}{2} + 7.$$

Find the "ideal" age of the bride if the groom's age is 18, 27, 32, 43, 58, and 14. What do you notice about age 14?

56. Calculus students use the formula

$$A = \left(\frac{b^3}{3} - \frac{b^2}{2} + 2b\right) - \left(\frac{a^3}{3} - \frac{a^2}{2} + 2a\right)$$

to find the irregularly shaped area A under a certain curve. Find A when:

a. $a = 1, b = 2$
b. $a = -1, b = 1$
c. $a = 0, b = 5$
d. $a = -1, b = 3$

57. The distance D traveled by an object traveling at rate R for time T is given by the formula

$$D = RT.$$

a. Find the distance traveled by a car going 55 miles per hour for 3 hours.
b. Two cars leave the same point at the same time but travel in *opposite* directions. One car goes 65 mph and the other goes 80 mph. Find their distance apart after two hours.
c. Now assume that these two cars were going in the *same* direction, and find their distance apart after four hours.
d. Car A leaves a point at 9:00 A.M. going 60 mph due east. Car B leaves from the same point at 10:30 A.M. going 75 mph due west. How far apart will these cars be at 2:00 P.M.?

58. This problem requires the simple interest formula

$$I = Prt$$

given in Exercise 1.3, Problem 38.

a. $15,000 is invested at 6% for 2 years, and $25,000 is invested at 11% for 3 years. Find the total interest earned.
b. On June 1, 1970, $20,000 was deposited into an account paying 7% interest. On June 1, 1974, $8000 was taken out of this account and deposited into another account paying 9.5%. On June 1, 1977, both accounts were closed. Find the total interest earned.

59. A $40,000 investment portfolio consists of $27,500 divided equally among a stock paying 12% for 2 years, a bond paying 8% for 3 years, and a real estate limited partnership that lost 15% in 1 year. Find the net gain on the investment.

Use the Pythagorean Theorem to decide which of the following sets of three numbers can form the sides of a right triangle:

60. 16, 12, 20

61. 15, 12, 9

62. 5, 7, 9

63. 11, 60, 61

64. 0.7, 2.4, 2.5

65. $\dfrac{1}{2}, \dfrac{2}{3}, \dfrac{5}{6}$

66. $1\dfrac{2}{3}, 4, 4\dfrac{1}{3}$

67. $1\dfrac{1}{3}, 2\dfrac{1}{2}, 2\dfrac{5}{6}$

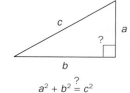

$a^2 + b^2 \stackrel{?}{=} c^2$

Find the shaded area in each figure. The units of measurement are unspecified.

68.

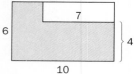

69.

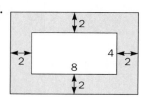

70.

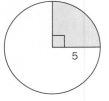

71.

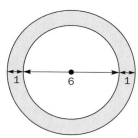

72.

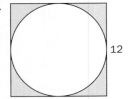

73.

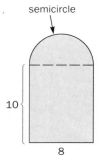

74. a. A stereo that normally sells for $750 is marked up 20%. Because of poor sales, the higher price is then marked down 20%. Find the new selling price. (Your answer, surprisingly, will *not* be $750.)

    b. Suppose the stereo is marked down 20% and then the reduced price is marked up 20%. Will the new selling price be the same as in part a?

1.4

Classify each number below as one or more of the following types: real, rational, irrational, integer, whole, natural, prime, or none of these.

75. $\dfrac{-2}{15}$ **76.** 0   **77.** 59   **78.** $1 + \sqrt{2}$ **79.** 0.25

80. $0.\overline{25}$ **81.** $\sqrt{0.25}$ **82.** 133   **83.** $\sqrt[3]{27}$ **84.** $\pi \cdot \dfrac{1}{\pi}$

85. $\sqrt{\dfrac{16}{25}}$ **86.** 143   **87.** $\dfrac{\sqrt{5}}{0}$ **88.** $3 + (\sqrt{3} - 3)$ **89.** $\sqrt{2809}$

90. 0.212121212 . . .   **91.** 0.2121121112 . . .

Simplify each polynomial expression:

92. $3(2x - 7) - (9 - x)$ **93.** $5(3a + 2b) - 9(-2a + 7b)$

94. $2(x^2 - 7x + 3) - 5(3x^2 + 2x - 4)$ **95.** $5(a^2 - 3ab - 5b^2) + 3(5ab - 4b^2 - a^2)$

96. $(y^2 - y - 1) - (2y + y^2 - 3) + 2(4 - y^2 - 4y)$ **97.** $8(t^2 - 2t + 2) + (3t - 5t^2 - 1) - 5(3 + 2t^2 - 5t)$

98. $2x - 3[3 - 2(2x - 5)]$ **99.** $5a + 2[5a - 4(2b - a)]$

100. $a - 4[a - 4(a - 4(a - 4))]$ **101.** $24 + 5[a - 7(3 - 6(2 - 5(1 - 2a))) + 3a]$

Translate into algebraic expressions, and then simplify:

102. Find twice the difference between $2x$ and five, decreased by the sum of seven and $-4x$.

103. Increase three times the sum of x and $-3y$ by four times the difference between $-2y$ and $-5x$.

104. Add a number to twice its square, then subtract twice the difference between the square of the number and the number itself.

105. From twice the sum of a number and seven, subtract the sum of twice the number and seven.

106. Decrease the difference between three times a number and five by three times the difference between the number and five.

107. Find nine less than twice the result of increasing six by five times a number.

108. From twice the difference between $2a$ and $-3b$, subtract the sum of $-3a$ and $2b$.

 Use the Distributive Property to obtain the following "quick products." *No calculators, please.* **See Exercise 1.4, Problems 47–54.**

109. 14(101)   **110.** 23(202)

111. 26(99)   **112.** 47(999)

113. 85(9998)   **114.** 127(2001)

115. Exhibit two prime numbers whose sum and difference are both prime.

116. Exhibit two prime numbers whose sum is prime but whose difference is not prime.

117. Exhibit two prime numbers whose difference is prime but whose sum is not prime.

118. Exhibit two prime numbers whose sum, difference, and product are not prime.

119. Is the sum of a rational and an irrational number rational or irrational? What about their product?

120. Is the "left-handed" Distributive Property of division:
$(ab) \div c = (a \div c)(b \div c)$ true or false? Experiment with numbers.

121. Is the Distributive Property of powers: $a^{(b+c)} = (a^b)(a^c)$ true or false? Experiment with numbers.

122. In Exercise 1.4, Problem 56, we used the fact that "any number times 0 equals 0." This is easy to prove for natural numbers, by using repeated addition, for example,

$$3 \cdot 0 = 0 + 0 + 0 = 0.$$

But repeated addition has no meaning in the case of other real numbers, such as

$$-8 \cdot 0, \quad \frac{2}{7} \cdot 0, \quad \text{and} \quad \sqrt{2} \cdot 0.$$

The proof below shows that $a \cdot 0 = 0$ for any real number a. *You* supply the correct property of real numbers.

	$0 + 0 = 0$	Why?
	$a \cdot (0 + 0) = a \cdot 0$	Multiply both sides by a (Section 2.1).
	$a \cdot 0 + a \cdot 0 = a \cdot 0$	Why?
We know that	$a \cdot 0 + 0 = a \cdot 0$	Why?
Therefore	$a \cdot 0 = 0$	Why?

CHAPTER 1 · TEST   *Answers, page A36*

Simplify:

1. $-5 - (-8) - 2$

2. $(-3)^2(-2)^3$

3. $\dfrac{3}{4} - \dfrac{5}{8} - \left(-\dfrac{1}{2}\right)$

4. $\dfrac{-2}{3} \cdot \dfrac{-15}{16}$

5. $5 - (3 - 8)^2$

6. $8 - 2[4 + 2(1 - 3)]$

7. $\dfrac{2}{5} \cdot \dfrac{-10}{12} \div \dfrac{-8}{25}$

8. $(2.6 - 4.3)(5.1 + 8.9)$

9. $\dfrac{7}{36} + \dfrac{5}{24} - \dfrac{11}{27}$

10. $3(2a - 9) - 4(a - 8) + 2(5 - a)$

11. $3(2x^2 - xy + y^2) - (3y^2 - yx + 6x^2)$

12. $5x - 2[3 - 4(x - 1)]$

13. $-3a + 4[a - 2(2a + 3(a - 4)) - 5a]$

Translate into algebraic expressions, and then simplify:

14. Find five more than three times the difference between twice a number and one.

15. Find twice the sum of a number and four, decreased by the sum of twice the number and four.

16. When a ball is thrown upward from the ground, its height h above the ground (in feet) after t seconds is given by the formula

$$h = -16t^2 + 80t.$$

   a. Find the height of the ball when $t = 0$, 1, 2, $\dfrac{5}{2}$, 3, 4, and 5.

   b. What is the maximum height of the ball, and when does this occur?
   c. When does the ball strike the ground?

17. A rectangular patio 28.3 feet by 11.8 feet is to be fenced off and tiled. The fencing costs $22.50 per foot, and the tile costs $36.00 per square foot. Find the total cost of the patio.

18. Which of these sets of three numbers can be the sides of a right triangle?

   a. 8, 10, 12   **b.** 13, 84, 85   **c.** $\dfrac{2}{3}, \dfrac{1}{2}, \dfrac{5}{6}$

Find the shaded areas. The units of measurement are unspecified.

19.

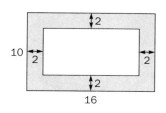

20.

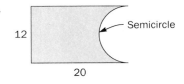

2 LINEAR EQUATIONS AND INEQUALITIES

2.1 ## First-Degree Equations

A **first-degree equation** in one variable is a statement about equality between first-degree polynomials in one variable and constants. Also called **linear equations,** they are illustrated by

$$2x = -30, \quad -3x + 5 = 26, \quad \text{and} \quad -5y - 11 = 5y + 4.$$

A **solution** of an equation is a value of the variable for which the equation is true. The equation $2x = -30$ has a solution $x = -15$, because $2(-15) \overset{\checkmark}{=} -30$; in fact, that is the only solution.

Equations can be solved by using the properties of equality listed below, where A, B, and C represent algebraic expressions or constants.

Properties of an Equation

The equation $A = B$ has the same solutions as the equations:

$$A + C = B + C$$
$$A - C = B - C$$
$$AC = BC, \quad C \neq 0$$
$$\frac{A}{C} = \frac{B}{C}, \quad C \neq 0$$

When we add, subtract, multiply, or divide both sides of the equation $A = B$, using appropriate choices of C, the result is an equation whose solution is the same as that of the original.

EXAMPLE 1

Solve $-3x + 5 = 26$.

Solution

$$
\begin{aligned}
-3x + 5 - 5 &= 26 - 5 \qquad \text{Subtract 5.}\\
-3x &= 21
\end{aligned}
$$

$$\frac{-3x}{-3} = \frac{21}{-3} \qquad \text{Divide by } -3.$$

$$x = -7. \qquad \text{Solution.}$$

Check: $-3(-7) + 5 = 26$, or $21 + 5 \overset{\checkmark}{=} 26$.

EXAMPLE 2 | Solve $-5y - 11 = 5y + 4$.

Solution

$$-5y - 11 + 5y = 5y + 4 + 5y \qquad \text{Add } 5y.$$
$$-11 \quad\;\; = 10y + 4$$
$$-11 - 4 = 10y + 4 - 4 \qquad \text{Subtract 4.}$$
$$-15 = 10y$$
$$\frac{-15}{10} = \frac{10y}{10} \qquad \text{Divide by 10.}$$
$$\frac{-3}{2} = y. \qquad \text{Solution.}$$

Check: $-5\left(\frac{-3}{2}\right) - 11 = 5\left(\frac{-3}{2}\right) + 4$

$$\frac{15}{2} - \frac{22}{2} = \frac{-15}{2} + \frac{8}{2}, \quad \text{or} \quad \frac{-7}{2} \overset{\checkmark}{=} \frac{-7}{2}.$$

Note: The answer $\frac{-3}{2} = y$ can also be written $y = \frac{-3}{2}$.

The strategy is to add and/or subtract appropriate terms and obtain an equation of the form $ax = b$. We then divide by a, the coefficient of x. In the next example, we **simplify each side separately** and then proceed as above.

EXAMPLE 3 | Solve $8(1 + 4z) - (2z - 5) = 3 - 2(3z - 5)$.

Solution

$$8 + 32z - 2z + 5 = 3 - 6z + 10 \qquad \text{Distributive Property.}$$
$$30z + 13 = 13 - 6z \qquad \text{Combine like terms.}$$
$$30z + 13 - 13 = 13 - 6z - 13 \qquad \text{Subtract 13.}$$
$$30z \quad\;\; = \quad -6z$$
$$30z + 6z = -6z + 6z \qquad \text{Add } 6z.$$
$$36z = 0$$
$$\frac{36z}{36} = \frac{0}{36} \qquad \text{Divide by 36.}$$
$$z = 0. \qquad \text{Solution.}$$

Check: You substitute $z = 0$ into the original equation and verify that it checks.

The next two examples have unexpected solutions, because the unknowns disappear.

EXAMPLE 4

Solve $14 + 7(t - 5) = 4t - 3(7 - t)$.

Solution

$$14 + 7t - 35 = 4t - 21 + 3t \qquad \text{Distributive Property.}$$
$$7t - 21 = 7t - 21 \qquad \text{Combine like terms.}$$
$$7t - 21 - 7t + 21 = 7t - 21 - 7t + 21 \qquad \text{Subtract 7t, add 21.}$$
$$0 = 0. \qquad \text{TRUE}$$

The unknown t has vanished, leaving the TRUE statement $0 = 0$. This means that substituting *any* value of t into the original equation leads to this true statement.

Answer: All real numbers.

The second line in this example, $7t - 21 = 7t - 21$, is identical, term for term, on both sides. Such an equation is called an **identity**, and so is the original equation from which it was derived. The solution of an identity consists of all real numbers for which every term is defined.

EXAMPLE 5

Solve $5(x - 8) + 27 = 5x + 7$.

Solution

$$5x - 40 + 27 = 5x + 7$$
$$5x - 13 = 5x + 7$$
$$5x - 13 - 5x + 13 = 5x + 7 - 5x + 13 \qquad \text{Subtract 5x, add 13.}$$
$$0 = 20. \qquad \text{FALSE}$$

The unknown x has vanished, leaving the FALSE statement $0 = 20$. This means that substituting *any* value of x into the original equation leads to this false statement.

Answer: No solution.

The next two examples are word problems. They make use of the phrases you learned in Section 1.4.

EXAMPLE 6

Taking five times the sum of a number and three is the same as decreasing one by twice the number. Find the number.

Solution

Let x = the number.

The phrase "is the same as" means the equals sign (=); this separates the problem into two sides of the equation. We translate the problem into an equation, consulting the Table of Phrases if necessary:

$$5(x + 3) = 1 - 2x \qquad \text{Translate into an equation.}$$

We now solve the equation:

$$5x + 15 = 1 - 2x$$
$$5x + 2x = 1 - 15 \qquad \text{Add } 2x, \text{ subtract } 15.$$
$$7x = -14$$
$$x = -2. \qquad \text{Divide by } 7.$$

Answer: The number is -2.

Check: $5(-2 + 3) = 1 - 2(-2)$, or $5 \overset{\checkmark}{=} 5$.

When checking the answer to a word problem, it is important to substitute the answer into the correct equation. If the equation itself were incorrect, the "correct" answer would also be incorrect.

EXAMPLE 7

One number is four more than another. Twice the larger is one more than three times the smaller. Find each number.

Solution

Let x = the smaller number; then $x + 4$ = the larger number.

Here, the word "is" represents the equals sign (=). The problem can be broken down into the equation

$$2(larger) = 3(smaller) + 1,$$
or
$$2(x + 4) = 3(x) + 1.$$

Solving in the usual manner gives

$$2x + 8 = 3x + 1$$
$$2x - 3x = -8 + 1 \qquad \text{Subtract } 3x, \text{ subtract } 8.$$
$$-x = -7$$
$$\frac{-x}{-1} = \frac{-7}{-1}$$
$$x = 7.$$

If $x = 7$ (the smaller), then $x + 4 = 7 + 4 = 11$ (the larger).

Answer: The numbers are 7 and 11.

Check: $2(11) = 3(7) + 1$, or $22 \overset{\checkmark}{=} 22$.

Here are the steps to follow in order to solve these and other word problems:

> **Helpful Steps in Solving a Word Problem**
>
> *Step 1*   **Carefully read** the problem.
>
> *Step 2*   Assign variables to **all** unknowns. (Example 7: x and $x + 4$)
>
> *Step 3*   Locate the phrase denoting the equals sign (=). This commonly appears as "is," "the result is," or "is the same as."
>
> *Step 4*   **Set up** the proper equation by translating into algebra the phrases *directly before* and *directly after* the equals sign (=). Use the Table of Phrases, if necessary.
>
> *Step 5*   Solve the equation obtained in Step 4.
>
> *Step 6*   If the problem asks for more than one number, such as a smaller and a larger number, go back to Step 2 to find all such numbers. (Example 7: 7 and 11)
>
> *Step 7*   **Check** your answer by substituting into the proper equation obtained in Step 4. *Warning!* If the original equation is incorrect, your answer(s) will also be incorrect!

EXERCISE 2.1   *Answers, page A36*

A

Solve and check, as in EXAMPLES 1 through 5:

1. $2x - 3 = 7$ **2.** $3y + 1 = -8$ **3.** $17 = 2 - 3z$ **4.** $-8 = 14 - 2t$

5. $5x - 9 = 7 - 3x$ **6.** $13 - 8n = n - 5$ **7.** $9 + 4m = 10m - 6$ **8.** $9r - 23 = r - 35$

9. $5x - 7 = 2x - 7$ **10.** $9 - 5w = 5w + 9$ **11.** $3x = x$ **12.** $-t = t$

13. $3(x - 7) = 5(2x + 4) + 1$ **14.** $2(3y - 2) = y - (2y - 3)$ **15.** $5z - 3(z - 9) = 1 + 2(z + 13)$

16. $8y - 3(2y - 8) = 4 + 2(10 + y)$ **17.** $4(2x - 1) + 7 = 2 + 8(1 + x)$ **18.** $7(3z - 3) + 6 = 8(2z + 1) - 2z$

19. $3(2x - 1) - 2(5 - x) = 5(1 - 4x) + 2(-3 - x)$ **20.** $4(3 + 5y) - (2y - 1) = 2(2y + 1) + 3(7 - 2y)$

21. $3y - 2[5y - 7(y - 1)] = 5[y - 4(1 - 3y)] - 4y$ **22.** $3[9z + 4(3z - 6)] + 22z = z - 6[4z + 3(5z - 7)]$

23. $5n + 10(3n - 2) + 25[2(3n - 2) + 4] = 1275$ **24.** $(2n + 1) + 5n + 10[3(2n + 1) - 4] = 728$

Set up and solve, as in EXAMPLE 6:

25. If twice a number is increased by seven, the result is 35. Find the number.

26. If four times a number is decreased by one, the result is 19. Find the number.

27. Five more than a number equals nine less than three times the number. What is the number?

28. Ten less than five times a number equals two more than three times the number. What is the number?

29. If seven is added to twice the difference between a number and six, the result is the same as three times the sum of the number and five. Find the number.

30. If three times the sum of a number and two is decreased by eight, the result is twelve increased by the number. What is the number?

31. Twice the difference between eight and a number equals the difference between twice the number and eight. Find the number.

32. If a number is increased by eight and the result is doubled, the result is five more than three times the result of decreasing two by the number. What is the number?

Set up and solve, as in EXAMPLE 7:

33. One number is five more than another. Twice the larger is one more than five times the smaller. Find each number.

34. One number is two less than another. Three times the smaller is two less than twice the larger. Find each number.

35. One number is three more than twice another. Twice the larger equals five times the smaller. What is each number?

36. One number is two more than three times another. Twice the larger is eight more than four times the smaller. What is each number?

B

37. Perform the following operations, in succession, to a certain number: add seven, multiply by two, subtract five, multiply by ten, and add six. The result is 136. Find the original number.

38. A cash register contains a certain amount of money. To this you add an equal amount of money and then remove $20. You match the current amount and then remove $20. Finally, you match the most current amount and then remove $20. There is $100 remaining. How much money was originally in the register?

39. Taking five more than three times the difference between four and twice a number is the same as subtracting one from twice the result of increasing the number by seven. Find the number.

40. One number is four more than another. A third number is twice as large as the larger of the first two numbers. Twice the largest number equals three times the sum of the two smaller numbers. Find each number.

41. Of four numbers, the second is twice the first; the third is four more than the second; and the fourth is one less than twice the third. The sum of the four numbers is 56. Find each number.

42. The sum of two numbers is 32. If twice the first is increased by three times the second, the result is 84. Find each number. *Hint:* the numbers are x and $32 - x$. See the Table of Phrases in Section 1.4.

C

43. Increase two by a certain number, and double the result; then increase the current result by the number, and double the new result; then increase the most current result by the number, and double the latest result; and finally, increase this latest result by the number. The result of all this is 31. Find the original number!

Solve:

44. $x + 2.6 = 0.2x - 3$ **45.** $x - 5.8 = 14.45 - 1.7x$ **46.** $6.1 - 3[4.1 - 2(3.7 - 1.4x)] = x - 6.5$

47. $21x - 2[8 - 3(2x + 5(x - 4))] = x + 3[8 + 2(3x - 2(x + 1)) - 6x]$

48. $2z + 3[2z + 3(2z + 3(2z + 3(2z + 3)))] = z - 2[z - 2(z - 2(z - 2(z - 2)))]$

2.2   Coin and Ticket Problems

Like problem 38 of Section 2.1, word problems often involve money. Suppose a collection of 32 coins consists of nickels and dimes, and suppose there are 13 nickels. Then there are $32 - 13 = 19$ dimes, and their total value is

$$5¢(13) + 10¢(19) = 65¢ + 190¢ = 255¢, \quad \text{or} \quad \$2.55.$$

More generally, if 32 coins consist of nickels and dimes, and if there are x nickels, then there are $32 - x$ dimes. Their total value is

$$5¢(x) + 10¢(32 - x).$$

We have used the fact that the value of any number of coins equals the face value of that coin times the number of coins.

EXAMPLE 1

A collection of 32 coins in nickels and dimes is worth $2.75. How many of each coin are there?

Solution

Let x = number of nickels; then $32 - x$ = number of dimes (see above). We can simplify the problem by using a chart:

	number of coins ·	face value =	value of coins
nickels	x	5¢	5¢(x)
dimes	$32 - x$	10¢	10¢$(32 - x)$

Total value = $2.75

Because the total value of the coins is $2.75, or 275¢, we have

$$5¢(x) + 10¢(32 - x) = 275¢$$
$$5x + 320 - 10x = 275$$
$$-5x = -45$$
$$x = 9.$$

If $x = 9$ (nickels), then $32 - x = 23$ (dimes).

Answer: 9 nickels, 23 dimes.
Check: $5¢(9) + 10¢(23) = 45¢ + 230¢ \overset{\checkmark}{=} 275¢$.

The next example involves ticket sales to a concert. The revenue generated from selling tickets at a particular price equals the price of the ticket times the number of tickets sold at that price.

EXAMPLE 2

Tickets to a concert cost $15, $25, and $40. There were 5000 more $40 tickets sold than $25 tickets, and the number of $15 tickets sold was twice the number of $40 tickets. The total revenue from ticket sales was $540,000. How many were sold at each price?

Solution

Let x = number of $25 tickets, because these were the fewest;
then $x + 5000$ = number of $40 tickets, (Why?)
and $2(x + 5000)$ = number of $15 tickets. (Why?)

We enter these data into a chart:

	number of tickets · price =		revenue
$25 tickets	x	$25	$25(x)$
$40 tickets	$x + 5000$	$40	$40(x + 5000)$
$15 tickets	$2(x + 5000)$	$15	$15(2x + 10,000)$

Total revenue = $540,000

Because the total revenue was $540,000, we add the last column:

$$\$25(x) + \$40(x + 5000) + \$15(2x + 10,000) = \$540,000$$
$$25x + 40x + 200,000 + 30x + 150,000 = 540,000$$
$$95x + 350,000 = 540,000$$
$$95x = 190,000$$
$$x = 2000.$$

If $x = 2000$, then $x + 5000 = 7000$ and $2(x + 5000) = 14,000$.

Answer: 2000 $25 tickets, 7000 $40 tickets, and 14,000 $15 tickets.
Check: $25(2000) + $40(7000) + $15(14,000)
    $= \$50,000 + \$280,000 + \$210,000 \overset{\checkmark}{=} \$540,000.$

EXERCISE 2.2   *Answers, page A36*

A

Solve as in EXAMPLE 1:

1. A collection of 23 coins in dimes and quarters is worth $3.35. How many of each coin are there?

2. Thirty-eight coins in pennies and nickels are worth $1.22. How many of each coin are there?

3. Tickets to Super Bowl I cost only $8 and $12! There were 62,000 sold, and the total revenue from ticket sales was $592,000. How many at each price were sold?

4. A Porsche/Audi dealership sells Porsches for $47,000 and Audis for $22,000. One month it sold 18 cars and grossed $646,000 in sales. How many of each car did it sell?

5. A cash register contains 35 bills in $5 bills and $20 bills. The value of the $5's equals the value of the $20's. How many of each bill are in the register?

6. A collection of 36 coins in dimes and quarters is such that the value of the dimes is twice the value of the quarters. How many of each coin are there?

Solve as in EXAMPLE 2:

7. In a collection of coins, there are seven more nickels than dimes and twice as many quarters as nickels. The total value is $9.05. How many of each coin are there?

8. In a collection of bills, there are twice as many $5's as $20's, and the number of $10's is seven more than three times the number of $5's. The total collection is worth $790. How many of each bill are there?

9. Tickets to the final day of track and field at the 1984 Olympics in Los Angeles cost $25, $35, and $50. The number of $35 tickets sold was twice the number of $25 tickets. The number of $50 tickets was 10,000 less than the number of $35 tickets. The total revenue was $3,400,000. How many of each were sold?

10. Tickets to gymnastics at the Olympics cost $40, $60, and $95. The number of $40 tickets sold was 1000 more than the number of $95 tickets. The number of $60 tickets was 1000 less than twice the number of $95 tickets. The total revenue was $745,000. How many of each were sold?

B

11. In one hour, the Burger Barn sold 95 hamburgers and 60 orders of fries. The total sales were $235.00. A burger costs $1.25 more than an order of fries. Find the cost of each item.

12. Reserved seats at a concert cost $8 more than general admission. There were 3000 reserved seats sold and 5000 general admission. The revenues from ticket sales at each price were equal. Find the cost of each ticket.

13. In a collection of 32 coins, there are equal numbers of nickels and dimes. The rest are quarters. The total value is $5.55. How many of each coin are there? *Hint:* Let x = number of nickels; then x = number of dimes, and ????? = number of quarters.

14. In a collection of 47 bills, there are twice as many $1's as $10's. The rest are $5's. If the total value is $199, how many of each bill are there? *Hint:* Let x = number of $10's; then $2x$ = number of $1's, and ????? = number of $5's.

C

15. Box A contains 22 coins in nickels and dimes. Box B contains as many nickels as A contains dimes and as many dimes as A contains nickels. Box B is worth 30¢ more than Box A. How many coins of each denomination are in each box? *Hint:* Let box A contain x nickels and $22 - x$ dimes; then what would be the composition of box B?

2.3   Investment Problems

Suppose you deposit $2000 into a bank account paying 6% annual interest. After one year, you will earn $2000(0.06) = $120. If this interest is withdrawn but the original principal of $2000 remains, then you will also earn $120 in interest after the second year, or a total of $240 interest in two years. By withdrawing the interest after each year, you are always earning interest on the original $2000 principal. This is called **simple annual interest,** and the formula that governs it is

Simple Annual Interest

$$I = Prt$$

where I = interest earned, P = principal, r = annual interest rate, and t = time in years.

For example, if P = $2000 is deposited at r = 6% = 0.06 for t = 5 years, then the interest earned is I = $2000(0.06)(5) = $600. (When you leave your interest in the account, you earn "interest on interest," and your money grows faster. This is called *compound interest,* which we will study in Section 10-1. In all problems in this section, "interest" will always mean "simple annual interest.")

EXAMPLE 1

Suppose $10,000 is invested, part in a stock paying 12% interest for 2 years, and the balance in a stock paying 7% interest for 3 years. If the total interest earned is $2340, how much is invested in each stock?

Solution

Let x = amount invested at 12%; then $10,000 − x$ = amount invested at 7%, because $x + (10,000 − x) = 10,000$.

	P · r · $t =$			I	
12% stock	x	0.12	2	0.12(2)(x)	Total interest
7% stock	$10,000 − x$	0.07	3	0.07(3)(10,000 − x)	= $2340

Because the total interest earned is $2340, we have

$$0.012(2)(x) + 0.07(3)(10,000 − x) = 2340,$$

or

$$0.24x + 0.21(10,000 − x) = 2340.$$

We can solve this equation as is, or we can clear the decimals by multiplying both sides by 100:

$$100(0.24x) + 100[0.21(10,000 − x)] = 100(2340)$$
$$24x + 21(10,000 − x) = 234,000$$
$$24x + 210,000 − 21x = 234,000$$
$$3x = 24,000$$
$$x = 8000.$$

If $x = 8000$, then $10,000 − x = 2000$.

Answer: $8000 at 12% for 2 years, $2000 at 7% for 3 years.

Check: $8000(0.12)(2) + $2000(0.07)(3) = $1920 + $420 $\overset{\checkmark}{=}$ $2340.

EXAMPLE 2

In planning a $50,000 investment portfolio, an investor decided to invest part in a stock paying 10% for 1 year and the balance in a bond paying $7\frac{1}{2}$% for 2 years. It turned out that the interest rates on the two investments were equal. How much was invested in each instrument?

Solution

Let x = amount invested at 10% = 0.10; then $50,000 − x$ = amount invested at $7\frac{1}{2}$% = 7.5% = 0.075.

	P	$\cdot$	r	$\cdot$	$t =$	I	
10% stock	x		0.10		1	$0.10(1)(x)$	
$7\frac{1}{2}$**% bond**	$50{,}000 - x$		0.075		2	$0.075(2)(50{,}000 - x)$	

These are equal.

Because the interest rates on the two investments are equal, we have

$$0.10(1)(x) = 0.075(2)(50{,}000 - x)$$
$$0.10x = 0.15(50{,}000 - x)$$
$$100(0.10x) = 100[0.15(50{,}000 - x)] \qquad \text{Multiply by } 100.$$
$$10x = 15(50{,}000 - x)$$
$$10x = 750{,}000 - 15x$$
$$25x = 750{,}000$$
$$x = 30{,}000.$$

If $x = 30{,}000$, then $50{,}000 - x = 20{,}000$.

Answer: \$30,000 at 10% for 1 year, \$20,000 at $7\frac{1}{2}$% for 2 years.

Check: \$30,000(0.10)(1) = \$20,000(0.075)(2), or \$3000 $\overset{\checkmark}{=}$ \$3000.

EXAMPLE 3

On July 1, 1980, a 6% account was opened with a \$16,000 deposit.
On July 1, 1981, a 12% account was opened with a certain deposit.
On July 1, 1983, both accounts were closed. The combined interest would have been equivalent to that earned on a single 10% account for 2 years. How much was deposited into the 12% account?

Solution

Let x = amount deposited at 12% for 2 years (1981 to 1983). Because \$16,000 was deposited at 6% for 3 years (1980 to 1983), there would have been a total deposit of \$16,000 + x at 10% for 2 years.

	P	$\cdot$	r	$\cdot$	$t =$	I	
6% account	\$16,000		0.06		3	$0.06(3)(16{,}000)$	
12% account	x		0.12		2	$0.12(2)(x)$	
10% account	\$16,000 + x		0.10		2	$0.10(2)(16{,}000 + x)$	

These two add up to the third.

(continued)

The combined interests on the first two accounts must total the interest on the single account:

$$0.06(3)(16{,}000) + 0.12(2)(x) = 0.10(2)(16{,}000 + x)$$
$$0.18(16{,}000) + 0.24x = 0.20(16{,}000 + x)$$
$$100[0.18(16{,}000)] + 100(0.24x) = 100[0.20(16{,}000 + x)] \qquad \text{Multiply by 100.}$$
$$18(16{,}000) + 24x = 20(16{,}000 + x)$$
$$288{,}000 + 24x = 320{,}000 + 20x$$
$$4x = 32{,}000$$
$$x = 8000.$$

Answer: $8000 was deposited at 12%.

Check: $16{,}000(0.06)(3) + \$8000(0.12)(2) \stackrel{?}{=} \$24{,}000(0.10)(2)$,
or $2880 + \$1920 \stackrel{\checkmark}{=} \$4800.$

EXERCISE 2.3   *Answers, pages A36–A37*

A

Solve as in EXAMPLE 1:

1. $50,000 is invested, part at 8% for 3 years and the balance at 11% for 2 years. How much is invested at each rate if the total interest is $11,400?

2. You invest $8000, part in a 2-year Apple bond paying 12% annual interest, and the rest in a 3-year Standard Oil bond paying $11\frac{1}{2}\%$. If you earn $2130 in interest, how much do you invest in each bond?

3. A low-risk municipal bond pays 6% interest, and a high-risk stock *claims* to yield 17%. You invest $20,000, part of which is in the bond and the rest in the stock. After 1 year, however, you find that the stock is *actually losing* 13%, so you sell it and take your loss. However, you continue with the bond for one more year. When all is said and done, you earn only $400 interest. How much did you invest in each instrument?

4. You deposit $10,000, part in Friendly Federal's insured 7% account for 3 years, and the balance in Unfriendly Federal's uninsured account. Unfortunately, you lose 15% in 1 year in the latter. For the whole venture, you end up losing $60. How much did you deposit in each bank?

Solve as in EXAMPLE 2:

5. You invested $35,000, part at 9% for 2 years, and the balance at 6% for 4 years. If both investments earned equal interest, how much did you invest at each rate? Find the total interest earned.

6. $10,000 is invested, part at $5\frac{1}{2}\%$ and the rest at $8\frac{1}{4}\%$. If both investments earn the same annual interest, how much is invested at each rate?

7. A certain amount of money is deposited into an 8% account. After 3 years, the account is closed and the interest is withdrawn. The original amount plus $3000 is then transferred into a $7\frac{1}{2}$% account for 2 years. If the same interest is earned at each rate, how much was invested at each rate? Find the total interest earned.

8. $50,000 is invested, part in stock A paying 8% for 4 years, and the balance in stock B paying 12% for 2 years. If the interest on stock A equals *twice* the interest on stock B, how much is invested in each stock? Find the total interest earned.

Solve as in EXAMPLE 3:

9. On January 1, 1970, Smith invested $10,000 at 16% plus a certain amount of money at 6%. At the same time, Jones matched Smith's entire investment, at 8%. Smith closed his first account on January 1, 1973 and his second account 3 years later. Jones closed her account on January 1, 1975. If Smith's investments earned the same interest as Jones's earned, how much did Smith invest at 6%? How much did Jones invest at 8%?

10. $6,000 has been invested at 8% for 1 year. How much money should one invest at 11% in order to obtain an annual return of 9% on the entire investment?

B

11. A total of $7000 is deposited, with equal amounts in $5\frac{1}{2}$% and 7% accounts, and the balance in an 8% account. After 2 years, the total interest is $980. How much was deposited into each account?

12. $14,500 is available for an investment portfolio. A certain amount is invested in a 3-year bond paying 6%. Twice that amount goes into a stock paying $9\frac{1}{2}$% for 2 years. The balance goes into a real estate limited partnership that *loses* $8\frac{1}{2}$% in 1 year. The three investments net $1620 in interest. How much was invested in each? *Hint:* Let x = amount in bond; then $2x$ = amount in stock and ????? = amount in partnership.

13. $5000 is invested at 6% for 2 years. A certain amount is invested at 7% for 3 years, and into a 9% investment for 2 years goes an amount of money $1000 greater than the amount invested at 7%. The total interest is $3510. How much is invested at each rate?

14. On May 1, 1975, a $5000 bank account paying 6% was opened. On May 1, 1977, a certain amount of money was added to this account, and an amount equal to this total was used to open another account paying 8%. On May 1, 1980, both accounts were closed. The total interest earned was $3120. How much money was added to the first account?

15. On September 1, 1965, $10,000 was deposited into a 7% account. On September 1, 1969, a certain amount was withdrawn and then deposited into an $8\frac{1}{2}\%$ account. On September 1, 1975, both accounts were closed. The total interest was $7360. How much was withdrawn?

16. Stock A pays 3% less annual interest than stock B. $5000 is invested in A for 2 years, and $3000 is invested in B for 3 years. These yield a total of $1790 in interest. What interest rates do both A and B pay? *Hint*: Let r = rate paid by B; then $r - 0.03$ = rate paid by A. Convert your decimal answer to % by moving the decimal point two places to the right.

17. $10,000 is deposited into an account paying $7\frac{1}{2}\%$, and $20,000 is deposited at 6%. After a certain number of years, the first account is closed, but the second account is left open for 2 more years. When all is said and done, the total interest earned is $6300. For how many years is each account open? *Hint*: Let t = time of $7\frac{1}{2}\%$ account; then $t + 2$ = time of 6% account.

18. $6000 is invested at 8% and $4000 is invested at 5%. What is the *effective interest rate* on the entire investment? That is, what interest rate on the entire principal would earn the same interest as the two investments earn? Assume each investment is for 1 year. *Hint*: Let r = interest rate on entire principal; your answer should be a decimal, so convert it to % by moving the decimal point two places to the right.

C

19. $40,000 is invested, part in a 6% bond and the balance in a real estate partnership paying 9%. After 2 years the bond is redeemed, but the real estate investment is continued for one more year, during which time it loses 30%. After all is said and done, the entire investment shows a net loss of $1200. How much is invested in the bond and in the partnership?

20. $15,000 was deposited into a 6% bank account. After a certain number of years, the account was closed and the principal was deposited, along with an additional $8000, into a 9% account. This account was closed 7 years after the first account was opened. The total interest earned on both accounts was $12,150. For how many years was each account open?

21. You open a 6% savings account with a certain amount of money. After 1 year, you close the account and are issued a check for $4770. This includes your original deposit plus interest. How much did you originally deposit? *Hint*: Let P = original principal; then principal + interest = amount of check.

22. A stereo normally selling at a certain price is discounted 15%. Its discounted price is $1700. Find the original price.

2.4   Distance-Rate-Time Problems

Suppose you drive your car with the cruise control set at 55 miles per hour for 3 hours. According to the formula

$$\text{rate} \cdot \text{time} = \text{distance}, \quad \text{or} \quad RT = D,$$

you will have traveled $55 \cdot 3 = 165$ miles. Even with cruise control, a car's rate (or speed) varies. For this reason, all rates henceforth given are to be interpreted as "average rate."

EXAMPLE 1

A Mercedes-Benz and a Porsche leave Beverly Hills, traveling in opposite directions. The Porsche goes 15 miles per hour faster than the Mercedes. In 3 hours they are 345 miles apart. How fast is each car going?

Solution

Let r = rate of Mercedes-Benz; then $r + 15$ = rate of Porsche.

	R	$\cdot\ T =$	D
Mercedes	r	3	$3r$
Porsche	$r + 15$	3	$3(r + 15)$

$\Big\}$ Total = 345

$D = 3r$ $D = 3(r + 15)$

Total = 345 miles

Because they are driving in opposite directions, we add their distances:

$$3r + 3(r + 15) = 345$$
$$3r + 3r + 45 = 345$$
$$6r = 300$$
$$r = 50.$$

If $r = 50$, then $r + 15 = 65$.

Answer: Mercedes goes 50 mph; Porsche goes 65 mph.

Check: $50 \cdot 3 + 65 \cdot 3 = 150 + 195 \overset{\checkmark}{=} 345.$

EXAMPLE 2

Two men rob a bank and make their getaway at 2:00 P.M., driving 90 mph. At 3:00 P.M., the police leave the scene of the crime, driving 110 mph, in hot pursuit of the robbers. How long will it take for the cops to catch the robbers? At what time of day and how far from the bank will this occur?

Solution

Let t = time required for cops to catch robbers, after 3:00 P.M.; then $t + 1$ = time driven by robbers, because of their 1-hour head start.

	R	$\cdot$ T	$=$ D
cops	110	t	$110t$
robbers	90	$t + 1$	$90(t + 1)$

$\left.\right\}$ Distances are equal.

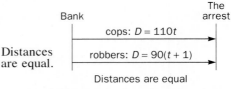

The cops will overtake the robbers when their distances from the bank are equal:

$$110t = 90(t + 1)$$
$$110t = 90t + 90$$
$$20t = 90$$
$$t = 4.5.$$

If $t = 4.5$ hours, then $D = 110(4.5) = 495$ miles. The police left the bank at 3:00 P.M., so $4\frac{1}{2}$ hours later it will be 7:30 P.M.

Answer: Cops catch robbers in $4\frac{1}{2}$ hours, at 7:30 P.M., 495 miles away.

Check: $110(4.5) = 90(5.5)$, or $495 \overset{\checkmark}{=} 495$.

EXAMPLE 3

Members of a track team are to warm up by jogging at the rate of 8 mph and then to run at 10 mph.* The workout is to last 1 hour and 15 minutes and is to cover $11\frac{1}{2}$ miles. How much time do they spend moving at each rate?

Find the distance covered at each rate.

Solution

We first convert the total time to hours: $1:15 = 1\frac{15}{60} = 1\frac{1}{4} = 1.25$ hours.

---

*For you runners, the formula $60 \div$ mph = *mile pace* converts speed (in mph) to mile pace. Thus, 10 mph is equivalent to $60 \div 10 = 6:00$ *mile pace*, and 8 mph is equivalent to $60 \div 8 = 7.5 = 7:30$ *mile pace*.

Now let t = time jogging; then $1.25 - t$ = time running, because
$t + (1.25 - t) = 1.25$.

	$R \cdot$	T =	D
jogging	8	t	$8t$
running	10	$1.25 - t$	$10(1.25 + t)$

$\left.\right\}$ Total $= 11\frac{1}{2}$

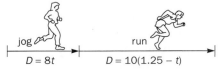

jog
run

$D = 8t$ $D = 10(1.25 - t)$

Total = 11.5 miles

Because the total distance covered is 11.5 miles, we add:

$$8t + 10(1.25 - t) = 11.5$$
$$8t + 12.5 - 10t = 11.5$$
$$-2t = -1.0$$
$$t = 0.5.$$

If $t = 0.5$, then $1.25 - t = 0.75$; $D = 8(0.5) = 4.0$ and $10(0.75) = 7.5$.

Answers: Jogging: 0.5 hours, 4.0 miles;   running: 0.75 hours, 7.5 miles.

Check: $4.0 + 7.5 \overset{\checkmark}{=} 11.5$ miles.

EXERCISE 2.4   *Answers, page A37*

A

Solve as in EXAMPLE 1:

1. A Toyota Supra and a Nissan 300 ZX leave New York and travel in opposite directions. The ZX travels 5 miles per hour faster than the Supra. After 4 hours they are 500 miles apart. Find the rate of each car.

2. A freight train and a passenger train are 380 kilometers apart and are headed toward each other. (Not to worry, they are on different tracks.) The freight train goes 30 kilometers per hour slower than the passenger train. In 2 hours they pass each other. Find the rate of each train.

3. Two planes flying at speeds of 150 and 180 mph, respectively, are 495 miles apart and headed toward each other. How long will it take for them to pass each other?

4. A Corvette and a BMW are 555 miles apart and headed toward each other on the same highway. The Corvette left at 9:00 A.M., traveling 75 mph; the BMW left at 10:00 A.M., traveling 85 mph. At what time of day will they meet?

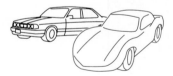

Solve as in EXAMPLE 2:

5. A Ferrari leaves Rome at 3:00 P.M., doing 120 kilometers per hour. At 4:00 P.M. a Maserati, traveling 150 kph, leaves the same point in pursuit of the Ferrari. How long does it take for the Maserati to overtake the Ferrari? At what time of day and how far from Rome does this occur?

6. Runner A, who can run 8 meters per second, starts running from a point. Two seconds later, runner B, who can run 10 mps, starts from the same point. In how many seconds does B overtake A? How far from the start does this occur?

7. A Mazda RX7 leaves St. Louis at 11:00 A.M., headed west. At 2:00 P.M. an Audi, traveling 15 miles per hour faster than the Mazda, leaves from the same point, also headed west. The Audi passes the Mazda at 4:00 A.M. the next morning. Find the speed of each car. Find the distance from St. Louis traveled by each car.

8. Runner C, who is 2 meters per second slower than runner D, is given a 3-second head start in a race. D catches C 12 seconds from the time that C started. Find the speed of each runner. Find the distance from the start where D catches C.

Solve as in EXAMPLE 3:

9. Suppose you jog at the rate of 6 miles per hour and then run at 10 mph. You cover 23 miles in 2 hours 30 minutes. Find your times for both jogging and running. Also find your distance for each.

10. You drive 55 miles per hour on the highway, but city traffic reduces your speed to 40 mph. You cover 170 miles in $3\frac{1}{2}$ hours. How much time did you drive at each speed? How much distance did you cover at each speed?

11. On a 220-mile trip, a driver spent 1 hour in city traffic but then increased her speed by 20 miles per hour on the highway. After 3 hours on the highway, she reached her destination. Find both city and highway speeds. Also find the distance she covered at each speed.

12. You drive at a certain speed for 3 hours and then double your speed for another 2 hours. You cover a total of 560 kilometers. Find your speed in each case. Also find your distance in each case.

B

13. You cycle to work at an average speed of 20 miles per hour, and you return home at an average speed of 15 mph. Your total travel time is 3 hours 30 minutes. How far is it from home to work? *Hint:* Let t = time going to work.

14. A family going on vacation arrived at their destination in 5 hours. On the drive home, they reduced their speed by 10 miles per hour, and consequently the return trip took 1 hour longer. Find the distance from home to the vacation spot. *Hint:* Let r = rate going.

15. In a triathlon, a competitor swam 2 miles per hour, then cycled 20 mph, and then ran 10 mph. He cycled 2 hours more than he swam, and he ran for 1 less hour than he cycled. He covered a total distance of 82 miles. How much time did he spend on each leg of the triathlon? Find the distance covered on each leg.

16. You drive 1 hour to the airport, board a plane that flies for 5 hours, and then bus 2 hours to your hotel. The total distance traveled is 2390 miles. Your car went 5 miles per hour faster than the bus, and the plane went nine times as fast as your car. Find the speed of each vehicle. Find the distance covered by each.

17. In a race for vintage sports cars, an MGB averaged 73 miles per hour and an Austin-Healy averaged 79 mph. How long did it take for the Austin-Healy to be 21 miles ahead of the MGB?

18. In a biathlon, a competitor ran 7.5 miles per hour and then cycled 20 mph. The distance she rode was 37.5 miles greater than the distance she ran. Her total time was 2:20. Find her time for each event. Then find her distance for each event.

19. On a 40-mile trip, you average 50 miles per hour in city traffic and 60 mph on the highway. Your driving time in the city exceeds your highway time by 15 minutes. Find your time in the city and on the highway. Find your distance for each. *Hint:* Convert 15 minutes to hours.

C

20. In an endurance race, car A averaged 80 kilometers per hour, while car B averaged 100 kph. Car B traveled 20 more kilometers than A did, in 12 minutes less time. Find the time of each car. Find the distance of each car. *Hint:* Convert 12 minutes to hours.

21. Driving to work you average 60 mph, and returning home you average 40 mph. Find your average speed for the whole round trip. (The answer is *not* 50 mph.) *Hint:* This problem does not depend on the distance each way, so use any convenient distance.

2.5	Linear Inequalities

Not all numbers are created equal. We use the following symbols to denote inequalities between numbers.

Symbol	Meaning	Examples
$a < b$	a is less than b	$-2 < 1, \quad 3 < 7$
$a > b$	a is greater than b	$1 > -2, \quad -\dfrac{1}{4} > -\dfrac{1}{2}$
$a \leq b$	a is less than or equal to b	$-6 \leq -4, \; 0 \leq 0$
$a \geq b$	a is greater than or equal to b	$\sqrt{3} \geq \sqrt{2}, \; \dfrac{2}{3} \geq \dfrac{4}{6}$
$a \neq b$	a is not equal to b	$\dfrac{5}{8} \neq \dfrac{3}{4}, \; 5\sqrt{2} \neq 4\sqrt{3}$

Geometrically, $-2 < 1$ because -2 lies to the left of 1 on the real-number line. Equivalently, $1 > -2$ because 1 lies to the right of -2.

-2 < 1 is equivalent to 1 > -2

A **linear inequality** in one variable is a statement about inequality between first-degree polynomials in one variable and constants. They are illustrated by

$$2x - 1 < 5, \quad 4 - 3x \leq 10 \quad \text{and} \quad 4x + 2(x + 3) > 14x - 4(x - 2).$$

Before solving them, we need some properties of inequalities similar to those for equations. The fourth property that follows is somewhat unexpected. Here A, B, and C represent algebraic expressions or constants.

Properties of an Inequality

The inequality $A < B$ has the same solutions as

the inequalities
$$A + C < B + C$$
$$A - C < B - C$$

$$AC < BC \text{ and } \frac{A}{C} < \frac{B}{C}, \quad C \text{ positive}$$

$$AC > BC \text{ and } \frac{A}{C} > \frac{B}{C}, \quad C \text{ negative}$$

All of these properties are true if $<$ is replaced by any of the symbols $>$, $\leq$, and $\geq$. The last property says that **multiplication** and **division** of an inequality by a **negative** quantity **reverses** the direction of the original inequality, whereas it remains the same in all other cases. We can illustrate these properties using numbers. Let us start with

$$6 < 8.$$

Then adding and subtracting 7, say, gives

$$6 + 7 < 8 + 7, \quad \text{or} \quad 13 < 15$$
$$6 - 7 < 8 - 7, \quad \text{or} \quad -1 < 1.$$

Multiplying and dividing by the positive number 2 gives

$$6 \cdot 2 < 8 \cdot 2, \quad \text{or} \quad 12 < 16$$
$$\frac{6}{2} < \frac{8}{2}, \quad \text{or} \quad 3 < 4.$$

Finally, multiplying and dividing by the *negative* number -2 *reverses* the direction of the original inequality; that is, $<$ becomes $>$:

$$6 < 8$$
$$6(-2) > 8(-2), \quad \text{or} \quad -12 > -16$$
$$\frac{6}{-2} > \frac{8}{-2}, \quad \text{or} \quad -3 > -4.$$

Using these properties, we can solve the inequality

$$2x - 1 < 5$$

by first adding 1 and then dividing by the positive number 2:

$$2x - 1 + 1 < 5 + 1 \qquad \text{Add 1.}$$
$$2x \qquad\quad < 6$$
$$\frac{2x}{2} < \frac{6}{2} \qquad \text{Divide by 2.}$$
$$x < 3. \qquad \text{Solution.}$$

The solution $x < 3$ consists of all real numbers less than 3. This is also written in set notation:

$\{x | x < 3\}$, which is read "the set of all numbers x such that $x < 3$."

For a more complete treatment of sets, see Selected Topic A: "Sets," page A1. We can graph this infinite set of numbers by shading in the portion of the real-number line to the left of, but not including, 3.

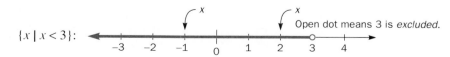

$\{x \mid x < 3\}$:

We use an open dot to show that the point $x = 3$ is excluded from the solution set. To "spot check" our solution, we can choose any number less than 3, say $x = 2$ or $x = -1$, and substitute it into the original inequality $2x - 1 < 5$.

Check $x = 2$: $2(2) - 1 < 5$, or $3 \overset{\checkmark}{<} 5$.

Check $x = -1$: $2(-1) - 1 < 5$, or $-3 \overset{\checkmark}{<} 5$.

EXAMPLE 1

Solve and graph $4 - 3x \leq 10$.

Solution

We proceed as follows:

$$4 - 3x - 4 \leq 10 - 4 \qquad \text{Subtract 4.}$$
$$-3x \quad\;\; \leq 6$$

$$\frac{-3x}{-3} \geq \frac{6}{-3} \qquad \begin{array}{l}\text{Divide by } -3, \\ \text{change } \leq \text{ to } \geq .\end{array}$$

$$x \geq -2. \qquad \text{Solution.}$$

Note that division by the negative number -3 requires that the inequality $\leq$ be *reversed* to $\geq$. The solution set is the shaded portion of the number line to the right of, and including, -2.

$$\{x \mid x \geq -2\}:$$

Closed dot means -2 is *included.*

We use a closed dot to include the point $x = -2$ in the solution set. Let us "spot check" as above, using $x = 1$.

Check $x = 1$: $4 - 3(1) \leq 10$,   or $1 \overset{\checkmark}{\leq} 10$.

EXAMPLE 2

Solve and graph $4x + 2(x + 3) > 14x - 4(x - 2)$.

Solution

We simplify each side separately, and then proceed as above:

$$4x + 2x + 6 > 14x - 4x + 8$$
$$6x + 6 > 10x + 8$$
$$-2 > 4x \qquad\qquad \text{Subtract } 6x, \text{ subtract 8.}$$
$$\frac{-2}{4} > \frac{4x}{4}$$

$$-\frac{1}{2} > x. \qquad\qquad \text{Solution.}$$

The solution $-\frac{1}{2} > x$ is equivalent to $x < -\frac{1}{2}$. This consists of all numbers to the left of, but not including, $-\frac{1}{2}$.

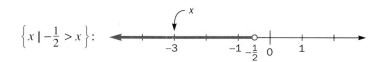

$\left\{ x \mid -\frac{1}{2} > x \right\}$:

Check: You substitute $x = -3$, say, into the original inequality.

The **double** (or **compound**) inequality

$$1 < 3 < 4$$

means that both $1 < 3$ and $3 < 4$. More simply, 3 lies **between** 1 and 4. In the next example we solve such an inequality. We do so by extending the properties used above on two sides of an inequality to the three "sides" of a double inequality.

EXAMPLE 3

Solve and graph $-5 < 3 - 2x < 7$.

Solution
The strategy here is to isolate x in the middle.

$$-5 - 3 < 3 - 2x - 3 < 7 - 3 \qquad \text{Subtract 3.}$$
$$-8 < \quad -2x \quad < 4$$
$$\frac{-8}{-2} > \frac{-2x}{-2} > \frac{4}{-2} \qquad \begin{array}{l}\text{Divide by } -2,\\ \text{change} < \text{to} >.\end{array}$$
$$4 > x > -2. \qquad \text{Solution.}$$

The solution $4 > x > -2$ is equivalent to $-2 < x < 4$. This consists of all numbers between -2 and 4, but not including -2 or 4.

$\{x \mid 4 > x > -2\}$:

Check $x = 1$: $-5 < 3 - 2(1) < 7$,   or $-5 \overset{\checkmark}{<} 1 \overset{\checkmark}{<} 7$.

EXAMPLE 4 | What combinations of 12 coins in dimes and quarters are worth less than $1.95?

Solution
Let x = number of dimes; then $12 - x$ = number of quarters.

We set this up as in Section 2.2, but the statement of the problem leads to an inequality.

	number of coins · face value = value of coins		
dimes	x	10¢	10¢(x)
quarters	$12 - x$	25¢	25¢($12 - x$)

} Total < $1.95

Because the total worth of the coins is less than $1.95, or 195¢, we have the inequality

$$10¢(x) + 25¢(12 - x) < 195¢$$
$$10x + 300 - 25x < 195$$
$$-15x < -105$$
$$\frac{-15x}{-15} > \frac{-105}{-15} \qquad \text{Divide by } -15,$$
$$\text{change } < \text{ to } >.$$
$$x > 7.$$

The number of dimes must be greater than 7. Because there are 12 coins in all, the only possible **answers** are as follows:

dimes	8	9	10	11	12
quarters	4	3	2	1	0
totals	$1.80	$1.65	$1.50	$1.35	$1.20

Note that all totals are less than $1.95.

EXERCISE 2.5   *Answers, pages A37–A38*

A

Solve and graph, as in **EXAMPLES 1, 2, and 3:**

1. $2x + 5 < 1$ **2.** $3x - 2 \leq 7$ **3.** $9 \geq 5x - 1$ **4.** $1 > 4x + 8$

5. $7 - 2x \leq 1$ **6.** $-4 < 8 - 3x$ **7.** $5x - 1 > 2x + 8$ **8.** $2 + 3x \geq 10 - x$

9. $8 - 3x < 2 + 5x$ 10. $2 - 7x > x + 6$ 11. $3x > x$ 12. $3x \le 5x$

13. $-4x \ge 8x$ 14. $-x < 2x$ 15. $-x - 9 > x - 9$ 16. $x + 2 \ge 2 - x$

17. $2x - 3 < x - 2(3x - 2)$ 18. $5(4 + 2x) > 3(x - 7) - 1$

19. $5(1 - 4x) - 2(x - 5) \ge 2(x + 3) + 3(1 + 2x)$ 20. $2(1 + 2x) - (1 - 2x) \le 3(2x - 7) + 4(5x + 3)$

21. $5x - 3[2x - 3(x - 1)] > 4 + 2[x - 2(3 + x)]$ 22. $1 + 5[3 - 2(1 - x)] \le 3x - 2[x - 3(2x - 1)]$

23. $0.7x + 1.1 > -1.3 - 0.5x$ 24. $0.2x - 1.5 \le 1.1x + 0.3$

25. $x + 4.7 \le 0.2x + 2.3$ 26. $1.5x - 2 < 0.25x + 0.2$

27. $5 < 2x - 1 < 9$ 28. $-3 < 2x + 3 < 7$

29. $-15 \le 2x + 1 \le 5$ 30. $14 \le 3x - 2 \le 28$

31. $3 < 1 - 2x \le 7$ 32. $-5 \le 3 - 2x < 1$

33. $7 > 1 - 3x \ge 1$ 34. $8 \ge 5x + 8 > -2$

Set up and solve, as in EXAMPLE 4:

35. What combinations of 9 coins in nickels and dimes are worth more than $0.70?

36. What combinations of 13 coins in nickels and quarters are worth between, but not equal to, $1.65 and $2.65?

37. $10,000 is to be invested, part at 8% for 2 years, and the balance at 7% for 3 years. For tax purposes, it has been determined that the total interest earned must be between $1800 and $1950. Within what range of money must the investment be at 8%?

38. $7000 is invested, part at 5% for 4 years, and the balance at $7\frac{1}{2}\%$ for 2 years. If the interest on the 5% investment is greater than the interest on the $7\frac{1}{2}\%$ investment, how much is in the first investment?

39. A Ford and a Chevy leave town at 1:00 P.M. traveling 50 and 60 mph, respectively, in opposite directions. Between what times of day will they be 55 and 220 miles apart?

40. In a business, profit is made only when revenue exceeds cost:

$$\text{revenue} > \text{cost}$$

A company manufacturing sneeze pins earns $15x$ for selling x pins at $15 per pin. The cost to manufacture x pins is $5x + 20{,}000$.
   a. What is the least number of pins that the company must sell in order to realize a profit?
   b. What is the "break even" point?

41. The length of a rectangle is 2 more than the width. Within what range must the width lie if the perimeter lies between 12 and 26, inclusive?

B

42. Your algebra professor awards an A in the course if your total points are 90% or more, of the total possible. Suppose your scores going into the final exam are 95, 97, 88, 100, 92, and 93 (each out of 100 points). You are shooting for an A in the course but are afraid of "choking" on the final, which is worth 200 points. What is the lowest score you can get on the final and still get an A in the course?

43. Going into the final exam your scores are 78, 61, 88, 93, 70, 52, and 83 (each out of 100 points). You are hoping for a B in the course, which requires that your total points be 80% or more, of the total possible, but less than 90%. The final is worth 200 points, and you are allowed to drop your worst previous score. Within what range must your final exam score fall in order for you to receive a B in the course?

Solve:

44. $4 + 3x < 5x < 3x + 8$ *Hint:* First subtract $3x$.   45. $-6 + 5x \le 2x \le 3 + 5x$ *Hint:* First subtract $5x$.

46. $1 - 2x \le x - 2 \le 7 - 2x$ 47. $1 + 2x < 3 < 7 + 2x$ 48. $5 + 2x \ge 3 - x \ge 2x - 7$

The *Trichotomy Law* states that
    "for any real numbers a and b, exactly one of the following options is true:

$$a < b \text{ or } a = b \text{ or } a > b.\text{"}$$

In essence, this law says that any two real numbers can be compared by size. This is obvious, but deciding the correct option can be challenging *without using a calculator.* For example, to compare

$$\frac{3}{4} \quad \text{and} \quad \frac{11}{15},$$

we write each fraction using the L.C.D. = 60:

$$\frac{3}{4} = \frac{3 \cdot 15}{4 \cdot 15} = \frac{45}{60} \quad \text{and} \quad \frac{11}{15} = \frac{11 \cdot 4}{15 \cdot 4} = \frac{44}{60}.$$

Because $\dfrac{45}{60} > \dfrac{44}{60}$, we conclude that $\dfrac{3}{4} > \dfrac{11}{15}$.

Calculator check: $\dfrac{3}{4} = 0.75 \overset{\checkmark}{>} 0.7333 \ldots = \dfrac{11}{15}.$

In like manner, compare the following pairs of numbers. Use a calculator *only to* check your results.

49. $\dfrac{5}{12}$ and $\dfrac{7}{18}$ 50. $\dfrac{2}{5}$ and $\dfrac{11}{25}$ 51. $\dfrac{7}{16}$ and $\dfrac{11}{24}$ 52. $\dfrac{11}{36}$ and $\dfrac{7}{24}$

53. $\dfrac{-5}{8}$ and $-\dfrac{3}{4}$ 54. $\dfrac{-17}{24}$ and $\dfrac{-13}{18}$ 55. $\dfrac{27}{47}$ and $\dfrac{4}{7}$ 56. $\dfrac{-8}{29}$ and $\dfrac{-3}{11}$

57. $\dfrac{26}{39}$ and $\dfrac{34}{51}$ 58. $\dfrac{-119}{68}$ and $\dfrac{-133}{78}$ 59. 0.12 and $0.\overline{12}$ 60. $0.1\overline{2}$ and $0.\overline{12}$

The *Triangle Inequality* states that
   "the sum of any two sides of a triangle must be greater than the third side."
The following illustrates this:

$$3 + 5 > 7$$
$$3 + 7 > 5$$
$$5 + 7 > 3$$

To illustrate further, look at these "triangles":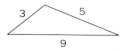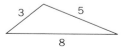

These are *not triangles* at all, because the two shorter sides are *too short*!

$$3 + 5 < 9 \qquad\qquad 3 + 5 = 8$$

Use the Triangle Inequality to decide which of the following sets of numbers can be the sides of a triangle. Assume a and b are positive.

61. 6, 7, 12

62. 3, 11, 8

63. 23, 65, 41

64. 279, 515, 137

65. $\dfrac{1}{5}, \dfrac{1}{3}, \dfrac{1}{4}$ *Hint:* L.C.D.

66. $\dfrac{1}{2}, \dfrac{1}{3}, \dfrac{1}{4}$

67. $\dfrac{3}{4}, \dfrac{1}{6}, \dfrac{2}{3}$

68. $\dfrac{2}{3}, \dfrac{29}{30}, \dfrac{4}{15}$

69. $a, a + 2, 2a + 1$

70. $a, 3a + 1, 2a + 3$

71. $2a + 2, a + 1, 3a + 3$

72. $a + b, 3a + 4b, 2a + 2b$

 C

For triangles with variable sides, the Triangle Inequality leads to a range of values for the variable. For example, in the triangle

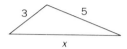

We must have
$$3 + 5 > x \;\longrightarrow\; 8 > x \;\left.\right\}$$
$$3 + x > 5 \;\longrightarrow\; x > 2 \;\left.\right\} \longrightarrow\; 8 > x > 2$$
$$5 + x > 3 \;\longrightarrow\; x > -2 \quad \text{is always true.}$$

Thus we must always have $2 < x < 8$.
In like manner, find the range of values for x in the triangles with the given sides. Assume x is positive.

73. $x, 5, 8$

74. $4, x, 7$

75. $x, x + 1, 9$

76. $x, x + 2, 12$

77. $5, 2x + 1, 8$

78. $x, 2x + 1, 2x + 4$

2.6   Absolute Value Equations and Inequalities

The **absolute value** of a number x is the distance from x to 0 on the number line. This distance is denoted by the symbol $|x|$. For example, the picture shows us that $|3| = 3$ and $|-3| = 3$.

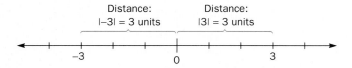

Distance:
$|-3| = 3$ units

Distance:
$|3| = 3$ units

Because "distance" is a measure of length, the absolute value of any number can never be negative. This leads to an equivalent definition of absolute value:

$$|x| = \begin{cases} x & \text{if } x \ge 0 \quad (case\ 1) \\ -x & \text{if } x < 0 \quad (case\ 2) \end{cases}$$

Thus $|3| = 3$ (*case 1*), $|0| = 0$ (*case 1*), and $|-3| = -(-3) = 3$ (*case 2*). In all cases, the absolute value of a number is always non-negative. Clearly, taking the absolute value makes negative numbers positive but leaves positive numbers alone.

EXAMPLE 1

Evaluate these expressions:

a. $|-8| = -(-8)$
$= 8.$

b. $|2 - 7| = |-5|$ Evaluate the inside first.
$= 5.$

c. $\left|\dfrac{-5}{8} - \dfrac{11}{12}\right| = \left|\dfrac{-15}{24} - \dfrac{22}{24}\right|$ L.C.D. = 24

$= \left|\dfrac{-37}{24}\right|$

$= \dfrac{37}{24}.$

d. $|-2| - |-9| = 2 - (9)$ WARNING! $|-2| - |-9| \ne |-2| + |+9|$
$= -7.$

e. $-|5 - |-8|| = -|5 - (8)|$
$= -|-3|$
$= -(3)$ WARNING! $-|-3| \ne 3$
$= -3.$

The equation $|x| = 3$ has solutions $x = 3$ and $x = -3$, as can be easily checked: $|3| = |-3| \overset{\checkmark}{=} 3$. This prototype leads to the following generalization:

> The **absolute value equation**
>
> $$|ax + b| = c, \quad c \text{ positive}$$
>
> is equivalent to the *two equations*
>
> $$ax + b = c, \quad ax + b = -c.$$

EXAMPLE 2

Solve $|3 - 2x| = 9$.

Solution
This is equivalent to the two equations

$$3 - 2x = 9 \qquad 3 - 2x = -9 \qquad \text{Write without } |\ | \text{ symbol.}$$
$$-2x = 6 \qquad -2x = -12$$
$$x = -3 \qquad x = 6$$

Answers: $x = -3, 6$.

Check $x = -3$: $|3 - 2(-3)| = 9$, or $|9| \overset{\checkmark}{=} 9$.
$\phantom{\textbf{Check }}x = 6$: $|3 - 2(6)| = 9$, or $|-9| \overset{\checkmark}{=} 9$.

The inequality $|x| < 3$ consists of all numbers x whose distance from 0 is less than 3 units. As shown below, this consists of all numbers between -3 and 3.

All points are less than 3 units from 0.

Spot check: $|-2| = 2 \overset{\checkmark}{<} 3$.

Thus $|x| < 3$ has solutions given by $-3 < x < 3$. We can generalize:

> The **absolute value inequality,**
>
> $$|ax + b| < c, \quad c \text{ positive}$$
>
> is equivalent to the *double inequality*
>
> $$-c < ax + b < c.$$

This is also true for $\leq$ instead of $<$.

EXAMPLE 3 Solve and graph $|2x + 3| < 7$.

Solution
This is equivalent to the double inequality

$$-7 < 2x + 3 < 7 \qquad \text{Write without } |\ | \text{ symbol.}$$
$$-10 < \quad 2x \quad < 4 \qquad \text{Subtract 3.}$$
$$\frac{-10}{2} < \frac{2x}{2} < \frac{4}{2} \qquad \text{Divide by 2.}$$
$$-5 < x < 2. \qquad \text{Solution.}$$

Written in set notation, the solution is

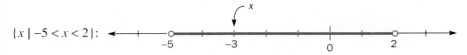

$\{x \mid -5 < x < 2\}$:

Spot check: $x = -3$: $|2(-3) + 3| < 7$, or $|-3| \overset{\checkmark}{<} 7$.

Finally, the inequality $|x| > 3$ consists of all numbers x whose distance from 0 is more than 3 units. This consists of two parts: all points to the right of 3 and all points to the left of -3.

All points $x < -3$ are more than 3 units from 0.

All points $x > 3$ are more than 3 units from 0.

Spot check: $|-5| = 5 \overset{\checkmark}{>} 3$.   **Spot check:** $|5| = 5 \overset{\checkmark}{>} 3$.

Thus $|x| > 3$ has solutions given by $x > 3$ or $x < -3$. We write this as the *union* of two sets:

$$\{x \mid x > 3\} \cup \{x \mid x < -3\}.$$

The symbol $\cup$ denotes the union of sets. See Selected Topic A for a more thorough discussion of sets. We can now generalize:

> The **absolute value inequality**
> $$|ax + b| > c, \quad c \text{ positive}$$
> is equivalent to the *two inequalities*
> $$ax + b > c \quad \text{or} \quad ax + b < -c.$$

As usual, this is true when $\geq$ appears in place of $>$.

EXAMPLE 4

Solve and graph $|7 - 2x| > 3$.

Solution

This is equivalent to the two inequalities:

$7 - 2x > 3$	or	$7 - 2x < -3$	Write without $|\ |$ symbol.
$-2x > -4$		$-2x < -10$	Subtract 7.
$\dfrac{-2x}{-2} < \dfrac{-4}{-2}$		$\dfrac{-2x}{-2} > \dfrac{-10}{-2}$	Divide by -2, reverse direction of $>$ and $<$.
$x < 2$	or	$x > 5$.	

We write the solution as the union of two sets:

$\{x \mid x < 2\} \cup \{x \mid x > 5\}$:

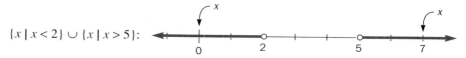

Spot check $x = 0$: $|7 - 2(0)| > 3$, or $|5| \overset{\checkmark}{>} 3$.

$x = 7$: $|7 - 2(7)| > 3$, or $|-7| = 7 \overset{\checkmark}{>} 3$.

WARNING 1! In Example 4,

$$|7 - 2x| > 3 \quad \text{is } not \text{ equivalent to} \quad -3 > 7 - 2x > 3.$$

In fact, there are no numbers at all that satisfy this compound inequality! If there were such a number x, it would mean that $-3 > 7 - 2x$ and $7 - 2x > 3$. According to the **Transitive Law of Inequalities*,** this statement would imply that $-3 > 3$, which is false. Remember,

$$|7 - 2x| > 3 \quad must \text{ be written as} \quad 7 - 2x > 3 \ or \ 7 - 2x < -3.$$

WARNING 2! The solution to Example 4,

$$x < 2 \text{ or } x > 5 \quad \text{is } not \text{ equivalent to} \quad 5 < x < 2.$$

As in Warning 1, this would imply that $5 < 2$, which is false. Remember,

$$x < 2 \text{ or } x > 5 \quad must \text{ be written as} \quad \{x|x < 2\} \cup \{x|x > 5\}.$$

---

*The **Transitive Law of Inequalities** states that if a $>$ b and b $>$ c, then a $>$ c.

Here is a useful summary, in which c is a positive constant.

To solve:	Remove $\| \|$ by solving:	Type of solution:
$\|ax + b\| = c$	$ax + b = c, \quad ax + b = -c$	$x = p, q$
$\|ax + b\| < c$	$-c < ax + b < c$	$\{x \mid p < x < q\}$: ○———○ $p \qquad q$
$\|ax + b\| > c$	$ax + b > c \quad \text{or} \quad ax + b < -c$	$\{x \mid x < p\} \cup \{x \mid x > q\}$: ←——○ ○——→ $p \qquad q$

Again, this is valid for $\leq$ in place of $<$ and for $\geq$ in place of $>$.

EXERCISE 2.6   *Answers, pages A38–A39*

A

Evaluate these expressions, as in EXAMPLE 1:

1. $|9 - 3|$ **2.** $|18 - 5|$ **3.** $|4 - 7|$

4. $|0 - 6|$ **5.** $|3 - 16 - 4|$ **6.** $|-27 - 8 + 35|$

7. $\left| \dfrac{1}{2} - \dfrac{3}{4} \right|$ **8.** $\left| \dfrac{5}{6} - \dfrac{2}{3} \right|$ **9.** $\left| \dfrac{-7}{12} - \dfrac{5}{9} \right|$

10. $\left| \dfrac{13}{10} - \dfrac{9}{25} \right|$ **11.** $\left| \dfrac{11}{12} - \dfrac{7}{8} - \dfrac{15}{16} \right|$ **12.** $\left| -\dfrac{4}{5} - \dfrac{13}{15} + \dfrac{17}{20} \right|$

13. $\left| 2 - \dfrac{5}{18} + \dfrac{3}{10} - \dfrac{1}{4} \right|$ **14.** $\left| \dfrac{3}{8} - 3 + \dfrac{5}{6} - \dfrac{17}{20} \right|$ **15.** $|-3| - |-8|$

16. $-|-7| - |-8|$ **17.** $-|5 - 9|$ **18.** $-|-3 - 4|$

19. $\big| 3 - |-7| \big|$ **20.** $\big| -12 + |-3| \big|$ **21.** $-\big| -7 - |-2| \big|$

22. $-\big| -|-3| - |-5| \big|$ **23.** $-\big| -|2 - 5| + |3 - 9| \big|$ **24.** $\big| -2 - |2 - 6| \big|$

Solve and check, as in EXAMPLE 2:

25. $|x| = 5$ **26.** $|-x| = 10$ **27.** $|-2y| = 12$ **28.** $|3z| = 15$

29. $|x - 1| = 6$ **30.** $|y + 2| = 9$ **31.** $|3t + 2| = 17$ **32.** $|2 + 3t| = 11$

33. $15 = |3 - 3x|$ **34.** $13 = |3 - 2z|$ **35.** $|2x - 8| = 0$ **36.** $0 = |-9 - 2x|$

37. $|1 - 2x| = 4$ **38.** $|4x + 3| = 3$ **39.** $|x - 7| = 0.1$ **40.** $|x + 6| = 0.01$

41. $|2x + 3| = 0.01$ **42.** $|6 - 2t| = 0.2$ **43.** $|2x - 1| + 3 = 8$ *Hint:* First subtract 3.

44. $|3 - 4x| - 6 = 1$ **45.** $15 = 3|3 - 2x|$ *Hint:* First divide by 3.

46. $2|2x + 3| = 8$ **47.** $5|2x + 1| - 3 = 7$ **48.** $4 = 13 - 3|3 - 2x|$

Solve and graph, as in EXAMPLE 3:

49. $|x| < 4$ **50.** $|x| \le 2$ **51.** $|-2x| \le 8$

52. $|-3x| < 12$ **53.** $|x - 1| < 3$ **54.** $|x + 2| \le 1$

55. $|2x - 3| \le 5$ **56.** $|2x + 1| < 1$ **57.** $|3 - x| < 1$

58. $|5 - x| < 2$ **59.** $2 \ge |5 - 2x|$ *Hint:* This is the same as $|5 - 2x| \le 2.$

60. $4 > |1 + 3x|$ **61.** $5 > |-2 - 3x|$ **62.** $3 \ge |3 - 6x|$

63. $|x - 5| < 0.1$ **64.** $|x + 5| < 0.2$ **65.** $|5 + 2x| \le 0.02$

66. $|3 - 2x| \le 0.2$ **67.** $|x - 3| < -1$ **68.** $|x + 2| < 0$

Solve and graph, as in EXAMPLE 4:

69. $|x| \ge 2$ **70.** $|x| > 4$ **71.** $|x - 2| > 2$

72. $|x - 3| \ge 4$ **73.** $|-3x| > 9$ **74.** $|-x| \ge 3$

75. $|5 - 2x| > 1$ **76.** $|4 - 3x| \ge 2$ **77.** $|x - 2| > 0.1$

78. $|x + 3| \ge 0.02$ **79.** $|-7 - 2x| \ge 0.01$ **80.** $|5 - 2x| > 0.2$

B

81. $3 < |3 - 2x|$ *Hint:* This is the same as $|3 - 2x| > 3.$

82. $4 \le |4 + x|$ **83.** $6 \le 1 + |x - 5|$ **84.** $10 < 5|3 - 2x|$

85. $|x - 2| > 0$ **86.** $|x + 3| \ge 0$

The equation $|x| = 2x - 3$ was not treated in the text, because the right side is not a positive constant. We still solve it as before, but we *must check* our answers.

$$
\begin{array}{ll}
x = 2x - 3 & x = -(2x - 3) \\
-x = \quad -3 & x = -2x + 3 \\
x = 3. & 3x = \quad 3 \\
& x = 1.
\end{array}
$$

Check $x = 3$: $|3| = 2(3) - 3$, or $3 \overset{\checkmark}{=} 3.$

            $x = 1$: $|1| = 2(1) - 3$, or $1 \ne -1.$

We see that $x = 3$ satisfies the original equation $|x| = 2x - 3$. However, $x = 1$ does not satisfy it and is called an *extraneous solution*. Extraneous solutions will also appear in Sections 4.5 and 6.5.

Answer: $x = 3$ ($x = 1$ is extraneous).

Solve these equations, and *be sure* to check your answers:

87. $|x| = 2x + 6$ **88.** $|-x| = 3x - 12$ **89.** $|-3x| = x - 12$ **90.** $|4x| = 2x + 8$

91. $|x - 2| = -2$ **92.** $|2x + 1| = -3$ **93.** $|z| = 16 - 3z$ **94.** $|x| = 2x - 30$

C

95. $|y + 9| = y - 1$ **96.** $|8 - x| = x + 4$ **97.** $|x + 8| = |x - 2|$ *Hint: $x + 8 = x - 2$ or $-(x - 2)$.*

98. $|2 + y| = |2y - 7|$ **99.** $|x - 6| = |12 - 2x|$ **100.** $|5 - x| = |2x + 10|$ **101.** $\left|1 + |2 - x|\right| = 7$

102. $\left||x + 3| - 2\right| = 5$ **103.** $\left||x + 3| - 6\right| = 1$ **104.** $\left||x - 4| - 5\right| = 2$

105. The expressions

$$\frac{a + b + |a - b|}{2} \quad \text{and} \quad \frac{a + b - |a - b|}{2}$$

always equal the *larger* and *smaller,* respectively, of the two numbers a and b. For example, if $a = 5$ and $b = 8$, then

$$\frac{5 + 8 + |5 - 8|}{2} = \frac{13 + |-3|}{2} = \frac{13 + 3}{2} = 8 \quad \text{(the larger)}$$

and

$$\frac{5 + 8 - |5 - 8|}{2} = \frac{13 - |-3|}{2} = \frac{13 - 3}{2} = 5 \quad \text{(the smaller)}.$$

In like manner, verify these for the following pairs of numbers: $a = 9$, $b = 3$; $a = -1$, $b = 4$; $a = -2$, $b = -7$; and $a = 5$, $b = 5$.

106. It is a fact that for any two numbers a and b,

$$|a + b| \le |a| + |b|.$$

a. Pick any two positive numbers and verify that $|a + b| = |a| + |b|$.
b. Pick any two negative numbers and verify that $|a + b| = |a| + |b|$.
c. Pick any two numbers with opposite signs and verify that $|a + b| < |a| + |b|$.

107. In the text, $|a|$ was defined to be the distance between a and 0 on the number line. We can extend this as follows:

$$|a - b| = \text{the } \textbf{distance} \text{ between any two points } a \text{ and } b.$$

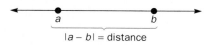

For example, the distance between $a = -1$ and $b = 4$ is 5 units.

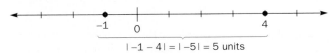

Find the distance between these pairs of points using absolute value:

$a = 2$, $b = 7$; $a = 3$, $b = -4$; $a = -5$, $b = 0$; $a = \dfrac{-3}{4}$, $b = \dfrac{-7}{8}$; and

$a = \dfrac{-5}{6}$, $b = -\dfrac{5}{6}$.

CHAPTER 2 · SUMMARY

2.1 First-Degree Equations

Example:

"Three less than five times a number is twice the sum of the number and six. Find the number."

Bring equation to the form

$$ax = b$$

and then divide by a.

Solution:

$$\text{"}5x - 3 = 2(x + 6)\text{"}$$
$$5x - 3 = 2x + 12$$
$$5x - 2x = 3 + 12$$
$$3x = 15$$
$$x = 5.$$

2.2 Coin and Ticket Problems

Example:

"Eighteen coins in nickels and dimes are worth $1.25. How many of each coin are there?"

Solution:

	number of coins	·	face value	=	value of coins
nickels	x		5¢		$5x$¢
dimes	$18 - x$		10¢		$10(18 - x)$¢

$$5x + 10(18 - x) = 125 \qquad (\$1.25 = 125¢)$$
$$\vdots$$
$$x = 11 \text{ nickels}$$
$$18 - x = 7 \text{ dimes.}$$

2.3 Investment Problems

Example:

"$5000 is invested, part at 6% for 2 years and the rest at 11% for 3 years. If the total interest is $915, how much is invested at each rate?"

Solution:

	P	·	r	·	t	=	I
6% inv.	x		0.06		2		$0.06(2)x$
11% inv.	$5000 - x$		0.11		3		$0.11(3)(5000 - x)$

$$0.12x + 0.33(5000 - x) = 915$$
$$12x + 33(5000 - x) = 91500$$
$$\vdots$$
$$x = \$3500 \text{ at } 6\%$$
$$5000 - x = \$1500 \text{ at } 11\%.$$

2.4 Distance-Rate-Time Problems

Example:

"A Ford and a Chevy leave town in opposite directions. The Ford goes 20 mph faster than the Chevy. After 3 hours, they are 360 miles apart. Find the rate of each car."

Solution:

	R	·	T	=	D
Chevy	x		3		$3x$
Ford	$x + 20$		3		$3(x + 20)$

$$3x + 3(x + 20) = 360$$
$$\vdots$$
$$x = 50 \text{ mph} \quad \text{(Chevy)}$$
$$x + 20 = 70 \text{ mph} \quad \text{(Ford)}$$

CHAPTER 2 ▪ SUMMARY

2.5 Linear Inequalities

When dividing or multiplying an inequality by a *negative* number, *reverse* the direction of the inequality symbol.

Example:

$$1 - 2x < 7$$
$$-2x < 6$$
$$\frac{-2x}{-2} > \frac{6}{-2} \quad \text{Change} < \text{to} >.$$
$$x > -3$$

$\{x \mid x > -3\}$:

2.6 Absolute Value Equations and Inequalities

$$|x| = \begin{cases} x & \text{if} \quad x \geq 0 \\ -x & \text{if} \quad x < 0 \end{cases}$$

Example:

$$|6 - 2| = |4| = 4$$
$$|-3 - 4| = |-7| = 7$$

$|ax + b| = c$ is equivalent to

$$ax + b = c, \quad ax + b = -c$$

Example:

$|x - 3| = 5$ is equivalent to

$$x - 3 = 5, \quad x - 3 = -5$$
$$x = 8, \quad\quad x = -2$$

$|ax + b| < c$ is equivalent to

$$-c < ax + b < c$$

Example:

$|x + 2| \leq 1$ is equivalent to

$$-1 \leq x + 2 \leq 1$$
$$-3 \leq x \leq -1$$

$\{x \mid -3 \leq x \leq -1\}$:

$|ax + b| > c$ is equivalent to

$$ax + b > c \quad \text{or} \quad ax + b < -c$$

Example:

$|x - 4| > 1$ is equivalent to

$$x - 4 > 1 \quad \text{or} \quad x - 4 < -1$$
$$x > 5 \quad \text{or} \quad x < 3$$

$\{x \mid x < 3\} \cup \{x \mid x > 5\}$:

CHAPTER 2 · REVIEW EXERCISES   *Answers, pages A39–A40*

2.1

Solve:

1. $5x - 11 = 5 + x$
2. $19 - 3y = 1 + y$
3. $5(z + 4) = 2(z + 10)$
4. $8x - 2(3 - x) = 5(2x + 3)$
5. $2 + 4(1 + 3x) = 6(2x + 1)$
6. $w = 2w$
7. $2.1x - 1.2 = 0.5x - 0.4$
8. $4z - 2[8z - 5(1 - z)] = 2[z - 2(3z - 1)]$

9. Decreasing five times a number by six is the same as increasing nine by twice the number. Find the number.

10. Nine more than twice the sum of a number and five equals seven less than three times the difference between seven and the number. What is the number?

11. The difference between nine and three times a number equals three times the difference between the number and nine. What is the number?

12. One number is two more than another. Twice the larger is one less than three times the smaller. Find each number.

13. A certain number is increased by three; the result is doubled and then decreased by four; this result is tripled and then increased by twice the number. The result is one less than the number. Find the number.

14. Of three numbers, the second is three more than the first, and the third is one less than the sum of the first two numbers. The sum of the three numbers is 25. Find each number.

15. The sum of two numbers is 10. Three times the smaller equals twice the larger. What are the numbers?

16. Increasing twice the difference between a number and five by six is the same as decreasing three times the sum of twice the number and five by three. Find the number!

2.2

Solve:

17. A collection of 21 coins in dimes and quarters is worth $4.50. How many of each coin are there?

18. Tickets to a concert cost $20 and $35. If there were 13,500 sold, and the revenue was $330,000, how many tickets were sold at each price?

19. A collection of 27 coins in nickels and dimes is such that the total value of all the nickels equals the total value of all the dimes. How many of each coin are there?

20. At a football game, there were 60,000 general admission tickets sold and 15,000 reserved tickets. The reserved seats cost $10 more than the general admission, but the revenue from the cheaper seats was twice as much as from the expensive seats. Find the cost of each ticket.

21. In a collection of coins, there are two more nickels than quarters and nine more dimes than quarters. The total value of the nickels and quarters equals the value of the dimes. How many of each coin are there?

22. In a collection of 25 coins, there are an equal number of nickels and quarters, the rest being dimes. The total value of the coins is $3.20. How many of each coin are there?

23. A cash register contains two more $5 bills than $10's and twice as many $20's as $5's. The total value of the bills is $585. How many of each bill are there?

24. Tickets to a concert cost $15, $25, and $40. The promoters project that approximately 2000 more $40 seats will be sold than $15 seats, and three times as many $25 seats as $40 seats. Their projected revenue from ticket sales is approximately $620,000. How many at each price do they expect to sell?

25. A retailer ordered 40 pairs of jeans, specifying a certain number of women's jeans costing $12 apiece wholesale and the rest men's jeans costing $10 apiece. By mistake, she was charged $10 for the women's jeans and $12 for the men's. She didn't complain, because she paid $20 less than expected for the order. How many of each pair did she order?

2.3

Solve:

26. $20,000 is invested, part at 11% for 2 years and the balance at 9% for 3 years. The total interest earned is $5025. How much is invested at each rate?

27. A certain amount of money was deposited into a 6% account. After a year, the account was closed and the principal, together with $2000, was invested in a stock paying 8% for 2 years. The total interest earned was $1420. How much money was invested at each rate?

28. $19,000 was invested. A certain amount went into a municipal bond paying 5% for 3 years. An equal amount plus $1000 went into a computer stock paying 8% for 2 years. The balance went into a real estate partnership, which lost 14% and was dissolved after only 1 year. The entire investment *broke even*. How much went into each investment?

29. $36,000 is invested, part at 8% for 5 years and the rest at 5% for 10 years. Both investments earn the same amount of interest. How much is invested at each rate?

30. $15,000 is invested, part at 5%, twice that amount at 7%, and the balance at 4%. The investments are terminated after 3, 1, and 4 years, respectively. The interest on the 4% investment is $90 more than the total interest on the other two investments. How much is invested at each rate?

31. $5000, $8000, and $10,000 are deposited into accounts paying 6%, 9%, and 7%, respectively. After a certain number of years, the first account is closed but the others are kept open. One year later the second account is closed but the third account is kept open. Two years after that, it is closed. The total interest earned is $7980. For how many years is each account open?

2.4

Solve:

32. A Ford and a Chevy leave town at the same time but travel in opposite directions. The Ford moves 10 miles per hour slower than the Chevy. After 3 hours they are 360 miles apart. Find the rate of each car.

33. Two men rob a bank and make their getaway at 11:00 A.M., traveling 85 miles per hour. At 1:00 P.M., the police leave the bank, traveling 105 mph in pursuit of the robbers. How long does it take for the cops to catch the robbers? At what time of day does this happen, and how far from the bank is the arrest made?

34. On a 405-kilometer trip, a driver averaged 90 kilometers per hour in the city and 135 kph on the highway. The trip took 3 hours 30 minutes. Find her time spent in each case and the distance at each rate.

35. You drive to work averaging 60 miles per hour, but traffic forces you to reduce your speed by 10 mph on the drive home. As a result, the trip home takes 15 minutes longer. Find the distance from home to work.

36. Two men rob a bank and make their getaway at 9:00 A.M. At 10:00 A.M., the police leave the bank in pursuit of the crooks. The cops drive 20 miles per hour faster than the robbers, and at 2:00 P.M. they catch the robbers. Find the speed of each.

37. You drive to the airport at 60 miles per hour, board a plane that travels 475 mph, and then take a bus to your hotel averaging 30 mph. You spend equal time driving and in the bus, and you cover 2420 total miles in 6 hours. Find the time you spent, and the distance you traveled, in each of these three vehicles.

2.5

Solve and graph:

38. $3x - 1 < 8$ 39. $2 + 5x \geq x - 10$ 40. $7 - 2x > 1$ 41. $9 - x \leq 5x - 3$

42. $4x \leq 2x$ 43. $3(2x - 1) - 2(x - 4) < 7 + 2(3 + 4x)$ 44. $-3 < 2x - 1 < 5$

45. $0 \leq 3 - 2x \leq 7$ 46. $2 \leq 2 - 3x < 8$ 47. $3 + x < 4x < x + 9$ *Hint:* First subtract x.

48. $-4 + 3x \leq x \leq 6 + 3x$ 49. $7 - x < 3x - 1 < 11 - x$

50. What combinations of 14 coins in nickels and dimes are worth less than $1.00?

51. What combinations of 12 coins in dimes and quarters are worth between $1.50 and $2.25, inclusive?

52. \$20,000 is to be invested, part at 6% for 3 years, and the balance at $7\frac{1}{2}\%$ for 2 years. How much should be invested at 6% if the total interest earned must be more than \$3360?

53. Cars A and B, traveling 55 and 65 miles per hour, respectively, are 240 miles apart and headed toward each other. Between what times are they within 60 miles of each other? Assume that they stop when they meet.

54. Going into the final exam, your algebra exam scores are 81, 93, 89, 85, and 91 (each is out of 100 points). To receive a B in the course, you must get 80%, or more, of the total possible points, but less than 90%. If the final exam is worth 200 points, within what range must your score fall in order for you to wind up with a B?

55. A company manufacturing fniffney rods finds that it costs $8x + 35,000$ to produce x rods. Its revenue for selling x rods at \$12 per rod is $12x$. What is the least number of rods that it can sell in order to make a profit? What is the "break even" point? *Hint:* Profit occurs when revenue exceeds cost:

$$\text{revenue} > \text{cost}$$

Using one of the symbols <, =, or >, compare the following pairs of numbers by size, *without using a calculator.* Hint: Use L.C.D.'s.

56. $\dfrac{11}{32}$ and $\dfrac{17}{48}$ **57.** $\dfrac{7}{24}$ and $\dfrac{5}{16}$ **58.** $-\dfrac{7}{9}$ and $\dfrac{-5}{6}$

59. $0.2\overline{1}$ and $0.\overline{21}$ **60.** $0.12\overline{1}$ and $0.\overline{121}$ **61.** $0.\overline{12}$ and $0.\overline{1212}$

Use the Triangle Inequality to decide which of the following numbers can be the sides of a triangle. Assume a and b are positive.

62. 5, 12, 14   **63.** 19, 85, 104   **64.** 32, 129, 162

65. $\dfrac{1}{2}, \dfrac{1}{3}, \dfrac{1}{6}$ **66.** $1, \dfrac{1}{2}, \dfrac{3}{4}$ **67.** $\dfrac{1}{6}, \dfrac{3}{4}, \dfrac{5}{12}$

68. $a, a + 3, 2a + 2$ **69.** $a + b, a + 2b, 2a + 3b$ **70.** $b + 1, 2b + 4, b + 2$

Use the Triangle Inequality to find the range of values for x in the triangles shown. Assume x is positive.

71. 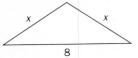   **72.** 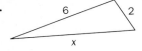   **73.**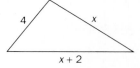

2.6

Evaluate these expressions:

74. $|2 - 8|$ **75.** $\big|3 - |-5|\big|$ **76.** $\left|\dfrac{1}{6} - \dfrac{3}{4} - \dfrac{5}{8}\right|$

77. $|-4| - |-7|$ **78.** $-\Big||2 - 7| - |-1 - 3|\Big|$ **79.** $\Big|-\big|2 - |3 - 6|\big|\Big|$

Solve and check:

80. $|x-2| = 7$

81. $|3 - 2x| = 11$

82. $10 = |1 + 3x|$

83. $|-2x - 1| - 3 = 8$

84. $12 = 2|3x - 2|$

85. $3|2x + 1| + 1 = 10$

86. $|-3x| = x + 12$

87. $|2x + 1| = x - 5$

88. $|2x - 1| = |5 - x|$

Solve and graph:

89. $|x + 1| < 4$

90. $|3 - 2x| \le 1$

91. $|x - 5| < 0.1$

92. $5 \ge |1 - x|$

93. $|2x - 1| > 7$

94. $|2 + 3x| \ge 5$

95. $|-3x| + 1 > 10$

96. $|-4 - 2x| \ge 0.02$

97. $2 < |2 - x|$

98. True or false: $|a - b| = |b - a|$? Experiment with numbers with the same sign and with opposite signs.

CHAPTER 2 · TEST   *Answers, pages A40–A41*

Evaluate:

1. $-|-3| - |2 - 7|$

2. $\left| \frac{3}{8} - \frac{5}{6} - 2 + \frac{7}{12} \right|$

3. $\dfrac{-6 + 2 + |-6 - 2|}{2}$

4. $\dfrac{-6 + 2 - |-6 - 2|}{2}$

Solve and check:

5. $6 - 2x = 2x - 6$

6. $6 + 2x = 6 - 2x$

7. $6 + 2x = 2x - 6$

8. $6 + 2x = 2x + 6$

9. $5x - 2(x - 3) = 3(2x + 1)$

10. $|2x - 1| = 13$

11. $|6 - x| = 0.1$

12. $|x + 9| = 2x$

13. $|x - 4| = 2x + 1$

Solve and graph:

14. $3 - 2x < 7$

15. $|3x + 1| \le 4$

16. $|1 - 2x| > 5$

17. In a collection of coins, there are three more nickels than dimes and twice as many quarters as nickels. The total value is $6.85. How many of each coin are there?

18. $10,000 is to be invested, with equal amounts at 6% and $9\frac{1}{2}$% and the balance at 7%. The first investment lasts 3 years, and the others last 2 years each. The total interest is $1670. How much is invested at each rate?

19. A Nissan 300 ZX and a Toyota Supra, traveling 65 and 70 miles per hour, respectively, left town in opposite directions. The Nissan left at 11:00 A.M. and the Toyota left at 1:00 P.M. At what time of day were they 400 miles apart?

20. Three times a number lies strictly between the sum of the number and four and the sum of the number and twelve. The number must range between what two other numbers?

3 POLYNOMIALS

3.1 Multiplying Polynomials

Polynomials were introduced in Section 1.4. There, we said that $x^3 + x^4$ cannot be further combined because of unlike terms. We can, however, multiply them as follows: $x^3 \cdot x^4 = xxx \cdot xxxx = x^7$. See Exercise 1.4, Problems 45 and 46. Likewise, $y^2 \cdot y^3 = yy \cdot yyy = y^5$. In each case, we see that when powers with the same base are multiplied, the result is a power with the same base and with an exponent equal to the *sum* of the original exponents.

> **Product Rule of Exponents**
> $$a^m a^n = a^{m+n}$$

EXAMPLE 1

a. $x^5 x^7 = x^{5+7}$
$\qquad = x^{12}$.

b. $a^3 a^9 a = a^{3+9+1}$ Note: $a = a^1$
$\qquad = a^{13}$.

c. $(a + b)^8 (a + b)^{12} = (a + b)^{20}$ **WARNING!**
$\qquad\qquad\qquad\qquad\qquad\qquad (a + b)^{20} \neq a^{20} + b^{20}$

d. $10^7 \cdot 10^{11} = 10^{18}$.   **WARNING!**
$\qquad\qquad\qquad\qquad\qquad 10^7 \cdot 10^{11} \neq 100^{18}$

e. $2x^3 y^2 \cdot 5xy^7 = 2 \cdot 5 \cdot x^3 \cdot x \cdot y^2 \cdot y^7$ Commutative Property.
$\qquad\qquad = 10x^4 y^9$.

f. $(-2x)(-3xy)(-y^2) = (-2x)(-3xy)(-1y^2)$ Note: $-y^2 = -1y^2$
$\qquad\qquad\qquad = (-2)(-3)(-1) \cdot x \cdot x \cdot y \cdot y^2$ Commutative Property.
$\qquad\qquad\qquad = -6x^2 y^3$.

g. $(2x + 1)^5 (x - 2)^4 \cdot (2x + 1)^7 (x - 2)^6$ Note binomials used as
$\qquad\qquad\qquad = (2x + 1)^{12}(x - 2)^{10}$.   a base.

h. $a^{3m} b^{n+1} \cdot a^{2m} b^{4n+3} = a^{3m+2m} b^{n+1+4n+3}$ Note variable
$\qquad\qquad\qquad = a^{5m} b^{5n+4}$.   exponents.

We obtain the product of a monomial and a polynomial by using the Distributive Property.

EXAMPLE 2

a. $5x^2(6x^2 + 7x) = 5x^2 \cdot 6x^2 + 5x^2 \cdot 7x$
$$= 30x^4 + 35x^3.$$

b. $4ab(7a^2 - 2ab + b^2) = 28a^3b - 8a^2b^2 + 4ab^3.$ Distributive Property

c. $-x^3y(x^3 - 3x^2y + 5x) = -x^6y + 3x^5y^2 - 5x^4y.$

d. $2x(3x^2 - 7x + 1) - 5x^2(4 - 2x)$
$$= 6x^3 - 14x^2 + 2x - 20x^2 + 10x^3$$
$$= 16x^3 - 34x^2 + 2x.$$ Combine like terms.

e. $a^n(3a^{n+2} - 7a^{n+1} + 5a^n)$
$$= 3a^{n+n+2} - 7a^{n+n+1} + 5a^{n+n}$$
$$= 3a^{2n+2} - 7a^{2n+1} + 5a^{2n}.$$

f. $7x^3(y + 1)^2[5x^4(y + 1)^9 - 10x^6(y + 1)^5]$
$$= 35x^{3+4}(y + 1)^{2+9} - 70x^{3+6}(y + 1)^{2+5}$$
$$= 35x^7(y + 1)^{11} - 70x^9(y + 1)^7.$$

We can multiply two binomials by repeated use of the Distributive Property, for example

$$(2x - 3)(x + 8) = 2x(x + 8) - 3(x + 8)$$ Distributive Property
$$= 2x^2 + 16x - 3x - 24$$ Distributive Property again
$$= 2x^2 + 13x - 24.$$

This type of product occurs so frequently that it is worth both a second look and a shortcut. Notice that in the second step there are four terms:

$2x^2 = 2x \cdot x$ is the product of the two *first* terms in each binomial.
$16x = 2x \cdot 8$ is the product of the two *outer* terms.
$-3x = -3 \cdot x$ is the product of the two *inner* terms.
$-24 = -3 \cdot 8$ is the product of the two *last* terms.

We can abbreviate and call this the FOIL pattern, which stands for "**F**irst, **O**uter, **I**nner, **L**ast":

$$(2x - 3) \cdot (x + 8) = 2x^2 + 16x - 3x - 24$$
$$= 2x^2 + 13x - 24.$$

Likewise, $(5a + 3b)(2a + 7b) = 10a^2 + 35ab + 6ab + 21b^2$
$$= 10a^2 + 41ab + 21b^2.$$

In a vast majority of cases, the two middle terms (Outer and Inner) are like terms and, therefore, can be combined. With practice, you should eventually be able to write the answer in one step, by mentally combining these terms.

EXAMPLE 3

a. $(2x + 7)(x + 3) = 2x^2 + 6x + 7x + 21$
$$= 2x^2 + 13x + 21.$$

b. $(5a - b)(2a - 3b) = 10a^2 - 15ab - 2ab + 3b^2$
$$= 10a^2 - 17ab + 3b^2.$$

c. $(x^3 + 5)(x^3 - 7) = x^{3+3} - 7x^3 + 5x^3 - 35$
$$= x^6 - 2x^3 - 35.$$

d. $(2x^m - 3y^n)(x^m + 4y^n) = 2x^{m+m} + 8x^my^n - 3x^my^n - 12y^{n+n}$
$$= 2x^{2m} + 5x^my^n - 12y^{2n}.$$

e. $(5x - 3)(2x - 1) = 10x^2 - 11x + 3.$ Mentally:
$-5x - 6x = -11x$

f. $2ab(3a - b)(a + 9b) = 2ab(3a^2 + 26ab - 9b^2)$ FOIL first.
$$= 6a^3b + 52a^2b^2 - 18ab^3.$$ Distributive Property.

g. $3(x - 3)(x + 11) - (2x + 5)(3x - 10)$
$$= 3(x^2 + 8x - 33) - (6x^2 - 5x - 50)$$ Keep parentheses!
$$= 3x^2 + 24x - 99 - 6x^2 + 5x + 50$$ Note sign changes!
$$= -3x^2 + 29x - 49.$$

h. $(2a + 3)(4a - b) = 8a^2 - 2ab + 12a - 3b.$ Middle terms are unlike.

The square of a binomial, such as $(5x + 3)^2$, can be obtained by FOIL:

$$(5x + 3)^2 = (5x + 3)(5x + 3) = 25x^2 + 15x + 15x + 9 = 25x^2 + 30x + 9.$$

This is a special case of the general form:

$$(A + B)^2 = (A + B)(A + B) = A^2 + AB + AB + B^2 = A^2 + 2AB + B^2.$$

Note that the middle term, $2AB$, is *twice* the product of the terms inside the parentheses. Thus we can write directly without using FOIL:

$$(5x + 3)^2 = 25x^2 + 2 \cdot 5x \cdot 3 + 9 = 25x^2 + 30x + 9.$$

A similar formula holds for $(A - B)^2$. We list both of these *special products*, which you should memorize:

Square of a Binomial

$$(A + B)^2 = A^2 + 2AB + B^2$$
$$(A - B)^2 = A^2 - 2AB + B^2$$

EXAMPLE 4

a. $(4x + 3)^2 = 16x^2 + 2 \cdot 4x \cdot 3 + 9$
$= 16x^2 + 24x + 9.$

b. $(a + 8b)^2 = a^2 + 2 \cdot a \cdot 8b + 64b^2$
$= a^2 + 16ab + 64b^2.$

c. $(x - 7y)^2 = x^2 - 2 \cdot x \cdot 7y + 49y^2$
$= x^2 - 14xy + 49y^2.$

d. $(2x^2 - 11y^3)^2 = 4x^4 - 2 \cdot 2x^2 \cdot 11y^3 + 121y^6$ $x^2x^2 = x^4, \quad y^3y^3 = y^6$
$= 4x^4 - 44x^2y^3 + 121y^6.$

e. $(x^n + y^n)^2 = x^{2n} + 2x^ny^n + y^{2n}.$ $x^nx^n = x^{n+n} = x^{2n}$

f. $-5a^2b(2a - 3b)^2 = -5a^2b(4a^2 - 12ab + 9b^2)$ Square the binomial.
$= -20a^4b + 60a^3b^2 - 45a^2b^3.$ Distributive Property.

g. $2(x + 3)^2 - 3(2x - 5)^2$
$= 2(x^2 + 6x + 9) - 3(4x^2 - 20x + 25)$ Square each binomial.
$= 2x^2 + 12x + 18 - 12x^2 + 60x - 75$ Distributive Property.
$= -10x^2 + 72x - 57.$ Combine like terms.

h. $[(x - y) + 3]^2 = (x - y)^2 + 2 \cdot 3 \cdot (x - y) + 9$ Square the binomial.
$= x^2 - 2xy + y^2 + 6x - 6y + 9.$ Now square $(x - y)$.

Our next special product is illustrated by

$$(x + 5)(x - 5) = x^2 - 5x + 5x + 25 = x^2 - 25.$$

Note that only two terms remain, because $-5x + 5x = 0$. This is a special case of the general form

$$(A + B)(A - B) = A^2 - AB + AB + B^2 = A^2 - B^2.$$

Thus multiplying the sum and the difference of the same two terms results in the difference of their squares:

Product of a Sum and a Difference

$$(A + B)(A - B) = A^2 - B^2$$

EXAMPLE 5

a. $(x + 7)(x - 7) = x^2 - 7^2$
$= x^2 - 49.$

b. $(3a + 5b)(3a - 5b) = (3a)^2 - (5b)^2$
$= 9a^2 - 25b^2.$

(continued)

c. $(4 - 11x^2)(4 + 11x^2) = 16 - 121x^4$.

d. $(x^3 - 8y^3)(x^3 + 8y^3) = x^6 - 64y^6$. $x^3x^3 = x^6$

e. $(x^n + 3y^m)(x^n - 3y^m) = x^{2n} - 9y^{2m}$. $x^n x^n = x^{2n}$

f. $(x^2 + 4)(x + 2)(x - 2) = (x^2 + 4)(x^2 - 4)$ $(x + 2)(x - 2)$ first.
                                 $= x^4 - 16$.

g. $3xy(2x - 9y)(2x + 9y) = 3xy(4x^2 - 81y^2)$ Parentheses first.
                                 $= 12x^3y - 243xy^3$.   Distributive Property

h. $(x + 5)(x - 5) - (x - 3)(x + 3)$
                  $= (x^2 - 25) - (x^2 - 9)$ Keep parentheses!
                  $= x^2 - 25 - x^2 + 9$ Note sign changes!
                  $= -16$.

i. $[(x + y) + 8][(x + y) - 8]$
                  $= (x + y)^2 - 64$ $(A + B)(A - B) = A^2 - B^2$
                  $= x^2 + 2xy + y^2 - 64$. $(A + B)^2 = A^2 + 2AB + B^2$

Finally, products involving trinomials can be obtained by repeated use of the Distributive Property.

EXAMPLE 6

a. $(x - 3)^3 = (x - 3)(x - 3)^2$
             $= (x - 3)(x^2 - 6x + 9)$
             $= x(x^2 - 6x + 9) - 3(x^2 - 6x + 9)$ Distributive Property
             $= x^3 - 6x^2 + 9x - 3x^2 + 18x - 27$ Distributive Property again
             $= x^3 - 9x^2 + 27x - 27$.

b. $(a^2 - 3a + 4)(a^2 + 2a - 1)$.

We can proceed as above, by using the Distributive Property three times. In this case, however, a vertical format is more convenient.

$$
\begin{array}{r}
a^2 - 3a + 4 \\
a^2 + 2a - 1 \\
\hline
\end{array}
$$

$$
\begin{array}{ll}
a^2(a^2 - 3a + 4) \longrightarrow & a^4 - 3a^3 + 4a^2 \\
2a(a^2 - 3a + 4) \longrightarrow & \quad\;\; 2a^3 - 6a^2 + 8a \\
-1(a^2 - 3a + 4) \longrightarrow & \quad\quad\quad\;\; -\; a^2 + 3a - 4 \\
\hline
& a^4 -\;\;\; a^3 - 3a^2 + 11a - 4.
\end{array}
$$

EXERCISE 3.1 *Answers, page A41*

A

Multiply as in EXAMPLES 1 AND 2:

1. $a^3 a^8$

2. $x^4 x^{13}$

3. $bb^7 b^9$

4. $y^6 y y^2$

5. $(a - b)^7 (a - b)^9$

6. $(x + 3)^6 (x + 3)$

7. $3^5 \cdot 3^7$

8. $10^4 \cdot 10^{13}$

9. $2a^3 b^4 \cdot 7ab^6$

10. $4xy^5 z^2 \cdot 9x^3 y^2 z^5$

11. $(-3m^2 n)(-m^4 n^2)(-5n^4)$

12. $(-p^2 q^3)(5p^4)(-7p^8 q^3)$

13. $(x + 1)^8 (y - 2)^3 \cdot (x + 1)(y - 2)^4$

14. $(a - 2b)^4 (a + 3b)^2 \cdot (a + 3b)(a - 2b)^9$

15. $2x^3 (x - 1)^4 (x + 2)^2 \cdot 3x^7 (x - 1)(x + 2)^8$

16. $5a^2 b(a + b)^4 (a - 2b)^6 \cdot 2ab^4 (a + b)^3 (a - 2b)^8$

17. $a^{3n} b^{5n} \cdot a^{7n} b^n$

18. $x^m y^{3m} \cdot x^{10m} y^{8m}$

19. $-3x^{2n} y^{3m+1} \cdot 8x^{4n-3} y^{m+6}$

20. $2a^{3m} b^{2m} c^{m+1} \cdot 5a^2 b^{m-2} c$

21. $2a(4a^2 - 7a - 11)$

22. $-5x^2 (x^2 + 9x - 7)$

23. $-xy(3x^2 + xy - 3y^2)$

24. $a^2 b(7a^2 - 2ab + 6b^2)$

25. $x^2 y(2x^3 + 7x^2 y - xy^2 - 2y^3)$

26. $-2c^2 d^3 (d^3 - 3c^2 d + 5cd^2 - 11d^3)$

27. $2a^2 (a + b)^3 [5a^3 (a + b)^7 - 7a^4 (a + b)^6]$

28. $x^3 (x - 1)[4x^2 (x - 1)^8 + 7x(x - 1)^9]$

29. $3x(2x^2 - 7x - 8) - 5x(x^2 + 8x - 6)$

30. $a^2 (a^2 - 2ab - b^2) - b^2 (2a^2 + ab - b^2)$

Multiply using FOIL or the Special Products, as in EXAMPLES 3, 4, and 5:

31. $(x + 8)(x + 2)$

32. $(a + 7)(a + 9)$

33. $(2a - b)(a - 5b)$

34. $(3x - 7y)(x - 3y)$

35. $(5y - 2z)(2y + 11z)$

36. $(4m + 7n)(3m - 8n)$

37. $(4x^2 y + 9z^3)(2x^2 y - 5z^3)$

38. $(cd^3 - 11d^2)(3cd^3 + 12d^2)$

39. $(2ab + 13c)(7c - 5ab)$

40. $(12x^2 - 10a^2)(9a^2 - 10x^2)$

41. $(3x + 7y)(3x - 8z)$

42. $(13a + 11)(10a + 11b)$

43. $(2x^n - 7y^m)(3x^n - 8y^m)$

44. $(a^m + 19)(3a^m - 21)$

45. $(x + 9)^2$

46. $(a + 11)^2$

47. $(2a - 5b)^2$

48. $(3x - 7y)^2$

49. $(4x^2 - 7y^3)^2$

50. $(5ab^2 + 3c^3)^2$

51. $(2x^n + 3y^n)^2$

52. $(3a^n - 13b^m)^2$

53. $(2x + 9)(2x - 9)$

54. $(3a + 7b)(3a - 7b)$

55. $(6y - 13b)(6y + 13b)$

56. $(11x - 14a)(11x + 14a)$

57. $(1 + 19xy)(1 - 19xy)$

58. $(3 + 22a^2)(3 - 22a^2)$

59. $(5x^3 - 17y^2)(17y^2 + 5x^3)$

60. $(7a^2 b - 13c^3)(13c^3 + 7a^2 b)$

61. $(2x^n + 3y^m)(2x^n - 3y^m)$

62. $(a^m - 7x^m)(a^m + 7x^m)$

63. $(5x^{2n} - 9y^{3m})(5x^{2n} + 9y^{3m})$

64. $(7a^n b^{2n} + b^{3n})(7a^n b^{2n} - b^{3n})$

65. $3x(2x + 7)(3x - 1)$

66. $4ab(a - 9b)(3a - 11b)$

67. $-2a^2 b(3a - 8b)^2$

68. $-xy(4x + 7y)^2$

69. $3xy(x - 5y)(x + 5y)$

70. $ax(2a - 5x)(5x + 2a)$

71. $2a(2a - 5b)(3a + b) - b(a - 3b)^2$

72. $5(2x + 3y)(2x - 3y) - (x - 5y)^2$

73. $6xy(x - 2y)^2 + 5(xy - 2)^2$

74. $2(3x - 7)(3x - 8) - (x + 9)(x - 9)$

75. $(a^2 + b^2)(a + b)(a - b)$

76. $(x^2 + 9)(x + 3)(x - 3)$

77. $(x - 1)(x + 1)(x^2 + 1)(x^4 + 1)$

78. $(a - 2b)(a + 2b)(a^2 + 4b^2) \cdot (a^4 + 16b^4)$

79. $(x + 3)(x - 3)(2x + 5)(2x - 5)$

80. $(3x - 7)(3x + 7)(2x + 1)(2x - 1)$

81. $[(a - b) + 7]^2$

82. $[(x + 2y) + 5]^2$

83. $[x + 3y - 2z]^2$

84. $[a - 4b + 3c]^2$

85. $[(x - y) + 11][(x - y) - 11]$

86. $[(a + 3) + 5b][(a + 3) - 5b]$

87. $[a + 2b - 7c][a + 2b + 7c]$

88. $[u - v - 5w][u - v + 5w]$

Multiply as in EXAMPLE 6:

89. $(x + 2)(3x^2 - 7x - 13)$

90. $(x - 5)(4x^2 + 5x - 1)$

91. $(a - 2b)(4a^2 + 9ab - 7b^2)$

92. $(2y + z)(8y^2 - 11yz - z^2)$

93. $(2x - 3y)(4x^2 + 6xy + 9y^2)$

94. $(3a + 5b)(9a^2 - 15ab + 25b^2)$

95. $(x + h)^3$

96. $(2x - y)^3$

97. $(x - 1)(x^3 + x^2 + x + 1)$

98. $(a - b)(a^3 + a^2b + ab^2 + b^3)$

99. $(3m^2 + m - 2)(m^2 - 2m + 7)$

100. $(x^2 + 2xy - 3y^2)(4x^2 - 7xy + y^2)$

101. $(x^2 + x + 1)(x^2 - x + 1)$

102. $(a^2 - 2ab + 4b^2)(a^2 + 2ab + 4b^2)$

103. $(x + 3)(2x - 7)(x + 5)$

104. $(2a - b)(3a + b)^2$

B

105. $[(x - 1)(x^2 + x + 1)][(x + 1)(x^2 - x + 1)]$

106. $[(a + 3)(a^2 - 3a + 9)][(a - 4)(a^2 + 4a + 16)]$

107. $(a - 2)^2(a + 2)^2$ *Hint:* Rearrange factors.

108. $(x - 3y)^2(x + 3y)^2$

109. $(x + 1)(x^4 - x^3 + x^2 - x + 1)$

110. $(a - b)(a^4 + a^3b + a^2b^2 + ab^3 + b^4)$

111. $(x + h)^2 - 2(x + h) + 5 - (x^2 - 2x + 5)$

112. $3(x + h)^2 + 4(x + h) - 7 - (3x^2 + 4x - 7)$

C

113. $(x + h)^3 + 2(x + h)^2 - 9(x + h) - 1 - (x^3 + 2x^2 - 9x - 1)$

114. $2(x + h)^3 - (x + h)^2 + 3(x + h) + 9 - (2x^3 - x^2 + 3x + 9)$

 The next set of problems are similar to Exercise 1.3, Problems 47–54. Certain numerical products can be performed quickly using the special product $(A + B)(A - B) = A^2 - B^2$, *without using a calculator.* For example,

$$101 \cdot 99 = (100 + 1)(100 - 1) = 10,000 - 1 = 9999$$
$$82 \cdot 78 = (80 + 2)(80 - 2) = 6400 - 4 = 6396$$

 In like manner, compute the following "quick products" *without using a calculator.*

115. $81 \cdot 79$

116. $51 \cdot 49$

117. $201 \cdot 199$

118. $301 \cdot 299$

119. $62 \cdot 58$

120. $42 \cdot 38$

121. $97 \cdot 103$

122. $196 \cdot 204$

The special products $(A + B)^2 = A^2 + 2AB + B^2$ and $(A - B)^2 = A^2 - 2AB + B^2$ can also be used to find the squares of certain numbers quickly. For example,

$$61^2 = (60 + 1)^2 = 60^2 + 2 \cdot 60 \cdot 1 + 1^2 = 3600 + 120 + 1 = 3721$$
$$48^2 = (50 - 2)^2 = 2500 - 200 + 4 = 2304$$

Compute these "quick products" *without using a calculator.*

123. 51^2 **124.** 31^2 **125.** 101^2 **126.** 201^2 **127.** 89^2 **128.** 39^2

129. 99^2 **130.** 199^2 **131.** 62^2 **132.** 102^2 **133.** 58^2 **134.** 98^2

135. The product of two binomials, $(a + b)(c + d) = ac + ad + bc + bd$, obtained by FOIL, has a geometric interpretation similar to Exercise 1.4, Problem 55. Give such an interpretation based on the following picture:

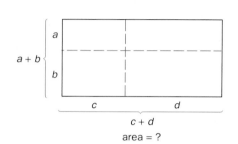 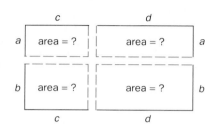

136. In like manner, give a geometric interpretation of the formula for the square of a binomial, $(a + b)^2 = a^2 + 2ab + b^2$, based on the following picture:

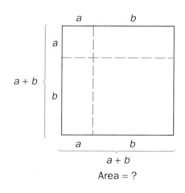 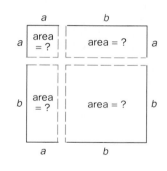

3.2   Common Factors and Grouping

The Distributive Law allows us to multiply:

$$3x(5x + 2) = 3x \cdot 5x + 3x \cdot 2 = 15x^2 + 6x.$$

Written in reverse, this is

$$15x^2 + 6x = 3x \cdot 5x + 3x \cdot 2 = 3x(5x + 2).$$

We have **factored** the polynomial sum $15x^2 + 6x$ into the product $3x(5x + 2)$, and $3x$ is the **common factor**.

Likewise, we can factor

$$8a^4 + 12a^3 - 4a^2 = 4a^2(2a^2 + 3a - 1),$$

as you can easily check by multiplying out. Alternatively, we could have factored $8a^4 + 12a^3 - 4a^2$ as $2a(4a^3 + 6a^2 - 2a)$, $4a(2a^3 + 3a^2 - a)$, $4(2a^4 + 3a^3 - a^2)$, or $\frac{1}{2}(16a^4 + 24a^3 - 8a^2)$. However, it is preferable to factor out the **greatest common factor,** which is the monomial of largest degree and largest integer factor that is a factor of each term in the original polynomial. Thus $4a^2$ is the greatest common factor (G.C.F.)

The **greatest common factor** of a polynomial can be found by these steps:

Step 1 Determine the largest integer that is a factor of each coefficient.
Step 2 If a variable (or variable expression) occurs in each term, determine the *smallest* power of that variable.
Step 3 The G.C.F. is the product of the results of steps 1 and 2.

Once you have found the G.C.F., factor it out and determine what remains in parentheses. Finally, check your answer by using the Distributive Law.

EXAMPLE 1 Factor out the greatest common factor:

a. $20x^3 - 15x^2 + 5x = 5x(4x^2 - 3x + 1).$ x is smallest power.

b. $30a^4b^3 + 45a^2b^4 = 15a^2b^3(2a^2 + 3b).$ a^2 and b^3 are smallest powers.

c. $x^2y - x^3y^2 + x^3 = x^2(y - xy^2 + x).$ y is not in last term.

d. $3x^{n+2} - 9x^{n+1} + 18x^n = 3x^n(x^2 - 3x + 6).$ x^n is smallest power; $x^n x^2 = x^{n+2}$, $x^n x = x^{n+1}$

In a polynomial, if the term of highest degree is negative, it is desirable that the common factor also be negative. For example,

$$-2x^3 - 8x^2 + 16x = -2x(x^2 + 4x - 8),$$

as you can check by multiplying out. Note that *factoring out a negative* monomial *changes the sign* of each term, leaving a polynomial inside the parentheses with a positive first term.

EXAMPLE 2 Factor out the negative greatest common factor:

a. $-5x^3 + 15x^2 - 25x = -5x(x^2 - 3x + 5).$ Note sign changes.

b. $-8a^3b^2 + 30ab^3 = -2ab^2(4a^2 - 15b).$

c. $-x^2 - x - 3 = -1(x^2 + x + 3).$ No other common factor.

The two preceding examples involve common factors that are monomials. We now extend these to common factors that are binomials.

EXAMPLE 3

Factor out the greatest common binomial factor:

a. $7x(y + 9) + 3(y + 9) = (y + 9)(7x + 3).$ $(y + 9)$ is the G.C.F.

b. $2(a - b)^5 + 18(a - b)^4 - 6(a - b)^3 =$
$\qquad 2(a - b)^3[(a - b)^2 + 9(a - b) - 3].$ $2(a - b)^3$ is the G.C.F.

Common factoring can be applied repeatedly to a polynomial containing four terms, such as

$$2x^2 + 2xy + 3x + 3y.$$

We start by factoring terms 1 and 2 and terms 3 and 4:

$$2x(x + y) + 3(x + y).$$

We factor out the common binomial factor $(x + y)$ as in Example 3:

$$(x + y)(2x + 3).$$

Thus $2x^2 + 2xy + 3x + 3y = (x + y)(2x + 3).$

We can check this by FOIL: $(x + y)(2x + 3) \overset{\checkmark}{=} 2x^2 + 3x + 2xy + 3y$. This method is called **factoring by grouping,** because we start by factoring groups of two terms.

EXAMPLE 4

Factor by grouping:

a. $x^2 + 5x + ax + 5a = x(x + 5) + a(x + 5)$ Factor terms 1 and 2,
$\qquad\qquad\qquad\quad = (x + 5)(x + a).$ 3 and 4.
$\qquad\qquad\qquad\qquad\qquad\qquad\qquad$ $(x + 5)$ is a
$\qquad\qquad\qquad\qquad\qquad\qquad\qquad$ common factor.

Check by FOIL: $(x + 5)(x + a) \overset{\checkmark}{=} x^2 + ax + 5x + 5a.$
WARNING! $x(x + 5) + a(x + 5) \neq (x + 5)(x + 5)(x + a).$

b. $7ab - 14ac + 3bx - 6cx = 7a(b - 2c) + 3x(b - 2c)$
$\qquad\qquad\qquad\qquad\quad = (b - 2c)(7a + 3x).$

c. $6x^2 - 8xy - 15x + 20y = 2x(3x - 4y) - 5(3x - 4y)$ Factor -5 from terms 3 and 4; change signs.
$\qquad\qquad\qquad\qquad\quad = (3x - 4y)(2x - 5).$

d. $5x^2 - 45ax - x + 9a = 5x(x - 9a) - 1(x - 9a)$ Note -1 as common factor.
$\qquad\qquad\qquad\qquad\quad = (x - 9a)(5x - 1).$

Now let us try factoring $10x^2 - 12a + 15ax - 8x$ by grouping.

$$10x^2 - 12a + 15ax - 8x = 2(5x^2 - 6a) + x(15a - 8).$$

Unfortunately, we cannot proceed further with common binomial factors, because the binomials inside the parentheses are different. Let us interchange the terms $-12a$ and $15ax$ and see what happens.

$$10x^2 - 12a + 15ax - 8x = 10x^2 + 15ax - 12a - 8x$$
$$= 5x(2x + 3a) - 4(3a + 2x) \quad \left.\right\} \text{ Note: } 3a + 2x =$$
$$= (2x + 3a)(5x - 4). \qquad\qquad 2x + 3a$$

We could also have interchanged the terms $-12a$ and $-8x$:

$$10x^2 - 12a + 15ax - 8x = 10x^2 - 8x + 15ax - 12a$$
$$= 2x(5x - 4) + 3a(5x - 4)$$
$$= (5x - 4)(2x + 3a).$$

This is the same answer as above.

EXAMPLE 5   Factor $3xy - 4z - 4y + 3xz$ by grouping, after *rearranging terms*. The first two terms have no common factor, so let us interchange the terms $-4z$ and $3xz$.

$$3xy - 4z - 4y + 3xz = 3xy + 3xz - 4y - 4z \qquad \text{Rearrange terms.}$$
$$= 3x(y + z) - 4(y + z) \qquad \text{Note sign changes.}$$
$$= (y + z)(3x - 4).$$

EXERCISE 3.2   *Answers, page A42*

A

Factor out the greatest common factor, as in EXAMPLE 1:

1. $7a^3 - 14a^2$
2. $15x^5 - 30x^6$
3. $2a^3b^2 + 2ab^3 - 4a^2b^2$
4. $6m^4n - 18m^3n^3 + 12mn^2$
5. $4x^2y + 6xy^3 - 10y^2$
6. $8ax^2 + 12a^2x - 16a$
7. $P + Prt$
8. $\pi r + \pi r^2$
9. $26x^3y^2z^3 + 52x^2y^3z^2 - 39x^2yz^4$
10. $34p^3q^2r - 51pq^3r^2 + 68p^2qr^2$
11. $99a^3b^2c^2 - 88a^2b^3c + 132a^2b^2$
12. $91m^3n^2p + 65mn^2p^2 + 104m^3p^3$
13. $5x^5 + 10x^{10} - 15x^{15} + 20x^{20}$
14. $10a^{10} - 20a^{20} - 30a^{30} + 40a^{40}$
15. $4x^{m+2} - 8x^{m+1} + 12x^m$
16. $12y^{2n+2} + 16y^{2n+1} - 8y^{2n}$
17. $a^{n+3} - a^{n+2} + a^{n+1}$
18. $b^{5m} + b^{4m} - b^{3m} - b^{2m}$
19. $x^{5n}y^{m+2} + x^{4n}y^{m+3} + x^{3n}y^{m+4}$
20. $u^{n+6}v^n - u^{n+4}v^{3n} - u^{n+8}v^{2n}$

Factor out the greatest negative common factor, as in EXAMPLE 2:

21. $-14ax - 21ay + 35az$
22. $-10cy^2 + 25cy - 5c$
23. $-6a^3 + 12a^2 - 18a$
24. $-2x^4 - 8x^2 - 4x$
25. $-8a^3b + 20ab^3$
26. $-27x^4y^2 - 36x^3y^3$
27. $-2x^2 - x + 7$
28. $-3a^2b - 6a - 4$
29. $-x^{n+2} - x^{n+1} + x^n$
30. $-a^{n+4} + 3a^{n+3} + a^{n+2}$

Factor out the greatest common binomial factor, as in EXAMPLE 3:

31. $2a(x + y) - 3(x + y)$

32. $7x(a - 2b) + 5y(a - 2b)$

33. $4x(m - n) + 3y(m - n) - z(m - n)$

34. $x^2(x^2 + 1) - x(x^2 + 1) + 3(x^2 + 1)$

35. $5(b - c)^3 - 15(b - c)^2 + 10(b - c)$

36. $24(x + y)^4 + 16(x + y)^3 - 12(x + y)^6$

Factor by grouping, as in EXAMPLE 4:

37. $x^2 + 2x + xy + 2y$

38. $2x^2 + 3x + 2xy + 3y$

39. $xz + yz + 3x + 3y$

40. $ab + ac + bd + cd$

41. $4x^2 - 3x + 4ax - 3a$

42. $5x^2 - ax + 25x - 5a$

43. $ax^2 - axy - 7x + 7y$

44. $2ax^2 - 6x - 9ax + 27$

45. $5abc + 35ab - 3c - 21$

46. $6xy + 4x - 21y - 14$

47. $10u^2v + 15u - 3 - 2uv$

48. $2mnc + mnd - d - 2c$

49. $2ax - bx + 2a - b$ *Hint:* 1 is a common factor of $2a - b$.

50. $3x^2 - 3xy + x - y$

First rearrange terms, then factor by grouping, as in EXAMPLE 5:

51. $ax^2 + bc + acx + bx$

52. $2x^2 + 3y + 6x + xy$

53. $b^2c - 6 + 2bc - 3b$

54. $7ab - 10c - 35 + 2abc$

55. $5b^2y - 30cy + 3bc - 50by^2$

56. $8a^2x + 28bx + 7ab + 32ax^2$

57. $ax^2 - 2yz - 2axz + xy$

58. $10a^2 - bc - 5ac + 2ab$

B

In the expression $8(x - 3)^7(2x + 1)^3 + 6(x - 3)^8(2x + 1)^2$, the greatest common factor of the coefficients is 2, and the lowest powers of the binomial factors are $(x - 3)^7$ and $(2x + 1)^2$. Thus the G.C.F. is $2(x - 3)^7(2x + 1)^2$. We now factor as in Example 1:

$$8(x - 3)^7(2x + 1)^3 + 6(x - 3)^8(2x + 1)^2 = 2(x - 3)^7(2x + 1)^2[4(2x + 1)^1 + 3(x - 3)^1] \quad \text{Factor out G.C.F.}$$
$$= 2(x - 3)^7(2x + 1)^2[8x + 4 + 3x - 9] \quad \text{Multiply inner parentheses.}$$
$$= 2(x - 3)^7(2x + 1)^2(11x - 5) \quad \text{Combine like terms.}$$

The answer is in its most simplified form. In like manner, factor the following expressions:

59. $8(x + 2)^7(x - 3)^6 + 6(x + 2)^8(x - 3)^5$

60. $6(x - 4)^5(x + 1)^3 + 3(x - 4)^6(x + 1)^2$

61. $10(2x - 1)^4(x + 4)^8 + 8(2x - 1)^5(x + 4)^7$

62. $6(x + 7)^5(3x + 1)^4 + 12(x + 7)^6(3x + 1)^3$

63. $6x^5(2x + 5)^4 + 8x^6(2x + 5)^3$

64. $3x^2(3x - 2)^5 + 15x^3(3x - 2)^4$

65. $10x^9(1 - 2x)^4 - 8x^{10}(1 - 2x)^3$

66. $2x(2 - x)^4 - 4x^2(2 - x)^3$

3.3 Factoring Trinomials

This section contains the most commonly used factoring in algebra, as well as the most challenging. As an example, the trinomal $x^2 + 10x + 21$ factors into the product of binomials:

$$x^2 + 10x + 21 = (x + 7)(x + 3).$$

Check this yourself by FOIL multiplication. Though seemingly pulled out of the hat, the two factors were obtained by methods you will now learn.

EXAMPLE 1

Factor $x^2 + 7x + 12$.

Solution

The first term x^2 factors as $x \cdot x$. Because all terms are positive, we must start with $(x + \, ? \,)(x + \, ? \,)$. The last term, 12, factors as $12 = 12 \cdot 1 = 6 \cdot 2 = 4 \cdot 3$, so we try all possibilities:

$$(x + 12)(x + 1) = x^2 + 13x + 12 \qquad \text{Wrong}$$
$$(x + 6)(x + 2) = x^2 + 8x + 12 \qquad \text{Wrong}$$
$$(x + 4)(x + 3) = x^2 + 7x + 12 \qquad \text{Right}$$

Thus $x^2 + 7x + 12 = (x + 4)(x + 3)$ is the correct factoring. You could have predicted the answer right away, because the choice $4 \cdot 3$ contains the only factors that add up to 7, the coefficient of the middle term, $7x$.

EXAMPLE 2

Factor $2x^2 - 11x + 5$.

Solution

The first term, $2x^2$, factors as $2x \cdot x$. In order to obtain the positive last term, $+5$, and the negative middle term, $-11x$, we must start with $(2x - \, ? \,)(x - \, ? \,)$. Fortunately, 5 factors only as $5 \cdot 1$, so the only choices are

$$(2x - 5)(x - 1) = 2x^2 - 7x + 5 \qquad \text{Wrong}$$
$$(2x - 1)(x - 5) = 2x^2 - 11x + 5 \qquad \text{Right}$$

Thus $2x^2 - 11x + 5 = (2x - 1)(x - 5)$ is the correct factoring.

EXAMPLE 3

Factor $3x^2 + 10x - 8$.

Solution

The first term, $3x^2$, factors as $3x \cdot x$. The negative last term, -8, requires the product of opposite signs. Also, $8 = 8 \cdot 1 = 4 \cdot 2$ gives two factorings for the last term. Here are some of the possibilities:

$$(3x + 8)(x - 1) = 3x^2 + 5x - 8 \qquad \text{Wrong}$$
$$(3x - 1)(x + 8) = 3x^2 + 23x - 8 \qquad \text{Wrong}$$
$$(3x + 4)(x - 2) = 3x^2 - 2x - 8 \qquad \text{Wrong}$$
$$(3x - 2)(x + 4) = 3x^2 + 10x - 8 \qquad \text{Right}$$

Thus $3x^2 + 10x - 8 = (3x - 2)(x + 4)$ is the correct factoring.

Note in each of these trial-and-error attempts that we always obtained the correct first and last terms, $3x^2$ and -8, but only the last attempt gave the correct middle term, $+10x$. This resulted from the proper combination of inner and outer terms in the FOIL multiplication:

$$(3x - 2)(x + 4) = \qquad + 12x - 2x \qquad = \qquad + 10x.$$

$$-2x$$

$$+12x$$

Picking the right combination, of course, is the key to factoring trinomials. As you do many of them, you will become skilled at choosing the combination of number and sign that produces the correct middle term.

In the next example, we show how to rule out certain combinations automatically and thus to limit the number of trial-and-error attempts.

EXAMPLE 4

Factor $4x^2 - 3x - 10$.

Solution

This is more challenging than Example 3, because both the first *and* the last terms have two factorings:

$$4x^2 = 4x \cdot x = 2x \cdot 2x \quad \text{and} \quad -10 = -(10 \cdot 1) = -(5 \cdot 2).$$

The middle term, $-3x$, has an *odd-number* coefficient, -3. Hence we can automatically *exclude* the choice $(2x \ ?)(2x \ ?)$, because in this case, the inner and outer terms would both have even numbers as coefficients; therefore, their sum would also have an even-number coefficient and could never be -3. Next we observe that $4x^2 - 3x - 10$ has no common factor. Hence we can *exclude* each of the following choices because they have common factors:

$$(4x \quad 2)(x \quad 5) = 2(2x \quad 1)(x \quad 5)$$
$$\text{and} \quad (4x \quad 10)(x \quad 1) = 2(2x \quad 5)(x \quad 1).$$

So far we have automatically excluded

$$
\begin{array}{lll}
4x^2 - 3x - 10 \neq (2x \quad ?)(2x \quad ?) & \text{Even terms can't produce the} \\
& \qquad \text{odd term } -3x. \\
\neq (4x \quad 2)(x \quad 5) & \text{Can't have a common factor.} \\
\neq (4x \quad 10)(x \quad 1). & \text{Can't have a common factor.}
\end{array}
$$

As in Example 3, we must have opposite signs to obtain the negative last term, -10. To yield the negative middle term, $-3x$, the larger product combination must be negative, and the smaller must be positive:

$$
\begin{array}{lll}
(4x + 1)(x - 10) = 4x^2 - 39x - 10 & \text{Wrong} \\
(4x + 5)(x - 2) \ = 4x^2 - 3x - 10 & \text{Right}
\end{array}
$$

Thus $4x^2 - 3x - 10 = (4x + 5)(x - 2)$ is the correct factoring.

For your reference, let us list the sign patterns based on these four examples and supply a summary of the discussion in Example 4.

Sign Patterns for Factoring Trinomials

$$ax^2 + bx + c = (\ + \ )(\ + \ )$$
$$ax^2 - bx + c = (\ - \ )(\ - \ )$$
$$\left.\begin{array}{l} ax^2 + bx - c \\ ax^2 - bx - c \end{array}\right\} = \begin{cases} (\ + \ )(\ - \ ) \\ (\ - \ )(\ + \ ) \end{cases} \text{or}$$

Tips for Factoring *Difficult* Trinomials

1. If the middle coefficient is an *odd* number, then *exclude*
   (even   )(even   ) and (even)(even).

   $$\textit{Example:} \quad 8x^2 + 5x - 4 \neq (4x \quad)(2x \quad)$$
   $$\underset{\text{odd}}{\uparrow} \qquad \neq (\quad 2)(\quad 2).$$

2. If the trinomial has *no common factor,* then *exclude* any binomial factors that have common factors themselves.

   $$\textit{Example:} \quad 8x^2 - 13x - 6 \neq (8x \quad 2)(x \quad 3)$$
   $$\neq (8x \quad 6)(x \quad 1).$$

3. If the trinomial has a *common factor,* factor out the G.C.F and then proceed. See Example 7.
4. If all else fails, try the *alternative method* (shown after Example 5).

EXAMPLE 5

Factor these trinomials:

a. $x^2 + 9x + 14 = (x + 7)(x + 2)$.

b. $a^2 - 8ab + 16b^2 = (a - 4b)(a - 4b)$ or $(a - 4b)^2$.

c. $5x^2 + 3xy - 8y^2 \neq (\quad 4y)(\quad 2y)$ 3 is odd.
    $= (5x \ + \ 8y)(x \ - \ y)$.

d. $4a^2 - 11ab - 3b^2 \neq (2a \quad)(2a \quad)$ -11 is odd.
    $= (4a \ + \ b)(a - 3b)$.

e. $12x^2 + 28x + 15 \neq (6x \quad 3)(2x \quad 5)$ No common factors.
    $= (6x \ + \ 5)(2x \ + \ 3)$.

The choices $(4x \quad)(3x \quad)$ and $(12x \quad)(x \quad)$ were unlikely, because the middle coefficient, 28, is an even number. Thus $(6x \quad)(2x \quad)$ is more likely in this case.

f. $x^4 - 8x^2 - 48 = (x^2 - 12)(x^2 + 4)$. $x^2 x^2 = x^4$

g. $16x^6 + 29x^3 y^3 - 6y^6 \neq (4x^3 \qquad)(4x^3 \qquad)$ ⎫
$\qquad\qquad\qquad\qquad \neq (8x^3 \qquad)(2x^3 \qquad)$ ⎬ 29 is odd.
$\qquad\qquad\qquad\qquad = (16x^3 - 3y^3)(x^3 + 2y^3)$. ⎭

An *alternative method* for factoring

$$ax^2 + bx + c$$

is based on factoring by grouping (Section 3.2, Example 4). In this method, we seek two numbers whose

$$\text{product} = ac \quad \text{and} \quad \text{sum} = b.$$

For example, let us factor

$$3x^2 + 26x + 35$$

by this method. Here,

$$a = 3, \ b = 26, \quad \text{and} \quad c = 35.$$

We seek two numbers whose

$$\text{product} = 3 \cdot 35 = 105 \quad \text{and} \quad \text{sum} = 26.$$

The two numbers are 5 and 21, because

$$5 \cdot 21 = 105 \quad \text{and} \quad 5 + 21 = 26.$$

We now write the middle term, $26x$, as the sum of two terms, using 5 and 21 as coefficients: $26x = 5x + 21x$. The resulting four-term polynomial is now factored by grouping:

$$3x^2 + 26x + 35 = 3x^2 + 5x + 21x + 35 \qquad 26x = 5x + 21x$$
$$= x(3x + 5) + 7(3x + 5) \quad \Big\rbrace \ \text{Factor by}$$
$$= (3x + 5)(x + 7). \qquad\qquad \text{grouping.}$$

EXAMPLE 6

Factor by the alternative method:

a. $6x^2 - 19x + 10$.

Solution
Here $a = 6$, $b = -19$, and $c = 10$. We seek two numbers whose

$$\text{product} = 6 \cdot 10 = 60 \quad \text{and} \quad \text{sum} = -19.$$

The numbers are -4 and -15, because $(-4)(-15) = 60$ and $-4 + (-15) = -19$. Proceeding as above, we get

$$6x^2 - 19x + 10 = 6x^2 - 4x - 15x + 10 \qquad -19x = -4x - 15x$$
$$= 2x(3x - 2) - 5(3x - 2) \quad \Big\rbrace$$
$$= (3x - 2)(2x - 5). \qquad\qquad \text{Factor by grouping.}$$

(continued)

b. $15x^2 + 19xy - 10y^2$.

Solution

Here $a = 15$, $b = 19$, and $c = -10$. We seek two numbers whose

$$\text{product} = 15(-10) = -150 \quad \text{and} \quad \text{sum} = 19.$$

The numbers are 25 and -6, because $25(-6) = -150$ and $25 + (-6) = 19$. It must be confessed that these numbers are not always easy to find. Proceeding as above, we get

$$
\begin{aligned}
15x^2 + 19xy - 10y^2 &= 15x^2 + 25xy - 6xy - 10y^2 \qquad \scriptstyle 19xy = 25xy - 6xy\\
&= 5x(3x + 5y) - 2y(3x + 5y) \quad \left.\right\} \; \text{Factor by}\\
&= (3x + 5y)(5x - 2y). \qquad\qquad\quad \text{grouping.}
\end{aligned}
$$

    In the next example, we first factor out the greatest common factor (G.C.F.) and then factor the resulting trinomial.

EXAMPLE 7   First factor out the G.C.F., then factor the resulting trinomial:

a. $\begin{aligned}[t] 3x^2 - 12x - 15 &= 3(x^2 - 4x - 5) &&\text{G.C.F.}\\ &= 3(x - 5)(x + 1). &&\text{Factor trinomial.} \end{aligned}$

b. $\begin{aligned}[t] 24x^3 + 28x^2 + 8x &= 4x(6x^2 + 7x + 2) &&\text{G.C.F.}\\ &= 4x(3x + 2)(2x + 1). &&\text{Factor trinomial.} \end{aligned}$

c. $\begin{aligned}[t] 10a^3b + 35a^2b^2 - 75ab^3 &= 5ab(2a^2 + 7ab - 15b^2)\\ &= 5ab(2a - 3b)(a + 5b). \end{aligned}$

d. $\begin{aligned}[t] -8x^3 + 24x^2y - 18xy^2 &= -2x(4x^2 - 12xy + 9y^2) &&\text{Negative G.C.F.}\\ &= -2x(2x - 3y)(2x - 3y)\\ &= -2x(2x - 3y)^2. \end{aligned}$

e. $\begin{aligned}[t] 6a^{n+2} + 11a^{n+1} - 21a^n &= a^n(6a^2 + 11a - 21)\\ &= a^n(6a - 7)(a + 3). \end{aligned}$

■ EXERCISE 3.3   *Answers, page A42*

A

Factor, as in EXAMPLES 1 through 6:

1. $x^2 + 6x + 5$ **2.** $x^2 + 12x + 11$ **3.** $a^2 - 8a + 7$ **4.** $y^2 - 24y + 23$

5. $x^2 + 18xy - 19y^2$ **6.** $a^2 + ab - 2b^2$ **7.** $a^2x^2 - 12ax - 13$ **8.** $x^2y^2 - 36xy - 37$

9. $x^2 + 8x + 16$ **10.** $t^2 + 12t + 36$ **11.** $x^2 - 10xy + 25y^2$ **12.** $b^2 - 14b + 49$

13. $u^2 + 7uv - 18v^2$ **14.** $w^2 + 8wz - 20z^2$ **15.** $x^4 - x^2 - 20$ **16.** $z^4 - 11z^2 - 26$

17. $a^4 + 26a^2b^2 + 169b^4$

18. $x^4 - 20x^2y^2 + 100y^4$

19. $m^6 + 16m^3 + 28$

20. $u^6 - 20u^3v^3 + 51v^6$

21. $x^4y^2 - 13x^2y - 68$

22. $a^2b^6 + 7ab^3 - 78$

23. $x^{2n} + 26x^n + 105$

24. $y^{4m} + 14y^{2m} - 95$

25. $x^2 + 72 - 18x$

26. $19ab + 48b^2 + a^2$

27. $x^2 + 24x + 135$

28. $x^2 - 24x + 143$

29. $a^2 - 34ab + 289b^2$

30. $x^2 - 24x + 144$

31. $2x^2 + 15x + 7$

32. $3x^2 - 14x + 11$

33. $5a^2 - 12ab - 17b^2$

34. $7m^2 + 6mn - 13n^2$

35. $11x^2 - 24xy + 4y^2$

36. $13u^2 - 4uv - 9v^2$

37. $4x^2 + 11x + 6$

38. $8a^2 - 21ab + 10b^2$

39. $6s^2 - 13st + 5t^2$

40. $6s^2 - 13st + 6t^2$

41. $6a^4 - 11a^2 - 35$

42. $9x^4 + 25x^2y^2 - 6y^4$

43. $4x^2 + 12x + 9$

44. $9a^2 - 30ab + 25b^2$

45. $16a^2x^2 - 8ax + 1$

46. $25b^2 + 10bcd + c^2d^2$

47. $6x^2 + 11x - 10$

48. $6x^2 - 17xy - 10y^2$

49. $6a^2 + 13ab - 15b^2$

50. $6x^2 + 5xy - 21y^2$

First factor out the greatest common factor, then factor the resulting trinomial, as in EXAMPLE 7:

51. $2x^2 - 32x + 78$

52. $3a^2 + 9a - 120$

53. $5a^3b - 70a^2b^2 + 240ab^3$

54. $4x^3y + 32x^2y^2 + 60xy^3$

55. $-3m^2 + 15mn + 42n^2$

56. $-2u^3 - 4u^2v + 96uv^2$

57. $5x^3 + 85x^2 - 300x$

58. $5x^3 - 85x^2 + 300x$

59. $3a^3 + 6a^2b - 360ab^2$

60. $5x^3 - 140x^2 + 660x$

61. $-x^3 - 5x^2 + 126x$

62. $-2ax^3 + 50ax^2 - 308ax$

63. $-x^2 + 9xy + 22y^2$

64. $-a^2b^2 - 9ab + 70$

65. $2x^2 - 140xy + 650y^2$

66. $3y^2 - 15y - 2250$

67. $12x^2 - 34x + 10$

68. $18x^2 - 21x - 15$

69. $20x^3y + 25x^2y^2 - 30xy^3$

70. $16a^3b - 100a^2b + 24ab$

71. $8x^3y - 28x^2y^2 - 60xy^3$

72. $6x^3 - 45x^2 + 81x$

73. $50x^2 - 140xy + 98y^2$

74. $18a^3b - 96a^2b^2 + 128ab^3$

75. $x^{n+2} - 7x^{n+1} - 18x^n$

76. $a^{n+2}b^m + 15a^{n+1}b^{m+1} + 26a^nb^{m+2}$

B

Factor completely:

77. $8x^2 + 29xy + 26y^2$

78. $8a^2 - 51a - 35$

79. $12x^4 - 23x^2 + 10$

80. $15m^6 - 31m^3n^3 - 12n^6$

81. $15x^4 - 11x^2 - 12$

82. $15x^2y^2 + 29xy + 12$

83. $8x^{2n} - 35x^n + 12$

84. $8x^{2m} + 95x^my^n - 12y^{2n}$

85. $24u^2 + 58uv + 35v^2$

86. $48z^2 - 22z - 15$

87. $36x^6 - 62x^4 + 12x^2$

88. $-32x^9 - 20x^6 - 42x^3$

89. $12x^8 + 116x^7 + 224x^6$

90. $15x^6 - 85x^5 - 280x^4$

91. $10x^{n+2} + 7x^{n+1} - 12x^n$

92. $30y^{m+3} - 57y^{m+2} - 36y^{m+1}$

C

93. $-12x^{3n}y^n - 14x^{2n}y^{2n} + 6x^ny^{3n}$

94. $-15a^{4m}b^n + 21a^{3m}b^{2n} + 18a^{2m}b^{3n}$

95. $48x^2 - 62x - 15$

96. $48x^2 + 175x + 50$

97. $36x^2y^2 - 65xy - 36$

98. $75a^2 - 65a + 14$

99. The **alternative method for factoring trinomials,** as given in the text, can be justified for the special case when $a = 1$: $x^2 + bx + c$. According to the method given, we seek two numbers m and n whose

$$\text{product } mn = 1 \cdot c = c \quad \text{and} \quad \text{sum } m + n = b.$$

By substitution, we get

$$x^2 + bx + c = x^2 + (m + n)x + mn$$
$$= x^2 + mx + nx + mn.$$

You finish the correct factoring.

3.4   Factoring Squares and Cubes

This section should offer a well-deserved break from the trial-and-error factoring problems that you just completed. Recall from Section 3.1 the special product $(A + B)(A - B) = A^2 - B^2$. Written in reverse, this becomes the formula for factoring the difference of two perfect-square terms into the product of their sum and their differences:

Factoring the Difference of Squares
$$A^2 - B^2 = (A + B)(A - B)$$

EXAMPLE 1   Factor the difference of squares:

a. $x^2 - 25 = x^2 - 5^2$ Check:
$= (x + 5)(x - 5)$. $x^2 - 5x + 5x - 25 \overset{\checkmark}{=} x^2 - 25$

b. $9a^2 - 16b^2 = (3a)^2 - (4b)^2$
$= (3a + 4b)(3a - 4b)$.

c. $36a^2x^2 - 121 = (6ax + 11)(6ax - 11)$.

d. $4x^6 - 49y^6 = (2x^3 + 7y^3)(2x^3 - 7y^3)$.

e. $z^{2n} - 1 = (z^n - 1)(z^n + 1)$.

f. $x^4 - 1 = (x^2 + 1)(x^2 - 1)$ Note
$= (x^2 + 1)(x + 1)(x - 1)$.   repeated
factoring.

g. $a^8 - 256 = (a^4 + 16)(a^4 - 16)$ $a^4a^4 = a^8,\ 16^2 = 256$

$\qquad = (a^4 + 16)(a^2 + 4)(a^2 - 4)$ $\left.\begin{array}{l}\text{Note}\\\text{repeated}\\\text{factoring.}\end{array}\right.$

$\qquad = (a^4 + 16)(a^2 + 4)(a + 2)(a - 2).$

h. $(a + b)^2 - c^2 = (a + b + c)(a + b - c).$ $(a + b)^2$ is a perfect square.

The repeated factoring in Example 1(f) is $(x^2 + 1)(x + 1)(x - 1)$. We cannot further factor $x^2 + 1$, because in general, $A^2 + B^2$ *does not factor.* (Try factoring it yourself.) We will now see that both the sum and the difference of perfect cubes factor. The sum of cubes, $A^3 + B^3$, factors into $(A + B)(A^2 - AB + B^2)$, as we can check by multiplying:

$$(A + B)(A^2 - AB + B^2) = A^3 - A^2B + AB^2 + A^2B - AB^2 + B^3 \stackrel{\checkmark}{=} A^3 + B^3.$$

The difference of cubes, $A^3 - B^3$, factors into $(A - B)(A^2 + AB + B^2)$, as you will be asked to verify in Problem 93. We list these factoring formulas, which you should memorize.

Factoring the Sum and Difference of Cubes

$$A^3 + B^3 = (A + B)(A^2 - AB + B^2)$$
$$A^3 - B^3 = (A - B)(A^2 + AB + B^2)$$

EXAMPLE 2

Factor the sum and difference of cubes:

a. $x^3 + 27 = x^3 + 3^3$

$\qquad = (x + 3)(x^2 - x \cdot 3 + 3^2)$ Check:

$\qquad = (x + 3)(x^2 - 3x + 9).$ $x^3 - 3x^2 + 9x + 3x^2 - 9x + 27 \stackrel{\checkmark}{=} x^3 + 27$

b. $125a^3 - 1 = (5a)^3 - 1^3$

$\qquad = (5a - 1)[(5a)^2 + 5a \cdot 1 + 1^2]$

$\qquad = (5a - 1)(25a^2 + 5a + 1).$ $25a^2 + 5a + 1$ does not factor.

c. $8x^3 + 343y^3 = (2x)^3 + (7y)^3$

$\qquad = (2x + 7y)(4x^2 - 14xy + 49y^2).$

d. $64a^3 - 729b^6c^9 = (4a)^3 - (9b^2c^3)^3$ $9b^2c^3 \cdot 9b^2c^3 \cdot 9b^2c^3 = 729b^6c^9$

$\qquad = (4a - 9b^2c^3)[(4a)^2 + 4a \cdot 9b^2c^3 + (9b^2c^3)^2]$

$\qquad = (4a - 9b^2c^3)(16a^2 + 36ab^2c^3 + 81b^4c^6).$

e. $(x - 2)^3 + 64 = (x - 2)^3 + 4^3$

$\qquad = [(x - 2) + 4][(x - 2)^2 - 4(x - 2) + 4^2]$ Multiply inner

$\qquad = [x - 2 + 4][x^2 - 4x + 4 - 4x + 8 + 16]$ parentheses.

$\qquad = (x + 2)(x^2 - 8x + 28).$

In the next example, we combine all of the factoring methods introduced so far. The flow chart suggests that you should always *first look for common factors.*

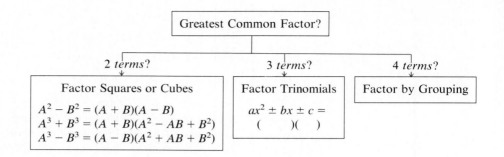

EXAMPLE 3 | Factor completely. First look for the greatest common factor.

a. $75x^3y - 12xy^3 = 3xy(25x^2 - 4y^2)$ G.C.F.
$$= 3xy(5x + 2y)(5x - 2y).$$ Squares

b. $2x^5 - 162x = 2x(x^4 - 81)$ G.C.F.
$$= 2x(x^2 + 9)(x^2 - 9)$$ Squares
$$= 2x(x^2 + 9)(x + 3)(x - 3).$$ Squares again

c. $54a^4 - 128ab^3 = 2a(27a^3 - 64b^3)$ G.C.F.
$$= 2a(3a - 4b)(9a^2 + 12ab + 16b^2).$$ Cubes

d. $-5x^6 + 65x^4 - 180x^2 = -5x^2(x^4 - 13x^2 + 36)$ Negative G.C.F.
$$= -5x^2(x^2 - 9)(x^2 - 4)$$ Trinomial
$$= -5x^2(x + 3)(x - 3)(x + 2)(x - 2).$$ Squares

e. $a^6 + 7a^3 - 8 = (a^3 + 8)(a^3 - 1)$ Trinomial
$$= (a + 2)(a^2 - 2a + 4)(a - 1)(a^2 + a + 1).$$ Cubes, cubes

f. $x^5 - 9x^3 - 8x^2 + 72 = x^3(x^2 - 9) - 8(x^2 - 9)$ ⎫
$$= (x^2 - 9)(x^3 - 8)$$ ⎬ Grouping
$$= (x + 3)(x - 3)(x - 2)(x^2 + 2x + 4).$$ ⎭   Squares, cubes

---

EXERCISE 3.4   *Answers, page A43*

A

Factor as in EXAMPLES 1 and 2. Consult the Table of Powers and Roots in the Appendix, if necessary.

1. $x^2 - 100$

2. $a^2 - 49$

3. $a^2b^2 - 36$

4. $x^2 - 81y^2$

5. $9u^2 - 121v^2w^2$

6. $121m^2 - 64n^2p^2$

7. $289a^4 - 400b^2$

8. $169x^2 - 324y^4$

9. $625x^2 - 361y^2$

10. $441a^2 - 676b^2$

11. $1369b^6 - 1849c^4$

12. $2601x^4y^2 - 10{,}000z^6$

13. $a^{2n} - b^{2m}$

14. $x^{2n}y^{2n} - 1$

15. $(x + y)^2 - 4z^2$

16. $(a - 2b)^2 - 9c^2$

17. $a^2 - (b + 2c)^2$

18. $4x^2 - (y - 3)^2$

19. $a^4 - b^4$

20. $x^4 - 16$

21. $x^8 - 256$ **22.** $a^8 - 100,000,000$ **23.** $x^{16} - 1$ **24.** $a^{16} - b^{16}$

25. $x^3 + 125$ **26.** $a^3 + 64$ **27.** $8a^3 - 27b^3$ **28.** $27x^3 - 1000$

29. $x^3y^6 - 216z^3$ **30.** $343a^6b^3 - 1$ **31.** $729a^3 + 1331b^6c^9$ **32.** $512x^6 + 2197y^3z^9$

33. $3375 - 64x^3$ **34.** $64 + 4913y^3$ **35.** $x^{3n} + y^{3n}$ **36.** $a^{3m}b^{3m} - 1$

37. $(x - 3)^3 + 125$ **38.** $(a + 4)^3 - 27$ **39.** $(y + 1)^3 - 8$ **40.** $(b - 2)^3 + 27$

41. $(x + h)^3 - x^3$ **42.** $(a + 2b)^3 - 8b^3$

Factor completely, using all of the factoring methods, as in EXAMPLE 3. First look for common factors.

43. $2x^2 - 50y^2$ **44.** $3a^2 - 48$ **45.** $3a^3b - 147ab^3$ **46.** $125x^3y - 5xy^3$

47. $12x^5y^2 - 675x^3y^4$ **48.** $32a^3b^3 - 98ab^5$ **49.** $3x^5y - 48xy^5$ **50.** $4a^5bc - 4ab^5c^5$

51. $x^{3n}y^n - x^ny^{3n}$ **52.** $a^{5n} - a^n$ **53.** $2x^3 + 16$ **54.** $3a^3 - 81$

55. $24a^4b - 81ab^4$ **56.** $128x^4y + 250xy^4$ **57.** $192x^4 - 3x$ **58.** $625x^5 + 5x^2$

59. $-2a^4b - 250ab$ **60.** $-24a^4b^7c + 81abc^{10}$ **61.** $x^4 - 26x^2 + 25$ **62.** $a^4 - 17a^2b^2 + 16b^4$

63. $4x^4 - 37x^2y^2 + 9y^4$ **64.** $9c^4 - 13c^2 + 4$ **65.** $x^6 - 7x^3y^3 - 8y^6$ **66.** $a^6 - 26a^3 - 27$

67. $x^6 - y^6$ **68.** $x^6 - 64$ **69.** $75x^5 - 87x^3 + 12x$ **70.** $72a^5 - 170a^3 + 18a$

71. $-2x^7 + 32x^4 - 128x$ **72.** $-x^5y - 24xy^3 + 25xy^5$ **73.** $x^4 - 9x^2 - 4x^2y^2 + 36y^2$

74. $4a^4 - a^2 - 36a^2b^2 + 9b^2$ **75.** $x^5 - x^3 + x^2y^3 - y^3$ **76.** $a^3x^2 - 25a^3y^2 + 8x^2 - 200y^2$

77. $x^6 - 8x^3 + x^3y^3 - 8y^3$ **78.** $8a^3b^3 - b^3 + 64a^3c^3 - 8c^3$ **79.** $x^5 - 9x^3 + 8x^2 - 72$

80. $x^5 + x^2 - 16x^3 - 16$ **81.** $32a^2x^3 - 4a^2y^3 - 72b^2x^3 + 9b^2y^3$ **82.** $72x^3u^2 + 9u^2y^3 - 32x^3v^2 - 4y^3v^2$

The polynomial $x^2 + 2x + 1 - 25y^2$ consists of a perfect-square trinomial and a perfect-square monomial. It can be factored as follows:

$$x^2 + 2x + 1 - 25y^2 = (x + 1)(x + 1) - 25y^2 \qquad \text{Factor trinomial.}$$
$$= (x + 1)^2 - (5y)^2 \qquad \text{Write as squares.}$$
$$= (x + 1 + 5y)(x + 1 - 5y) \qquad \text{Factor difference of squares.}$$

In like manner, factor each of these polynomials:

83. $x^2 + 10x + 25 - 4y^2$ **84.** $a^2 + 12a + 36 - 49b^2$ **85.** $x^2 - 6xy + 9y^2 - 25$

86. $a^2 + 4ab + 4b^2 - 9$ **87.** $4a^2 + 12ab + 9b^2 - 16c^2$ **88.** $9x^2 - 30xy + 25y^2 - 4z^2$

The binomial $x^6 - 1$ can be factored both as the difference of squares (Example 3f) and as the difference of cubes:

As difference of squares: $x^6 - 1 = (x^3)^2 - 1^2$ $x^3x^3 = x^6$

$\qquad\qquad\qquad\qquad\qquad = (x^3 + 1)(x^3 - 1)$ Factor squares.

$\qquad\qquad\qquad\qquad\qquad = (x + 1)(x^2 - x + 1)(x - 1)(x^2 + x + 1)$ Factor cubes.

As difference of cubes: $x^6 - 1 = (x^2)^3 - 1^3$ $x^2 x^2 x^2 = x^6$

$$= (x^2 - 1)[(x^2)^2 + x^2 \cdot 1 + 1^2] \quad \text{Factor cubes.}$$
$$= (x + 1)(x - 1)(x^4 + x^2 + 1). \quad \text{Factor squares.}$$

By comparing these two answers, we discover it must be true that

$$(x^2 - x + 1)(x^2 + x + 1) = x^4 + x^2 + 1.$$

It is easy to verify this fact by multiplying out as in Section 3.1. A more challenging task, however, is to *factor* the third trinomial into the product of the first two trinomials. We can do this by first adding and subtracting x^2 and then proceeding as in Problems 83–88.

$$x^4 + x^2 + 1 = x^4 + x^2 + x^2 + 1 - x^2 \quad \text{Add and subtract } x^2.$$
$$= x^4 + 2x^2 + 1 \quad\quad - x^2 \quad x^2 + x^2 = 2x^2$$
$$= (x^2 + 1)(x^2 + 1) - x^2 \quad \text{Factor trinomial.}$$
$$= (x^2 + 1)^2 - x^2 \quad \text{Write as square.}$$
$$= (x^2 + 1 + x)(x^2 + 1 - x) \quad \text{Factor squares.}$$
$$= (x^2 + x + 1)(x^2 - x + 1). \quad \text{Rearrange terms.}$$

We have obtained the factoring promised above. By adding and subtracting the indicated term, factor these polynomials in like manner:

89. $x^4 + 4x^2 + 16$; $4x^2$

90. $x^4 + 9x^2 + 81$; $9x^2$

91. $x^4 - 6x^2y^2 + 25y^4$; $16x^2y^2$

92. $x^4 - 13x^2y^2 + 4y^4$; $9x^2y^2$

93. Verify the factoring formula for cubes, $A^3 - B^3 = (A - B)(A^2 + AB + B^2)$, by multiplying out the parentheses.

94. The factoring formula $A^2 - B^2 = (A + B)(A - B)$ has a geometric interpretation. The first of the accompanying figures is a square of side A, with a smaller square of side B removed from it. If it is cut along the dotted line and then rectangle I is pasted onto rectangle II, as indicated, the new rectangle on the right is formed.

a. Find the area of the first figure. *Hint:* Big square minus little square.

b. Find the length and the width of the new rectangle.

c. Find the area of the new rectangle.

d. Equate your answers from parts a and c. What is the result?

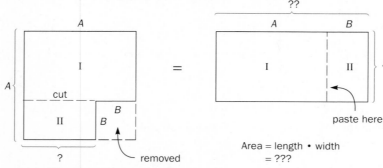

3.5   Factoring Second-Degree Equations

A **second-degree equation** in one variable is a statement about equality between single-variable polynomials of degree two and less. Also called **quadratic equations**, they are illustrated by

$$x^2 + 7x - 18 = 0, \quad 20 - 8y^2 = 12y, \quad \text{and} \quad 2z^2 = 8z.$$

In order to solve these equations, we need an important property of numbers. You will discover the property by first trying this: "Try to find two numbers, neither of which is zero, yet whose product is zero."

Time's up. If you said it's impossible, you are right. If you tried 5 and -5, for example, you are wrong, because their product equals -25. The truth of the matter is that if two numbers multiplied together give zero, then at least one of them must be zero. We now state the important rule:

> **Zero Product Rule**
> If $AB = 0$, then $A = 0$ or $B = 0$ (or both).

In order to solve a second-degree equation, we first put it into **standard form**:

$$ax^2 + bx + c = 0.$$

We then factor it, if possible*, and then set each factor equal to zero in accordance with the Zero Product Rule.

EXAMPLE 1   Solve $x^2 + 7x - 18 = 0$.

Solution

$$
\begin{array}{ll}
(x + 9)(x - 2) = 0 & \text{Factor.} \\
x + 9 = 0 \quad \text{or} \quad x - 2 = 0 & \text{Zero Product Rule.} \\
x = -9 \quad \text{or} \qquad x = 2. & \text{Solutions.}
\end{array}
$$

The word "or" might lead you to ask, "Which answer works?" They both work, as can be checked:

$$
\begin{array}{ll}
(-9)^2 + 7(-9) - 18 = 0 & \qquad 2^2 + 7 \cdot 2 - 18 = 0 \\
81 - 63 - 18 = 0 & \qquad 4 + 14 - 18 = 0 \\
0 \overset{\checkmark}{=} 0. & \qquad\qquad 0 \overset{\checkmark}{=} 0.
\end{array}
$$

---

*In Chapter 6, we will solve quadratic equations that don't factor.

EXAMPLE 2   Solve $20 - 8y^2 = 12y$.

Solution

We first put this into standard form, with positive second-degree term:

$$0 = 8y^2 + 12y - 20 \qquad \text{Standard form.}$$
$$0 = 4(2y^2 + 3y - 5) \qquad \text{Common factor.}$$
$$0 = 4(2y + 5)(y - 1) \qquad \text{Factor trinomial.}$$
$$2y + 5 = 0 \quad \text{or} \quad y - 1 = 0 \qquad \text{Zero Product Rule.}$$
$$2y = -5$$
$$y = \frac{-5}{2} \quad \text{or} \quad y = 1. \qquad \text{Solutions.}$$

Note: The common factor 4 *is not equal to* zero, so we can disregard it in the Zero Product Rule.

EXAMPLE 3   Solve $2z^2 = 8z$.

Solution

$$2z^2 - 8z = 0 \qquad \text{Standard form.}$$
$$2z(z - 4) = 0 \qquad \text{Common factor.}$$
$$2z = 0 \quad \text{or} \quad z - 4 = 0 \qquad \text{Zero Product Rule.}$$
$$\frac{2z}{2} = \frac{0}{2}$$
$$z = 0 \quad \text{or} \quad z = 4. \qquad \text{Solutions.}$$

Note: The common factor $2z$ *is equal to* zero, which gives the solution $z = 0$. Unlike in Example 2, we *cannot disregard* $2z$ here, because it contains a variable.

EXAMPLE 4   Solve $(2x - 1)^2 = (2x + 5)(x + 2) - x(2x + 13)$.

Solution

We simplify each side separately:

$$4x^2 - 4x + 1 = 2x^2 + 9x + 10 - 2x^2 - 13x$$
$$4x^2 - 4x + 1 = -4x + 10$$
$$4x^2 - 9 = 0 \qquad\qquad\qquad \text{Standard form.}$$
$$(2x + 3)(2x - 3) = 0$$
$$2x + 3 = 0 \quad \text{or} \quad 2x - 3 = 0$$
$$2x = -3 \qquad\qquad 2x = 3$$
$$x = \frac{-3}{2} \quad \text{or} \quad x = \frac{3}{2}. \qquad\qquad \text{Solutions.}$$

Translating word problems into algebra can often lead to quadratic equations. You should consult the Table of Phrases in Section 1.4, as necessary, in order to set them up.

EXAMPLE 5

When twice the square of a number is increased by three, the result equals seven times the sum of the number and one. Find the number.

Solution
Let x = the number. Then

$$2x^2 + 3 = 7(x + 1) \qquad \text{Translate into equation.}$$
$$2x^2 + 3 = 7x + 7$$
$$2x^2 - 7x - 4 = 0$$
$$(2x + 1)(x - 4) = 0 \qquad \text{Solve equation.}$$
$$x = -\frac{1}{2}, \; x = 4.$$

There are *two answers*: $-\dfrac{1}{2}$ and 4.

Check: $2\left(-\dfrac{1}{2}\right)^2 + 3 = 7\left(-\dfrac{1}{2} + 1\right)$ $\qquad$ $2(4)^2 + 3 = 7(4 + 1)$

$\qquad\qquad 2\left(\dfrac{1}{4}\right) + 3 = 7\left(\dfrac{1}{2}\right)$ $\qquad\qquad$ $2 \cdot 16 + 3 = 7(5)$

$\qquad\qquad\qquad \dfrac{1}{2} + 3 \overset{\checkmark}{=} \dfrac{7}{2}.$ $\qquad\qquad\qquad$ $35 \overset{\checkmark}{=} 35.$

EXAMPLE 6

One number is seven more than another. When the square of the larger number is decreased by twice the smaller, the result is 75 more than the smaller. Find the numbers.

Solution
Let x = the smaller number; then $x + 7$ = the larger number.

$$(x + 7)^2 - 2x = x + 75 \qquad \text{Translate into equation.}$$
$$x^2 + 14x + 49 - 2x = x + 75$$
$$x^2 + 11x - 26 = 0$$
$$(x - 2)(x + 13) = 0 \qquad \text{Solve equation.}$$
$$x = 2, \; x = -13.$$

If $x = 2$ (smaller), then $x + 7 = 9$ (larger).
If $x = -13$ (smaller), then $x + 7 = -6$ (larger).

There are *two sets of answers*: 2 and 9; -13 and -6.

Check: $9^2 - 2 \cdot 2 \overset{\checkmark}{=} 2 + 75$ $\qquad$ $(-6)^2 - 2(-13) \overset{\checkmark}{=} -13 + 75$
$\qquad\qquad 81 - 4 \overset{\checkmark}{=} 77.$ $\qquad\qquad\qquad$ $36 + 26 \overset{\checkmark}{=} 62.$

Problems involving consecutive integers occur very commonly in algebra. They come in three types:

consecutive integers: . . ., $-1, 0, 1, 2, 3, \ldots, n, n+1, n+2, \ldots$
consecutive *even* integers: . . ., $-2, 0, 2, 4, 6, \ldots, n, n+2, n+4, \ldots$
consecutive *odd* integers: . . ., $-1, 1, 3, 5, 7, \ldots, n, n+2, n+4, \ldots$

We see that consecutive integers increase by one: $n, n+1, n+2, \ldots$ Both consecutive even and consecutive odd integers increase by two: $n, n+2, n+4, \ldots$

Consecutive integer phrases	Algebraic expressions
sum of three consecutive integers	$n + (n+1) + (n+2)$
sum of three consecutive *even* or *odd* integers	$n + (n+2) + (n+4)$
sum of squares of three consecutive integers	$n^2 + (n+1)^2 + (n+2)^2$
sum of squares of three consecutive *even* or *odd* integers	$n^2 + (n+2)^2 + (n+4)^2$
square of the sum of two consecutive integers	$(n + n + 1)^2$, or $(2n+1)^2$
product of two consecutive *even* or *odd* integers	$n(n+2)$

EXAMPLE 7

The sum of the squares of three consecutive even integers is 116. Find the integers.

Solution
Let $n =$ first *even* integer, then
$n + 2 =$ second *even* integer, and
$n + 4 =$ third *even* integer.

$$n^2 + (n+2)^2 + (n+4)^2 = 116 \qquad \text{Translate into equation.}$$
$$n^2 + n^2 + 4n + 4 + n^2 + 8n + 16 = 116 \qquad (A+B)^2 = A^2 + 2AB + B^2$$
$$3n^2 + 12n - 96 = 0 \qquad \text{Standard form.}$$
$$3(n^2 + 4n - 32) = 0 \qquad \text{Common factor.}$$
$$3(n - 4)(n + 8) = 0$$
$$n = 4, n = -8.$$

If $n = 4$, then $n + 2 = 6$ and $n + 4 = 8$.
If $n = -8$, then $n + 2 = -6$ and $n + 4 = -4$.

There are *two sets of answers*: 4, 6, 8; and -8, -6, -4.

Check: $4^2 + 6^2 + 8^2 = 116$ $\qquad$ $(-8)^2 + (-6)^2 + (-4)^2 = 116$

$\qquad\qquad$ $16 + 36 + 64 \overset{\checkmark}{=} 116.$ $\qquad\qquad\qquad$ $64 + 36 + 16 \overset{\checkmark}{=} 116.$

EXERCISE 3.5 $\quad$ *Answers, pages A43–A44*

A

Solve by first putting into standard form and then factoring, as in EXAMPLES 1 through 4. First look for common factors.

1. $x^2 - 8x + 15 = 0$ $\qquad$ **2.** $y^2 + 12y + 35 = 0$ $\qquad$ **3.** $z^2 = 5z + 36$ $\qquad$ **4.** $w^2 - 48 = -8w$

5. $15 - 2x^2 = 7x$ $\qquad$ **6.** $15t - 2t^2 = 7$ $\qquad$ **7.** $15m - 9 - 4m^2 = 0$ $\qquad$ **8.** $31n - 6n^2 = 28$

9. $30 + 43x = 8x^2$ $\qquad$ **10.** $12 - 7r = 12r^2$ $\qquad$ **11.** $x^2 - 9 = 0$ $\qquad$ **12.** $4y^2 - 25 = 0$

13. $49 = 16z^2$ $\qquad$ **14.** $0 = 100 - 81u^2$ $\qquad$ **15.** $x^2 = 9x$ $\qquad$ **16.** $4y^2 = 25y$

17. $4x^2 + 12x + 9 = 0$ $\qquad$ **18.** $25y^2 + 20y + 4 = 0$ $\qquad$ **19.** $49n^2 = 84n - 36$ $\qquad$ **20.** $88x - 16x^2 = 121$

21. $4x^2 + 8x - 12 = 0$ $\qquad$ **22.** $5x^2 + 30x - 35 = 0$ $\qquad$ **23.** $-18x^2 + 33x - 9 = 0$ $\qquad$ **24.** $-18y^2 + 18y - 4 = 0$

25. $18y^2 + 24y + 8 = 0$ $\qquad$ **26.** $18z^2 = 60z - 50$ $\qquad$ **27.** $2x^2 = 32x$ $\qquad$ **28.** $8x^2 = 98x$

29. $2x^2 = 32$ $\qquad$ **30.** $8x^2 = 98$ $\qquad$ **31.** $-16t^2 + 64t + 80 = 0$ $\qquad$ **32.** $-16t^2 + 8t + 24 = 0$

33. $121 - 81z^2 = 0$ $\qquad$ **34.** $169 - 144y^2 = 0$ $\qquad$ **35.** $6z^2 - 9z = 0$ $\qquad$ **36.** $12x^2 - 16x = 0$

37. $t^2 = t$ $\qquad$ **38.** $t = 2t^2$ $\qquad$ **39.** $20n^2 + 33n - 27 = 0$ $\qquad$ **40.** $10m^2 - 39m + 27 = 0$

41. $x(1 - 2x) = -10$ $\qquad$ **42.** $y(2 - 3y) = -33$ $\qquad$ **43.** $20 = (y - 5)(y + 3)$ $\qquad$ **44.** $16 = (x + 4)(x - 2)$

45. $z^2 + (z + 7)^2 = 169$ $\qquad$ **46.** $u^2 + 576 = (4u - 3)^2$ $\qquad$ **47.** $(x + 2)(x + 6) = (2x - 3)(2x + 1)$

48. $2x(2x - 1) = (x - 2)(3x + 1)$ $\qquad$ **49.** $12 - (4x - 3)(x + 1) = 1 - (x + 2)(x - 2)$

50. $2x - (3x - 4)^2 = (5x + 1)(x - 2)$

Solve as in EXAMPLES 5, 6, and 7. Use the Table of Phrases as necessary.

51. Three times the square of a number is increased by the number itself; the result is the difference between five times the number and one. What is the number?

52. The square of a number is decreased by four times the number; the result is the same as subtracting five from twice the square of the number. Find the number.

53. What number equals its own square?

54. What number equals twice its own square?

55. The square of the sum of a number and three is decreased by one; the result is thirteen more than twice the square of the number. Find the number.

56. Twice the square of the difference between two and a number is decreased by two; the result is the difference between the square of the number and three times the number. What is the number?

57. One number is five more than another. When the square of the larger number is decreased by 17, the result is eight times the sum of the smaller number and two. Find the numbers.

58. One number is four less than another. When the square of the smaller number is increased by one, the result is one less than three times the larger number. Find the numbers.

59. One number is three more than twice another. Their product is 65. What are the numbers?

60. One number is five less than three times another. The sum of their squares is five. Find the numbers.

61. Find three consecutive integers whose sum is 87.

62. Find four consecutive even integers whose sum is 44.

63. Find four consecutive odd integers whose sum is 38 more than twice the second integer.

64. Find three consecutive even integers whose sum is 36 less than five times the second integer.

65. The sum of the squares of three consecutive integers is 50. Find them.

66. Find three consecutive even integers such that the sum of the squares of the first two equals the square of the third.

67. Find two consecutive odd integers whose product is 63.

68. Find two consecutive integers whose product is five more than their sum.

B

69. Find three consecutive even integers the sum of whose squares is four less than three times the square of the first integer.

70. The sum of the squares of two consecutive integers equals the product of the integers increased by the sum of the integers.

71. The square of the sum of two consecutive integers is twelve more than the sum of the squares of the integers. Find the integers.

72. To an integer is added one, and the result is squared. This is doubled and then decreased by the integer. The result is 29. Find the integer.

73. When a ball is thrown upward from the top of a building, its height h above the ground (in feet) after t seconds is given by the formula

$$h = -16t^2 + 48t + 64.$$

When does the ball strike the ground? *Hint:* Its height will be zero.

74. Repeat Problem 73, using the formula $h = -16t^2 + 56t$.

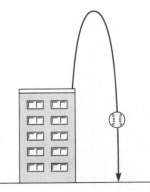

The absolute-value equation $\left|x^2 - 4x + 2\right| = 2$ can be solved using the methods of Section 2.6:

$$x^2 - 4x + 2 = 2 \quad \text{or} \quad x^2 - 4x + 2 = -2$$

$x^2 - 4x = 0$	$x^2 - 4x + 4 = 0$	Standard form
$x(x - 4) = 0$	$(x - 2)(x - 2) = 0$	Factor
$x = 0, \quad x = 4$	$x = 2$	Solutions

Check $x = 0$: $\ \left|0 - 0 + 2\right| \overset{?}{=} 2$ $\quad$ $x = 4$: $\ \left|16 - 16 + 2\right| \overset{?}{=} 2$ $\quad$ $x = 2$: $\ \left|4 - 8 + 2\right| \overset{?}{=} 2$
$$\left|2\right| \overset{\checkmark}{=} 2 \qquad\qquad\qquad \left|2\right| \overset{\checkmark}{=} 2 \qquad\qquad\qquad \left|-2\right| \overset{\checkmark}{=} 2$$

In like manner, solve and check these equations:

75. $\left|x^2 - 8x + 8\right| = 8$ $\qquad\qquad$ **76.** $\left|x^2 + 12x + 18\right| = 18$

77. $\left|x^2 + 2x - 25\right| = 10$ $\qquad\qquad$ **78.** $\left|x^2 - 8\right| = 8$

79. $\left|x^2 + 3x - 6\right| = 2x$ $\qquad\qquad$ **80.** $\left|x^2 - 2x - 6\right| = 3x$

By combining the factoring methods, as in Exercise 3.4, we can solve equations whose degree is higher than two. For example:

$3x^5 - 15x^3 + 12x = 0$	
$3x(x^4 - 5x^2 + 4) = 0$	Common factor
$3x(x^2 - 4)(x^2 - 1) = 0$	Factor trinomial
$3x(x + 2)(x - 2)(x + 1)(x - 1) = 0$	Difference of squares
$x = 0, x = -2, x = 2, x = -1, x = 1$	Solutions

Now solve these equations. First look for common factors.

81. $2x^3 - 18x^2 - 44x = 0$ $\qquad\qquad$ **82.** $5x^3 + 65x^2 - 150x = 0$

83. $-12x^4 + 72x^3 - 105x^2 = 0$ $\qquad\qquad$ **84.** $-24x^5 - 52x^4 + 20x^3 = 0$

85. $x^4 - 10x^2 + 9 = 0$ $\qquad\qquad$ **86.** $4x^4 - 17x^2 + 4 = 0$

87. $2x^5 - 34x^3 + 32x = 0$ $\qquad\qquad$ **88.** $18x^6 - 74x^4 + 8x^2 = 0$

89. $x^3 + 3x^2 - 4x - 12 = 0$ $\quad$ *Hint*: Factor by grouping. $\qquad$ **90.** $2x^3 - 2x + 3x^2 - 3 = 0$

C

91. $4x^3 - 9x + 20x^2 - 45 = 0$ $\qquad\qquad$ **92.** $9x^3 - 27x^2 - 4x + 12 = 0$

93. Find three consecutive integers such that their product equals their sum. *Hint:* After setting up the equation, get all terms on one side of the equation and zero on the other side; then solve by factoring by grouping, as in Problems 89–92. You should get three sets of answers.

94. The Zero Product Rule says that if $AB = 0$, then $A = 0$ or $B = 0$. You will prove it in this problem, by answering the questions below. Let $AB = 0$; then it is true that either $A = 0$ or $A \neq 0$.
Case I: $A = 0$. What can you say?
Case II: $A \neq 0$. Divide both sides of $AB = 0$ by A. What do you conclude?

95. What is wrong with this "proof" that $2 = 1$? Let a and b be any *non-zero* numbers such that

$$a = b.$$

Then $\quad a^2 = ab$	Multiply both sides by a ($\neq 0$).
$a^2 - b^2 = ab - b^2$	Subtract b^2.
$(a + b)(a - b) = b(a - b)$	Factor.
$a + b = b$	Divide by $a - b$.
$a + a = a$	Substitute a in place of b.
$2a = a$	
$2 = 1.$	Divide by a ($\neq 0$).

96. In the equation

$$(x - 5)(x + 3) = 20,$$

let us set each factor equal to 20:

$$x - 5 = 20 \quad \text{or} \quad x + 3 = 20$$
$$x = 25 \qquad\qquad x = 17.$$

Are these the correct solutions? If not, tell why the method is incorrect, and then solve the equation correctly.

3.6   **Geometry Problems**

In Section 1.3, you were introduced to rectangles, circles, squares, and triangles. In this section, problems involving them lead to both linear and quadratic equations.

EXAMPLE 1

The length of a rectangle is 1 meter more than twice the width. Find the width and the length, given each of the following conditions:
a. The perimeter is 50 meters.
b. The area is 36 square meters.

Solution
These are two different problems, but the following information and picture apply to both parts.

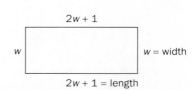

Let $w =$ width; then $2w + 1 =$ length.

a. We use the formula for perimeter, $P = 2W + 2L$.

$$2w + 2(2w + 1) = 50 \qquad 2W + 2L = P$$
$$2w + 4w + 2 = 50$$
$$6w = 48$$
$$w = 8.$$

If $w = 8$, then $2w + 1 = 2 \cdot 8 + 1 = 17$.

Answer: width = 8 meters, length = 17 meters.
Check: Perimeter = $2 \cdot 8 + 2 \cdot 17 = 16 + 34 \overset{\checkmark}{=} 50$.

b. We use the formula for area, $A = LW$.

$$w(2w + 1) = 36 \qquad LW = A$$
$$2w^2 + w - 36 = 0$$
$$(2w + 9)(w - 4) = 0$$
$$w = \frac{-9}{2}, \; w = 4.$$

We must *reject* $w = \dfrac{-9}{2}$, because the width cannot be negative.

If $w = 4$, then $2w + 1 = 2 \cdot 4 + 1 = 9$.

Answer: width = 4 meters, length = 9 meters.
Check: Area $= 4 \cdot 9 \overset{\checkmark}{=} 36$.

EXAMPLE 2

A square is 6 inches on a side. If each side is increased by a certain amount, a larger square is formed. Find this amount of increase, given each of the following conditions:

a. The perimeter of the large square is 12 inches more than the perimeter of the small square.

b. The area of the large square is 64 square inches more than the area of the small square.

Solution

As in Example 1, there are two problems, each reflected in the pictures shown.

Let x = amount each side is increased; then $6 + x$ = side of large square.

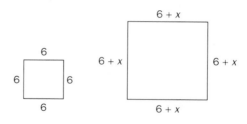

(continued)

a. The perimeter of a square with side s is $P = 4s$. Because the large perimeter is 12 more than the small perimeter, we have

$$\text{Large perimeter} = \text{small perimeter} + 12$$
$$4(6 + x) = 4(6) + 12 \qquad P = 4s$$
$$24 + 4x = 24 + 12$$
$$4x = 12$$
$$x = 3.$$

Answer: Each side of the small square must be increased by 3 inches.
Check: The large square is $6 + 3 = 9$ inches on a side.
Perimeters: $4 \cdot 9 = 4 \cdot 6 + 12$, or $36 \overset{\checkmark}{=} 24 + 12$.

b. The area of a square with side s is $A = s^2$. Because the large area is 64 more than the small area, we have

$$\text{Large area} = \text{small area} + 64$$
$$(6 + x)^2 = 6^2 + 64 \qquad A = s^2.$$
$$36 + 12x + x^2 = 36 + 64$$
$$x^2 + 12x - 64 = 0 \qquad \text{Standard form}$$
$$(x - 4)(x + 16) = 0$$
$$x = 4, \quad x = -16 \qquad \text{Reject } x = -16. \text{ Why?}$$

Answer: Each side of the small square must be increased by 4 inches.
Check: The large square is $6 + 4 = 10$ inches on a side.
Areas: $10^2 = 6^2 + 64$, or $100 \overset{\checkmark}{=} 36 + 64$.

EXAMPLE 3

A garden measuring 10 feet by 20 feet is surrounded by a path of uniform width. The combined area of garden and path is 600 square feet. Find the width of the path.

Solution
Let x = uniform width of path.

The picture shows that the length of the outer rectangle is $x + 20 + x$, or $20 + 2x$; likewise, the width of the outer rectangle is $x + 10 + x$, or $10 + 2x$. Because the combined area is the area of the outer rectangle, we have

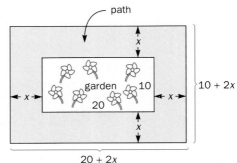

$$(20 + 2x)(10 + 2x) = 600$$
$$200 + 60x + 4x^2 = 600$$
$$4x^2 + 60x - 400 = 0 \qquad \text{Standard form}$$
$$4(x^2 + 15x - 100) = 0 \qquad \text{Common factor}$$
$$4(x - 5)(x + 20) = 0$$
$$x = 5, \quad x = -20 \qquad \text{Reject } x = -20. \text{ Why?}$$

Answer: The path is 5 feet wide.
Check: Outer length = $20 + 2 \cdot 5 = 30$; outer width = $10 + 2 \cdot 5 = 20$;
outer area = $30 \cdot 20 \overset{\checkmark}{=} 600$.

In Section 1.3, we introduced the **Pythagorean Theorem,*** which states that

"in any right triangle, the sum of the squares of
the legs equals the square of the hypotenuse."

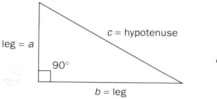

$$a^2 + b^2 = c^2$$

EXAMPLE 4

One leg of a right triangle is 2 inches more than twice the other leg. The hypotenuse is 13 inches. Find the lengths of the two legs.

Solution
Let x = length of short leg; then
$2x + 2$ = length of long leg.

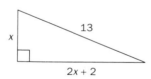

By the Pythagorean Theorem,

$$x^2 + (2x + 2)^2 = 13^2$$
$$x^2 + 4x^2 + 8x + 4 = 169$$
$$5x^2 + 8x - 165 = 0$$
$$(5x + 33)(x - 5) = 0$$
$$x = \frac{-33}{5}, \quad x = 5. \qquad \text{Reject } x = \frac{-33}{5}. \text{ Why?}$$

If $x = 5$, then $2x + 2 = 2 \cdot 5 + 2 = 12$.

Answer: short leg = 5 inches, long leg = 12 inches.
Check: $5^2 + 12^2 = 13^2$, or $25 + 144 \overset{\checkmark}{=} 169$.

---

*Pythagoras of Syracuse, 6th century B.C. Among the many people to have proven the Pythagorean Theorem is the American President James Garfield.

■ **EXERCISE 3.6**   *Answers, page A44*

 A

Solve as in **EXAMPLE 1:**

1. The length of a rectangle is 3 feet more than twice the width. Find the width and the length, given that:
   a. The perimeter is 48 feet.
   b. The area is 27 square feet.

2. The length of a rectangle is 1 inch less than three times the width. Find its dimensions, given that:
   a. Its perimeter is 30 inches.
   b. Its area is 140 square inches.

3. A garden's width is 8 meters less than its length. Find its dimensions, given that:
   a. 44 meters of fencing enclose it.
   b. It encloses 48 square meters.

4. The length of a rectangle is twice the width. Find its dimensions, given that:
   a. The perimeter is 45 centimeters.
   b. The area is 72 square centimeters.

Solve as in **EXAMPLE 2:**

5. A square is 8 inches on a side. When each side is increased by a certain amount, a larger square is formed. Find this amount of increase, given that:
   a. The perimeter of the large square is 16 inches more than the perimeter of the small square.
   b. The area of the large square is 132 square inches more than the area of the small square.

6. A square is 10 meters on a side. When each side is decreased by a certain amount, a smaller square is formed. Find this amount of decrease, given that:
   a. The smaller perimeter is 12 meters less than the larger perimeter.
   b. The smaller area is 36 square meters less than the larger area.

7. If the sides of a square are each increased by 4 feet, a new square is formed whose area is 56 square feet more than that of the original square. Find the side of each square.

8. A square of unknown sides is distorted into a rectangle by decreasing two of its opposite sides by 2 inches and increasing its other two sides by 3 inches. The rectangle has the same area as the square. Find the sides of the square and of the rectangle.

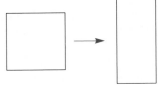

Solve as in EXAMPLE 3:

9. Surrounding a painting 10 inches by 16 inches is a frame of uniform width. The combined area of painting and frame is 280 square inches. Find the width of the frame.

10. A 12-foot by 20-foot room has wall-to-wall carpeting. A uniform strip of carpeting is removed along the edges and then replaced by tile. Find the width of the strip removed, if the remaining carpet is 180 square feet.

11. The length of a rectangle is 8 meters more than its width. When a uniform border one meter in width is added to the outside of the rectangle, a larger rectangle is formed whose area is 105 square meters. Find the dimensions of each rectangle.

12. The length of a garden is three times its width. When a path of a uniform width of 2 feet is installed around the garden, the combined area of garden and path is 220 square feet. Find the dimensions of the garden.

Use the Pythagorean Theorem to find the unknown sides of these right triangles. See EXAMPLE 4. Assume all measurements are in inches.

13.
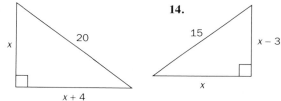

14.
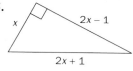

15.

16.
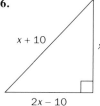

17.
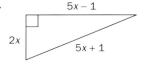

18.

19. One leg of a right triangle is 5 inches more than the other leg. The hypotenuse is 25 inches. Find the lengths of the two legs.

20. One leg of a right triangle is 4 feet less than the other leg. The hypotenuse is 4 feet greater than the longer of the two legs. Find the lengths of all three sides.

21. The three sides of a right triangle are consecutive even integers. Find them.

22. The length of a rectangle is 4 meters more than twice the width. The diagonal is 26 meters. Find the length and width. *Hint*: Draw a picture.

Solve using any method:

23. Surrounding a photograph 4 inches by 7 inches is a frame of uniform width. Find the width of the frame if the area of the frame alone is 26 square inches.

24. The length of a garden is 5 meters longer than twice its width. A path of a uniform width of 1 meter is installed around the garden. Find the dimensions of the garden, given that:
    a. The outer perimeter is 162 meters.
    b. The area of the path is 46 square meters.

25. The length of a rectangle is 3 feet more than twice the width. When all dimensions are doubled, a larger rectangle is formed. Find the dimensions of each rectangle, given that:
    a. The large perimeter is 42 feet more than the small perimeter.
    b. The large area is 42 square feet more than the small area.

26. Find the side of a square whose area is numerically equal to its perimeter.

27. The length of a rectangle is three more than the width. Find the sides of the rectangle if the area is numerically equal to the perimeter.

28. The picture shows a square 8 inches on a side surrounded by a shaded area (path) of uniform width. If the shaded area is 80 square inches, find the width of the path.

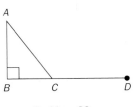

Problem 28

29. A rope is hanging from the top of a flagpole. The rope is 3 feet longer than the pole, so the excess rope is on the ground. When the rope is pulled taut, the end of the rope reaches a point on the ground 9 feet from the base of the pole. How high is the pole?

30. A ladder is resting vertically against a wall. When the bottom is pulled 12 feet away from the wall, the top of the ladder slides 4 feet down the wall. How long is the ladder?

31. While swinging from its highest point to its lowest, the end of a pendulum drops 2 feet vertically while moving 6 feet horizontally. How long is the pendulum? See the accompanying diagram.

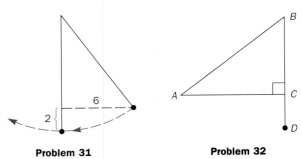

Problem 31   **Problem 32**   **Problem 33**

32. In the accompanying diagram, $AB = BD$, $AC = 15$, and $CD = 5$. Find AB.

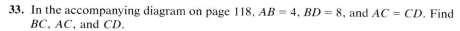

33. In the accompanying diagram on page 118, $AB = 4$, $BD = 8$, and $AC = CD$. Find BC, AC, and CD.

34. The side of one square is 5 inches longer than the side of another square. When the squares are attached at the dotted line as shown, a six-sided polygon is formed whose *outer* perimeter is 62 inches. Find the sides of each square.

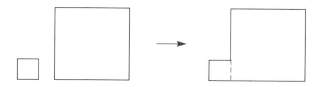

35. In the accompanying drawing, the middle square is 4 inches greater on a side than the small square. The large square is 6 inches greater on a side than the middle square. If the squares are attached at the dotted lines as shown, the eight-sided polygon formed has an outer perimeter of 72 inches. Find the sides of each square.

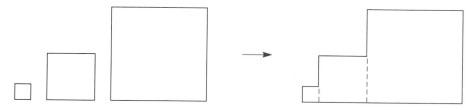

36. Right triangles whose sides are natural numbers can be obtained from the following picture. (See Exercise 1.3, Problem 43).

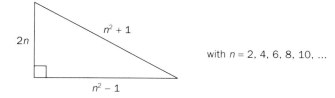

with $n = 2, 4, 6, 8, 10, \ldots$

For example, if $n = 2$, then the sides of a right triangle are

$$2 \cdot 2, \quad 2^2 - 1, \quad 2^2 + 1, \quad \text{or} \quad 4, 3, 5. \qquad \text{Check:} \quad 4^2 + 3^2 \overset{\checkmark}{=} 5^2$$
$$16 + 9 \overset{\checkmark}{=} 25$$

a. You obtain the three sides of a right triangle for each of the following numbers: $n = 4, 6, 8, 10, 12, 14,$ and 20. Check your answers in each case.

b. Using algebra, show that the triangle above satisfies

$$(2n)^2 + (n^2 - 1)^2 = (n^2 + 1)^2.$$

■ **CHAPTER 3 · SUMMARY**

3.1 Multiplying Polynomials

$a^m a^n = a^{m+n}$

Example:
$$(-2x^3y^2)(-5xy^4) = 10x^4y^6.$$

Distributive Property

Example:
$$3ab(2a - 5b) = 6a^2b - 15ab^2.$$

FOIL

Example:
$$(2x + 3)(x - 6) = 2x^2 - 9x - 18.$$

$(A \pm B)^2 = A^2 \pm 2AB + B^2$

Example:
$$(3a + 5)^2 = 9a^2 + 30a + 25.$$

$(A + B)(A - B) = A^2 - B^2$

Example:
$$(5x + 9)(5x - 9) = 25x^2 - 81.$$

3.2 Common Factors and Grouping

Example:
$$2x^2 - 4x + 3ax - 6a = 2x(x - 2) + 3a(x - 2)$$
$$= (x - 2)(2x + 3a).$$

3.3 Factoring Trinomials

Examples:
$$x^2 + 12x + 35 = (x + 7)(x + 5).$$
$$4a^2 - 5ab - 6b^2 = (4a + 3b)(a - 2b).$$

3.4 Factoring Squares and Cubes

$A^2 - B^2 = (A + B)(A - B)$

Example:
$$4x^2 - 25y^2 = (2x + 5y)(2x - 5y).$$

$A^3 + B^3 = (A + B)(A^2 - AB + B^2)$

Example:
$$x^3 + 27 = (x + 3)(x^2 - 3x + 9).$$

$A^3 - B^3 = (A - B)(A^2 + AB + B^2)$

Example:
$$8a^3 - 125 = (2a - 5)(4a^2 + 10a + 25).$$

Factoring Flow Chart:

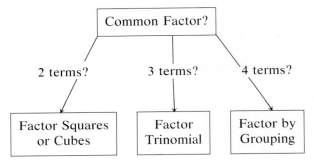

Example:
$$8x^5 - 26x^3 + 18x = 2x(4x^4 - 13x^2 + 9)$$
$$= 2x(4x^2 - 9)(x^2 - 1)$$
$$= 2x(2x + 3)(2x - 3)(x + 1)(x - 1).$$

3.5 Factoring Second-Degree Equations

Bring equation to the form
$$ax^2 + bx + c = 0,$$

then factor, set each factor equal to zero, and solve.

Example:
$$2x^2 + 9x - 5 = 0$$
$$(2x - 1)(x + 5) = 0$$
$$2x - 1 = 0, \quad x + 5 = 0$$
$$x = \frac{1}{2}, \quad x = -5.$$

CHAPTER 3 · SUMMARY

3.6 Geometry Problems

Rectangle: $P = 2L + 2W$, $A = LW$
Square: $P = 4s$, $A = s^2$
Pythagorean Theorem: $a^2 + b^2 = c^2$

Example:

"The length of a rectangle is 7 more than the width. The area is 60. Find the dimensions."

$$x(x + 7) = 60 \qquad LW = A$$
$$x^2 + 7x - 60 = 0$$
$$(x - 5)(x + 12) = 0$$
$$x = 5, \quad x = -12 \qquad \text{Reject } x = -12.$$
$$\text{width} = 5, \quad \text{length} = 5 + 7 = 12.$$

CHAPTER 3 · REVIEW EXERCISES   *Answers, pages A44–A46*

3.1

Multiply:

1. $(-3x^3y)(-5xy^2)$

2. $4^5 \cdot 4^7$

3. $(a + b)^8(a + b)^2(a + b)$

4. $x^{2m}y^{3n} \cdot x^{5m}y^{n+2}$

5. $(-4m^2n)(-3mn^3)(-mn)$

6. $3x(x^2 - 2x + 5) - x(2x^2 + 4x - 1)$

7. $(2x - 7)(x + 3)$

8. $(3a - 7b)(2a - 5b)$

9. $(a^2b + 2c)(2a^2b + 5c)$

10. $(2x + 3y)(2y - 5x)$

11. $(4a + 3b)(2a - 7)$

12. $(2x + 3y)^2$

13. $(5a - 2b)^2$

14. $(3a^2 - 7b^3)^2$

15. $(3x + 8)(3x - 8)$

16. $(13 - 17c^4)(13 + 17c^4)$

17. $3x(5x - 7)(x - 1)$

18. $-2ab(3a - 5b)^2$

19. $2(2a - 1)(a + 3) - 4(3a - 5)^2$

20. $(a^4 + b^4)(a^2 + b^2)(a + b)(a - b)$

21. $(a + b + 5)(a + b - 5)$

22. $(x - y - z)(x - y + z)$

23. $(x - 3)(3x^2 - 7x + 4)$

24. $(3a + 2)^3$

25. $(2a - 3b)(4a^2 + 6ab + 9b^2)$

26. $(x^2 + 2x + 4)(x^2 - 2x + 4)$

27. $(x - 1)(x^4 + x^3 + x^2 + x + 1)$

28. $(2x + 5)(2x - 5)(x + 3)(x - 3)$

29. $(x + 1)^2(x - 1)^2$

30. $(a - 2)(2a + 7)(a + 2)(2a - 7)$

31. $(x + h)^2 + 3(x + h) - 7 - (x^2 + 3x - 7)$

32. $(x + h)^3 + (x + h)^2 - 4(x + h) + 1 - (x^3 + x^2 - 4x + 1)$

3.2, 3.3, 3.4

Factor completely. First look for the greatest common factor.

33. $12x^3y - 4x^2y^3 - 6xy^2$

34. $-8ax - 12ay + 20az$

35. $5x^2 + 10xy + 3x + 6y$

36. $7a^2b - 21ab - 2ac + 6c$

37. $5x^2 + 2xy + 5x + 2y$

38. $ax + 14b - 7bx - 2a$

39. $x^2 + 13xy + 42y^2$

40. $a^2 - 16a + 48$

41. $2x^2 - x - 21$

42. $3a^2 + 2ab - 16b^2$

43. $4m^2 + 3mn - 10n^2$

44. $4x^2 - 36x + 45$

45. $4x^2 - 31xy - 45y^2$

46. $6u^2 + 17uv + 10v^2$

47. $25x^2 + 49 + 70x$

48. $9a^2 + 100b^2 - 60ab$

49. $2x^3 - 12x^2 - 144x$

50. $-3x^3 - 9x^2 + 162x$

51. $10m^3n + 36m^2n^2 - 16mn^3$

52. $-30a^3b + 35a^2b^2 + 100ab^3$

53. $16t^2 - 46t + 15$

54. $-24x^{n+2} - 26x^{n+1} + 28x^n$

55. $24x^2 + 55x - 24$

56. $24x^2 - 10xy - 21y^2$

57. $4x^2 - 9y^2$

58. $144x^2 - 225y^2z^2$

59. $x^4y^4 - 16z^4$

60. $3x^9 - 363x^7y^2$

61. $(x + y)^2 - 16z^2$

62. $a^7b^4c^3 - a^3b^8c$

63. $x^3 - 8y^3$

64. $a^3b^3 + 27c^3$

65. $64x^3 + 1331$

66. $125a^3b^6 - 729c^9$

67. $3a^4b - 3000ab^4$

68. $(x - y)^3 - 8y^3$

69. $9x^4 - 37x^2 + 4$

70. $x^6 + 16x^3y^3 + 64y^6$

71. $50x^5 + 182x^3 - 72x$

72. $64x^6 - y^6$

73. $2x^7y - 52x^4y^4 - 54xy^7$

74. $x^6 - 4x^4y^2 - 16x^2 + 64y^2$

75. $4x^5 + 4x^2 - 25x^3 - 25$

76. $x^2 + 6xy + 9y^2 - 4z^2$

3.5

Solve:

77. $x^2 - 9x - 22 = 0$

78. $2y^2 - 3y - 5 = 0$

79. $z^2 = 20z - 91$

80. $6t^2 = 14 + 17t$

81. $4v^2 = 9v$

82. $4v^2 = 9$

83. $27 = 3x^2$

84. $27x = 3x^2$

85. $-16t^2 + 64t + 80 = 0$

86. $-16t^2 + 80t = 0$

87. $18w - 5w^2 = 9$

88. $252 - y^2 = 4y$

89. $5 = 13x - 6x^2$

90. $9n^2 = 6 - 25n$

91. $11x = 21 - 6x^2$

92. $25y = 30 - 20y^2$

93. $x^2 - 26x - 87 = 0$

94. $0 = 57 + 22x + x^2$

95. $x^2 + 5x - 126 = 0$

96. $x^2 + 15x - 126 = 0$

97. $x^2 - 29x - 132 = 0$

98. $x^2 + 41x - 132 = 0$

99. $3x^2 - 50x - 5000 = 0$

100. $5x^2 + 16x - 660 = 0$

101. $12x^2 - 25x + 2 = 0$

102. $12x^2 - 25x + 12 = 0$

103. $(5t - 2)(2t + 3) = (t + 6)(3t - 1)$

104. $3x(2x - 7) = (2x + 3)^2 - 25$

105. $7y - (2y + 5)(2y - 5) = 13 + (3y + 2)(1 + 2y)$

106. $(z - 1)(z^2 + 3z - 2) = z(z - 3)(z - 6)$

107. $-24m^2 + 153m + 105 = 0$

108. $48x^2 = 22x + 15$

109. $|x^2 + 4x + 2| = 2$

110. $|x^2 - 18| = 18$

111. $|x^2 - 2x - 9| = 6$

112. $|x^2 + 4x - 24| = 6x$ *Hint*: Check your answers.

113. $2x^3 + 2x^2 - 12x = 0$

114. $4y^3 - 4y^2 - 15y = 0$

115. $x^4 - 13x^2 + 36 = 0$

116. $9y^5 - 37y^3 + 4y = 0$

117. $x^3 + 3x^2 - 25x - 75 = 0$

118. $8z^3 - 20z^2 - 18z + 45 = 0$

119. When twice the square of a number is decreased by the number, the result is three more than four times the number. Find the number.

120. The square of the difference between two and a number equals the difference between two and the square of the number. What is the number?

121. One number is one more than twice another. The sum of their squares is one more than four times the sum of the two numbers. Find the numbers.

122. Find two consecutive even integers such that the sum of their squares is 52.

123. The sum of the squares of the first, second, and third of four consecutive integers is two more than three times the square of the fourth. What are the integers?

3.6

Solve:

124. The length of a rectangle is 3 inches more than its width. Find the width and the length, given that:
   a. The perimeter is 26 inches.
   b. The area is 108 square inches.

125. If the sides of a square were each increased by 3 meters, a new square would be formed whose area is 51 square meters greater than the area of the original square. Find the side of each square.

126. The length of a rectangle is 5 feet more than the width. If the width is decreased by 1 foot and the length is increased by 4 feet, a new rectangle is formed with the same area as the original rectangle. Find the dimensions of each rectangle.

127. Surrounding a garden 10 feet by 18 feet is a path of uniform width. The combined area of garden and path is 240 square feet. Find the width of the path.

128. The length of a photograph is 1 inch less than twice its width. Surrounding the picture is a frame of uniform width 2 inches. The area of the frame is 72 square inches. Find the dimensions of the photograph.

Use the Pythagorean Theorem to find the unknown sides of these right triangles:

129.
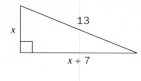

130.
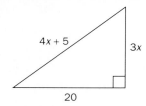

131.
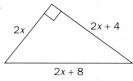

132. The length of a rectangle is 2 inches more than twice the width. The diagonal is 1 inch longer than the length. Find the width, the length, and the diagonal.

133. A ladder is resting vertically against a wall. When the bottom of the ladder is pulled horizontally 20 feet away from the wall, the top of the ladder slides 10 feet down the wall. How long is the ladder?

Find x in each figure. C represents the center of each circle. *Hints:* All radii in any circle are equal, and use the Pythagorean Theorem.

134.
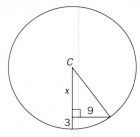

135.
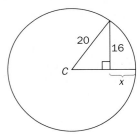

CHAPTER 3 · TEST   *Answers, page A46*

Multiply:

1. $(-4a^5b^6)(-2ab)(3b^2)$

2. $(2x - 7)(x + 1) - (x - 3)^2$

3. $(x - 2)(x^3 + 2x^2 + 4x + 8)$

4. $(2x - 3y)(2x + 3y)(5x + y)(5x - y)$

Factor completely:

5. $x^2 + 5x - 36$

6. $9x^2 - 169y^2$

7. $3a^2 - 11ab + 10b^2$

8. $6a^2 + 13a - 15$

9. $27x^3 + 1000y^3$

10. $2ax^2 - bx + 6abx - 3b^2$

11. $-12x^3y - 10x^2y^2 + 12xy^3$

12. $x^5 - 4x^3 + 8x^2 - 32$

Solve:

13. $3x^2 - 5x - 2 = 0$

14. $4z^2 = 25$

15. $4z^2 = 25z$

16. $6x^2 = 17x + 10$

17. $4x^4 - 101x^2 + 25 = 0$

18. When twice the square of a number is decreased by seven, the result is the same as adding one to four times the sum of the number and two. Find the number.

19. The length of a rectangle is 2 inches more than twice the width. The area is 144 square inches. Find the width and the length.

20. The sides of a right triangle are consecutive integers. Find them.

CHAPTERS 1—3 · CUMULATIVE REVIEW EXERCISES   *Answers, page A47*

Perform the indicated operations and simplify:

1. $\dfrac{5}{8} \cdot \dfrac{-16}{25} \div \dfrac{-4}{15}$

2. $|5 - 9|$

3. $(-4a^2b)(5ab^3)(-a^3)$

4. $\left|\dfrac{1}{4} - \dfrac{2}{3} - \dfrac{5}{8}\right|$

5. $(2x + 3y)(5x - y)$

6. $3a + 7[5a - 2(4 - 7a)]$

7. $(5y - 9z)(5y + 9z)$

8. $(7x + 11)^2$

9. $(-2)^3 - (-3)^2$

10. $(3a - 2)(9a^2 + 6a + 4)$

11. $(x^2 + 3x + 9)(x^2 - 3x + 9)$

12. $(2x + 3)(2x - 3)(x + 1)(x - 1)$

13. $(x + 4)^2 - (x - 2)(x + 3)$

14. $3xy(2x^2 - xy + y^2) - xy(x^2 - y^2 + 5xy)$

Use the Pythagorean Theorem to decide which of these sets of three numbers can be the sides of a right triangle:

15. 7, 11, 13

16. 9, 41, 40

17. 0.5, 1.2, 1.3

18. $1\dfrac{1}{4}, 1\dfrac{2}{3}, 2\dfrac{1}{12}$

Find the area of each figure. All measurements are in meters.

19.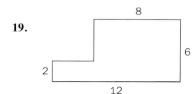

20.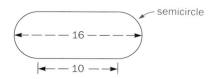

Factor completely:

21. $2a^2 - 5ab - 3b^2$

22. $2a^2 - 5ab + 3b^2$

23. $6x^2 - 19x + 15$

24. $6x^2 + 13x - 15$

25. $x^2 - 64$

26. $x^3 - 64$

27. $a^3 + 27b^3c^3$

28. $2x^3 + 12x^2 - 182x$

29. $3x^3y - 147xy^3$

30. $6x^2 + 10ax - 9x - 15a$

31. $x^4 - 16$

32. $x^5 - 9x^3 + 8x^2 - 72$

Solve each equation and solve and graph each inequality:

33. $5x - 2(x - 3) = 4(x - 7)$

34. $6(2x + 5) = 5(3x + 6)$

35. $3x^2 + x - 2 = 0$

36. $|2x - 1| = 7$

37. $3x^2 - 27 = 0$

38. $x - 4.1 = 0.2x + 5.5$

39. $3x - 1 < 7 - x$

40. $-5 \le 3 - 2x \le 7$

41. $x^2 = 2(x + 4)$

42. $6y^2 - 10y - 4 = 0$

43. $4z^2 - 25 = 0$

44. $4z^2 - 25z = 0$

45. $x(2 - x) = 1$

46. $x^2 + (x + 2)^2 = (x + 4)^2$

47. $|x + 4| > 1$

48. $|3 - 2x| \le 3$

49. $4t^4 - 37t^2 + 9 = 0$

50. $27w^3 + 18w^2 - 12w - 8 = 0$

Set up and solve:

51. Twice the sum of a number and four equals the difference between four and twice the number. Find the number.

52. One more than twice the square of a number equals four more than the number. What is the number?

53. The sum of the squares of three consecutive integers is 4 more than 5 times the second of these integers. Find the integers.

54. Thirty-seven coins in nickels and dimes is worth $2.60. How many of each coin are there?

55. In a collection of coins, there is one more dime than quarters, and twice as many nickels as dimes. The value of the quarters equals the total value of the nickels and dimes. How many of each coin are there?

56. $50,000 is invested, part at 7% for three years, an equal amount at 12% for two years, and the balance at $8\frac{1}{2}\%$ for one year. The total interest earned is $8450. How much is invested at each rate?

57. *A* leaves town at 1:00 p.m. travelling 60 m.p.h. *B* leaves town at 2:30 p.m. travelling 75 m.p.h., in pursuit of *A*. At what time does *B* overtake *A*? How far from town does this occur?

58. The length of a rectangle is three inches more than twice its width. Find its width, given that:
a. the perimeter is 48 inches;   **b.** the area is 90 square inches.

Use the Pythagorean Theorem to find the unknown sides of these right triangles:

59.

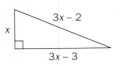

60.

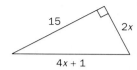

61. The sides of a right triangle are consecutive even integers. Find them.

4 RATIONAL EXPRESSIONS

4.1 Reducing, Multiplying, and Dividing Rational Expressions

A **rational expression** is the ratio of two polynomials. Examples of rational expressions include

$$\frac{-3z}{7w}, \quad \frac{5x}{2x + 8}, \quad \text{and} \quad \frac{7x + y}{y^2 - 2y - 3}.$$

Because division by zero is undefined (see Section 1.1), each of the foregoing expressions is defined only when its denominator is not zero. Thus

$$\frac{-3z}{7w} \text{ is defined, provided that } 7w \neq 0, \text{ or } w \neq 0.$$

The next example further illustrates how we place restrictions on variables in the denominator of a rational expression.

EXAMPLE 1 State any restrictions on the variables in each expression:

a. $\dfrac{5x}{2x + 8}.$

The denominator $2x + 8$ must *not* equal zero:

$$2x + 8 \neq 0, \quad \text{or} \quad x \neq -4.$$

b. $\dfrac{7x + y}{y^2 - 2y - 3}.$

The denominator $y^2 - 2y - 3$ must *not* equal zero:

$$(y - 3)(y + 1) \neq 0, \quad \text{or} \quad y \neq 3, -1.$$

The fraction $\dfrac{12}{18}$ can be reduced to lowest terms by factoring:

$$\frac{12}{18} = \frac{2 \cdot 6}{3 \cdot 6} = \frac{2}{3}.$$

This is a special instance of the more general rule for reducing rational expressions:

The Fundamental Principle of Fractions

$$\frac{AC}{BC} = \frac{A}{B}, \quad B, C \neq 0$$

We can use this rule to divide powers with the same base. For example, if $a \neq 0$,

$$\frac{a^5}{a^2} = \frac{\cancel{a}\cancel{a}aaa}{\cancel{a}\cancel{a}} = a^3 \quad \text{and} \quad \frac{a^2}{a^5} = \frac{\cancel{a}\cancel{a}}{\cancel{a}\cancel{a}aaa} = \frac{1}{a^3}.$$

Thus we obtain the quotient of two powers with the same base by *subtracting* exponents. If the exponent in the numerator is larger than the exponent in the denominator, the result is a positive power in the numerator; if the exponent in the denominator is larger than the exponent in the numerator, the result is a positive power in the denominator.

Quotient Rule of Exponents

$$\frac{a^m}{a^n} = \begin{cases} a^{m-n} & \text{if } m > n \\ \dfrac{1}{a^{n-m}} & \text{if } m < n \end{cases}$$

Question: What is $\frac{a^m}{a^n}$ if $m = n$?   *Hint:* What is $\frac{a^3}{a^3}$?

Our next item concerns signs of fractions. We saw in Section 1.1 that, by convention, the rational number $\frac{7}{-4}$ is written as either $\frac{-7}{4}$ or $-\frac{7}{4}$. More generally, the rational expression

$$\frac{A}{-B} \quad \text{should be written as} \quad \frac{-A}{B} \quad \text{or} \quad -\frac{A}{B}.$$

We now apply these facts to reduce the following expressions to lowest terms.

EXAMPLE 2 | Reduce to lowest terms. Assume the denominators do not equal zero.

a. $\dfrac{x^7}{x^4} = x^{7-4}$

$\qquad = x^3.$

b. $\dfrac{y^2}{y^8} = \dfrac{1}{y^{8-2}}$

$\qquad = \dfrac{1}{y^6}.$

c. $\dfrac{8a^3b^5c^6(x+y)^9}{-6a^4bc^6(x+y)^5} = \dfrac{-8b^{5-1}(x+y)^{9-5}}{6a^{4-3}}$ *Note:* $\dfrac{c^6}{c^6} = 1$

$\qquad\qquad = \dfrac{-4b^4(x+y)^4}{3a}.$

d. $\dfrac{x^{7n}y^{5m}z^{n+3}}{x^{2n}y^{6m}z^{n-1}} = \dfrac{x^{7n-2n}z^{(n+3)-(n-1)}}{y^{6m-5m}}$

$\qquad\qquad = \dfrac{x^{5n}z^4}{y^m}.$ $z^{(n+3)-(n-1)} = z^{n+3-n+1} = z^4$

Here we assume m and n are positive integers.

Rational expressions can be reduced to lowest terms by *first factoring* both numerator and denominator, if possible, and then applying the Fundamental Principle of Fractions.

EXAMPLE 3 | Reduce to lowest terms. Assume the denominators do not equal zero.

a. $\dfrac{3x^3}{6x^2 + 12x} = \dfrac{3x^3}{6x(x+2)}$ Common factor.

$\qquad\qquad = \dfrac{x^2}{2(x+2)}.$ Reduce.

b. $\dfrac{2a-4b}{4a-8b} = \dfrac{2(a-2b)}{4(a-2b)}$ Common factors.

$\qquad\qquad = \dfrac{1}{2}.$ Reduce.

c. $\dfrac{x^2 - 2x - 8}{x^3 + 8} = \dfrac{(x-4)(x+2)}{(x+2)(x^2 - 2x + 4)}$ Factor trinomial and cubes.

$\qquad\qquad = \dfrac{x-4}{x^2 - 2x + 4}.$ Reduce.

(continued)

d. $\dfrac{x^2 - 4}{-x - 2} = \dfrac{(x + 2)(x - 2)}{-1(x + 2)}$ Factor squares and a negative.

$\qquad = \dfrac{x - 2}{-1}$ Reduce.

$\qquad = \dfrac{-(x - 2)}{1}$ Move negative sign to numerator.

$\qquad = 2 - x.$ Change signs.

Before continuing, it is important to warn students that we must guard against the *incorrect* "canceling" of terms, illustrated by

$$\frac{a^2 + 3a + 2}{a^2 + 4a + 3} = \frac{\overset{1}{\cancel{a^2}} + 3\cancel{a} + 2}{\underset{1}{\cancel{a^2}} + 4\cancel{a} + 3} = \frac{6}{8} = \frac{3}{4}.$$

To see that this is *false*, we need only substitute $a = 2$ into the expression:

$$\frac{2^2 + 3 \cdot 2 + 2}{2^2 + 4 \cdot 2 + 3} = \frac{4 + 6 + 2}{4 + 8 + 3} = \frac{12}{15} = \frac{4}{5} \neq \frac{3}{4}.$$

The Fundamental Principle of Fractions allows rational expressions to be reduced only when both numerator and denominator are products of factors, not sums of terms. Thus $\dfrac{AC}{BC} = \dfrac{A}{B}$ is correctly reduced, but $\dfrac{A + C}{B + C}$ cannot be reduced. The moral of this story is that we *can reduce factors* but we *cannot "cancel" terms.* The rational expression illustrated above can be correctly reduced by *first factoring:*

$$\frac{a^2 + 3a + 2}{a^2 + 4a + 3} = \frac{(a + 2)(a + 1)}{(a + 3)(a + 1)} = \frac{a + 2}{a + 3}.$$

The product and the quotient of two rational expressions are the same as those for two rational numbers:

Product and Quotient of Rational Expressions

$$\frac{A}{B} \cdot \frac{C}{D} = \frac{AC}{BD} \quad \text{and} \quad \frac{A}{B} \div \frac{C}{D} = \frac{A}{B} \cdot \frac{D^*}{C}$$

The quotient is obtained by inverting the divisor and then multiplying.

---

*You will be asked to prove this fact in Problem 86 in the exercises.

EXAMPLE 4

Perform the indicated operations. Assume the denominators do not equal zero.

a. $\dfrac{14a^3}{11} \cdot \dfrac{10}{7a} = \dfrac{\overset{2a^2}{\cancel{14a^3}}}{11} \cdot \dfrac{10}{\underset{1}{\cancel{7a}}}$ Reduce.

$\qquad = \dfrac{20a^2}{11}.$

b. $\dfrac{5x}{9y} \div \dfrac{-20x^4}{21y^2} = \dfrac{5x}{9y} \cdot \dfrac{21y^2}{-20x^4}$ Invert divisor.

$\qquad = \dfrac{\overset{1}{\cancel{5x}}}{\underset{3}{\cancel{9y}}} \cdot \dfrac{\overset{7y}{\cancel{21y^2}}}{\underset{-4x^3}{\cancel{-20x^4}}}$ Reduce.

$\qquad = \dfrac{7y}{-12x^3}$

$\qquad = \dfrac{-7y}{12x^3}.$

c. $\dfrac{6a+8}{6a^2+6a} \cdot \dfrac{3a^2-a-4}{9a^2-16} = \dfrac{2(3a+4)}{6a(a+1)} \cdot \dfrac{(3a-4)(a+1)}{(3a-4)(3a+4)}$ Factor.

$\qquad = \dfrac{2}{6a}$ Binomials all reduce.

$\qquad = \dfrac{1}{3a}.$

d. $\dfrac{x^3-27}{x^2-16} \div \dfrac{x^2-9}{x^2-x-12}$

$\qquad = \dfrac{x^3-27}{x^2-16} \cdot \dfrac{x^2-x-12}{x^2-9}$ Invert divisor.

$\qquad = \dfrac{(x-3)(x^2+3x+9)}{(x+4)(x-4)} \cdot \dfrac{(x-4)(x+3)}{(x-3)(x+3)}$ Factor.

$\qquad = \dfrac{x^2+3x+9}{x+4}.$ Reduce.

e. $(a+5) \cdot \dfrac{2x+6}{a^2-25} \div \dfrac{x+3}{a-5} = \dfrac{a+5}{1} \cdot \dfrac{2(x+3)}{(a+5)(a-5)} \cdot \dfrac{a-5}{x+3}$ Invert divisor.

$\qquad = 2.$ Reduce.

In the next example, expressions of the form $\dfrac{A - B}{B - A}$ occur. It is a fact that

$$\frac{A - B}{B - A} = -1 \text{ because } A - B = -1(B - A), \text{ or } A - B \overset{\checkmark}{=} -B + A.$$

$$\frac{A - B}{B - A} = -1$$

For example, $\dfrac{x - 2}{2 - x} = -1$ and $\dfrac{a - 3b}{3b - a} = -1$. This useful reduction will be illustrated in the next example.

EXAMPLE 5 | Simplify. Assume the denominators do not equal zero.

a. $\dfrac{x^2 - 4}{6 - 2x} = \dfrac{(x + 2)\overset{-1}{\cancel{(x - 2)}}}{3\cancel{(2 - x)}}$ $\dfrac{x - 2}{2 - x} = -1$

$$= \frac{-(x + 2)}{3} \quad \text{or} \quad \frac{-x - 2}{3}.$$

b. $\dfrac{a^2 - 2ab - 3b^2}{7a + 7b} \cdot \dfrac{10a}{15ab - 5a^2} = \dfrac{(a + b)\overset{-1}{\cancel{(a - 3b)}}}{7(a + b)} \cdot \dfrac{\overset{2}{\cancel{10a}}}{\cancel{5a}(3b - a)}$

$$= \frac{-2}{7}.$$

EXERCISE 4.1   *Answers, pages A47–A48*

A

State any restrictions on the variables in each expression, as in EXAMPLE 1:

1. $\dfrac{12}{5x}$ **2.** $\dfrac{-4a}{3y}$ **3.** $\dfrac{2x}{15}$ **4.** $\dfrac{z - 2}{4}$

5. $\dfrac{8}{x - 3}$ **6.** $\dfrac{y + 1}{5 + y}$ **7.** $\dfrac{x + 3}{(x - 5)(x + 2)}$ **8.** $\dfrac{2y + 1}{(2y - 1)(y + 3)}$

9. $\dfrac{x - 4}{x^2 + x - 6}$ **10.** $\dfrac{4t}{3t^2 - t - 2}$ **11.** $\dfrac{6}{z^2 - 4z}$ **12.** $\dfrac{6}{z^2 - 4}$

Reduce to lowest terms, as in EXAMPLES 2 and 3. Assume the denominators do not equal zero. Here *m* and *n* are positive integers.

13. $\dfrac{a^8}{a^2}$

14. $\dfrac{x^{13}}{x^2}$

15. $\dfrac{x^3}{x^9}$

16. $\dfrac{y}{y^5}$

17. $\dfrac{4a^8b^2c^5}{-6ab^4c^5}$

18. $-\dfrac{27x^5(y+z)^3}{-18x^6(y+z)}$

19. $\dfrac{x^{3n}y^{m+1}}{x^{8n}y^{m-2}}$

20. $\dfrac{a^{7n}b^{2n+3}}{a^n b^{n+1}}$

21. $\dfrac{5a^2}{10a^3 + 15a}$

22. $\dfrac{x^2 - x}{x}$

23. $\dfrac{6x^4y - 4x^2y^2}{4x^3y^2}$

24. $\dfrac{3cd^4}{9c^2d^2 + 12cd^3}$

25. $\dfrac{2a + 10b}{4a + 20b}$

26. $\dfrac{6x - 24}{3x - 12}$

27. $\dfrac{b^2 - b - 6}{3b + 6}$

28. $\dfrac{4a - 16}{a^2 - a - 12}$

29. $\dfrac{x^2 - 9}{12x^2 - 18x - 54}$

30. $\dfrac{16c^2 - 56c + 48}{4c^2 - 9}$

31. $\dfrac{2x^2 + 13x + 21}{4x^2 + 12x - 7}$

32. $\dfrac{3a^2 - 26a + 35}{6a^2 - a - 15}$

33. $\dfrac{x^3 - 8y^3}{2x^2 - xy - 6y^2}$

34. $\dfrac{3a^2 + 4ab - 15b^2}{a^3 + 27b^3}$

35. $\dfrac{x^3 - 16x}{x^3 - 64}$

36. $\dfrac{a^3 - 1}{a^3 - a}$

37. $\dfrac{a^2 - 25b^2}{-2a - 10b}$

38. $\dfrac{-x^2 + 2x + 15}{-x^2 - x + 6}$

Perform the indicated operations, as in EXAMPLE 4. Assume the denominators do not equal zero.

39. $\dfrac{8x^4}{17y^3} \cdot \dfrac{34y}{24x}$

40. $\dfrac{10a}{27b^3} \cdot \dfrac{9b^4}{2a^3}$

41. $\dfrac{8x^3}{3y^2} \div \dfrac{-4x}{15y^5}$

42. $\dfrac{-26a^5}{9b} \div \dfrac{-39a^2}{18b^4}$

43. $-16ab \cdot \dfrac{-12a^2b^3}{15c^3} \div \dfrac{-8ab^5}{10c^5}$

44. $\dfrac{-25c^2d}{12e} \cdot (-16c^2de) \div \dfrac{-5c^2e^4}{8d}$

45. $5x^3 \div \dfrac{-15x}{8x^2} \cdot \dfrac{3}{4x}$

46. $-7a^4 \div \dfrac{14a^3}{3a} \cdot \dfrac{-2}{9a}$

47. $\dfrac{2x + 6y}{3x - 6y} \cdot \dfrac{6x - 12y}{8x + 24y}$

48. $\dfrac{10a - 5b}{12a + 36b} \cdot \dfrac{8a + 24b}{20a - 10b}$

49. $\dfrac{4x + 4}{10a - 10} \div \dfrac{6x + 6}{5a - 5}$

50. $\dfrac{4a - 3}{5a + 10} \div \dfrac{8a - 6}{15a + 30}$

51. $\dfrac{2a^2b - 10ab^2}{6a^3 + 3a^2b} \cdot \dfrac{18a^4b + 9a^3b^2}{4a^3b - 20a^2b^2}$

52. $\dfrac{5c^3d - 5c^2d^2}{24c^2 + 36cd} \div \dfrac{15c^2d^3 - 15cd^4}{16c^3 + 24c^2d}$

53. $\dfrac{x^2 - 2x - 3}{2x^2 + 3x + 1} \cdot \dfrac{2x^2 + 7x + 3}{x^2 - 9}$

54. $\dfrac{a^2 + a - 6}{a^2 - 2a - 15} \cdot \dfrac{a^2 - 25}{a^2 + 3a - 10}$

55. $\dfrac{x^2 + 6x + 9}{x^2 - 3x - 10} \div \dfrac{x^2 - 9}{x^2 - 8x + 15}$

56. $\dfrac{a^2 + 4ab + 4b^2}{a^2 - 16b^2} \div \dfrac{a^2 - 2ab - 8b^2}{a^2 - 8ab + 16b^2}$

57. $\dfrac{8x^3 - 27}{9x^2 - 4} \cdot \dfrac{9x^2 - 12x + 4}{6x^2 - 13x + 6}$

58. $\dfrac{25y^2 - 16}{64y^3 + 125} \div \dfrac{25y^2 + 40y + 16}{20y^2 + 9y - 20}$

Simplify using $\dfrac{A - B}{B - A} = -1$, as in EXAMPLE 5. Assume the denominators do not equal zero.

59. $\dfrac{x^2 - 9}{3 - x}$

60. $\dfrac{a^2 - 16b^2}{4b - a}$

61. $\dfrac{6a - 12b}{18b - 9a}$

62. $\dfrac{30 - 10x}{5x - 15}$

63. $\dfrac{9 - x^2}{x^3 - 27}$

64. $\dfrac{b^2 - 4a^2}{8a^3 - b^3}$

65. $\dfrac{3a - 6b}{4x - 7y} \cdot \dfrac{21y - 12x}{12b - 6a}$

66. $\dfrac{5x - 10}{8x - 4} \div \dfrac{28 - 14x}{6 - 12x}$

67. $\dfrac{-8x + 24}{-2x^2 + 7x - 5} \cdot \dfrac{2x - 5}{12 - 4x}$

68. $\dfrac{x^2 - 2x - 15}{1 - x^2} \cdot \dfrac{8 - 8x}{-2x - 6}$

B

Simplify. Assume the denominators do not equal zero.

69. $\dfrac{-2x^2 + 2xy + 12y^2}{12y - 4x}$

70. $\dfrac{40b - 16a}{-8a^2 + 44ab - 60b^2}$

71. $\dfrac{-5x^3 - 135}{-10x - 30}$

72. $\dfrac{-14x - 14y}{-7x^3 - 7y^3}$

73. $\dfrac{4a^3 - 16a^2 - 20a}{3a^2 - 9a - 30} \cdot \dfrac{6a^2 - 24}{5a^4 + 5a^3 - 30a^2}$

74. $\dfrac{5x^4 - 30x^3 + 40x^2}{8x^2 + 8x - 48} \div \dfrac{10x^3 + 30x^2 - 280x}{12x^2 - 108}$

75. $\dfrac{ax^2 + acx + 5x + 5c}{x^2 - c^2}$ *Hint:* Factor by grouping.

76. $\dfrac{x^2 - 2xy + 3x - 6y}{x^2 - 4y^2}$

77. $\dfrac{x^3 - 8y^3}{x^2 + 3xy - 5x - 15y} \cdot \dfrac{x^2 + 2xy - 5x - 10y}{x^2 - 4y^2}$

78. $\dfrac{a^2 - a + 3ab - 3b}{a^2 + ab - 6b^2} \div \dfrac{a^2 - a + 2ab - 2b}{a^2 - 4b^2}$

79. $\dfrac{6a^2 - a - 2}{2a^2 + 11a - 21} \cdot \dfrac{a^2 + 2a - 35}{3a^2 + 16a - 12} \cdot \dfrac{6a^2 - 13a + 6}{2a^2 - 9a - 5}$

80. $\dfrac{3x^2 + x - 4}{12x^2 + 25x + 12} \cdot \dfrac{10x^2 + x - 21}{2x^2 + x - 3} \div \dfrac{10x^2 + 11x - 35}{8x^2 + 26x + 15}$

C

81. $\dfrac{45x^3 - 78x^2 + 24x}{30x^4 + 65x^3 - 140x^2} \cdot \dfrac{80x^2 + 220x - 210}{180x^2 - 207x + 54}$

82. $\dfrac{28a^3 + 28a^2 - 245a}{108a^2 - 99a + 18} \div \dfrac{84a^3 + 350a^2 + 196a}{135a^2 - 60}$

83. $\dfrac{a}{b} \div \left(\dfrac{c}{d} \div \dfrac{e}{f} \right)$

84. $\left(\dfrac{a}{b} \div \dfrac{c}{d} \right) \div \left(\dfrac{e}{f} \div \dfrac{g}{h} \right)$

85. $\dfrac{a}{b} \div \left[\dfrac{c}{d} \div \left(\dfrac{e}{f} \div \dfrac{g}{h} \right) \right]$

86. Prove that the equation

$$\frac{A}{B} \div \frac{C}{D} = \frac{A}{B} \cdot \frac{D}{C}$$

is valid by multiplying the divisor $\frac{C}{D}$ by the quotient $\frac{A}{B} \cdot \frac{D}{C}$. You should get the dividend $\frac{A}{B}$.

87. In general, it is incorrect to "cancel" the C's in the fraction $\frac{A + C}{B + C}$ and get $\frac{A}{B}$; that is,

$$\frac{A + C}{B + C} \neq \frac{A}{B}.$$

For example, $\frac{2 + 3}{4 + 3} \neq \frac{2}{4}$, or $\frac{5}{7} \neq \frac{1}{2}$. In some instances, however, these expressions are equal. Can you discover specific values for A, B, and C (all non-zero) for which

$$\frac{A + C}{B + C} = \frac{A}{B} ?$$

Can you discover a relationship between the variables for which this will always be true?

88. As noted above, it is very important that you don't "cancel" terms in the numerator and denominator of a fraction. Curiously enough, however, the following "cancellations" do hold true:

$$\frac{16}{64} = \frac{1\!\!\!/6}{6\!\!\!/4} = \frac{1}{4} \quad \text{and} \quad \frac{19}{95} = \frac{1\!\!\!/9}{9\!\!\!/5} = \frac{1}{5}.$$

Can you discover other such cases using two-digit numbers?

89. If a chicken-and-a-half can lay an egg-and-a-half in a day-and-a-half, how many eggs can one chicken lay in one week? *Hint:* How many eggs can one chicken lay in one day?

4.2   Adding and Subtracting Rational Expressions

The sum and the difference of rational expressions with the same denominator are the same as those for rational numbers:

Sum and Difference of Rational Expressions

$$\frac{A}{C} + \frac{B}{C} = \frac{A + B}{C}$$

$$\frac{A}{C} - \frac{B}{C} = \frac{A - B}{C}$$

EXAMPLE 1 | Combine and simplify. Assume the denominators are not zero.

a. $\dfrac{7x}{5ab} + \dfrac{11x}{5ab} - \dfrac{3x}{5ab} = \dfrac{7x + 11x - 3x}{5ab}$ Combine numerators.

$= \dfrac{15x}{5ab}$

$= \dfrac{3x}{ab}.$ Reduce.

b. $\dfrac{2x - 3}{x + 2} - \dfrac{4x - 14}{2 + x} + \dfrac{7x - 1}{x + 2}.$

$= \dfrac{(2x - 3) - (4x - 14) + (7x - 1)}{x + 2}$ Keep parentheses!

$= \dfrac{2x - 3 - 4x + 14 + 7x - 1}{x + 2}$ Note signs!

$= \dfrac{5x + 10}{x + 2}$ Combine like terms.

$= \dfrac{5(x + 2)}{x + 2}$ Factor and reduce.

$= 5.$

Combining rational expressions with unlike denominators is similar to the method for rational numbers (see Exercise 1.1, Probs. 61–66). To obtain the L.C.D. (least common denominator), we factor each denominator and then take as factors of the L.C.D. the *highest power* of each base that occurs.

EXAMPLE 2 | Add: $\dfrac{5}{18x^2y} + \dfrac{7}{24xy^2}.$ Assume the denominators are not zero.

Solution
We first factor the denominators.

$$18x^2y = 2 \cdot 3^2 \cdot x^2 \cdot y$$
$$24xy^2 = 2^3 \cdot 3 \cdot x \cdot y^2$$ Take highest power of 2, 3, x, and y.
$$\text{L.C.D.} = 2^3 \cdot 3^2 \cdot x^2 \cdot y^2 = 72x^2y^2$$

The least common denominator is the product of the highest powers of all bases that occur.

$$2^3 \cdot 3^2 \cdot x^2 \cdot y^2 = 72x^2y^2$$

Using the Fundamental Principle of Fractions, we multiply both numerator and denominator of each fraction by a suitable *"building factor,"* producing equivalent fractions,* each with denominator $72x^2y^2$. We then add:

$$\frac{5}{18x^2y} + \frac{7}{24xy^2} = \frac{5 \cdot 4y}{18x^2y \cdot 4y} + \frac{7 \cdot 3x}{24xy^2 \cdot 3x} \qquad \text{Building factors.}$$

$$= \frac{20y}{72x^2y^2} + \frac{21x}{72x^2y^2} \longleftarrow \text{L.C.D.}$$

$$= \frac{20y + 21x}{72x^2y^2}.$$

Denominators containing binomials or trinomials must be *factored first*. The least common denominator is then obtained, and we proceed as above.

EXAMPLE 3

Combine: $\dfrac{7}{5x - 10} - \dfrac{9}{10x - 20} + \dfrac{8}{3x - 6}$. Assume the denominators are not zero.

Solution
We first factor the denominators.

$$\frac{7}{5(x - 2)} - \frac{9}{10(x - 2)} + \frac{8}{3(x - 2)}.$$

We see that the L.C.D. $= 30(x - 2)$. Multiplying by building factors gives

$$\frac{7}{5(x - 2)} - \frac{9}{10(x - 2)} + \frac{8}{3(x - 2)}$$

$$= \frac{7 \cdot 6}{5(x - 2) \cdot 6} - \frac{9 \cdot 3}{10(x - 2) \cdot 3} + \frac{8 \cdot 10}{3(x - 2) \cdot 10} \qquad \text{Building factors.}$$

$$= \frac{42}{30(x - 2)} - \frac{27}{30(x - 2)} + \frac{80}{30(x - 2)} \longleftarrow \text{L.C.D.}$$

$$= \frac{95}{30(x - 2)}$$

$$= \frac{19}{6(x - 2)} \qquad \text{Reduce.}$$

---

*Multiplying by a building factor does not change the value of a rational expression, because *each* building factor equals 1: $\frac{4y}{4y} = 1$ and $\frac{3x}{3x} = 1$.

EXAMPLE 4   Combine: $\dfrac{a+2}{a+1} - \dfrac{2a}{a-3} + \dfrac{2a^2+2}{a^2-2a-3}$. Assume the denominators are not zero.

Solution
We start by factoring denominators.

$$\frac{a+2}{a+1} - \frac{2a}{a-3} + \frac{2a^2+2}{(a+1)(a-3)}.$$

The L.C.D. $= (a+1)(a-3)$, so we multiply by building factors.

$$\frac{a+2}{a+1} - \frac{2a}{a-3} + \frac{2a^2+2}{(a+1)(a-3)}$$

$$= \frac{(a+2)\cdot(a-3)}{(a+1)\cdot(a-3)} - \frac{2a\cdot(a+1)}{(a-3)\cdot(a+1)} + \frac{2a^2+2}{(a+1)(a-3)} \qquad \text{Building factors.}$$

$$= \frac{(a+2)(a-3) - 2a(a+1) + 2a^2+2}{(a+1)(a-3)} \longleftarrow \qquad\qquad \text{L.C.D.}$$

$$= \frac{a^2-a-6-2a^2-2a+2a^2+2}{(a+1)(a-3)} \longleftarrow \quad \text{Remove parentheses.}$$

$$= \frac{a^2-3a-4}{(a+1)(a-3)} \qquad\qquad \text{Combine like terms.}$$

$$= \frac{(a+1)(a-4)}{(a+1)(a-3)} \qquad\qquad \text{Factor.}$$

$$= \frac{a-4}{a-3}. \qquad\qquad \text{Reduce.}$$

In our final example, the denominator of the second fraction, $2-x$, is "backwards." We first "turn it around," so that it becomes $x-2$, by using the building factor $\dfrac{-1}{-1}$.

EXAMPLE 5   Combine: $\dfrac{5x^2+x}{x^3-8} - \dfrac{3}{2-x}$. Assume the denominators are not zero.

Solution
As stated above, we multiply the second fraction by $\dfrac{-1}{-1}$:

$$\frac{5x^2+x}{x^3-8} - \frac{3\cdot(-1)}{(2-x)\cdot(-1)} = \frac{5x^2+x}{x^3-8} - \frac{-3}{x-2} \leftarrow (2-x)(-1)=-2+x$$

Now we factor the difference of cubes in the first fraction:

$$\frac{5x^2+x}{(x-2)(x^2+2x+4)} + \frac{3}{x-2} \qquad -\frac{-3}{x-2}=+\frac{3}{x-2}$$

We see that the L.C.D. $= (x - 2)(x^2 + 2x + 4)$, so we proceed as in the previous examples.

$$\frac{5x^2 + x}{(x - 2)(x^2 + 2x + 4)} + \frac{3 \cdot (x^2 + 2x + 4)}{(x - 2) \cdot (x^2 + 2x + 4)} \leftarrow \text{L.C.D.}$$

$$= \frac{5x^2 + x + 3x^2 + 6x + 12}{(x - 2)(x^2 + 2x + 4)} \longleftarrow \text{Remove parentheses.}$$

$$= \frac{8x^2 + 7x + 12}{(x - 2)(x^2 + 2x + 4)}.$$

EXERCISE 4.2   *Answers, pages A48–A49*

A

Combine and simplify, as in EXAMPLE 1. Assume the denominators are not zero.

1. $\dfrac{5a}{3x} - \dfrac{4a}{3x} + \dfrac{8a}{3x}$

2. $\dfrac{-8b}{7yz} + \dfrac{9b}{7yz} - \dfrac{22b}{7zy}$

3. $-\dfrac{4a}{5xy} + \dfrac{3b}{5yx} - \dfrac{7c}{5xy}$

4. $\dfrac{3u}{7t} - \dfrac{4y}{7t} + \dfrac{-8w}{7t}$

5. $\dfrac{2x + 3}{x + 1} + \dfrac{7x - 10}{x + 1} - \dfrac{4x - 12}{1 + x}$

6. $\dfrac{5y + 3}{y - 3} - \dfrac{2y + 9}{y - 3} + \dfrac{y - 6}{y - 3}$

7. $\dfrac{5c - 12}{c + 7} - \dfrac{9c - 1}{7 + c} + \dfrac{c - 10}{c + 7}$

8. $\dfrac{2a - 6b}{2b + a} + \dfrac{a - 3b}{a + 2b} - \dfrac{7a - b}{a + 2b}$

Combine as in EXAMPLES 2, 3, and 4. Assume the denominators are not zero.

9. $\dfrac{11}{8a^2} + \dfrac{2}{3ab} - \dfrac{5}{4b^2}$

10. $\dfrac{2}{7x^2} - \dfrac{7}{2xy} + \dfrac{13}{21y^2}$

11. $\dfrac{3}{5x^3} - \dfrac{3}{2xy^2} + \dfrac{7}{4y^2}$

12. $\dfrac{7}{12a^3} + \dfrac{2}{5a^2b} + \dfrac{3}{4ab^2}$

13. $2 + \dfrac{1}{a} - \dfrac{1}{a^2}$ *Hint:* $2 = \dfrac{2}{1}$

14. $x - \dfrac{1}{x} + \dfrac{1}{x^3}$

15. $x + 1 - \dfrac{1}{x - 1}$

16. $2a - 1 + \dfrac{2}{a + 1}$

17. $\dfrac{1}{x + h} - \dfrac{1}{x}$ *Hint:* L.C.D. $= x(x + h)$

18. $\dfrac{1}{x} + \dfrac{1}{1 - x}$

19. $\dfrac{x}{x - 1} - \dfrac{x}{x + 1}$

20. $\dfrac{a}{a + 1} + \dfrac{1}{a - 1}$

21. $\dfrac{a - b}{a + b} + \dfrac{a + b}{a - b}$

22. $\dfrac{x + 3}{x - 3} - \dfrac{x - 3}{x + 3}$

23. $\dfrac{3}{2x+6} + \dfrac{4}{3x+9} - \dfrac{1}{6x+18}$

24. $\dfrac{2}{5a-15b} - \dfrac{3}{10a-30b} + \dfrac{5}{4a-12b}$

25. $\dfrac{2x}{x+2} + \dfrac{16}{x^2-4} - \dfrac{2x}{x-2}$

26. $\dfrac{a}{a-b} - \dfrac{b}{a+b} + \dfrac{2ab}{a^2-b^2}$

27. $\dfrac{7}{x-8} + \dfrac{5}{x+3} - \dfrac{77}{x^2-5x-24}$

28. $\dfrac{3}{y+7} - \dfrac{9}{y^2+5y-14} + \dfrac{1}{y-2}$

29. $\dfrac{2x}{x-4y} - \dfrac{2y}{x-y} - \dfrac{24y^2}{x^2-5xy+4y^2}$

30. $\dfrac{a}{a-2b} - \dfrac{3ab}{a^2-ab-2b^2} - \dfrac{b}{a+b}$

31. $\dfrac{3x^2+12}{x^3-8} + \dfrac{2}{x^2+2x+4} - \dfrac{2}{x-2}$

32. $\dfrac{x^2+1}{x^3+1} - \dfrac{1}{x+1} + \dfrac{x}{x^2-x-1}$

33. $1 - \dfrac{x}{x+1} - \dfrac{2}{x^2+2x+1}$

34. $2 + \dfrac{a}{a-3} - \dfrac{18}{a^2-6a+9}$

35. $\dfrac{x-1}{2x+2} - \dfrac{x+1}{x^2+2x+1}$

36. $\dfrac{a-b}{a^2-10ab+25b^2} + \dfrac{a+b}{4a-20b}$

37. $\dfrac{3a}{2a-2} + \dfrac{5}{6a+12} - \dfrac{a^2}{4a^2+4a-8}$

38. $\dfrac{7x}{5x-10a} - \dfrac{2a}{3x^2-3ax-6a^2} + \dfrac{3}{4x+4a}$

Combine by first using the building factor $\dfrac{-1}{-1}$, as in **EXAMPLE 5:**

39. $\dfrac{6}{x-y} + \dfrac{8}{y-x}$

40. $\dfrac{3a}{2a-b} - \dfrac{5a}{b-2a}$

41. $\dfrac{-2b}{b-a} - \dfrac{2a}{a-b}$

42. $\dfrac{3}{3-x} + \dfrac{x}{x-3}$

43. $\dfrac{7}{-x-y} + \dfrac{8}{x+y}$

44. $\dfrac{4}{x+3a} - \dfrac{6}{-3a-x}$

45. $\dfrac{x-a}{x-2a} - \dfrac{a+x}{4a-2x}$

46. $\dfrac{a+b}{2a-7b} + \dfrac{a-2b}{21b-6a}$

47. $\dfrac{2}{x^2-1} - \dfrac{2x}{1-x^2}$

48. $\dfrac{3x}{x^2-a^2} - \dfrac{3a}{a^2-x^2}$

49. $\dfrac{2x}{x^2-9} - \dfrac{2}{3-x} + \dfrac{1}{x+3}$

50. $\dfrac{b}{a-5b} - \dfrac{a}{a+5b} + \dfrac{10b^2-a^2}{25b^2-a^2}$

51. $2 + \dfrac{3}{1-x} + \dfrac{6}{x^2-1}$

52. $1 + \dfrac{2x-1}{9-4x^2} - \dfrac{3}{2x+3}$

53. $\dfrac{2x^2+3x}{x^3-27} + \dfrac{1}{3-x}$

54. $\dfrac{3a^2+3b^2}{a^3-b^3} + \dfrac{2}{b-a}$

B

Combine and simplify:

55. $\dfrac{2x+y}{5x-10y} + \dfrac{x+y}{4y-2x} - \dfrac{xy}{x^2-4y^2}$

56. $\dfrac{a-2}{2a-6} + \dfrac{a-1}{9-3a} - \dfrac{a}{a^2-9}$

57. $\dfrac{3a}{27a^3 + 64} - \dfrac{1}{9a^2 - 16}$

58. $\dfrac{1}{4x^2 - 25a^2} - \dfrac{2x}{8x^3 - 125a^3}$

59. $\dfrac{2}{y^2 + y - 6} - \dfrac{y - 1}{y^2 + 3y - 10} + \dfrac{3}{y^2 + 8y + 15}$

60. $\dfrac{x - a}{2x^2 + 5ax + 2a^2} + \dfrac{3x}{2x^2 - 5ax - 3a^2} - \dfrac{2a}{x^2 - ax - 6a^2}$

61. $\dfrac{x + y}{6x^2 - xy - 2y^2} - \dfrac{x - y}{3x^2 - 5xy + 2y^2} + \dfrac{3x}{2x^2 - xy - y^2}$

62. $\dfrac{a - 1}{8a^2 + 2a - 3} + \dfrac{2a + 1}{12a^2 + a - 6} - \dfrac{3a}{6a^2 - 7a + 2}$

63. $\dfrac{2x}{x^2 - 9} + \dfrac{x}{x^2 - 6x + 9} - \dfrac{x}{x^2 + 6x + 9}$

64. $\dfrac{a}{a^2 - 4ab + 4b^2} - \dfrac{2a}{a^2 - 4b^2} + \dfrac{b}{a^2 + 4ab + 4b^2}$

C

65. $4 + \dfrac{x}{x^2 - 2x + 1} - \dfrac{2}{3x^2 - 3x} + \dfrac{1}{2x}$

66. $1 - \dfrac{2}{a^2 - 4} + \dfrac{1}{2a^2 - 8a + 8} - \dfrac{2}{a + 2}$

67. $\dfrac{3a}{x^2 + ax + 2x + 2a} - \dfrac{2x}{x^2 - 3ax + 2x - 6a}$ *Hint:* Factor by grouping.

68. $\dfrac{a}{a^2 - ad + ac - cd} + \dfrac{2d}{2a^2 - 3ad + 2ac - 3cd}$

69. To *multiply* fractions, we do *not* need a least common denominator. We simply multiply numerator and denominator:

$$\frac{A}{B} \cdot \frac{C}{D} = \frac{A \cdot C}{B \cdot D}$$

Why, then, all the fuss about L.C.D.'s when *adding* fractions? Why not just add numerator and denominator:

$$\frac{A}{B} + \frac{C}{D} = \frac{A + C}{B + D}?$$

Give a specific numerical example to show that this is *false*. Then combine $\dfrac{A}{B} + \dfrac{C}{D}$ correctly.

70. We know that fractions with the *same denominator* can be added by simply adding their numerators:

$$\frac{A}{C} + \frac{B}{C} = \frac{A + B}{C}$$

What if they have the *same numerator*? Can we just add their denominators:

$$\frac{A}{B} + \frac{A}{C} = \frac{A}{B + C}?$$

Give a specific numerical example to show that this is *false*. Then combine $\dfrac{A}{B} + \dfrac{A}{C}$ correctly.

71. This problem is similar to Exercise 1.1, Prob. 55. A three-volume book is standing on a bookshelf, with volume I on the left, volume II in the middle, and volume III on the right. Each hard-cover flap is $\frac{1}{16}$ inch thick; the pages of volume I comprise $\frac{7}{8}$ inch, those of volume II comprise $\frac{5}{6}$ inch, and those of volume III comprise $\frac{9}{16}$ inch. A bookworm eats its way from the first page of volume I to the last page of volume III. How far does it travel horizontally from left to right: Neglect any blank pages at the beginning and end of each volume.

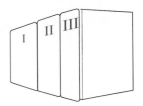

4.3   Complex Fractions

Rational expressions whose numerators and/or denominators themselves contain fractions are called **complex fractions.** Examples of complex fractions include

$$\frac{\dfrac{3}{8}}{\dfrac{1}{4}}, \qquad \frac{\dfrac{2x^4}{3y^2}}{\dfrac{5x^2}{6y}}, \qquad \frac{1 + \dfrac{7}{15}}{3 - \dfrac{2}{5}}, \qquad \text{and} \qquad 2 - \frac{3}{4 + \dfrac{1}{x}}.$$

They can be simplified via every technique you have learned in this chapter, which makes this section an *excellent opportunity for review.* Because the denominator of a fraction is a divisor, the first expression above,

$$\frac{\dfrac{3}{8}}{\dfrac{1}{4}}, \qquad \text{means} \qquad \frac{3}{8} \div \frac{1}{4} = \frac{3}{8} \cdot \frac{4}{1} = \frac{3}{2}.$$

A complex fraction, therefore, can be simplified by *inverting the denominator and then multiplying.* In the following examples, assume that the denominators are not zero.

EXAMPLE 1

a. $\dfrac{\dfrac{2x^4}{3y^2}}{\dfrac{5x^2}{6y}} = \dfrac{2x^4}{3y^2} \cdot \dfrac{6y}{5x^2}$ Invert denominator.

$= \dfrac{12x^4y}{15x^2y^2}$ Multiply.

$= \dfrac{4x^2}{5y}.$ Reduce.

b. $\dfrac{\dfrac{6ab}{-3a^2}}{5b} = \dfrac{6ab}{1} \cdot \dfrac{5b}{-3a^2}$ Invert denominator; $6ab = \dfrac{6ab}{1}$

$\qquad\qquad = \dfrac{-10b^2}{a}.$

c. $\dfrac{\dfrac{x^2 + 5x + 6}{2x}}{\dfrac{6x + 12}{6x^2}} = \dfrac{x^2 + 5x + 6}{2x} \cdot \dfrac{6x^2}{6x + 12}$ Invert denominator.

$\qquad\qquad = \dfrac{(x + 3)(x + 2)}{2x} \cdot \dfrac{6x^2}{6(x + 2)}$ Factor.

$\qquad\qquad = \dfrac{x(x + 3)}{2}.$ Reduce.

In the next example, we must *first combine numerators and denominators into single fractions,* and then invert the denominator.

EXAMPLE 2

a. $\dfrac{1 + \dfrac{7}{15}}{3 - \dfrac{2}{5}} = \dfrac{\dfrac{1 \cdot 15}{1 \cdot 15} + \dfrac{7}{15}}{\dfrac{3 \cdot 5}{1 \cdot 5} - \dfrac{2}{5}}$ Combine numerator and denominator into single fractions, using L.C.D.'s.

$\qquad\qquad = \dfrac{\dfrac{22}{15}}{\dfrac{13}{5}}$ Single fractions.

$\qquad\qquad = \dfrac{22}{15} \cdot \dfrac{5}{13}$ Invert denominator.

$\qquad\qquad = \dfrac{22}{39}.$

b. $\dfrac{1 - \dfrac{1}{4x^2}}{1 + \dfrac{1}{8x^3}} = \dfrac{\dfrac{1 \cdot 4x^2}{1 \cdot 4x^2} - \dfrac{1}{4x^2}}{\dfrac{1 \cdot 8x^3}{1 \cdot 8x^3} + \dfrac{1}{8x^3}}$ Combine numerator and denominator into single fractions, using L.C.D.'s.

$\qquad\qquad = \dfrac{\dfrac{4x^2 - 1}{4x^2}}{\dfrac{8x^3 + 1}{8x^3}}$ Single fractions.

(continued)

$$= \frac{4x^2 - 1}{4x^2} \cdot \frac{8x^3}{8x^3 + 1}$$ Invert denominator.

$$= \frac{(2x - 1)\cancel{(2x + 1)}}{\cancel{4x^2}} \cdot \frac{\overset{2x}{\cancel{8x^3}}}{\cancel{(2x + 1)}(4x^2 - 2x + 1)}$$ Factor and reduce.

$$= \frac{2x(2x - 1)}{4x^2 - 2x + 1}.$$

c. $$\dfrac{1 - \dfrac{72}{x^2 - 9}}{\dfrac{2}{x + 3} - \dfrac{1}{x - 3}} = \dfrac{\dfrac{1 \cdot (x^2 - 9)}{1 \cdot (x^2 - 9)} - \dfrac{72}{x^2 - 9}}{\dfrac{2 \cdot (x - 3)}{(x + 3) \cdot (x - 3)} - \dfrac{1 \cdot (x + 3)}{(x - 3) \cdot (x + 3)}}$$ $\leftarrow$ $\begin{aligned} x^2 - 9 - 72 \\ = x^2 - 81 \end{aligned}$

                                                  $\leftarrow$ $\begin{aligned} 2x - 6 - x - 3 \\ = x - 9 \end{aligned}$

$$= \frac{\dfrac{x^2 - 81}{x^2 - 9}}{\dfrac{x - 9}{(x + 3)(x - 3)}}$$ Single fractions.

$$= \frac{x^2 - 81}{x^2 - 9} \cdot \frac{(x + 3)(x - 3)}{x - 9}$$ Invert denominator.

$$= \frac{(x + 9)(x - 9)}{(x + 3)(x - 3)} \cdot \frac{(x + 3)(x - 3)}{x - 9}$$ Factor.

$$= x + 9.$$ Reduce.

In the next example, we simplify the complex fraction on the bottom and then work our way upward.

EXAMPLE 3

a. $$2 - \frac{3}{4 + \dfrac{1}{x}} = 2 - \frac{3}{\dfrac{4 \cdot x}{1 \cdot x} + \dfrac{1}{x}}$$ $\leftarrow$ L.C.D.

$$= 2 - \frac{\dfrac{3}{1}}{\dfrac{4x + 1}{x}}$$

$$= 2 - \frac{3x}{4x + 1}$$ Invert denominator: $\dfrac{3}{1} \cdot \dfrac{x}{4x + 1}$

$$= \frac{2 \cdot (4x + 1)}{1 \cdot (4x + 1)} - \frac{3x}{4x + 1}$$ $\leftarrow$ L.C.D.

$$= \frac{5x + 2}{4x + 1}.$$ $\leftarrow$ $8x + 2 - 3x = 5x + 2$

b. $10 + \dfrac{4}{2 - \dfrac{1}{5 + \dfrac{2}{3}}} = 10 + \dfrac{4}{2 - \dfrac{1}{\dfrac{5 \cdot 3}{1 \cdot 3} + \dfrac{2}{3}}}$ Start on lower right.

$\longleftarrow$ L.C.D.

$= 10 + \dfrac{4}{2 - \dfrac{1}{\dfrac{17}{3}}}$ $\longleftarrow$ $5 \cdot 3 + 2 = 17$

$= 10 + \dfrac{4}{2 - \dfrac{3}{17}}$ Invert: $\dfrac{1}{1} \cdot \dfrac{3}{17} = \dfrac{3}{17}$

$= 10 + \dfrac{4}{\dfrac{2 \cdot 17}{1 \cdot 17} - \dfrac{3}{17}}$ $\longleftarrow$ L.C.D.

$= 10 + \dfrac{4}{\dfrac{1}{\dfrac{31}{17}}}$ $\longleftarrow$ $2 \cdot 17 - 3 = 31$

$= 10 + \dfrac{68}{31}$ Invert: $\dfrac{4}{1} \cdot \dfrac{17}{31} = \dfrac{68}{31}$

$= \dfrac{10 \cdot 31}{1 \cdot 31} + \dfrac{68}{31}$ $\longleftarrow$ L.C.D.

$= \dfrac{378}{31}.$ $\longleftarrow$ $10 \cdot 31 + 68 = 378$

EXERCISE 4.3 *Answers, pages A49–A50*

Simplify as in EXAMPLE 1:

1. $\dfrac{\dfrac{2a^2}{7b}}{\dfrac{4a^3}{21b^3}}$

2. $\dfrac{\dfrac{3x^3}{8y^2}}{\dfrac{15x}{16y^5}}$

3. $\dfrac{\dfrac{3x^3y^5}{7y}}{\dfrac{-6xy^4}{35y^2}}$

4. $\dfrac{\dfrac{-8ab^2}{3a^3b}}{\dfrac{-16a^2b}{9ab^4}}$

5. $\dfrac{\dfrac{3z^2}{6z^5}}{\dfrac{5}{}}$

6. $\dfrac{\dfrac{11m^2n}{-33m^4}}{\dfrac{6mn}{}}$

7. $\dfrac{\dfrac{4p^2q}{3rp}}{-8pq^2}$

8. $\dfrac{-\dfrac{15cd^3}{2e^2}}{30cde}$

9. $\dfrac{\dfrac{3x-6}{5xy}}{\dfrac{9x-18}{10x^2y^2}}$

10. $\dfrac{\dfrac{3a+9b}{7a^2b}}{\dfrac{6a+18b}{14ab^2}}$

11. $\dfrac{\dfrac{b^2-a^2}{a^2b^2}}{\dfrac{b^3-a^3}{a^3b^3}}$

12. $\dfrac{\dfrac{8x^3-27y^3}{12x^3y}}{\dfrac{4x^2-9y^2}{3xy^2}}$

13. $\dfrac{\dfrac{x^2+x-6}{x^2-4x-5}}{\dfrac{x^2-3x+2}{x^2-6x+5}}$

14. $\dfrac{\dfrac{a^2-a-20}{a^2-3a-10}}{\dfrac{a^2+2a-8}{a^2-5a-14}}$

15. $\dfrac{\dfrac{-4x^2-4x+8}{3x-15}}{\dfrac{2x^2+14x+20}{25-x^2}}$

16. $\dfrac{\dfrac{2a-4b}{a^2-b^2}}{\dfrac{8b-4a}{a^3+b^3}}$

17. $\dfrac{\dfrac{8+2x-x^2}{9-x^2}}{\dfrac{x^2-x-12}{x^2+2x-15}}$

18. $\dfrac{\dfrac{2b^2-3ab+a^2}{a^2+2ab-3b^2}}{\dfrac{9b^2-a^2}{}}$

Simplify as in EXAMPLE 2:

19. $\dfrac{8+\dfrac{1}{3}}{5-\dfrac{1}{6}}$

20. $\dfrac{2-\dfrac{2}{5}}{4+\dfrac{3}{10}}$

21. $\dfrac{\dfrac{1}{2}+\dfrac{2}{3}}{\dfrac{5}{6}-\dfrac{5}{2}}$

22. $\dfrac{\dfrac{4}{5}+\dfrac{1}{2}}{\dfrac{3}{10}-\dfrac{7}{5}}$

23. $\dfrac{1-\dfrac{1}{2}+\dfrac{2}{3}}{\dfrac{3}{4}+\dfrac{4}{5}-\dfrac{5}{6}}$

24. $\dfrac{1+\dfrac{1}{2}-\dfrac{1}{4}}{\dfrac{1}{8}-\dfrac{1}{16}+\dfrac{1}{32}}$

25. $\dfrac{\dfrac{1}{x}-\dfrac{1}{x^2}}{2-\dfrac{2}{x^2}}$

26. $\dfrac{\dfrac{1}{a}+\dfrac{3}{a^2}}{1-\dfrac{9}{a^2}}$

27. $\dfrac{\dfrac{a}{b}-\dfrac{b}{a}}{\dfrac{1}{b}+\dfrac{1}{a}}$

28. $\dfrac{\dfrac{1}{5}+\dfrac{1}{x}}{\dfrac{x}{5}-\dfrac{5}{x}}$

29. $\dfrac{1+\dfrac{4b}{a}-\dfrac{5b^2}{a^2}}{1-\dfrac{2b}{a}+\dfrac{b^2}{a^2}}$

30. $\dfrac{1-\dfrac{y}{2x}-\dfrac{3y^2}{2x^2}}{1-\dfrac{7y}{2x}+\dfrac{3y^2}{2x^2}}$

31. $\dfrac{\dfrac{1}{x^2}-\dfrac{1}{y^2}}{\dfrac{1}{y}-\dfrac{1}{x}}$

32. $\dfrac{\dfrac{1}{2b}+\dfrac{1}{3a}}{\dfrac{1}{9a^2}-\dfrac{1}{4b^2}}$

33. $\dfrac{\dfrac{1}{x^3}-\dfrac{1}{8y^3}}{\dfrac{1}{x^2}-\dfrac{1}{4y^2}}$

34. $\dfrac{\dfrac{1}{9b^2}-\dfrac{1}{a^2}}{\dfrac{1}{27b^3}+\dfrac{1}{a^3}}$

35. $\dfrac{1-\dfrac{1}{25x^2}}{1+\dfrac{1}{125x^3}}$

36. $\dfrac{1-\dfrac{64}{x^3}}{1-\dfrac{16}{x^2}}$

37. $\dfrac{x-y}{x-\dfrac{y^2}{x}}$

38. $\dfrac{\dfrac{a^2}{b}-b}{a+b}$

39. $\dfrac{\dfrac{1}{x+h}-\dfrac{1}{x}}{h}$

40. $\dfrac{\dfrac{3}{5+h}-\dfrac{3}{5}}{h}$

41. $\dfrac{\dfrac{2}{3x}-\dfrac{2}{3a}}{x-a}$

42. $\dfrac{\dfrac{5}{x^2}-\dfrac{5}{a^2}}{x-a}$

43. $\dfrac{\dfrac{2}{x-4}-\dfrac{2}{x+4}}{\dfrac{6}{x^2-16}}$

44. $\dfrac{\dfrac{10}{x^2-25}}{\dfrac{1}{x+5}-\dfrac{1}{x-5}}$

45. $\dfrac{\dfrac{x+1}{x^2+4x-5}}{\dfrac{2}{x+5}+\dfrac{1}{x-1}}$

46. $\dfrac{\dfrac{2}{a-4}+\dfrac{4}{a+3}}{\dfrac{3a-5}{a^2-a-12}}$

47. $\dfrac{\dfrac{2}{x-5}-\dfrac{1}{x+5}}{1-\dfrac{200}{x^2-25}}$

48. $\dfrac{1-\dfrac{128}{a^2-16}}{\dfrac{2}{a+4}-\dfrac{1}{a-4}}$

Simplify as in EXAMPLE 3:

49. $2+\dfrac{3}{2+\dfrac{1}{3}}$

50. $1+\dfrac{2}{3-\dfrac{1}{2}}$

51. $4-\dfrac{5}{7+\dfrac{3}{5-\dfrac{1}{2}}}$

52. $8-\dfrac{6}{2-\dfrac{4}{3+\dfrac{1}{3}}}$

53. $3+\dfrac{2}{2-\dfrac{1}{x}}$

54. $2-\dfrac{1}{4-\dfrac{2}{a}}$

55. $x + \dfrac{1}{x + \dfrac{1}{x}}$

56. $a + \dfrac{1}{2a - \dfrac{1}{a}}$

B

Simplify:

57. $\dfrac{\dfrac{x}{y^2} - \dfrac{y}{x^2}}{\dfrac{1}{x^2} - \dfrac{1}{y^2}}$

58. $\dfrac{\dfrac{b}{a} - \dfrac{a}{b}}{\dfrac{a}{b^2} - \dfrac{b}{a^2}}$

59. $\dfrac{2D}{\dfrac{D}{R} + \dfrac{D}{r}}$

60. $\dfrac{e^2 - \dfrac{1}{e^2}}{2}$

61. $\dfrac{x - \dfrac{2}{x - 1}}{x + 2 + \dfrac{1}{x}}$

62. $\dfrac{a + 2 - \dfrac{5}{a - 2}}{a + 3 - \dfrac{6}{a - 2}}$

63. $\dfrac{\dfrac{y}{r}}{1 - \dfrac{x}{r}} - \dfrac{\dfrac{y}{r}}{1 + \dfrac{x}{r}}$

64. $\dfrac{\dfrac{x}{r}}{\dfrac{r}{x} + 1} + \dfrac{\dfrac{x}{r}}{\dfrac{r}{x} - 1}$

65. $\dfrac{\dfrac{2}{x + a} - \dfrac{1}{x - a}}{1 - \dfrac{8a^2}{x^2 - a^2}}$

66. $\dfrac{\dfrac{1}{(x + h)^2} - \dfrac{1}{x^2}}{h}$

67. $2 + \dfrac{1}{2 - \dfrac{1}{2 + \dfrac{1}{2 - \dfrac{1}{2}}}}$

68. $3 - \dfrac{1}{3 + \dfrac{1}{3 - \dfrac{1}{3 - \dfrac{1}{3}}}}$

C

69. $\dfrac{1 - \dfrac{2}{x - \dfrac{3}{x}}}{1 + \dfrac{2}{x + \dfrac{1}{x}}}$

70. $\dfrac{1 + \dfrac{m}{1 - \dfrac{1}{m}}}{1 - \dfrac{m}{1 + \dfrac{1}{m}}}$

71. $\dfrac{1}{\dfrac{3}{\dfrac{2}{\dfrac{1}{4} + 1} + 1} + 2} + 1$

For you
left-handers .

72. $\dfrac{2}{\dfrac{2}{\dfrac{5}{1-\dfrac{1}{2}}-1}+1}-2$

73. $1-\dfrac{1}{1-\dfrac{1}{1-\dfrac{1}{1-x}}}$

74. $\dfrac{\dfrac{\dfrac{a}{b}}{\dfrac{c}{d}}}{\dfrac{\dfrac{e}{f}}{\dfrac{g}{h}}}$

TRUE or FALSE? (*Hint:* Simplify each side.)

75. $\dfrac{\dfrac{a}{b}}{c}=\dfrac{a}{\dfrac{b}{c}}$

76. $\dfrac{\dfrac{a}{b}}{c}=\dfrac{a}{bc}$

77. $\dfrac{\dfrac{a}{b}}{\dfrac{c}{d}}=\dfrac{a}{\dfrac{\dfrac{b}{c}}{d}}$

78. $\dfrac{\dfrac{a}{b}}{\dfrac{c}{d}}=\dfrac{\dfrac{a}{c}}{\dfrac{d}{b}}$

Because complex fractions represent quotients of fractions, we can express them using the division operation. For example, we can write

$$\dfrac{x-\dfrac{1}{y}}{x+\dfrac{1}{y}} \quad \text{as} \quad (x-1\div y)\div(x+1\div y).$$

In like manner, express the following complex fractions in terms of the division operation. Do *not* simplify.

79. $\dfrac{\dfrac{1}{x}}{1+\dfrac{1}{x}}$

80. $\dfrac{\dfrac{a}{b-c}}{1+\dfrac{a}{b}}$

81. $\dfrac{\dfrac{1}{a}+\dfrac{1}{b}}{\dfrac{1}{a^2}-\dfrac{1}{b^2}}$

82. $1+\dfrac{a}{b-\dfrac{1}{a+b}}$

83. $x+\dfrac{1}{x-\dfrac{1}{x+\dfrac{1}{x+1}}}$

84. $\dfrac{a}{\dfrac{a}{\dfrac{a}{a-2}+3}-1}+2$

Conversely, we can express $a\div(1+b\div c)$ **as** $\dfrac{a}{1+\dfrac{b}{c}}$.

Express each of the following as complex fractions. Do *not* simplify.

85. $a \div (b \div c)$

86. $(a \div b) \div (c \div d)$

87. $(1 + a \div b) \div (1 - a^2 \div b^2)$

88. $(x - 1 \div x) \div [x + 1 \div (2 - 3 \div x)]$

89. $1 + x \div [1 - x \div (1 + x)]$

90. $[a \div (b - 1 \div b)] \div [a \div (1 + 2 \div (1 - a \div b))]$

4.4 Dividing Polynomials

In Section 4.1 we saw how rational expressions can be reduced to lowest terms by factoring. For example, if $x \neq 0$, then

$$\frac{6x^2 + 14x}{2x} = \frac{2x(3x + 7)}{2x} = 3x + 7.$$

This quotient of polynomials can also be obtained by using the rule for adding fractions with the same denominator, but in reverse order:

$$\frac{A + B}{C} = \frac{A}{C} + \frac{B}{C}.$$

Thus

$$\frac{6x^2 + 14x}{2x} = \frac{6x^2}{2x} + \frac{14x}{2x} = 3x + 7.$$

In the event that the quotient cannot be obtained by factoring, we *must* use this new method. For example, if $a \neq 0$, then

$$\frac{18a^2 - 9a + 4}{3a} = \frac{18a^2}{3a} - \frac{9a}{3a} + \frac{4}{3a} = 6a - 3 + \frac{4}{3a}.$$

Note that our answer is not expressed in terms of a least common denominator but instead has the "remainder term" $\frac{4}{3a}$. Although this is contrary to normal procedure, the answer is often more desirable in this form.

EXAMPLE 1 Divide, assuming no denominator equals zero:

a. $\dfrac{15x^3 - 18x^2 + 3x}{3x} = \dfrac{15x^3}{3x} - \dfrac{18x^2}{3x} + \dfrac{3x}{3x}$

$\qquad\qquad\qquad\quad = 5x^2 - 6x + 1$

b. $\dfrac{30a^3b^2 + 20a^2b - 5a}{10ab} = \dfrac{30a^3b^2}{10ab} + \dfrac{20a^2b}{10ab} - \dfrac{5a}{10ab}$

$\qquad\qquad\qquad\qquad = 3a^2b + 2a - \dfrac{1}{2b}.$

Quotients whose divisor (denominator) is a binomial can also be obtained by factoring. For example, if $x \neq -3$, then

$$\frac{2x^2 + 5x - 3}{x + 3} = \frac{(2x - 1)(x + 3)}{x + 3} = 2x - 1.$$

In the next example, the numerator is not factorable, so we must use a method similar to the "long division" of whole numbers.

EXAMPLE 2

Divide: $\dfrac{5x^2 - 8x + 1}{x - 1}$

Step 1 Writing this as $x - 1 \overline{)5x^2 - 8x + 1}$, we divide x into $5x^2$ to get $5x$, which is then used to multiply the divisor.

$$\begin{array}{r} 5x \quad\longleftarrow \\ x - 1 \overline{)5x^2 - 8x + 1} \end{array} \qquad \frac{5x^2}{x} = 5x$$

Multiply: $5x(x - 1) \longrightarrow 5x^2 - 5x$

Step 2 Subtract by *changing the sign* of each term in the bottom row and then adding. Then bring down the next term.

$$\begin{array}{r} 5x \\ x - 1 \overline{)\ 5x^2 - 8x + 1} \end{array}$$

Change signs. $\longrightarrow -5x^2 \not{+} 5x \quad \downarrow$

Add. $\longrightarrow \quad -3x + 1$

Step 3 Divide x into $-3x$ to get -3, and then multiply out as in Step 1.

$$\begin{array}{r} 5x - 3 \quad \longleftarrow \\ x - 1 \overline{)\ 5x^2 - 8x + 1} \\ -5x^2 + 5x \\ \hline -3x + 1 \end{array} \qquad \frac{-3x}{x} = -3$$

Multiply: $-3(x - 1) \longrightarrow -3x + 3$

Step 4 *Change the sign* of each term in the bottom row and then add. This gives a remainder of -2.

$$\begin{array}{r} 5x - 3 \\ x - 1 \overline{)\ 5x^2 - 8x + 1} \\ -5x^2 + 5x \\ \hline -3x + 1 \end{array}$$

Change signs. $\longrightarrow \not{+}3x \not{+} 3$

Add. $\longrightarrow -2 \ = \text{remainder}$

(continued)

In algebra, we write the remainder in fractional form, so the answer is

$$\frac{5x^2 - 8x + 1}{x - 1} = 5x - 3 + \frac{-2}{x - 1}, \quad \text{or} \quad 5x - 3 - \frac{2}{x - 1}.$$

We can check by expressing our answer as a single fraction.

$$5x - 3 - \frac{2}{x - 1} = \frac{(5x - 3) \cdot (x - 1)}{1 \cdot (x - 1)} - \frac{2}{x - 1}$$

$$= \frac{5x^2 - 8x + 3 - 2}{x - 1} \stackrel{\checkmark}{=} \frac{5x^2 - 8x + 1}{x - 1}.$$

EXAMPLE 3

Divide: $\dfrac{4x^3 - 3x + 26}{2x + 3}$

Solution

First we must add the term $0x^2$ to the dividend as a *place holder*. We then proceed as in the foregoing example:

Multiply: $2x^2(2x + 3)$,
then change signs.

$$
\begin{array}{r}
2x^2 - 3x + 3 \\
2x + 3 \overline{)\ 4x^3 + 0x^2 - 3x + 26} \\
-4x^3 \not\mp 6x^2 \\
\hline
-6x^2 - 3x \\
\not\mp 6x^2 \not\mp 9x \\
\hline
6x + 26 \\
\not\mp 6x \not\mp 9 \\
\hline
17 = \text{remainder}
\end{array}
$$

Answer: $\dfrac{4x^3 - 3x + 26}{2x + 3} = 2x^2 - 3x + 3 + \dfrac{17}{2x + 3}.$

You should check by expressing the answer as a single fraction.

EXERCISE 4.4   *Answers, page A50*

Divide as in EXAMPLE 1:

1. $\dfrac{8x^3 - 4x^2 + 16x}{4x}$

2. $\dfrac{5a^3 - 10a^2 - 15a}{5a}$

3. $\dfrac{P + Prt}{P}$

4. $\dfrac{\pi r^2 + 2\pi r}{\pi r}$

5. $\dfrac{3x^2h + 3xh^2 + h^3}{h}$

6. $\dfrac{4x^3h + 6x^2h^2 + 4xh^3 + h^4}{h}$

7. $\dfrac{16a^4 + 24a^3 - 8a^2 + 4a}{8a^2}$

8. $\dfrac{15m^3 - 30m^2 - 45m - 10}{15m}$

9. $\dfrac{-14x^2y^3 + 21xy^2 - 35xy}{-7xy^2}$

10. $\dfrac{-9a^3b^2 - 15a^2b + 27ab^2}{-3a^2b^2}$

11. $\dfrac{4a^2b^6 - 8ab^3 + 6a^2b^3 - 9ab^2}{4a^2b^2}$

12. $\dfrac{22uv^3 - 16u^2v^3 - 33uv^4 + 12u^2v^2}{24u^2v^2}$

Divide as in EXAMPLE 2:

13. $\dfrac{2x^2 + x - 15}{x + 3}$

14. $\dfrac{5x^2 - 13x + 8}{x - 1}$

15. $\dfrac{6x^2 - 7x - 5}{2x - 3}$

16. $\dfrac{12x^2 - 11x + 2}{3x + 1}$

17. $\dfrac{2x - 3}{x + 5}$

18. $\dfrac{6x + 7}{3x - 2}$

19. $\dfrac{6x^2 + 3x - 2}{2x + 1}$

20. $\dfrac{10x^2 - 4x + 9}{5x - 2}$

21. $\dfrac{20x^2 + 13x - 1}{2x + 1}$ *Hint:* Use decimals.

22. $\dfrac{8x^2 - 17x + 3}{2x - 3}$

23. $\dfrac{5x + 7}{2x - 5}$

24. $\dfrac{10x - 3}{4x + 3}$

25. $\dfrac{-8x^2 + 3x - 2}{x + 5}$

26. $\dfrac{-3x^2 - 7x + 9}{x - 3}$

27. $\dfrac{2x^3 + 7x^2 - 4x - 21}{x + 3}$

28. $\dfrac{2x^3 - 9x^2 + 13x - 6}{x - 2}$

29. $\dfrac{9x^3 + 9x^2 - 28x + 9}{3x - 2}$

30. $\dfrac{8x^3 - 22x^2 + 3x + 6}{4x + 3}$

31. $\dfrac{5x^3 + 10x^2 + 6x + 10}{x + 2}$

32. $\dfrac{4x^3 - 10x^2 - 6x + 3}{x - 3}$

33. $\dfrac{8x^3 - 2x^2 + 4x + 1}{2x - 1}$

34. $\dfrac{6x^3 - x^2 + x - 2}{2x + 1}$

35. $\dfrac{5x^4 - 7x^3 + 8x^2 + x - 2}{x + 3}$

36. $\dfrac{3x^4 + 5x^3 + x^2 - 2x - 1}{x - 1}$

Divide by first adding place holders (0, $0x$, $0x^2$, $0x^3$), as in EXAMPLE 3:

37. $\dfrac{2x^3 + 5x - 7}{x - 1}$

38. $\dfrac{5x^3 + x + 6}{x + 1}$

39. $\dfrac{7x^3 + 4x^2 - 2}{x + 2}$

40. $\dfrac{x^3 + 2x^2 - 3}{x - 2}$

41. $\dfrac{x^3 - 8}{x - 2}$

42. $\dfrac{8x^3 + 27}{2x + 3}$

43. $\dfrac{x}{x + 1}$

44. $\dfrac{x^2}{x - 1}$

45. $\dfrac{x^4 + 2x^2 - 9}{x - 3}$

46. $\dfrac{5x^4 + x^3 - 7}{x + 2}$

47. $\dfrac{4x^3 - 6x^2 - 2x}{2x + 1}$

48. $\dfrac{4x^3 + 4x^2 - x}{2x - 1}$

B

49. $\dfrac{x^4 - 16}{x - 2}$

50. $\dfrac{x^5 + 32}{x + 2}$

Divide. You do *not* need to add place holders.

51. $\dfrac{8x^6 + 7x^4 - 2x^2 + 9}{x^2 + 2}$

52. $\dfrac{a^6 - 9a^4 + 2a^2 + 3}{a^2 - 7}$

53. $\dfrac{2x^4 + x^3 - 10x^2 - 6x + 7}{x^2 + x - 1}$

54. $\dfrac{3x^4 - 5x^3 - x^2 + 5x - 2}{x^2 - 2x + 1}$

55. $\dfrac{a^5 + 2a^4 - 5a^3 + a^2 + a - 1}{a^2 - 2a - 1}$

56. $\dfrac{2a^5 + a^4 + a^3 - 3a^2 - a + 2}{a^2 + a - 2}$

4.5   **Fractional Equations and Inequalities**

One of the properties of an equation (Section 2.1) is that if A, B, and C are expressions and $C \neq 0$, then the fractional equation $\dfrac{A}{C} = \dfrac{B}{C}$ has the same solutions as the equation $A = B$. For example, if $x + 5 \neq 0$, then the fractional equation

$$\frac{3x - 1}{x + 5} = \frac{x + 7}{x + 5} \quad \text{has the same solution as} \quad 3x - 1 = x + 7.$$

Extending this to several terms, we see that if $y^2 - 9 \neq 0$, then the equations

$$\frac{2y^2 - y}{y^2 - 9} - \frac{2y - 3}{y^2 - 9} = \frac{y^2 + 7}{y^2 - 9} \quad \text{and} \quad 2y^2 - y - (2y - 3) = y^2 + 7$$

have the same solutions.

In the examples that follow, the denominators are not equal. By using building factors, we can express all fractions in terms of the least common denominator, as in Section 4.2. We then equate numerators as above, and solve.

EXAMPLE 1   Solve: $\dfrac{x}{x + 1} = \dfrac{3x + 5}{x^2 - 1}$

Solution
We first factor the denominator and then specify restrictions on the variable to avoid zero denominators (see Section 4.1, Example 1):

$$\frac{x}{x + 1} = \frac{3x + 5}{(x + 1)(x - 1)}, \quad x \neq -1, 1.$$

We now use a building factor on the left side to obtain the L.C.D. = $(x + 1)(x - 1)$:

$$\frac{x \cdot (x - 1)}{(x + 1) \cdot (x - 1)} = \frac{3x + 5}{(x + 1)(x - 1)}. \leftarrow \text{L.C.D.}$$

Because the denominators are equal, we equate the numerators and solve:

$$
\begin{aligned}
x(x - 1) &= = 3x + 5 \quad &&\text{Equate numerators.} \\
x^2 - x &= 3x + 5 \\
x^2 - 4x - 5 &= 0 \\
(x - 5)(x + 1) &= 0 \quad &&\text{Solve.} \\
x = 5, \quad x &= -1.
\end{aligned}
$$

We must reject $x = -1$ because of the restriction imposed. A solution that arises algebraically, but violates any restrictions imposed on the variable, is called an **extraneous solution.** Extraneous solutions often occur in fractional equations, so be on the lookout in this section. We write the answer as follows:

Answer: $x = 5$ ($x = -1$ is extraneous).

EXAMPLE 2

Solve: $\dfrac{6}{x^2 + 2x - 8} - \dfrac{3x}{x + 4} = \dfrac{x}{x - 2}$

Solution
We first factor and specify restrictions on the variable:

$$\frac{6}{(x + 4)(x - 2)} - \frac{3x}{x + 4} = \frac{x}{x - 2}, \quad x \neq -4, 2.$$

We now use building factors to obtain the L.C.D. = $(x + 4)(x - 2)$:

$$\frac{6}{(x + 4)(x - 2)} - \frac{3x \cdot (x - 2)}{(x + 4) \cdot (x - 2)} = \frac{x \cdot (x + 4)}{(x - 2) \cdot (x + 4)} \leftarrow \text{L.C.D.}$$

$$6 - 3x(x - 2) = x(x + 4) \quad \text{Equate numerators}$$
$$6 - 3x^2 + 6x = x^2 + 4x \quad \text{and solve.}$$
$$0 = 4x^2 - 2x - 6$$

Note sign.

$$0 = 2(2x^2 - x - 3)$$
$$0 = 2(2x - 3)(x + 1)$$
$$x = \frac{3}{2}, \quad x = -1. \quad \text{Answers}$$

In our next example, the denominator in the second fraction is "backwards." We "turn it around" by multiplying by $\frac{-1}{-1}$, as in Section 4.2, Example 5.

EXAMPLE 3

Solve: $\dfrac{7}{3z + 6} + \dfrac{11}{10 - 5z} = 0$

Solution

$$\frac{7}{3z + 6} + \frac{11 \cdot (-1)}{(10 - 5z) \cdot (-1)} = 0$$

$$\frac{7}{3z + 6} + \frac{-11}{5z - 10} = 0$$

$$\frac{7}{3(z + 2)} + \frac{-11}{5(z - 2)} = \frac{0}{1}, \quad z \ne -2, 2 \qquad \text{Factor.}$$

$$\frac{7 \cdot 5(z - 2)}{3(z + 2) \cdot 5(z - 2)} + \frac{-11 \cdot 3(z + 2)}{5(z - 2) \cdot 3(z + 2)} = \frac{0 \cdot 15(z + 2)(z - 2)}{1 \cdot 15(z + 2)(z - 2)} \leftarrow \text{L.C.D.}$$

$$35(z - 2) - 33(z + 2) = 0(z + 2)(z - 2) \qquad \text{Equate}$$
$$35z - 70 - 33z - 66 = 0 \qquad\qquad\qquad \text{numerators}$$
$$2z = 136 \qquad\qquad\qquad\qquad \text{and solve.}$$
$$z = 68 \qquad\qquad\qquad\qquad\qquad \text{Answer}$$

Linear inequalities containing fractions can be solved in a manner similar to that of fractional equations. One of the properties of an inequality (Section 2.5) states that if A, B, and C are expressions and C is positive, then the fractional inequality $\dfrac{A}{C} < \dfrac{B}{C}$ has the same solutions as the inequality $A < B$. For example, because 7 is positive,

$$\frac{5x - 2}{7} < \frac{x - 10}{7} \qquad \text{has the same solution as } 5x - 2 < x - 10.$$

This is also true if $<$ is replaced by the symbol $>$, $\le$, or $\ge$. In the next example, this idea will be extended to several terms.

EXAMPLE 4

Solve and graph: $\dfrac{3x + 1}{4} - \dfrac{4x - 5}{5} < 2$

Solution
The L.C.D. $= 20$, so we use building factors.

$$\frac{(3x + 1) \cdot 5}{4 \cdot 5} - \frac{(4x - 5) \cdot 4}{5 \cdot 4} < \frac{2 \cdot 20}{1 \cdot 20} \leftarrow \text{L.C.D.}$$

$$5(3x + 1) - 4(4x - 5) < 2 \cdot 20 \qquad \text{Solve numerators.}$$
$$15x + 5 - 16x + 20 < 40 \qquad \text{Note signs!}$$
$$-x + 25 < 40$$
$$-x < 15$$
$$\frac{-x}{-1} > \frac{15}{-1} \qquad \begin{array}{l}\text{Divide by } -1, \\ \text{change} < \text{to} > .\end{array}$$
$$x > -15. \qquad \text{Answer}$$

$\{x \mid x > -15\}:$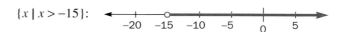

EXERCISE 4.5   *Answers, page A51*

A

Solve as in EXAMPLES 1 and 2. Watch for extraneous solutions.

1. $\dfrac{-3x}{4} = \dfrac{15}{8}$

2. $\dfrac{8y}{5} = -\dfrac{4}{7}$

3. $\dfrac{2z - 3}{5} = 7$ *Hint:* $7 = \dfrac{7}{1}$

4. $\dfrac{5 - 3t}{6} = -2$

5. $\dfrac{x^2 + 2x - 8}{2} = 0$

6. $\dfrac{w^2 - 2w - 3}{3} = 0$

7. $\dfrac{z^2}{9} = \dfrac{1}{4}$

8. $\dfrac{1}{y} = 9y$

9. $\dfrac{x}{3} - \dfrac{2x}{5} = 4$

10. $\dfrac{3y}{7} + \dfrac{5y}{4} = -1$

11. $\dfrac{t - 3}{8} - \dfrac{2t - 7}{6} = \dfrac{t + 5}{4}$

12. $\dfrac{z + 5}{12} - \dfrac{4z - 5}{8} = \dfrac{2z - 7}{9}$

13. $\dfrac{x}{x - 4} = \dfrac{4}{x - 4} + 3$

14. $\dfrac{2y}{y - 3} = \dfrac{6}{y - 3} - 6$

15. $x = 2 - \dfrac{1}{x}$

16. $\dfrac{3}{y} = 4 - y$

17. $\dfrac{2}{3x} = \dfrac{1}{6x^2} + \dfrac{1}{2}$

18. $\dfrac{1}{x^2} = \dfrac{1}{2} + \dfrac{7}{4x}$

19. $\dfrac{x + 5}{x - 4} = \dfrac{x + 4}{x - 2}$

20. $\dfrac{y - 1}{y + 2} = \dfrac{y - 7}{y + 4}$

21. $\dfrac{x + 2}{2x - 3} = \dfrac{2x + 1}{x + 6}$

22. $\dfrac{2z - 1}{z - 2} = \dfrac{3z + 1}{2z}$

23. $\dfrac{2t}{t + 2} = \dfrac{t^2 + 5t + 10}{t^2 + 3t + 2}$

24. $\dfrac{7 - 4y}{y^2 + y - 2} = \dfrac{y}{y - 1}$

25. $\dfrac{3}{z - 2} + \dfrac{7}{z + 2} = 2$

26. $\dfrac{9}{x + 1} - \dfrac{1}{x - 1} = 2$

27. $\dfrac{2r + 1}{3r - 6} + \dfrac{3r - 4}{5r - 10} = \dfrac{r + 7}{2r - 4}$

28. $\dfrac{n - 1}{2n + 6} - \dfrac{2n - 5}{3n + 9} = \dfrac{5n}{4n + 12}$

29. $\dfrac{x}{x + 2} = \dfrac{8}{x^2 - 4}$

30. $\dfrac{21 - y}{y^2 - 9} = \dfrac{y}{y - 3}$

31. $\dfrac{3z}{z + 2} - \dfrac{z}{z - 3} = \dfrac{z^2 + 26}{z^2 - z - 6}$

32. $\dfrac{2}{2w + 1} = \dfrac{1 - 3w}{2w^2 - 9w - 5} - \dfrac{4}{w - 5}$

33. $\dfrac{x - 1}{x + 2} + \dfrac{x - 2}{x - 3} = \dfrac{x^2 + 4}{x^2 - x - 6}$

34. $\dfrac{3y + 1}{2y + 3} - \dfrac{y - 3}{2y + 1} = \dfrac{y^2 - y + 4}{4y^2 + 8y + 3}$

Solve by first using the building factor $\dfrac{-1}{-1}$, as in EXAMPLE 3:

35. $\dfrac{7}{x + 2} - \dfrac{3}{2 - x} = \dfrac{2}{x^2 - 4}$

36. $\dfrac{9}{y + 1} + \dfrac{1}{1 - y} = \dfrac{2}{y^2 - 1}$

37. $\dfrac{5}{z - 3} + \dfrac{1}{4 - z} = \dfrac{3}{z^2 - 7z + 12}$

38. $\dfrac{3}{w + 4} + \dfrac{2}{2 - w} = \dfrac{-3}{w^2 + 2w - 8}$

39. $\dfrac{2}{x + 1} + \dfrac{3}{1 - x} = \dfrac{4x + 5}{x^2 - 1}$

40. $\dfrac{5}{2y + 3} - \dfrac{2y + 1}{9 - 4y^2} = \dfrac{2}{2y - 3}$

Solve and graph, as in EXAMPLE 4:

41. $\dfrac{x}{10} + \dfrac{2x}{5} > x - \dfrac{5}{2}$

42. $\dfrac{x}{6} - 2 \le \dfrac{2x}{3} - \dfrac{1}{2}$

43. $\dfrac{2x + 5}{9} \ge \dfrac{x + 3}{4}$

44. $\dfrac{5x + 1}{3} < \dfrac{x - 4}{2}$

45. $\dfrac{2x}{3} - \dfrac{x - 1}{4} < \dfrac{3}{2}$

46. $\dfrac{3x + 4}{8} - \dfrac{2 - x}{6} \ge -2$

47. $\dfrac{x}{5} > \dfrac{x}{8}$

48. $\dfrac{x}{2} \ge 2x$

49. $-3 < \dfrac{2x + 1}{5} < 1$

50. $2 \le \dfrac{3x - 2}{7} \le 4$

51. $6 \le \dfrac{4 - 3x}{2} < 12$

52. $-3 < \dfrac{3 - 2x}{3} \le 5$

53. $6 > \dfrac{5x + 1}{2} > 3$

54. $1 \ge \dfrac{1 - x}{3} \ge -2$

B

Solve:

55. $\dfrac{2z + 3}{6z + 7} = \dfrac{z - 4}{3z - 2}$

56. $\dfrac{8t + 1}{4t + 1} = \dfrac{4t - 1}{2t - 1}$

57. $\dfrac{2x}{5x - 15} - 7 = \dfrac{x + 1}{9 - 3x}$

58. $\dfrac{3v + 1}{2v - 4} - 5 = \dfrac{3 + v}{16 - 8v}$

59. $\dfrac{3}{2y} + \dfrac{5}{y + 3} = \dfrac{2}{3y + 9}$

60. $\dfrac{3}{z - 5} - \dfrac{5}{3z} = \dfrac{1}{4z - 20}$

61. $\dfrac{x - 1}{x^2 - 9} + \dfrac{3}{4x + 12} = \dfrac{2}{2x - 6}$

62. $\dfrac{5}{3x - 15} = \dfrac{3}{2x + 10} - \dfrac{x - 2}{x^2 - 25}$

63. $\dfrac{x^2 - 8x}{x^3 - 8} = \dfrac{2x + 5}{x^2 + 2x + 4}$

64. $\dfrac{y - 5}{y^2 - 5y + 25} = \dfrac{5y - 1}{y^3 + 125}$

65. $\dfrac{x+4}{x^2+x-6} + \dfrac{x-1}{x^2-2x-15} = \dfrac{x}{x^2-7x+10}$

66. $\dfrac{y+3}{2y^2-3y+1} = \dfrac{y-2}{2y^2+y-1} - \dfrac{3}{y^2-1}$

67. $x = \dfrac{\dfrac{2}{x}}{1+\dfrac{1}{x}}$

68. $\dfrac{x-\dfrac{2}{x}}{1+\dfrac{3}{x}} = 2$

Use the methods of Section 2.6 to solve the following:

69. $\left|2x - \dfrac{3}{4}\right| = \dfrac{1}{2}$

70. $\left|\dfrac{2}{3} - \dfrac{3x}{4}\right| = 2$

71. $\left|\dfrac{x+1}{x+8}\right| = \dfrac{2}{3}$

72. $\left|\dfrac{x}{2} - \dfrac{3}{5}\right| = \left|\dfrac{2x}{3} + \dfrac{1}{10}\right|$

73. $\left|\dfrac{1-2x}{3}\right| < 5$

74. $\left|\dfrac{5x}{6} + \dfrac{1}{2}\right| \le \dfrac{1}{3}$

75. $\left|\dfrac{x}{2} - 1\right| \ge \dfrac{3}{4}$

76. $\left|2 + \dfrac{2x}{3}\right| > \dfrac{5}{6}$

Students often confuse rational *expressions* and rational *equations*. For example,

$$\dfrac{2x}{3} + \dfrac{x}{4} - \dfrac{11}{2} \quad \text{is an **expression**.} \qquad \text{(Section 4.2)}$$

$$\dfrac{2x}{3} + \dfrac{x}{4} = \dfrac{11}{2} \quad \text{is an **equation**.} \qquad \text{(Section 4.5)}$$

We *simplify* the expression as follows:

$$\dfrac{2x \cdot 4}{3 \cdot 4} + \dfrac{x \cdot 3}{4 \cdot 3} - \dfrac{11 \cdot 6}{2 \cdot 6} = \dfrac{8x + 3x - 66}{12}$$

$$= \dfrac{11x - 66}{12}, \quad \text{which is an **expression**.}$$

We *solve* the equation as follows:

$$\dfrac{2x \cdot 4}{3 \cdot 4} + \dfrac{x \cdot 3}{4 \cdot 3} = \dfrac{11 \cdot 6}{2 \cdot 6}$$
$$8x + 3x = 66$$
$$11x = 66$$
$$x = 6, \quad \text{which is a **solution**.}$$

Simplify the following expressions and solve the equations:

77. $\dfrac{2x}{3} - \dfrac{1}{2} = \dfrac{x}{6} - 2$

78. $\dfrac{3x}{4} + \dfrac{2}{3} - \dfrac{x}{2} + 4$

79. $\dfrac{2x}{5} - \dfrac{x-1}{4} + 1$

80. $\dfrac{x+3}{2} - 1 = \dfrac{x+5}{4}$

81. $\dfrac{2x}{5} = \dfrac{x-1}{4} + 1$

82. $\dfrac{x+3}{2} - 1 - \dfrac{x+5}{4}$

83. $\dfrac{3x}{4} + \dfrac{x}{2} - \dfrac{1}{3} - 0$

84. $\dfrac{3x}{4} + \dfrac{x}{2} - \dfrac{1}{3} = 0$

4.6   Word Problems

In Sections 2.1 and 3.5, you translated verbal statements into linear and quadratic equations, respectively. Continuing in the same spirit, these statements will now lead to fractional equations. As before, a list of commonly used fractional phrases is helpful.

Fractional phrases	Algebraic expressions
two-thirds of a number	$\frac{2}{3}x$, or $\frac{2}{3} \cdot \frac{x}{1}$, or $\frac{2x}{3}$
one-half of the sum of a number and three	$\frac{1}{2}(x + 3)$, or $\frac{x + 3}{2}$
four-fifths of the difference between two and a number	$\frac{4}{5}(2 - x)$, or $\frac{4(2 - x)}{5}$
half the result of decreasing a number by one	$\frac{1}{2}(x - 1)$, or $\frac{x - 1}{2}$
six less than one-third of the result of increasing eight by a number	$\frac{1}{3}(8 + x) - 6$, or $\frac{8 + x}{3} - 6$
a fraction whose denominator is four more than its numerator	$\frac{x}{x + 4}$
a fraction such that the sum of its numerator and denominator is 20	$\frac{x}{20 - x}$
the *reciprocal* of x; of $2x$; of $\frac{x}{y}$	$\frac{1}{x}$; $\frac{1}{2x}$; $\frac{y}{x}$
twice the reciprocal of x	$2\left(\frac{1}{x}\right)$, or $\frac{2}{x}$
the sum of the reciprocals of two consecutive integers	$\frac{1}{n} + \frac{1}{n + 1}$
the *ratio* of x to y; of y to x	$\frac{x}{y}$; $\frac{y}{x}$

EXAMPLE 1

If a number is increased by seven, then one-half of the result equals five less than four-fifths of the difference between the number and one. What is the number?

Solution

Let x = the number; then this "tongue-twister" translates into

$$\frac{1}{2}(x + 7) = \frac{4}{5}(x - 1) - 5, \quad \text{or}$$

$$\frac{x + 7}{2} = \frac{4(x - 1)}{5} - 5$$

$$\frac{(x + 7) \cdot 5}{2 \cdot 5} = \frac{(4x - 4) \cdot 2}{5 \cdot 2} - \frac{5 \cdot 10}{1 \cdot 10} \leftarrow \text{L.C.D.} = 10$$

$$5x + 35 = 8x - 8 - 50 \qquad \text{Equate numerators and solve.}$$
$$93 = 3x$$
$$31 = x.$$

Answer: The number is 31.

Check: $\dfrac{31 + 7}{2} = \dfrac{4(31 - 1)}{5} - 5, \quad \text{or} \quad \dfrac{38}{2} = \dfrac{4(30)}{5} - 5, \quad \text{or} \quad 19 \overset{\checkmark}{=} 24 - 5.$

EXAMPLE 2

The denominator of a fraction is four more than the numerator. If both numerator and denominator are increased by one, a new fraction is formed whose value is $\frac{3}{4}$. Find the original fraction and the new fraction.

Solution

Let $\dfrac{x}{x + 4}$ = original fraction; then $\dfrac{x + 1}{x + 4 + 1}$ = new fraction.

We have

$$\frac{x + 1}{x + 5} = \frac{3}{4} \qquad \text{New fraction} = \frac{3}{4}$$

$$\frac{(x + 1) \cdot 4}{(x + 5) \cdot 4} = \frac{3 \cdot (x + 5)}{4 \cdot (x + 5)} \leftarrow \text{L.C.D.} = 4(x + 5)$$

$$4x + 4 = 3x + 15$$
$$x = 11$$

Answer: original fraction = $\dfrac{11}{11 + 4} = \dfrac{11}{15}$;   new fraction = $\dfrac{11 + 1}{15 + 1} = \dfrac{12}{16}$.

Check: new fraction $\dfrac{12}{16} \overset{\checkmark}{=} \dfrac{3}{4}$.

EXAMPLE 3 | The sum of the reciprocals of two consecutive even integers is $\frac{5}{12}$. What are the integers?

Solution

Let n = first even integer; then $n + 2$ = second even integer. The sum of their reciprocals is

$$\frac{1}{n} + \frac{1}{n + 2} = \frac{5}{12}$$

$$\frac{1 \cdot 12(n + 2)}{n \cdot 12(n + 2)} + \frac{1 \cdot 12n}{(n + 2) \cdot 12n} = \frac{5 \cdot n(n + 2)}{12 \cdot n(n + 2)} \longleftarrow \text{L.C.D.} = 12n(n + 2)$$

$$12n + 24 + 12n = 5n^2 + 10n$$
$$0 = 5n^2 - 14n - 24$$
$$0 = (5n + 6)(n - 4)$$
$$n = \frac{-6}{5}, \; n = 4.$$

We reject $\frac{-6}{5}$ because it is not an integer, so $n = 4$ and $n + 2 = 6$.

Answer: The integers are 4 and 6.

Check: The sum of their reciprocals is $\frac{1}{4} + \frac{1}{6} = \frac{3}{12} + \frac{2}{12} \stackrel{\checkmark}{=} \frac{5}{12}$.

EXERCISE 4.6   *Answers, page A51*

A

Solve as in EXAMPLES 1, 2, and 3. Consult the Table of Phrases if necessary.

1. One-half of the sum of a number and five equals two-thirds of the difference between the number and seven. Find the number.

2. When twice a number is decreased by seven, one-third of the result equals three-fifths of the sum of the number and eight. What is the number?

3. When two-thirds of the difference between seven and a number is increased by 14, the result is the same as one-fourth of the difference between the number and six. What is the number?

4. Five-sixths of the sum of four and twice a number equals the result of adding ten-thirds to twice the difference between the number and one. What is the number?

5. The denominator of a fraction is six more than the numerator. When four is added to both numerator and denominator, a new fraction is formed whose value is $\frac{3}{5}$. Find both the original fraction and the new fraction.

6. The numerator of a fraction is seven less than the denominator. When numerator and denominator are each increased by twelve, the new fraction formed equals $\frac{2}{3}$. Find both the original and the new fraction.

7. The numerator of a fraction is two less than the denominator. When the numerator is decreased by one, and the denominator is increased by three, the newly formed fraction equals $\frac{1}{4}$.

8. The denominator of a fraction is one more than twice the numerator. When the denominator is decreased by three, and the numerator is increased by two, the new fraction equals $\frac{3}{4}$. Find both fractions.

9. What number must be added to both the numerator and the denominator of the fraction $\frac{2}{7}$ to make a new fraction equal to $\frac{1}{2}$?

10. By what number must both the numerator and the denominator of the fraction $\frac{11}{15}$ be decreased to make a new fraction equal to $\frac{2}{3}$?

11. The sum of the numerator and the denominator of a fraction is 40. The fraction equals $\frac{2}{3}$. Find the fraction.

12. Find two numbers whose sum is 24 and whose ratio is $\frac{5}{7}$.

13. When three consecutive integers are divided by 2, 3, and 4, respectively, the sum of the quotients is 16. Find the integers.

14. When four consecutive *even* integers are divided by 3, 4, 6, and 8, respectively, the sum of the quotients is $\frac{32}{3}$. Find the integers.

15. The sum of the reciprocals of two consecutive even integers is $\frac{7}{24}$. Find the integers.

16. The difference between the reciprocals of two consecutive integers is $\frac{1}{30}$. Find the integers.

17. The sum of a number and its reciprocal is $\frac{25}{12}$. Find the number.

18. What number equals its own reciprocal?

B

19. When twice the reciprocal of a number is added to the reciprocal of twice the number, the result is $\frac{5}{12}$. What is the number?

20. One-half of the reciprocal of an even integer, increased by one-third of the reciprocal of the next even integer, equals $\frac{1}{3}$. Find the integers.

21. The reciprocal of the square of a number is two less than three times the reciprocal of the number. Find the number.

22. The sum of the reciprocals of two consecutive integers is two-thirds more than the product of their reciprocals. Find the integers.

23. What number must be added to both the numerators and the denominators of the fractions $\frac{3}{8}$ and $\frac{5}{12}$ so that the new fractions that are formed are equal to each other? What are the new fractions?

24. The numerator of a fraction is four less than the denominator. When two is added to both numerator and denominator, a new fraction is formed that is $\frac{1}{6}$ more than the original fraction. Find the original fraction and the new fraction.

25. The rectangle shown below has perimeter 30 inches. Find the width and the length. *Formula*: $P = 2L + 2W$.

26. The rectangle shown below has area 90 square meters. Find the width and the length. *Formula*: $A = LW$.

$\frac{x}{3} + 7$

Problems 25 and 26

Problem 27

27. Find the three sides of the right triangle shown above. *Hint*: $a^2 + b^2 = c^2$.

28. The length of a rectangle is 3 feet more than one-half of the width. The area is 20 square feet. Find the width and the length.

29. The length of a rectangle is 11 inches more than one-third of the width. The perimeter is 38 inches. Find the width and the length.

30. One leg of a right triangle is 1 inch more than one-half of the hypotenuse. The other leg is 2 inches longer than the first leg. Find each side.

The area of a triangle with base b and height h is given by the formula

$$A = \frac{bh}{2}$$

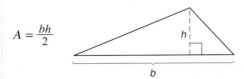

Use this formula to solve the following problems.

31. The base of a triangle is 5 meters longer than the height. The area is 12 square meters. Find the height and the base.

32. The height of a triangle is three times the base. The area is 54 square inches. Find the height and the base.

33. The height of a triangle is $\frac{1}{2}$ inch less than the base. The area is $3\frac{3}{4}$ inches. Find the base and the height. *Suggestion*: Express the area as an improper fraction.

34. One leg of a right triangle is 2 feet shorter than the other leg. The area is 24 square feet. Find the two legs.

35. An estate is divided among three children. The eldest gets half, the next eldest gets one-fourth of the balance, and the youngest gets two-thirds of the remaining balance. There is $125,000 left over, which is donated to charity. Find the original estate and each child's share.

36. A real estate partnership buys a piece of property. Smith puts up $\frac{1}{3}$ of the cash and therefore controls $\frac{1}{3}$ of the partnership. Jones puts up $\frac{1}{4}$ of the cash, and therefore controls $\frac{1}{4}$ of the partnership. Smith then buys Jones's control for $20,000. The property is eventually sold for $120,000. Smith's net profit is $30,000. Find the original price of the property. *Hint*: Smith's profit = his share of the proceeds from the sale of the property minus his total cash outlay.

4.7 **Literal Equations and Formulas**

An equation having letters as terms or as coefficients of the unknown is called a **literal equation.** For example, $5ax + 7b = ax - 9b$ is a literal equation with unknown x, though a and b could just as well be considered unknowns. In this section, we will be solving for one of the variables in terms of the others.

EXAMPLE 1

Solve for x: $5ax + 7b = ax - 9b$

Solution
We get all terms containing x to one side of the equation and all other terms to the other side:

$$5ax - ax = -7b - 9b \qquad \text{All } x\text{-terms to one side.}$$
$$4ax = -16b \qquad \text{Combine like terms.}$$
$$\frac{4ax}{4a} = \frac{-16b}{4a} \qquad \text{Divide by coefficient of } x.$$
$$x = \frac{-4b}{a}. \qquad \text{Answer}$$

In the third line of Example 1, we divided by $4a$. Here we assume $a \neq 0$ in order to make this division possible. In the *next two examples* and *throughout the exercises* we will make similar assumptions about the variables so that division is possible and, hence, each problem has a solution.

EXAMPLE 2

Solve for y: $\dfrac{m}{y + n} = \dfrac{n}{m - y}$

Solution

$$\frac{m \cdot (m - y)}{(y + n) \cdot (m - y)} = \frac{n \cdot (y + n)}{(m - y) \cdot (y + n)} \longleftarrow \text{L.C.D.}$$

$$m^2 - my = ny + n^2 \qquad \text{Equate numerators and multiply parentheses.}$$

$$m^2 - n^2 = ny + my \qquad \text{All } y \text{ terms to one side.}$$

$$(m + n)(m - n) = y(n + m) \qquad \text{Factor.}$$

$$\frac{(m + n)(m - n)}{n + m} = \frac{y(n + m)}{n + m} \qquad \text{Divide by coefficient of } y.$$

$$m - n = y. \qquad \text{Answer}$$

These examples illustrate the steps taken in order to solve for a specific variable, say x:

To Solve a Literal Equation for x

Step 1 If fractions are present, use L.C.D. to rewrite without denominator.

Step 2 Multiply all parentheses and combine like terms, if any.

Step 3 By adding and subtracting, get all terms containing x to *one side* of the equation, all terms without x to other side.

Step 4 Combine like terms, if any.

Step 5 If either side contains two or more terms, *factor* if possible.

Step 6 Divide by the entire coefficient of x.

EXAMPLE 3

If R_1 and R_2 are resistors in a parallel circuit, and R is the total resistance, then they are related by the formula

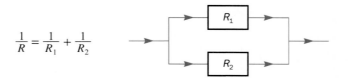

$$\frac{1}{R} = \frac{1}{R_1} + \frac{1}{R_2}$$

Solve this formula for R_1.

Note: R_1 and R_2 are read "R sub one" and "R sub two," respectively. They are *not* powers R^1 and R^2.

Solution

$$\frac{1 \cdot R_1R_2}{\cdot R_1R_2} = \frac{1 \cdot RR_2}{R_1 \cdot RR_2} + \frac{1 \cdot RR_1}{R_2 \cdot RR_1} \Bigg\} \quad \text{Step 1}$$

$$R_1R_2 = RR_2 + RR_1$$

$$R_1R_2 - RR_1 = RR_2 \qquad\qquad\qquad \text{Step 3:} \quad \text{All } R_1\text{-terms to left side.}$$

$$R_1(R_2 - R) = RR_2 \qquad\qquad\qquad \text{Step 5:} \quad \text{Factor.}$$

$$R_1 = \frac{RR_2}{R_2 - R}. \qquad\qquad\qquad\qquad \text{Step 6:} \quad \text{Divide.}$$

EXERCISE 4.7   *Answers, page A52*

A

Solve for the indicated variable, as in EXAMPLES 1, 2, and 3:

1. $V = LWH$; for L (Volume of a box)

2. $PV = nrT$; for r (Ideal Gas Law)

3. $A = \dfrac{1}{2}\,bh$; for h (Area of a triangle)

4. $s = \dfrac{gt^2}{2}$; for g (Law of gravity)

5. $S = \dfrac{200WH^2}{L}$; for W (Stress on a beam)

6. $F = \dfrac{km_1m_2}{d^2}$; for m_2 (Universal Law of Gravitation)

7. $3ab - 7bx = bx - 13ab$; for x

8. $3ab - 7bx = bx - 13ab$; for a

9. $A = P + Prt$; for P

10. $A = P + Prt$; for t (Amount = principal + interest)

11. $2x - 3y - 6 = 0$; for y

12. $ax + by + c = 0$; for y (Equation of a straight line)

13. $L = a + (n - 1)d$; for n

14. $L = a + (n - 1)d$; for d (Last term in a sequence)

15. $a(9y - 4b) = 3(4a^2 - by)$; for y

16. $2c^2(3x + 8d^2) = 4cd(2c^2 + 3x)$; for x

17. $\dfrac{y - 2}{x + 5} = -3$; for y

18. $\dfrac{y + 1}{x - 7} = \dfrac{2}{5}$; for x

19. $\dfrac{a}{x - b} = \dfrac{4b}{2x - a}$; for x

20. $\dfrac{2c}{9d} = \dfrac{y - d}{3y - c}$; for y

21. $\dfrac{2x + a}{2x - a} = \dfrac{x + b}{x - 2b}$; for x

22. $\dfrac{2y + 3b}{5b + 6y} = \dfrac{y - 2b}{3y - b}$; for y

23. $x = \dfrac{y}{y + 1}$; for y

24. $x = \dfrac{y - 2}{2y}$; for y

25. $F = \dfrac{9}{5} C + 32$;   for C (Celsius/Fahrenheit relationship)

26. $C = \dfrac{5}{9} (F - 32)$;   for F

27. $\dfrac{1}{c} = \dfrac{1}{c_1} + \dfrac{1}{c_2}$;   for c_2 (Capacitors in a series circuit)

28. $\dfrac{1}{c} = \dfrac{1}{c_1} + \dfrac{1}{c_2}$;   for c

29. $\dfrac{x}{b} + \dfrac{b}{a} = \dfrac{x}{a} + \dfrac{a}{b}$;   for x

30. $\dfrac{x}{18d} + \dfrac{d}{4c} = \dfrac{c}{9d} - \dfrac{x}{12d}$;   for x

31. $A = \dfrac{h(a + b)}{2}$;   for a (Area of a trapezoid)

32. $S = \dfrac{n(A + L)}{2}$;   for n (Sum of a sequence)

B

33. $a(y - a) = b(2a + b - y)$;   for y

34. $n(x + m + 2n) = m(m - x)$;   for x

35. $b(b^2 - x) = a(a^2 - x)$;   for x

36. $\dfrac{a}{2b} = \dfrac{4b^2 - y}{y - a^2}$;   for y

37. $(x - b)(x + 2b) = 2(x + b)(x - 3b) - x(x + 4b)$;   for x

38. $(3x - c)(x + 2c) - (x - 4c)^2 = (x - c)(x + c) + (x - 3c)^2$;   for x

39. $S = \dfrac{a(1 - rn)}{1 - r}$;   for r

40. $S = \dfrac{n[2a + (n - 1)d]}{2}$;   for d (Sum of a sequence)

41. $\dfrac{y}{ab^2} - \dfrac{b}{a^2} = \dfrac{a}{b^2} - \dfrac{y}{a^2b}$;   for y

42. $(x - a)^2 = (x - b)^2 - 2b(b - a)$;   for x

C

43. $\dfrac{1}{R} = \dfrac{1}{R_1} + \dfrac{1}{R_2} + \dfrac{1}{R_3}$;   for R_3

44. $\dfrac{1}{R} = \dfrac{1}{R_1} + \dfrac{1}{R_2} + \dfrac{1}{R_3}$;   for R

45. $\dfrac{w}{1 + \dfrac{w}{a}} = 2$;   for a

46. $\dfrac{w}{1 + \dfrac{w}{a}} = 2$;   for w

47. $E = \dfrac{1}{\dfrac{1}{m} + \dfrac{1}{M}}$;   for m

48. $\dfrac{B}{A} = \dfrac{1}{1 + \dfrac{A}{C}}$;   for A

CHAPTER 4 · SUMMARY

4.1 Reducing, Multiplying, and Dividing Rational Expressions

$$\frac{a^m}{a^n} = \begin{cases} a^{m-n} & \text{if } m > n \\ \dfrac{1}{a^{n-m}} & \text{if } m < n \end{cases}$$

Example:

$$\frac{6a^5b}{-8a^2b^7} = \frac{-3a^3}{4b^6}.$$

Divide by 2

To divide fractions, invert divisor and multiply.
Use reduction $\dfrac{A-B}{B-A} = -1$ when applicable.

Example:

$$\frac{2x-6}{4x+4} \div \frac{9-x^2}{x^2+4x+3}$$

Invert.

$$= \frac{2(x-3)}{4(x+1)} \cdot \frac{(x+1)(x+3)}{(3-x)(3+x)}$$

$$= \frac{-1}{2}.$$

4.2 Adding and Subtracting Rational Expressions

Factor to get L.C.D.; then use building factors.

Example:

$$\frac{3}{x+2} + \frac{2}{x^2-4}$$

$$= \frac{3 \cdot (x-2)}{(x+2) \cdot (x-2)} + \frac{2}{(x+2)(x-2)}$$

$$= \frac{3x-4}{(x+2)(x-2)}. \leftarrow \text{L.C.D.}$$

4.3 Complex Fractions

Write numerator and denominator as single fractions; then invert and multiply.

Example:

$$\frac{\dfrac{1}{x} - \dfrac{1}{x^2}}{1 - \dfrac{1}{x^2}} = \frac{\dfrac{1 \cdot x}{x \cdot x} - \dfrac{1}{x^2}}{\dfrac{1 \cdot x^2}{1 \cdot x^2} - \dfrac{1}{x^2}}$$

$$= \frac{\dfrac{x-1}{x^2}}{\dfrac{x^2-1}{x^2}} \begin{array}{l}\leftarrow \text{Single fraction} \\ \\ \leftarrow \text{Single fraction}\end{array}$$

$$= \frac{x-1}{x^2} \cdot \frac{x^2}{(x-1)(x+1)} \quad \text{Invert.}$$

$$= \frac{1}{x+1}.$$

4.4 Dividing Polynomials

Division by a monomial:

Example:

$$\frac{8x^3 - 4x^2 + 3x}{4x} = \frac{8x^3}{4x} - \frac{4x^2}{4x} + \frac{3x}{4x}$$

$$= 2x^2 - x + \frac{3}{4}.$$

Division by a binomial:

Example:

$$\begin{array}{r} 5x - 3 + \dfrac{-2}{x-1} \\ x-1 \overline{)\, 5x^2 - 8x + 1} \end{array}$$

$5x(x-1) \longrightarrow$
Change signs and add.

$$-5x^2 \not{+} 5x \quad \downarrow$$
$$-3x+1$$
$$\not{+}3x \not{+} 3$$
$$-2 = \text{remainder}.$$

■ **CHAPTER 4 ▪ SUMMARY**

4.5 Fractional Equations and Inequalities

Express all fractions in terms of the L.C.D.; then equate numerators and solve. *Watch for extraneous solutions.*

Example:

$$\text{Solve: } 1 - \frac{2}{x} - \frac{8}{x^2} = 0, \quad x \neq 0$$

$$\frac{1 \cdot x^2}{1 \cdot x^2} - \frac{2 \cdot x}{x \cdot x} - \frac{8}{x^2} = \frac{0 \cdot x^2}{1 \cdot x^2} \leftarrow \text{L.C.D.}$$

$$x^2 - 2x - 8 = 0$$
$$(x - 4)(x + 2) = 0$$
$$x = 4, \quad x = -2.$$

4.6 Word Problems

"Two-thirds of the sum of a number and five equals one-half of the difference between three times the number and five. Find the number."

Use the Table of Phrases if necessary.

Example:

$$\frac{2}{3}(x + 5) = \frac{1}{2}(3x - 5)$$

$$\frac{(2x + 10) \cdot 2}{3 \cdot 2} = \frac{(3x - 5) \cdot 3}{2 \cdot 3}$$

$$4x + 20 = 9x - 15$$
$$35 = 5x$$
$$7 = x.$$

4.7 Literal Equations and Formulas

Step 1   Use the L.C.D. to rewrite fraction without denominator.

Step 2   Multiply parentheses.

Step 3   Isolate all x terms to one side.

Step 4   Combine like terms, if any.

Step 5   Factor, if possible.

Step 6   Divide by coefficient of x.

Example:

$$\text{Solve for } x: \quad \frac{x}{a} = \frac{x + b}{b}$$

$$\frac{x \cdot b}{a \cdot b} = \frac{(x + b) \cdot a}{b \cdot a}$$

$$bx = ax + ab$$

$$bx - ax = \qquad ab$$

$$x(b - a) = ab$$

$$x = \frac{ab}{b - a}.$$

CHAPTER 4 · REVIEW EXERCISES   *Answers, pages A52–A53*

4.1

State any restrictions on the variables in each expression:

1. $\dfrac{19}{2x}$

2. $\dfrac{x + 7}{x - 1}$

3. $\dfrac{5y}{2y + 6}$

4. $\dfrac{x + 1}{(x + 7)(x - 2)}$

5. $\dfrac{3z}{z^2 + 2z - 8}$

6. $\dfrac{5t + 4}{6t^2 - 11t - 21}$

Reduce to lowest terms. Assume the denominators are not zero.

7. $\dfrac{-12x^5y^4z^3}{-16x^3yz^4}$

8. $\dfrac{5a - 10b}{8a - 16b}$

9. $\dfrac{10x^3y - 25x^2y^2}{20x^2y^2 - 50xy^3}$

10. $\dfrac{x^2 - 7x - 8}{64 - x^2}$

11. $\dfrac{8a^3 - 27b^3}{2a^2 - 9ab + 9b^2}$

12. $\dfrac{x^2 - 2x + ax - 2a}{2x^2 - 3bx + 2ax - 3ab}$

Multiply or divide, as indicated. Assume the denominators are not zero.

13. $\dfrac{18a^2b^5}{5ac^2} \cdot \dfrac{15a^3c}{24abc}$

14. $\dfrac{4mn^2}{14mp^3} \div \dfrac{24mn^3p}{-21m^2}$

15. $\dfrac{x^2 - 81}{4x + 12} \cdot \dfrac{x^2 + 4x + 3}{x^2 + 10x + 9}$

16. $\dfrac{12y - 4x}{12x - 30y} \cdot \dfrac{-4x^2 + 14xy - 10y^2}{8x - 24y}$

17. $\dfrac{4x^2 - 16x + 15}{8x^3 + 12x^2} \div \dfrac{6x^2 - x - 35}{24x^2y + 56xy}$

18. $\dfrac{18a^3 - 2a}{2a^2 - 9a - 35} \cdot \dfrac{6a^2 - 13a - 5}{12a^3 + 4a^2} \div \dfrac{4a^2 - 20a + 25}{16a^2 + 40a}$

4.2

Combine and reduce to lowest terms. Assume the denominators are not zero.

19. $\dfrac{5x - 2}{2x - 1} + \dfrac{4x - 5}{2x - 1} - \dfrac{x - 3}{2x - 1}$

20. $\dfrac{5b - 3a}{a - 2b} - \dfrac{-a + 3b}{2b - a}$

21. $\dfrac{5}{12x^2} - \dfrac{3}{8xy} + \dfrac{7}{36y^2}$

22. $\dfrac{4}{x - y} + \dfrac{3}{x - 4y} - \dfrac{x - 13y}{x^2 - 5xy + 4y^2}$

23. $\dfrac{3}{x + 2} + \dfrac{20}{x^2 - 4} - \dfrac{5}{x - 2}$

24. $\dfrac{5}{2a - 5} - \dfrac{2}{5 - a} - \dfrac{3a - 5}{2a^2 - 15a + 25}$

25. $\dfrac{5}{x^2 - 6x + 9} + \dfrac{2}{2x^2 - 18} + \dfrac{3}{4x + 12}$

26. $\dfrac{2}{x + h} - \dfrac{2}{x} + \dfrac{2}{x - h}$

27. $\dfrac{2}{a - 2b} - \dfrac{2a}{a^2 + 2ab + 4b^2} - \dfrac{24b^2}{a^3 - 8b^3}$

28. $\dfrac{x + 2}{x^2 - 1} - \dfrac{x - 1}{x^2 + 3x + 2} + \dfrac{x + 1}{x^2 + x - 2}$

4.3

Simplify these complex fractions:

29. $\dfrac{\dfrac{-3a^2b}{5c}}{\dfrac{9a}{10bc^2}}$

30. $\dfrac{\dfrac{2x + 6y}{-5xy}}{4x + 12y}$

31. $\dfrac{1 + \dfrac{a}{b}}{1 - \dfrac{a^2}{b^2}}$

32. $\dfrac{\dfrac{x}{y^3} - \dfrac{1}{x^2}}{\dfrac{1}{x} - \dfrac{x}{y^2}}$

33. $\dfrac{1 - \dfrac{8}{x^2 - 1}}{\dfrac{2}{x - 1} - \dfrac{1}{x + 1}}$

34. $\dfrac{x + 1 + \dfrac{1}{x - 1}}{2x - 1 - \dfrac{1}{x - 1}}$

35. $1 + \dfrac{1}{1 - \dfrac{1}{1 + \dfrac{1}{x}}}$

4.4

Divide:

36. $\dfrac{20x^3 - 10x^2 + 15x}{5x}$

37. $\dfrac{8a^3b^2 - 12ab^2 + 2ab}{-4ab^2}$

38. $\dfrac{4x - 7}{2x + 3}$

39. $\dfrac{x}{x + 1}$

40. $\dfrac{3x^2 - 8x - 5}{x - 2}$

41. $\dfrac{8x^2 - 2x + 5}{2x + 1}$

42. $\dfrac{5x^3 - 7x^2 + 8x - 6}{x - 1}$

43. $\dfrac{2x^4 + 7x^2 - 9x + 3}{x + 2}$

44. $\dfrac{2x^5 + x^4 + 2x^3 - 7x^2 + x - 5}{x^2 - x + 1}$

4.5

Solve. Watch for extraneous solutions.

45. $\dfrac{3x - 5}{4} = \dfrac{5 + 2x}{6}$

46. $\dfrac{2}{3}(2x + 1) - \dfrac{1}{4}(x - 5) = -10$

47. $\dfrac{x}{x - 5} = \dfrac{5}{x - 5} - 8$

48. $\dfrac{y}{5y + 10} + \dfrac{y - 3}{2y + 4} = \dfrac{y + 1}{3y + 6}$

49. $\dfrac{5}{x - 2} - \dfrac{7}{x + 3} = \dfrac{x - 1}{x^2 + x - 6}$

50. $\dfrac{4}{z + 5} + \dfrac{2}{5 - z} = \dfrac{5 - 3z}{z^2 - 25}$

51. $\dfrac{1}{n} + \dfrac{1}{n + 2} = \dfrac{9}{40}$

52. $\dfrac{2x - 2}{5x - 7} = \dfrac{7x - 2}{10x - 2}$

53. $\dfrac{4x}{x - 1} + \dfrac{2x}{x + 2} = \dfrac{72}{x^2 + x - 2}$

54. $\dfrac{4}{4 - x} - \dfrac{6}{4 + x} = 1$

55. $\dfrac{3x}{2x - 1} - \dfrac{2}{x + 1} = \dfrac{9x^2 + 4x - 4}{2x^2 + x - 1}$

56. $\dfrac{2u - 3}{5u + 10} - 1 = \dfrac{2 - 3u}{8 + 4u}$

57. $\dfrac{1}{25} = \dfrac{x^2}{16}$

58. $\dfrac{25}{t} + \dfrac{150}{t+100} = \dfrac{3}{2}$

59. $\dfrac{x+1}{3} = 3x - 7 + \dfrac{5x-2}{4}$

60. $\dfrac{x-1}{6} + \dfrac{2x-3}{5} = \dfrac{3}{2}$

61. $\dfrac{x^2+15}{x^3-1} = \dfrac{2x}{x^2+x+1}$

62. $\dfrac{1+\dfrac{3}{x}}{1-\dfrac{1}{x}} = x$

63. $\left|3x - \dfrac{5}{6}\right| = \dfrac{2}{3}$

64. $\left|\dfrac{1}{2} + \dfrac{z}{5}\right| = 3$

65. $\left|\dfrac{x-5}{x+1}\right| = \dfrac{3}{4}$

66. $\left|\dfrac{3}{4} - \dfrac{x}{6}\right| = \left|\dfrac{x}{4} - \dfrac{1}{2}\right|$

Solve and graph:

67. $\dfrac{x}{3} - \dfrac{1}{2} < \dfrac{x}{12} + \dfrac{1}{4}$

68. $\dfrac{2x}{3} - \dfrac{5x-3}{7} \geq 2$

69. $1 \leq \dfrac{5-2x}{3} \leq 5$

70. $\dfrac{x}{7} > \dfrac{x}{5}$

71. $\left|x - \dfrac{1}{2}\right| < 3$

72. $\left|\dfrac{1}{6} + \dfrac{x}{3}\right| \geq \dfrac{1}{2}$

4.6

Solve:

73. When a number is increased by three, one-half of the result equals five less than two-thirds of the sum of the number and five. Find the number.

74. The denominator of a fraction is one more than twice the numerator. When both numerator and denominator are decreased by three, a new fraction is formed that equals $\dfrac{1}{4}$. Find the original fraction and the new fraction.

75. What number must be added to both the numerator and the denominator of the fraction $\dfrac{5}{8}$ in order to obtain a new fraction equal to $\dfrac{3}{4}$?

76. Find two numbers whose sum is 40 and whose ratio is $\dfrac{3}{7}$.

77. What number must be added to both the numerators and the denominators of the fractions $\dfrac{5}{7}$ and $\dfrac{8}{11}$ so that the new fractions that are formed are equal to each other? What are the new fractions?

78. The sum of a number and its reciprocal is $\frac{17}{4}$. Find the number.

79. The sum of the reciprocals of two consecutive integers is $\frac{11}{30}$. What are the integers?

80. When three times the reciprocal of a number is added to the reciprocal of three times the number, the result is $\frac{5}{6}$. What is the number?

81. The largest temperature change in a 24-hour period was recorded in Browning, Montana, where the temperature dropped from 44°F to −56°F. Using the formula $F = \frac{9}{5}C + 32$, express this range in degrees Celsius. *Hint:* $-56° \le F \le 44°$, and substitute the foregoing formula.

82. The lowest temperature ever recorded on the earth's surface was −88.3°C in Antarctica (1960); the highest was 57.8°C in Libya (1922): $-88.3° \le C \le 57.8°$. Use the formula $C = \frac{5}{9}(F - 32)$ to express this range in degrees Fahrenheit.

83. The rectangle shown has perimeter 25 inches. Find the width and the length.

84. The rectangle shown has area 7 square inches. Find the width and the length.

85. If the rectangle shown were actually a square, what would be the length of a side?

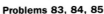

Problems 83, 84, 85

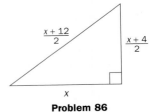

Problem 86

86. Find the sides of the right triangle shown.

87. The base of a triangle is 5 inches more than twice the height. The area is 26 square inches. Find the height and the base. *Formula:* $A = \frac{1}{2}bh$.

4.7

Solve for the indicated variable:

88. $E = mc^2$; for m (Einstein's famous equation)

89. $P = 2L + 2W$; for L

90. $\frac{P_1V_1}{T_1} = \frac{P_2V_2}{T_2}$; for T_2 (Ideal Gas Law)

91. $\frac{1}{f} = \frac{1}{d_i} + \frac{1}{d_o}$; for d_i (Focal length equation)

92. $x = \frac{y}{y - 2}$; for y

93. $\frac{y - 2}{x + 5} = \frac{-2}{3}$; for y

94. $\frac{x}{5b} - \frac{2}{a} = \frac{1}{b} - \frac{2x}{5a}$; for x

95. $a(x - a) = 3(x - 3)$; for x

96. $R_1 = \frac{RR_2}{R + R_2}$; for R_2

97. $2a(x - a) = b(x + a) - b^2$; for x

98. $\frac{m^2}{n^2} = \frac{y - n}{y - m}$; for y

99. $I = \dfrac{E}{1 + \dfrac{R}{r}}$; for r

CHAPTER 4 · TEST   *Answers, page A54*

Simplify according to the methods of this chapter. Assume the denominators are not zero.

1. $\dfrac{3a^3b^2}{14a} \div \dfrac{-9ab^4}{21a^2b^3}$

2. $\dfrac{2x^4y - x^3y^2}{2x^2y^2 - xy^3}$

3. $\dfrac{x^3 + 8}{x^2 - 4}$

4. $\dfrac{6a^2 - ab - 2b^2}{8b^3 - 27a^3}$

5. $\dfrac{x^2 + x - 6}{x^2 - 9} \cdot \dfrac{x^2 + 2x - 15}{2x^2 + 7x - 15}$

6. $\dfrac{2a^3 - 3a^2}{a^2 - 2a + 1} \div \dfrac{9a - 4a^3}{2a^2 + a - 3}$

7. $\dfrac{x + \dfrac{1}{x^2}}{x - \dfrac{1}{x}}$

8. $\dfrac{\dfrac{4}{x + h} - \dfrac{4}{x}}{h}$

Simplify the following expressions (assume the denominators are not zero), solve the equations (watch for extraneous solutions), and solve and graph the inequalities.

9. $\dfrac{5x + 3}{2x + 5} + \dfrac{3x - 1}{2x + 5} - \dfrac{2x - 13}{5 + 2x}$

10. $\dfrac{2x + 3}{5} - \dfrac{3 - 2x}{3} = 6$

11. $\dfrac{x}{x - 3} = \dfrac{18}{x^2 - 9}$

12. $\dfrac{5}{x - 2} - \dfrac{2}{x + 3} + \dfrac{x - 7}{x^2 + x - 6}$

13. $\dfrac{1}{8x^2} + \dfrac{5}{12xy} - \dfrac{3}{16y^2}$

14. $\dfrac{3x - 4}{3} + \dfrac{5 - 7x}{2} < \dfrac{3}{4}$

Divide:

15. $\dfrac{25a^3b^2 + 15ab^3 - 3a^2b}{5ab^2}$

16. $\dfrac{5x^3 - 2x + 3}{x + 2}$

Solve for the indicated variable:

17. $\dfrac{1}{c} = \dfrac{1}{y} - \dfrac{1}{d}$; for y

18. $\dfrac{x - a}{x - b} = \dfrac{b}{a}$; for x

Solve:

19. The numerator of a fraction is three less than the denominator. When two is subtracted from the numerator, and one is added to the denominator, the new fraction equals $\dfrac{1}{3}$. Find the original fraction and the new fraction.

20. The difference between the reciprocals of two consecutive even integers is $\dfrac{1}{24}$. Find the integers.

5 EXPONENTS, ROOTS, AND RADICALS

Integer Exponents

The product or quotient of powers with the same base can be obtained by adding or subtracting exponents, respectively. These rules were covered in Sections 3.1 and 4.1:

$$\textbf{Product Rule of Exponents:} \quad a^m a^n = a^{m+n}$$

$$\textbf{Quotient Rule of Exponents:} \quad \frac{a^m}{a^n} = \begin{cases} a^{m-n} & \text{if} \quad m > n \\ \dfrac{1}{a^{n-m}} & \text{if} \quad m < n \end{cases}$$

Examples include $x^5 x^{12} = x^{17}$, $\dfrac{y^8}{y^2} = y^6$, $\dfrac{z^3}{z^8} = \dfrac{1}{z^5}$, and $\dfrac{a^2 b^8}{a^5 b} = \dfrac{b^7}{a^3}$.

We now see what happens when powers are raised to another power. The following examples should provide the clue:

$$(x^5)^4 = x^5 x^5 x^5 x^5 = x^{20}.$$

$$(4a^3 b^7)^2 = 4a^3 b^7 \cdot 4a^3 b^7 = 16a^6 b^{14}.$$

$$\left(\frac{xy^2}{z^4}\right)^3 = \frac{xy^2}{z^4} \cdot \frac{xy^2}{z^4} \cdot \frac{xy^2}{z^4} = \frac{x^3 y^6}{z^{12}}.$$

In each case, we could have obtained the answers by multiplying exponents. Thus $(x^5)^4 = x^{5 \cdot 4} = x^{20}$ and $(4a^3 b^7)^2 = 4^2 a^{3 \cdot 2} b^{7 \cdot 2} = 16a^6 b^{14}$. In general, when a power, or a product or quotient of powers, is raised to another power, the result can be obtained by *multiplying exponents:*

Generalized Power Rule of Exponents

$$\left(\frac{a^m b^n}{c^p}\right)^k = \frac{a^{mk} b^{nk}}{c^{pk}}$$

EXAMPLE 1

a. $(a^5)^9 = a^{5 \cdot 9}$
$= a^{45}.$

b. $(xy)^7 = (x^1 y^1)^7$
$= x^7 y^7.$

c. $\left(\dfrac{2ab^3}{3c^5}\right)^2 = \dfrac{2^2a^2b^6}{3^2c^{10}}$

$\qquad = \dfrac{4a^2b^6}{9c^{10}}.$

d. $(-3x^5)^4 = (-3)^4x^{20}$

$\qquad\quad = 81x^{20}.$

e. $\left(\dfrac{-a^5}{b}\right)^7 = \left(\dfrac{-1 \cdot a^5}{b}\right)^7 \qquad -a^5 = -1 \cdot a^5$

$\qquad\quad = \dfrac{(-1)^7a^{35}}{b^7}$

$\qquad\quad = \dfrac{-a^{35}}{b^7}. \qquad\qquad (-1)^7 = -1$

f. $(x^ny^{m+1})^3 = x^{3n}y^{3m+3}$

g. $[(a^4)^2]^3 = [a^8]^3 \qquad\qquad$ Inside first.

$\qquad\quad = a^{24}.$

WARNING! The Generalized Power Rule is *false* for sums and differences. Thus

$$(x^3 + y^3)^2 \neq x^6 + y^6.$$

The *correct* result must be obtained by FOIL:

$$(x^3 + y^3)^2 = (x^3 + y^3)(x^3 + y^3) = x^6 + 2x^3y^3 + y^6.$$

Up to now, we have used powers with only natural-number exponents, such as a^2 and a^5. We now wish to extend the definition of powers to include all integer exponents, including a^0 and a^{-3}. Because powers denote repeated multiplication, we might wonder how a can appear as a factor zero or a negative number of times! The answer is that it can't, so we must assign meanings to these exponents in a different manner.

Let a be any non-zero number, and let n be any positive integer. Because any non-zero quantity divided by itself is 1, we know that

$$\frac{a^n}{a^n} = 1, \quad a \neq 0.$$

To be consistent with the Quotient Rule of Exponents, we can subtract exponents:

$$\frac{a^n}{a^n} = a^{n-n} = a^0.$$

Comparing these two results leads us to conclude that $a^0 = 1$. Note that 0^0 is undefined, because $\dfrac{0^n}{0^n}$ is undefined.

To interpret negative exponents, let us form the product of a^{-n} and a^n, by adding exponents:

$$a^{-n} \cdot a^n = a^{-n+n} = a^0 = 1, \quad a \neq 0.$$

If we divide both sides of this resulting equation, $a^{-n}a^n = 1$, by a^n, we get

$$\frac{a^{-n}a^n}{a^n} = \frac{1}{a^n}, \quad \text{or} \quad a^{-n} = \frac{1}{a^n}.$$

We see that a power with a negative exponent is the reciprocal of the positive power. Let us summarize our results in a definition:

Zero and Negative Exponents

$$a^0 = 1 \quad \text{and} \quad a^{-n} = \frac{1}{a^n}, \quad \text{provided } a \neq 0.$$

For example, if $x \neq 0$, then

$$x^0 = 1, \quad x^{-3} = \frac{1}{x^3}, \quad (x^2 + 1)^0 = 1, \quad (-4)^0 = 1, \quad \text{and} \quad 5^{-2} = \frac{1}{5^2} = \frac{1}{25}.$$

As noted above, 0^0 is undefined.

It is interesting to illustrate the "transition" from positive exponents to zero and negative exponents, by means of the following pattern:

$$a^3 = aaa$$
$$\Big\rangle \div a$$
$$a^2 = aa$$
$$\Big\rangle \div a$$
$$a^1 = a$$
$$\Big\rangle \div a$$
$$a^0 = 1$$
$$\Big\rangle \div a$$
$$a^{-1} = \frac{1}{a}$$
$$\Big\rangle \div a$$
$$a^{-2} = \frac{1}{a^2}$$

Reducing each exponent by 1 is equivalent to dividing each power by a.

Negative exponents can produce complex fractions. For example,

$$\frac{1}{a^{-n}} = \frac{1}{\dfrac{1}{a^n}} = \frac{1}{1} \cdot \frac{a^n}{1} = a^n;$$

and

$$\frac{a^{-n}}{b^{-m}} = \frac{\dfrac{1}{a^n}}{\dfrac{1}{b^m}} = \frac{1}{a^n} \cdot \frac{b^m}{1} = \frac{b^m}{a^n}.$$

Notice that a negative exponent in the denominator becomes a positive exponent in the numerator, and vice versa. This "flip-flopping" is a valid shortcut and will be useful in the next example.

Shortcuts for Negative Exponents

$$\frac{1}{a^{-n}} = a^n \quad \text{and} \quad \frac{a^{-n}}{b^{-m}} = \frac{b^m}{a^n}$$

EXAMPLE 2

a. $(7x)^0 + 7x^0 = 1 + 7 \cdot 1$ *Note!* $(7x)^0 = 1$ but $7x^0 = 7 \cdot 1 = 7.$
$= 8.$

b. $a^{-5} = \dfrac{1}{a^5}.$ Assume $a \neq 0.$

c. $4^{-3} = \dfrac{1}{4^3}$

$= \dfrac{1}{64}.$

d. $(-2)^{-5} = \dfrac{1}{(-2)^5}$

$= \dfrac{1}{-32}$

$= \dfrac{-1}{32}.$ WARNING! $(-2)^{-5} \neq +32$ or $10.$

e. $\dfrac{1}{x^{-7}} = x^7.$ Shortcut for negative exponents.

f. $\dfrac{a^{-3}}{b^{-2}} = \dfrac{b^2}{a^3}.$ Shortcut for negative exponents.

g. $\dfrac{a^{-5}b^0}{c^{-2}d^4} = \dfrac{c^2}{a^5 d^4}.$ Shortcut: $b^0 = 1.$

The next example combines the Product, Quotient, and Power Rules. It can be shown that these are valid for negative, as well as positive, exponents.

EXAMPLE 3 Simplify in terms of positive exponents:

a. $(x^2x^5)^3 = (x^7)^3$ Inside parentheses first.

$\quad\quad\quad = x^{21}.$ Or $(x^2x^5)^3 = x^6x^{15} = x^{21}$

b. $\left(\dfrac{a^7b}{a^2b^3}\right)^4 = \left(\dfrac{a^5}{b^2}\right)^4$ Inside parentheses first.

$\quad\quad\quad = \dfrac{a^{20}}{b^8}.$ Or $\left(\dfrac{a^7b}{a^2b^3}\right)^4 = \dfrac{a^{28}b^4}{a^8b^{12}} = \dfrac{a^{20}}{b^8}$

c. $x^{-5}x^2 = x^{-5+2}$ Add exponents.

$\quad\quad\quad = x^{-3}$

$\quad\quad\quad = \dfrac{1}{x^3}.$ Or $x^{-5}x^2 = \dfrac{1}{x^5}\cdot\dfrac{x^2}{1} = \dfrac{1}{x^3}$

d. $\dfrac{a^{-5}b}{a^{-2}b^{-3}} = \dfrac{b^{1-(-3)}}{a^{-2-(-5)}}$ Subtract exponents: largest $-$ smallest.

$\quad\quad\quad = \dfrac{b^4}{a^3}.$ Or $\dfrac{a^{-5}b}{a^{-2}b^{-3}} = \dfrac{a^2b^3b}{a^5} = \dfrac{b^4}{a^3}$

e. $\left(\dfrac{2x}{5y}\right)^{-3} = \dfrac{2^{-3}x^{-3}}{5^{-3}y^{-3}}$ Generalized Power Rule.

$\quad\quad\quad = \dfrac{5^3y^3}{2^3x^3}$ Shortcut.

$\quad\quad\quad = \dfrac{125y^3}{8x^3}.$

f. $\left(\dfrac{x^{-2}y^3}{z^{-5}}\right)^{-4} = \dfrac{x^{-2\cdot-4}y^{3\cdot-4}}{z^{-5\cdot-4}}$ Multiply exponents (Power Rule).

$\quad\quad\quad = \dfrac{x^8y^{-12}}{z^{20}}$

$\quad\quad\quad = \dfrac{x^8}{y^{12}z^{20}}.$

WARNING! The shortcut method we have used on negative exponents applies only to quotients containing products, not to quotients containing sums and differences. Thus

$$\frac{a^{-2}b^{-2}}{a^{-1}b^{-1}} = \frac{a^1 b^1}{a^2 b^2} \quad \text{but} \quad \frac{a^{-2} - b^{-2}}{a^{-1} + b^{-1}} \neq \frac{a^1 - b^1}{a^2 + b^2}.$$

Instead, we are lead to a *complex fraction* that is simplified in the next example.

EXAMPLE 4

a. $xy^{-1} + (xy)^{-1} = x \cdot \dfrac{1}{y} + \dfrac{1}{xy}$ *Note:* $xy^{-1} \neq (xy)^{-1}$

$$= \frac{x}{y} + \frac{1}{xy} \qquad\qquad x \cdot \frac{1}{y} = \frac{x}{1} \cdot \frac{1}{y} = \frac{x}{y}$$

$$= \frac{x \cdot x}{y \cdot x} + \frac{1}{xy} \longleftarrow \text{L.C.D.} = xy$$

$$= \frac{x^2 + 1}{xy}.$$

b. $\dfrac{a^{-2} - b^{-2}}{a^{-1} + b^{-1}} = \dfrac{\dfrac{1}{a^2} - \dfrac{1}{b^2}}{\dfrac{1}{a} + \dfrac{1}{b}}$ Now simplify this complex fraction.

$$= \frac{\dfrac{1 \cdot b^2}{a^2 \cdot b^2} - \dfrac{1 \cdot a^2}{b^2 \cdot a^2}}{\dfrac{1 \cdot b}{a \cdot b} + \dfrac{1 \cdot a}{b \cdot a}} \begin{array}{l} \longleftarrow \text{L.C.D.} = a^2 b^2 \\ \\ \longleftarrow \text{L.C.D.} = ab \end{array}$$

$$= \frac{\dfrac{b^2 - a^2}{a^2 b^2}}{\dfrac{b + a}{ab}}$$

$$= \frac{(b - a)(b + a)}{a^2 b^2} \cdot \frac{ab}{b + a} \qquad \text{Invert.}$$

$$= \frac{b - a}{ab}. \qquad\qquad\qquad \text{Reduce.}$$

c. $\dfrac{1 - x^{-3}}{1 - x^{-2}} = \dfrac{1 - \dfrac{1}{x^3}}{1 - \dfrac{1}{x^2}}$ Now simplify this complex fraction.

$$= \frac{\dfrac{x^3 - 1}{x^3}}{\dfrac{x^2 - 1}{x^2}} \longleftarrow \qquad \frac{1 \cdot x^3}{1 \cdot x^3} - \frac{1}{x^3} = \frac{x^3 - 1}{x^3}$$

(continued)

$$= \frac{(x-1)(x^2+x+1)}{x^3} \cdot \frac{x^2}{(x-1)(x+1)} \qquad \text{Invert.}$$

$$= \frac{x^2+x+1}{x(x+1)}. \qquad \text{Reduce.}$$

Very large numbers, often cumbersome to write or too large to display on a calculator, can be expressed in a more compact form using powers of 10. For example,

$$5,000,000 = 5 \times 1,000,000 = 5 \times 10^6$$

$$10^6 = 10 \cdot 10 \cdot 10 \cdot 10 \cdot 10 \cdot 10$$
$$= 1,000,000$$

$$430,000,000 = 4.3 \times 100,000,000 = 4.3 \times 10^8$$

Very small numbers can be similarly expressed:

$$0.007 = \frac{7}{1000} = \frac{7}{10^3} = 7 \times 10^{-3}$$

$$0.0000039 = \frac{3.9}{1,000,000} = 3.9 \times 10^{-6}$$

Our four numbers have been written in **scientific notation,** which is the product of a number between 1 and 10 and an integer power of 10:

$$a \times 10^n$$
$$\text{where } 1 \le a < 10 \quad \text{and } n = \pm1, \pm2, \pm3, \ldots$$

To express a number in scientific notation, move the existing decimal point to the position just to the *right of the first non-zero digit*. The number of places that the decimal point moved now becomes the power of 10—a positive power if the decimal point was moved to the left, a negative power if it was moved to the right.

EXAMPLE 5   Convert to scientific notation:

a. $80,000 = 8.\underbrace{0000}_{\text{4 places}} = 8 \times 10^4$

b. $53,000,000,000 = 5.\underbrace{3000000000}_{\text{10 places}} = 5.3 \times 10^{10}$

c. $0.00068 = 0\underbrace{\,0006}_{\text{4 places}}.8 = 6.8 \times 10^{-4}$

d. $0.0000000403 = 0\underbrace{\,00000004}_{\text{8 places}}.03 = 4.03 \times 10^{-8}$

This process is reversible, and we reverse it when we transform a number from scientific notation back to decimal notation. We move the decimal point the number of places indicated by the exponent—to the right if the exponent is positive, to the left if it is negative.

EXAMPLE 6 | Convert to decimal notation:

a. $3 \times 10^5 = 3\underbrace{00000.}_{5 \text{ places}} = 300{,}000$

b. $2.9 \times 10^7 = 2\underbrace{9000000.}_{7 \text{ places}} = 29{,}000{,}000$

c. $9 \times 10^{-4} = \underbrace{.0009}_{4 \text{ places}} = 0.0009$

d. $1.07 \times 10^{-6} = \underbrace{.000001}_{6 \text{ places}} 07 = 0.00000107$

A number exceeding the display capability of a scientific calculator (usually 8 or 10 digits) can be entered in scientific notation. On a Casio or Sharp scientific calculator, the EXP button is used; on other brands one uses an EE or SCI button. The following chart shows how this is done, using the results from Example 5:

Number	Press	Display
$80{,}000 = 8 \times 10^4$	8 EXP 4	8. 04
$53{,}000{,}000{,}000 = 5.3 \times 10^{10}$	5.3 EXP 10	5.3 10
$0.00068 = 6.8 \times 10^{-4}$	6.8 EXP 4 +/−	6.8 − 04
$0.0000000403 = 4.03 \times 10^{-8}$	4.03 EXP 8 +/−	4.03 − 08

EXERCISE 5.1 *Answers, pages A54–A55*

A

Simplify as in EXAMPLE 1. Assume no denominator is zero.

1. $(x^4)^9$
2. $(a^{12})^8$
3. $(ab^2)^7$
4. $(x^3y^2)^5$
5. $(-3c^4d^7)^4$

6. $(4a^7x)^5$
7. $(-x^4)^3$
8. $(-x^3)^4$
9. $(10^3)^2$
10. $(5^4)^3$

11. $\left(\dfrac{3xy^2}{4z^3}\right)^3$
12. $\left(\dfrac{5abc^2}{3d^7}\right)^3$
13. $\left(\dfrac{-x^8}{y^2}\right)^9$
14. $\left(-\dfrac{a^2b}{2c^4}\right)^{11}$
15. $[(a-b)^8]^9$

16. $[x^3(y+1)^4]^6$
17. $(2x^{3m}y^{n+1})^4$
18. $(a^2b^3c^{3n})^n$
19. $[(x^2y^3)^4]^5$
20. $[(2ab^2)^3]^2$

Simplify in terms of positive exponents, as in EXAMPLES 2 and 3. Assume no variable is zero.

21. $(5a)^0$ **22.** $(-2xy^2)^0$ **23.** $4^0 + 0^4$ **24.** $7^0 + 0^7$ **25.** $(-6x)^0 - 6x^0$

26. $-4y^0 - (-4y)^0$ **27.** $(4 \cdot 8 - 2^5)^0$ **28.** $(4^3 - 8^2)^0$ **29.** x^{-8} **30.** y^{-11}

31. 2^{-6} **32.** 7^{-3} **33.** $(-3)^{-5}$ **34.** $(-4)^{-3}$ **35.** $(-1)^{-8}$

36. $(-2)^{-6}$ **37.** $\dfrac{1}{a^{-4}}$ **38.** $\dfrac{1}{x^{-9}}$ **39.** $\dfrac{x^{-3}}{2^{-5}}$ **40.** $\dfrac{4^{-2}}{a^{-8}}$

41. $\dfrac{x^{-5}y^{-2}}{z^{-7}w^4}$ **42.** $\dfrac{a^{-7}b^2c^{-2}}{d^{-6}}$ **43.** $\dfrac{2^{-4}x^{-3}y^0}{4^{-2}z^{-3}w}$ **44.** $\dfrac{3^2a^{-4}b^{-2}}{2^{-3}cd^0}$ **45.** $(a^3a^4)^5$

46. $(x^5x)^7$ **47.** $\left(\dfrac{b^7}{b}\right)^8$ **48.** $\left(\dfrac{a^2}{a^9}\right)^{10}$ **49.** $\left(\dfrac{x^7y^2}{xy^{13}}\right)^5$ **50.** $\left(\dfrac{ab^9}{a^7b^2}\right)^5$

51. $(x^5)^2(x^2)^7$ **52.** $(a^3)^8(a^4)^5$ **53.** $\left(\dfrac{a^2}{b^7}\right)^3\left(\dfrac{b^4}{a}\right)^8$ **54.** $\left(\dfrac{x^5}{y^2}\right)^3 \div \left(\dfrac{x^2}{y^3}\right)^4$ **55.** $(-2x^3y)^3(-3xy^5)^2$

56. $(-a^2b^5)^3(-2a^4b)^5$ **57.** $(x^m)^2(x^3)^{4m}$ **58.** $(a^5)^n(a^{2n})^7$ **59.** $x^{-12}x^2x^3$ **60.** $aa^{-8}a^2$

61. $\dfrac{a^{-5}}{a^{-7}}$ **62.** $\dfrac{b^{-3}}{b^4}$ **63.** $\dfrac{x^{-4}y^{-7}}{x^{-1}y^2}$ **64.** $\dfrac{a^{-5}b^{-6}}{a^{-2}b^{-1}}$ **65.** $(x^{-5})^4$

66. $(a^3)^{-7}$ **67.** $\left(\dfrac{3}{4}\right)^{-2}$ **68.** $\left(\dfrac{2}{3}\right)^{-4}$ **69.** $\left(\dfrac{x^{-3}}{y^4}\right)^{-5}$ **70.** $\left(\dfrac{z^7}{w^{-2}}\right)^{-3}$

71. $(8x^{-3})^{-2}$ **72.** $(4x^3)^{-3}$ **73.** $\left(\dfrac{a^{-5}b^{-7}}{a^{-1}b^2}\right)^{-4}$ **74.** $\left(\dfrac{x^{-2}y^{-7}}{xy^{-3}}\right)^{-3}$

Simplify in terms of positive exponents, as in EXAMPLE 4. Assume no variable is zero.

75. $a^{-1} + b^{-1}$ **76.** $x^{-2} + y^{-2}$ **77.** $(xy)^{-2} - xy^{-2}$ **78.** $\dfrac{x^{-1}}{y^{-1}} - \dfrac{y^{-1}}{x^{-1}}$ **79.** $\dfrac{x^{-1} + y^{-1}}{x^{-2} - y^{-2}}$

80. $\dfrac{a^{-3} + b^{-3}}{a^{-2} - b^{-2}}$ **81.** $\dfrac{1 + a^{-1}}{a - a^{-1}}$ **82.** $\dfrac{8 - x^{-3}}{2 - x^{-1}}$

Convert to scientific notation, as in EXAMPLE 5:

83. 600,000   **84.** 80,000,000   **85.** 49,000,000,000   **86.** 9,030,000,000,000   **87.** 0.00003

88. 0.00000087   **89.** 0.430   **90.** 0.000000000602

91. 93,000,000 miles = the distance from the earth to the sun

92. One trillion, two hundred forty billion dollars = the estimated United States national debt in 1988

93. 0.0000075 meter, the average diameter of a human red blood cell

94. 5.2 microseconds; the prefix *micro* means "millionth"

95. a "googol," the largest number with a name: 1 followed by 100 zeros.

96. 320 megabytes of memory; the prefix *mega* means "million"

Convert to decimal notation, as in EXAMPLE 6:

97. 2×10^5 **98.** 4×10^6 **99.** 7.3×10^8 **100.** 9.07×10^7 **101.** 3×10^{-4}

102. 8.2×10^{-7} **103.** 9.025×10^{-6} **104.** 4.130×10^{-1}

105. 4.8×10^8 miles = the distance from the earth to Jupiter

106. 6.023×10^{23} = *Avogadro's number*, the number of atoms in 1 mole of a chemical molecule

107. 1.53×10^{-6} = the probability of being dealt a royal flush in poker

108. 4.5×10^9 years = the estimated age of the earth

B

Simplify in terms of positive exponents. Assume no variable is zero.

109. $[((x^2)^3)^4]^5$ **110.** $[(a^{-3})^2]^{-5}$ **111.** $[(-xy^3)^7]^2$ **112.** $[(-a^4b^2)^3]^5$

113. $\left(\dfrac{c^{m+5}d^{2m}}{c^2d^{3m+1}}\right)^2$, $m > 0$ **114.** $\left(\dfrac{x^{n+1}y^{5n+2}}{x^{2n+2}y^{n-1}}\right)^4$, $n > 0$ **115.** $\left(\dfrac{2x^3y}{3z^4}\right)^3\left(\dfrac{3xz^2}{4y^5}\right)^2$ **116.** $\left(\dfrac{2a^{-1}b^2}{3a^0b^{-1}}\right)^{-2} \div \left(\dfrac{4ab^{-3}}{3a^{-2}b}\right)^2$

117. $\dfrac{a^{-1} - b^{-1}}{a^{-1}b^{-1}}$ **118.** $\dfrac{e^2 - e^{-2}}{2}$ **119.** $\dfrac{2^{-1} + 3^{-2}}{3^{-1} - 4^{-2}}$ **120.** $\dfrac{1 - 2^{-3}}{1 + 4^{-2}}$

121. $\dfrac{xy^{-1} - x^{-1}y}{y^{-1} - x^{-1}}$ **122.** $\dfrac{ab^{-2} - (ab)^{-2}}{ab^{-1} - (ab)^{-1}}$

 Without using a calculator, we can employ scientific notation to compute the following expression:

$$\frac{(6000)(800{,}000)}{0.00004} = \frac{(6 \times 10^3)(8 \times 10^5)}{4 \times 10^{-5}} \qquad \text{Write in scientific notation.}$$

$$= \frac{48 \times 10^8}{4 \times 10^{-5}} \qquad 10^{3+5} = 10^8$$

$$= 12 \times 10^{13} \qquad 10^{8-(-5)} = 10^{13}$$
$$= (1.2 \times 10^1) \times 10^{13} \qquad \text{Write } 12 = 1.2 \times 10^1$$
$$= 1.2 \times 10^{14} \qquad \text{Answer.}$$

 In like manner, compute the following expressions *without using a calculator*:

123. $(200{,}000)(0.0004)$ **124.** $(0.000003)(3000)$ **125.** $\dfrac{80{,}000}{0.002}$ **126.** $\dfrac{0.0000009}{4500}$ **127.** $\dfrac{(30{,}000)(400)}{0.0002}$

128. $\dfrac{(7500)(4000)}{(0.08)(0.15)}$ **129.** $(0.000003)^2$ **130.** $(50{,}000)^3$

131. A *light-year* is the distance light travels in 1 year. The speed of light is 186,000 miles per second. Multiply this number by the number of seconds in 1 year to find the number of miles in 1 light-year.

5.2   Roots

Recall from beginning algebra that both 5 and -5 are square roots (2nd roots) of 25, because $5^2 = (-5)^2 = 25$. The symbol $\sqrt{25}$ (or $\sqrt[2]{25}$) denotes the positive square root, 5; and $-\sqrt{25}$ denotes the negative square root, -5. Thus

$$\sqrt{25} = 5 \quad \text{and} \quad -\sqrt{25} = -5.$$

Also, -25 has no real square root, because the square of any real number can never be negative; thus

$$\sqrt{-25} \text{ is not a real number.*}$$

More generally, for $n = 2, 3, 4, 5, \ldots$, we say that

$$b \text{ is an } n^{\text{th}} \text{ root of } a, \text{ provided } b^n = a.$$

Thus 4 is a cube root (3rd root) of 64 because $4^3 = 64$; likewise, -4 is a cube root of -64 because $(-4)^3 = -64$. These symbols denote cube roots:

$$\sqrt[3]{64} = 4 \quad \text{and} \quad \sqrt[3]{-64} = -4.$$

We now define the symbol $\sqrt[n]{a}$, called the **principal n^{th} root of a**:

Principal n^{th} root of a: $\sqrt[n]{a}$, a positive

For all values of n,

   $\sqrt[n]{a}$ is the unique *positive* real number such that $(\sqrt[n]{a})^n = a$.
   $\sqrt[n]{0} = 0$.

For $n = 2, 4, 6, \ldots$,

   $-\sqrt[n]{a}$ is the unique *negative* real number such that $(-\sqrt[n]{a})^n = a$.
   $\pm\sqrt[n]{a}$ denotes both $\sqrt[n]{a}$ and $-\sqrt[n]{a}$.
   $\sqrt[n]{-a}$ is *not* a real number.

For $n = 3, 5, 7, \ldots$,

   $\sqrt[n]{-a}$ is the unique *negative* real number such that $(\sqrt[n]{-a})^n = -a$.

This symbol consists of three parts, as shown:

$$\text{index} \longrightarrow \sqrt[n]{a} \quad \begin{matrix} \nearrow \text{radical} \\ \searrow \text{radicand} \end{matrix}$$

A radical without an index denotes the square root. Thus $\sqrt{a}$ means $\sqrt[2]{a}$.

---

*In Section 6.1 we will use imaginary numbers to express square roots of negative numbers.

EXAMPLE 1 The following roots illustrate the previous definitions.

Even index	Check	Odd index	Check
$\sqrt{36} = 6$	$6^2 \stackrel{\checkmark}{=} 36$	$\sqrt[3]{27} = 3$	$3^3 \stackrel{\checkmark}{=} 27$
$-\sqrt[2]{9} = -3$	$(-3)^2 = 9$	$\sqrt[3]{-125} = -5$	$(-5)^3 = -125$
$\pm\sqrt{\dfrac{16}{49}} = \pm\dfrac{4}{7}$	$\left(\pm\dfrac{4}{7}\right)^2 = \dfrac{16}{49}$	$\sqrt[5]{32} = 2$	$2^5 = 32$
$\sqrt[4]{81} = 3$	$3^4 = 81$	$\sqrt[7]{-1} = -1$	$(-1)^7 = -1$
$-\sqrt[6]{64} = -2$	$(-2)^6 = 64$	$\sqrt[3]{0} = 0$	$0^3 = 0$
$\sqrt{-9}$ is not a real number			

Expressions containing roots can be simplified by starting with the inner-most quantities.

EXAMPLE 2

a. $\sqrt[3]{\sqrt{625} + \sqrt[4]{16}} = \sqrt[3]{25 + 2}$ $\qquad \sqrt{625} = 25, \sqrt[4]{16} = 2$

$\qquad\qquad\qquad\qquad = \sqrt[3]{27}$

$\qquad\qquad\qquad\qquad = 3.$

b. $\dfrac{-(-5) \pm \sqrt{(-5)^2 - 4(2)(-3)}}{2(2)} = \dfrac{5 \pm \sqrt{25 + 24}}{4}$

$\qquad\qquad\qquad\qquad\qquad = \dfrac{5 \pm \sqrt{49}}{4}$

$\qquad\qquad\qquad\qquad\qquad = \dfrac{5 \pm 7}{4}$

$\qquad\qquad\qquad\qquad\qquad = \dfrac{12}{4}, \dfrac{-2}{4} \qquad\qquad \dfrac{5+7}{4} = \dfrac{12}{4}, \dfrac{5-7}{4} = \dfrac{-2}{4}$

$\qquad\qquad\qquad\qquad\qquad = 3, \dfrac{-1}{2}.$ $\qquad\qquad$ Note two answers.

c. $\sqrt{(4 - (-2))^2 + (-3 - 5)^2} = \sqrt{(6)^2 + (-8)^2}$

$\qquad\qquad\qquad\qquad\qquad = \sqrt{36 + 64}$

$\qquad\qquad\qquad\qquad\qquad = \sqrt{100}$

$\qquad\qquad\qquad\qquad\qquad = 10.$

The rest of this section is devoted to roots of powers. For example,

$$\sqrt[3]{a^3} = a \qquad \text{because} \quad aaa \stackrel{\checkmark}{=} a^3$$
$$\sqrt[5]{a^5 b^{10}} = ab^2 \qquad \text{because} \quad (ab^2)^5 \stackrel{\checkmark}{=} a^5 b^{10}.$$

These results can be obtained by dividing the power by the index:

$$\sqrt[3]{a^3} = a^{3 \div 3} = a^1$$
$$\sqrt[5]{a^5 b^{10}} = a^{5 \div 5} b^{10 \div 5} = a^1 b^2$$

Can we make a similar remark for square roots:

$$\sqrt{a^2} = a?$$

The answer is complicated by the two possibilities:

$$\sqrt{a^2} = a \quad \text{or} \quad -a \quad \text{because} \quad aa \overset{\checkmark}{=} (-a)(-a) \overset{\checkmark}{=} a^2.$$

To select the correct choice, a or $-a$, we must remember that the principal square root, $\sqrt{a^2}$, must always be positive (or zero). We look at two cases:

Case 1: $a \geq 0$; then $\sqrt{a^2} = a$, which is positive (or zero).

Case 2: $a < 0$; then $-a > 0$, so $\sqrt{a^2} = -a$, which is positive.

In summary, therefore, we have

$$\sqrt{a^2} = \begin{cases} a & \text{if } a \geq 0 \ (\textit{Case 1}) \\ -a & \text{if } a < 0 \ (\textit{Case 2}) \end{cases}$$

You, of course, recognize this as the definition of absolute value (Section 2.6), which answers the problem posed earlier:

$$\sqrt{a^2} = |a| \quad \text{for all values of } a.$$

A similar discussion extends this to all roots with even indices. We summarize once more:

Let a be *any* real number; then
$$\sqrt[n]{a^n} = |a|, \quad n = 2, 4, 6, \ldots$$
$$\sqrt[n]{a^n} = a, \quad n = 3, 5, 7, \ldots$$

EXAMPLE 3

Obtain the indicated roots. Assume the variables represent *any* real numbers.

a. $\sqrt{49x^2} = |7x|$
    $= 7|x|.$ $\qquad\qquad |7x| = |7||x| = 7|x|$

b. $\sqrt[3]{-125x^3 y^6} = -5x^{3 \div 3} y^{6 \div 3}$
    $= -5xy^2.$ $\qquad$ **Check:** $(-5xy^2)^3 \overset{\checkmark}{=} -125x^3 y^6$

c. $\sqrt[4]{81a^4 b^8 c^{12}} = |3a^{4 \div 4} b^{8 \div 4} c^{12 \div 4}|$
    $= 3|ab^2 c^3|$
    $= 3b^2 |ac^3|.$ $\qquad |b^2| = b^2 \text{ because } b^2 \geq 0$

d. $\sqrt[5]{\dfrac{32a^5b^{10}}{c^{15}}} = \dfrac{2ab^2}{c^3},\ c \neq 0.$ Check: $\left(\dfrac{2ab^2}{c^3}\right)^5 \overset{\checkmark}{=} \dfrac{32a^5b^{10}}{c^{15}}$

e. $\sqrt{(x+3)^2} = |x+3|.$

f. $\sqrt{a^2 - 2ab + b^2} = \sqrt{(a-b)^2}$ Factor first!

$\qquad\qquad\qquad\quad = |a-b|.$

The trinomial must *first be factored* into the perfect square binomial $(a-b)^2$, and then the square root is taken.

Square roots and cube roots can be obtained by referring to the Table of Roots and Powers on page A30 or by using a calculator equipped with the functions $\boxed{\sqrt{\ }}$ and $\boxed{\sqrt[3]{\ }}$. Other roots can be obtained on a scientific calculator, using the $\boxed{\sqrt[x]{y}}$ or $\boxed{y^{1/x}}$ button. For example,

$$\sqrt{289} = 17:\quad 289\ \boxed{\sqrt{\ }}\ \boxed{=}$$

$$\sqrt[5]{32} = 2:\quad 32\ \boxed{\sqrt[x]{y}}\ 5\ \boxed{=}\quad \text{or}\quad 32\ \underbrace{\boxed{\text{INV}}\ \boxed{y^x}}_{\boxed{y^{1/x}}}\ 5\ \boxed{=}$$

EXERCISE 5.2   *Answers, page A55*

Obtain the indicated roots and simplify where applicable, as in EXAMPLES 1, 2, and 3. Assume the variables represent any real numbers.

1. $\sqrt{64}$

2. $-\sqrt{81}$

3. $-\sqrt{\dfrac{25}{49}}$

4. $\pm\sqrt{\dfrac{121}{169}}$

5. $\sqrt{\dfrac{324}{361}}$

6. $\pm\sqrt{\dfrac{144}{289}}$

7. $\sqrt[3]{125}$

8. $-\sqrt[3]{512}$

9. $\sqrt[3]{-729}$

10. $\sqrt[3]{-1331}$

11. $\sqrt[3]{\dfrac{-343}{216}}$

12. $\sqrt[3]{-\dfrac{8}{2197}}$

13. $\sqrt[4]{\dfrac{81}{16}}$

14. $\sqrt[4]{10000}$

15. $\sqrt[5]{\dfrac{-1}{32}}$

16. $\sqrt[5]{\dfrac{243}{100000}}$

17. $\sqrt[6]{\dfrac{64}{729}}$

18. $\sqrt[6]{0}$

19. $-\sqrt[3]{-64}$

20. $-\sqrt{-64}$

21. $\sqrt{\sqrt{36} + \sqrt{9}}$

22. $\sqrt{\sqrt[5]{32} + \sqrt[3]{343}}$

23. $\sqrt[3]{\sqrt{100} - \sqrt[3]{8}}$

24. $\sqrt[3]{\sqrt{4} - \sqrt[3]{1000}}$

25. $\sqrt{\sqrt[3]{200} + 16 - \sqrt{5-1}}$

26. $\sqrt{\sqrt{401-1} - \sqrt[3]{2-127}}$

27. $\dfrac{-2 \pm \sqrt{2^2 - 4(1)(-15)}}{2(1)}$

28. $\dfrac{-(-10) \pm \sqrt{(-10)^2 - 4\cdot 1\cdot 21}}{2\cdot 1}$

29. $\dfrac{-(-13) \pm \sqrt{(-13)^2 - 4\cdot 3(-10)}}{2\cdot 3}$

30. $\dfrac{-1 \pm \sqrt{1^2 - 4 \cdot 6(-2)}}{2(6)}$

31. $\dfrac{-(-9) \pm \sqrt{(-9)^2 - 4 \cdot 4 \cdot 0}}{2 \cdot 4}$

32. $\dfrac{-0 \pm \sqrt{0^2 - 4(4)(-9)}}{2 \cdot 4}$

33. $\sqrt{(1 - 4)^2 + (2 - (-2))^2}$

34. $\sqrt{(-7 - 5)^2 + (-8 - (-3))^2}$

35. $\sqrt{(14 - 2)^2 + (-2 - 7)^2}$

36. $\sqrt{(25 - 1)^2 + (5 - (-2))^2}$

37. $\sqrt{4x^2}$

38. $\sqrt{9a^2b^2}$

39. $\sqrt{16a^{16}}$

40. $\sqrt{100x^2y^6z^{16}}$

41. $\sqrt[3]{-27x^3y^{27}}$

42. $\sqrt[3]{64a^6b^{12}}$

43. $\sqrt[4]{\dfrac{16x^4y^8}{81z^{12}}}, z \neq 0$

44. $\sqrt[4]{\dfrac{a^4b^{20}}{10000}}$

45. $\sqrt[5]{32x^{10}y^{15}}$

46. $\sqrt[5]{\dfrac{-243a^5b^{25}}{c^{10}}}, c \neq 0$

47. $\sqrt[6]{64x^6y^{12}}$

48. $\sqrt[7]{-128c^7d^{21}}$

49. $\sqrt{(x - 1)^2}$

50. $\sqrt{(2a + b)^2}$

51. $\sqrt[3]{(x + y)^3}$

52. $\sqrt[4]{(a - 3b)^8}$

53. $\sqrt{x^2 + 2x + 1}$ *Hint:* Factor.

54. $\sqrt{a^2 - 2ab + b^2}$

55. $\sqrt{4a^4 + 4a^2 + 1}$

56. $\sqrt{x^4 + 4x^2 + 4}$

57. $\sqrt{9x^4 + 24x^2y^2 + 16y^4}$

58. $\sqrt{16a^4 + 56a^2b^2 + 49b^4}$

B

59. $\sqrt{x^6 + 2x^5y + x^4y^2}$ *Hint:* Common factor.

60. $\sqrt{a^4b^2 - 10a^3b^3 + 25a^2b^4}$

61. $\sqrt{4x^6y^4 - 12x^4y^6 + 9x^2y^8}$

62. $\sqrt{16a^8 + 80a^6b^2 + 100a^4b^4}$

63. $\sqrt[3]{\sqrt{\sqrt{169} + \sqrt{9}} + \sqrt{\sqrt[3]{1000} + \sqrt[3]{216}}}$

64. $\sqrt{\sqrt{\sqrt{1600} + \sqrt{81}} - \sqrt[3]{\sqrt[3]{512} + \sqrt{361}}}$

65. $\sqrt{\left(\dfrac{5}{6} - \dfrac{1}{6}\right)^2 + \left(\dfrac{1}{3} - \left(\dfrac{-1}{6}\right)\right)^2}$ *Hint:* L.C.D.

66. $\sqrt{\left(\dfrac{5}{12} - \left(\dfrac{-1}{4}\right)\right)^2 + \left(\dfrac{-1}{6} - \dfrac{1}{3}\right)^2}$

Roots of large numbers not appearing on the Table of Roots and Powers can be obtained by prime factorization (Section 1.1, Example 5). For example,

$$\sqrt{11025} = \sqrt{3^2 \cdot 5^2 \cdot 7^2} = 3 \cdot 5 \cdot 7 = 105.$$

 In like manner, find these roots *without using a calculator:*

67. $\sqrt{22500}$

68. $\sqrt{1764}$

69. $\sqrt{27225}$

70. $\sqrt{53361}$

71. $\sqrt{99225}$

72. $\sqrt{148225}$

73. $\sqrt[3]{74088}$

74. $\sqrt[3]{1157625}$

75. $\sqrt[3]{287496}$

76. $\sqrt[3]{12326391}$

77. Suppose on a test you were asked to simplify both $\sqrt{x^4}$ and $\sqrt{x^6}$, with x any real number. In the heat of the exam, *you forgot* the absolute-value symbol required for roots with even indices and wrote

$$\sqrt{x^4} = x^2 \quad \text{and} \quad \sqrt{x^6} = x^3.$$

Your instructor marked these incorrect, saying that the correct answers should be

$$\sqrt{x^4} = |x^2| \quad \text{and} \quad \sqrt{x^6} = |x^3|.$$

Tell why *your* first answer above is actually *correct,* but not your second answer.

5.3 Simplifying and Combining Radicals

Not all numbers have rational roots, as the numbers we worked with in the previous section did. For example, $\sqrt{2}$, $\sqrt{3}$, and $\sqrt[3]{7}$ are not "perfect roots." As discussed in Section 1.4, these are irrational numbers and can be expressed as non-terminating, non-repeating decimals:

$$\sqrt{2} = 1.414213562 \ldots \approx 1.414 \quad \textbf{Check:}\ 1.414^2 = 1.999396 \overset{\checkmark}{\approx} 2$$

$$\sqrt{3} = 1.732050808 \ldots \approx 1.732$$

$$\sqrt[3]{7} = 1.912931183 \ldots \approx 1.913$$

The approximations ($\approx$) to three decimal places were taken from the Table of Roots and Powers page A30. This table gives roots only up to $\sqrt{50}$. How might we find larger roots, such as $\sqrt{200}$, without using a calculator? The answer lies in the following simple calculations:

$$\sqrt{4 \cdot 25} = \sqrt{100} = 10 \quad \text{and} \quad \sqrt{4}\sqrt{25} = 2 \cdot 5 = 10.$$

From this we see that $\sqrt{4 \cdot 25} = \sqrt{4}\sqrt{25}$, which is an instance of the more general rule:

Product Rule for Roots

$$\sqrt[n]{ab} = \sqrt[n]{a}\sqrt[n]{b}$$

provided all roots are real numbers.*

This rule says that "the root of a product of two numbers equals the product of their roots." See Problem 91 in the exercises for a proof of this fact.

Returning to the problem posed earlier, we write 200 in factored form, $200 = 100 \cdot 2$, in which 100 is the *largest perfect-square factor* of 200. Then

$$\sqrt{200} = \sqrt{100 \cdot 2}$$
$$= \sqrt{100}\sqrt{2} \quad \text{Product Rule.}$$
$$= 10\sqrt{2}$$
$$\approx 10(1.414)$$
$$\approx 14.14.$$

This decimal approximation is not so important as the method used to obtain it. We have simplified $\sqrt{200}$ as $10\sqrt{2}$, meaning that this last radical cannot be

---

*In Exercise 6.1, Problem 85, we will see that this rule leads to a contradiction for imaginary numbers.

further broken down into a smaller root. In like fashion,

$$\sqrt{18} = \sqrt{9 \cdot 2}$$
$$= \sqrt{9}\sqrt{2}$$
$$= 3\sqrt{2}.$$

Note that the factoring $\sqrt{18}$ as $\sqrt{6 \cdot 3}$ is useless, because neither 6 nor 3 is a perfect square.

EXAMPLE 1 | Simplify these radicals:

a. $\sqrt{50} = \sqrt{25 \cdot 2}$
$\qquad = \sqrt{25}\sqrt{2}$
$\qquad = 5\sqrt{2}.$

b. $\sqrt{48} = \sqrt{16 \cdot 3}$ $\qquad$ or $\quad \sqrt{48} = \sqrt{4 \cdot 12} = \sqrt{4 \cdot 4 \cdot 3}$
$\qquad = \sqrt{16}\sqrt{3}$ $\qquad\qquad\qquad = \sqrt{4}\sqrt{4}\sqrt{3} = 2 \cdot 2\sqrt{3}$
$\qquad = 4\sqrt{3}.$ $\qquad\qquad\qquad = 4\sqrt{3}.$

c. $\sqrt{45} = \sqrt{9 \cdot 5}$
$\qquad = \sqrt{9}\sqrt{5}$
$\qquad = 3\sqrt{5}.$

d. $\sqrt[3]{16} = \sqrt[3]{8 \cdot 2}$ $\qquad$ 8 is a perfect cube.
$\qquad = \sqrt[3]{8}\sqrt[3]{2}$
$\qquad = 2\sqrt[3]{2}.$

e. $\sqrt[3]{-375} = \sqrt[3]{-125 \cdot 3}$
$\qquad\quad = \sqrt[3]{-125}\sqrt[3]{3}$
$\qquad\quad = -5\sqrt[3]{3}.$

f. $\sqrt[4]{32} = \sqrt[4]{16 \cdot 2}$ $\qquad$ 16 is a perfect 4th power.
$\qquad = \sqrt[4]{16}\sqrt[4]{2}$
$\qquad = 2\sqrt[4]{2}.$

Radicals containing powers of variables are *simplified* when the exponents under the radical are less than the index. For example,

$$\sqrt[2]{x^3} = \sqrt[2]{x^2 \cdot x}$$
$$= \sqrt[2]{x^2}\,\sqrt[2]{x}$$
$$= |x|\sqrt[2]{x}.$$

Note the absolute value, as discussed in the previous section. This annoying feature is needed in case x is negative. Other instances of even indices can occur where absolute value is needed, but we can avoid this problem altogether by assuming throughout the remainder of this book that *all variables under the radical are positive* (unless otherwise stated). Thus absolute value will not be needed.

In the following example, we factor the powers so that one of the exponents is the largest number divisible by the index.

EXAMPLE 2   Simplify these radicals. Assume all variables are positive.

a. $\sqrt{27a^3b^7} = \sqrt{9 \cdot 3 \cdot a^2 \cdot a \cdot b^6 \cdot b}$ —————— Exponents divisible by 2.

$= \sqrt{9a^2b^6}\sqrt{3ab}$

$= 3ab^3\sqrt{3ab}$.

b. $\sqrt[3]{24x^4y^9z^{17}} = \sqrt[3]{8 \cdot 3 \cdot x^3 \cdot x \cdot y^9 \cdot z^{15} \cdot z^2}$ —————— Exponents divisible by 3.

$= \sqrt[3]{8x^3y^9z^{15}}\sqrt[3]{3xz^2}$

$= 2xy^3z^5\sqrt[3]{3xz^2}$.

c. $\sqrt[4]{162a^5b^{22}c^{11}} = \sqrt[4]{81 \cdot 2 \cdot a^4 \cdot a \cdot b^{20} \cdot b^2 \cdot c^8 \cdot c^3}$ —————— Exponents divisible by 4.

$= \sqrt[4]{81a^4b^{20}c^8}\sqrt[4]{2ab^2c^3}$

$= 3ab^5c^2\sqrt[4]{2ab^2c^3}$.

d. $\sqrt{x^4 + x^2} = \sqrt{x^2(x^2 + 1)}$ Common Factor.

$= \sqrt{x^2}\sqrt{x^2 + 1}$

$= x\sqrt{x^2 + 1}$.

WARNING! $\sqrt{x^4 + x^2} \neq x^2 + x$;   in general, $\sqrt[n]{a + b} \neq \sqrt[n]{a} + \sqrt[n]{b}$.

See Problems 83–90 in the exercises.

In our next example, we first simplify the radical, then factor out a common factor, and, finally, reduce to lowest terms.

EXAMPLE 3

$$\frac{-2 \pm \sqrt{2^2 - 4(1)(-4)}}{2(1)} = \frac{-2 \pm \sqrt{4 + 16}}{2}$$

$$= \frac{-2 \pm \sqrt{20}}{2}$$

$$= \frac{-2 \pm \sqrt{4 \cdot 5}}{2}$$

$$= \frac{-2 \pm 2\sqrt{5}}{2} \qquad \sqrt{4 \cdot 5} = \sqrt{4}\sqrt{5} = 2\sqrt{5}$$

(continued)

$$= \frac{2(-1 \pm \sqrt{5})}{2} \qquad \text{Common factor.}$$

$$= -1 \pm \sqrt{5}. \qquad \text{Reduce.}$$

$-1 \pm \sqrt{5}$ denotes the irrational numbers $-1 + \sqrt{5}$ and $-1 - \sqrt{5}$.

The Product Rule for Roots enables us to multiply radicals with the same index. For example,

$$\sqrt{9}\sqrt{16} = \sqrt{9 \cdot 16}, \quad \text{because} \quad 3 \cdot 4 = \sqrt{144}, \quad \text{or} \quad 12 \overset{\checkmark}{=} 12.$$

However, we cannot add these radicals; that is,

$$\sqrt{9} + \sqrt{16} \neq \sqrt{9 + 16}, \quad \text{because} \quad 3 + 4 \neq \sqrt{25}, \quad \text{or} \quad 7 \neq 5.$$

Likewise,

$$\sqrt{8} + \sqrt{18} \neq \sqrt{26}, \quad \text{because} \quad 2.828 + 4.243 \neq 5.009.$$

We can, however, simplify these radicals and then combine like terms:

$$\sqrt{8} + \sqrt{18} = 2\sqrt{2} + 3\sqrt{2}$$
$$= (2 + 3)\sqrt{2}$$
$$= 5\sqrt{2}.$$

EXAMPLE 4

Simplify radicals and then combine:

a. $\sqrt{50} + \sqrt{32} - \sqrt{2} = \sqrt{25 \cdot 2} + \sqrt{16 \cdot 2} - \sqrt{2}$
$$= 5\sqrt{2} + 4\sqrt{2} - 1\sqrt{2}$$
$$= 8\sqrt{2}.$$

b. $\sqrt{98} + 2\sqrt{27} - 4\sqrt{200} - \sqrt{48} = \sqrt{49 \cdot 2} + 2\sqrt{9 \cdot 3} - 4\sqrt{100 \cdot 2}$
$$- \sqrt{16 \cdot 3}$$
$$= 7\sqrt{2} + 2 \cdot 3\sqrt{3} - 4 \cdot 10\sqrt{2} - 4\sqrt{3}$$
$$= 7\sqrt{2} + 6\sqrt{3} - 40\sqrt{2} - 4\sqrt{3}$$
$$= -33\sqrt{2} + 2\sqrt{3} \quad \text{or} \quad 2\sqrt{3} - 33\sqrt{2}.$$

These unlike radicals cannot be further combined.

c. $\sqrt[3]{-40} + 2\sqrt[3]{135} - \sqrt[3]{5000} = \sqrt[3]{-8 \cdot 5} + 2\sqrt[3]{27 \cdot 5} - \sqrt[3]{1000 \cdot 5}$
$$= -2\sqrt[3]{5} + 2 \cdot 3\sqrt[3]{5} - 10\sqrt[3]{5}$$
$$= -6\sqrt[3]{5}.$$

d. $\sqrt{20x^3} - \sqrt{125x^3} - x\sqrt{80x} = \sqrt{4x^2 \cdot 5x} - \sqrt{25x^2 \cdot 5x} - x\sqrt{16 \cdot 5x}$
$$= 2x\sqrt{5x} - 5x\sqrt{5x} - 4x\sqrt{5x}$$
$$= -7x\sqrt{5x}.$$

In Section 5.2, Example 1, we found roots of fractions. For example,

$$\sqrt{\frac{16}{49}} = \frac{4}{7} \quad \text{because} \quad \left(\frac{4}{7}\right)^2 = \frac{16}{49}; \quad \text{but} \quad \frac{\sqrt{16}}{\sqrt{49}} = \frac{4}{7} \quad \text{also.}$$

From this we see that $\sqrt{\frac{16}{49}} = \frac{\sqrt{16}}{\sqrt{49}}$, which is an instance of a more general rule:

Quotient Rule for Roots

$$\sqrt[n]{\frac{a}{b}} = \frac{\sqrt[n]{a}}{\sqrt[n]{b}}$$

provided all roots are real numbers.*

This rule says, "the root of a quotient equals the quotient of the roots." See Problem 91 in the exercises for a proof of this fact. For example,

$$\sqrt{\frac{18}{25}} = \frac{\sqrt{18}}{\sqrt{25}} = \frac{\sqrt{9 \cdot 2}}{5} = \frac{3\sqrt{2}}{5}.$$

In the next example, we simplify radicals such as these and then combine terms, using the least common denominator.

EXAMPLE 5

$$\sqrt{\frac{45}{16}} + \sqrt{\frac{20}{9}} - \sqrt{\frac{245}{36}} = \frac{\sqrt{9 \cdot 5}}{\sqrt{16}} + \frac{\sqrt{4 \cdot 5}}{\sqrt{9}} - \frac{\sqrt{49 \cdot 5}}{\sqrt{36}} \quad \left. \right\} \text{Simplify radicals.}$$

$$= \frac{3\sqrt{5}}{4} + \frac{2\sqrt{5}}{3} - \frac{7\sqrt{5}}{6}$$

$$= \frac{3 \cdot 3\sqrt{5}}{3 \cdot 4} + \frac{4 \cdot 2\sqrt{5}}{4 \cdot 3} - \frac{2 \cdot 7\sqrt{5}}{2 \cdot 6}$$

$$= \frac{9\sqrt{5} + 8\sqrt{5} - 14\sqrt{5}}{12} \quad \left. \right\} \text{L.C.D.} = 12$$

$$= \frac{3\sqrt{5}}{12}$$

$$= \frac{\sqrt{5}}{4}. \quad \text{Reduce.}$$

---

*In Exercise 6.1, problem 86, we will see that this leads to a contradiction for imaginary numbers.

EXERCISE 5.3   *Answers, pages A55–A56*

A

Simplify these radicals, as in EXAMPLES 1 and 2. All variables in this exercise are positive.

1. $\sqrt{24}$ **2.** $\sqrt{20}$ **3.** $-\sqrt{27}$ **4.** $\pm\sqrt{18}$

5. $\sqrt{80}$ **6.** $\sqrt{72}$ **7.** $\sqrt{150}$ **8.** $\sqrt{108}$

9. $\sqrt[3]{24}$ **10.** $\sqrt[3]{-81}$ **11.** $\sqrt[3]{-108}$ **12.** $\sqrt[3]{128}$

13. $\sqrt[4]{80}$ **14.** $\sqrt[4]{48}$ **15.** $\sqrt{8a^3b^9}$ **16.** $\sqrt{28x^5y^{13}}$

17. $-\sqrt{72x^4y^8z^9}$ **18.** $\pm\sqrt{99a^{15}b^{12}c^{25}}$ **19.** $-\sqrt[3]{72x^4y^8z^9}$ **20.** $\sqrt[3]{-81x^{11}y^{13}}$

21. $\sqrt[3]{-5c^3d^{10}e^7}$ *Hint:* $-5 = -1 \cdot 5$ **22.** $\sqrt[3]{-40x^7y^{14}z^{15}}$ **23.** $\sqrt[4]{32x^4y^5z^{10}w^{23}}$ **24.** $\sqrt[5]{64c^{10}d^{11}e^7}$

25. $\sqrt{8x^3 + 4x^2}$ *Hint:* Factor.   **26.** $\sqrt{9a^6 + 18a^4}$ **27.** $-\sqrt{9x^2 + 9}$ **28.** $\pm\sqrt{100 + 4x^2}$

Simplify as in EXAMPLE 3:

29. $\dfrac{3 + \sqrt{18}}{3}$ **30.** $\dfrac{-2 \pm \sqrt{32}}{2}$ **31.** $\dfrac{-5 - \sqrt{75}}{10}$ **32.** $\dfrac{6 - \sqrt{48}}{12}$ **33.** $\dfrac{\sqrt{8} + \sqrt{12}}{4}$ **34.** $\dfrac{\sqrt{75} - \sqrt{200}}{10}$

35. $\dfrac{-(-2) \pm \sqrt{(-2)^2 - 4(1)(-4)}}{2(1)}$ **36.** $\dfrac{-4 \pm \sqrt{4^2 - 4(1)(-8)}}{2 \cdot 1}$ **37.** $\dfrac{-3 \pm \sqrt{3^2 - 4 \cdot 1 \cdot (-9)}}{2 \cdot 1}$

38. $\dfrac{-(-8) \pm \sqrt{(-8)^2 - 4 \cdot 1 \cdot 4}}{2(1)}$ **39.** $\sqrt{(5 - 1)^2 + (-2 - 6)^2}$ **40.** $\sqrt{(3 - (-1))^2 + (-4 - 0)^2}$

41. $\sqrt{(-2 - 3)^2 + (-1 - (-11))^2}$ **42.** $\sqrt{(-1 - 2)^2 + (5 - (-1))^2}$

Simplify radicals and then combine, as in EXAMPLES 4 and 5:

43. $\sqrt{50} + \sqrt{18} - \sqrt{2}$ **44.** $-4\sqrt{7} + \sqrt{63} - \sqrt{28}$

45. $2\sqrt{12} - \sqrt{75} + \sqrt{27}$ **46.** $2\sqrt{5} + \sqrt{125} - 3\sqrt{80}$

47. $-2\sqrt{8} + 4\sqrt{27} + 4\sqrt{32} - \sqrt{243}$ **48.** $2\sqrt{125} + 5\sqrt{8} - \sqrt{3125} - \sqrt{128}$

49. $\sqrt[3]{16} - 3\sqrt[3]{2} - 2\sqrt[3]{-54}$ **50.** $\sqrt[3]{3000} - \sqrt[3]{-81} + 2\sqrt[3]{375}$

51. $\sqrt[3]{40} + \sqrt[3]{-81} + \sqrt[3]{135} + \sqrt[3]{192}$ **52.** $\sqrt[3]{-250} - \sqrt[3]{320} + \sqrt[3]{625} - \sqrt[3]{128}$

53. $\sqrt{8x^2} - 2\sqrt{18x^2} + 5\sqrt{50x^2}$ **54.** $4\sqrt{75x^4} + \sqrt{27x^4} + \sqrt{48x^4}$

55. $\sqrt{12a^3} + a\sqrt{27a} - 10\sqrt{3a^3}$ **56.** $2\sqrt{125b^5} - b^2\sqrt{45b} - 12\sqrt{5b^5}$

57. $\sqrt[3]{16x^4} + 2\sqrt[3]{2x^4} + x\sqrt[3]{-54x}$ **58.** $\sqrt[3]{-3a^5} + a\sqrt[3]{81a^2} - 6a^2\sqrt[3]{24a}$ *Hint:* $-3 = -1 \cdot 3$

59. $\sqrt{\dfrac{8}{9}} - \sqrt{\dfrac{18}{25}} + \sqrt{\dfrac{50}{49}}$ **60.** $\sqrt{\dfrac{147}{16}} + \sqrt{\dfrac{75}{4}} - \sqrt{\dfrac{363}{64}}$

61. $\sqrt{75} + \sqrt{\dfrac{12}{25}} - \sqrt{\dfrac{27}{100}}$ **62.** $\sqrt{\dfrac{8}{49}} + \sqrt{\dfrac{8}{9}} - \sqrt{98}$

B

63. $\sqrt{\dfrac{3}{4}} + \sqrt{\dfrac{2}{9}} - \sqrt{\dfrac{75}{36}} + \sqrt{\dfrac{18}{16}}$

64. $\sqrt{\dfrac{5}{9}} - \sqrt{\dfrac{12}{25}} + \sqrt{\dfrac{27}{4}} - \sqrt{\dfrac{125}{36}}$

65. $\sqrt{72} + 2\sqrt{98} - \sqrt{338} + \sqrt{242}$

66. $\sqrt{147} + 5\sqrt{363} + \sqrt{507} - 2\sqrt{1875}$

67. $\sqrt[3]{-24a^8} + 3a\sqrt[3]{81a^5} - a^2\sqrt[3]{375a^2}$

68. $\sqrt[3]{16x^7} - x\sqrt[3]{-2x^4} + 4x^2\sqrt[3]{54x}$ *Hint:* $-2 = -1 \cdot 2$

Simplify by first factoring:

69. $\sqrt{\dfrac{4x^2 + 4}{9}}$

70. $\pm\sqrt{\dfrac{16y^2 + 64}{25}}$

71. $-\sqrt{\dfrac{36 + 4x^2}{36}}$

72. $\sqrt{\dfrac{25 + 100x^2}{100}}$

73. Suppose you get off a roller coaster and are totally disoriented. You are leaning against a lamppole and then stagger away from it in such a manner that each step is 1 foot long, but in a random direction. According to **The Principle of Random Walk** from statistics, after n steps you will be approximately $\sqrt{n}$ feet away from the pole. How many feet from the lamppole will you be after 25 steps? 20 steps? 50 steps? 300 steps? Give your answers to the nearest tenth of a foot.

74. Hero's Formula from geometry gives the area of any triangle in terms of its sides. If the triangle has sides a, b, and c, then its area is

$$A = \frac{1}{4}\sqrt{(a + b + c)(a + b - c)(a - b + c)(-a + b + c)}.$$

Find the area of each of the following triangles:

a.

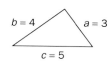

b.

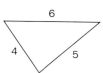

In Exercise 5.2, Problems 67–76, we showed how to simplify roots of large numbers using prime factorization. Continuing in the same spirit, we can simplify $\sqrt{1008}$ as follows:

$$\sqrt{1008} = \sqrt{2^4 \cdot 3^2 \cdot 7} = 2^2 \cdot 3\sqrt{7} = 12\sqrt{7}.$$

In like manner, simplify these radicals:

75. $\sqrt{1620}$

76. $\sqrt{52500}$

77. $\sqrt{2352}$

78. $\sqrt{3528}$

C

79. $\sqrt{5292}$

80. $\sqrt{8820}$

81. $\sqrt[3]{2592}$

82. $\sqrt[3]{60750}$

Certain statements about roots are true and some are false. For example,

$$\sqrt[3]{a^3b^3} = ab \quad \text{is } \textit{true} \text{ because} \quad (ab)^3 \overset{\checkmark}{=} a^3b^3;$$

but $\sqrt{a^2 + b^2} = a + b$ is *false* because $(a + b)^2 = a^2 + 2ab + b^2 \neq a^2 + b^2.$

(continued)

Tell whether each of the following statements is *true or false*, and give reasons to justify your answers. Assume all variables are positive. Use specific numbers and a calculator if need be.

83. $\sqrt[3]{a^3 + b^3} = a + b$ **84.** $\sqrt{a^2} + \sqrt{b^2} = a + b$ **85.** $\sqrt{36a^{36}} = 6a^6$ **86.** $\sqrt{a + b} = \sqrt{a} + \sqrt{b}$

87. $\sqrt[3]{2^9} = 8$ **88.** $\sqrt[5]{5^{25}} = 25$ **89.** $\sqrt[3]{2} + \sqrt[2]{2} = \sqrt[5]{2}$ **90.** $\sqrt{5} + \sqrt{5} = \sqrt{10}$

91. The *Product Rule for Roots* can be proven as follows:

$$\sqrt[n]{ab} \overset{?}{=} \sqrt[n]{a}\,\sqrt[n]{b} \qquad \text{To be proven.}$$
$$(\sqrt[n]{ab})^n \overset{?}{=} (\sqrt[n]{a}\,\sqrt[n]{b})^n \qquad \text{Raise both sides to the } n\text{th power.}$$
$$(\sqrt[n]{ab})^n \overset{?}{=} (\sqrt[n]{a})^n(\sqrt[n]{b})^n \qquad \text{Generalized Power Law of Exponents.}$$
$$ab \overset{\checkmark}{=} (a)(b) \qquad \text{Definition of principal } n\text{th root.}$$

These steps are reversible. Therefore, because the last line is true, the first line must be true: $\sqrt[n]{ab} \overset{\checkmark}{=} \sqrt[n]{a}\sqrt[n]{b}$. You prove the *Quotient Rule for Roots* in the same manner, starting with

$$\sqrt[n]{\frac{a}{b}} \overset{?}{=} \frac{\sqrt[n]{a}}{\sqrt[n]{b}}.$$

92. Find the perimeter of this rectangle:

$\sqrt{24} - \sqrt{18}$ [rectangle]

$\sqrt{32} + \sqrt{54}$

5.4   Multiplying Radicals

The Product Rule for Roots, applied in reverse, can be used to multiply radicals with the same index.*

EXAMPLE 1

a. $\sqrt{3}\sqrt{27} = \sqrt{3 \cdot 27}$
$= \sqrt{81}$
$= 9.$

b. $\sqrt[3]{-5x^2}\sqrt[3]{25x} = \sqrt[3]{-125x^3}$
$= -5x.$

c. $6\sqrt{10} \cdot 5\sqrt{10} = 6 \cdot 5(\sqrt{10})^2$ Commutative Property.
$= 30(10)$ $(\sqrt{10})^2 = 10$ by definition of $\sqrt{}$
$= 300.$

d. $(3\sqrt{x + 1})^2 = 3^2(\sqrt{x + 1})^2$
$= 9(x + 1)$ $(\sqrt{x + 1})^2 = x + 1$
$= 9x + 9.$

---

*In Selected Topic C, we will multiply radicals with different indices.

e. $\sqrt{6x^3}\sqrt{12x^6} = \sqrt{72x^9}$

$\qquad\qquad = \sqrt{36 \cdot 2 \cdot x^8 \cdot x}$ Simplify radical.

$\qquad\qquad = 6x^4\sqrt{2x}.$

f. $(\sqrt{3})^5 = \underbrace{\sqrt{3}\sqrt{3}}_{3}\underbrace{\sqrt{3}\sqrt{3}}_{3}\sqrt{3}$ $\sqrt{3}\sqrt{3} = (\sqrt{3})^2 = 3$

$\qquad\quad = 9\sqrt{3}.$

Radical expressions containing sums and differences can be multiplied by using the Distributive Property, FOIL, or one of the special products.

EXAMPLE 2

a. $3\sqrt{2}(5\sqrt{8} - 7\sqrt{6} + 4)$

$\qquad = 3\sqrt{2} \cdot 5\sqrt{8} - 3\sqrt{2} \cdot 7\sqrt{6} + 3\sqrt{2} \cdot 4$ Distributive Property.

$\qquad = 15\sqrt{16} - 21\sqrt{12} + 12\sqrt{2}$

$\qquad = 15 \cdot 4 - 21 \cdot 2\sqrt{3} + 12\sqrt{2}$

$\qquad = 60 - 42\sqrt{3} + 12\sqrt{2}.$

b. $(5\sqrt{2} + 3)(\sqrt{2} - 4)$

$\qquad = 5\sqrt{2}\sqrt{2} - 20\sqrt{2} + 3\sqrt{2} - 12$ FOIL.

$\qquad = 5 \cdot 2 - 17\sqrt{2} - 12$ Combine like radicals.

$\qquad = -2 - 17\sqrt{2}.$ $10 - 12 = -2$

WARNING! $-2 - 17\sqrt{2} \neq -19\sqrt{2}.$ Why?

c. $(5\sqrt{6} + \sqrt{2})^2$

$\qquad = (5\sqrt{6})^2 + 2 \cdot 5\sqrt{6}\sqrt{2} + (\sqrt{2})^2$ $(A + B)^2 = A^2 + 2AB + B^2$

$\qquad = 25 \cdot 6 + 10\sqrt{12} + 2$

$\qquad = 150 + 10 \cdot 2\sqrt{3} + 2$ $\sqrt{12} = \sqrt{4 \cdot 3} = 2\sqrt{3}$

$\qquad = 152 + 20\sqrt{3}.$

d. $(3 - \sqrt{x + 3})^2$

$\qquad = 9 - 2 \cdot 3\sqrt{x + 3} + (\sqrt{x + 3})^2$ $(A - B)^2 = A^2 - 2AB + B^2$

$\qquad = 9 - 6\sqrt{x + 3} + x + 3$

$\qquad = 12 - 6\sqrt{x + 3} + x.$ $9 + 3 = 12$

e. $(\sqrt{x + h} + \sqrt{x})(\sqrt{x + h} - \sqrt{x})$

$\qquad = (\sqrt{x + h})^2 - (\sqrt{x})^2$ $(A + B)(A - B) = A^2 - B^2$

$\qquad = x + h - x$

$\qquad = h.$

EXERCISE 5.4   *Answers, page A56*

A

Multiply and simplify, as in EXAMPLES 1 AND 2:

1. $\sqrt{2}\sqrt{72}$ 2. $\sqrt{5}\sqrt{45}$ 3. $\sqrt[3]{-9}\sqrt[3]{3}$ 4. $\sqrt[3]{7}\sqrt[3]{49}$

5. $\sqrt{3x^3}\sqrt{27x^7}$ 6. $\sqrt{8y^5}\sqrt{8y^{11}}$ 7. $\sqrt[3]{2a^2}\sqrt[3]{4a}$ 8. $\sqrt[3]{-5b^4}\sqrt[3]{25b^2}$

9. $\sqrt[4]{4x^7y^{-2}}\sqrt[4]{4x^{-3}y^2}$ 10. $\sqrt[5]{-8a^{-3}b^9}\sqrt[5]{4a^3b^{-4}}$ 11. $3\sqrt{6}\cdot5\sqrt{24}$ 12. $2\sqrt{10}\cdot3\sqrt{40}$

13. $(\sqrt{2})^6$ 14. $(\sqrt{7})^4$ 15. $(\sqrt[3]{5})^6$ 16. $(\sqrt[3]{5})^9$

17. $(5\sqrt{x-1})^2$ 18. $(2a\sqrt{a+2})^2$ 19. $(\sqrt{x}\sqrt{x+2})^2$ 20. $(2y\sqrt[3]{y})^3$

21. $\sqrt{5}\sqrt{40}$ 22. $\sqrt{12}\sqrt{24}$ 23. $\sqrt{20x^3}\sqrt{2x^6}$ 24. $\sqrt{30a^7}\sqrt{6a^{12}}$

25. $\sqrt[3]{25x^4}\sqrt[3]{-10x}$ 26. $\sqrt[3]{-50y^7}\sqrt[3]{4y^4}$ 27. $(\sqrt{3})^7$ 28. $(\sqrt{5})^5$

29. $(\sqrt{8})^5$ 30. $(\sqrt[3]{6})^8$

31. $5\sqrt{2}(2\sqrt{3}-4\sqrt{2}+\sqrt{6})$ 32. $-2\sqrt{3}(4\sqrt{3}-\sqrt{8}+3\sqrt{7})$

33. $\sqrt{5x}(2\sqrt{2x}-3\sqrt{5x^3}+\sqrt{15x^2}+2)$ 34. $2\sqrt{3y}(\sqrt{3y}-2\sqrt{12y^3}-\sqrt{8y}-7)$

35. $(2\sqrt{3}+7)(3\sqrt{3}-5)$ 36. $(5\sqrt{2}-6)(\sqrt{2}-4)$

37. $(5\sqrt{6}+2\sqrt{2})(2\sqrt{6}+3\sqrt{2})$ 38. $(2\sqrt{15}-\sqrt{5})(\sqrt{15}+4\sqrt{5})$

39. $(2\sqrt{6}-\sqrt{3})(5\sqrt{2}-5\sqrt{12})$ 40. $(\sqrt{10}-3\sqrt{8})(2\sqrt{5}+\sqrt{20})$

41. $(\sqrt{5}+\sqrt{3})^2$ 42. $(2\sqrt{3}+7)^2$

43. $(3\sqrt{6}-\sqrt{3})^2$ 44. $(\sqrt{6}-\sqrt{2})^2$

45. $(4+\sqrt{x})^2$ 46. $(\sqrt{2a}-3)^2$

47. $(\sqrt{x+1}-2)^2$ 48. $(\sqrt{x-2}+4)^2$

49. $(\sqrt{x-1}+\sqrt{x})^2$ 50. $(\sqrt{x+1}+\sqrt{x-1})^2$

51. $(\sqrt{5}+\sqrt{3})(\sqrt{5}-\sqrt{3})$ 52. $(\sqrt{3}+1)(\sqrt{3}-1)$

53. $(2\sqrt{7}-4)(2\sqrt{7}+4)$ 54. $(3\sqrt{2}-2\sqrt{3})(3\sqrt{2}+2\sqrt{3})$

55. $(\sqrt{x}+\sqrt{a})(\sqrt{x}-\sqrt{a})$ 56. $(\sqrt{9+h}+3)(\sqrt{9+h}-3)$

57. $(\sqrt{x^2+h}+x)(\sqrt{x^2+h}-x)$ 58. $(\sqrt{2x+2h}-\sqrt{2x})(\sqrt{2x+2h}+\sqrt{2x})$

B

Find the area and perimeter of each rectangle:

59.

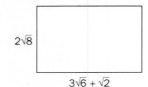

$2\sqrt{8}$

$3\sqrt{6}+\sqrt{2}$

60.

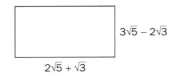

$3\sqrt{5}-2\sqrt{3}$

$2\sqrt{5}+\sqrt{3}$

Show that each of the following triangles is a right triangle by verifying that its sides satisfy the Pythagorean Theorem: $a^2 + b^2 = c^2$. Then find each area, using the formula $A = \frac{1}{2}bh$.

61.

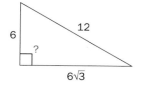

62.

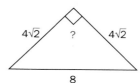

63.

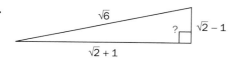

64.

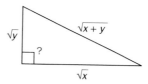

The product of a binomial and a trinomial containing radicals can be obtained by long multiplication, as follows:

$$(\sqrt[3]{a} + \sqrt[3]{b})(\sqrt[3]{a^2} - \sqrt[3]{ab} + \sqrt[3]{b^2})$$
$$= \sqrt[3]{a^3} \underbrace{- \sqrt[3]{a^2b} + \sqrt[3]{ab^2} + \sqrt[3]{ab^2} - \sqrt[3]{ab^2}}_{\text{these} = 0} + \sqrt[3]{b^3}$$
$$= a + b.$$

In like manner, obtain the following products:

65. $(\sqrt[3]{x} - 2)(\sqrt[3]{x^2} + 2\sqrt[3]{x} + 4)$

66. $(\sqrt[3]{x} + \sqrt[3]{a})(\sqrt[3]{x^2} - \sqrt[3]{ax} + \sqrt[3]{a^2})$

67. $(\sqrt[4]{x} - 1)(\sqrt[4]{x^3} + \sqrt[4]{x^2} + \sqrt[4]{x} + 1)$

68. $(\sqrt[5]{x} + 1)(\sqrt[5]{x^4} - \sqrt[5]{x^3} + \sqrt[5]{x^2} - \sqrt[5]{x} + 1)$

The absolute value of the difference of terms depends on their sizes:

$$|a - b| = \begin{cases} a - b & \text{if} \quad a \geq b \\ b - a & \text{if} \quad a < b \end{cases}$$

Thus

$$|\sqrt{5} - \sqrt{2}| = \sqrt{5} - \sqrt{2} \quad \text{because} \quad \sqrt{5} > \sqrt{2}$$
$$\text{but} \quad |\sqrt{3} - \sqrt{7}| = \sqrt{7} - \sqrt{3} \quad \text{because} \quad \sqrt{3} < \sqrt{7}.$$

Not so easy is $|4\sqrt{3} - 5\sqrt{2}|$. To decide whether $4\sqrt{3}$ or $5\sqrt{2}$ is larger, we square each of them:

$$(4\sqrt{3})^2 = 16 \cdot 3 = 48 \quad \text{and} \quad (5\sqrt{2})^2 = 25 \cdot 2 = 50.$$

(continued)

Because $(4\sqrt{3})^2 = 48 < 50 = (5\sqrt{2})^2$, we must conclude that $4\sqrt{3} < 5\sqrt{2}$.* Thus

$$|4\sqrt{3} - 5\sqrt{2}| = 5\sqrt{2} - 4\sqrt{3} \quad \text{because} \quad 4\sqrt{3} < 5\sqrt{2}.$$

 In like manner, find these absolute-value expressions *without using a calculator*:

69. $|\sqrt{13} - \sqrt{10}|$ **70.** $|\sqrt{2} - \sqrt{6}|$ **71.** $|3\sqrt{5} - 2\sqrt{10}|$ **72.** $|5\sqrt{2} - 2\sqrt{5}|$ **73.** $|5\sqrt{3} - 4\sqrt{5}|$

74. $|9\sqrt{3} - 10\sqrt{2}|$ **75.** $|5\sqrt{6} - 6\sqrt{5}|$ **76.** $|3\sqrt{7} - 2\sqrt{17}|$ **77.** $\left|\sqrt{3} - \dfrac{9}{5}\right|$ **78.** $\left|\dfrac{7}{2} - 2\sqrt{3}\right|$

C

The binomial shown as follows contains "radicals within radicals." It's square can be obtained by using the special product $(A + B)^2 = A^2 + 2AB + B^2$:

$$(\sqrt{2 + \sqrt{3}} + \sqrt{2 - \sqrt{3}})^2 = (\sqrt{2 + \sqrt{3}})^2 + 2\sqrt{2 + \sqrt{3}}\,\sqrt{2 - \sqrt{3}} + (\sqrt{2 - \sqrt{3}})^2 \qquad (A + B)^2 = A^2 + 2AB + B^2$$

$$= 2 + \cancel{\sqrt{3}} + 2\sqrt{(2 + \sqrt{3})(2 - \sqrt{3})} + 2 - \cancel{\sqrt{3}} \qquad (A + B)(A - B) = A^2 - B^2$$

$$= 2 + 2 + 2\sqrt{4 - (\sqrt{3})^2}$$

$$= 4 + 2\underbrace{\sqrt{4 - 3}}_{1}$$

$$= 6$$

In like manner, obtain the following squares of binomials:

79. $(\sqrt{4 + \sqrt{7}} + \sqrt{4 - \sqrt{7}})^2$ **80.** $(\sqrt{3 + \sqrt{5}} + \sqrt{3 - \sqrt{5}})^2$

81. $(\sqrt{7 + \sqrt{24}} - \sqrt{7 - \sqrt{24}})^2$ **82.** $(\sqrt{6 - \sqrt{11}} - \sqrt{6 + \sqrt{11}})^2$

Multiply:

83. $(\sqrt{2} + \sqrt{3} + \sqrt{5})(\sqrt{2} + \sqrt{3} - \sqrt{5})$ **84.** $(\sqrt{2} + \sqrt{3} + \sqrt{5})^2$

85. $(1 + \sqrt{2} - \sqrt{6})^2$ **86.** $(\sqrt{6} + \sqrt{3} - 3)(\sqrt{6} + \sqrt{3} + 3)$

5.5   Dividing Radicals

The Quotient Rule for Roots, applied in reverse, can be used to divide radicals with the same index.**

EXAMPLE 1   **a.** $\dfrac{\sqrt{72}}{\sqrt{6}} = \sqrt{\dfrac{72}{6}}$

---

*We use the fact that if a and b are positive and $a^2 < b^2$, then $a < b$.

**In Selected Topic C, we will divide radicals with different indices.

$$= \sqrt{12} \qquad \text{Divide.}$$
$$= 2\sqrt{3}. \qquad \text{Simplify radical.}$$

b. $\dfrac{\sqrt{45x^5}}{\sqrt{5x}} = \sqrt{\dfrac{45x^5}{5x}}$

$$= \sqrt{9x^4} \qquad \text{Divide.}$$
$$= 3x^2.$$

c. $\dfrac{\sqrt[3]{54a^4b}}{\sqrt[3]{16ab^7}} = \sqrt[3]{\dfrac{54a^4b}{16ab^7}}$

$$= \sqrt[3]{\dfrac{27a^3}{8b^6}} \qquad \text{Reduce fraction.}$$

$$= \dfrac{3a}{2b^2}.$$

d. $\dfrac{\sqrt{6xy} + \sqrt{12x}}{\sqrt{2x}} = \dfrac{\sqrt{6xy}}{\sqrt{2x}} + \dfrac{\sqrt{12x}}{\sqrt{2x}} \qquad \text{Write as two fractions (Section 4.4).}$

$$= \sqrt{\dfrac{6xy}{2x}} + \sqrt{\dfrac{12x}{2x}}$$

$$= \sqrt{3y} + \sqrt{6}.$$

We have just seen how quotients of radicals can be reduced and then simplified. However, $\dfrac{\sqrt{5}}{\sqrt{2}} = \sqrt{\dfrac{5}{2}}$ cannot even be reduced. The decimal approximation

$$\dfrac{\sqrt{5}}{\sqrt{2}} \approx \dfrac{2.236}{1.414}$$

requires a lengthy division in the absence of a calculator. This fraction can, however, be expressed in such a way that the radical (decimal) does not appear in the denominator. We multiply numerator and denominator by $\sqrt{2}$ and get

$$\dfrac{\sqrt{5}}{\sqrt{2}} = \dfrac{\sqrt{5} \cdot \sqrt{2}}{\sqrt{2} \cdot \sqrt{2}} = \dfrac{\sqrt{10}}{2}$$

$$\approx \dfrac{3.162}{2} = 1.581.$$

Long division by a decimal has been converted to the much easier division by a whole number. This process of rewriting the denominator without a radical is called **rationalizing the denominator**.

EXAMPLE 2 | Rationalize the denominators and then simplify:

a. $\dfrac{\sqrt{7}}{\sqrt{3}} = \dfrac{\sqrt{7} \cdot \sqrt{3}}{\sqrt{3} \cdot \sqrt{3}}$

$\quad = \dfrac{\sqrt{21}}{3}.$ $\qquad\qquad\qquad\qquad$ $\sqrt{3}\sqrt{3} = 3$

b. $\dfrac{10}{\sqrt{5}} = \dfrac{10 \cdot \sqrt{5}}{\sqrt{5} \cdot \sqrt{5}}$

$\quad = \dfrac{10\sqrt{5}}{5}$

$\quad = 2\sqrt{5}.$ $\qquad\qquad\qquad\qquad$ Reduce.

c. $\sqrt{\dfrac{5x}{18y^3}} = \dfrac{\sqrt{5x}}{3y\sqrt{2y}} \longleftarrow$ $\qquad$ First simplify: $\sqrt{18y^3} = \sqrt{9 \cdot 2 \cdot y^2 \cdot y}$
$\qquad\qquad\qquad\qquad\qquad\qquad\qquad\qquad\qquad\qquad = 3y\sqrt{2y}$

$\quad = \dfrac{\sqrt{5x} \cdot \sqrt{2y}}{3y\sqrt{2y} \cdot \sqrt{2y}}$ $\qquad$ Now rationalize denominator.

$\quad = \dfrac{\sqrt{10xy}}{3y(2y)}$

$\quad = \dfrac{\sqrt{10xy}}{6y^2}.$

d. $\dfrac{\sqrt{2} + \sqrt{3}}{\sqrt{6}} = \dfrac{(\sqrt{2} + \sqrt{3}) \cdot \sqrt{6}}{\sqrt{6} \cdot \sqrt{6}}$

$\quad = \dfrac{\sqrt{12} + \sqrt{18}}{6} \longleftarrow$ $\qquad$ Distributive Property.

$\quad = \dfrac{2\sqrt{3} + 3\sqrt{2}}{6}.$ $\qquad$ Simplify radicals.

e. $\sqrt{\dfrac{2}{3}} - \sqrt{\dfrac{1}{5}} + \sqrt{6} = \dfrac{\sqrt{2} \cdot \sqrt{3}}{\sqrt{3} \cdot \sqrt{3}} - \dfrac{1}{\sqrt{5}} \dfrac{\sqrt{5}}{\cdot \sqrt{5}} + \sqrt{6}$ $\qquad$ Rationalize denominators.

$\quad = \dfrac{\sqrt{6}}{3} - \dfrac{\sqrt{5}}{5} + \dfrac{\sqrt{6}}{1}$

$\quad = \dfrac{5 \cdot \sqrt{6}}{5 \cdot 3} - \dfrac{3 \cdot \sqrt{5}}{3 \cdot 5} + \dfrac{15 \cdot \sqrt{6}}{15 \cdot 1} \longleftarrow$ L.C.D. = 15

$\quad = \dfrac{20\sqrt{6} - 3\sqrt{5}}{15}.$ $\qquad$ Combine like radicals.

$\qquad$ Rationalizing denominators containing other roots requires a slightly different approach. If, for example, we tried the previous method on $\dfrac{1}{\sqrt[3]{2}}$, we would get

$$\frac{1 \cdot \sqrt[3]{2}}{\sqrt[3]{2} \cdot \sqrt[3]{2}} = \frac{\sqrt[3]{2}}{\sqrt[3]{4}}.$$

Unfortunately, $\sqrt[3]{4}$ is *not a perfect root,* so we are no better off than before. Instead, multiply by $\sqrt[3]{2^2}$, which converts the denominator into $\sqrt[3]{2^3} = 2$:

$$\frac{1}{\sqrt[3]{2}} = \frac{1 \cdot \sqrt[3]{2^2}}{\sqrt[3]{2} \cdot \sqrt[3]{2^2}} = \frac{\sqrt[3]{2^2}}{\sqrt[3]{2^3}} = \frac{\sqrt[3]{4}}{2}.$$

EXAMPLE 3

Rationalize the denominators and then simplify:

a. $\dfrac{15}{\sqrt[3]{5}} = \dfrac{15 \cdot \sqrt[3]{5^2}}{\sqrt[3]{5} \cdot \sqrt[3]{5^2}}$

$\phantom{\dfrac{15}{\sqrt[3]{5}}} = \dfrac{15 \cdot \sqrt[3]{5^2}}{\sqrt[3]{5^3}}$

$\phantom{\dfrac{15}{\sqrt[3]{5}}} = \dfrac{15\sqrt[3]{25}}{5}$

$\phantom{\dfrac{15}{\sqrt[3]{5}}} = 3\sqrt[3]{25}.$ Reduce.

b. $\sqrt[3]{\dfrac{4y}{9x}} = \dfrac{\sqrt[3]{4y} \cdot \sqrt[3]{3x^2}}{\sqrt[3]{3^2x} \cdot \sqrt[3]{3x^2}}$

              First write $9x = 3^2x$, then rationalize denominator.

$\phantom{\sqrt[3]{\dfrac{4y}{9x}}} = \dfrac{\sqrt[3]{4y \cdot 3x^2}}{\sqrt[3]{3^3x^3}}$

$\phantom{\sqrt[3]{\dfrac{4y}{9x}}} = \dfrac{\sqrt[3]{12x^2y}}{3x}.$

c. $\dfrac{8a}{\sqrt[4]{4a^3}} = \dfrac{8a \cdot \sqrt[4]{2^2a}}{\sqrt[4]{2^2a^3} \cdot \sqrt[4]{2^2a}}$

              First write $4a^3 = 2^2a^3$, then rationalize denominator.

$\phantom{\dfrac{8a}{\sqrt[4]{4a^3}}} = \dfrac{8a\sqrt[4]{2^2a}}{\sqrt[4]{2^4a^4}}$

$\phantom{\dfrac{8a}{\sqrt[4]{4a^3}}} = \dfrac{8a\sqrt[4]{4a}}{2a}$

$\phantom{\dfrac{8a}{\sqrt[4]{4a^3}}} = 4\sqrt[4]{4a}.$ Reduce.

Rationalizing a denominator that contains two terms, such as $\dfrac{12}{\sqrt{5} + \sqrt{2}}$, is more involved. We first need a definition. The **conjugate** of the binomial sum $A + B$ is the difference $A - B$; the conjugate of $A - B$ is $A + B$. When a binomial containing square-root radicals is multiplied by its conjugate, the result is *free of radicals*. For example,

$$(\sqrt{5} + \sqrt{2})(\sqrt{5} - \sqrt{2}) = \underbrace{(\sqrt{5})^2}_{5} - \sqrt{10} + \sqrt{10} - \underbrace{(\sqrt{2})^2}_{-\ 2}$$
$$= 5 \qquad\qquad -\ 2$$
$$= 3.$$

This is just a special case of $(A + B)(A - B) = A^2 - B^2$, in which there is no middle term. We can, therefore, rationalize such denominators by multiplying by the conjugate of the denominator.

EXAMPLE 4 | Rationalize the denominators and then simplify:

a. $\dfrac{12}{\sqrt{5} + \sqrt{2}} = \dfrac{12 \cdot (\sqrt{5} - \sqrt{2})}{(\sqrt{5} + \sqrt{2}) \cdot (\sqrt{5} - \sqrt{2})}$ Multiply by conjugate.

$$\underset{\text{conjugates}}{\llcorner\qquad\lrcorner}$$

$$= \dfrac{12(\sqrt{5} - \sqrt{2})}{(\sqrt{5})^2 - (\sqrt{2})^2}$$

$$= \dfrac{12(\sqrt{5} - \sqrt{2})}{3} \qquad\qquad 5 - 2 = 3$$

$$= 4(\sqrt{5} - \sqrt{2}). \qquad\qquad \text{Reduce.}$$

b. $\dfrac{\sqrt{12}}{\sqrt{6} - 4} = \dfrac{2\sqrt{3} \cdot (\sqrt{6} + 4)}{(\sqrt{6} - 4) \cdot (\sqrt{6} + 4)}$ Note: $\sqrt{12} = 2\sqrt{3}$

$$\underset{\text{conjugates}}{\llcorner\qquad\lrcorner}$$

$$= \dfrac{2\sqrt{18} + 8\sqrt{3}}{(\sqrt{6})^2 - 4^2}$$

$$= \dfrac{2 \cdot 3\sqrt{2} + 8\sqrt{3}}{-10} \qquad\qquad 6 - 16 = -10$$

$$= \dfrac{\overset{1}{\cancel{2}}(3\sqrt{2} + 4\sqrt{3})}{\underset{-5}{\cancel{-10}}} \qquad\qquad \text{Factor and reduce.}$$

$$= \dfrac{-(3\sqrt{2} + 4\sqrt{3})}{5}. \qquad\qquad \begin{array}{l}\text{Negative sign must be}\\ \text{in the numerator.}\end{array}$$

c. $\dfrac{h}{\sqrt{x + h} - \sqrt{x}} = \dfrac{h \cdot (\sqrt{x + h} + \sqrt{x})}{(\sqrt{x + h} - \sqrt{x}) \cdot (\sqrt{x + h} + \sqrt{x})}$

$$\underset{\text{conjugates}}{\llcorner\qquad\lrcorner}$$

$$= \dfrac{h(\sqrt{x + h} + \sqrt{x})}{(\sqrt{x + h})^2 - (\sqrt{x})^2}$$

$$= \dfrac{h(\sqrt{x + h} + \sqrt{x})}{x + h - x}$$

$$= \frac{h(\sqrt{x+h} + \sqrt{x})}{h}$$

$$= \sqrt{x+h} + \sqrt{x}. \qquad \text{Reduce.}$$

d. $\dfrac{3\sqrt{6} - 2\sqrt{2}}{3\sqrt{6} + 2\sqrt{2}} = \dfrac{(3\sqrt{6} - 2\sqrt{2}) \cdot (3\sqrt{6} - 2\sqrt{2})}{(3\sqrt{6} + 2\sqrt{2}) \cdot (3\sqrt{6} - 2\sqrt{2})}$ Multiply by conjugate.

$$= \frac{(3\sqrt{6})^2 - 12\sqrt{12} + (2\sqrt{2})^2}{(3\sqrt{6})^2 - (2\sqrt{2})^2} \longleftarrow \begin{array}{l}(A-B)(A-B) = \\ A^2 - 2AB + B^2\end{array}$$

$$= \frac{9 \cdot 6 - 12 \cdot 2\sqrt{3} + 4 \cdot 2}{9 \cdot 6 - 4 \cdot 2} \longleftarrow \sqrt{12} = 2\sqrt{3}$$

$$= \frac{62 - 24\sqrt{3}}{46} \qquad 9 \cdot 6 + 4 \cdot 2 = 54 + 8 = 62$$

$$= \frac{2(31 - 12\sqrt{3})}{46} \qquad \text{Factor.}$$

$$= \frac{31 - 12\sqrt{3}}{23}. \qquad \text{Reduce.}$$

e. $\dfrac{a-x}{\sqrt{x} - \sqrt{a}} = \dfrac{(a-x) \cdot (\sqrt{x} + \sqrt{a})}{(\sqrt{x} - \sqrt{a}) \cdot (\sqrt{x} + \sqrt{a})}$

$$= \frac{(a-x)(\sqrt{x} + \sqrt{a})}{(\sqrt{x})^2 - (\sqrt{a})^2} \longleftarrow \text{Do } not \text{ multiply out.}$$

$$= \frac{\overset{-1}{\cancel{(a-x)}}(\sqrt{x} + \sqrt{a})}{\underset{x-a}{\cancel{x-a}}} \qquad \text{Reduce: } \frac{a-x}{x-a} = -1$$

$$= -(\sqrt{x} + \sqrt{a}).$$

By *not* multiplying out the numerator, we are able to reduce while still in factored form.

EXERCISE 5.5 *Answers, pages A56–A57*

A

Divide and then simplify, as in EXAMPLE 1. All variables are positive.

1. $\dfrac{\sqrt{48}}{\sqrt{3}}$ **2.** $\dfrac{\sqrt{125}}{\sqrt{5}}$ **3.** $\dfrac{\sqrt{27x^3}}{\sqrt{3x}}$ **4.** $\dfrac{\sqrt{98a^9}}{\sqrt{2a^3}}$

5. $\dfrac{\sqrt[3]{24a^7}}{\sqrt[3]{-3a}}$ **6.** $\dfrac{\sqrt[3]{-81x^{11}}}{\sqrt[3]{3x^2}}$ **7.** $\dfrac{\sqrt{50x^3y}}{\sqrt{8xy^5}}$ **8.** $\dfrac{\sqrt[3]{2000a^7b^4}}{\sqrt[3]{54ab^4}}$

9. $\dfrac{\sqrt{60}}{\sqrt{5}}$ **10.** $\dfrac{\sqrt[3]{-48}}{\sqrt[3]{3}}$ **11.** $\dfrac{\sqrt{24x^{11}y^7z^9}}{\sqrt{3x^2y^6z^0}}$ **12.** $\dfrac{\sqrt[3]{243a^7b^5c^8}}{\sqrt[3]{3a^2bc^0}}$

13. $\dfrac{\sqrt[3]{3a^2}\,\sqrt[3]{81a^5}}{\sqrt[3]{9a}}$

14. $\dfrac{\sqrt[3]{2x}\,\sqrt[3]{16x^7}}{\sqrt[3]{4x^2}}$

15. $\dfrac{\sqrt{18a^7}}{\sqrt{2a^{-1}}}$

16. $\dfrac{\sqrt[3]{2x^5y^{-2}}}{\sqrt[3]{54x^2y^4}}$

17. $\dfrac{\sqrt{40}+\sqrt{125}}{\sqrt{5}}$

18. $\dfrac{\sqrt[3]{-81}-\sqrt[3]{48}}{\sqrt[3]{3}}$

19. $\dfrac{\sqrt{8ab}-\sqrt{24a}+\sqrt{40ac}}{\sqrt{2a}}$

20. $\dfrac{\sqrt{60xy}+\sqrt{20x}+\sqrt{10x}}{\sqrt{5x}}$

Rationalize the denominators and then simplify, as in EXAMPLE 2:

21. $\dfrac{\sqrt{2}}{\sqrt{7}}$

22. $\dfrac{\sqrt{5}}{\sqrt{3}}$

23. $\dfrac{4}{\sqrt{2}}$

24. $\dfrac{15}{\sqrt{5}}$

25. $\sqrt{\dfrac{8}{5}}$

26. $\sqrt{\dfrac{12}{7}}$

27. $\dfrac{6}{\sqrt{8}}$

28. $\dfrac{8}{\sqrt{12}}$

29. $\dfrac{9x^2}{\sqrt{3x}}$

30. $\dfrac{10a}{\sqrt{5a}}$

31. $\dfrac{6x}{\sqrt{12x^3}}$

32. $\dfrac{16y}{\sqrt{8y^3}}$

33. $\sqrt{\dfrac{1}{x^7y^5}}$

34. $\sqrt{\dfrac{2}{a^3b}}$

35. $\dfrac{1}{\sqrt{2x+y}}$

36. $\sqrt{\dfrac{1}{a+b}}$

37. $\dfrac{\sqrt{3}+\sqrt{6}-\sqrt{8}}{\sqrt{2}}$

38. $\dfrac{\sqrt{15}-\sqrt{20}+\sqrt{12}}{\sqrt{6}}$

39. $\dfrac{\sqrt{xy}+\sqrt{y}}{\sqrt{x}}$

40. $\dfrac{\sqrt{6a}-\sqrt{3b}}{\sqrt{2ab}}$

41. $\dfrac{5\sqrt{6}-4\sqrt{12}+3\sqrt{7}}{2\sqrt{3}}$

42. $\dfrac{5\sqrt{18}+\sqrt{24}-2\sqrt{3}}{3\sqrt{6}}$

43. $\sqrt{\dfrac{2}{5}}-\sqrt{\dfrac{1}{2}}+\sqrt{\dfrac{2}{3}}$

44. $\sqrt{\dfrac{4}{3}}+\sqrt{\dfrac{3}{2}}-\sqrt{\dfrac{1}{6}}$

45. $\sqrt{2}-\sqrt{\dfrac{1}{2}}+\sqrt{\dfrac{1}{8}}$

46. $\sqrt{\dfrac{1}{3}}+\sqrt{3}-\sqrt{\dfrac{1}{27}}$

Rationalize the denominators and then simplify, as in EXAMPLE 3:

47. $\dfrac{1}{\sqrt[3]{2}}$

48. $\dfrac{1}{\sqrt[3]{3}}$

49. $\dfrac{10}{\sqrt[3]{25}}$

50. $\dfrac{8}{\sqrt[3]{4}}$

51. $\sqrt[3]{\dfrac{2x}{9y}}$

52. $\sqrt[3]{\dfrac{3a}{2b^2}}$

53. $\dfrac{12x}{\sqrt[3]{4x^2}}$

54. $\dfrac{15a^2}{\sqrt[3]{25a}}$

55. $\dfrac{25ab}{\sqrt[3]{5a^2b}}$

56. $\dfrac{3xy}{\sqrt[3]{9xy^2}}$

57. $\dfrac{4a^2}{\sqrt[4]{8a^2}}$

58. $\dfrac{25x}{\sqrt[4]{125x^3}}$

59. $\sqrt[5]{\dfrac{1}{27x^2y^3}}$

60. $\dfrac{2ab}{\sqrt[5]{4ab^4}}$

Rationalize the denominators and then simplify, as in EXAMPLE 4:

61. $\dfrac{8}{\sqrt{6}+\sqrt{2}}$

62. $\dfrac{14}{\sqrt{5}+\sqrt{3}}$

63. $\dfrac{-6}{\sqrt{6}-3}$

64. $\dfrac{22}{4-\sqrt{5}}$

65. $\dfrac{50}{3\sqrt{3}-\sqrt{2}}$

66. $\dfrac{58}{\sqrt{3}+4\sqrt{2}}$

67. $\dfrac{\sqrt{21}}{\sqrt{3}-\sqrt{7}}$

68. $\dfrac{\sqrt{12}}{\sqrt{2}+\sqrt{6}}$

69. $\dfrac{\sqrt{xy}}{\sqrt{x}+\sqrt{y}}$

70. $\dfrac{\sqrt{2ab}}{\sqrt{a}-\sqrt{b}}$

71. $\dfrac{h}{\sqrt{25 + h} - 5}$

72. $\dfrac{h}{\sqrt{7 + h} - \sqrt{7}}$

73. $\dfrac{\sqrt{3} - 1}{\sqrt{3} + 1}$

74. $\dfrac{2 + \sqrt{2}}{2 - \sqrt{2}}$

75. $\dfrac{\sqrt{5} + \sqrt{3}}{\sqrt{5} - \sqrt{3}}$

76. $\dfrac{\sqrt{2} + \sqrt{5}}{\sqrt{2} - \sqrt{5}}$

77. $\dfrac{2\sqrt{6} + \sqrt{3}}{\sqrt{6} - 2\sqrt{3}}$

78. $\dfrac{\sqrt{6} - 4\sqrt{2}}{3\sqrt{6} + \sqrt{2}}$

79. $\dfrac{a - b}{\sqrt{a} - \sqrt{b}}$

80. $\dfrac{x - 3}{\sqrt{x} - \sqrt{3}}$

81. $\dfrac{y - x^2}{x - \sqrt{y}}$

82. $\dfrac{1 - x^2}{\sqrt{x} - 1}$

B

83. $\dfrac{\sqrt{6} + \sqrt{2}}{\sqrt{5} - \sqrt{3}}$

84. $\dfrac{\sqrt{7} - \sqrt{5}}{\sqrt{5} + \sqrt{3}}$

85. $\dfrac{\dfrac{3}{\sqrt{3}} + 1}{\dfrac{3}{\sqrt{3}} - 1}$ *Hint:* First rationalize each denominator.

86. $\dfrac{\sqrt{x} - \dfrac{1}{\sqrt{x}}}{x}$

87. $\dfrac{\dfrac{1}{\sqrt{6}} - \dfrac{1}{\sqrt{2}}}{\dfrac{1}{\sqrt{6}} + \dfrac{1}{\sqrt{2}}}$

88. $\dfrac{\dfrac{1}{\sqrt{3}} + \dfrac{1}{\sqrt{5}}}{\dfrac{1}{\sqrt{2}} - \dfrac{1}{\sqrt{6}}}$

89. $\dfrac{h}{\dfrac{1}{\sqrt{x + h}} - \dfrac{1}{\sqrt{x}}}$

90. $\dfrac{h}{\dfrac{1}{\sqrt{4 + h}} - \dfrac{1}{2}}$

Calculus students are required to rationalize the *numerators* of the following expressions. You do likewise, and be sure to simplify your answers.

91. $\dfrac{\sqrt{x + h} - \sqrt{x}}{h}$

92. $\dfrac{\sqrt{4 + h} - 2}{h}$

93. $\dfrac{\sqrt{2x + 2h} - \sqrt{2x}}{h}$

94. $\dfrac{\sqrt{x + h + 1} - \sqrt{x + 1}}{h}$

95. $\dfrac{\sqrt{x} - \sqrt{a}}{x - a}$

96. $\dfrac{\sqrt{x} - 1}{x - 1}$

In Exercise 5.4, preceding Problems 65–68, we multiplied a binomial and a trinomial:

$$(\sqrt[3]{a} + \sqrt[3]{b})(\sqrt[3]{a^2} - \sqrt[3]{ab} + \sqrt[3]{b^2}) = a + b.$$

We can use this fact to rationalize the denominator in the expression $\dfrac{a + b}{\sqrt[3]{a} + \sqrt[3]{b}}$ by taking $\sqrt[3]{a^2} - \sqrt[3]{ab} + \sqrt[3]{b^2}$ as the *conjugate* of $\sqrt[3]{a} + \sqrt[3]{b}$:

$$\frac{a + b}{\sqrt[3]{a} + \sqrt[3]{b}} = \frac{(a + b) \cdot (\sqrt[3]{a^2} - \sqrt[3]{ab} + \sqrt[3]{b^2})}{\underbrace{(\sqrt[3]{a} + \sqrt[3]{b}) \cdot (\sqrt[3]{a^2} - \sqrt[3]{ab} + \sqrt[3]{b^2})}_{\text{conjugates}}}$$

$$= \frac{(a + b)(\sqrt[3]{a^2} - \sqrt[3]{ab} + \sqrt[3]{b^2})}{a + b} \quad \longleftarrow \text{From above.}$$

$$= \sqrt[3]{a^2} - \sqrt[3]{ab} + \sqrt[3]{b^2}.$$

In like fashion, rationalize the denominators and then simplify, using the conjugates provided after each expression:

97. $\dfrac{x - y}{\sqrt[3]{x} - \sqrt[3]{y}}$; $\sqrt[3]{x^2} + \sqrt[3]{xy} + \sqrt[3]{y^2}$

98. $\dfrac{a + 8}{\sqrt[3]{a} + 2}$; $\sqrt[3]{a^2} - 2\sqrt[3]{a} + 4$

C

99. $\dfrac{x - 1}{\sqrt[4]{x} - 1}$; $\sqrt[4]{x^3} + \sqrt[4]{x^2} + \sqrt[4]{x} + 1$

100. $\dfrac{x + 1}{\sqrt[5]{x} + 1}$; $\sqrt[5]{x^4} - \sqrt[5]{x^3} + \sqrt[5]{x^2} - \sqrt[5]{x} + 1$

101. $\dfrac{1}{\sqrt{2} + \sqrt{3} + \sqrt{5}}$; $\sqrt{2} + \sqrt{3} - \sqrt{5}$

102. $\dfrac{1}{\sqrt{6} + \sqrt{3} - 3}$; $\sqrt{6} + \sqrt{3} + 3$

Rational Exponents

In Section 5.2, we obtained roots of powers by dividing the exponent by the index. For example,

$$\sqrt[3]{a^6} = a^{6 \div 3} = a^2.$$

By extending this to the case in which the division is not exact, we get fractional exponents:

$$\sqrt[3]{a^1} = a^{1 \div 3} = a^{1/3} \quad \text{and} \quad \sqrt[3]{a^4} = a^{4 \div 3} = a^{4/3}.$$

We now *define* these fractional exponents in terms of cube roots:

$$a^{1/3} = \sqrt[3]{a} \quad \text{and} \quad a^{4/3} = \sqrt[3]{a^4}.$$

But are these valid definitions? That is, are they consistent with the properties of exponents and roots? *Yes* they are, as we now demonstrate by cubing each side of the first definition:

$$(a^{1/3})^3 = a^{3/3} = a \qquad \text{Power Rule: Multiply exponents.}$$

$$\text{and} \quad (\sqrt[3]{a})^3 = a. \qquad \text{Principal } n\text{th root: } (\sqrt[n]{a})^n = a$$

Comparing these two lines, we see that $a^{1/3} = \sqrt[3]{a}$ *is* a valid definition. The second definition given is also valid by the Power Rule:

$$a^{4/3} = (a^4)^{1/3} = \sqrt[3]{a^4}. \qquad (\quad)^{1/3} \text{ now means } \sqrt[3]{}$$

The Power Rule gives us an alternative definition:

$$a^{4/3} = (a^{1/3})^4 = \left(\sqrt[3]{a}\right)^4.$$

Let us generalize all of these facts with the following definitions:

Definition of Rational Exponents

$$a^{1/n} = \sqrt[n]{a}$$
$$a^{m/n} = (\sqrt[n]{a})^m = \sqrt[n]{a^m}$$

provided all roots are real numbers* and m/n is in lowest terms.

EXAMPLE 1

Express in terms of roots and then simplify. All variables are positive.

a. $9^{1/2} = \sqrt[2]{9}$
    $= 3.$

b. $(-125x^3y^6)^{1/3} = \sqrt[3]{-125x^3y^6}$
    $= -5xy^2.$

c. $16^{3/2} = (\sqrt[2]{16})^3$ Or $\sqrt[2]{16^3} = \sqrt{4096} = 64$
    $= 4^3$ This method is harder!
    $= 64.$

d. $27^{-1/3} = (\sqrt[3]{27})^{-1}$
    $= (3)^{-1}$
    $= \dfrac{1}{3}.$

e. $(-32a^5)^{-3/5} = (\sqrt[5]{-32a^5})^{-3}$ **WARNING!** $(-32a^5)^{-3/5} \neq (32a^5)^{3/5} = 8a^3$
    $= (-2a)^{-3}$

    $= \dfrac{1}{(-2a)^3}$

    $= \dfrac{1}{-8a^3}$

    $= \dfrac{-1}{8a^3}.$ Move negative sign to numerator.

f. $\left(\dfrac{4x^2}{9y^4}\right)^{-3/2} = \left(\sqrt[2]{\dfrac{4x^2}{9y^4}}\right)^{-3}$

    $= \left(\dfrac{2x}{3y^2}\right)^{-3}$

(continued)

---

*In the case of imaginary numbers, this second definition leads to a mathematical contradiction.
See Exercise 6.1, Problem 87.

$$= \frac{1}{\left(\dfrac{2x}{3y^2}\right)^3}$$

$$= \frac{1}{\dfrac{8x^3}{27y^6}}$$

$$= \frac{27y^6}{8x^3}.$$

Invert: $\dfrac{1}{1} \cdot \dfrac{27y^6}{8x^3}$

In the next example, rational exponents lead to roots that must be simplified as in the previous section.

EXAMPLE 2

Express in terms of roots and then simplify:

a. $(250x^4)^{1/3} = \sqrt[3]{250x^4}$
$= \sqrt[3]{125 \cdot 2 \cdot x^3 \cdot x}$
$= 5x\sqrt[3]{2x}.$

b. $8^{3/2} = (\sqrt{8})^3$
$= (2\sqrt{2})^3$
$= 2 \cdot 2 \cdot 2 \cdot \underset{2}{\underbrace{\sqrt{2}\sqrt{2}}}\sqrt{2}$
$= 16\sqrt{2}.$

c. $12^{-1/2} = (\sqrt{12})^{-1}$
$= \dfrac{1}{\sqrt{12}}$
$= \dfrac{1}{2\sqrt{3}}$ Simplify radical.
$= \dfrac{1 \cdot \sqrt{3}}{2\sqrt{3} \cdot \sqrt{3}}$ Rationalize denominator.
$= \dfrac{\sqrt{3}}{6}.$

The definition of rational exponents that we have given was validated by using the Power Rule, in which we multiplied exponents. We will assume that the rational exponents also satisfy the Product and Quotient Rules of exponents. Because these calculations require the addition and subtraction of fractional exponents, we will need least common denominators to perform them.

EXAMPLE 3   Perform the indicated operations and express the answers in terms of positive exponents:

a. $aa^{1/2}a^{3/4} = a^{1+1/2+3/4}$ Add exponents.

$\qquad = a^{4/4+2/4+3/4}$ ⟵——————————————— L.C.D. = 4

$\qquad = a^{9/4}.$

b. $\dfrac{x^{5/6}y^{-3/4}}{x^{2/3}y^2} = \dfrac{x^{5/6-2/3}}{y^{2-(-3/4)}}$ Subtract exponents.

$\qquad = \dfrac{x^{5/6-4/6}}{y^{8/4+3/4}}$ ⟵——————————— L.C.D. = 6
$\qquad\qquad\qquad\qquad$ ⟵——————————— L.C.D. = 4

$\qquad = \dfrac{x^{1/6}}{y^{11/4}}.$

c. $\left(\dfrac{a^{-2/5}b^{3/4}}{c^{-1/10}}\right)^{-20} = \dfrac{a^{40/5}b^{-60/4}}{c^{20/10}}$ Multiply exponents, L.C.D. not needed.

$\qquad = \dfrac{a^8 b^{-15}}{c^2}$

$\qquad = \dfrac{a^8}{b^{15}c^2}.$

d. $x^{1/2}(7x^{3/2} - 5x + 3x^{-1/2})$

$\qquad = 7x^{3/2}x^{1/2} - 5x^1x^{1/2} + 3x^{-1/2}x^{1/2}$ Distributive Property.

$\qquad = 7x^{4/2} - 5x^{2/2}x^{1/2} + 3x^0$

$\qquad = 7x^2 - 5x^{3/2} + 3.$ $3x^0 = 3(1) = 3$

e. $(a^{1/2} + a^{-1/2})^2 = (a^{1/2})^2 + 2a^{1/2}a^{-1/2} + (a^{-1/2})^2$ $(A - B)^2 =$
$\qquad\qquad\qquad\qquad\qquad\qquad\qquad\qquad\qquad\qquad\qquad A^2 - 2AB + B^2$

$\qquad = a^{2/2} + 2a^0 + a^{-2/2}$

$\qquad = a^1 + 2(1) + a^{-1}$

$\qquad = a + 2 + \dfrac{1}{a}$

$\qquad = \dfrac{a^2 + 2a + 1}{a}$ $\dfrac{a^2}{a} + \dfrac{2a}{a} + \dfrac{1}{a}$

$\qquad = \dfrac{(a + 1)^2}{a}.$

EXERCISE 5.6   *Answers, pages A57–A58*

Express in terms of roots and then simplify, as in **EXAMPLES 1 and 2.** All variables are positive.

1. $25^{1/2}$ **2.** $(-27)^{1/3}$ **3.** $(8x^6)^{1/3}$ **4.** $(16a^2b^4)^{1/2}$ **5.** $36^{3/2}$

6. $81^{3/4}$ **7.** $(-64x^3y^9)^{2/3}$ **8.** $(9x^4y^8)^{5/2}$ **9.** $144^{-1/2}$ **10.** $125^{-2/3}$

11. $(81x^4)^{-3/4}$ **12.** $(32x^5y^{10})^{-3/5}$ **13.** $(-27)^{-1/3}$ **14.** $(-1000)^{-1/3}$ **15.** $(-8y^6)^{-2/3}$

16. $(-x^{10})^{-4/5}$ **17.** $\left(\dfrac{27}{64}\right)^{1/3}$ **18.** $\left(\dfrac{25}{81}\right)^{1/2}$ **19.** $\left(\dfrac{9x^2}{16y^4}\right)^{3/2}$ **20.** $\left(\dfrac{-27a^3}{125b^9}\right)^{4/3}$

21. $\left(\dfrac{16}{25}\right)^{-1/2}$ **22.** $\left(\dfrac{8}{27}\right)^{-1/3}$ **23.** $\left(\dfrac{8x^{12}}{27y^9}\right)^{-2/3}$ **24.** $\left(\dfrac{9a^4}{4b^8}\right)^{-3/2}$ **25.** $12^{1/2}$

26. $20^{1/3}$ **27.** $(8x^3)^{1/2}$ **28.** $(50a^5b^6)^{1/2}$ **29.** $(54a^3b^7)^{1/3}$ **30.** $(-16x^4y^8)^{1/3}$

31. $20^{3/2}$ **32.** $27^{3/2}$ **33.** $(18x^3)^{3/2}$ **34.** $(50a^5)^{3/2}$ **35.** $8^{-1/2}$

36. $(12a^3)^{-1/2}$ **37.** $(16x^4)^{-1/3}$ **38.** $9^{-1/3}$

Perform the indicated operations, and express your answers in terms of positive exponents, as in EXAMPLE 3:

39. $x^{2/3}x^{1/2}x^{1/4}$ **40.** $a^{5/8}a^{3/4}a^{1/2}$ **41.** $a^{1/3}b^{2/5} \cdot a^{3/8}b$ **42.** $x^2y^{5/6} \cdot x^{3/4}y^{1/3}$ **43.** $\dfrac{x^{2/3}}{x^{1/6}}$

44. $\dfrac{y^{1/5}}{y^{7/10}}$ **45.** $\dfrac{a^{3/4}b^{-1/3}}{a^{1/2}b^{5/6}}$ **46.** $\dfrac{x^{1/3}y^{7/8}}{x^{-3}y^{3/4}}$ **47.** $(x^{-2/3}y^{5/6})^{-12}$ **48.** $(a^{3/10}b^{-2/5})^{20}$

49. $\left(\dfrac{a^{-8}}{b^{12}}\right)^{-3/4}$ **50.** $\left(\dfrac{x^{-6}}{y^{-9}}\right)^{2/3}$

51. $x^{1/3}(2x - 5x^{2/3} + 4x^{-1/3})$ **52.** $a^{1/4}(3a^{3/4} - 2a + 3a^{-1/4})$ **53.** $3a^{-2/5}(4a^2 + 2a^{2/5} - a^{7/5})$

54. $2y^{-2/3}(4y + y^{2/3} - 7y^3)$ **55.** $(a^{1/2} + b^{1/2})(a^{1/2} - b^{1/2})$ *Hint:* FOIL.   **56.** $(x^{1/2} - 5y^{1/2})(x^{1/2} + 5y^{1/2})$

57. $(3x^{1/3} - y^{2/3})(x^{1/3} + 2y^{2/3})$ **58.** $(5a^{2/3} - b^{1/3})(2a^{2/3} + 3b^{1/3})$ **59.** $(x^{1/2} - x^{-1/2})^2$

60. $(2a^{3/2} - a^{-3/2})^2$

B

61. $(x^{1/3} + y^{1/3})(x^{2/3} - x^{1/3}y^{1/3} + y^{2/3})$ **62.** $(a^{1/4} - b^{1/4})(a^{3/4} + a^{1/2}b^{1/4} + a^{1/4}b^{1/2} + b^{3/4})$

The expression $\dfrac{x^{1/3} - 5x^{-2/3}}{x}$ **can be simplified as a complex fraction:**

$$\frac{x^{1/3} - 5x^{-2/3}}{x} = \frac{x^{1/3} - \dfrac{5}{x^{2/3}}}{x} = \frac{\dfrac{x^{1/3} \cdot x^{2/3}}{1 \cdot x^{2/3}} - \dfrac{5}{x^{2/3}}}{x} \longleftarrow \text{L.C.D.} = x^{2/3}$$

$$= \frac{\dfrac{x^{3/3} - 5}{x^{2/3}}}{\dfrac{x}{1}} = \frac{x - 5}{x^{2/3}} \cdot \frac{1}{x} \qquad \text{Invert.}$$

$$= \frac{x - 5}{x^{5/3}}. \qquad \longleftarrow \qquad x^{2/3}x = x^{2/3}x^{3/3} = x^{5/3}$$

In like manner, simplify the following expressions:

63. $x^{1/2} + x^{-1/2}$

64. $a^{2/5} - 4a^{-3/5}$

65. $\dfrac{x^{1/3} + 2x^{-2/3}}{x}$

66. $\dfrac{x^{1/4} + 3x^{-3/4}}{x}$

67. $\dfrac{(x + 1)^{1/2} - (x + 1)^{-1/2}}{(x + 1)}$

68. $\dfrac{(a - 1)^{1/6} + (a - 1)^{-5/6}}{(a - 1)}$

Calculus students find that expressions containing radicals are more useful when written with positive and negative rational exponents. For example,

$$3\sqrt{x} - 5\sqrt[3]{x^2} + \frac{7}{(\sqrt[4]{x})^3} = 3x^{1/2} - 5x^{2/3} + 7x^{-3/4}.$$

Similarly write the following expressions:

69. $5\sqrt[3]{x} - (\sqrt{x})^3 + \dfrac{3}{\sqrt[5]{x^2}}$

70. $\sqrt[4]{x^3} + 2(\sqrt{x})^2 - \dfrac{1}{(\sqrt[5]{x})^3}$

71. $\sqrt{2x} + 2\sqrt{x} - (\sqrt[3]{x + 1})^2 + \dfrac{3}{x^3}$

72. $\sqrt[3]{(x - 1)^2} + 3\sqrt{x - 1} + \sqrt[4]{3x} + \dfrac{1}{(x + 2)^2}$

True or False? Give reasons to justify your answers. All variables are positive.

73. $x^{1/2}x^{1/3} = x^{5/6}$

74. $(x^{1/2})^{1/3} = x^{1/6}$

75. $x^{1/2}x^{3/2} = x^{3/4}$

76. $\dfrac{x^{1/3}}{x^{1/4}} = x^{4/3}$

77. $(x^{1/2} + y^{1/2})^2 = x + y$

78. $\sqrt[6]{x^2} = \sqrt[3]{x}$ *Hint:* Use fractional exponents.

79. $\sqrt[6]{x^5} = (\sqrt[6]{x})^5$

80. $(x^{1/2} + x^{-1/2})(x^{1/2} - x^{-1/2}) = \dfrac{(x + 1)(x - 1)}{x}$

C

81. We proved the **Product Rule for Roots,** $\sqrt[n]{a}\,\sqrt[n]{b} = \sqrt[n]{ab}$, in Exercise 5.3, Problem 91, by raising both sides to the nth power. It can also be proven using rational exponents, in conjunction with the Power Rule for Exponents:

$$\sqrt[n]{a}\,\sqrt[n]{b} = a^{1/n}b^{1/n} = (ab)^{1/n} \stackrel{\checkmark}{=} \sqrt[n]{ab}$$

You prove the **Quotient Rule for Roots,** $\dfrac{\sqrt[n]{a}}{\sqrt[n]{b}} = \sqrt[n]{\dfrac{a}{b}}$, in the same manner.

■ **CHAPTER 5 · SUMMARY**

5.1 Further Properties of Exponents

$$\left(\frac{a^m b^n}{c^p}\right)^k = \frac{a^{mk} b^{nk}}{c^{pk}}$$

Example:

$$\left(\frac{-2ab^3}{3c^5}\right)^4 = \frac{16a^4 b^{12}}{81c^{20}}$$

$a^0 = 1$

Example:

$$(3x^4)^0 = 1$$

$a^{-n} = \dfrac{1}{a^n}$

Example:

$$(3x^4)^{-2} = \frac{1}{(3x^4)^2} = \frac{1}{9x^8}$$

$\dfrac{1}{a^{-n}} = a^n$ and $\dfrac{a^{-n}}{b^{-m}} = \dfrac{b^m}{a^n}$

Example:

$$\frac{a^{-3} b^4}{c^{-5}} = \frac{b^4 c^5}{a^3}$$

Scientific Notation

$a \times 10^n$, $n = \pm 1, \pm 2, \pm 3, \ldots$
$1 \le a < 10$

Examples:

$$800,000 = 8 \times 10^5$$
$$0.00092 = 9.2 \times 10^{-4}$$

5.2 Roots

$\sqrt[n]{a}$ = principal nth root of a, satisfies
$$(\sqrt[n]{a})^n = a$$
$\sqrt[2]{a}, \sqrt[4]{a}, \sqrt[6]{a}, \ldots$ are ≥ 0 for $a \ge 0$

Examples:

$$\sqrt{25x^4 y^8} = 5x^2 y^4$$
$$\sqrt[3]{-27a^3 b^6} = -3ab^2$$

5.3 Simplifying and Combining Radicals

Factor radicand, one factor being the largest perfect root.

Example:

$$\sqrt[3]{-250a^5} = \sqrt[3]{-125 \cdot 2 \cdot a^3 \cdot a^2}$$
$$= -5a\sqrt[3]{2a^2}.$$

To combine radicals, simplify and then combine like terms.

Example:

$$\sqrt{12} + \sqrt{48} = \sqrt{4 \cdot 3} + \sqrt{16 \cdot 3}$$
$$= 2\sqrt{3} + 4\sqrt{3}$$
$$= 6\sqrt{3}.$$

5.4 Multiplying Radicals

$\sqrt[n]{a} \; \sqrt[n]{b} = \sqrt[n]{ab}$

Example:

$$\sqrt[3]{5x^2}\sqrt[3]{25x} = \sqrt[3]{125x^3}$$
$$= 5x.$$

Binomials with radicals can be multiplied using FOIL.

Example:

$$(2\sqrt{3} + 1)(\sqrt{3} - 4) = 2(\sqrt{3})^2 - 8\sqrt{3} + 1\sqrt{3} - 4$$
$$= 2(3) - 7\sqrt{3} - 4$$
$$= 2 - 7\sqrt{3}.$$

CHAPTER 5 · SUMMARY

5.5 Dividing Radicals

$$\frac{\sqrt[n]{a}}{\sqrt[n]{b}} = \sqrt[n]{\frac{a}{b}}$$

Example:

$$\frac{\sqrt{96}}{\sqrt{12}} = \sqrt{\frac{96}{12}} = \sqrt{8} = 2\sqrt{2}.$$

Rationalize denominators:
Remove radicals from denominator by multiplying by a suitable factor.

Examples:

$$\frac{10}{\sqrt{5}} = \frac{10 \cdot \sqrt{5}}{\sqrt{5} \cdot \sqrt{5}} = \frac{10\sqrt{5}}{5} = 2\sqrt{5}.$$

$$\frac{6}{\sqrt{2} + 1} = \frac{6 \cdot (\sqrt{2} - 1)}{(\sqrt{2} + 1) \cdot (\sqrt{2} - 1)}$$
$$\underset{\text{conjugates}}{\underbrace{\qquad\qquad}}$$
$$= \frac{6(\sqrt{2} - 1)}{2 - 1}$$
$$= 6(\sqrt{2} - 1).$$

5.6 Rational Exponents

$$a^{1/n} = \sqrt[n]{a}$$

Example:

$$25^{1/2} = \sqrt{25} = 5.$$

$$a^{m/n} = (\sqrt[n]{a})^m \quad \text{or} \quad \sqrt[n]{a^m}$$

Examples:

$$(-64a^3)^{2/3} = (\sqrt[3]{-64a^3})^2$$
$$= (-4a)^2$$
$$= 16a^2.$$

$$81^{-3/4} = (\sqrt[4]{81})^{-3}$$
$$= (3)^{-3}$$
$$= \frac{1}{3^3}$$
$$= \frac{1}{27}.$$

Powers with rational exponents satisfy the same rules as those for integer exponents. Use L.C.D. to add and subtract exponents.

Example:

$$\frac{a^2 a^{1/2}}{a^{3/4}} = a^{2 + 1/2 - 3/4}$$
$$= a^{8/4 + 2/4 - 3/4}$$
$$= a^{7/4}.$$

CHAPTER 5 · REVIEW EXERCISES   *Answers, pages A58–A59*

5.1

Simplify in terms of positive exponents:

1. $(a^3)^9$

2. $\left(\dfrac{-2x^3y}{3z^2}\right)^3$

3. $\left(\dfrac{b^5c^2}{bc^7}\right)^4$

4. $\left(\dfrac{x^2}{y}\right)^3\left(\dfrac{y^8}{x^4}\right)^2$

5. $(-c^2d^3)^3(-2cd^2)^2$

6. $(x^{n-1}x^{2n})^5$

7. $(-x^4)^3(-x^3)^4$

8. $(7x)^0 - 7x^0$

9. 5^{-4}

10. $(-6x^2)^{-3}$

11. $\dfrac{1}{(2a^2)^{-3}}$

12. $\dfrac{8^{-2}}{2^{-3}}$

13. $\dfrac{a^{-8}b^{-3}c^0}{a^{-2}b^6c^{-2}}$

14. $(x^2y^{-7})^4$

15. $\left(\dfrac{c^{-2}d^{-7}}{c^{-9}d^2}\right)^{-3}$

16. $[(a^3b^{-4})^{-5}]^{-3}$

17. $\left(\dfrac{3x^2}{5y^3}\right)^{-2}$

18. $(-2x^{-4}y^3)^{-4}$

19. $\dfrac{a^{-1}+b^{-1}}{a^{-2}-b^{-2}}$

20. $\dfrac{4-x^{-2}}{8+x^{-3}}$

Convert to scientific notation:

21. 4,000,000

22. 535,000,000,000

23. 0.00002

24. 0.0000000062

25. five billion, two hundred thirty million

26. 7 nanoseconds; the prefix *nano* means "billionth"

Convert to decimal notation:

27. 7×10^5

28. 8.20×10^9

29. 1.02×10^{-3}

30. 3.270×10^{-6}

31. 6.6×10^{21} tons, the estimated weight of the earth

32. 7.86×10^{-2} pounds, the weight of 1 cubic foot of air

Compute using scientific notation, *without using a calculator:*

33. $(30{,}000)(2{,}000{,}000)$

34. $\dfrac{900{,}000}{0.003}$

35. $\dfrac{(3000)(0.005)}{0.000002}$

36. $(200{,}000)^3$

37. In the text, we showed that $a^0 = 1$ by subtracting exponents. It can also be done by adding exponents in the following product:

$$a^0a^n = a^{0+n} = a^n.$$

Why does this imply that $a^0 = 1$? *Hint*: Use the Identity Property for Real Numbers in Section 1.4.

38. It was also proven in the text that $\dfrac{1}{a^{-n}} = a^n$ by simplifying a complex fraction.

You prove it differently, by dividing the equation $a^{-n}a^n = 1$ by a^{-n}.

True or False? If necessary, experiment with specific values of the variables.

39. $5^3 \cdot 2^3 = 10^3$ **40.** $2^4 \cdot 3^3 = 6^7$ **41.** $\dfrac{6^4}{2^4} = 3^4$ **42.** $\left(\dfrac{-3}{4}\right)^{-3} = \dfrac{27}{64}$ **43.** $\left(\dfrac{1}{2}\right)^{-6} = 64$

44. $(x^m)^n = (x^n)^m$ **45.** $x^2 y^3 = (xy)^5$ **46.** $\dfrac{x^{2n+2}}{x^{n+1}} = x^2$ **47.** $(-x)^{-n} = x^n$ **48.** $(x^3 + y^3)^2 = x^6 + y^6$

5.2, 5.3

Simplify these expressions. All variables are positive.

49. $\sqrt{121a^4 b^6}$

50. $\sqrt[3]{\dfrac{-125x^6 y^{12}}{216z^9}}$

51. $\sqrt{\sqrt{50-1} + \sqrt{6-2}}$

52. $\dfrac{-(-2) \pm \sqrt{(-2)^2 - 4(3)(-5)}}{2(3)}$

53. $\sqrt{(2-(-7))^2 + (-4-8)^2}$

54. $\sqrt{x^2 + 8x + 16}$ *Hint*: Factor.

55. $\sqrt{98}$

56. $\sqrt[3]{-81}$

57. $\sqrt{12x^3 y^6 z^9}$

58. $\sqrt[3]{16a^3 b^4 c^8}$

59. $\sqrt{9x^4 + 36x^2}$ *Hint*: Factor.

60. $\sqrt{\dfrac{9x^2 + 9}{16}}$

61. $\dfrac{2 - \sqrt{12}}{2}$

62. $\dfrac{-4 + \sqrt{72}}{8}$

63. $\dfrac{10x - \sqrt{50x^3}}{5x}$

64. $\dfrac{\sqrt{32a^3} + \sqrt[3]{32a^5}}{2a}$

65. $\sqrt{(3-1)^2 + (-8-(-2))^2}$

66. $\dfrac{-4 \pm \sqrt{4^2 - 4(1)(-16)}}{2 \cdot 1}$

67. $\sqrt{8} - \sqrt{50} + 3\sqrt{200}$

68. $2\sqrt[3]{81} + \sqrt[3]{-24} + \sqrt[3]{375}$

69. $\sqrt{125} - \sqrt{27} + 2\sqrt{500} + 3\sqrt{12}$

70. $\sqrt{50x^3} - 2x\sqrt{98x} + \sqrt{2x^3}$

71. $\sqrt{\dfrac{18}{25}} + \sqrt{\dfrac{125}{36}} - \sqrt{\dfrac{5}{16}} + \sqrt{8}$

72. $\sqrt{8820}$ *Hint*: Factor into prime numbers.

5.4

Multiply and then simplify:

73. $\sqrt{3}\sqrt{27}$ **74.** $\sqrt[3]{3x^2}\sqrt[3]{-9x^7}$ **75.** $5\sqrt{6} \cdot 3\sqrt{24}$ **76.** $(3\sqrt{5})^2$ **77.** $(4\sqrt{x+1})^2$

78. $\sqrt{15}\sqrt{6}$ **79.** $\sqrt{5x^3}\sqrt{10x^8}$ **80.** $\sqrt[3]{-4x^2}\sqrt[3]{12x^5}$ **81.** $(\sqrt{5})^5$

82. $5\sqrt{2}(\sqrt{8} - 3\sqrt{6} + 2\sqrt{12})$ **83.** $(2\sqrt{5} + 5)(\sqrt{5} - 7)$ **84.** $(3\sqrt{6} - 4\sqrt{2})(\sqrt{6} + 7\sqrt{2})$

85. $(2\sqrt{3} + \sqrt{7})(2\sqrt{3} - \sqrt{7})$ **86.** $(\sqrt{6} + \sqrt{3})^2$ **87.** $(2 - \sqrt{x+1})^2$

88. $(\sqrt{1-x} + \sqrt{x})^2$ **89.** $(\sqrt{9+h} + 3)(\sqrt{9+h} - 3)$ **90.** $(\sqrt{6+\sqrt{32}} + \sqrt{6-\sqrt{32}})^2$

91. $(\sqrt[3]{x} - \sqrt[3]{a})(\sqrt[3]{x^2} + \sqrt[3]{ax} + \sqrt[3]{a^2})$

Find the perimeter and area of each of the rectangles shown:

92.

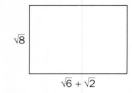

93.

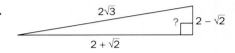

Use the Pythagorean Theorem to show that each of the triangles shown is a *right triangle*.

Then find each area, using the formula $A = \frac{1}{2}bh$.

94.

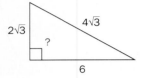

95.

$$\begin{array}{c} 2\sqrt{3} \\ ?\,\lrcorner\, 2 - \sqrt{2} \\ 2 + \sqrt{2} \end{array}$$

5.5

Divide and then simplify:

96. $\dfrac{\sqrt{54x^3}}{\sqrt{6x}}$ **97.** $\dfrac{\sqrt[3]{-54x^5}}{\sqrt[3]{2x^2}}$ **98.** $\dfrac{\sqrt{a^3b^5}}{\sqrt{ab^9}}$ **99.** $\dfrac{\sqrt[3]{72a^4b^5}}{\sqrt[3]{9ab^{-1}}}$ **100.** $\dfrac{\sqrt{80x^5}}{\sqrt{2x^2}}$

101. $\dfrac{\sqrt[3]{48a^5b^7}}{\sqrt[3]{2ab^2}}$ **102.** $\dfrac{\sqrt{72}-\sqrt{108}}{\sqrt{6}}$ **103.** $\dfrac{\sqrt[3]{-64}+\sqrt[3]{216}}{\sqrt[3]{4}}$ **104.** $\dfrac{\sqrt{8xy}-\sqrt{24x}+\sqrt{10x}}{\sqrt{2x}}$

Rationalize the denominators and then simplify:

105. $\dfrac{26}{\sqrt{13}}$ **106.** $\dfrac{\sqrt{5}}{\sqrt{8}}$ **107.** $\sqrt{\dfrac{3}{7}}$ **108.** $\sqrt{\dfrac{2x}{3y}}$

109. $\dfrac{2x^2}{\sqrt[3]{8x^3y}}$ **110.** $\sqrt{\dfrac{1}{x+2}}$ **111.** $\dfrac{6}{\sqrt[3]{9}}$ **112.** $\sqrt{\dfrac{1}{5xy^2}}$

113. $\dfrac{\sqrt{5}+\sqrt{30}-\sqrt{15}}{\sqrt{10}}$ **114.** $\sqrt{\dfrac{2}{3}}+\sqrt{\dfrac{1}{5}}-\sqrt{\dfrac{3}{2}}+\sqrt{5}$ **115.** $\dfrac{10}{\sqrt{7}-\sqrt{2}}$ **116.** $\dfrac{h}{\sqrt{2+h}-\sqrt{2}}$

117. $\dfrac{x-16}{\sqrt{x}-4}$ **118.** $\dfrac{\sqrt{2}-2}{\sqrt{2}+2}$ **119.** $\dfrac{\sqrt{12}+\sqrt{2}}{\sqrt{12}-\sqrt{2}}$ **120.** $\dfrac{\sqrt{6}-2\sqrt{2}}{3\sqrt{6}+\sqrt{2}}$

121. $\dfrac{\sqrt{8}}{\sqrt{6}-\sqrt{2}}$ **122.** $\dfrac{\dfrac{2}{\sqrt{2}}-1}{\dfrac{2}{\sqrt{2}}+1}$

5.6

Express in terms of radicals and then simplify:

123. $(-125)^{1/3}$ **124.** $(9x^4)^{1/2}$ **125.** $32^{3/5}$ **126.** $(25x^2y^6)^{3/2}$ **127.** $64^{-1/2}$ **128.** $(49x^6)^{-3/2}$

129. $(-27x^6)^{-2/3}$ **130.** $\left(\dfrac{4}{9}\right)^{3/2}$ **131.** $\left(\dfrac{64}{125}\right)^{-1/3}$ **132.** $\left(\dfrac{25a^2}{16b^6}\right)^{-3/2}$ **133.** $50^{1/2}$ **134.** $(-250)^{1/3}$

135. $12^{3/2}$ **136.** $(-16)^{2/3}$ **137.** $8^{-1/2}$ **138.** $(4x)^{-1/3}$

Perform the indicated operations, and express your answers in terms of positive exponents:

139. $x^{1/2}x^{3/4}x^2$ **140.** $\dfrac{a^{2/3}b^{3/5}}{a^{1/6}b^{7/10}}$ **141.** $\dfrac{x^3y^{-1/2}}{x^{-2/3}y^{-3/4}}$

142. $\left(\dfrac{a^{-2/5}b^{-3/10}}{c^{1/2}}\right)^{-2}$ **143.** $2x^{1/2}(3x^{1/2} - 5x^{1/4} + 7x^{-1/2})$ **144.** $a^{-1/3}(a^3 + a^{4/3} - a^{1/3})$

145. $(x^{1/2} - y^{1/2})(x^{1/2} + y^{1/2})$ **146.** $(a^{-1/2} - a^{1/2})^2$ **147.** $(a^{1/3} - b^{1/3})(a^{2/3} + a^{1/3}b^{1/3} + b^{2/3})$

148. $x^{1/3} - x^{-2/3}$ **149.** $\dfrac{x^{1/2} + x^{-1/2}}{x}$

CHAPTER 5 · TEST   *Answers, pages A59–A60*

1. Express 602,000,000 in scientific notation.

2. Express 5.2×10^{-6} in decimal notation.

Simplify according to the methods of this chapter. All variables are positive.

3. $\left(\dfrac{2a^3b^0}{3c^4}\right)^2$ **4.** $\dfrac{a^2b^{-3}c^4}{a^{-5}b^3c^0}$ **5.** $(3x^4)^{-2}$ **6.** $\left(\dfrac{5x}{3y}\right)^{-3}$ **7.** $\sqrt{25x^4y^{16}}$

8. $(-27a^6b^9)^{2/3}$ **9.** $\sqrt{x^2 + 4x + 4}$ **10.** $\left(\dfrac{27x^3}{64y^6}\right)^{-1/3}$ **11.** $\sqrt{27x^3}$ **12.** $\sqrt{8} + \sqrt{50} - \sqrt{98}$

13. $a^2a^{1/2}a^{-3/4}$ **14.** $\dfrac{12}{\sqrt{6}}$ **15.** $\dfrac{10x^2}{\sqrt[3]{25x}}$ **16.** $\dfrac{a^{-1} + b^{-1}}{a^{-2} - b^{-2}}$ **17.** $(a^{1/2} + a^{-1/2})^2$

18. $\dfrac{\sqrt{10} + \sqrt{2}}{\sqrt{10} - \sqrt{2}}$ **19.** $\dfrac{\sqrt{12}}{\sqrt{6} + \sqrt{3}}$ **20.** $\dfrac{h}{\sqrt{x + h} - \sqrt{x}}$

6 QUADRATIC EQUATIONS AND INEQUALITIES

Imaginary and Complex Numbers

In Section 3.5, we solved the quadratic equation $x^2 + 7x - 18 = 0$ by factoring: $(x + 9)(x - 2) = 0$, or $x = -9, 2$. In this chapter we will solve equations that are not factorable, such as $x^2 + 8x - 18 = 0$ and $x^2 + 9 = 0$. In many cases, their solutions involve square roots of negative numbers, such as $\sqrt{-9}$, which Section 5.2 told us are not real numbers. Because we will need to incorporate them into algebra, we start by defining the symbol i:

$$i = \sqrt{-1}, \quad \text{with the property} \quad i^2 = -1$$

This is consistent with the property of square roots: $i^2 = (\sqrt{-1})^2 = -1$. The symbol i is called the **imaginary unit** and is used to define the square root of a negative number:

$$\sqrt{-a} = i\sqrt{a}, \quad \text{where} \quad a > 0$$

EXAMPLE 1

a. $\sqrt{-9} = i\sqrt{9}$
$\qquad = i3 \quad \text{or} \quad 3i.$ **Check:** $(3i)^2 = 9i^2 = 9(-1) \overset{\checkmark}{=} -9$

b. $-\sqrt{-25} = -i\sqrt{25}$
$\qquad = -5i.$ WARNING! $-\sqrt{-25} \neq \sqrt{25} = 5$

c. $\pm\sqrt{-12} = \pm i\sqrt{12}$
$\qquad = \pm i \cdot 2\sqrt{3} \quad \text{or} \quad \pm 2i\sqrt{3}.$

d. $\sqrt{-\dfrac{1}{18}} = i\sqrt{\dfrac{1}{18}}$ First "pull out" the i.

$\qquad = \dfrac{i}{1} \cdot \dfrac{1}{3\sqrt{2}}$ Simplify $\sqrt{18} = \sqrt{9 \cdot 2} = 3\sqrt{2}$.

$\qquad = \dfrac{i \cdot \sqrt{2}}{3\sqrt{2} \cdot \sqrt{2}}$ Rationalize denominator.

$\qquad = \dfrac{i\sqrt{2}}{6}.$

All of the foregoing answers are of the form bi, where $b \neq 0$ is a real number, and are called **imaginary numbers.** * They can be added as like terms: $2i + 3i = 5i$. However, $2 + 3i$ cannot be further combined because of unlike terms. More important, the sum of a real number and an imaginary number is neither real nor imaginary. It represents a special kind of number, according to the following definition. A **complex number** is any number of the form $a + bi$, where a and b are real numbers. Thus $2 + 3i$ is a complex number. Other examples of complex numbers are

$$5 - 7i, \quad \sqrt{2} + 3\pi i, \quad \text{and} \quad -\frac{1}{2} + 4i\sqrt{3}.$$

Real and imaginary numbers are also considered complex numbers, because they can be written in this form. For example,

$$\text{the real number} \quad 8 = 8 + 0i$$
$$\text{and the imaginary number} \quad 5i = 0 + 5i$$

are both complex numbers.

Complex numbers will be assumed to obey the usual laws of algebra that you have learned so far, with one important provision: $i^2 = -1$. In the next example, we show how to add, subtract, and multiply complex numbers, using familiar properties from Chapter 3.

EXAMPLE 2

Express in terms of i and then perform the indicated operation:

a. $(2 + \sqrt{-9}) + (5 - \sqrt{-49}) - (4 - \sqrt{-81})$
$$= (2 + 3i) + (5 - 7i) - (4 - 9i)$$
$$= 2 + 3i + 5 - 7i - 4 + 9i$$
$$= 3 + 5i.$$

b. $6(8 - 3i) - 5i(9 + 2i) = 48 - 18i - 45i - 10i^2$
$$= 48 - 63i - 10(-1) \qquad i^2 = -1$$
$$= 58 - 63i. \qquad 48 - 63i + 10$$

c. $(2 - i)(5 - 7i) = 10 - 19i + 7i^2 \qquad \text{FOIL}$
$$= 10 - 19i + 7(-1) \qquad i^2 = -1$$
$$= 3 - 19i.$$

d. $(3 + 4i)(3 - 4i) = 9 - 16i^2 \qquad (A + B)(A - B) = A^2 - B^2$
$$= 9 - 16(-1)$$
$$= 25.$$

e. $(7 - \sqrt{-8})(3 + \sqrt{-50})$
$$= (7 - i\sqrt{8})(3 + i\sqrt{50}) \qquad \text{"Pull out" the } i\text{'s.}$$
$$= (7 - 2i\sqrt{2})(3 + 5i\sqrt{2}) \qquad \text{Simplify radicals.}$$
$$= 21 + 35i\sqrt{2} - 6i\sqrt{2} - 10i^2 \cdot 2 \qquad \text{FOIL}$$
$$= 21 + 29i\sqrt{2} - 20(-1) \qquad i^2 = -1$$
$$= 41 + 29i\sqrt{2}.$$

---

*They are also called **pure imaginary numbers**.

(continued)

f. $\dfrac{-(-2) + \sqrt{(-2)^2 - 4 \cdot 1 \cdot 4}}{2 \cdot 1} = \dfrac{2 \pm \sqrt{4 - 16}}{2}$

$$= \dfrac{2 \pm \sqrt{-12}}{2}$$

$$= \dfrac{2 \pm 2i\sqrt{3}}{2} \qquad\qquad \sqrt{-12} = i\sqrt{12} = 2i\sqrt{3}$$

$$= \dfrac{2(1 \pm i\sqrt{3})}{2} \qquad\qquad \text{Factor.}$$

$$= 1 \pm i\sqrt{3}. \qquad\qquad \text{Reduce.}$$

g. $(2 + 3i)^3 = (2 + 3i)(2 + 3i)^2$
$\qquad\qquad = (2 + 3i)(4 + 12i + 9i^2) \qquad\quad (A + B)^2 = A^2 + 2AB + B^2$
$\qquad\qquad = (2 + 3i)(4 + 12i - 9) \qquad\qquad 9i^2 = 9(-1) = -9$
$\qquad\qquad = (2 + 3i)(-5 + 12i)$
$\qquad\qquad = -10 + 24i - 15i + 36i^2 \qquad\quad \text{FOIL again!}$
$\qquad\qquad = -10 + 9i + 36(-1)$
$\qquad\qquad = -46 + 9i.$

Because $i = \sqrt{-1}$, any denominator containing i must be rationalized in order to remove this radical. The methods used in Section 5.5 can be used here. Note in Example 2d, the product is

$$(3 + 4i)(3 - 4i) = 9 - 16i^2 = 25,$$

which is a real number. This illustrates the fact that the product of a complex number $a + bi$ and its **conjugate** $a - bi$ is a real number. We will use this idea in parts c and d of the next example.

EXAMPLE 3   Rationalize the denominators, and then simplify:

a. $\dfrac{2}{i} = \dfrac{2 \cdot i}{i \cdot i}$ Multiply numerator and denominator by i.

$\qquad = \dfrac{2i}{i^2}$

$\qquad = \dfrac{2i}{-1}$

$\qquad = -2i.$

b. $\dfrac{6}{i\sqrt{3}} = \dfrac{6 \cdot i\sqrt{3}}{i\sqrt{3} \cdot i\sqrt{3}}$

$$= \frac{6i\sqrt{3}}{i^2 \cdot 3}$$

$$= \frac{6i\sqrt{3}}{-3}$$

$$= -2i\sqrt{3}. \qquad\qquad i^2 \cdot 3 = (-1) \cdot 3 = -3$$

Reduce.

c. $\dfrac{50}{3 + 4i} = \dfrac{50 \cdot (3 - 4i)}{(3 + 4i) \cdot (3 - 4i)}$ Multiply by conjugate.

conjugates

$$= \frac{50(3 - 4i)}{9 - 16i^2}$$

$$= \frac{50(3 - 4i)}{25} \longleftarrow\qquad 9 - 16(-1) = 9 + 16$$

$$= 2(3 - 4i) \quad \text{or} \quad 6 - 8i. \qquad \text{Reduce.}$$

d. $\dfrac{5 + 10i}{1 - 3i} = \dfrac{(5 + 10i) \cdot (1 + 3i)}{(1 - 3i) \cdot (1 + 3i)}$

conjugates

$$= \frac{5 + 25i + 30i^2}{1 - 9i^2} \qquad\qquad \text{FOIL}$$

$$= \frac{-25 + 25i}{10} \begin{array}{l} \longleftarrow \quad 5 + 30i^2 = 5 + 30(-1) = -25 \\ \longleftarrow \quad 1 - 9i^2 = 1 - 9(-1) = 10 \end{array}$$

$$= \frac{\overset{5}{\cancel{25}}(-1 + i)}{\underset{2}{\cancel{10}}} \qquad\qquad \text{Factor and reduce.}$$

$$= \frac{-5 + 5i}{2} \quad \text{or} \quad \frac{-5}{2} + \frac{5i}{2}.$$

At this point, you should be reminded to first "pull out" the i's before performing any operation in the exercises. This is especially critical when multiplying, as evidenced by the following product:

$$\sqrt{-2}\sqrt{-8} = i\sqrt{2} \cdot i\sqrt{8} = i^2\sqrt{16} = -4.$$

Be aware that the Product Rule for Roots (Section 5.3) does *not apply* to imaginary numbers; thus

$$\sqrt{-2}\sqrt{-8} \neq \sqrt{(-2)(-8)} = \sqrt{16} = 4.$$

EXERCISE 6.1   *Answers, page A60*

A

Express in terms of the imaginary unit *i*, as in EXAMPLE 1:

1. $\sqrt{-49}$ **2.** $\sqrt{-81}$ **3.** $-\sqrt{-144}$ **4.** $\pm\sqrt{-169}$ **5.** $\sqrt{-\dfrac{16}{25}}$ **6.** $\sqrt{-\dfrac{121}{289}}$

7. $\sqrt{-50}$ **8.** $-\sqrt{-20}$ **9.** $-\sqrt{-75}$ **10.** $\sqrt{-108}$ **11.** $\sqrt{-\dfrac{1}{8}}$ **12.** $\pm\sqrt{-\dfrac{1}{72}}$

13. $\sqrt{-\dfrac{25}{3}}$ **14.** $-\sqrt{-\dfrac{49}{12}}$ **15.** $\pm\sqrt{\dfrac{-5}{48}}$ **16.** $\sqrt{\dfrac{-3}{28}}$

Express in terms of *i*, if necessary, and then perform the indicated operation, as in EXAMPLE 2:

17. $(6 - 2i) - (7 - 9i) + (3 + 5i)$ **18.** $(-4 - i) + (6 - 5i) - (3 - i)$ **19.** $3i(2 - 7i) - 5(8 - i) + i(4 + 3i)$

20. $9(7 - i) - 2i(-3 - 7i) + 6i(1 + i)$ **21.** $(2 + 3i)(4 - 5i)$ **22.** $(6 + i)(3 - 2i)$

23. $(5 - \sqrt{-4})(2 + \sqrt{-9})$ **24.** $(3 + \sqrt{-1})(2 - \sqrt{-25})$ **25.** $(3 + 5i\sqrt{2})(4 + 3i\sqrt{2})$

26. $(-4 - 3i\sqrt{3})(4 + 2i\sqrt{3})$ **27.** $(6 - \sqrt{-20})(3 + \sqrt{-45})$ **28.** $(1 + \sqrt{-28})(2 + \sqrt{-63})$

29. $\dfrac{-(-4) \pm \sqrt{(-4)^2 - 4 \cdot 1 \cdot 8}}{2 \cdot 1}$ **30.** $\dfrac{-6 \pm \sqrt{6^2 - 4 \cdot 1 \cdot 18}}{2 \cdot 1}$ **31.** $\dfrac{-3 \pm \sqrt{3^2 - 4 \cdot 1 \cdot 9}}{2 \cdot 1}$

32. $\dfrac{-0 \pm \sqrt{0^2 - 4 \cdot 3 \cdot 25}}{2 \cdot 3}$ **33.** $(2 - 5i)^2$ **34.** $(1 + 4i)^2$

35. $(3 - 2i)^3$ **36.** $(2 + i)^3$

Rationalize the denominators and then simplify, as in EXAMPLE 3:

37. $\dfrac{5}{i}$ **38.** $\dfrac{-7}{i}$ **39.** $\dfrac{1 + 3i}{2i}$ **40.** $\dfrac{2 - 7i}{3i}$ **41.** $\dfrac{10}{i\sqrt{5}}$ **42.** $\dfrac{8}{i\sqrt{2}}$

43. $\dfrac{16}{\sqrt{-12}}$ **44.** $\dfrac{12}{\sqrt{-8}}$ **45.** $\dfrac{26}{2 + 3i}$ **46.** $\dfrac{20}{3 + i}$ **47.** $\dfrac{15i}{3 - 4i}$ **48.** $\dfrac{87i}{2 - 5i}$

49. $\dfrac{3 - 2i}{2 + i}$ **50.** $\dfrac{10 - 5i}{1 - 2i}$ **51.** $\dfrac{4 - \sqrt{-18}}{3 + \sqrt{-8}}$ **52.** $\dfrac{1 + \sqrt{-12}}{2 + \sqrt{-27}}$

B

The property $i^2 = -1$ and the laws of exponents allow us to simplify higher powers of *i*. For example,

$$i^3 = i^2 i = (-1)i = -i \quad \text{and} \quad i^4 = (i^2)^2 = (-1)^2 = 1.$$

Even powers of i can be expressed in terms of i^2:

$$i^{56} = (i^2)^{28} = (-1)^{28} = 1.$$
$$i^{70} = (i^2)^{35} = (-1)^{35} = -1.$$

Odd powers of i can be expressed in terms of i^2 and i:

$$i^{85} = i^{84}i = (i^2)^{42}i = (-1)^{42}i = i.$$
$$i^{31} = i^{30}i = (i^2)^{15}i = (-1)^{15}i = -i.$$

Now express each of these powers as either 1, -1, i, or $-i$:

53. i^{16} **54.** i^{60} **55.** i^{26} **56.** i^{80} **57.** i^{75}

58. i^{59} **59.** i^{37} **60.** i^{77} **61.** i^{103} **62.** i^{153}

Negative powers of i can be handled by first inverting, then using the "even/odd" method above, and, finally, rationalizing the denominator when necessary.

$$i^{-26} = \frac{1}{i^{26}} = \frac{1}{(i^2)^{13}} = \frac{1}{(-1)^{13}} = \frac{1}{-1} = -1.$$

$$i^{-67} = \frac{1}{i^{67}} = \frac{1}{i^{66}i} = \frac{1}{(i^2)^{33}i} = \frac{1}{(-1)^{33}i} = \frac{1}{-i}.$$

$$= \frac{1 \cdot i}{-i \cdot i} = \frac{i}{-i^2} = \frac{i}{-(-1)} = i. \qquad \text{Now rationalize denominator.}$$

Now express each of these powers as either 1, -1, i, or $-i$:

63. i^{-14} **64.** i^{-34} **65.** i^{-20} **66.** i^{-44} **67.** i^{-21} **68.** i^{-65} **69.** i^{-35} **70.** i^{-59}

True or false?

71. $4 - i\sqrt{2}$ is a complex number.

72. 2 is a complex number.

73. Every irrational number is a complex number.

74. $i\sqrt{2}$ is irrational.

75. $i = -1$

76. $\sqrt[3]{-8} = 2i$

77. $\sqrt{-4}\sqrt{-9} = 6$

78. $\sqrt{-2}\sqrt{-8} = -4$

79. $i^{-1} = -i$

80. The sum of two imaginary numbers is always an imaginary number.

81. The product of two imaginary numbers is always a real number.

82. The sum of a complex number $a + bi$, with $b \neq 0$, and its conjugate $a - bi$ is always an imaginary number.

83. The difference between a complex number $a + bi$, with $b \neq 0$, and its conjugate $a - bi$ is always an imaginary number.

84. The product of a complex number and its conjugate is always a real number.

C

85. Find the incorrect step in this "proof" that $1 = -1$:

$$\sqrt{(-1)(-1)} = \sqrt{-1}\sqrt{-1}$$
$$\sqrt{1} = i \cdot i$$
$$1 = i^2$$
$$1 = -1.$$

86. Find the incorrect step in this "proof" that $1 = -1$:

$$\sqrt{\frac{1}{-1}} = \sqrt{\frac{-1}{1}}$$

$$\frac{\sqrt{1}}{\sqrt{-1}} = \frac{\sqrt{-1}}{\sqrt{1}}$$

$$\frac{1}{i} = \frac{i}{1}$$

$$\frac{1}{i} = \frac{i \cdot i}{1 \cdot i} \longleftarrow \text{L.C.D.} = i$$

$$1 = i^2 \longleftarrow \text{Equate numerators.}$$
$$1 = -1.$$

87. In Section 5.6, we defined the fractional exponent by

$$a^{m/n} = (\sqrt[n]{a})^m = \sqrt[n]{a^m}.$$

This would mean that

$$(-1)^{3/2} = (\sqrt[2]{-1})^3 = \sqrt[2]{(-1)^3}.$$

a. Show that $(\sqrt[2]{-1})^3 = -i$ but $\sqrt[2]{(-1)^3} = i.$

b. The results of part **a.** tell us that $-i = i$, which is absurd! Find the fallacy in this whole line of reasoning.

| **6.2** | **Extracting Roots** |

In Section 3.5, we learned how to solve the equation $x^2 = 9$ by factoring:

$$x^2 = 9$$
$$x^2 - 9 = 0$$
$$(x + 3)(x - 3) = 0$$
$$x = -3, x = 3, \quad \text{or} \quad x = \pm 3.$$

In like fashion, we can solve the more general equation $x^2 = c$ by factoring, provided we use radicals:

$$x^2 = c$$
$$x^2 - c = 0$$

$$(x + \sqrt{c})(x - \sqrt{c}) = 0 \qquad \text{Check: } x^2 - (\sqrt{c})^2 \overset{\checkmark}{=} x^2 - c$$
$$x = -\sqrt{c}, \ x = \sqrt{c}, \quad \text{or} \quad x = \pm\sqrt{c}.$$

Thus the equation $x^2 = c$ has solutions $x = \pm\sqrt{c}$, which means we could have solved it by simply taking the positive and negative square roots of c. For this reason, we have the extracting-roots method:

Extracting-Roots Method

$$\text{If} \qquad x^2 = c,$$
$$\text{then} \qquad x = \pm\sqrt{c}.$$

EXAMPLE 1

Solve by extracting roots.

a. $x^2 = 20$

$$x = \pm\sqrt{20} \qquad\qquad \text{Extract roots.}$$
$$x = \pm 2\sqrt{5}. \qquad\qquad \textbf{Check: } (\pm 2\sqrt{5})^2 = 4 \cdot 5 \overset{\checkmark}{=} 20$$

b. $x^2 - 9 = 0$

$$x^2 = 9 \qquad\qquad \text{Isolate } x^2.$$
$$x = \pm\sqrt{9} \qquad\qquad \text{Extract roots.}$$
$$x = \pm 3.$$

c. $4y^2 + 27 = 0$

$$4y^2 = -27$$
$$y^2 = \frac{-27}{4} \qquad\qquad \Bigg\} \quad \text{First isolate } y^2.$$

$$y = \pm\sqrt{\frac{-27}{4}} \qquad\qquad \text{Now extract roots.}$$

$$y = \pm\frac{3i\sqrt{3}}{2}. \qquad\qquad \sqrt{-27} = i\sqrt{27}$$

d. $\dfrac{x^2}{3} - \dfrac{1}{5} = 0$

$$\frac{x^2}{3} = \frac{1}{5}$$
$$\frac{x^2 \cdot 5}{3 \cdot 5} = \frac{1 \cdot 3}{5 \cdot 3} \qquad\qquad \Bigg\} \quad \text{First isolate } x^2.$$
$$5x^2 = 3$$
$$x^2 = \frac{3}{5}$$

(continued)

$$x = \pm\sqrt{\frac{3}{5}}$$ Now extract roots.

$$x = \pm\frac{\sqrt{3} \cdot \sqrt{5}}{\sqrt{5} \cdot \sqrt{5}}$$ Rationalize denominator.

$$x = \pm\frac{\sqrt{15}}{5}.$$

e. $(x + 1)^2 = 25$

$$x + 1 = \pm\sqrt{25}$$ Extract roots, leaving $x + 1$.
$$x = -1 \pm 5$$ Subtract 1.
$$x = -1 + 5, \; -1 - 5$$ -1 ± 5 means $-1 + 5$ or $-1 - 5$.
$$x = 4, \; -6.$$

f. $(x - 3)^2 - 50 = 0$

$$(x - 3)^2 = 50$$ Isolate $(x - 3)^2$.
$$x - 3 = \pm\sqrt{50}$$ Extract roots, leaving $x - 3$.
$$x = 3 \pm 5\sqrt{2}.$$ Add 3.

Note: $x = 3 \pm 5\sqrt{2}$ means $x = 3 + 5\sqrt{2}$ or $x = 3 - 5\sqrt{2}$.

In the next example, we use the Pythagorean Theorem, $a^2 + b^2 = c^2$, to find the sides of the right triangles (see Sections 1.3 and 3.6).

EXAMPLE 2

In the following right triangles, a and b are the legs and c is the hypotenuse. In each case, find the unknown side or sides.

a. $a = 2, b = 6, c = x$

$$x^2 = 2^2 + 6^2$$
$$x^2 = 40$$
$$x = \sqrt{40}$$ $-\sqrt{40}$ has no meaning here.
$$x = 2\sqrt{10}.$$

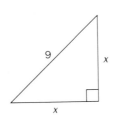

b. $a = b = x, c = 9$

$$x^2 + x^2 = 9^2$$
$$2x^2 = 81$$
$$x^2 = \frac{81}{2}$$
$$x = \sqrt{\frac{81}{2}}$$
$$x = \frac{9 \cdot \sqrt{2}}{\sqrt{2} \cdot \sqrt{2}}$$ Rationalize denominator.
$$x = \frac{9\sqrt{2}}{2}.$$

c. $a = x$, $b = 6$, $c = 2x$

$$(2x)^2 = x^2 + 6^2$$

$(2x)^2 \neq 2x^2 \longrightarrow 4x^2 = x^2 + 36$

$$3x^2 = 36$$
$$x^2 = 12$$
$$x = \sqrt{12}$$
$$x = 2\sqrt{3}.$$

Then $2x = 4\sqrt{3}.$

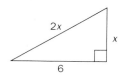

In the next example, we solve for the indicated variable in terms of the other variables. This was the subject of Section 4.7, except that now we solve by extracting roots.

EXAMPLE 3   Solve for the indicated variable.

a. $s = \dfrac{gt^2}{2}$;   for t,   where $t \geq 0$

$$\frac{s \cdot 2}{1 \cdot 2} = \frac{gt^2}{2}$$

$$2s = gt^2$$

$$\frac{2s}{g} = t^2$$

First isolate t^2.

$$\pm\sqrt{\frac{2s}{g}} = t.$$ Now extract roots.

The stipulation $t \geq 0$ means that we can use only the *positive* root. Rationalizing the denominator gives

$$t = \frac{\sqrt{2s} \cdot \sqrt{g}}{\sqrt{g} \cdot \sqrt{g}}$$

$$t = \frac{\sqrt{2sg}}{g}.$$

b. $\dfrac{x^2}{4} + \dfrac{y^2}{9} = 1$;   for y

$$\frac{x^2 \cdot 9}{4 \cdot 9} + \frac{y^2 \cdot 4}{9 \cdot 4} = \frac{1 \cdot 36}{1 \cdot 36}$$

$$9x^2 + 4y^2 = 36$$
$$4y^2 = 36 - 9x^2$$

First isolate y^2.

$$y^2 = \frac{36 - 9x^2}{4}$$

(continued)

$$y = \pm\sqrt{\frac{36 - 9x^2}{4}} \qquad \text{Now extract roots.}$$

$$y = \pm\sqrt{\frac{9(4 - x^2)}{4}} \longleftarrow \text{Factor.}$$

$$y = \pm\frac{3\sqrt{4 - x^2}}{2}. \qquad \longleftarrow \frac{\sqrt{9}\sqrt{4 - x^2}}{\sqrt{4}}$$

WARNING! $\sqrt{4 - x^2} \neq 2 - x$

EXERCISE 6.2   *Answers, pages A60–A61*

A

Solve by extracting roots, as in EXAMPLE 1:

1. $x^2 = 54$

2. $18 = y^2$

3. $y^2 - 25 = 0$

4. $49 - z^2 = 0$

5. $12 + w^2 = 0$

6. $t^2 + 50 = 0$

7. $4z^2 - 121 = 0$

8. $169 = x^2$

9. $81 - 36x^2 = 0$

10. $-16t^2 + 36 = 0$

11. $3y^2 = 96$

12. $100 + 5w^2 = 0$

13. $2x^2 - 9 = 0$

14. $16 = 3x^2$

15. $\dfrac{z}{6} = \dfrac{8}{z}$

16. $\dfrac{3}{t} = \dfrac{t}{9}$

17. $\dfrac{5x^2}{7} - \dfrac{3}{4} = 0$

18. $\dfrac{7}{8} + \dfrac{2x^2}{3} = 0$

19. $9y = \dfrac{1}{y}$

20. $\dfrac{4}{x} = x$

21. $(x + 3)^2 = 36$

22. $(x - 5)^2 = 100$

23. $(y - 2)^2 - 12 = 0$

24. $(z + 1)^2 + 8 = 0$

25. $\left(x - \dfrac{1}{2}\right)^2 = \dfrac{25}{4}$

26. $\left(x + \dfrac{1}{3}\right)^2 = \dfrac{20}{9}$

27. $(2x + 1)^2 - 18 = 0$

28. $(4x - 3)^2 = \dfrac{49}{16}$

29. $\dfrac{3x - 2}{x + 2} = \dfrac{x + 8}{x + 4}$

30. $\dfrac{1}{x} + \dfrac{1}{x + 2} = 1$

In the following right triangles, *a* and *b* are the legs and *c* is the hypotenuse. In each case, find the unknown side or sides, as in EXAMPLE 2.

31. $a = 4$, $b = 8$, $c = x$

32. $a = 5$, $b = 9$, $c = x$

33. $a = 5$, $b = x$, $c = 13$

34. $a = x$, $b = 9$, $c = 15$

35. $a = \sqrt{2}$, $b = \sqrt{6}$, $c = x$

36. $a = \sqrt{5}$, $b = x$, $c = \sqrt{17}$

37. $a = \dfrac{1}{3}$, $b = \dfrac{1}{4}$, $c = x$

38. $a = x$, $b = \dfrac{1}{2}$, $c = 1$

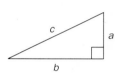

39. $a = b = 5$, $c = x$

40. $a = b = 8$, $c = x$

41. $a = b = x$, $c = 10$

42. $a = b = x$, $c = 6$

43. $a = b = x$, $c = 3$

44. $a = b = x$, $c = 1$

45. $a = x$, $b = 3$, $c = 2x$

46. $a = x$, $b = 9$, $c = 2x$

47. $a = x$, $b = 4$, $c = 2x$

48. $a = x$, $b = 2$, $c = 2x$

49. $a = \sqrt{2} - 1$, $b = \sqrt{2} + 1$, $c = x$

50. $a = 3 - \sqrt{3}$, $b = 3 + \sqrt{3}$, $c = x$

Solve for the indicated variable, as in EXAMPLE 3:

51. $a^2x^2 - b^2 = 0$;   for x

52. $4c^2 = 9d^2y^2$;   for y

53. $E = mc^2$;   for c, $c > 0$

54. $F = \dfrac{Gm_1m_2}{d^2}$;   for d, $d > 0$

55. $\dfrac{mx}{n} = \dfrac{n}{x}$;   for x

56. $\dfrac{ay}{b} - \dfrac{b}{y} = 0$;   for y

57. $g = \dfrac{4\pi^2 L}{T^2}$;   for T, $T > 0$

58. $V = \dfrac{1}{3}\pi r^2 h$;   for r, $r \geq 0$

59. $x^2 + y^2 = 1$;   for x

60. $x^2 - y^2 = 9$;   for y

61. $\dfrac{x^2}{25} + \dfrac{y^2}{16} = 1$;   for y

62. $\dfrac{x^2}{64} - \dfrac{y^2}{16} = 1$;   for y

 B

63. $\dfrac{x^2}{a^2} - \dfrac{y^2}{b^2} = 1$;   for y

64. $V = h(R^2 + r^2)$;   for R, $R \geq 0$

Solve for x in terms of the other letters:

65.

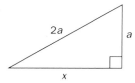

66.

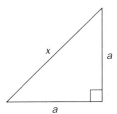

67.

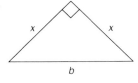

68.

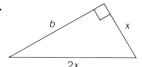

69.

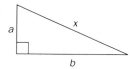

70.

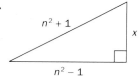

71.

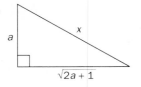

72.

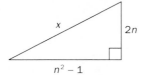

73. Find w, x, y, and then z.

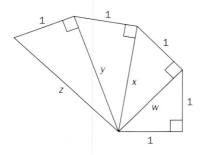

74. Find the unknown sides.

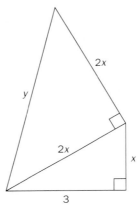

75. Find x. *Hint*: First find y and z.

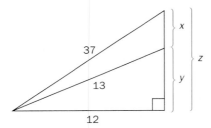

76. Find x and then y.

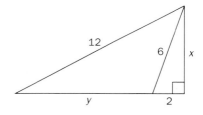

77. If the sides of a square are each increased by 2, a new square is formed having area 12. Find the side of the original square. Then find both the area and the perimeter of the original square.

78. A baseball diamond is actually a square with the bases 90 feet apart.
   a. Find the distance from home plate to second base.
   b. A shortstop fields a ball half-way between second base and third base and throws it to first base. How far does he throw the ball?
   (Express both answers to the nearest tenth of a foot.)

C

79. A square has diagonal of length a.
  a. Find x (the length of one side) in terms of a.
  b. Express the area of the square in terms of a.

80. Who says a square peg won't fit into a round hole? The square shown has side a.
  a. Find the radius of the circle, r, in terms of a.
  b. Show that the circle has area $\dfrac{\pi a^2}{2}$.

  Formula: $A = \pi r^2$.

81. A rectangle with sides 4 and 6 is inscribed in a circle.
  a. Find the radius r of the circle.
  b. Find the shaded area inside the circle but outside the rectangle.
  Hint: Shaded area = area of circle − area of rectangle. (Express your answer both in terms of π and to the nearest hundredth.)

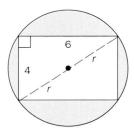

Problem 81

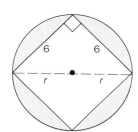

Problem 82

82. Repeat Problem 81, for the square shown in the diagram.

83. A ball is dropped, and its height h above the ground (in feet) after t seconds is given by the formula

$$h = -16t^2 + 144.$$

When will the ball hit the ground?

Solve, using the methods of Section 2.6 and of Exercise 3.5, Problems 75–80:

84. $|x^2 - 5| = 4$ **85.** $|x^2 - 10| = 2$ **86.** $|x^2 + 2| = 6$ **87.** $|x^2 + 6| = 2$

88. Here is a sneaky way to find out a woman's (or man's) age. Ask her to multiply her age one year ago by her age one year from now and to give you the answer. You simply add 1 to the result and then take the square root. The result is her present age. For example, if she gives you the product 624, you add 1 and take the square root:

$$\sqrt{624 + 1} = \sqrt{625} = 25.$$

(continued)

She is 25 years old. **Check:** $(25 - 1)(25 + 1) = 24 \cdot 26 \overset{\checkmark}{=} 624$.

a. Find the present ages of people who give you these products: 960, 360, 728, 1295, 1848, and 2808.

b. Let P = someone's present age; then $P - 1$ = age 1 year ago, and $P + 1$ = age 1 year from now. Write a formula, similar to the foregoing one, that gives that person's present age. Verify that the formula, when simplified, equals P.

6.3   Completing the Square

In Section 6.2, Example 1f, we solved the equation $(x - 3)^2 = 50$ by extracting roots: $x = 3 \pm 5\sqrt{2}$. After multiplying, this equation becomes $x^2 - 6x + 9 = 50$, or $x^2 - 6x - 41 = 0$. We have solved a nonfactorable equation! The obvious question now is "How can $x^2 - 6x - 41 = 0$ be converted back to $(x - 3)^2 = 50$, which we can solve by extracting roots?"

To answer this, we first look at the following trinomial, which factors into the square of a binomial:

$$x^2 + bx + \left(\frac{b}{2}\right)^2 = \left(x + \frac{b}{2}\right)^2.$$

To check this, we expand the right side, using $(A + B)^2 = A^2 + 2AB + B^2$:

$$\left(x + \frac{b}{2}\right)^2 = x^2 + 2 \cdot \frac{b}{2} \cdot x + \left(\frac{b}{2}\right)^2 \overset{\checkmark}{=} x^2 + bx + \left(\frac{b}{2}\right)^2.$$

In each of these "perfect-square trinomials," the third term $\left(\frac{b}{2}\right)^2$ can be obtained by taking one-half of the middle coefficient b and then squaring the result.

Now let us use this idea to solve

$$x^2 - 6x - 41 = 0.$$

We start by isolating the constant to the right side of the equation:

$$x^2 - 6x \qquad = 41.$$

Now we take one-half of the middle coefficient, -6, and square the result: $\left(\frac{-6}{2}\right)^2 = (-3)^2 = 9$. This is added to both sides of the equation, creating a perfect square trinomial:

$$x^2 - 6x + 9 = 41 + 9$$

or

$$(x - 3)^2 = 50.$$

This is precisely the equation in Example 1f (mentioned in Section 6.2), which we solved by extracting roots:

$$x - 3 = \pm\sqrt{50}$$
$$x = 3 \pm 5\sqrt{2}.$$

By adding 9 to both sides of the equation $x^2 - 6x = 41$, we created the perfect-square trinomial $x^2 - 6x + 9$, or $(x - 3)^2$. Appropriately enough, this method of solving quadratic equations is called **completing the square.**

EXAMPLE 1 | Solve by completing the square.

a. $x^2 + 10x - 7 = 0$

$$\begin{aligned} x^2 + 10x &= 7 \\ x^2 + 10x + 25 &= 7 + 25 \\ (x + 5)^2 &= 32 \\ x + 5 &= \pm\sqrt{32} \\ x &= -5 \pm 4\sqrt{2}. \end{aligned}$$

Isolate constant.
Add $\left(\frac{10}{2}\right)^2 = 25$ to both sides to complete the square. Then factor.
Extract roots.
Subtract 5.

b. $y^2 - 8y + 28 = 0$

$$\begin{aligned} y^2 - 8y &= -28 \\ y^2 - 8y + 16 &= -28 + 16 \\ (y - 4)^2 &= -12 \\ y - 4 &= \pm\sqrt{-12} \\ y &= 4 \pm 2i\sqrt{3}. \end{aligned}$$

Isolate constant.
Add $\left(\frac{-8}{2}\right)^2 = 16$ to both sides.
Factor.
Extract roots.
Add 4.

In the next example, the middle coefficient, -1, is an odd number. This creates a fraction as the term to be added to both sides in order to complete the square.

EXAMPLE 2 | Solve $x^2 - x - 3 = 0$ by completing the square.

Solution

$$x^2 - 1x = 3 \qquad \text{Isolate constant.}$$

$$x^2 - 1x + \frac{1}{4} = 3 + \frac{1}{4} \qquad \text{Add } \left(\frac{-1}{2}\right)^2 = \frac{1}{4} \text{ to both sides.}$$

$$\left(x - \frac{1}{2}\right)^2 = \frac{3 \cdot 4}{1 \cdot 4} + \frac{1}{4} \qquad \text{Factor left side; L.C.D.} = 4$$

$$\left(x - \frac{1}{2}\right)^2 = \frac{13}{4} \qquad \frac{12}{4} + \frac{1}{4} = \frac{13}{4}$$

$$x - \frac{1}{2} = \pm\sqrt{\frac{13}{4}} \qquad \text{Extract roots.}$$

(continued)

$$x = \frac{1}{2} \pm \frac{\sqrt{13}}{2} \qquad \text{Add } \frac{1}{2}.$$

$$x = \frac{1 \pm \sqrt{13}}{2}. \qquad \text{Write as single fraction.}$$

Alternative method: Instead of factoring after the second step, we can first rewrite the equation *without denominators:*

$$x^2 - 1x + \frac{1}{4} = 3 + \frac{1}{4} \qquad \text{Second step as before.}$$

$$\frac{x^2 \cdot 4}{1 \cdot 4} - \frac{1x \cdot 4}{1 \cdot 4} + \frac{1}{4} = \frac{3 \cdot 4}{1 \cdot 4} + \frac{1}{4} \longleftarrow \text{L.C.D.} = 4$$

$$4x^2 - 4x + 1 = 12 + 1 \qquad \text{Equate numerators.}$$

$$(2x - 1)^2 = 13 \qquad \text{Factor.}$$

$$2x - 1 = \pm \sqrt{13} \qquad \text{Extract roots.}$$

$$2x = 1 \pm \sqrt{13} \qquad \text{Add 1.}$$

$$x = \frac{1 \pm \sqrt{13}}{2}. \qquad \text{Divide by 2.}$$

In Examples 1 and 2, the first term in each equation was x^2 or y^2. In the next example, the first term is $2x^2$. Completing the square works if the first term is x^2, so we will need to divide each term of the equation by 2, resulting in an equation with first term x^2. We then continue as in the two preceding examples.

EXAMPLE 3

Solve $2x^2 + 3x - 5 = 0$ by completing the square.

Solution
As mentioned above, we start by dividing each term by 2:

$$\frac{2x^2}{2} + \frac{3x}{2} - \frac{5}{2} = \frac{0}{2} \qquad \text{Divide by first coefficient 2.}$$

$$x^2 + \frac{3x}{2} = \frac{5}{2} \qquad \text{Isolate constant.}$$

$$x^2 + \frac{3x}{2} + \frac{9}{16} = \frac{5}{2} + \frac{9}{16} \qquad \text{Add } \left(\frac{\frac{3}{2}}{2} \right)^2 = \left(\frac{3}{2} \cdot \frac{1}{2} \right)^2 = \left(\frac{3}{4} \right)^2 = \frac{9}{16}$$

$$\left(x + \frac{3}{4} \right)^2 = \frac{5 \cdot 8}{2 \cdot 8} + \frac{9}{16} \qquad \text{Factor left side; L.C.D.} = 16$$

$$\left(x + \frac{3}{4} \right)^2 = \frac{49}{16} \qquad \frac{40}{16} + \frac{9}{16} = \frac{49}{16}$$

$$x + \frac{3}{4} = \pm \sqrt{\frac{49}{16}}$$ Extract roots.

$$x = -\frac{3}{4} \pm \frac{7}{4}$$ Subtract $\frac{3}{4}$.

$$x = \frac{-3 \pm 7}{4}$$ Combine fractions.

$$x = \frac{-3 + 7}{4}, \frac{-3 - 7}{4}$$

$$x = 1, \frac{-5}{2}.$$ $\frac{4}{4} = 1, \frac{-10}{4} = \frac{-5}{2}$

Alternative method: Instead of factoring after the third step, we first rewrite the equation without denominators:

$$x^2 + \frac{3x}{2} + \frac{9}{16} = \frac{5}{2} + \frac{9}{16}$$ Third step as before.

$$\frac{x^2 \cdot 16}{1 \cdot 16} + \frac{3x \cdot 8}{2 \cdot 8} + \frac{9}{16} = \frac{5 \cdot 8}{2 \cdot 8} + \frac{9}{16}$$ L.C.D. = 16

$$16x^2 + 24x + 9 = 40 + 9$$ Equate numerators.
$$(4x + 3)^2 = 49$$ Factor.
$$4x + 3 = \pm \sqrt{49}$$ Extract roots.
$$4x = -3 \pm 7$$ Subtract 3.

$$x = \frac{-3 \pm 7}{4}$$ Divide by 4.

$$x = 1, \frac{-5}{2}.$$ See above.

EXAMPLE 4

The length of a rectangle is 4 inches longer than the width. The area is 8 square inches. Find the width and the length, both in terms of radicals and approximated to the nearest hundredth.

Solution

Let x = the width; then $x + 4$ = the length.

The formula $A = LW$ gives us

$$x(x + 4) = 8.$$

Solving by completing the square gives

$$x^2 + 4x = 8$$
$$x^2 + 4x + 4 = 8 + 4$$ Add $\left(\frac{4}{2}\right)^2 = 2^2 = 4$.   *(continued)*

$$(x + 2)^2 = 12$$
$$x + 2 = \pm \sqrt{12}$$
$$x = -2 + 2\sqrt{3}. \quad -2 - 2\sqrt{3} \text{ is meaningless here. Why?}$$

If $x = -2 + 2\sqrt{3}$, then $x + 4 = (-2 + 2\sqrt{3}) + 4 = 2 + 2\sqrt{3}$.
On a calculator, we get

$$-2 + 2\sqrt{3} \approx 1.46 \quad \text{using} \quad 2 \boxed{+/-} \boxed{+} 2 \boxed{\times} 3 \boxed{\sqrt{}} \boxed{=} 1.4641016$$
$$2 + 2\sqrt{3} \approx 5.46.$$

Answer: Width $= -2 + 2\sqrt{3} \approx 1.46$ inches,
Length $= 2 + 2\sqrt{3} \approx 5.46$ inches.
Check: Area $= (-2 + 2\sqrt{3})(2 + 2\sqrt{3}) = -4 - 4\sqrt{3} + 4\sqrt{3} + 4 \cdot 3 \overset{\checkmark}{=} 8$.

EXERCISE 6.3   *Answers, page A61*

Solve by completing the square, as in EXAMPLE 1:

1. $x^2 - 4x - 4 = 0$ **2.** $x^2 + 2x - 2 = 0$ **3.** $y^2 + 6y + 21 = 0$ **4.** $y^2 - 8y + 24 = 0$

5. $z^2 + 10z - 24 = 0$ **6.** $z^2 - 12z + 20 = 0$ **7.** $y^2 - 26y + 168 = 0$ **8.** $z^2 + 14z - 120 = 0$

9. $t^2 = 2(t + 2)$ **10.** $3 = u(u + 6)$ **11.** $\dfrac{x}{x - 2} = \dfrac{4}{x}$ **12.** $2y(y + 1) = (y - 2)^2$

13. $\dfrac{1}{x - 1} + \dfrac{1}{x + 1} = 1$ **14.** $\dfrac{2x + 1}{3x - 1} = \dfrac{x - 5}{2x + 1}$

Solve by completing the square, as in EXAMPLE 2:

15. $x^2 - 3x - 9 = 0$ **16.** $y^2 - 5y - 5 = 0$ **17.** $z^2 - z + 2 = 0$ **18.** $w^2 + w - 3 = 0$

19. $x^2 + 7x + 12 = 0$ **20.** $x^2 + 5x - 14 = 0$ **21.** $u^2 = 3(3u + 2)$ **22.** $z(5 - z) = 25$

23. $x = \dfrac{1}{x + 1}$ **24.** $(2x + 1)(x - 2) = (x + 2)^2$

Solve by completing the square. Start by dividing by the coefficient of x^2, as in EXAMPLE 3.

25. $2x^2 + 7x - 15 = 0$ **26.** $3x^2 - 5x - 2 = 0$ **27.** $4y^2 - 2y - 7 = 0$ **28.** $5x^2 + 3x - 1 = 0$

29. $6 = z(3z + 1)$ **30.** $3(x^2 - 3) = x$

Solve as in EXAMPLE 4:

31. The length of a rectangle is 6 meters longer than the width. The area is 11 square meters. Find the width and the length, both in terms of radicals and approximated to the nearest hundredth.

32. The width of a rectangle is two feet shorter than the length. The area is seven square feet. Find the width and the length, both in terms of radicals and approximated to the nearest hundredth.

33. The square of a number is twice the sum of the number and 13. Find the number.

34. The sum of a number and its reciprocal is four. Find the number.

B

Find the unknown sides of these right triangles. Give answers both in terms of radicals and approximated to the nearest tenth.

35.

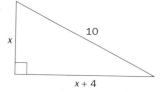

36.

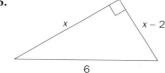

37. When a ball is thrown upward from the top of a building, its height h above the ground (in feet) after t seconds is given by the formula

$$h = -16t^2 + 64t + 64.$$

When does the ball strike the ground? *Hint*: See the hint for Exercise 3.5, Problem 73. Give your answer both expressed as a radical and approximated to the nearest tenth of a second.

38. The base of a triangle is four inches more than the height. If the triangle has area four square inches, find the height and the base. Give answers both expressed in terms of radicals and approximated to the nearest tenth. *Hint*: $A = \dfrac{bh}{2}$.

39. We saw in the text that if $\left(\dfrac{b}{2}\right)^2$ is added to $x^2 + bx +$,

the result is a perfect square trinomial:

$$x^2 + bx + \left(\frac{b}{2}\right)^2 = \left(x + \frac{b}{2}\right)^2.$$

a. Verify this by multiplying out the right side of the equation. The picture consists of a square, two rectangles, and an "unknown square." By multiplying the outer dimensions, we see that the entire square has area $\left(x + \dfrac{b}{2}\right)^2$. But the sum of its parts is $x^2 + 2\left(\dfrac{bx}{2}\right) + \,?$,   or $x^2 + bx + \,?$ Thus $x^2 + bx + \,? = \left(x + \dfrac{b}{2}\right)^2$.   *(continued next page)*

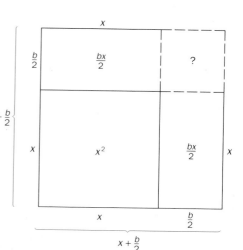

b. What are the sides of the unknown square?

c. What is its area?

d. What must therefore be added to the above equation to "complete the square"?

6.4   The Quadratic Formula

Completing the square can be used to solve second-degree equations, whether they are factorable or not. One of its drawbacks occurred in Examples 2 and 3 of the previous section, where we encountered fractions. In this section, we will derive a formula for solving any quadratic equation, thus eliminating the need for least common denominators. The formula itself is derived by completing the square, using the methods of Example 3, where we first divided by the coefficient of the first term.

Here is the plan. We start with the standard second-degree equation

$$ax^2 + bx + c = 0, \quad a \neq 0,$$

and solve it for x in terms of the coefficients a, b, and c.

$$\frac{ax^2}{a} + \frac{bx}{a} + \frac{c}{a} = \frac{0}{a} \qquad \text{Divide by } a.$$

$$x^2 + \frac{bx}{a} \qquad = \frac{-c}{a} \qquad \text{Isolate constant.}$$

$$x^2 + \frac{bx}{a} + \frac{b^2}{4a^2} = \frac{-c}{a} + \frac{b^2}{4a^2} \qquad \text{Add } \left(\frac{\frac{b}{a}}{2}\right)^2 = \left(\frac{b}{2a}\right)^2 = \frac{b^2}{4a^2}.$$

$$\frac{x^2 \cdot 4a^2}{1 \cdot 4a^2} + \frac{bx \cdot 4a}{a \cdot 4a} + \frac{b^2}{4a^2} = \frac{-c \cdot 4a}{a \cdot 4a} + \frac{b^2}{4a^2} \longleftarrow \text{L.C.D.} = 4a^2$$

$$4a^2x^2 + 4abx + b^2 = -4ac + b^2 \qquad \text{Equate numerators.}$$

$$(2ax + b)^2 = b^2 - 4ac \qquad \text{Factor.}$$

$$2ax + b = \pm\sqrt{b^2 - 4ac} \qquad \text{Extract roots.}$$

$$2ax = -b \pm \sqrt{b^2 - 4ac} \qquad \text{Subtract } b.$$

$$x = \frac{-b \pm \sqrt{b^2 - 4ac}}{2a}. \qquad \text{Divide by } 2a.$$

The last line is known as the **Quadratic Formula.** Let us summarize:

The Quadratic Formula

The second-degree ("quadratic") equation

$$ax^2 + bx + c = 0, \quad a \neq 0,$$

has solutions given by the Quadratic Formula:

$$x = \frac{-b \pm \sqrt{b^2 - 4ac}}{2a}.$$

EXAMPLE 1

Solve $2x^2 + 3x - 5 = 0$ by using the Quadratic Formula.

Solution

The coefficients are $a = 2$, $b = 3$, and $c = -5$. Substituting into the formula gives

$$x = \frac{-3 \pm \sqrt{3^2 - 4(2)(-5)}}{2(2)}$$

$$= \frac{-3 \pm \sqrt{49}}{4} \longleftarrow \quad 9 + 40 = 49$$

$$= \frac{-3 \pm 7}{4}$$

$$= \frac{-3 + 7}{4}, \frac{-3 - 7}{4}$$

$$x = 1, \frac{-5}{2}. \longleftarrow \quad \frac{4}{4} = 1, \frac{-10}{4} = \frac{-5}{2}$$

EXAMPLE 2

Solve $4x^2 - 12x + 9 = 0$ by using the Quadratic Formula.

Solution

$$x = \frac{-(-12) \pm \sqrt{(-12)^2 - 4 \cdot 4 \cdot 9}}{2 \cdot 4} \qquad a = 4, b = -12, c = 9$$

$$= \frac{12 \pm \sqrt{0}}{8} \longleftarrow \quad 144 - 144 = 0$$

$$x = \frac{3}{2}.$$

In the next two examples we must first put the equations into the standard form $ax^2 + bx + c = 0$ before applying the Quadratic Formula.

EXAMPLE 3

Solve $(2y - 1)(y + 4) = y(y + 9)$ by using the Quadratic Formula.

Solution

$2y^2 + 7y - 4 = y^2 + 9y$	Multiply out.
$y^2 - 2y - 4 = 0$	Standard form.
$y = \dfrac{-(-2) \pm \sqrt{(-2)^2 - 4(1)(-4)}}{2 \cdot 1}$	$a = 1,\ b = -2,\ c = -4$
$= \dfrac{2 \pm \sqrt{20}}{2}$ ⟵	$4 + 16 = 20$
$= \dfrac{2 \pm 2\sqrt{5}}{2}$ ⟵	$\sqrt{20} = 2\sqrt{5}$
$= \dfrac{2(1 \pm \sqrt{5})}{2}$	Factor.
$y = 1 \pm \sqrt{5}.$	Answers.

EXAMPLE 4

Solve $\dfrac{2}{z} - \dfrac{3}{z - 1} = 5$ by using the Quadratic Formula.

Solution

$\dfrac{2 \cdot (z - 1)}{z \cdot (z - 1)} - \dfrac{3 \cdot z}{(z - 1) \cdot z} = \dfrac{5 \cdot z(z - 1)}{1 \cdot z(z - 1)}$ ⟵	L.C.D. $= z(z - 1)$
$2z - 2 - 3z = 5z^2 - 5z$	Equate numerators.
$0 = 5z^2 - 4z + 2$	Standard form.
$z = \dfrac{-(-4) \pm \sqrt{(-4)^2 - 4 \cdot 5 \cdot 2}}{2 \cdot 5}$	$a = 5,\ b = -4,\ c = 2$
$= \dfrac{4 \pm \sqrt{-24}}{10}$ ⟵	$16 - 40 = -24$
$= \dfrac{4 \pm 2i\sqrt{6}}{10}$ ⟵	$i\sqrt{24} = 2i\sqrt{6}$
$= \dfrac{\cancel{2}(2 \pm i\sqrt{6})}{\cancel{10}_5}$	Factor.
$z = \dfrac{2 \pm i\sqrt{6}}{5}.$	Answers.

The **discriminant** of the equation $ax^2 + bx + c = 0$ is the value of $b^2 - 4ac$. This is just the expression under the radical in the Quadratic Formula:

$$x = \dfrac{-b \pm \sqrt{b^2 - 4ac}}{2a} \longleftarrow \text{Discriminant}$$

We illustrate with Examples 1, 2, 3, and 4:

Equation	Discriminant $b^2 - 4ac$	Factorable?	Types of solutions
$2x^2 + 3x - 5 = 0$	$3^2 - 4(2)(-5) = 49$	Yes: $(2x + 5)(x - 1) = 0$	2 rationals: $x = -5/2,\ 1$
$4x^2 - 12x + 9 = 0$	$(-12)^2 - 4 \cdot 4 \cdot 9 = 0$	Yes: $(2x - 3)(2x - 3) = 0$	1 rational: $x = 3/2$
$y^2 - 2y - 4 = 0$	$(-2)^2 - 4(1)(-4) = 20$	No	2 irrationals: $y = 1 \pm \sqrt{5}$
$5z^2 - 4z + 2 = 0$	$(-4)^2 - 4 \cdot 5 \cdot 2 = -24$	No	2 complex: $z = \dfrac{2 \pm i\sqrt{6}}{5}$

The value of the discriminant tells us whether or not the equation, with *integer coefficients,* is solvable by factoring, and what type of solutions it will have. Using these examples as prototypes, we can make the following generalizations:

Value of Discriminant $b^2 - 4ac$	Factorable?	Types of Solutions
1. Positive, perfect square	Yes	2 rationals
2. Zero	Yes	1 rational
3. Positive, not a perfect square	No	2 irrationals
4. Negative	No	2 complex

EXAMPLE 5 Compute the discriminant of each of the following equations. Then tell whether or not the equation is factorable, and what type of solutions it will have. Do *not* solve the equations.

a. $3x^2 - 7x - 9 = 0$

$$\text{The discriminant} = (-7)^2 - 4(3)(-9)$$
$$= 49 + 108$$
$$= 157.$$

Because 157 is not a perfect square, category 3 says that this equation is *not factorable* and has *2 irrational* solutions.

(continued)

b. $6x^2 + 37x + 56 = 0$

$$\text{The discriminant} = 37^2 - 4 \cdot 6 \cdot 56$$
$$= 1369 - 1344$$
$$= 25.$$

Because 25 is a positive, perfect square, category 1 says that this equation *is factorable* and has *2 rational* solutions. (Can you factor it?)

EXERCISE 6.4   *Answers, pages A61–A62*

Solve by using the Quadratic Formula, as in **EXAMPLES 1–4:**

1. $x^2 - 3x - 18 = 0$ **2.** $x^2 - 5x - 14 = 0$ **3.** $2y^2 = 3 - 5y$

4. $4 - 7z = 2z^2$ **5.** $w = 12 - 6w^2$ **6.** $6 = 19y - 10y^2$

7. $25u^2 = 10u - 1$ **8.** $9v^2 + 6v + 1 = 0$ **9.** $3x^2 + 2x = 0$ *Hint:* $3x^2 + 2x + 0 = 0$

10. $5y^2 = 3y$ **11.** $4z^2 - 9 = 0$ *Hint:* $4z^2 + 0z - 9 = 0$ **12.** $16x^2 - 25 = 0$

13. $3x^2 + 16 = 0$ **14.** $5y^2 - 9 = 0$ **15.** $x^2 + x - 12 = 0$

16. $y^2 - y - 2 = 0$ **17.** $z^2 = 4z - 1$ **18.** $y^2 = 6y - 7$

19. $9x^2 + 3x + 1 = 0$ **20.** $y^2 - 2y + 4 = 0$ **21.** $z^2 - 6z - 3 = 0$

22. $w^2 + 4w - 4 = 0$ **23.** $2y + 1 = 2y^2$ **24.** $5 - 3z^2 = 4z$

25. $(4x + 3)(2x + 7) = (x + 20)(x + 2)$

26. $(5y + 7)(1 - 2y) = (1 - 3y)(y + 5)$

27. $5x + (3x - 2)^2 = x^2 - (2x - 7)(3x + 1)$

28. $x^2 - (2 - 5x)^2 = (x + 3)(7 - 2x) - 9(x - 3)$

29. $\dfrac{2x + 4}{5x - 7} = \dfrac{6x + 4}{7x - 8}$ **30.** $\dfrac{2y - 1}{y - 2} = \dfrac{7y - 11}{5y - 12}$ **31.** $\dfrac{15}{t - 1} - \dfrac{10}{t + 2} = \dfrac{2}{3}$ **32.** $\dfrac{8}{x - 2} + \dfrac{7}{x + 3} = \dfrac{1}{5}$

33. $\dfrac{25}{r} + \dfrac{30}{r + 10} = 1$ **34.** $\dfrac{15}{r} + \dfrac{30}{r + 15} = \dfrac{7}{6}$

Compute the discriminant of each the following equations. Then tell whether or not the equation is factorable, what type of solution it has, and how many solutions. Do *not* solve the equations.

35. $x^2 - 6x - 9 = 0$ **36.** $4x^2 + x - 8 = 0$ **37.** $4y^2 - 20y + 25 = 0$ **38.** $16z^2 + 24z + 9 = 0$

39. $w^2 + 6w + 10 = 0$ **40.** $2t^2 - 3t + 5 = 0$ **41.** $18x^2 - 39x + 20 = 0$ **42.** $15y^2 + 53y + 42 = 0$

43. $25t^2 - 131t + 162 = 0$ **44.** $25t^2 + 130t + 162 = 0$

B

Solve by the Quadratic Formula:

45. $x^2 + 2\sqrt{2}\,x - 7 = 0$ *Hint:* $b = 2\sqrt{2}$

46. $x^2 - 2\sqrt{3}\,x - 8 = 0$

47. $\sqrt{2}\,x^2 + 4x - 2\sqrt{2} = 0$

48. $\sqrt{3}\,y^2 - 9y + 3\sqrt{3} = 0$

49. $x^2 - 3ix + 4 = 0$ *Hint:* $b = -3i$

50. $ix^2 + 4x - 2i = 0$

51. $40x^2 - 123x + 77 = 0$

52. $18y^2 + 79y + 56 = 0$

53. $-36x^2 - 19x + 220 = 0$

54. $-16t^2 + 96t + 48 = 0$

55. $12x^2 + 83x + 126 = 0$

56. $25x^2 - 20x + 53 = 0$

57. $2 = \dfrac{3}{x} + \dfrac{9}{2x^2}$

58. $25 = 5x^{-1} - x^{-2}$

Set up and then solve by the Quadratic Formula:

59. The difference between a number and its reciprocal is one. Find the number.

60. The square of a number is three times the difference between two and the number. Find the number.

61. When a ball is thrown upward, its height h above the ground (in feet) after t seconds is given by the formula

$$h = -16t^2 + 48t + 192.$$

When does the ball strike the ground? *Hint:* See the hint Exercise 3.5, Problem 73. Give your answer both expressed as a radical and approximated to the nearest tenth of a second. (See Section 6.3, Example 4, for the calculator sequence.)

C

62. The distance traveled by a falling rock is given by

$$d = 16t^2,$$

where d is measured in feet and t is measured in seconds. A person drops a rock into a well, and the sound of the splash reaches the observer 3 seconds after the rock is dropped. The distance (in feet) traveled by the sound is given by

$$d = (1080)(\text{time of the sound}).$$

How deep is the well? *Hint:* Let t = time for rock to hit bottom of well; then $3 - t$ = time for sound to reach observer. Express your answer to the nearest hundredth.

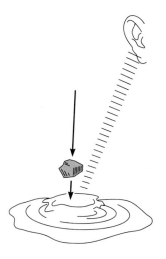

63. The **Golden Rectangle,** according to the ancient Greeks, is the rectangle whose shape, or length-to-width ratio, is the most pleasing to the eye. Let G = its length and 1 = its width; then $\frac{G}{1} = G$ is called the "Golden ratio."

We will see below that G is approximately 1.6. Artists and photographers try to frame their works so that the length-to-width ratio is 1.6, a number that is in harmony with the distance our eyes are set apart horizontally. The important property of the Golden Rectangle is that if a square is cut away from it, as shown below, the small rectangle remaining has the *same* length-to-width ratio as the original rectangle.

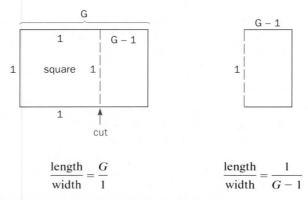

$$\frac{\text{length}}{\text{width}} = \frac{G}{1} \qquad\qquad \frac{\text{length}}{\text{width}} = \frac{1}{G-1}$$

Use the fact that these ratios are equal to show that the

$$\text{Golden Ratio } G = \frac{1 + \sqrt{5}}{2} \approx 1.618.$$

64. In the derivation of the Quadratic Formula from the equation $ax^2 + bx + c = 0$, the first three steps given in the text are

$$\frac{ax^2}{a} + \frac{bx}{a} + \frac{c}{a} = \frac{0}{a}$$

$$x^2 + \frac{bx}{a} = \frac{-c}{a}$$

$$x^2 + \frac{bx}{a} + \frac{b^2}{4a^2} = \frac{-c}{a} + \frac{b^2}{4a^2}.$$

In the text we then used the L.C.D. to rewrite the equation without the denominator. This can also be accomplished by now factoring the left side of the equation and combining the right side:

$$\left(x + \frac{b}{2a}\right)^2 = \frac{-c \cdot 4a}{a \cdot 4a} + \frac{b^2}{4a^2}.$$

You finish the derivation of the Quadratic Formula by solving this last step for x, using the extraction of roots method.

65. The equation $ax^2 + bx + c = 0$ has two solutions given by the Quadratic Formula:

$$x_1 = \frac{-b + \sqrt{b^2 - 4ac}}{2a} \quad \text{and} \quad x_2 = \frac{-b - \sqrt{b^2 - 4ac}}{2a}.$$

a. By adding these expressions, show that $x_1 + x_2 = \frac{-b}{a}$.

b. By multiplying these expressions, show that $x_1 \cdot x_2 = \frac{c}{a}$.   *Hint:* FOIL.

66. The Table of Discriminants given in the text is valid for equations $ax^2 + bx + c = 0$, with a, b, and c integers (or rational numbers). It is false in other cases.

a. Show that the discriminant of $x^2 - 2\sqrt{2}x - 7 = 0$ is a positive, perfect square (category 1 on the Table).

b. Solve this equation, and show that its solutions are irrational, in apparent violation of category 1.

67. Repeat Problem 66 for the equation $x^2 + 2ix - 2 = 0$.

68. The equation $ax^2 + bx + c = 0$ has solutions given by the Quadratic Formula:

$$x = \frac{-b \pm \sqrt{b^2 - 4ac}}{2a}, \quad a \neq 0.$$

Tell what happens in the case $a = 0$.

6.5   Equations Containing Radicals

The title of this section suggests square roots and cube roots. Indeed so, and we can eliminate them by squaring or cubing both sides of the equation.

EXAMPLE 1   Solve $\sqrt[3]{3x - 5} = 4$.

Solution

$$(\sqrt[3]{3x - 5})^3 = 4^3 \qquad \text{Cube both sides.}$$
$$3x - 5 = 64 \qquad (\sqrt[3]{A})^3 = A$$
$$3x = 69$$
$$x = 23.$$

Check: $\sqrt[3]{3 \cdot 23 - 5} \overset{?}{=} 4$ **Answer:** $x = 23$.

$$\sqrt[3]{64} \overset{\checkmark}{=} 4.$$

EXAMPLE 2 | Solve $2\sqrt{y-1} + 4 = y$.

Solution

We first isolate the radical term, and then square both sides:

$$2\sqrt{y-1} = y - 4 \qquad \text{Isolate radical term.}$$
$$(2\sqrt{y-1})^2 = (y-4)^2 \qquad \text{Square both sides.}$$
$$4(y-1) = y^2 - 8y + 16$$
$$4y - 4 = y^2 - 8y + 16$$
$$0 = y^2 - 12y + 20$$
$$0 = (y-10)(y-2)$$
$$y = 10, \quad y = 2.$$

Check $y = 10$: $2\sqrt{10-1} + 4 \overset{?}{=} 10$ **Check** $y = 2$: $2\sqrt{2-1} + 4 \overset{?}{=} 2$

$\qquad\qquad\qquad 2\sqrt{9} + 4 \overset{?}{=} 10 \qquad\qquad\qquad\qquad\quad 2\sqrt{1} + 4 \overset{?}{=} 2$

$\qquad\qquad\qquad\quad 6 + 4 \overset{\checkmark}{=} 10. \qquad\qquad\qquad\qquad\qquad\quad 2 + 4 \neq 2.$

Answer: $y = 10$ ($y = 2$ is extraneous).

The extraneous solution* $y = 2$ arose when we squared both sides in the second step. It is a solution of the "squared equation" $4(y-1) = y^2 - 8y + 16$, but not of the original equation. This is typical of equations containing radicals, so you will have to check all of your answers in this section. Let us summarize:

> Solutions to the equation
>
> $$A(x) = B(x),$$
>
> if any, are among solutions to the equation
>
> $$[A(x)]^n = [B(x)]^n,$$
>
> where $n = 2, 3, 4,\ldots$. However, solutions to the latter *might not* be solutions to the former. These are called *extraneous* solutions. You *must check* all answers by substituting into the original equation.

A clear-cut case is illustrated by the equation $2x = 6$, whose only solution is $x = 3$. However, squaring both sides gives $4x^2 = 36$, or $x^2 = 9$, or $x = \pm 3$. So $x = -3$ satisfies the second equation but not the original equation. Thus $x = -3$ is an extraneous solution.

---

*Extraneous solutions to equations were discussed in Section 4.5.

EXAMPLE 3 | Solve $\sqrt{2x-1}+\sqrt{x+3}=3$ and check all answers:

Solution

This equation is best solved by first separating the radicals and then squaring both sides:

$$\sqrt{2x-1}=3-\sqrt{x+3}$$ Separate radicals.

$$(\sqrt{2x-1})^2=(3-\sqrt{x+3})^2$$ Square both sides.
$(A-B)^2=A^2-2AB+B^2$

$$2x-1=9\underline{-6\sqrt{x+3}}+x+3$$ Note middle term!

$$2x-1-9-x-3=\ -6\sqrt{x+3}$$ Isolate radical.

$$x-13=\ -6\sqrt{x+3}$$ Combine like terms.

$$(x-13)^2=(-6\sqrt{x+3})^2$$ Square both sides again!

$$x^2-26x+169=36(x+3)$$

$$x^2-26x+169=36x+108$$

$$x^2-62x+61=0$$

$$(x-61)(x-1)=0$$

$$x=61,\quad x=1.$$

Check $x=61$: $\sqrt{2\cdot61-1}+\sqrt{61+3}\overset{?}{=}3$ **Check** $x=1$: $\sqrt{2\cdot1-1}+\sqrt{1+3}\overset{?}{=}3$

$\qquad\qquad\quad\sqrt{121}\ +\ \sqrt{64}\ =3\qquad\qquad\qquad\quad\sqrt{1}\ +\ \sqrt{4}\ =3$

$\qquad\qquad\qquad 11\ \ +\ \ 8\quad\neq3.\qquad\qquad\qquad\qquad\ \ 1\ +\ 2\ \ \overset{\checkmark}{=}3.$

Answer: $x=1$ ($x=61$ is extraneous).

WARNING! The second and third lines of Example 3 are

$$(\sqrt{2x-1})^2=(3-\sqrt{x+3})^2$$

$$2x-1=9-6\sqrt{x+3}+x+3.$$

The right side was obtained from $(A-B)^2=A^2-2AB+B^2$. See also Section 5.4, Example 2d. Students often *incorrectly* omit the middle term, $-6\sqrt{x+3}$, and simply square each term inside the parentheses. Thus

$$(3-\sqrt{x+3})^2\neq9+x+3.$$

Don't you make this mistake!

EXERCISE 6.5   *Answers, page A62*

A

Solve as in EXAMPLES 1 and 2, by first isolating the radical term, if necessary. Check your answers. Be sure to list any extraneous solutions.

1. $\sqrt[3]{2x+3}=-5$

2. $\sqrt[3]{2x-4}=6$

3. $\sqrt{2x-1}=9$

4. $\sqrt{3-2x}=5$

5. $\sqrt{y^2-y+2}=2$

6. $\sqrt[3]{y^2-y+2}=2$

7. $2\sqrt{u+4}=u-4$

8. $3\sqrt{2v-1}=v+4$

9. $x+2=(8x+1)^{1/2}$

10. $(2x - 6)^{1/2} = x - 3$ *Hint:* $(\quad)^{1/2} = \sqrt{\quad}$ **11.** $w\sqrt{3} = 3\sqrt{w}$ **12.** $x = \sqrt{x}$

13. $\sqrt{x}\sqrt{x - 5} = 6$ **14.** $\sqrt{x}\sqrt{2x + 3} = 3$ **15.** $\sqrt{4 - x}\sqrt{2 - x} = 2\sqrt{6}$

16. $\sqrt{2 - 3t}\sqrt{2 - t} = 4\sqrt{2}$ **17.** $\sqrt{x + 2} - x = 0$ **18.** $\sqrt{3 - x} - 2x = 0$

19. $\sqrt{5x + 1} + x = 7$ **20.** $\sqrt{3y + 1} + 3 = y$ **21.** $\dfrac{\sqrt{27 + x^2}}{2x} = 1$

22. $\dfrac{x}{\sqrt{16 - x^2}} - 1 = 0$ **23.** $\sqrt{y + 3} - \dfrac{1}{\sqrt{y + 3}} = 0$ **24.** $(2z - 1)^{1/2} - \dfrac{5}{(2z - 1)^{1/2}} = 0$

Solve as in EXAMPLE 3, by first separating radicals, if necessary. Check your answers. Be sure to list any extraneous solutions.

25. $\sqrt{x + 4} = 5 - \sqrt{x - 1}$ **26.** $\sqrt{x + 3} = 1 + \sqrt{x - 2}$ **27.** $\sqrt{z + 7} - \sqrt{z} = 1$

28. $\sqrt{w - 7} + \sqrt{w} = 9$ **29.** $\sqrt{2y - 1} - \sqrt{y - 4} = 2$ **30.** $\sqrt{2z + 1} + \sqrt{z - 3} = 4$

31. $(3x + 1)^{1/2} + (x - 1)^{1/2} = 6$ **32.** $(y + 1)^{1/2} - (y - 4)^{1/2} = 1$

B

Solve using any method, and be sure to check your answers. List any extraneous solutions.

33. $m - 5\sqrt{m} - 14 = 0$ **34.** $n - 7\sqrt{n} + 6 = 0$

35. $v - 3 = \sqrt{\dfrac{3v - 7}{2}}$ **36.** $6 - u = \dfrac{\sqrt{3u + 4}}{2}$

37. $\sqrt{7 + \sqrt{x + 2}} = 3$ *Hint:* Square outer radical, then isolate inner radical.   **38.** $\sqrt{6 - \sqrt{x - 1}} = 2$

39. $\sqrt{x - \sqrt{x - 9}} = 3$ **40.** $\sqrt{x + \sqrt{x - 7}} = 3$

41. $\sqrt[3]{7 + \sqrt{y - 1}} = 2$ **42.** $\sqrt[3]{\sqrt{y + 7} - 30} = -3$

43. $\sqrt{x - 3} + \sqrt{x} = \sqrt{4x - 5}$ **44.** $\sqrt{x - 5} + \sqrt{x} = \sqrt{5x - 5}$

45. $\sqrt{3w + 1} - \sqrt{w - 4} = \sqrt{w + 4}$ **46.** $\sqrt{2z + 4} - \sqrt{z + 3} = \sqrt{z - 5}$

Solve for the indicated variable:

47. $r = \sqrt{\dfrac{A}{\pi}}$;   for A **48.** $r = \sqrt[3]{\dfrac{3V}{4\pi}}$;   for V **49.** $d = \sqrt{\dfrac{Gm_1m_2}{F}}$;   for F

50. $T = 2\pi\sqrt{\dfrac{L}{g}}$;   for $L, L > 0$ **51.** $y = \sqrt{r^2 - x^2}$;   for x **52.** $x = \dfrac{3\sqrt{y^2 + 4}}{2}$;   for y

The following word problem contains phrases with radicals.

> ''The square root of the sum of a number and seven is one more than the square root of the number. Find the number.''

This translates into

$$\sqrt{x + 7} = \sqrt{x} + 1.$$

(continued on next page)

By squaring both sides and then solving, we get $x = 9$.

Check: $\sqrt{9+7} = \sqrt{9} + 1,$ or $\sqrt{16} \overset{\checkmark}{=} 3 + 1.$

Now set up and solve these word problems, and check your answers.

53. The square root of a number is six less than the number. Find the number.

54. What number is two more than its own square root?

55. Twice the square root of a number is three less than the number. Find the number.

56. The square root of the sum of a number and 3 is three less than the number. Find the number.

57. When the square root of the sum of a number and 21 is decreased by one, the result is the number itself. What is the number?

58. What number is twice its own square root?

59. When one is added to the square root of a number, the result is the same as the square root of the sum of the number and seven. Find the number.

60. When the square root of the sum of a number and five is decreased by the square root of the difference between the number and 3, the result is the square root of the number itself. Find the number!

In Problems 61 and 62, use the Pythagorean Theorem to solve for x in the right triangles shown. Then use your answers to find the lengths of the three sides:

61.

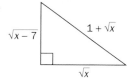

62.

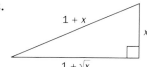

C

63. The picture contains unknowns x, y, and z.
   a. Express y as a radical in terms of x, using the Pythagorean Theorem.
   b. Express z in terms of x, using the length 10 that is shown.
   c. Given that
$$y + z = 14,$$
   write an equation involving only x by substituting your results from parts a and b.
   d. Solve this radical equation for x.
   e. Use your value of x to find y and z.

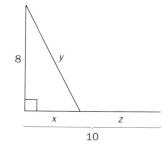

64. The picture contains unknowns x, y, and z.
  a. Express y as a radical in terms of x, and similarly express z in terms of x, using the Pythagorean Theorem.
  b. Given that
$$y + z = 28,$$
write an equation involving only x by substituting your results from part a.
  c. Solve this radical equation for x.
  d. Use your value of x to find y and z.

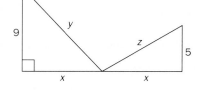

65. What is wrong with this "proof" that $-1 = 1$?

$$\begin{array}{lll} \text{Let} & x = 1 & \\ \text{Then} & x^2 = 1^2 & \text{Square both sides.} \\ & x^2 - 1 = 0 & \text{Subtract 1.} \\ & (x+1)(x-1) = 0 & \text{Factor.} \\ & x = -1, \quad x = 1 & \text{Solutions.} \end{array}$$

Now substitute the solution $x = -1$ into the original equation $x = 1$. This gives $-1 = 1$.

6.6   **Combined Methods of Solving Equations**

In this section, we will use all the techniques we have learned for solving equations. In our first example, we start with the most important method of all—factoring.

EXAMPLE 1   Solve $2x^5 - 8x^3 - 64x = 0$.

Solution
This is similar to Exercise 3.5, Problems 81–92. We begin by factoring:

$$\begin{array}{lll} 2x(x^4 - 4x^2 - 32) = 0 & & \text{Common factor.} \\ 2x(x^2 - 8)(x^2 + 4) = 0 & & \text{Factor trinomial.} \end{array}$$

$$\begin{array}{llll} 2x = 0 & x^2 = 8 & x^2 = -4 & \\ & x = \pm\sqrt{8} & x = \pm\sqrt{-4} & \text{Extract roots.} \\ x = 0, & x = \pm 2\sqrt{2}, & x = \pm 2i & \text{Answers.} \end{array}$$

EXAMPLE 2   Solve $x^6 - 7x^3 - 8 = 0$.

Solution
We start by factoring the trinomial:

$$(x^3 - 8)(x^3 + 1) = 0.$$

Next we factor the difference and the sum of cubes, as in Section 3.4:

$$(x - 2)(x^2 + 2x + 4)(x + 1)(x^2 - x + 1) = 0.$$
$$x = 2, \quad \text{see below}; \quad x = -1, \quad \text{see below}$$

$x^2 + 2x + 4 = 0$ is best solved by completing the square, because the middle coefficient, 2, is an even integer:

$$x^2 + 2x \qquad = -4$$
$$x^2 + 2x + 1 = -4 + 1 \qquad \text{Add } \left(\frac{2}{2}\right)^2 = 1$$
$$(x + 1)^2 = -3$$
$$x + 1 = \pm\sqrt{-3}$$
$$x = -1 \pm i\sqrt{3}.$$

$x^2 - x + 1 = 0$ is best solved by using the Quadratic Formula, because the middle coefficient, -1, is an odd integer:

$$x = \frac{-(-1) \pm \sqrt{(-1)^2 - 4 \cdot 1 \cdot 1}}{2 \cdot 1} \qquad a = 1, b = -1, c = 1$$
$$x = \frac{1 \pm \sqrt{-3}}{2}$$
$$x = \frac{1 \pm i\sqrt{3}}{2}.$$

Answers: $x = 2, \quad -1, \quad -1 \pm i\sqrt{3}, \quad \dfrac{1 \pm i\sqrt{3}}{2}.$

WARNING! In the first step of the preceding,

$$(x^3 - 8)(x^3 + 1) = 0.$$

it is tempting to solve by "extracting cube roots":

$$x^3 = 8 \qquad \text{or} \quad x^3 = -1$$
$$x = \sqrt[3]{8} \quad \text{or} \quad x = \sqrt[3]{-1} \qquad \textit{Note:} \ x \neq \pm\sqrt[3]{8} = \pm 2$$
$$x = 2 \quad \text{or} \quad x = -1. \qquad \text{because } (-2)^3 \neq 8$$

Unfortunately, this "method" *omits* the four complex solutions obtained above, $x = -1 \pm i\sqrt{3}$ and $x = \dfrac{1 \pm i\sqrt{3}}{2}$. Therefore, the correct method begins with the *complete* factorization of these cubes:

$$(x^3 - 8)(x^3 + 1) = 0$$
$$(x - 2)(x^2 + 2x + 4)(x + 1)(x^2 - x + 1) = 0$$

The *six* solutions are then obtained as in the preceding example.

EXAMPLE 3 | Solve $8x^5 - 16x^3 + 27x^2 - 54 = 0$.

Solution

We start by factoring by grouping, as in Section 3.2, Example 4:

$$8x^3(x^2 - 2) + 27(x^2 - 2) = 0$$
$$(x^2 - 2)(8x^3 + 27) = 0$$

Next we factor the sum of cubes and then solve:

$$(x^2 - 2)(2x + 3)(4x^2 - 6x + 9) = 0$$
$$x^2 = 2 \qquad 2x = -3$$
$$x = \pm\sqrt{2}, \quad x = \frac{-3}{2}, \quad \text{see below}$$

$4x^2 - 6x + 9 = 0$ is best solved by using the Quadratic Formula:

$$x = \frac{-(-6) \pm \sqrt{(-6)^2 - 4 \cdot 4 \cdot 9}}{2 \cdot 4}$$

$$x = \frac{3 \pm 3i\sqrt{3}}{4}.$$

You fill in the details.

Answers: $x = \pm\sqrt{2}, \quad \dfrac{-3}{2}, \quad \dfrac{3 \pm 3i\sqrt{3}}{4}$.

EXAMPLE 4 | Solve $x^{2/3} + 5x^{1/3} - 14 = 0$.

Solution

This is not a polynomial equation, but it can be solved by first factoring it as a trinomial:

$$(x^{1/3} + 7)(x^{1/3} - 2) = 0$$
$$x^{1/3} = -7, \quad x^{1/3} = 2.$$

Check by FOIL:
$$x^{1/3}x^{1/3} - 2x^{1/3} + 7x^{1/3} - 14$$
$$\overset{\checkmark}{=} x^{2/3} + 5x^{1/3} - 14$$

We have solved for $x^{1/3}$; in order to solve for x, we cube both sides of these last two equations:

$$(x^{1/3})^3 = (-7)^3, \quad (x^{1/3})^3 = 2^3$$
$$x = -343, \qquad\qquad x = 8. \qquad (x^{1/3})^3 = x^{3/3} = x$$

Answers: $x = -343, \quad 8$.

EXERCISE 6.6 *Answers, pages A62–A63*

Solve by first factoring, as in **EXAMPLES 1 and 2**:

1. $x^4 - 5x^2 - 36 = 0$ **2.** $x^4 - 12x^2 + 32 = 0$ **3.** $4y^4 + 5y^2 - 6 = 0$

4. $9y^4 + 11y^2 + 2 = 0$ **5.** $2x^5 - 18x^3 - 72x = 0$ **6.** $3z^6 + 12z^4 - 135z^2 = 0$

7. $t^3 + 2t^2 - 11t = 0$ **8.** $5v^3 - 20v^2 - 20v = 0$ **9.** $x^6 + 7x^3 - 8 = 0$

10. $x^6 - 26x^3 - 27 = 0$ **11.** $x^6 - 1 = 0$ *Hint:* $(x^3 + 1)(x^3 - 1)$ **12.** $x^6 - 64 = 0$

Solve by first factoring by grouping, as in **EXAMPLE 3**:

13. $2x^3 + 5x^2 - 16x - 40 = 0$ **14.** $3y^3 - y^2 + 36y - 12 = 0$

15. $z^4 - 5z^3 + 8z - 40 = 0$ **16.** $2w^4 + 3w^3 - 16w - 24 = 0$

17. $3t^5 + 4t^4 - 3t - 4 = 0$ **18.** $2v^5 + 3v^4 - 162v - 243 = 0$

Solve by first factoring, as in **EXAMPLE 4**:

19. $x^{2/3} + 2x^{1/3} - 8 = 0$ **20.** $y^{2/3} + 10y^{1/3} + 21 = 0$

21. $z^{2/3} - 12z^{1/3} + 35 = 0$ **22.** $w^{2/3} - w^{1/3} - 6 = 0$

23. $x^{2/3} - 25 = 0$ **24.** $u^{2/3} - 4 = 0$

25. $x - 5x^{1/2} + 6 = 0$ *Hint:* $(x^{1/2} - ?)(x^{1/2} - ?)$ **26.** $x - 6x^{1/2} + 8 = 0$

B

Solve by any method:

27. $\dfrac{1}{x^2} + \dfrac{6}{x} - 27 = 0$ **28.** $\dfrac{2}{x^2} - \dfrac{13}{x} + 20 = 0$

29. $y^{-4} - 4y^{-2} + 3 = 0$ **30.** $3z^{-4} + 4z^{-2} - 32 = 0$

31. $16x^4 - 81 = 0$ **32.** $3x^5 - 3x = 0$

33. $3z^6 + 23z^4 = 36z^2$ **34.** $2w^5 = 3w^3 + 27w$

35. $24y^7 + 105y^4 + 81y = 0$ **36.** $16y^8 - 38y^5 - 54y^2 = 0$

37. $z^9 - 17z^5 + 16z = 0$ **38.** $2w^{11} - 4w^7 + 2w^3 = 0$

39. $x^5 + 4x^3 - x^2 - 4 = 0$ **40.** $3y^5 - 25y^3 + 24y^2 - 200 = 0$

41. $z^6 + z^4 - z^2 - 1 = 0$ **42.** $w^6 + 2w^4 - 81w^2 - 162 = 0$

43. $3x^{2/3} - x^{1/3} - 4 = 0$ **44.** $2y^{2/3} + 5y^{1/3} - 12 = 0$

45. $x^{-2/3} + 5x^{-1/3} - 14 = 0$ **46.** $v^{-2/3} - 10v^{-1/3} + 16 = 0$

47. $x^2 - 8x^{-2} + 7 = 0$ **48.** $x - 5x^{-1} - 36x^{-3} = 0$

49. $x\sqrt{x^2 - 7} = 3\sqrt{2}$ Check your answers.

50. $\dfrac{4}{x} = \sqrt{x^2 - 6}$

51. $\sqrt{9 + x^2} = \dfrac{2x^2}{\sqrt{9 + x^2}}$ Check your answers.

52. $\sqrt{6 - x^2} - \dfrac{2x^2}{\sqrt{6 - x^2}} = 0$

Certain equations can be solved by a *substitution method*. **For example,**

$$(x^2 - 2x)^2 - 11(x^2 - 2x) + 24 = 0$$

can be solved by first letting

$$U = x^2 - 2x.$$

Substituting this into the first equation gives the simpler equation

$$U^2 - 11U + 24 = 0.$$

We now solve for U by factoring:

$$(U - 8)(U - 3) = 0$$
$$U = 8, \quad U = 3.$$

We now *resubstitute* $x^2 - 2x$ for U and then solve for x:

$$
\begin{array}{ll}
x^2 - 2x = 8 & x^2 - 2x = 3 \\
x^2 - 2x - 8 = 0 & x^2 - 2x - 3 = 0 \\
(x - 4)(x + 2) = 0 & (x - 3)(x + 1) = 0 \\
x = 4, \quad x = -2, & x = 3, \quad x = -1. \qquad \text{Answers}
\end{array}
$$

Solve these equations using the substitution method:

53. $(x^2 - x)^2 - 14(x^2 - x) + 24 = 0$ **54.** $(x^2 + 2x)^2 - 7(x^2 + 2x) - 8 = 0$ **55.** $(y^2 - 2y)^2 + (y^2 - 2y) - 12 = 0$

56. $(z^2 + 4z)^2 + 3(z^2 + 4z) - 40 = 0$ **57.** $(t^2 + 2)^2 + 4(t^2 + 2) - 21 = 0$ **58.** $(w^2 - 3)^2 + (w^2 - 3) - 42 = 0$

C

59. $2(x^2 + 2x)^2 - 3(x^2 + 2x) - 5 = 0$

60. $6(y^2 - 2)^2 + (y^2 - 2) - 2 = 0$

61. $\left(\dfrac{x^2 - 8}{x}\right)^2 + 5\left(\dfrac{x^2 - 8}{x}\right) - 14 = 0$

62. $\left(\dfrac{x^2 - 6}{x}\right)^2 - 4\left(\dfrac{x^2 - 6}{x}\right) - 5 = 0$

63. $\left(r - \dfrac{10}{r}\right)^2 + 6\left(r - \dfrac{10}{r}\right) - 27 = 0$

64. $\left(s - \dfrac{12}{s}\right)^2 + 7\left(s - \dfrac{12}{s}\right) - 44 = 0$

65. $\left(\dfrac{x}{x + 1}\right)^2 + 3\left(\dfrac{x}{x + 1}\right) - 10 = 0$

66. $\left(\dfrac{2y}{y - 1}\right)^2 - 7\left(\dfrac{2y}{y - 1}\right) + 12 = 0$

Use the Pythagorean Theorem to solve for x. Then find the other unknown sides of the triangles.

67.

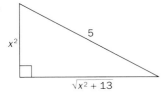

68.

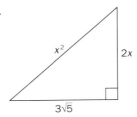

69.

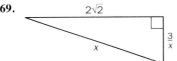

70.

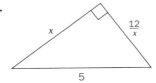

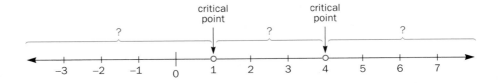

6.7   Quadratic and Rational Inequalities

In Sections 2.5 and 2.6, we saw that the solutions of linear and absolute-value inequalities consist of entire sets of numbers that were graphed on the real-number line. We now look at the quadratic inequality

$$x^2 - 5x + 4 > 0$$

and see how to find the correct sets. First, let us regard this as an equation and solve it:

$$x^2 - 5x + 4 = 0$$
$$(x - 1)(x - 4) = 0$$
$$x = 1, \quad x = 4.$$

These solutions are called **critical points.** Since the original inequality involved the symbol $>$, we plot them as open dots. This divides the number line into three intervals:

An easy way to decide which of these intervals is the correct solution set is to pick a **test point** from each and substitute it into the *original inequality*. If the result is TRUE, we take that interval as part of the solution; if it is FALSE, we do not take it. For the three intervals shown, let us select as test points $x = 0$, $x = 2$, and $x = 5$. These are chosen for convenience, though any other points will work equally well, *except* the critical points $x = 1$ and $x = 4$.

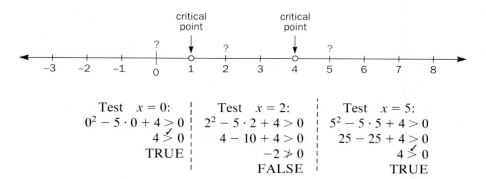

The answer consists of the TRUE intervals:

$\{x \mid x < 1\} \cup \{x \mid x > 4\}$:

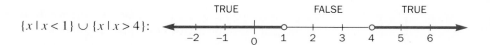

The nice feature of this method is that it obtains the answer and checks it at the same time. It can be shown that if a particular test point checks out TRUE, then any other test point chosen in that interval will also check out TRUE; a similar remark holds if it checks out FALSE. As mentioned above, *never* use a critical point as a test point. It is not "inside" any interval, so it would not give any information about an interval.

WARNING! The original inequality,

$$x^2 - 5x + 4 > 0,$$
$$\text{or} \quad (x - 1)(x - 4) > 0,$$

does *not* have the solution

$$x > 1, \quad x > 4.$$

You must use test points as we have described, in order to obtain the correct solution.

EXAMPLE 1

Solve and graph: $2x^2 - x \le 6$.

Solution

We first solve the equation, plot the critical points as closed dots and then use test points:

$$2x^2 - x - 6 = 0 \qquad \text{Solve the equation.}$$
$$(2x + 3)(x - 2) = 0$$

$$x = \frac{-3}{2}, \quad x = 2 \qquad \text{Critical points: plot them as closed dots. Why?}$$

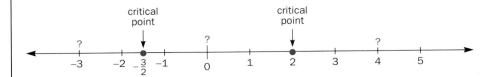

Test $x = -3$:	Test $x = 0$:	Test $x = 4$:
$2(-3)^2 - (-3) - 6 \le 0$	$2 \cdot 0^2 - 0 - 6 \le 0$	$2 \cdot 4^2 - 4 - 6 \le 0$
$18 + 3 - 6 \le 0$	$-6 \overset{\checkmark}{\le} 0$	$32 - 4 - 6 \le 0$
$15 \not\le 0$	TRUE	$22 \not\le 0$
FALSE		FALSE

The answer is the TRUE interval:

Answer: $\{x \mid -\frac{3}{2} \le x \le 2\}$:

The rational inequality $\dfrac{x - 3}{x + 1} > 0$ can be solved in a similar manner.

The *critical points* of this inequality are the values of x that make the numerator and the denominator equal to zero. We plot them and use test points as before.

EXAMPLE 2

Solve and graph: $\dfrac{x - 3}{x + 1} > 0, \quad x \ne -1$.

Solution

$$x - 3 = 0, \quad x + 1 = 0 \qquad \text{Set numerator and denominator} = 0$$
$$x = 3, \qquad x = -1. \qquad \text{Critical points: plot them as open dots.}$$

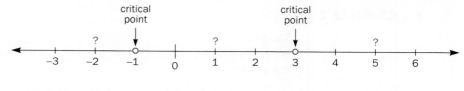

Test $x = -2$:

$$\frac{-2 - 3}{-2 + 1} > 0$$

$$\frac{-5}{-1} > 0$$

$$5 \overset{\checkmark}{>} 0$$

TRUE

Test $x = 1$:

$$\frac{1 - 3}{1 + 1} > 0$$

$$\frac{-2}{2} > 0$$

$$-1 \not> 0$$

FALSE

Test $x = 5$:

$$\frac{5 - 3}{5 + 1} > 0$$

$$\frac{2}{6} \overset{\checkmark}{>} 0$$

TRUE

Answer: $\{x \mid x < -1\} \cup \{x \mid x > 3\}$:

TRUE   FALSE   TRUE

EXAMPLE 3

Solve and graph: $\dfrac{x + 2}{x - 1} \geq 2, \quad x \neq 1$.

Solution

We write this as a single fraction with zero on one side:

$$\frac{x + 2}{x - 1} - 2 \geq 0$$

$$\frac{x + 2}{x - 1} - \frac{2 \cdot (x - 1)}{1 \cdot (x - 1)} \geq 0 \qquad \text{L.C.D.}$$

$$\frac{x + 2 - 2x + 2}{x - 1} \geq 0$$

$$\frac{4 - x}{x - 1} \geq 0 \qquad \text{Single fraction.}$$

$$\begin{aligned} 4 - x = 0, \quad & x - 1 = 0 \qquad \text{Set numerator and denominator} = 0. \\ 4 = x, \quad & x \quad = 1 \qquad \text{Critical points.} \end{aligned}$$

We plot $x = 4$ as a closed dot (included), because the numerator can equal zero. However, we must plot $x = 1$ as an *open* dot (excluded), because the denominator would . equal zero when $x = 1$, making the fraction undefined.

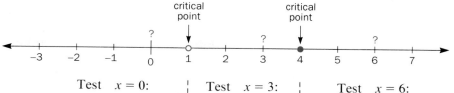

Test $x = 0$:   Test $x = 3$:   Test $x = 6$:

$$\frac{0 + 2}{0 - 1} \geq 2$$ $$\frac{3 + 3}{3 - 1} \geq 2$$ $$\frac{6 + 2}{6 - 1} \geq 2$$

$$-2 \not\geq 2$$ $$\frac{5}{2} \overset{\checkmark}{\geq} 2$$ $$\frac{8}{5} \not\geq 2$$

FALSE   TRUE   FALSE

Answer: $\{x \mid 1 < x \leq 4\}$:

EXERCISE 6.7   *Answers, pages A63–A64*

Solve and graph, as in EXAMPLE 1:

1. $x^2 - 5x - 6 > 0$ **2.** $x^2 - x - 12 \geq 0$ **3.** $x^2 \leq 3x + 10$ **4.** $x^2 < 6x - 5$

5. $2x^2 + 5x < 12$ **6.** $3x^2 - 3 \leq 8x$ **7.** $4 - x^2 \geq 0$ **8.** $1 - x^2 < 0$

9. $x^2 - 9 > 0$ **10.** $x^2 \leq 25$ **11.** $x^2 - 4x + 4 > 0$ **12.** $x^2 + 6x + 9 \geq 0$

13. $20x \geq 4x^2 + 25$ **14.** $12x > 4 + 9x^2$ **15.** $3x^2 > 3x$ **16.** $2x^2 > 6x$

17. $1 - 2x - 3x^2 \geq 0$ **18.** $6 - x - 2x^2 \leq 0$ **19.** $20 < (x + 3)(x - 5)$ **20.** $16 > (x + 4)(x - 2)$

Solve and graph, as in EXAMPLE 2:

21. $\dfrac{x + 2}{x - 5} > 0$ **22.** $\dfrac{x - 4}{x - 1} > 0$ **23.** $\dfrac{2x - 1}{x + 4} < 0$ **24.** $\dfrac{x - 2}{2x + 3} < 0$

25. $\dfrac{x - 3}{x} \geq 0$ **26.** $\dfrac{2x}{x + 2} \geq 0$ **27.** $0 > \dfrac{3}{3 + x}$ **28.** $0 < \dfrac{-2}{1 - x}$

Solve and graph, as in EXAMPLE 3:

29. $\dfrac{x + 3}{x - 3} > 2$ **30.** $\dfrac{2x + 3}{x - 2} > 1$ **31.** $\dfrac{x + 4}{x - 1} \leq -2$ **32.** $\dfrac{x + 2}{x + 1} \leq 3$

B

33. $\dfrac{1}{x} < \dfrac{1}{x-2}$

34. $\dfrac{1}{x-1} > \dfrac{1}{x+3}$

35. As stated in Exercise 2.5, Problem 40, a company earns a profit when revenue exceeds cost:

$$\text{revenue} > \text{cost}.$$

The Blue Sky-Hook company finds that it earns \$20x for selling x sky-hooks at \$20 each, whereas it costs \$36 + 29x - x^2 to manufacture the same number of hooks. What is the least number of hooks that the company must sell in order to make a profit? What is the "break even" point?

36. When a ball is thrown upward from the ground, its height (in feet) after t seconds is given by the formula

$$h = -16t^2 + 96t.$$

When is the ball 128 feet or more above the ground?

37. For what values of the side of a square does the area of the square exceed its perimeter numerically?

38. The length of a rectangle is two more than its width. For what values of the width is the area numerically greater than the perimeter?

39. The length of a rectangle is three inches longer than the width. For what values of the width is the area of the rectangle less than 18 square inches?

40. A circle has radius r. For what values of the radius does the circumference exceed the area numerically? *Formulas:* $C = 2\pi r$, $A = \pi r^2$.

41. The denominator of a fraction is three more than the numerator. Find all such positive fractions that are less then $\dfrac{3}{4}$. *Hint:* Let $\dfrac{n}{n+3}$ be such a fraction; you are interested only in positive integers n satisfying a rational inequality.

42. The sum of the numerator and denominator of a fraction is 15. Find all such positive fractions that are less than or equal to $\dfrac{2}{3}$. Consult the Tables of Phrases (Section 4.6), if necessary.

43. A positive integer is added to both the numerator and the denominator of the fraction $\dfrac{1}{2}$ such that the newly formed fraction is more than $\dfrac{9}{10}$. Find all such integers.

44. The sum of the squares of two consecutive integers is less than or equal to 25. Find all such pairs of integers.

The method in the text can be extended to higher-degree inequalities. For example,

$$(x + 2)(x - 1)(x - 4) > 0$$

has critical points $x = -2$, $x = 1$, and $x = 4$. We plot them as open dots and then use test points. You should be able to fill in the details.

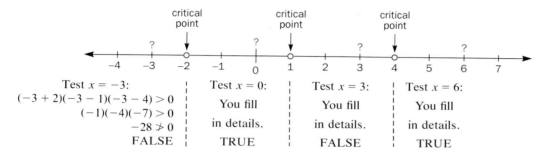

Test $x = -3$:
$(-3 + 2)(-3 - 1)(-3 - 4) > 0$
$(-1)(-4)(-7) > 0$
$-28 \not> 0$
FALSE

Test $x = 0$:
You fill in details.
TRUE

Test $x = 3$:
You fill in details.
FALSE

Test $x = 6$:
You fill in details.
TRUE

FALSE   TRUE   FALSE   TRUE

Answer: $\{x \mid -2 < x < 1\} \cup \{x \mid x > 4\}$:

In like manner, solve and graph these inequalities:

45. $(x + 3)(x - 2)(x - 5) > 0$

46. $(x + 4)(x + 1)(x - 3) \geq 0$

47. $x(x - 3)(x + 4) \leq 0$

48. $x(x + 5)(x - 2) < 0$

49. $(x + 1)^2(x - 3) \geq 0$

50. $(x + 4)(x - 2)^2 \leq 0$

51. $x^3 + x^2 - 6x < 0$

52. $2x^3 - 5x^2 - 12x > 0$

53. $\dfrac{(x + 3)(x - 2)}{x + 1} > 0$

54. $\dfrac{x - 3}{(x + 2)(x + 5)} < 0$

55. $\dfrac{2x}{x^2 - 9} < 0$

56. $\dfrac{x^2 + 5x - 6}{2x + 6} > 0$

C

57. $\dfrac{2x}{x - 1} \geq x$

58. $\dfrac{3x}{x + 2} \leq x$

59. $x^4 - 10x^2 + 9 < 0$

60. $x^3 + x^2 - 12x \geq 0$

61. $x^3 + 4x^2 - 4x - 16 \geq 0$ *Hint*: Factor by grouping.

62. $x^5 - 29x^3 + 100x > 0$

For polynomials that are not factorable, the critical points must be found by other methods. For example,

$$x^2 - 2x - 1 < 0$$

has critical points found by completing the square or using the Quadratic Formula:

$$x = 1 \pm \sqrt{2} \approx 1 + 1.414, \quad 1 - 1.414$$
$$x \approx 2.414, \quad x \approx -0.414$$

(continued)

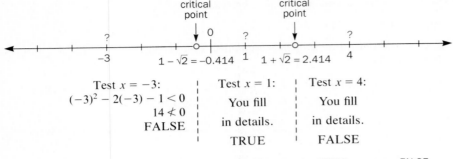

Test $x = -3$:

$(-3)^2 - 2(-3) - 1 < 0$

$14 \not< 0$

FALSE

Test $x = 1$:

You fill
in details.

TRUE

Test $x = 4$:

You fill
in details.

FALSE

Answer: $\{x \mid 1 - \sqrt{2} < x < 1 + \sqrt{2}\}$:

In like manner, solve and graph these inequalities:

63. $x^2 + 2x - 2 > 0$

64. $x^2 - 4x - 4 \geq 0$

65. $x^2 - 6x + 1 \leq 0$

66. $x^2 < 9 - 3x$

67. $x^2 - 2 \geq 0$

68. $x^2 > 8$

69. $3x^2 < 25$

70. $2x^2 \leq 9$

CHAPTER 6 · SUMMARY

6.1 Imaginary Numbers

$i = \sqrt{-1}$ such that $i^2 = -1$

$\sqrt{-a} = i\sqrt{a}$ if $a > 0$

Example:

$$\sqrt{-9} = i\sqrt{9}$$
$$= 3i.$$

Complex Numbers

$a + bi$, a and b real numbers

Example:

$$3 + \sqrt{-8} = 3 + 2i\sqrt{2}.$$

Addition of complex numbers

Example:

$(3 + 5i) + (7 - 9i) = 10 - 4i.$

Multiplication of complex numbers

Example:

$$(2 + 3i)(4 - i) = 8 + 10i - 3i^2$$
$$= 8 + 10i - 3(-1)$$
$$= 11 + 10i.$$

6.2 Extracting Roots

If $x^2 = c$,
then $x = \pm\sqrt{c}$.

Example:

$$4x^2 - 45 = 0$$

$$x^2 = \frac{45}{4}$$

$$x = \frac{\pm 3\sqrt{5}}{2}.$$

CHAPTER 6 · SUMMARY

6.3 Completing the Square

If $\quad x^2 + bx + c = 0,$

then $\quad x^2 + bx + \left(\dfrac{b}{2}\right)^2 = -c + \left(\dfrac{b}{2}\right)^2$

Example:

$$x^2 + 6x - 3 = 0$$
$$x^2 + 6x + 9 = 3 + 9 \qquad \text{Add } \left(\dfrac{6}{2}\right)^2 = 9.$$
$$(x + 3)^2 = 12$$
$$x + 3 \;= \pm\sqrt{12}$$
$$x = -3 \pm 2\sqrt{3}.$$

6.4 The Quadratic Formula

If $\quad ax^2 + bx + c = 0$

then $\quad x = \dfrac{-b \pm \sqrt{b^2 - 4ac}}{2a}.$

Example:

$$x^2 - 5x + 2 = 0$$
$$x = \dfrac{-(-5) \pm \sqrt{(-5)^2 - 4 \cdot 1 \cdot 2}}{2 \cdot 1}$$
$$x = \dfrac{5 \pm \sqrt{17}}{2}.$$

6.5 Equations Containing Radicals

Square both sides to remove radicals. *Check* all answers to find any extraneous solutions.

Example:

$$\sqrt{x + 1} = x - 5$$
$$(\sqrt{x + 1})^2 = (x - 5)^2$$
$$x + 1 = x^2 - 10x + 25$$
$$0 = x^2 - 11x + 24$$
$$0 = (x - 8)(x - 3)$$
$$x = 8 \quad (x = 3 \text{ is extraneous}).$$

6.6 Combined Methods of Solving Equations

Start by factoring *completely*. Then solve by techniques learned in this chapter.

Example:

$$x^5 + 2x^3 - 8x = 0$$
$$x(x^4 + 2x^2 - 8) = 0$$
$$x(x^2 + 4)(x^2 - 2) = 0$$
$$x = 0, \quad x^2 = -4, \quad x^2 = 2$$
$$x = 0, \quad x = \pm 2i, \quad x = \pm\sqrt{2}.$$

6.7 Quadratic and Rational Inequalities

Solve equation to get *critical points*. Then substitute *test points* into the original inequality to get TRUE intervals.

Example:

$$\text{Solve: } x^2 - 3x - 4 > 0$$
$$(x + 1)(x - 4) = 0$$
$$x = -1, \quad x = 4 \qquad \text{Critical points.}$$

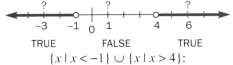

$$\{x \mid x < -1\} \cup \{x \mid x > 4\}:$$

CHAPTER 6 · REVIEW EXERCISES   *Answers, pages A64–A66*

6.1

Express in terms of the imaginary unit i:

1. $\sqrt{-64}$

2. $\sqrt{-40}$

3. $\sqrt{-\dfrac{49}{3}}$

4. $-\sqrt{\dfrac{-63}{50}}$

Express in the form $a + bi$, and then perform the indicated operation:

5. $(6 - \sqrt{-9}) + (4 + \sqrt{-16}) - (-2 - \sqrt{-4})$

6. $\sqrt{-4}(3 - \sqrt{-25}) - 5(7 + \sqrt{-1})$

7. $(5 - 4i)(2 + 7i)$

8. $(3 - 2i)^2$

9. $(1 - i)^3$

10. $(4 + \sqrt{-8})(2 - \sqrt{-18})$

Rationalize denominators:

11. $\dfrac{-1}{2i}$

12. $\dfrac{4}{i\sqrt{2}}$

13. $\dfrac{3 - 5i}{2i}$

14. $\dfrac{18}{\sqrt{-12}}$

15. $\dfrac{50}{3 + 4i}$

16. $\dfrac{14i}{2 - i\sqrt{3}}$

17. $\dfrac{5 + 10i}{2 - i}$

18. $\dfrac{3 - i\sqrt{5}}{5 + 2i\sqrt{5}}$

6.2

Solve by extracting roots:

19. $x^2 = 25$

20. $x^2 - 12 = 0$

21. $3y^2 + 72 = 0$

22. $49 - 4z^2 = 0$

23. $8 = 5x^2$

24. $\dfrac{x}{3} = \dfrac{3}{2x}$

25. $(x - 2)^2 = 18$

26. $(3y + 1)^2 - 49 = 0$

27. $\dfrac{2bx}{a} = \dfrac{5a}{2bx}$; for x

28. $\dfrac{v^2}{2a} = s$; for v

29. $\dfrac{x^2}{4} + \dfrac{y^2}{9} = 1$; for y

30. $F = \dfrac{kwh^2}{L}$; for h, $h > 0$

Solve for x in terms of the other numbers or letters:

31.
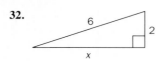

32.

33.
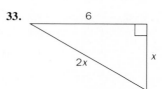

34.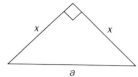

35.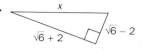

36.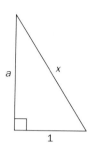

37. Find x and then y:

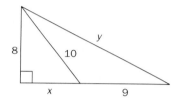

38. The rectangle has sides as shown.
  a. Find the radius r of the circle.
  b. Find the shaded area inside the circle but outside the rectangle.

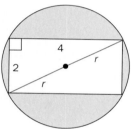

39. If the sides of a square are each decreased by 1, a new square is formed having area 8. Find the side of the original square. Then find both the area and the perimeter of the original square.

40. Twice a number equals nine times the reciprocal of the number. Find the number.

6.3

Solve by completing the square:

41. $x^2 - 6x - 16 = 0$ **42.** $y^2 + 4y - 8 = 0$ **43.** $z^2 + 10z + 33 = 0$

44. $w^2 + 3w - 1 = 0$ **45.** $2x^2 = 3 - 5x$ **46.** $6y = 1 + 2y^2$

47. The length of a rectangle is 4 inches longer than the width. The area is 16 square inches. Find the width and the length, both expressed in terms of radicals and approximated to the nearest hundredth of an inch.

48. When a ball is thrown vertically upward, its height h above the ground (in feet) after t seconds is given by the formula

$$h = -16t^2 + 96t + 80.$$

At what time is the ball 128 feet above the ground? Give your answers both expressed as a radical and approximated to the nearest tenth of a second.

Find the unknown sides of these right triangles. Give your answers in terms of radicals, and approximate them to the nearest tenth.

49.

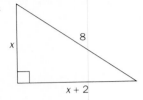

50.

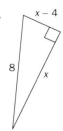

6.4

Solve by using the Quadratic Formula:

51. $3x^2 - x - 4 = 0$

52. $x^2 = 2 - 2x$

53. $4z^2 + 9 = 0$

54. $4z^2 + 9z = 0$

55. $t^2 - 6t + 17 = 0$

56. $x^2 + 30x - 259 = 0$

57. $x = \dfrac{1}{x - 1}$

58. $\dfrac{2}{y} - \dfrac{3}{2y - 1} = 2$

59. $x(3x + 2) = (x - 1)^2$

60. $x^2 + 2\sqrt{2}\,x - 6 = 0$

61. $2x^2 - 5ix + 3 = 0$

62. $\sqrt{3}\,x^2 - 3x - 6\sqrt{3} = 0$

Compute the discriminant of each of the following equations. Then tell whether or not the equation is factorable, what type of solution it has, and how many solutions it has. Do *not* solve the equations.

63. $x^2 + 2x - 10 = 0$

64. $4x^2 - 20x + 25 = 0$

65. $2y^2 - y + 5 = 0$

66. $9x^2 + 25 = 0$

67. $9x^2 + 25x = 0$

68. $x^2 - 2x - 288 = 0$

69. $6x^2 - 11x - 72 = 0$

70. $6x^2 - 11x + 72 = 0$

Set up and then solve by using the Quadratic Formula:

71. The square of a number is five times the sum of the number and two. Find the number.

72. When a number is decreased by its own reciprocal, the result is three. What is the number?

Solve the following equations by using the Quadratic Formula, and then give the solutions to three decimal places:

73. $x^2 - 2x - 4 = 0$

74. $x^2 + x - 1 = 0$

75. $2y^2 - y - 2 = 0$

6.5

Solve and check (be sure to list any extraneous solutions):

76. $\sqrt{x^2 - 8x} = 3$

77. $\sqrt{2y + 1} + 1 = y$

78. $\sqrt[3]{1 - 2x} = -4$

79. $\sqrt{z + 9} + \sqrt{z - 6} = 5$

80. $\sqrt{11 - \sqrt{w - 2}} = 3$

81. $(x + 2)^{1/2} - (x - 3)^{1/2} = 1$

82. $\sqrt{v + 1} + \sqrt{v - 4} = \sqrt{3v + 1}$

83. $x = \dfrac{\sqrt{x^2 + 36}}{2}$

84. When the square root of the sum of a number and four is increased by two, the result is the number itself. What is the number?

85. When the square root of the sum of a number and seven is decreased by the square root of the number itself, the result is one. Find the number.

86. Use the Pythagorean Theorem to solve for x in the right triangle shown. Then find the lengths of the three sides.

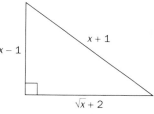

Problem 86

87. The figure contains unknowns x and y.
   a. Express y in terms of x, using the length 8 that is shown.
   b. Express y in terms of x, using the Pythagorean Theorem.
   c. By equating your results from parts a and b, write a radical equation involving only x.
   d. Solve this radical equation for x.
   e. Use your value of x to find y.

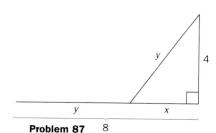

Problem 87

88. The figure contains unknowns x, y, and z.
   a. Express y as a radical in terms of x, and similarly express z in terms of x, using the Pythagorean Theorem.
   b. Given that
   $$y + z = 5,$$
   write a radical equation involving only x by substituting your results from parts a and b.
   c. Solve this equation for x.
   d. Use your value of x to find y and z.

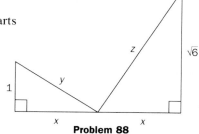

Problem 88

6.6

Solve using combined methods:

89. $2x^5 + 2x^3 - 144x = 0$

90. $3y^4 - 31y^2 + 36 = 0$

91. $z^6 - 16z^3 + 64 = 0$

92. $w^5 - 8w^3 + 8w^2 - 64 = 0$

93. $2t^5 + 3t^4 - 32t - 48 = 0$

94. $2x^{2/3} - 3x^{1/3} - 9 = 0$

95. $2x^{-2} + 5x^{-1} - 3 = 0$

96. $x^{-2/3} + 7x^{-1/3} - 18 = 0$

97. $(x^2 - 3x)^2 - 2(x^2 - 3x) - 8 = 0$

98. $z^6 - 2z^4 - 16z^2 + 32 = 0$

99. $\dfrac{x^2}{4} = \dfrac{6}{x^2 + 2}$

In Problems 100 and 101, solve for x in the right triangles. Then find the unknown sides.

100.

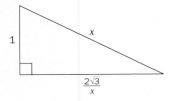

101.

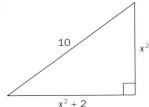

6.7

Solve and graph:

102. $x^2 - x - 6 < 0$

103. $2x^2 + 3x \geq 5$

104. $x^2 - 4 \leq 0$

105. $x^2 - 9x > 0$

106. $x^2 - 9 \geq 0$

107. $\dfrac{x - 2}{x + 3} > 0$

108. $\dfrac{2x + 1}{x - 4} < 0$

109. $\dfrac{x - 1}{x} \leq 0$

110. $\dfrac{2x + 1}{x - 4} > 2$

111. $5x + 15 < 10x^2$

112. $\dfrac{4}{x} > x$

113. $(x - 2)(x - 4)(x + 1) > 0$

114. $\dfrac{x^2 - 2x - 3}{2x + 6} < 0$

115. $x^2 - 5 \geq 0$

116. $x^2 - 2x - 2 < 0$

117. The formula $h = -16t^2 + 64t$ gives the height h (in feet) of a ball t seconds after it is thrown vertically upward. When will the ball be higher than 48 feet?

118. Acme Frammis Company finds that it costs $\$57 + 63x - 2x^2$ to manufacture x framisses. If each frammis sells for \$50, what is the least number that must be sold in order for the company to earn a profit? *Hint:* Revenue = \$50x.

119. A square has side x. For what values of x is the area numerically less than the perimeter?

120. The numerator of a fraction is four less than the denominator. Find all such positive fractions that are less than $\frac{2}{3}$.

CHAPTER 6 · TEST   *Answers, pages A66–A67*

---

1. Write in terms of i and then multiply: $(2 - \sqrt{-9})(3 + \sqrt{-4})$

Solve by extracting roots:

2. $x^2 - 12 = 0$

3. $3y^2 + 50 = 0$

4. Solve by completing the square: $z^2 - 8z - 4 = 0$

5. Solve by using the Quadratic Formula: $9w^2 - 3w + 1 = 0$

Solve and check:

6. $\sqrt{2r - 1} + 2 = r$

7. $\sqrt{t + 3} - \sqrt{t - 2} = 1$

Solve by first factoring:

8. $3x^5 - 12x^3 - 96x = 0$

9. $x^{2/3} + 3x^{1/3} - 18 = 0$

Solve by any method:

10. $\dfrac{2x}{7} = \dfrac{4}{3x}$

11. $\dfrac{1}{x} + \dfrac{1}{x - 1} = 2$

12. $\dfrac{\sqrt{40 - x^2}}{3} = x$

13. Solve for r: $A = 4\pi r^2,\ r > 0$

Solve and graph your answers:

14. $x^2 + 3x - 4 \geq 0$

15. $\dfrac{x}{x + 3} < 0$

16. The length of a rectangle is 4 more than the width. The area is 8. Find the dimensions, both expressed in terms of radicals and approximated to the nearest tenth.

17. The square root of the sum of a number and three is three less than the number. Find the number.

18. Find the unknown sides:

19. Solve for x in terms of a:

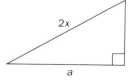

20. Find x and then find y:

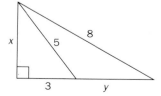

CHAPTERS 1—6 · CUMULATIVE REVIEW EXERCISES   *Answers, pages A67–A68*

Simplify each expression, solve each equation, and solve and graph each inequality:

1. $19 + 5[2 - 6(1 - 9)]$

2. $13 - (-6)^2$

3. $3(x - 3) = -2(x - 8)$

4. $(2x - 5)(2x + 3) - (4x - 1)(x - 2)$

5. $\dfrac{(-3x^3y^5)(-6xy^4)}{-2x^3y^{13}}$

6. $1 - t = 1 + t$

7. $\left(\dfrac{-2a^5b^7}{3ab^9}\right)^4$

8. $\dfrac{a^3 + 1}{a^2 + 2a + 1} \cdot \dfrac{a^2 + 3a + 2}{a^2 + a - 2}$

9. $\dfrac{2z + 1}{5} - \dfrac{3z - 2}{3} = 8$

10. $(4x^3)^{-2}$

11. $\left(\dfrac{a^{-8}b^{-3}c^0}{a^{-2}b^3c^{-5}}\right)^{-4}$

12. $6y^2 + 5y - 6 = 0$

13. $\dfrac{x - \dfrac{1}{x^2}}{x - \dfrac{1}{x}}$

14. $-|2 - 7|$

15. $5x - 3 < 9 - x$

16. $\dfrac{x - 5}{x^2 - x - 2} - \dfrac{2}{x + 1} + \dfrac{5}{x - 2}$

17. $5\sqrt{8} - 5\sqrt{50} + 9\sqrt{2}$

18. $|x + 1| = 7$

19. $(2 + 3i)(4 - 7i)$

20. $16x^2 = 25$

21. $\left(\dfrac{27x^3}{64y^6}\right)^{4/3}$

22. $\dfrac{2x}{3} - \dfrac{5x - 3}{7} \geq 2$

23. $16x^2 = 25x$

24. $(9x^6)^{-1/2}$

25. $\dfrac{3x^2 - 8}{2x^2 - 7x + 6} = \dfrac{5}{2x - 3} + \dfrac{2x}{x - 2}$

26. $\dfrac{2x - 6x^2}{6x^2 - x - 2} \div \dfrac{12x^3 - 4x^2}{3x^2 + x - 2}$

27. $|2x - 3| < 7$

28. $x^{1/2}x^{2/3}x^{3/4}$

29. $\dfrac{1 - \dfrac{8}{x^2 - 1}}{\dfrac{2}{x + 1} - \dfrac{1}{x - 1}}$

30. $z^2 - 6z - 3 = 0$

31. $1 \leq \dfrac{5 - 2x}{3} \leq 5$

32. $w^2 - 8 = 0$

33. $(x^{1/2} + x^{-1/2})^2$

34. $\dfrac{6a^2 - 13a + 6}{8a^2 + 26a + 21} \div \dfrac{10a^2 - 15a}{40a^4 + 270a^3 + 350a^2}$

35. $3v^2 + 25 = 0$

36. $|1 - 2x| \geq 5$

37. $\dfrac{a - a^{-1}}{a^{-1} + a^{-2}}$

38. $\dfrac{2}{x - 2} - \dfrac{4}{x^2 + 2x + 4} - \dfrac{x^2 + 20}{x^3 - 8}$

39. $\sqrt{-50}$

40. $w^2 + 3w + 9 = 0$

41. $x^2 + 2x - 8 \geq 0$

42. $\dfrac{(x + h)^2 - 2(x + h) + 5 - (x^2 - 2x + 5)}{h}$

43. $x^5 - 8x^3 - 48x = 0$

44. $\dfrac{x + 3}{x - 1} < 0$

45. $\sqrt{2x + 7} - x = 2$ (Check your answers.)

46. $\dfrac{1}{x} = \dfrac{x}{x + 1}$

47. $3x^8 - 21x^5 - 24x^2 = 0$

48. $\dfrac{2}{x - 1} - \dfrac{1}{x^2 - 4x + 3} - \dfrac{4}{x - 3} - \dfrac{3}{x^2 - 2x + 1}$

49. $x^5 - 4x^3 - 27x^2 + 108 = 0$

50. $\sqrt{y - 3} + \sqrt{y + 5} = 4$ (Check your answers.)

51. $\dfrac{2x}{x + 2} \geq 1$

52. $z^{2/3} + 5z^{1/3} - 14 = 0$

53. $x = 6x^{-1} + 27x^{-3}$

54. $\sqrt{2x - 1} - \sqrt{x - 1} = \sqrt{x - 4}$ (Check your answers.)

55. $|x^2 - 2x - 16| = 8$

56. $(x^2 - 2x)^2 - 34(x^2 - 2x) - 35 = 0$ *Hint*: Let $U = x^2 - 2x$.

57. Express in scientific notation: **a.** 513,000,000   **b.** 0.0000302

58. Express in decimal notation: **a.** 6.7×10^{-4} **b.** 7.901×10^8

Divide:

59. $\dfrac{2x^2 - 6x + 8}{2x}$

60. $\dfrac{8x^3 - 16x^2 + 20x - 21}{2x - 3}$

61. $\dfrac{2x^4 + 3x^3 - 7x + 9}{x - 2}$

62. $\dfrac{x}{x + 1}$

Rationalize denominators and simplify:

63. $\dfrac{8x^2}{\sqrt{2x}}$

64. $\dfrac{\sqrt{6} - \sqrt{2}}{\sqrt{6} + \sqrt{2}}$

65. $\dfrac{5i}{3 + 4i}$

66. $\dfrac{9x}{\sqrt[3]{3x^2}}$

67. $\dfrac{h}{\sqrt{4 + h} - 2}$

68. $\dfrac{-15}{2i\sqrt{5}}$

69. $\dfrac{5 - 10i}{2 + 4i}$

70. $\dfrac{x - a}{\sqrt{x} - \sqrt{a}}$

Solve for the indicated variable:

71. $\dfrac{1}{A} - \dfrac{1}{B} = \dfrac{1}{B} + \dfrac{1}{C}$;   for B

72. $\dfrac{a}{b} + \dfrac{x}{a} = \dfrac{b}{a} + \dfrac{x}{b}$;   for x

73. $2x^2 - ax - 6a^2 = 0$;   for x

74. $x^2 + y^2 = r^2$;   for y

75. $\dfrac{x^2}{36} - \dfrac{y^2}{16} = 1$;   for x

76. $P = 2\pi\sqrt{\dfrac{L}{g}}$;   for g

77. $E = mc^2$;   for c, $c > 0$

78. $\dfrac{1}{c} = \dfrac{1}{c_1} + \dfrac{1}{c_2} + \dfrac{1}{c_3}$;   for c_2

79. $\dfrac{3x}{a} = \dfrac{9a}{x}$;   for x

80. $\dfrac{x-a}{b} + \dfrac{2b}{a} = \dfrac{2x-a}{a};$ for x

81. Twice the difference between a number and three is five more than three times the sum of the number and eight. Find the number.

82. Find three consecutive odd integers whose sum is eight more than the largest.

83. Find four consecutive integers such that the sum of their squares is ten less than four times the product of the first and second integers.

84. When twice a number is decreased by three, one-half of the result equals six more than two-thirds of the sum of the number and eight. Find the number.

85. The sum of a number and its reciprocal is $\dfrac{25}{12}$. Find the number.

86. In a collection of 23 coins worth $3.30, there are equal numbers of nickels and dimes, and the rest are quarters. How many of each coin are there?

87. A collection of 18 coins consists of nickels and dimes. How many of each coin are there if the value of the nickels equals the value of the dimes?

88. A certain amount of money is invested in a stock paying 9% simple annual interest for 3 years. An equal amount of money is invested in a bond paying 8% for 4 years. An amount of money equal to $2000 more than the total already invested is put into a high-risk commodity that *loses* $7\frac{1}{2}\%$ annually for 2 years. The net profit on these investments is $1150. How much is invested in each instrument?

89. $10,000 is invested at 8% for 2 years. How much money should be invested at 11% for the same time so that the combined investment is equivalent to a 9% investment?

90. A family left home averaging 50 mph on its way to the mountains. On the trip home, it saved one hour by increasing its speed by 10 mph. Find the distance in each direction. *Hint*: Let t = time going.

91. A car drives from A to B averaging 60 mph but returns to A averaging 40 mph. If the total driving time is 5 hours, find the time in each direction. Then find the distance from A to B.

92. A man drives 50 mph a certain distance in city traffic and then drives 60 mph on the highway. He covers a total of 45 miles in 50 minutes. How far did he travel in each case? *Hint*: First convert 50 minutes to hours.

93. The length of a rectangle is 3 inches longer than the width. Find its dimensions, given that:
   a. The perimeter is 34 inches.
   b. The area is 108 square inches.
   c. The diagonal is 15 inches.
      (*Note*: These are three separate problems.)

94. A square of unknown sides is made into a rectangle of equal area by increasing two of its sides by 6 inches each and decreasing each of its other sides by 4 inches. Find the dimensions of each figure.

95. When the sides of a square are each increased by 4 meters, a new square is formed whose area is 56 square meters greater than that of the original square. Find the sides of each square.

96. A ladder is resting vertically against a wall. When the base of the ladder is pulled 7 feet away from the wall, the top of the ladder slides down 1 foot vertically along the wall. How long is the ladder?

97. One leg of a right triangle is 2 more than the other leg, and the hypotenuse is 10. Find the two legs.

98. The length of a rectangle is twice the width. The area is 16. Find the dimensions of the rectangle.

Solve for x in these right triangles:

99.

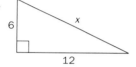

100.

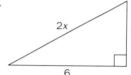

101.

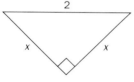

102.

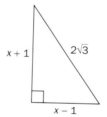

103.

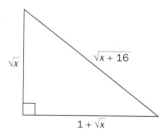

104.

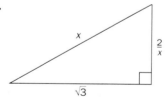

7 LINES AND LINEAR SYSTEMS

Graphing Points and Lines

The inequalities of Sections 2.5, 2.6, 4.5, and 6.7 had solutions with values of x on the one-dimensional real-number line. In this chapter we move up a dimension to points (x,y) in the two-dimensional plane. This plane is obtained by placing two number lines, called the x- and y-axes, perpendicular to each other:

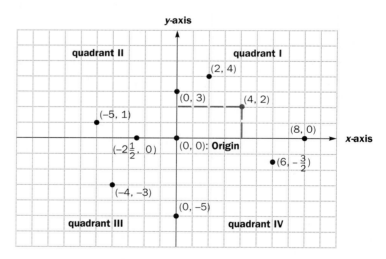

Points in the plane are represented by ordered pairs of numbers, or coordinates. The point $(4,2)$ indicates 4 units along the positive x-axis and 2 units along the positive y-axis. Note that the point $(4,2)$ is different from the point $(2,4)$. The "xy-plane" is also called the **Cartesian coordinate system.** Its inventor, René Descartes, was a Renaissance mathematician and philosopher who founded analytic geometry, the study of geometry with coordinates. (This is in contrast to Euclidean, or plane, geometry, which has no coordinate system.) The plane is divided into four quadrants, numbered counterclockwise as shown. In the midst of it all lies the *origin,* with coordinates $(0,0)$.

Points in the Cartesian coordinate system can be plotted randomly, as shown on the previous graph, or, more importantly, as the solution to an equation in two variables. Recall that an equation in the single variable x, such as $-2x + 7 = 1$, has the solution $x = 3$. An equation in two variables, such as

$$-2x + y = 1$$

has infinitely many solutions. For example, if $x = 0$, then $-2 \cdot 0 + y = 1$, or $y = 1$; thus $(0,1)$ is a solution of $-2x + y = 1$. Again, if $x = 1$, then $-2 \cdot 1 + y = 1$, or $y = 3$; thus $(1,3)$ is also a solution. This can be repeated indefinitely, which generates an infinite set of solutions: $(0,1)$, $(1,3)$, ... When these points are plotted in the plane, the "solution graph" of the equation is obtained.

A more systematic method for graphing the equation $-2x + y = 1$ is first to solve it for y:

$$y = 2x + 1.$$

We next prepare a table of values for x and y, plot them, and then connect the points.

EXAMPLE 1

Graph $-2x + y = 1$.

Solution
We first solve for y: $y = 2x + 1.$

Now we choose convenient values of x, (say $x = 0$, 1, 2, -1, and -2) and substitute into this equation to obtain the corresponding values of y.

x	y
0	1
1	3
2	5 $\leftarrow y = 2 \cdot 2 + 1$
-1	-1
-2	-3 $\leftarrow y = 2(-2) + 1$

The graph of $-2x + y = 1$,
                        or
$y = 2x + 1$, is a *straight line*.

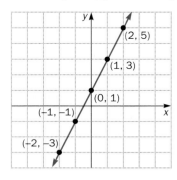

This example, $-2x + y = 1$, leads to an observation: The *line*ar equation

$$ax + by = c$$

has a graph that is a straight line. This equation is called the *standard form* of a straight line.

In the next example,

$$2x - 5y = -10,$$

solving for y leads to a fraction: $2x + 10 = 5y,$ or $y = \frac{2}{5}x + 2.$ Preparing a table as in Example 1 will also lead to fractions, so we take a simpler approach. We first

let $x = 0$; then $2 \cdot 0 - 5y = -10,$ or $-5y = -10,$ or $y = 2.$

Next, we

let $y = 0$; then $2x - 5 \cdot 0 = -10,$ or $2x = -10,$ or $x = -5.$

The point $(0,2)$ is called the y-**intercept,** because it is where the line crosses the y-axis (see the graph accompanying Example 2). The point $(-5,0)$ is called the x-**intercept** because the graph crosses the x-axis at this point. Geometry tells us that two points determine a line, so these two intercepts are sufficient for us to draw the graph. As a check, we include a third point, using $x = 5$ in the table in Example 2. It has the *opposite sign* from the $x = -5$ we used before and results in $y = 4,$ a value that is not a fraction. The foregoing is called the **intercept method** of graphing a straight line.

EXAMPLE 2

Graph $2x - 5y = -10$ by the intercept method.

Solution

Let $x = 0$; then $y = 2$ (see above).
Let $y = 0$; then $x = -5$ (see above).

x	y	
0	2	y-intercept
-5	0	x-intercept
5	4	$\leftarrow 2 \cdot 5 - 5y = -10$
		$-5y = -20$
		$y = 4.$

Change sign. $\longrightarrow$

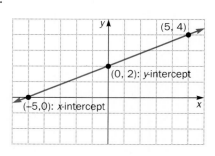

(5, 4)
(0, 2): *y*-intercept
(-5,0): *x*-intercept

EXAMPLE 3

Graph $3x + 4y - 12 = 0$ by the intercept method.

Solution

Let $x = 0$; then $3 \cdot 0 + 4y - 12 = 0$
$$4y = 12$$
$$y = 3.$$

Let $y = 0$;   then $3x + 4 \cdot 0 - 12 = 0$

$$3x \qquad\qquad = 12$$
$$x = 4.$$

x	y
0	3 y-intercept
4	0 x-intercept
−4	6 ← $3(−4) + 4y − 12 = 0$

Change sign. ⟶ −4

$$4y \qquad = 24$$
$$y = 6.$$

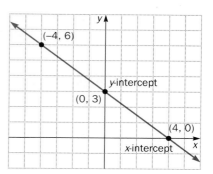

Our next examples are missing a variable. This makes things simple, because the lines turn out to be either horizontal or vertical.

EXAMPLE 4

Graph $y = 2$.

Solution

Here x is missing. So no matter what x-values we choose, y is always 2.

x	y
0	2
1	2
2	2
−1	2
57	2

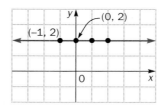

The graph of $y = 2$ is a *horizontal* line.

EXAMPLE 5

Graph $2x + 6 = 0$.

Solution

This is $2x = −6$,   or $x = −3$.
Here y is missing, so regardless of its value, x is always −3.

x	y
−3	0
−3	1
−3	−2
−3	19

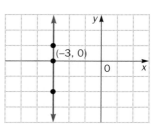

The graph of $x = −3$ is a *vertical* line.

Using these two examples as prototypes, we can graph the following horizontal and vertical lines "by inspection."

EXAMPLE 6

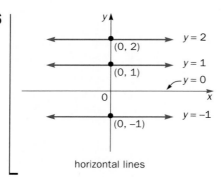

horizontal lines

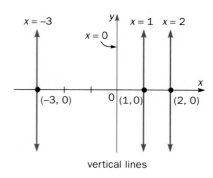

vertical lines

EXERCISE 7.1   *Answers, pages A69–A70*

A

Graph these lines by first preparing a table, as in EXAMPLE 1. First solve for *y* if necessary.

1. $y = 2x - 3$ 　　　　**2.** $y = 3x - 2$ 　　　　**3.** $y = 4 - 3x$ 　　　　**4.** $y = 6 - x$

5. $x + y - 4 = 0$ 　　**6.** $2x - y + 1 = 0$ 　　**7.** $x + y = 0$ 　　　**8.** $2x - y = 0$

Graph these lines by the intercept method, as in EXAMPLES 2 and 3:

9. $3x - 2y = 6$ 　　**10.** $3x - 5y = -15$ 　　**11.** $2x + 7y = 14$ 　　**12.** $8x - 3y = 24$

13. $5x + 4y - 20 = 0$ 　**14.** $7x - 3y + 21 = 0$ 　**15.** $2x - 3y + 9 = 0$ 　**16.** $5x + 2y = 15$

17. $\dfrac{x}{2} + \dfrac{y}{5} = 1$ 　　**18.** $\dfrac{x}{7} - \dfrac{y}{2} = 1$ 　　**19.** $\dfrac{2x}{3} - \dfrac{3y}{2} = 6$ 　　**20.** $\dfrac{3x}{4} + \dfrac{2y}{3} = 3$

Graph these horizontal or vertical lines by inspection. See EXAMPLES 4, 5, and 6.

21. $y = 5$ 　　　　**22.** $y = -3$ 　　　　**23.** $x = -4$ 　　　　**24.** $x = 1$

25. $y + 4 = 0$ 　　**26.** $y - 7 = 0$ 　　**27.** $2x + 3 = 0$ 　　**28.** $5 - 3x = 0$

29. $5 - 2y = 0$ 　　**30.** $-4y + 3 = 0$

B

The line $x - 2y = -5$ and the *x*- and *y*-axes form a triangle as shown on the next page.

Its *x*- and *y*-intercepts are $(-5, 0)$ and $(0, \frac{5}{2})$, respectively, which we obtained by using the method of Examples 2 and 3. Using the formula $A = \frac{1}{2}bh$, we find that the shaded triangle has area

$$A = \frac{1}{2}(5)\left(\frac{5}{2}\right) = \frac{25}{4}.$$

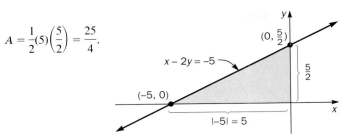

In like manner, find the *x*- and *y*-intercepts of these lines. Then find the areas of the triangles formed by the lines and the *x*- and *y*-axes.

31. $2x + y = 4$

32. $3x - y = 6$

33. $3x - 5y - 15 = 0$

34. $7y - 2x - 14 = 0$

35. $y = -2x - 2$

36. $\dfrac{x}{4} + \dfrac{y}{6} = 1$

The point $(2,-3)$ lies on the line $5x - 7y = 31$ because it satisfies the equation:

$$5(2) - 7(-3) = 31, \quad \text{or} \quad 10 + 21 \overset{\checkmark}{=} 31.$$

Conversely, the point $(-1,6)$ does *not* lie on this line because it does not satisfy the equation:

$$5(-1) - 7(6) \neq 31, \quad \text{or} \quad -5 - 42 \neq 31.$$

For each of the following lines, decide whether or not the given point lies on the given line:

37. $x + 2y = 7$; $(1,3)$

38. $2x - 3y = 5$; $(1,-1)$

39. $2x - 5y + 17 = 0$; $(-3,2)$

40. $-4x + 3y + 11 = 0$; $(2,-3)$

41. $y = 3x - 1$; $(0,-1)$

42. $y = 9 - 2x$; $(1,8)$

43. $2x + 5y = 4$; $\left(\dfrac{3}{4},\dfrac{1}{2}\right)$

44. $6x - 2y = 1$; $\left(\dfrac{1}{3},\dfrac{3}{2}\right)$

45. $x\sqrt{5} - y\sqrt{3} = \sqrt{2}$; $(\sqrt{10},\sqrt{6})$

46. $x\sqrt{3} - y\sqrt{10} = \sqrt{5}$; $(\sqrt{15},\sqrt{2})$

By connecting the following sequence of points, you will form a recognizable "stick figure." After each semicolon (;), take your pencil off the paper, and then connect the next sequence of points.

47. $(1,5)$ to $(2,4)$ to $(6,4)$ to $(6,2)$ to $(2,2)$ to $(2,4)$; then $(3,2)$ to $(2,1)$ to $(3,1)$; then $(5,2)$ to $(4,1)$ to $(5,1)$; then $(6,2)$ to $(8,3)$ to $(6,4)$; then $(7,3)$; then $(8,5)$ to $(9,4)$ to $(13,4)$ to $(13,2)$ to $(9,2)$ to $(9,4)$; then $(10,2)$ to $(9,1)$ to $(10,1)$; then $(12,2)$ to $(11,1)$ to $(12,1)$; then $(13,2)$ to $(11,3)$ to $(13,4)$; then $(12,3)$.

48. $(1,1)$ to $(1,3)$ to $(2\frac{1}{2},3)$; then $(1,2)$ to $(2,2)$; then $(3,1)$ to $(3,3)$ to $(5,3)$ to $(5,1)$ to $(3,1)$; then $(6,3)$ to $(6,1)$; then $(7,3)$ to $(7,1)$ to $(9,1)$.

7.2   Slope

The activity of a stock over a 12-month period from January 1 to January 1 is shown on the accompanying graph.

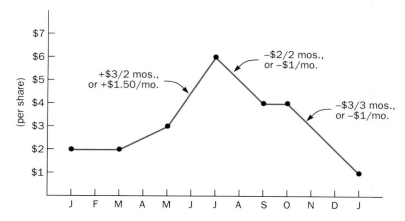

When did it perform the best? From May 1 to July 1, where it gained $3, or +$1.50 per month. Its worst performance was July 1 to September 1, where it lost $2, or −$1 per month, and from October 1 to January 1, where it lost $3, again −$1 per month. In each case, the "steepness" of the graph measures the stock's performance. This brief sojourn into the financial world leads to a more precise definition, in the mathematics world, of slope.

To define the slope of a line, we look at the line through the points $P_1(-1,2)$ and $P_2(5,6)$:

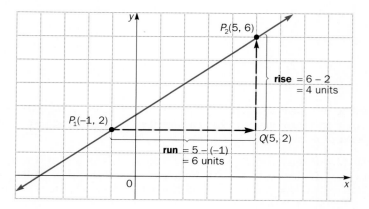

We form a right triangle with point $Q(5,2)$. The "rise" of the line is the vertical distance QP_2, which can be obtained by subtracting the y components: $6 - 2 = 4$. The "run" of the line is the horizontal distance P_1Q, which we obtain by subtracting the x components: $5 - (-1) = 6$. Finally, the **slope** of the line through P_1 and P_2 is defined as **the rise divided by the run:**

$$\text{Slope} = \frac{\text{rise}}{\text{run}} = \frac{6 - 2}{5 - (-1)} = \frac{4}{6} = \frac{2}{3}.$$

A slope of $\frac{2}{3}$ means that for every 3 units the line "runs" horizontally to the right, it "rises" 2 units vertically.

More generally, suppose P_1 and P_2 are any two points in the plane. As before, we define the *slope* of the line through these points to be *the rise divided by the run.*

$$\text{Slope} = \frac{\text{rise}}{\text{run}} = \frac{y_2 - y_1}{x_2 - x_1}$$

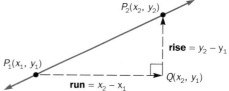

In mathematics, the letter m is traditionally used to denote slope,* so from the picture we get

The Slope Formula

The slope of the line through points $P_1(x_1, y_1)$ and $P_2(x_2, y_2)$ is

$$\text{Slope } m = \frac{y_2 - y_1}{x_2 - x_1}.$$

It can be shown that the slope of a line is the same regardless of the two points selected to compute it. And once the points are chosen, it does not matter which is labeled P_1 and which is labeled P_2. See Problems 63 and 64 in the exercises.

EXAMPLE 1

Find the slope of the line through the given pairs of points:

a. $(1, -2)$ and $(3, 2)$

Labeling the points P_1 and P_2 as shown, we have

$$\text{Slope } m = \frac{2 - (-2)}{3 - 1} = \frac{4}{2} = 2.$$

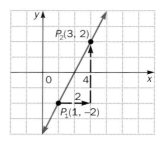

(continued)

*The letter m is from the French *monter*, which means "to climb."

b. $(-2,-1)$ and $(3,-2)$

$$\text{Slope } m = \frac{-2-(-1)}{3-(-2)} = \frac{-1}{5}$$

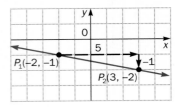

c. $(-2,1)$ and $(3,1)$

$$\text{Slope } m = \frac{1-1}{3-(-2)} = \frac{0}{5} = 0$$

The slope of a horizontal line is 0.

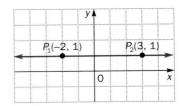

d. $(2,-1)$ and $(2,3)$

$$\text{Slope } m = \frac{3-(-1)}{2-2} = \frac{4}{0} \text{ (undefined)}$$

The slope of a vertical line is *undefined*.

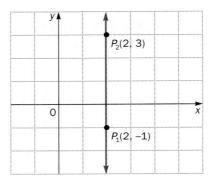

We can summarize these four results as follows:

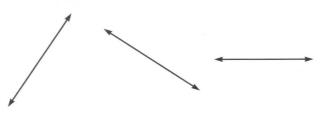

positive slope: negative slope: horizontal line: vertical line:
$m > 0$ $m < 0$ $m = 0$ m is undefined

Let us return to the line we graphed in Section 7.1, Example 1: $y = 2x + 1$.

The picture shows that the slope is 2 and the y-intercept is 1, which happen to be the constants in the equation $y = 2x + 1$. This equation is said to be in **slope–intercept form,** because it reveals directly the line's slope and y-intercept. This leads to a generalization:

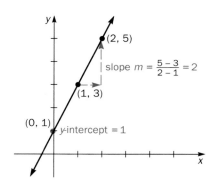

slope $m = \frac{5-3}{2-1} = 2$

y-intercept = 1

Slope–Intercept Form of a Line

$$y = mx + b$$

has slope $= m$, y-intercept $= b$.

In the next example, we show how the slope and y-intercept can be used to graph each line, without using a table as in Section 7.1.

EXAMPLE 2

Put each equation into the slope–intercept form of a line. Give the slope and y-intercept, and use them to graph each line.

a. $2x - 3y = 3$.

Solving for y gives

$$-3y = -2x + 3$$

$$y = \frac{2}{3}x - 1. \longleftarrow \text{ Slope–intercept form}$$

We see that the slope $= \frac{2}{3}$ and the y-intercept $= -1$. Starting from the point $(0, -1)$, we proceed 3 units to the right and 2 units up, arriving at $(3, 1)$. The line is drawn through both points.

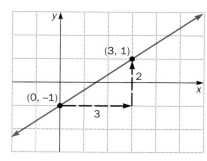

(continued)

b. $y = -3x + 2$.

This is already in slope–intercept form: slope $= -3 = \dfrac{-3}{1}$ and y-intercept $= 2$. Starting from $(0,2)$, we proceed 1 unit to the right and 3 units down, arriving at $(1,-1)$.

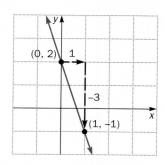

In geometry, we say that two lines in the same plane are **parallel** provided they do not intersect. In order not to intersect, they must have the same steepness, or slope: $m_1 = m_2$.

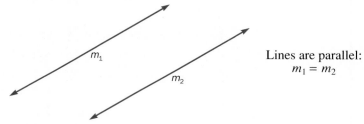

Lines are parallel:
$m_1 = m_2$

Two lines in the same plane are **perpendicular** if they intersect at right angles, or 90°. Not so obvious, and certainly not easy to visualize, is the relationship between the slopes m_1 and m_2 of perpendicular lines. The following discussion will show that their product is *negative one*:

$$m_1 m_2 = -1.$$

In the picture, we assume (to make the calculations easy) that the line with slope m_1 goes through the origin $(0,0)$ and some other point (a,b). By rotating this line 90°, we get a second line, with slope m_2, perpendicular to the first line and going through $(-b,a)$. We then compute the slope of each line.

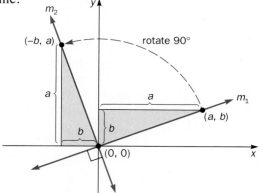

Compute slopes:

$$m_1 = \frac{b-0}{a-0} = \frac{b}{a}$$

$$m_2 = \frac{a-0}{-b-0} = \frac{a}{-b} = \frac{-a}{b}$$

Their product is:

$$m_1 m_2 = \frac{b}{a} \cdot \frac{-a}{b} = \frac{-ab}{ab} \overset{\checkmark}{=} -1$$

Let us summarize these two important facts:

Non-vertical lines with slopes m_1 and m_2 are

1. **parallel** if and only if $m_1 = m_2$.

2. **perpendicular** if and only if $m_1 m_2 = -1$, or $m_1 = \dfrac{-1}{m_2}$.

These properties of slopes apply only to "non-vertical lines." What about a vertical line, whose slope is undefined? It is easy to see that two vertical lines are parallel and that a vertical line is perpendicular to a horizontal line. One more comment. The last equation in the box, $m_1 = \dfrac{-1}{m_2}$, says that the slopes of perpendicular lines are *negative reciprocals* of each other.

EXAMPLE 3

Which of the following pairs of lines are parallel, perpendicular, or neither?

a. Line 1 goes through (1,2) and (7,5); line 2 goes through (2,0) and (6,2).

Compute slopes:

$$m_1 = \frac{5-2}{7-1} = \frac{3}{6} = \frac{1}{2}$$

$$m_2 = \frac{2-0}{6-2} = \frac{2}{4} = \frac{1}{2}$$

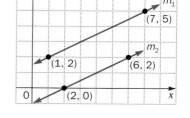

The lines are *parallel*, because $m_1 = m_2$.

b. Line 1 goes through (1,2) and (4,7); line 2 goes through (3,3) and (8,0).

Compute slopes:

$$m_1 = \frac{7-2}{4-1} = \frac{5}{3}$$

$$m_2 = \frac{0-3}{8-3} = \frac{-3}{5}$$

$$m_1 m_2 = \frac{5}{3} \cdot \frac{-3}{5} = \frac{-15}{15} = -1$$

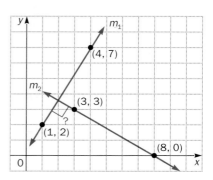

The lines are *perpendicular*, because $m_1 m_2 = -1$.

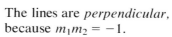
(continued)

c. Line 1: $2x + 3y - 7 = 0$; line 2: $4x - 6y + 3 = 0$.
We put both equations into slope–intercept form:

$$2x + 3y - 7 = 0 \qquad\qquad 4x - 6y + 3 = 0$$
$$3y = -2x + 7 \qquad\qquad 4x + 3 = 6y$$
$$y = \frac{-2}{3}x + \frac{7}{3} \qquad\qquad \frac{2}{3}x + \frac{1}{2} = y$$
$$\text{Slope } m_1 = \frac{-2}{3} \qquad\qquad \text{Slope } m_2 = \frac{2}{3}$$

The lines are *neither* parallel nor perpendicular, because $m_1 \neq m_2$ and $m_1 m_2 = \frac{-2}{3} \cdot \frac{2}{3} = \frac{-4}{9} \neq -1$.

EXERCISE 7.2   *Answers, pages A70–A71*

A

Find the slope of the line through the given pairs of points, as in EXAMPLE 1:

1. (1,3) and (3,7)

2. (2,1) and (8,5)

3. (−1,4) and (3,−2)

4. (6,−2) and (−3,−8)

5. (6,−2) and the origin

6. the origin and (−6,8)

7. (13,33) and (−21,−18)

8. (−10,42) and (−102,−27)

9. (−3,−1) and (2,−1)

10. (4,2) and (0,2)

11. (−2,1) and (−2,4)

12. (5,0) and (5,−3)

Put each equation into the slope–intercept form of a line, if it is not already in this form. Give the slope and *y*-intercept, and use them to graph each line. See EXAMPLE 2.

13. $y = -2x + 3$

14. $y = 3x - 2$

15. $y = \frac{x}{2} - 2$

16. $y = \frac{3x}{2}$

17. $3x - y = 1$

18. $x + y = 3$

19. $3x + 4y - 12 = 0$

20. $5x - 7y + 35 = 0$

21. $2x + 7 - 3y = 0$

22. $9y - 6 + 2x = 0$

23. $2x + 5y = 0$

24. $x - 2y = 4$

25. $\frac{x}{2} + \frac{y}{6} = 1$

26. $2y + 5 = 0$

Which of the following pairs of lines are parallel, which are perpendicular, and which are neither? See EXAMPLE 3.

27. Line 1: $m_1 = \frac{5}{8}$; line 2: $m_2 = \frac{-8}{5}$

28. Line 1: $m_1 = \frac{-17}{19}$; line 2: $m_2 = \frac{59}{51}$

29. Line 1 goes through (1,−2) and (5,4); line 2 goes through (−7,4) and (1,16).

30. Line 1 goes through (−3,4) and (5,2); line 2 goes through (0,7) and (4,6).

31. Line 1 goes through (3,4) and (5,−2); line 2 goes through (5,−2) and (−4,−5).

32. Line 1 goes through $(4,0)$ and $(-2,3)$; line 2 goes through $(-7,-2)$ and $(-3,6)$.

33. Line 1: $3x + 5y = 7$; line 2: $6x + 10y = 1$

34. Line 1: $2x - 7y = 5$; line 2: $4x - 14y + 9 = 0$

35. Line 1: $3x - 6 + 2y = 0$; line 2: $4x = 6y + 5$

36. Line 1: $x + 2y = 0$; line 2: $2x = y - 9$

37. Line 1: $3x + 4y - 1 = 0$; line 2: $6x = 8y + 11$

38. Line 1: $x + 5y = 0$; line 2: $5x + y = 0$

B

39. Line 1: $m_1 = 6\sqrt{2}$; line 2: $m_2 = 3\sqrt{8}$

40. Line 1: $m_1 = \dfrac{\sqrt{3}}{3}$; line 2: $m_2 = \dfrac{1}{\sqrt{3}}$

41. Line 1: $m_1 = 0$; line 2: m_2 is undefined

42. Line 1: m_1 is undefined; line 2: m_2 is undefined

43. Line 1: $y - 3 = 0$; line 2: $x + 2 = 0$

44. Line 1: $m_1 = a - b$; line 2: $m_2 = \dfrac{1}{b - a}$

45. Line 1: $m_1 = \sqrt{2} + 1$; line 2: $m_2 = 1 - \sqrt{2}$

46. Line 1: $m_1 = \dfrac{\sqrt{7} + \sqrt{3}}{2}$; line 2: $m_2 = \dfrac{\sqrt{3} - \sqrt{7}}{2}$

47. Line 1: $m_1 = \dfrac{2}{\sqrt{5} - \sqrt{3}}$; line 2: $m_2 = \sqrt{5} + \sqrt{3}$

48. Line 1: $m_1 = \dfrac{1}{\sqrt{2} + 1}$; line 2: $m_2 = \sqrt{2} - 1$

The slope of the line through $\left(\frac{1}{2},\frac{2}{3}\right)$ and $\left(\frac{3}{4},\frac{1}{6}\right)$ involves a complex fraction (Section 4.3):

$$m = \frac{\dfrac{1}{6} - \dfrac{2}{3}}{\dfrac{3}{4} - \dfrac{1}{2}} = \frac{\dfrac{1}{6} - \dfrac{4}{6}}{\dfrac{3}{4} - \dfrac{2}{4}} = \frac{\dfrac{-3}{6}}{\dfrac{1}{4}} = \frac{-3}{6} \cdot \frac{4}{1} = -2.$$

L.C.D. Invert and multiply.

In like manner, find the slope of the lines through the given points:

49. $\left(1,\dfrac{-1}{3}\right)$ and $\left(\dfrac{1}{2},\dfrac{3}{4}\right)$

50. $\left(\dfrac{-3}{2},\dfrac{3}{4}\right)$ and $\left(-2,\dfrac{1}{2}\right)$

51. $\left(\dfrac{1}{3},\dfrac{1}{2}\right)$ and $\left(\dfrac{1}{6},\dfrac{-2}{3}\right)$

52. $\left(\dfrac{1}{2},\dfrac{-1}{6}\right)$ and $\left(\dfrac{1}{3},\dfrac{1}{5}\right)$

53. $\left(-2,\dfrac{1}{2}\right)$ and $\left(\dfrac{5}{3},-\dfrac{1}{4}\right)$

54. $\left(\dfrac{5}{8},\dfrac{1}{6}\right)$ and $\left(\dfrac{-3}{4},0\right)$

Finding the slope of the line through $(-1,1)$ and $(\sqrt{2},\sqrt{2})$ requires you to rationalize the denominator (Section 5.5):

$$m = \frac{\sqrt{2} - 1}{\sqrt{2} - (-1)} = \frac{(\sqrt{2} - 1) \cdot (\sqrt{2} - 1)}{(\sqrt{2} + 1) \cdot (\sqrt{2} - 1)} = \frac{2 - 2\sqrt{2} + 1}{2 - 1} = 3 - 2\sqrt{2}.$$

conjugates.

In like manner, find the slope of the lines through the given points:

55. $(-2,2)$ and $(\sqrt{2},\sqrt{2})$

56. $(\sqrt{3},\sqrt{3})$ and $(3,-3)$

57. $(0,0)$ and $(\sqrt{2},3\sqrt{8})$

58. $(4\sqrt{27},2\sqrt{3})$ and $(0,0)$

59. $(0,0)$ and $(\sqrt{3},-6)$

60. $(0,0)$ and $(-\sqrt{5},20)$

61. $(\sqrt{2},-1)$ and $(\sqrt{7},4)$

62. $(-1,2)$ and $(\sqrt{5},6)$

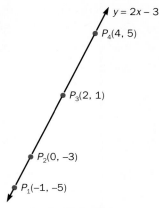

Problem 63

63. In the text it was mentioned that the slope of a line is the same, regardless of the two points selected to compute it. In this problem, the line in question will be $y = 2x - 3$.
   a. Verify that every point shown lies on this line by substituting each of them into the equation.
   b. Compute the slope through P_1 and P_2 and the slope between P_3 and P_4. They should be equal.

64. It was also mentioned in the text that it does not matter which of two points is labeled $P_1(x_1, y_1)$ and which is labeled $P_2(x_2, y_2)$, because the slope will be the same in either case. Verify this fact by showing that

$$\frac{y_2 - y_1}{x_2 - x_1} = \frac{y_1 - y_2}{x_1 - x_2}.$$

Hint: Multiply the second expression by $\frac{-1}{-1}$.

65. A **parallelogram** is a four-sided polygon in which both pairs of opposite sides are **parallel**. Show that the given figure is a parallelogram by first computing the slopes of the sides.

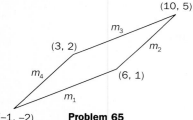

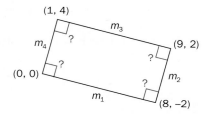

Problem 65

Problem 66

66. A **rectangle** is a four-sided polygon in which all angles are 90°. Show that the given figure is a rectangle by first computing the four slopes and then showing that the sides are **perpendicular.**

67. A **right triangle** has two sides **perpendicular.** Show that the given figure is a right triangle by first computing the three slopes and then showing that the product of two of them is -1. Warning: The triangle is not drawn to scale!

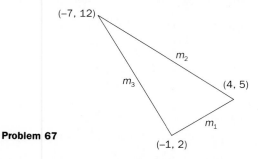

Problem 67

68. Below are monthly prices of various stocks. Calculate their performances on a "per month" gain or loss basis. Then rank them from best to worst.

Stock A: Jan. 1 = \$2, Mar. 1 = \$7   Stock B: Mar. 1 = \$8, June 1 = \$20
Stock C: July 1 = \$10, Sept. 1 = \$2   Stock D: Sept. 1 = \$3, Dec. 1 = \$18
Stock E: Feb. 1 = \$5, July 1 = \$5   Stock F: Jan. 1 = \$17, Apr. 1 = \$9
Stock G: Feb. 1 = \$21, Aug. 1 = \$5   Stock H: May 1 = \$27, Oct. 1 = \$42

Finding the slope of the line through (b,b^2) **and** (a,a^2) **involves factoring:**

$$m = \frac{a^2 - b^2}{a - b} = \frac{(a + b)(a - b)}{a - b} = a + b.$$

In like manner, find the slope of the lines through the given points:

69. $(3,9)$ and (x,x^2)

70. $(-5,25)$ and (x,x^2)

71. $(a,2a)$ and $(x,2x)$

72. $(4b,6b)$ and $(8y,12y)$

73. $(a,2b)$ and $(-5a,-2b)$

74. $(-2c,3d)$ and $(4c,-5d)$

75. $(2,8)$ and (x,x^3)

76. $(-3,-27)$ and (x,x^3)

77. $\left(a,\dfrac{1}{a}\right)$ and $\left(x,\dfrac{1}{x}\right)$

78. $\left(x,\dfrac{1}{x}\right)$ and $\left(x + h,\dfrac{1}{x + h}\right)$

79. $\left(2,\dfrac{1}{4}\right)$ and $\left(x,\dfrac{1}{x^2}\right)$

80. (x,x^2) and $(x + h,(x + h)^2)$

Here is a problem involving slope that leads to an equation: "Find k such that the line through $(-2,k)$ and $(2,1)$ has slope $\dfrac{3}{2}$." The solution is obtained as follows:

$$\frac{1 - k}{2 - (-2)} = \frac{3}{2}, \quad \text{or} \quad \frac{1 - k}{4} = \frac{3}{2}$$

Solving as in Section 4.5 gives $1 - k = 6$, **or** $-5 = k$.
In a like manner, solve the following problems:

81. Find k such that the line through $(1,k)$ and $(3,4)$ has slope $\dfrac{2}{3}$.

82. Find k such that the line through $(-1,2)$ and $(k,3k)$ has slope $\dfrac{-3}{4}$.

83. Find k such that the line through $(3,-2)$ and $(2k,k)$ is parallel to the line through $(-6,-1)$ and $(k,2k)$. *Hint*: The slopes are equal; you will need to solve a second-degree equation.

84. Find k such that the line through $(-2,-10)$ and $(7,k)$ is perpendicular to the line through $(-3,4)$ and $(2,k)$. *Hint*: Product of slopes = -1.

7.3   Deriving Equations of Lines

In Section 7.1 you were given the equation of a line and asked to graph it. Now you will do the reverse: Given a geometric description of a line, derive its equation. To accomplish this, we need an important fact. "Given a fixed point (x_1, y_1) and a fixed slope m, there is a unique line going through the point with the given slope." In a sense, these conditions "fix" the line.

To find the equation, we let (x, y) be a variable point on the line. We then compute its slope.

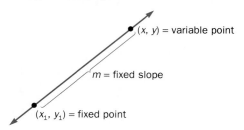

By the slope formula,

$$\frac{y - y_1}{x - x_1} = m.$$

Equivalently, this is

$$y - y_1 = m(x - x_1).$$

This last equation is called the **point–slope** form of a straight line. If we are given one point and the slope of the line, we can determine the equation of the line.

Point–Slope Form of a Line

$$y - y_1 = m(x - x_1)$$

is the equation of the line with slope m, going through (x_1, y_1).

EXAMPLE 1

Find the equation of the line satisfying the given conditions. Express your answer in slope–intercept form: $y = mx + b$.

a. The slope $m = -2$, and the line goes through $(4, -3)$.

Substituting into the point–slope form gives

$$y - (-3) = -2(x - 4)$$
$$y + = -2x + 8$$
$$y = 2x + 5. \qquad \text{Slope–intercept form.}$$

b. The line goes through $(3, -2)$ and $(9, 6)$.

We first compute the slope: $m = \dfrac{6 - (-2)}{9 - 3} = \dfrac{8}{6} = \dfrac{4}{3}$.

We now substitute *either* of the two given points—say $(9, 6)$—into the point–slope form:

$$y - 6 = \frac{4}{3}(x - 9) \qquad \text{Point–slope form.}$$

$$\frac{(y - 6) \cdot 3}{1 \cdot 3} = \frac{4(x - 9)}{3} \qquad \text{L.C.D. = 3}$$

$$3y - 18 = 4x - 36 \qquad \text{Equate numerators.}$$
$$3y \quad\;\; = 4x - 18$$

$$y = \frac{4x}{3} - 6. \qquad \text{Divide by 3.}$$

c. The line goes through $(-1,5)$ and $(7,5)$.

First compute the slope: $m = \dfrac{5 - 5}{7 - (-1)} = \dfrac{0}{8} = 0.$

Now substitute $(-1,5)$ into the point–slope form:

$$y - 5 = 0[x - (-1)]$$
$$y - 5 = 0$$
$$y = 5.$$

You should recognize this as a *horizontal* line.

d. The line goes through $(2,-1)$ and $(2,3)$.

Compute the slope: $m = \dfrac{3 - (-1)}{2 - 2} = \dfrac{4}{0}$ is *undefined*.

We cannot use the point–slope form, but you should recognize this as the *vertical* line in Example 1d of Section 7.2. This line has the equation

$$x = 2.$$

Two axioms from plane geometry give us further criteria for deriving equations of lines. The **Parallel Axiom** states that

> "Given a line L_1 and a point P (not on L_1), there is a unique line L_2 going through P that is *parallel* to L_1."

The **Perpendicular Axiom** states that

> "Given a line L_1 and a point P (which may or may not be on L_1), there is a unique line L_2 going through P that is *perpendicular* to L_1."

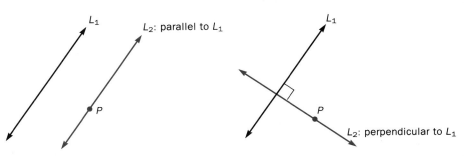

In the next example, we show how to find the equations of the lines guaranteed by these axioms.

EXAMPLE 2

Given the line $3x - 4y + 16 = 0$ and the point $(1,3)$, find the equation of the line through this point that is
a. *parallel* to the given line.
b. *perpendicular* to the given line.

Solution
We first put the given line into slope–intercept form:

$$-4y = -3x - 16, \quad \text{or} \quad y = \frac{3}{4}x + 4.$$

Thus the given line has slope $m = \frac{3}{4}$.

a. Because the lines are to be *parallel*, our new line must also have slope $m = \frac{3}{4}$. The point–slope form gives

$$y - 3 = \frac{3}{4}(x - 1)$$

$$\frac{(y - 3) \cdot 4}{1 \cdot 4} = \frac{3(x - 1)}{4}$$

$$4y - 12 = 3x - 3$$
$$4y = 3x + 9$$

$$y = \frac{3}{4}x + \frac{9}{4}.$$

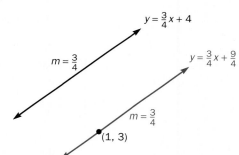

b. Because the lines are to be *perpendicular*, our new line must have slope $m = \frac{-4}{3}$, which is the negative reciprocal of the given slope $m = \frac{3}{4}$. The point–slope form gives

$$y - 3 = \frac{-4}{3}(x - 1)$$

$$\frac{(y - 3) \cdot 3}{1 \cdot 3} = \frac{-4(x - 1)}{3}$$

$$3y - 9 = -4x + 4$$
$$3y = -4x + 13$$

$$y = \frac{-4}{3}x + \frac{13}{3}.$$

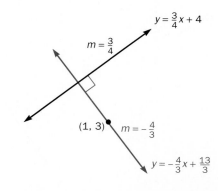

EXERCISE 7.3   *Answers, pages A71–A72*

A

Find the equation of the line satisfying the given conditions. Express your answer in slope–intercept form: $y = mx + b$. See EXAMPLE 1.

1. $m = -3$, goes through $(-5,2)$

2. $m = 3$, goes through $(6,-1)$

3. $m = \dfrac{2}{5}$, goes through $(15,-10)$

4. $m = \dfrac{-1}{2}$, goes through $(8,-3)$

5. $m = 0$, goes through $(2,-1)$

6. m is undefined, goes through $(3,-4)$

7. $m = -2$, goes through the origin

8. $m = \dfrac{2}{3}$, goes through the origin

9. $m = 4$, y-intercept $= 5$ *Hint*: $(0,5)$

10. $m = -5$, x-intercept $= 7$ *Hint*: $(7,0)$

11. Line goes through $(2,-1)$ and $(4,-7)$

12. Line goes through $(0,3)$ and $(5,-7)$

13. Line goes through $(-4,5)$ and $(2,1)$

14. Line goes through $(0,-3)$ and $(-8,1)$

15. Line goes through $(-3,3)$ and $(4,3)$

16. Line goes through $(4,-3)$ and $(4,5)$

17. x-intercept $= 2$, y-intercept $= 5$

18. x-intercept $= -3$, y-intercept $= 4$

19. $m = -3$, goes through $\left(\dfrac{5}{6},\dfrac{2}{3}\right)$

20. $m = 2$, goes through $\left(\dfrac{1}{2},\dfrac{-3}{4}\right)$

In the following problems, you will be given a line and a point. In each case, find the equation of the line through this point that is
a. *parallel* to the given line.   **b.** *perpendicular* to the given line.
See EXAMPLE 2. Express your answers in slope-intercept form.

21. Line: $y = -2x + 7$, point: $(4,-6)$

22. Line: $y = 3x - 2$, point: $(-1,9)$

23. Line: $2x + 3y = 5$, point: $(4,2)$

24. Line: $3x - 4y + 1 = 0$, point: $(2,0)$

25. Line: goes through $(-3,2)$ and $(5,4)$, point: $(1,-12)$
Hint: Find the slope of the given line.

26. Line: goes through $(2,0)$ and $(-6,12)$, point: origin

B

27. Fahrenheit and Celsius temperatures are related by a linear equation in two variables (C,F). Find this equation for F in terms of C, given that the freezing point of water is $(0°C,32°F)$ and the boiling point of water is $(100°C,212°F)$. Then use your equation to find F when $C = 25°$.

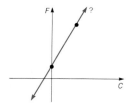

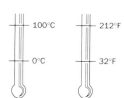

28. Two lesser-known temperature scales are the Suzy and the Mary scales, which are also related by a linear equation in two variables (M,S). On these scales, water freezes at $-40°$M and $-10°$S, and it boils at $50°$M and $125°$S. Find this equation for S in terms of M. Then use your equation to find S when $M = 20°$.

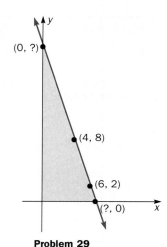

C

29. A line goes through the points (4,8) and (6,2).
   a. Find the equation of this line.
   b. Use your equation to find its x- and y-intercepts.
   c. Find the area of the shaded triangle in the first quadrant formed by this line, the x-axis, and the y-axis.

Problem 29

7.4 **Linear Systems in Two Variables**

We begin this section by graphing the lines $2x + y = 5$ and $4x - y = 7$ on the same set of axes and locating their point of intersection.

$2x + y = 5$

x	y
0	5
1	3
2	1
3	-1

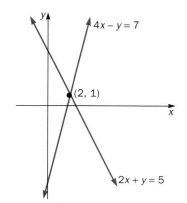

$4x - y = 7$

x	y
0	-7
1	-3
2	1
3	5

We see that the lines intersect at (2,1). We were lucky! Two lines might just as easily intersect at a point that does not readily appear on the table or graph, because its coordinates are very large or are fractions. Even worse, the lines might be parallel and not intersect at all, as Example 4 will illustrate. The point here is that graphical methods for locating points of intersections are inefficient. Algebraically, however, it is simple. By adding the two equations, we eliminate one of the variables and solve for the other:

$$
\begin{array}{rl}
2x + y &= 5 \\
4x - y &= 7 \\
\hline
6x &= 12 \\
x &= 2.
\end{array}
$$

Now substitute $x = 2$ into either of the equations—say the first:

$$
\begin{aligned}
2 \cdot 2 + y &= 5 \\
y &= 1.
\end{aligned}
$$

The lines intersect at $x = 2$, $y = 1$, or the point $(2,1)$.

In the preceding process, two equations were added in order to eliminate one of the unknowns. To justify this, we have the following axiom for algebraic expressions A, B, C, and D:

Addition Property of Equations

$$
\begin{array}{rl}
\text{If} & A = B \\
\text{and} & C = D, \\
\hline
\text{then} & A + C = B + D.
\end{array}
$$

This says that "equals added to equals are equal." It is an extension of the first property of an equation given at the beginning of Chapter 2, in which $A + C = B + C$. The method used to solve such a system of linear equations is appropriately called the **addition method**.

EXAMPLE 1

Solve this system by the addition method:

$$
\begin{aligned}
3x + 5y &= -23 \\
-x + 7y &= -53
\end{aligned}
$$

Solution

Adding the two equations results in the equation $2x + 12y = -76$, which is useless because neither variable is eliminated. Instead, we can eliminate x by multiplying both sides of the second equation by 3* and then adding:

$$
\begin{array}{l}
3x + 5y = -23 \xrightarrow{\hspace{3cm}} 3x + 5y = -23 \\
-x + 7y = -53 \xrightarrow{\text{Multiply by 3.}} -3x + 21y = -159 \\
\hline
\hspace{5cm} 26y = -182 \\
\hspace{5cm} y = -7.
\end{array}
$$

---

*Multiplying both sides of an equation by 3 is valid by the second property of an equation given at the beginning of Chapter 2.

(continued)

Now substitute $y = -7$ into the second equation, because it is the easiest:

$$-x + 7(-7) = -53$$
$$-x - 49 = -53$$
$$4 = x.$$

Answer: $x = 4$, $y = -7$ or $(4, -7)$.

Check by substituting into *both* equations:

$$3(4) + 5(-7) = -23, \quad \text{or} \quad 12 - 35 \stackrel{\checkmark}{=} -23$$
$$-(4) + 7(-7) = -53, \quad \text{or} \quad -4 - 49 \stackrel{\checkmark}{=} -53.$$

EXAMPLE 2   Solve this system by the addition method:

$$6x + 4y = -5$$
$$14x + 3y = 1$$

Solution

In order to eliminate y, say, we need to make its coefficients *equal* and *opposite in sign:* 12 and -12. To do this, we multiply as shown:

$$
\begin{array}{lll}
6x + 4y = -5 & \xrightarrow{\text{Multiply by 3.}} & 18x + 12y = -15 \\
14x + 3y = 1 & \xrightarrow{\text{Multiply by } -4.} & -56x - 12y = -4 \\
& & \overline{\quad -38x \qquad\quad = -19} \\
& & \qquad\qquad x = \dfrac{1}{2}.
\end{array}
$$

Substitute $x = \dfrac{1}{2}$ into the first equation:

$$6\left(\frac{1}{2}\right) + 4y = -5$$
$$3 + 4y = -5$$
$$4y = -8$$
$$y = -2.$$

Answer: $x = \dfrac{1}{2}$, $y = -2$ or $(\dfrac{1}{2}, -2)$.

Check: You should verify these answers in *both* equations.

The addition method necessitates multiplying one or both of the original equations by suitable numbers and then adding in order to eliminate one of the unknowns. Another technique, called the **substitution method,** is easier when one of the variables is isolated. For example, the system

$$x = 2y + 10$$
$$3x + 7y = -9$$

can be solved by substituting $x = 2y + 10$ into the second equation:

$$3(2y + 10) + 7y = -9$$
$$6y + 30 + 7y = -9$$
$$13y = -39$$
$$y = -3.$$

Now substitute $y = -3$ into the first equation:

$$x = 2(-3) + 10$$
$$x = 4.$$

The **answer** is $x = 4$, $y = -3$ or $(4, -3)$.

EXAMPLE 3

Solve this system by the substitution method:

$$2x - 3y = 2$$
$$5x - y = 5$$

Solution

It is easiest to isolate y in the second equation:

$$5x - 5 = y$$

Now substitute this expression into the first equation:

$$2x - 3(5x - 5) = 2$$
$$2x - 15x + 15 = 2 \qquad \text{Pay careful attention to signs!}$$
$$-13x = -13$$
$$x = 1.$$

Then substitute $x = 1$ into the expression for y:

$$5(1) - 5 = y$$
$$0 = y.$$

Answer: $x = 1$, $y = 0$ or $(1, 0)$.

Check: You should verify these answers in both equations.

Because the solution to a linear system of two equations in two unknowns represents the intersection of two lines, it is natural to ask what happens in the case of parallel lines. We shall see in the next example that the attempt to eliminate one of the unknowns results in the elimination of both!

EXAMPLE 4

Solve this system:

$$5x - 2y = 7$$
$$-10x + 4y = -8$$

Solution
This is best solved by the addition method:

$$
\begin{array}{l}
5x - 2y = 7 \xrightarrow{\text{Multiply by 2.}} 10x - 4y = 14 \\
-10x + 4y = -8 \xrightarrow{\phantom{\text{Multiply by 2.}}} \underline{-10x + 4y = -8} \\
\phantom{-10x + 4y = -8 \xrightarrow{}} 0 = 6. \qquad \text{FALSE}
\end{array}
$$

Both variables have vanished, leaving the FALSE statement $0 = 6$.

Answer: No solution

An explanation of the foregoing result lies in the slope–intercept forms of these lines: $y = \frac{5}{2}x - \frac{7}{2}$ and $y = \frac{5}{2}x - 2$. We see that the lines are *parallel* because they have the same slope $m = \frac{5}{2}$. However, they have different y-intercepts and therefore no points of intersection. This system is called *inconsistent*. By contrast, Examples 1, 2, and 3 contain systems that have a single point of intersection as a solution. These systems are called *consistent*.

EXAMPLE 5

Solve this system:

$$2x - 5y = 10$$
$$6x - 15y = 30$$

Solution
We solve by the addition method:

$$
\begin{array}{l}
2x - 5y = 10 \xrightarrow{\text{Multiply by } -3.} -6x + 15y = -30 \\
6x - 15y = 30 \xrightarrow{\phantom{\text{Multiply by } -3.}} \underline{6x - 15y = 30} \\
\phantom{6x - 15y = 30 \xrightarrow{}} 0 = 0. \qquad \text{TRUE}
\end{array}
$$

Both unknowns have been eliminated, but certainly $0 = 0$ is TRUE. A quick inspection of the original system reveals that the second equation is exactly 3 times the first equation, term for term. Thus both equations represent the same line, which, in slope–intercept form, is $y = \frac{2}{5}x - 2$. The solution of the system consists of all points on this line, which contains infinitely many points. Such a system is called *dependent*.

Answer: Infinitely many solutions

Let us now summarize the three possible solutions for a system of two linear equations in two unknowns:

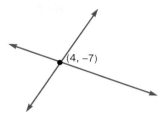   

One solution
The lines intersect in one point.
The system is "consistent."
(Examples 1, 2, 3)

No solution
(get 0 = 6)
The lines are parallel.
The system is "inconsistent."
(Example 4)

Infinitely many solutions
(get 0 = 0)
Only one line
The system is "dependent."
(Example 5)

The *standard form* of a linear system in unknowns x and y is

$$a_1x + b_1y = c_1$$
$$a_2x + b_2y = c_2$$

where the a's, b's, and c's are constant coefficients. Quite often, algebraic maneuvering is needed to put complicated systems into this form.

EXAMPLE 6

Solve this system:

$$\frac{x + 3}{6} + \frac{y - 1}{3} = \frac{1}{3}$$

$$\frac{x - 1}{2} - \frac{y - 2}{5} = \frac{-1}{5}$$

Solution
We first rewrite each equation without fractions, using L.C.D.'s:

$$\frac{x + 3}{6} + \frac{(y - 1) \cdot 2}{3 \cdot 2} = \frac{1 \cdot 2}{3 \cdot 2} \qquad \frac{(x - 1) \cdot 5}{2 \cdot 5} - \frac{(y - 2) \cdot 2}{5 \cdot 2} = \frac{-1 \cdot 2}{5 \cdot 2}$$

$$x + 3 + 2y - 2 = 2 \qquad\qquad 5x - 5 - 2y + 4 = -2 \qquad \text{Note signs!}$$

$$x + 2y = 1. \qquad\qquad\qquad 5x - 2y = -1.$$

Our original system is in standard form, which we now solve:

$$\begin{array}{r} x + 2y = 1 \\ 5x - 2y = -1 \\ \hline 6x \qquad = 0 \\ x = 0. \end{array}$$

(continued)

Substitute $x = 0$ into the first of these equations:

$$0 + 2y = 1$$

$$y = \frac{1}{2}.$$

Answer: $x = 0$, $y = \frac{1}{2}$ or $(0, \frac{1}{2})$.

EXERCISE 7.4   *Answers, page A72*

A

Solve by the addition method, as in EXAMPLES 1 and 2:

1. $x + y = 11$
$x - y = 3$

2. $x + y = 19$
$x - y = -7$

3. $3x + 5y = 8$
$7x - y = 6$

4. $4x - 3y = -2$
$2x + y = 4$

5. $2u + 3v = 8$
$5u - 7v = 20$

6. $9w + 2z = -6$
$5w - 3z = 9$

7. $2s + 3t = 0$
$3s + 5t = 0$

8. $3r + 5t = 0$
$4r + 3t = 0$

9. $3x + 6 = -5y$
$7y = 5x - 13$

10. $4x + 7y - 5 = 0$
$7 - 6x = 9y$

11. $9r + 11s = 73$
$7r = 59 - 9s$

12. $8s = 6t - 54$
$9t = 30 - 5s$

13. $14m + 7n + 19 = 0$
$5n + 18 = 21m$

14. $11u - 12v = -8$
$5u = 12 - 18v$

Solve by the substitution method, as in EXAMPLE 3:

15. $2x - 3y = 20$
$y = 12 - 2x$

16. $y = 19 - 3x$
$5x - 4y = 60$

17. $x + 2y = 17$
$3x + 5y = 43$

18. $-3x + 5y = -31$
$x - y = 9$

19. $3z - w = -1$
$5z - 9w = -20$

20. $9r + 5t = 3$
$7r - t = -5$

21. $u = 2v - 15$
$u = 106 - 9v$

22. $s = 4t + 17$
$s = -4 - 4t$

23. $x = 10 - 3y$
$y = 5x - 18$

24. $w = 5z - 3$
$z = 6w + 19$

25. $x - 3y = 0$
$5x = 8y$

26. $7u = 6v$
$5 + v = 2u$

Solve as in EXAMPLES 4 and 5, using either method:

27. $2x - 3y = 11$
$4x - 6y = 13$

28. $-5x + 8y = -4$
$10x - 16y = 7$

29. $3x - 4y = -7$
$-6x + 8y = 14$

30. $6x + 3y = 9$
$8x + 4y = 12$

31. $6x - 3y = 15$
$y = 2x - 5$

32. $7 + 5x = y$
$10x = 2y - 13$

Solve as in EXAMPLE 6:

33. $\dfrac{x}{2} + \dfrac{2y}{3} = 3$

$\dfrac{x}{5} - \dfrac{y}{15} = \dfrac{1}{5}$

34. $\dfrac{x}{4} + \dfrac{5y}{3} = \dfrac{19}{4}$

$\dfrac{2x}{7} - \dfrac{5y}{6} = \dfrac{-39}{14}$

35. $\dfrac{x+2}{4} - \dfrac{y-3}{3} = \dfrac{5}{3}$

$\dfrac{x-4}{5} + \dfrac{1-2y}{6} = \dfrac{-7}{30}$

36. $\dfrac{2x}{5} + 3y = \dfrac{-24}{5}$

$\dfrac{x-3}{4} - \dfrac{y}{2} = 1$

37. $\dfrac{2x-y}{3} = \dfrac{x-2y}{2}$

$x - \dfrac{y+2}{3} = \dfrac{x-4y}{2}$

38. $\dfrac{3z+2w}{5} = \dfrac{2z-w}{3}$

$z - \dfrac{2-w}{5} = \dfrac{z+11w}{2}$

B

39. $\dfrac{x+2}{y+5} = \dfrac{x+1}{y-3}$

$(x+3)(y-7) = (x-9)(y+13)$

40. $(u+1)(v-2) = (u-4)(v+4)$

$\dfrac{u-5}{u+7} = \dfrac{v-3}{v+4}$

The system

$$\frac{5}{x} + \frac{4}{y} = -2$$

$$\frac{7}{x} - \frac{3}{y} = 23$$

can be solved *without* using the L.C.D. $= xy$, as follows:

$$\frac{5}{x} + \frac{4}{y} = -2 \quad \xrightarrow{\text{Multiply by 3.}} \quad \frac{15}{x} + \frac{12}{y} = -6$$

$$\frac{7}{x} - \frac{3}{y} = 23 \quad \xrightarrow{\text{Multiply by 4.}} \quad \frac{28}{x} - \frac{12}{y} = 92$$

$$\overline{\qquad\qquad}$$

$$\frac{43}{x} = 86$$

$$43 = 86x$$

$$\frac{1}{2} = x.$$

Substituting $x = \dfrac{1}{2}$ into the first equation gives

$$\frac{5}{\frac{1}{2}} + \frac{4}{y} = -2$$

$$10 + \frac{4}{y} = -2 \qquad \text{Recall: } \frac{5}{\frac{1}{2}} = \frac{5}{1} \cdot \frac{2}{1} = 10$$

$$\frac{4}{y} = -12$$

$$4 = -12y \qquad \textit{(continued)}$$

$$\frac{-1}{3} = y.$$

Answer: $x = \dfrac{1}{2}$, $y = \dfrac{-1}{3}$ or $\left(\dfrac{1}{2}, \dfrac{-1}{3}\right)$.

In like manner, solve the following systems:

41. $\dfrac{3}{x} + \dfrac{2}{y} = -2$

$\dfrac{5}{x} - \dfrac{2}{y} = 18$

42. $\dfrac{5}{x} + \dfrac{7}{y} = 16$

$\dfrac{3}{x} - \dfrac{7}{y} = -24$

43. $\dfrac{5}{x} - \dfrac{2}{y} = 4$

$\dfrac{4}{x} + \dfrac{3}{y} = 17$

44. $\dfrac{2}{x} + \dfrac{3}{y} = 11$

$\dfrac{5}{x} + \dfrac{4}{y} = 17$

45. $\dfrac{3}{x+1} + \dfrac{2}{y-1} = 0$

$\dfrac{5}{x+1} - \dfrac{3}{y-1} = -19$

46. $\dfrac{4}{x+2} - \dfrac{5}{y-3} = -11$

$\dfrac{3}{x+2} + \dfrac{2}{y-3} = 9$

The system $\quad ax - by = a^2 - b^2$

$\qquad\qquad\quad x + \ \ y = 4a + 4b$

with literal coefficients a and b can be solved by the methods of this section. This is similar to solving literal equations in Section 4.7.

$$ax - by = a^2 - b^2 \xrightarrow{} ax - by = a^2 - b^2$$
$$x + \ \ y = 4a + 4b \xrightarrow{\text{Multiply by } b.} bx + by = 4ab + 4b^2$$

$$\overline{ \ ax + bx \quad\ \ = a^2 + 4ab + 3b^2} \quad \text{Add.}$$
$$x(a + b) = (a + b)(a + 3b) \quad \text{Factor.}$$
$$x = a + 3b. \quad \text{Divide.}$$

Substituting $x = a + 3b$ into the second equation gives

$$a + 3b + y = 4a + 4b$$
$$y = 3a + b.$$

Answer: $x = a + 3b$, $y = 3a + b$.

In like manner, solve the following systems for x and y:

47. $2x + y = 4a + b$
$5x - y = 3a + 6b$

48. $5x + 3y = 12c + d$
$2x + 5y = c + 8d$

49. $2mx - 3ny = -2$
$mx + \ \ ny = 4$

50. $3px - 5qy = 4pq$
$5px + 3qy = \ \ pq$

51. $ax + y = a^2$
$bx + y = b^2$

52. $x - ay = a^2$
$x + by = b^2$

C

53. $x + cy = c^3$
$\quad x + dy = d^3$

54. $\quad cx + y = c^3$
$-dx + y = -d^3$

55. $\dfrac{3x}{a} - \dfrac{4y}{b} = 1$

$\dfrac{4x}{a} - \dfrac{3y}{b} = 6$

56. $\dfrac{3x}{4c} + \dfrac{y}{4d} = 1$

$\quad\;\; \dfrac{x}{2c} - \dfrac{3y}{2d} = 4$

57. $rx - sy = r^2 - s^2$

$\quad\;\; x - y = s - r$

58. $2mx - 3ny = 2m^2 + 3n^2$

$\quad\;\; x + y = 3m + 2n$

59. $ax + by = 2a^2 - 2b^2$

$\quad\;\; y - x = -a - b$

60. $cx + dy = c^2 + d^2$

$\quad\;\; y + 2x = 3c - d$

61. $\dfrac{ax - by}{a + b} = a - b$

$\quad\;\; \dfrac{y + x}{4} = a + b$

62. $\dfrac{ax - by}{a^2 + b^2} = 1$

$\quad\;\; \dfrac{x}{a} + \dfrac{y}{b} = \dfrac{a^2 + b^2}{ab}$

Solve for *x* and *y*, using your knowledge of radicals and complex numbers:

63. $5\sqrt{2}\,x + 2\sqrt{3}\,y = 22$

$\quad\;\; \sqrt{2}\,x - 3\sqrt{3}\,y = -16$

64. $2x\sqrt{5} - 3y\sqrt{2} = 36$

$\quad\;\; 3x\sqrt{5} + 4y\sqrt{2} = 37$

65. $ix + 3y = -4$

$\quad\;\; 3ix - 2y = -1$

66. $5x + 7iy = 29$

$\quad\;\; -x + 8iy = 13$

67. $x + y = 2i$

$\quad\;\; ix - y = 0$

68. $ix + 2y = -1 + 4i$

$\quad\;\; 2x - y = 0$

| | 7.5 | **Word Problems in Two Variables** |

In previous chapters we used a single variable to set up and solve word problems. Now we show how two unknowns can often make the task easier.

EXAMPLE 1

A collection of 51 coins in nickels and dimes is worth $4.20. How many of each coin are there?

Solution

Let n = number of nickels and d = number of dimes.

We set up a chart as in Section 2.2:

	number of coins ·	face value =	value of coins
nickels	n	5¢	$5n$¢
dimes	d	10¢	$10d$¢
total	51		$4.20, or 420¢

The two equations are

$$n + d = 51 \qquad \text{Number of coins.}$$
$$5n + 10d = 420 \qquad \text{Value of coins (in cents).}$$

(continued)

Solving by the addition method gives

$$n + d = 51 \xrightarrow{\text{Multiply by } -5.} \begin{array}{r} -5n - 5d = -255 \\ 5n + 10d = 420 \\ \hline 5d = 165 \\ d = 33. \end{array}$$

Substituting $d = 33$ into the first equation gives

$$n + 33 = 51$$
$$n = 18.$$

Answer: 18 nickels, 33 dimes.
Check: $5¢(18) + 10¢(33) = 90¢ + 330¢ \overset{\checkmark}{=} 420¢.$

EXAMPLE 2

$16,000 is invested, part at 11% for 2 years and the balance at 6% for 5 years. How much is invested at each rate if the combined interest earned is equivalent to the interest earned by investing the total amount at 7% for 4 years?

Solution
Let x = amount invested at 11% and y = amount invested at 6%.
We set up a chart as in Section 2.3:

	P ·	r ·	t	= I
11% investment	x	0.11	2	$2(0.11)x = 0.22x$
6% investment	y	0.06	5	$5(0.06)y = 0.30y$
7% investment	$16,000	0.07	4	$4(0.07)(16,000) = \$4480$

The two equations are

$$x + y = 16,000$$
$$0.22x + 0.30y = 4480$$

Solving by the addition method gives

$$\begin{array}{l} x + y = 16,000 \\ 0.22x + 0.30y = 4480 \end{array} \xrightarrow[\text{Multiply by 100.}]{\text{Multiply by } -22.} \begin{array}{r} -22x - 22y = -352,000 \\ 22x + 30y = 448,000 \\ \hline 8y = 96,000 \\ y = 12,000. \end{array}$$

Substituting $y = 12,000$ into the first equation gives

$$x + 12,000 = 16,000$$
$$x = 4000.$$

Answer: $4000 at 11%, $12,000 at 6%.
Check: You should check these figures in the second equation.

Our next example requires a short lesson from chemistry. An acid solution consists of pure acid diluted with water. Thus a 40% sulfuric acid (H_2SO_4) solution consists of 40% pure H_2SO_4 and 60% water. Then 30 liters of this solution would consist of

$$30(0.40) = 12 \text{ liters of pure } H_2SO_4$$

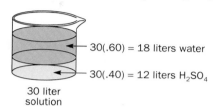

30 liter
solution

and $30(0.60) = 18$ liters of water. We will be concerned only with the acid, not with the water, so we will be using the formula

$$\begin{array}{c}\text{volume of} \\ \text{solution}\end{array} \times \begin{array}{c}\text{strength} \\ \text{of acid}\end{array} = \begin{array}{c}\text{volume of} \\ \text{pure acid}\end{array}$$

In our next example, we mix a 40% H_2SO_4 solution with an 80% H_2SO_4 solution to produce a 50% H_2SO_4 solution. In general, mixing chemical solutions of two different strengths yields a solution that is stronger than the weaker of the two but weaker than the stronger of the two.

EXAMPLE 3

How many liters of an 80% H_2SO_4 solution should be mixed with 30 liters of a 40% H_2SO_4 solution to produce a 50% solution? How many liters will be produced?

Solution

Let x = liters of 80% solution required
and y = liters of 50% solution produced.

	liters solution	× strength of acid	= liters of pure acid
80% solution	x	0.80	$0.80x$
40% solution	30	0.40	$0.40(30) = 12$
50% solution	y	0.50	$0.50y$

(continued)

Because the liters of the two component solutions must equal the total liters in the mixture, we get

$$x + 30 = y.$$

And because the liters of pure acid in the two component solutions must equal those in the mixture, we get

$$0.80x + 12 = 0.50y.$$

We now solve this system by the substitution method:

$$
\begin{aligned}
0.80x + 12 &= 0.50(x + 30) &&\text{Substitute } y = x + 30.\\
8x + 120 &= 5(x + 30) &&\text{Multiply by 10.}\\
8x + 120 &= 5x + 150\\
3x &= 30\\
x &= 10.
\end{aligned}
$$

Now substitute $x = 10$ into the first equation:

$$
\begin{aligned}
10 + 30 &= y\\
40 &= y.
\end{aligned}
$$

Answer: 10 liters of 80% H_2SO_4 are required, and 40 liters of 50% H_2SO_4 are produced.

The next example uses the formula $RT = D$ from Section 2.4, but with a slight wrinkle. Suppose you own a power boat that has a top speed of 60 miles per hour on a lake with no current. If you took the boat to a river that had a current of 10 miles per hour, then its net speeds would be

Downstream: 60 + 10 = 70 mph   (with current)
Upstream: 60 − 10 = 50 mph   (against current)

In general, if r = rate of boat in still water, and c = rate of current, then its net rates are

Downstream: $r + c$ (with current)
Upstream: $r - c$ (against current)

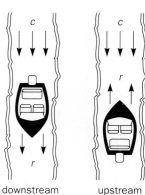

downstream
$r + c$

upstream
$r - c$

EXAMPLE 4

A power boat can go a certain speed on a lake with no current. When taken to a river, it can travel 90 miles downstream in 2 hours and 105 miles upstream in 3 hours. Find the rate of the boat on the lake and the rate of the current.

Solution

Let r = rate of boat on lake and c = rate of current.
We set up a table as in Section 2.4, using the foregoing discussion.

	R · T = D		
downstream (with current)	$r + c$	2	$2(r + c) = 90$
upstream (against current)	$r - c$	3	$3(r - c) = 105$

Our system of two equations is

$$2(r + c) = 90$$
$$3(r - c) = 105$$

Multiply the first equation by $\frac{1}{2}$, multiply the second by $\frac{1}{3}$, and then add:

$$2(r + c) = 90 \xrightarrow{\text{Multiply by } \frac{1}{2}.} r + c = 45$$

$$3(r - c) = 105 \xrightarrow{\text{Multiply by } \frac{1}{3}.} \underline{r - c = 35}$$
$$2r \quad = 80$$
$$r = 40.$$

Substituting $r = 40$ into the equation $r + c = 45$ gives

$$40 + c = 45$$
$$c = 5.$$

Answer: Rate of boat on lake = 40 mph, and rate of current = 5 mph.

EXERCISE 7.5   *Answers, pages A72–A73*

A

Solve using two unknowns, as in Example 1:

1. A collection of 43 coins in dimes and quarters is worth $6.70. How many of each coin are there?

2. A collection of 20 bills consists of $5's and $20's. How many of each are there if their total values are equal?

3. In a collection of 37 coins, there are an equal number of nickels and quarters, the rest being dimes. If the collection is worth $4.50, how many of each coin are there? *Hint:* Let n = number of nickels = number of quarters.

4. Tickets to a concert cost $15, $20, and $30. There were 20,000 sold, with equal numbers of the two lower-priced tickets. If the total revenue from sales was $400,000, how many at each price were sold?

Solve using two unknowns, as in EXAMPLE 2:

5. $24,000 is invested, part at 12% for 3 years and the balance at 6% for 5 years. How much is invested at each rate if the combined interest earned is equivalent to the interest earned by investing the total amount at 8% for 4 years?

6. Friendly Federal Savings pays 6% annual interest on insured savings, whereas Unfriendly Federal pays 9% on uninsured savings. If you have $5000, how much should you deposit for 2 years at each place to obtain a total of $810 interest?

7. You have $9000 invested in stock A, paying 10% interest, plus a certain amount in stock B, paying 15%. How much is invested in B if the entire investment has an equivalent return of 12%? Find the total amount invested. Assume 1 year's time.

8. $12,000 is invested, part at 8% for 3 years, and the balance at 6% for 2 years. How much is invested at each rate if the interest from the 8% investment equals the interest from the 6% investment?

Solve using two unknowns, as in EXAMPLE 3:

9. How many liters of a 20% hydrochloric acid (HCl) solution should be mixed with 105 liters of 70% HCl to obtain a 55% HCl solution? Find the volume of the 55% acid.

10. How many grams of a 15% copper (Cu) alloy should be melted with 20 grams of a 70% Cu alloy to obtain a 25% Cu alloy? Find the weight of the 25% alloy.

11. How many cubic centimeters of water should be added to 50 cubic centimeters of a 40% saline (NaCl) solution to obtain a 16% NaCl solution? Find the total volume that results. *Hint:* Water is 0% NaCl.

12. How many quarts of water should be added to 9 quarts of pure alcohol to obtain 45% alcohol? Find the total volume of alcohol that results. *Hint:* Water is 0% alcohol; pure alcohol is 100% alcohol.

13. How many liters of 25% sulfuric acid (H_2SO_4) and how many liters of 80% H_2SO_4 should be mixed to obtain 110 liters of 65% acid?

14. How many grams each of a 15% silver (Ag) alloy and of a 50% Ag alloy must be melted together to form 70 grams of a 20% Ag alloy?

15. How many ounces of beef that is 40% fat and how many ounces of cheese that is 60% fat are used to make Burger Barn's half-pound Triple-Bypass Cheeseburger that is 44.5% fat? *Doctors recommend a low-fat diet to prevent heart disease.

16. A Toyota radiator contains 8 quarts of antifreeze solution, which is pure ethylene glycol (100% E.G.) mixed with water (0% E.G.). In mild climates, a 40% E.G. mixture is recommended by the manufacturer. How many quarts of pure E.G. and of water must be used to fill this radiator?

Solve using two unknowns, as in EXAMPLE 4:

17. A boat can go a certain speed on a lake with no current. On a river, it can travel 40 miles downstream in 1 hour and 90 miles upstream in 3 hours. Find the rate of the boat on the lake and the rate of the current.

18. A plane can fly a certain speed with no wind. Flying with a certain tailwind, it flew 420 miles in 3 hours. Flying into the same headwind, it flew 400 miles in 4 hours. Find the rate of the plane in still air and the rate of the wind.

19. A bicyclist can average 20 miles per hour if there is no wind. One day she rode for 3 hours with the wind at her back, then turned around and returned to her starting point in 5 hours, while fighting the same headwind. Find the speed of the wind and the distance traveled in each direction.

20. A boat can go a certain speed in still water. On a river with a 3-mph current, it goes a certain distance downstream in 2 hours, then turns around and returns upstream in 3 hours to the starting point. Find the speed of the boat in still water and the distance traveled each way on the river.

Solve using two unknowns:

21. Three hamburgers and two orders of french fries cost $7.50, whereas four hamburgers and five orders of fries cost $11.75. Find the cost of each item.

22. In 1876, Samuel Tilden won the popular vote for the United States presidency by 250,807 votes, out of a total of 8,318,707 votes cast for him and Rutherford B. Hayes. How many votes did each candidate receive? (*Historical note:* Tilden lost the election because Hayes won the electoral vote, 185 to 184!)

23. Solution A is 25% acid, B is 15%, C is 80%, and D is 70%. Equal amounts of A and B are added to 100 liters of C to obtain D. How many liters each of A and B are needed? How many liters of D will result?

24. Solution E is 20% alcohol, F is 60%, G is 90%, and H is 50%. Equal amounts of E and F are added to a certain amount of G to obtain 100 cc of H. How many cubic centimeters each of E, F, and G must be used?

25. How many pounds of nuts selling for $1.50 per pound and how many pounds of nuts selling for $2.00 per pound should be mixed in order to make a 10-pound mix selling for $1.85 per pound?

26. The price of tea in China is 57¢ per kilogram, whereas the price of tea in India is 48¢ per kilogram. How many kilograms of the Chinese tea must be added to 14 kilograms of the Indian tea to produce a blend selling for 50¢ per kilogram?

27. You invest $12,000 at 9% for a certain number of years, after which you remove the principal and invest it at 11% for a few more years. If your money was tied up for 9 years and earned $11,160 in interest, for how many years was your money in each investment?

28. Crummy Granola contains 3% protein. Unbeknownst to most folk, the box it comes in contains a surprising 7% protein (a new source of food value!). The entire package has a gross weight of 20 ounces, and chemical analysis reveals it (granola and box) to have an average protein value of 4%. Find the net weight of the Crummy in the box and of the box itself.

29. A 50-question test has two parts. On part A a student got 80% correct, and on part B he got only 20% correct. If his overall score was 62% correct, how many questions were on each part?

30. When reduced to lowest terms, a fraction equals $\frac{4}{5}$. If both numerator and denominator of the fraction are decreased by 8, the new fraction that is formed equals $\frac{2}{3}$. Find the original fraction. *Hint:* Let $\frac{x}{y}$ = the original fraction.

31. If both numerator and denominator of a fraction are increased by 3, a new fraction is formed that equals $\frac{3}{4}$. If both numerator and denominator of the original fraction are decreased by 3, a new fraction is formed that equals $\frac{2}{3}$. Find the original fraction.

32. If the width of a rectangle is decreased by 1 inch and the length is increased by 3 inches, a second rectangle is formed with the same area as the first rectangle. If the width of the first rectangle is increased by 2 inches and the length is decreased by 3 inches, a third rectangle is formed with the same area as the first rectangle. Find the length and width of each rectangle. *Formula: A = LW.*

first

second

third

33. The perimeter of a rectangle is 26 inches. If the length is increased by 3 inches and the width is decreased by 1 inch, a new rectangle is formed whose area equals that of the original rectangle. Find the width and the length of the original rectangle. *Formulas: P = 2L + 2W, A = LW.*

34. The combined perimeters of the two squares shown below is 52 inches. When they are placed together, the third figure shown is formed, which has an *outer* perimeter of 42 inches. Find the sides of the two squares.

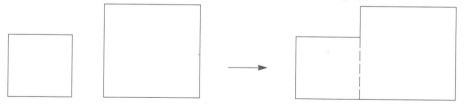

35. The perimeter of a rectangle is 40 inches. If the width is divided by two and the length is divided by three, a new rectangle is formed that is actually a square. Find the dimensions of the original rectangle. Then find the side of the new square.

C

36. The shaded triangle is bounded by the line $y = 2x - 6$, the line $x + 2y = 4$, and the y-axis.
   a. Find the lines' point of intersection.
   b. Find the y-intercept of each line.
   c. Use your results to find the area of the triangle. *Formula:*
   $$A = \frac{base \cdot height}{2}$$

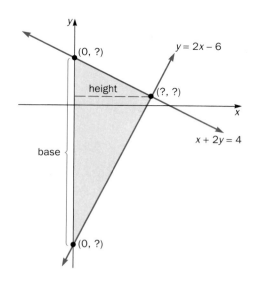

37. Line 1 and line 2 go through the points shown.
   a. Use the method of Section 7.3, Example 1b, to find the equation of each line.
   b. Use these equations to find the point of intersection of these lines.

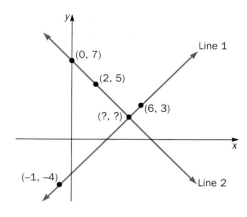

7.6	**Linear Systems in Three Variables**

In Section 7.4 we saw that the solution to a system of two linear equations in x and y represents the intersection of lines in the two-dimensional plane. By the same token, a system of three linear equations in x, y, and z represents three planes in three-dimensional space, and its solution is the intersection of these planes. Graphing planes in "3-space" is a topic for second-year calculus students to worry about, so we will content ourselves only with the algebraic aspects. To solve such a system, we must *eliminate one of the variables twice* in order to arrive at a **reduced system** of two equations in two unknowns. We then solve this system by our earlier methods.

EXAMPLE 1 Solve this system:

$$\begin{array}{ll} x + y + z = 7 & (1) \\ 2x - y + 2z = 8 & (2) \\ 4x + y - 3z = -6 & (3) \end{array}$$

Solution
The easiest variable to eliminate twice is y, which we can eliminate by adding equations (1) and (2) and then adding equations (2) and (3):

$$\begin{array}{ll} x + y + \ z = 7 & (1) \\ 2x - y + 2z = 8 & (2) \\ \hline 3x \quad\ \ + 3z = 15 & (4) \end{array} \qquad \begin{array}{ll} 2x - y + 2z = 8 & (2) \\ 4x + y - 3z = -6 & (3) \\ \hline 6x \quad\ \ - \ z = 2 & (5) \end{array}$$

Our **reduced system,** consisting of equations (4) and (5), can now be solved:

$$\begin{array}{l} 3x + 3z = 15 \xrightarrow{\text{Multiply by } \frac{1}{3}.} x + z = 5 \\ 6x - \ z = 2 \xrightarrow{\hspace{3cm}} \underline{6x - z = 2} \\ \hspace{5.5cm} 7x \quad\ \ = 7 \\ \hspace{6cm} x = 1. \end{array}$$

Now substitute $x = 1$ back into the equation $x + z = 5$:

$$\begin{array}{l} 1 + z = 5 \\ \quad\ \ z = 4. \end{array}$$

Working backwards, we substitute $x = 1$, $z = 4$ into one of the original equations, the simplest of which is equation (1):

$$\begin{array}{l} 1 + y + 4 = 7 \\ \hspace{1.3cm} y = 2. \end{array}$$

Answer: $x = 1$, $y = 2$, $z = 4$ or   (1,2,4)
Check: You should check this answer in *all three* equations.

EXAMPLE 2   Solve this system:

$$2x - 4y + 3z = 0 \qquad (1)$$
$$5x + 3y - 2z = 19 \qquad (2)$$
$$-x + 2y + 5z = -13 \qquad (3)$$

Solution

It is easiest to eliminate x. To do this, we multiply equation (3) by 2 and add it to equation (1), and then multiply equation (3) by 5 and add it to equation (2):

$$
\begin{array}{ll}
2x - 4y + 3z = 0 & (1) \\
-2x + 4y + 10z = -26 \quad 2\times(3) \\
\hline
13z = -26 & (4)
\end{array}
\qquad
\begin{array}{ll}
5x + 3y - 2z = 19 & (2) \\
-5x + 10y + 25z = -65 \quad 5\times(3) \\
\hline
13y + 23z = -46 & (5)
\end{array}
$$

Equation (4) in our **reduced system** contains only one unknown, z. We solve for z immediately and then substitute its value into equation (5):

$$13z = -26 \xrightarrow{\text{Solve for } z.} z = -2.$$

$$13y + 23z = -46 \xrightarrow[\text{into (5).}]{\text{Substitute}} \begin{array}{l} 13y + 23(-2) = -46 \\ 13y - 46 = -46 \\ 13y = 0 \\ y = 0. \end{array}$$

Now substitute $y = 0$, $z = -2$ into one of the original equations, say (3):

$$
\begin{array}{l}
-x + 2(0) + 5(-2) = -13 \\
-x + 0 - 10 = -13 \\
-x = -3 \\
x = 3.
\end{array}
$$

Answer: $x = 3$, $y = 0$, $z = -2$ or $(3, 0, -2)$.
Check: You should check this answer in *all three* equations.

In our next example, variables are missing in each equation. This makes things easier, because one of the original equations can be used in the reduced system.

EXAMPLE 3   Solve this system:

$$
\begin{array}{ll}
3x + 2y = 7 & (1) \\
2x + 7z = -26 & (2) \\
5y + z = 6 & (3)
\end{array}
$$

Solution

We first eliminate z using equations (2) and (3):   *(continued)*

$$
\begin{array}{rll}
2x \phantom{{}-35y} + 7z &= -26 & (2) \\
-35y - 7z &= -42 & -7 \times (3) \\
\hline
2x - 35y \phantom{{}+7z} &= -68 & (4)
\end{array}
$$

We now use equation (1), along with equation (4), to form our **reduced system;** it is then solved as shown:

$$
\begin{array}{lll}
3x + 2y = 7 & \xrightarrow{\text{Multiply by 2.}} & 6x + 4y = 14 \\
2x - 35y = -68 & \xrightarrow{\text{Multiply by } -3.} & -6x + 105y = 204 \\
& & \hline \\[-1.2em]
& & 109y = 218 \\
& & y = 2.
\end{array}
$$

Now substitute $y = 2$ into both equations $3x + 2y = 7$ and $5y + z = 6$:

$$
\begin{array}{ll}
3x + 2(2) = 7 & 5(2) + z = 6 \\
3x = 3 & z = -4. \\
x = 1. &
\end{array}
$$

Answer: $x = 1$, $y = 2$, $z = -4$ or $(1, 2, -4)$.

EXAMPLE 4

A collection of 23 coins in nickels, dimes, and quarters is worth \$3.20. There are three more dimes than nickels. How many of each coin are there?

Solution
Let n = number of nickels, d = number of dimes, and q = number of quarters. We use a chart as in Sections 2.2 and 7.5:

	number of coins	× face value	= value of coins
nickels	n	5¢	$5n$¢
dimes	d	10¢	$10d$¢
quarters	q	25¢	$25q$¢
total	23		\$3.20, or 320¢

Our system of equations is

$$
\begin{array}{lll}
n + d + q = 23 & \text{Number of coins.} & (1) \\
5n + 10d + 25q = 320 & \text{Total value (in cents).} & (2) \\
d = n + 3. & \text{3 more dimes than nickels.} & (3)
\end{array}
$$

This is best solved by substituting equation (3), $d = n + 3$, into both equations (1) and (2):

into (1):

$$
\begin{array}{l}
n + (n + 3) + q = 23 \\
2n + q = 20 \qquad (4)
\end{array}
$$

into (2):

$$
\begin{array}{l}
5n + 10(n + 3) + 25q = 320 \\
5n + 10n + 30 + 25q = 320 \\
15n + 25q = 290 \qquad (5)
\end{array}
$$

Our **reduced system** (4) and (5) can now be solved by either the addition or the substitution method; we choose the latter:

$$2n + q = 20 \xrightarrow{\text{Solve for } q.} q = 20 - 2n$$

$$15n + 25q = 290 \xrightarrow{\text{Substitute.}} 15n + 25(20 - 2n) = 290$$
$$15n + 500 - 50n = 290$$
$$-35n = -210$$
$$n = 6.$$

Now substitute $n = 6$ into both $2n + q = 20$ and $d = n + 3$:

$$2(6) + q = 20 \qquad d = 6 + 3$$
$$q = 8. \qquad d = 9.$$

Answer: 6 nickels, 9 dimes, and 8 quarters.

EXERCISE 7.6   *Answers, page A73*

A

Solve as in **EXAMPLES 1 and 2**:

1.
$$x + y + z = 3$$
$$2x - y + 7z = 8$$
$$4x + y + 8z = 13$$

2.
$$x + y + z = 10$$
$$3x - 4y - z = 4$$
$$2x + 3y - z = 13$$

3.
$$2x - 3y + 2z = 0$$
$$x + 2y - z = 2$$
$$2x + y + 3z = -1$$

4.
$$3x - y + 2z = 0$$
$$2x + 2y + z = 0$$
$$x - 5y + 2z = 0$$

5.
$$5x + 2y + 3z = -3$$
$$2x - y - 2z = -9$$
$$6x - 3y + 4z = 3$$

6.
$$2x + 2y + z = 0$$
$$4x + 3y - 2z = 4$$
$$2x + 10y + 5z = -4$$

Solve as in **EXAMPLE 3**:

7.
$$x + 2y = 9$$
$$x - 3z = 8$$
$$y + 2z = 0$$

8.
$$2x - y = 5$$
$$2y - 3z = 4$$
$$2x + z = 2$$

9.
$$3x - 5y + 2z = -2$$
$$5x + 7y - 5z = 2$$
$$x - y = 2$$

10.
$$4x - y + 3z = -4$$
$$3y + z = -1$$
$$3x + 6y + 2z = -4$$

11.
$$2x + y - 3z = 0$$
$$x + 2z = 0$$
$$2y - z = 0$$

12.
$$3x + 2y + z = 2$$
$$3y + 2z = -8$$
$$2x - y = 4$$

Solve using three unknowns, as in **EXAMPLE 4**:

13. The sum of three numbers is 10. The third number equals the sum of the first two numbers. Twice the second is one less than the sum of the first and third. Find the numbers.

14. The sum of the three angles of a triangle is 180°. The first angle is 10° less than the sum of the second and third angles. Twice the second angle equals the sum of the first and third angles. Find each angle.

15. Thirty-three coins in pennies, nickels, and dimes are worth $1.98. The total number of pennies and nickels is seven more than the number of dimes. How many of each coin are there?

16. Tickets to a rock concert cost $10, $15, and $25. There were 20,000 tickets sold, and revenue from sales was $285,000. The total number of $10 and $25 tickets was three times the number of $15 tickets. How many of each were sold?

B

17. How many liters each of 10% alcohol, 40% alcohol, and 60% alcohol should be mixed to form 9 liters of 50% alcohol, if the volume of 60% alcohol is twice the total volume of the 10% and 40% alcohols?

18. $10,000 is invested, part in a stock paying 6% interest, part in a bond paying 9%, and the balance in a money market certificate paying 11%. After one year, the total interest is $810, at which time the stock is sold. The bond is held for 2 more years, and the money market certificate for 3 more. These two yield an additional $1350 in interest. How much was originally invested in each instrument?

Solve by first simplifying these systems:

19. $\dfrac{x}{2} + \dfrac{3y}{2} - \dfrac{z}{4} = \dfrac{15}{4}$

$x + y - z = 4$

$\dfrac{x}{3} + \dfrac{y}{2} + \dfrac{z}{3} = 1$

20. $\dfrac{2x}{3} + \dfrac{3y}{2} + \dfrac{z}{6} = \dfrac{7}{3}$

$\dfrac{x}{2} + \dfrac{y}{4} - z = \dfrac{-1}{4}$

$x - y + z = 1$

21. $\dfrac{z}{x + y} = 1$

$\dfrac{x + 3}{y + 4} = \dfrac{2}{3}$

$\dfrac{z - 1}{2x + y} = \dfrac{1}{2}$

22. $\dfrac{x + 2y}{z + 3} = 0$

$\dfrac{z - 2x}{y - z} = 4$

$\dfrac{y + 1}{x - 3} = \dfrac{-2}{5}$

23. $(x + 3)(y - 2) = xy - 5$

$(y + 1)(z - 3) = y(z - 1) - (x + y)$

$\dfrac{x - 2}{z} = \dfrac{x + 1}{z + 3}$

24. $(y + z)(y - 1) = y(z - 4) + (y^2 - 3x)$

$(x - y)(x + 2) = x(x - y) + 3z$

$(x + z)(x - 2) = z(x - 4) + x(x + 1)$

Solve for x, y, and z. See Exercise 7.4, Problems 47–62.

C

25. $5x - 2y + 2z = 3a$
$2x + 3y - z = 7a$
$x - y + z = 0$

26. $x + y + z = 3a$
$x + 2y - 3z = 4a - 5b$
$2x - 4y + 3z = 7b$

27. $2x + y + 3z = 3c + 7d$
$2y + z = 2c$
$x = y + z$

28. This problem can be done with or without algebra. What fifty United States coins total $1.00? There are two possible answers. (U.S. coins are pennies nickels, dimes, quarters, half-dollars, and silver dollars.)

7.7 Determinants and Cramer's Rule

You have learned both the addition and substitution methods for solving a linear system of equations. In this section, you will learn a third method. We start with a definition. The symbol

$$\begin{vmatrix} a_1 & b_1 \\ a_2 & b_2 \end{vmatrix}$$

is called a **2 × 2 determinant.** The value of this determinant is defined by

$$\begin{vmatrix} a_1 & b_1 \\ a_2 & b_2 \end{vmatrix} = a_1 b_2 - a_2 b_1.$$

Note that the value of a 2×2 determinant is simply the difference between its "cross products."

EXAMPLE 1

a. $\begin{vmatrix} 2 & 3 \\ 1 & 7 \end{vmatrix} = 2 \cdot 7 - 1 \cdot 3 = 14 - 3 = 11.$

b. $\begin{vmatrix} -4 & 2 \\ -5 & 0 \end{vmatrix} = -4 \cdot 0 - (-5) \cdot 2 = 0 + 10 = 10.$

To obtain the value of a 3×3 determinant, we first define the **minor** of any of its terms to be the 2×2 determinant obtained by *deleting* the row and column occupied by that term. For example,

in $\begin{vmatrix} a_1 & b_1 & c_1 \\ a_2 & b_2 & c_2 \\ a_3 & b_3 & c_3 \end{vmatrix}$, the **minor of** a_1 is $\begin{vmatrix} b_2 & c_2 \\ b_3 & c_3 \end{vmatrix}$

and in $\begin{vmatrix} a_1 & b_1 & c_1 \\ a_2 & b_2 & c_2 \\ a_3 & b_3 & c_3 \end{vmatrix}$, the **minor of** b_2 is $\begin{vmatrix} a_1 & c_1 \\ a_3 & c_3 \end{vmatrix}.$

(continued)

We define the 3×3 determinant, **expanded by minors about column 1,** to be

$$
\begin{array}{c}
\text{Column 1} \\
\downarrow
\end{array}
$$

$$
\begin{vmatrix} a_1 & b_1 & c_1 \\ a_2 & b_2 & c_2 \\ a_3 & b_3 & c_3 \end{vmatrix} = a_1 \underbrace{\begin{vmatrix} b_2 & c_2 \\ b_3 & c_3 \end{vmatrix}}_{\text{minor of } a_1} - a_2 \underbrace{\begin{vmatrix} b_1 & c_1 \\ b_3 & c_3 \end{vmatrix}}_{\text{minor of } a_2} + a_3 \underbrace{\begin{vmatrix} b_1 & c_1 \\ b_2 & c_2 \end{vmatrix}}_{\text{minor of } a_3}.
$$

For example, expanding by minors about column 1 in the following determinant gives

$$
\begin{array}{c}
\text{Column 1} \\
\downarrow
\end{array}
$$

$$
\begin{vmatrix} 2 & 3 & 1 \\ -1 & 0 & -4 \\ -5 & 6 & 8 \end{vmatrix} = 2\begin{vmatrix} 0 & -4 \\ 6 & 8 \end{vmatrix} - (-1)\begin{vmatrix} 3 & 1 \\ 6 & 8 \end{vmatrix} + (-5)\begin{vmatrix} 3 & 1 \\ 0 & -4 \end{vmatrix}
$$

$$
\begin{aligned}
&= 2[0 - (-24)] + 1(24 - 6) - 5(-12 - 0) \\
&= 2(24) \qquad\quad + 1(18) \qquad\;\; - 5(-12) \\
&= 48 + 18 + 60 \\
&= 126.
\end{aligned}
$$

A 3×3 determinant can also be obtained by expanding about any other row or column, using the following **sign array:**

$$
\begin{array}{ccc}
+ & - & + \\
- & + & - \\
+ & - & +
\end{array}
$$

In the next example, we start by expanding the preceding determinant about row 2, using this sign pattern. The result is also 126, which illustrates that the value of a 3×3 determinant is the same regardless of the row or column about which it is expanded.

EXAMPLE 2

a.

$$
\text{Row 2} \rightarrow \begin{vmatrix} 2 & 3 & 1 \\ -1 & 0 & -4 \\ -5 & 6 & 8 \end{vmatrix} = -(-1)\begin{vmatrix} 3 & 1 \\ 6 & 8 \end{vmatrix} + 0\begin{vmatrix} 2 & 1 \\ -5 & 8 \end{vmatrix} - (-4)\begin{vmatrix} 2 & 3 \\ -5 & 6 \end{vmatrix}
$$

$$
\begin{aligned}
&= 1(24 - 6) \qquad + 0 \qquad\qquad + 4[12 - (-15)] \\
&= 18 \qquad\qquad\qquad\qquad\qquad + 108 \\
&= 126.
\end{aligned}
$$

Note here that 0 times the second determinant equals 0. Expanding about any row or column containing 0 will make your work easier.

b. $\begin{vmatrix} 2 & 5 & 4 \\ 3 & -4 & 6 \\ -3 & 1 & -2 \end{vmatrix} = 4\begin{vmatrix} 3 & -4 \\ -3 & 1 \end{vmatrix} - 6\begin{vmatrix} 2 & 5 \\ -3 & 1 \end{vmatrix} + (-2)\begin{vmatrix} 2 & 5 \\ 3 & -4 \end{vmatrix}$

Column 3 ↓

$$= 4(3 - 12) - 6[2 - (-15)] - 2(-8 - 15)$$
$$= 4(-9) \qquad - 6(17) \qquad - 2(-23)$$
$$= -36 \qquad - 102 \qquad + 46$$
$$= -92.$$

A 3×3 determinant can also be evaluated by a **shortcut method.** Place columns 1 and 2 outside the determinant, creating columns 4 and 5. *Add* the products along the arrows going diagonally *downward;* then *subtract* the products along the arrows going diagonally *upward.* We illustrate by computing the determinants in Example 2, using this method. The results are the same as those obtained by expanding by minors.

EXAMPLE 3

a. $\begin{vmatrix} 2 & 3 & 1 \\ -1 & 0 & -4 \\ -5 & 6 & 8 \end{vmatrix} = \begin{vmatrix} 2 & 3 & 1 & 2 & 3 \\ -1 & 0 & -4 & 1 & 0 \\ -5 & 6 & 8 & -5 & 6 \end{vmatrix}$

Column 1 ⌐Column 2

$$= 2 \cdot 0 \cdot 8 + 3(-4)(-5) + 1(-1) \cdot 6 - (-5) \cdot 0 \cdot 1 - 6(-4) \cdot 2 - 8(-1) \cdot 3$$
$$= 0 \qquad + 60 \qquad - 6 \qquad - 0 \qquad + 48 \qquad + 24$$
$$= 126, \quad \text{the same result as Example 2a.}$$

b. $\begin{vmatrix} 2 & 5 & 4 \\ 3 & -4 & 6 \\ -3 & 1 & -2 \end{vmatrix} = \begin{vmatrix} 2 & 5 & 4 & 2 & 5 \\ 3 & -4 & 6 & 3 & -4 \\ -3 & 1 & -2 & -3 & 1 \end{vmatrix}$

$$= 2(-4)(-2) + 5 \cdot 6(-3) + 4 \cdot 3 \cdot 1 - (-3)(-4) \cdot 4 - 1 \cdot 6 \cdot 2 - (-2) \cdot 3 \cdot 5$$
$$= 16 \qquad - 90 \qquad + 12 \qquad - 48 \qquad - 12 \qquad + 30$$
$$= -92, \quad \text{the same result as Example 2b.}$$

We are now ready to solve the linear system

$$a_1x + b_1y = c_1$$
$$a_2x + b_2y = c_2$$

by determinants. Using the addition method, we first solve for x:

$a_1x + b_1y = c_1$ ⎯⎯ Multiply by b_2. ⎯⎯→ $a_1b_2x + b_1b_2y = c_1b_2$

$a_2x + b_2y = c_2$ ⎯⎯ Multiply by $-b_1$. ⎯⎯→ $-a_2b_1x - b_1b_2y = -c_2b_1$

$$a_1b_2x - a_2b_1x = c_1b_2 - c_2b_1 \qquad \text{Add.}$$
$$x(a_1b_2 - a_2b_1) = c_1b_2 - c_2b_1 \qquad \text{Factor.}$$
$$x = \frac{c_1b_2 - c_2b_1}{a_1b_2 - a_2b_1}. \qquad \text{Divide.}$$

In like manner, we can solve for y:

$$a_1x + b_1y = c_1 \xrightarrow{\text{Multiply by } -a_2.} -a_1a_2x - a_2b_1y = -a_2c_1$$

$$a_2x + b_2y = c_2 \xrightarrow{\text{Multiply by } a_1.} a_1a_2x + a_1b_2y = a_1c_2$$

$$a_1b_2y - a_2b_1y = a_1c_2 - a_2c_1 \qquad \text{Add.}$$

$$y(a_1b_2 - a_2b_1) = a_1c_2 - a_2c_1 \qquad \text{Factor.}$$

$$y = \frac{a_1c_2 - a_2c_1}{a_1b_2 - a_2b_1}. \qquad \text{Divide.}$$

Interestingly enough, our answers can be expressed using determinants:

$$x = \frac{\begin{vmatrix} c_1 & b_1 \\ c_2 & b_2 \end{vmatrix}}{\begin{vmatrix} a_1 & b_1 \\ a_2 & b_2 \end{vmatrix}} \quad \text{and} \quad y = \frac{\begin{vmatrix} a_1 & c_1 \\ a_2 & c_2 \end{vmatrix}}{\begin{vmatrix} a_1 & b_1 \\ a_2 & b_2 \end{vmatrix}}.$$

We now summarize our results, known as **Cramer's Rule,** which were first discovered by eighteenth-century Swiss mathematician Gabriel Cramer:

Cramer's Rule

The linear system

$$a_1x + b_1y = c_1$$
$$a_2x + b_2y = c_2$$

can be solved by first forming the determinants

$$D = \begin{vmatrix} a_1 & b_1 \\ a_2 & b_2 \end{vmatrix}, \quad D_x = \begin{vmatrix} c_1 & b_1 \\ c_2 & b_2 \end{vmatrix}, \quad D_y = \begin{vmatrix} a_1 & c_1 \\ a_2 & c_2 \end{vmatrix}.$$

Then the solutions are given by

$$x = \frac{D_x}{D}, \quad y = \frac{D_y}{D}, \quad D \neq 0.$$

Note that D is the determinant consisting of the coefficients a_1, b_1, a_2, b_2 appearing on the left sides of the two equations. D_x is the determinant formed by replacing column 1 of D with the constants c_1 and c_2; D_y is formed by replacing column 2 of D with c_1 and c_2. Note also the requirement that $D \neq 0$. (See Problems 55–58 in the exercises.)

EXAMPLE 4 Solve by Cramer's Rule:

$$2x + 5y = 1$$
$$7x + 3y = -4$$

We form the determinants described above:

$$D = \begin{vmatrix} 2 & 5 \\ 7 & 3 \end{vmatrix} = 6 - 35 \qquad D_x = \begin{vmatrix} 1 & 5 \\ -4 & 3 \end{vmatrix} = 3 - (-20) \qquad D_y = \begin{vmatrix} 2 & 1 \\ 7 & -4 \end{vmatrix} = -8 - 7$$
$$= -29. \qquad\qquad = 23. \qquad\qquad = -15.$$

By Cramer's Rule, the solutions are given by

$$x = \frac{D_x}{D} = \frac{23}{-29} = \frac{-23}{29}, \qquad y = \frac{D_y}{D} = \frac{-15}{-29} = \frac{15}{29}.$$

Answer: $x = \dfrac{-23}{29}$, $y = \dfrac{15}{29}$ or $\left(\dfrac{-23}{29}, \dfrac{15}{29}\right)$.

Cramer's Rule for solving a linear system of equations in three unknowns is similar to the above Rule for two unknowns. We now state it:

Cramer's Rule

The linear system

$$a_1 x + b_1 y + c_1 z = d_1$$
$$a_2 x + b_2 y + c_2 z = d_2$$
$$a_3 x + b_3 y + c_3 z = d_3$$

can be solved by first forming the determinants

$$D = \begin{vmatrix} a_1 & b_1 & c_1 \\ a_2 & b_2 & c_2 \\ a_3 & b_3 & c_3 \end{vmatrix}, \quad D_x = \begin{vmatrix} d_1 & b_1 & c_1 \\ d_2 & b_2 & c_2 \\ d_3 & b_3 & c_3 \end{vmatrix}, \quad D_y = \begin{vmatrix} a_1 & d_1 & c_1 \\ a_2 & d_2 & c_2 \\ a_3 & d_3 & c_3 \end{vmatrix}, \quad D_z = \begin{vmatrix} a_1 & b_1 & d_1 \\ a_2 & b_2 & d_2 \\ a_3 & b_3 & d_3 \end{vmatrix}.$$

Then the solutions are given by

$$x = \frac{D_x}{D}, \quad y = \frac{D_y}{D}, \quad z = \frac{D_z}{D}, \quad D \neq 0.$$

Here, as above, D is the determinant consisting of the coefficients a_1, b_1, c_1, a_2, ... , c_3 appearing on the left sides of the three equations. D_x is the determinant formed by replacing column 1 of D with the constants d_1, d_2, and d_3; D_y and D_z are similarly formed.

EXAMPLE 5 | Solve by Cramer's Rule:

$$2x + 3y - z = 0$$
$$3y + 2z = -6$$
$$x - 4y + 5z = 11$$

We form the determinants described above:

$$D = \begin{vmatrix} 2 & 3 & -1 \\ 0 & 3 & 2 \\ 1 & -4 & 5 \end{vmatrix} = \begin{vmatrix} 2 & 3 & -1 \\ 0 & 3 & 2 \\ 1 & -4 & 5 \end{vmatrix}\begin{matrix} 2 & 3 \\ 0 & 3 \\ 1 & -4 \end{matrix} = 30 + 6 + 0 - (-3) - (-16) - 0 = 55.$$

D was computed by the shortcut method of Example 3. The other determinants can be computed either by using the shortcut method or by expanding by minors, preferably about a row or column containing 0. Their values, as you should verify, are

$$D_x = \begin{vmatrix} 0 & 3 & -1 \\ -6 & 3 & 2 \\ 11 & -4 & 5 \end{vmatrix} = 165, \quad D_y = \begin{vmatrix} 2 & 0 & -1 \\ 0 & -6 & 2 \\ 1 & 11 & 5 \end{vmatrix} = -110, \quad D_z = \begin{vmatrix} 2 & 3 & 0 \\ 0 & 3 & -6 \\ 1 & -4 & 11 \end{vmatrix} = 0.$$

According to Mr. Cramer, the solutions are given by

$$x = \frac{D_x}{D} = \frac{165}{55} = 3, \quad y = \frac{D_y}{D} = \frac{-110}{55} = -2, \quad z = \frac{D_z}{D} = \frac{0}{55} = 0.$$

Answer: $x = 3, \quad y = -2, \quad z = 0$ or $(3, -2, 0)$.

Shortcut: Once x and y are solved, you may substitute them back into one of the original equations and solve for z. This avoids the need to compute the determinant D_z.

EXERCISE 7.7 *Answers, page A73*

Compute these determinants, as in **EXAMPLE 1, EXAMPLE 2 (by minors)**, or **EXAMPLE 3 (by the shortcut method)**:

1. $\begin{vmatrix} 4 & 2 \\ -1 & 3 \end{vmatrix}$ 　　　 **2.** $\begin{vmatrix} -5 & -1 \\ 6 & -7 \end{vmatrix}$ 　　　 **3.** $\begin{vmatrix} 1 & 7 \\ 2 & 14 \end{vmatrix}$ 　　　 **4.** $\begin{vmatrix} 2 & -5 \\ -4 & 10 \end{vmatrix}$

5. $\begin{vmatrix} 0 & -8 \\ 17 & -9 \end{vmatrix}$ 　　　 **6.** $\begin{vmatrix} 13 & -6 \\ 0 & 17 \end{vmatrix}$ 　　　 **7.** $\begin{vmatrix} 3 & 5 & 0 \\ -2 & 2 & 3 \\ 2 & -1 & 0 \end{vmatrix}$ 　　　 **8.** $\begin{vmatrix} -1 & 2 & -4 \\ 5 & 0 & 0 \\ 3 & -7 & 8 \end{vmatrix}$

9. $\begin{vmatrix} 2 & 3 & -1 \\ 4 & 0 & 2 \\ 3 & -1 & 6 \end{vmatrix}$ 　　 **10.** $\begin{vmatrix} 4 & 1 & 2 \\ 2 & -1 & 0 \\ 1 & -2 & 3 \end{vmatrix}$ 　　 **11.** $\begin{vmatrix} 1 & 2 & -3 \\ -4 & -5 & 6 \\ 7 & -8 & 9 \end{vmatrix}$ 　　 **12.** $\begin{vmatrix} 4 & -5 & 2 \\ -1 & 12 & 8 \\ -3 & 5 & 10 \end{vmatrix}$

Solve by Cramer's Rule, as in EXAMPLES 4 and 5:

13. $2x + 3y = 0$
$5x - 2y = -1$

14. $6x + 7y = -9$
$4x + 6y = 0$

15. $2x + y = 3$
$5x - 3y = 2$

16. $7x + 2y = 5$
$5x - y = 6$

17. $13x + 17y = -26$
$-9x - 11y = 18$

18. $15x + 11y = 7$
$13x - 17y = -47$

19. $2x - 7y - z = 35$
$x + y = 1$
$2y + z = -5$

20. $5x - 3y + 3z = 0$
$2x + z = 1$
$3x + 4y = -1$

21. $3x + 5y - 2z = 0$
$x + 2y = 3$
$2x - y - 4z = -15$

22. $2x - 3y + 4z = 8$
$3x - 2y - 3z = -10$
$y - z = -1$

23. $x + y + z = 3$
$2x - y + 7z = 8$
$4x + y + 8z = 13$

24. $x + y + z = 10$
$3x - 4y - z = 4$
$2x + 3y - z = 13$

25. $2x + 5y + 3z = 0$
$3x + y + z = 1$
$5x + 2y - 4z = 6$

26. $5x + y + 2z = 4$
$x - 2y + 6z = 1$
$3x + y - 2z = 2$

B

How much do you remember about fractions, radicals, complex numbers, and FOIL?
Find out by expanding these determinants:

27. $\begin{vmatrix} \frac{1}{2} & \frac{3}{2} \\ \frac{1}{4} & \frac{3}{8} \end{vmatrix}$

28. $\begin{vmatrix} -3 & \frac{1}{3} \\ \frac{1}{4} & \frac{5}{6} \end{vmatrix}$

29. $\begin{vmatrix} \sqrt{6} & 2\sqrt{8} \\ \sqrt{5} & 3\sqrt{2} \end{vmatrix}$

30. $\begin{vmatrix} 2\sqrt{6} & -5 \\ 4\sqrt{3} & 3\sqrt{2} \end{vmatrix}$

31. $\begin{vmatrix} 1 & i \\ i & 1 \end{vmatrix}$

32. $\begin{vmatrix} -i & 2 \\ 1 & 2i \end{vmatrix}$

33. $\begin{vmatrix} 1 + \sqrt{2} & 1 - \sqrt{2} \\ 2 + \sqrt{2} & 2 - \sqrt{2} \end{vmatrix}$

34. $\begin{vmatrix} 1 + i & 2 - i \\ 3 + i & 4 + i \end{vmatrix}$

35. $\begin{vmatrix} a + b & a - b \\ a - b & a + b \end{vmatrix}$

36. $\begin{vmatrix} a & ka + c \\ b & kb + d \end{vmatrix}$

37. $\begin{vmatrix} \frac{1}{2} & 1 & -\frac{1}{2} \\ 3 & \frac{1}{4} & -1 \\ -\frac{1}{2} & -\frac{1}{4} & 2 \end{vmatrix}$

38. $\begin{vmatrix} i & 1 & 0 \\ -1 & -i & 1 \\ 2 & i & -i \end{vmatrix}$

39. $\begin{vmatrix} a & b & c \\ d & e & f \\ a & b & c \end{vmatrix}$

40. $\begin{vmatrix} 1 & x & x^2 \\ x^2 & 1 & x \\ x & x^2 & 1 \end{vmatrix}$ (Expand, simplify, and factor completely!)

Using the method of Section 7.3, Example 1b, one can show that the equation of the line through the points

$$(-2,3) \text{ and } (4,-9) \quad \text{is} \quad y = -2x - 1.$$

Interestingly enough, this equation can be obtained by inserting these points into the following determinant, which is set equal to 0:

$$\begin{vmatrix} x & y & 1 \\ -2 & 3 & 1 \\ 4 & -9 & 1 \end{vmatrix} = 0.$$

Expanding this determinant by the shortcut method gives

$$3x + 4y + 18 - 12 - (-9x) - (-2y) = 0$$
$$12x + 6y + 6 = 0$$
$$6y = -12x - 6$$
$$y = -2x - 1, \qquad \text{the same equation as above.}$$

Find the equation of the line through the following pairs of points:

a. By the determinant method shown above.

b. By the method of Section 7.3, Example 1b.
In each case, your results should be identical.

41. $(2,5)$ and $(4,-3)$ **42.** $(-3,6)$ and $(3,2)$ **43.** $(0,b)$ and $(1,m + b)$

The following equations are given in both standard form and slope–intercept form of a straight line (see Section 7.2, Example 2a).

Standard form	Slope–intercept form
$2x - 5y = 3$	$y = \dfrac{2}{5}x - \dfrac{3}{5}$
$-4x + 10y = 1$	$y = \dfrac{2}{5}x + \dfrac{1}{10}$

Because each line has slope $m = \dfrac{2}{5}$, the lines are *parallel*. But the determinant consisting of the coefficients of x and y is 0:

$$\begin{vmatrix} 2 & -5 \\ -4 & 10 \end{vmatrix} = 20 - 20 = 0.$$

This illustrates the fact that non-vertical lines

$$\begin{cases} ax + by = e \\ cx + dy = f \end{cases} \quad \text{are } \textit{parallel} \text{ if and only if} \quad \begin{vmatrix} a & b \\ c & d \end{vmatrix} = 0.$$

Use this result to decide which of the following pairs of lines are parallel:

44. $-3x + 4y = 8$
$\quad\;\; 6x - 8y = 3$

45. $\quad ax + by = c$
$\quad kax + kby = d$

46. $\dfrac{1}{2}x - \dfrac{1}{4}y = 17$
$\quad -2x + 4y = -3$

c

Solve these systems by Cramer's Rule:

47. $5\sqrt{2}x + 2\sqrt{3}y = 22$
$\quad \sqrt{2}x - 3\sqrt{3}y = -16$

48. $x + iy = -1$
$\quad ix + y = 5i$

49. $ix + 3y = -4$
$\quad 3ix - 2y = -1$

50. $x + y = 2i$
$\quad ix - y = 0$

51. $ax + y = a^2$
$\quad bx + y = b^2$

52. $x + ay = a^3$
$\quad x + by = b^3$

53. $x - cy = c^2$
$\quad x + dy = d^2$

54. $cx + dy = c^2$
$\quad dx + cy = d^2$

Cramer's Rule does not work if $D = 0$, because $\dfrac{D_x}{0}$ and $\dfrac{D_y}{0}$ are undefined. In the following problems,

a. Show that $D = 0$.

b. Use the methods of Section 7.4, Examples 4 and 5, to show that they have either *no solution* or *infinitely many solutions*.

55. $-3x + 4y = 8$
$\quad\;\; 6x - 8y = 3$

56. $5x + 2y = \dfrac{1}{2}$
$\quad 10x + 4y = 3$

57. $2x - y = 3$
$\quad 4x - 2y = 6$

58. $-2x + 3y = 7$
$\quad 4x - 6y = -14$

59. The examples in the text illustrated that the value of a 3×3 determinant is the same regardless of the method used to obtain it. Expanding the following determinant by minors about *column 1* gives

$$\begin{vmatrix} a_1 & b_1 & c_1 \\ a_2 & b_2 & c_2 \\ a_3 & b_3 & c_3 \end{vmatrix} = a_1 \begin{vmatrix} b_2 & c_2 \\ b_3 & c_3 \end{vmatrix} - a_2 \begin{vmatrix} b_1 & c_1 \\ b_3 & c_3 \end{vmatrix} + a_3 \begin{vmatrix} b_1 & c_1 \\ b_2 & c_2 \end{vmatrix}$$

$$= a_1(b_2c_3 - b_3c_2) - a_2(b_1c_3 - b_3c_1) + a_3(b_1c_2 - b_2c_1)$$

$$= a_1b_2c_3 - a_1b_3c_2 - a_2b_1c_3 + a_2b_3c_1 + a_3b_1c_2 - a_3b_2c_1$$

Expand this determinant:

    a. By minors about *row 2*.
    b. By the shortcut method.

Your results should be equivalent in all three cases.

CHAPTER 7 · SUMMARY

7.1 Graphing Points and Lines

$y = mx + b$: Make up a table.
$ax + by = c$: Use the intercept method.

Example:

Graph $2x + 3y = 6$ by the intercept method.

x	y
0	2
3	0
-3	4

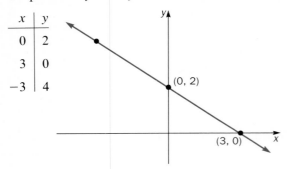

7.2 Slope

Slope $m = \dfrac{y_2 - y_1}{x_2 - x_1}$

Example:

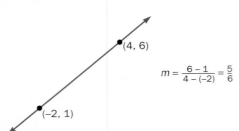

$m = \dfrac{6 - 1}{4 - (-2)} = \dfrac{5}{6}$

Parallel lines: $m_1 = m_2$

Example:

$y = -2x + 7$ is parallel to
$y = -2x - 9$ because $-2 = -2$.

Perpendicular lines: $m_1 m_2 = -1$

Slope–intercept form of a line: $y = mx + b$

7.3 Deriving Equations of Lines

Point–slope form of a line:

$$y - y_1 = m(x - x_1)$$

Example:

The equation of the line that has slope $m = -4$ and goes through $(7,5)$ is

$$y - 5 = -4(x - 7), \text{ or}$$
$$y = -4x + 33.$$

7.4 Linear Systems in Two Variables

Example:

Solve by the **addition method**:

$2x + 3y = 3 \quad \underrightarrow{\text{Mult. by 4.}} \quad 8x + 12y = 12$

$5x + 4y = 11 \quad \underrightarrow{\text{Mult. by } -3} \quad \dfrac{-15x - 12y - 33}{-7x \qquad = -21,}$

$$\text{or} \quad x = 3.$$

Substitute: $2(3) + 3y = 3$, or $y = -1$.

Answer: $x = 3$, $y = -1$ or $(3, -1)$.

7.5 Word Problems in Two Variables

Example:

"How many liters each of 15% acid and of 70% acid should be mixed to obtain 22 liters of 30% acid?"

	liters	× strength =	pure acid
15%	x	0.15	$0.15x$
70%	y	0.70	$0.70y$
30%	22	0.30	6.6

CHAPTER 7 · SUMMARY

Solve by the **substitution method**:

$$x + y = 22$$

Solve for y. $\quad y = 22 - x$

$$0.15x + 0.70y = 6.6$$

Substitute. $\quad 0.15x + 0.70(22 - x) = 6.6$
$$15x + 1540 - 70x = 660$$
$$-55x = -880$$
$$x = 16 \text{ liters.}$$
Substitute: $y = 22 - 16,$ or $y = 6 \text{ liters.}$

7.6 Linear Systems in Three Variables

Example:

$$\left. \begin{matrix} x + y + z = -1 \\ 3x - y + 2z = -3 \\ x + y - 4z = 19 \end{matrix} \right\} \begin{matrix} \xrightarrow{\text{Add.}} 4x + 3z = -4 \\ \xrightarrow{\text{Add.}} 4x - 2z = 16 \end{matrix}$$

Reduced system

$$4x + 3z = -4 \longrightarrow 4x + 3z = -4$$
$$4x - 2z = 16 \xrightarrow[\text{by } -1.]{\text{Mult.}} -4x + 2z = -16$$
$$5z = -20$$
$$z = -4.$$

Substitute: $4x + 3(-4) = -4,$
$\qquad$ or $4x = 8,$ or $x = 2.$
Substitute: $2 + y + (-4) = -1,$ or $y = 1.$

Answer: $x = 2,$ $y = 1,$ $z = -4$ or $(2,1,-4).$

7.7 Determinants

2×2 Determinants

Example:

$$\begin{vmatrix} 5 & -3 \\ 7 \end{vmatrix} = 5 \cdot 7 - 4(-3) = 47.$$

3×3 Determinants
Expand by minors about column 1.

Example:

$$\begin{vmatrix} 2 & 1 & 4 \\ -3 & 5 & -8 \\ 0 & 7 & 6 \end{vmatrix} = 2 \begin{vmatrix} 5 & -8 \\ 7 & 6 \end{vmatrix} - (-3) \begin{vmatrix} 1 & 4 \\ 7 & 6 \end{vmatrix} + 0 \begin{vmatrix} 1 & 4 \\ 5 & -8 \end{vmatrix}$$

$$= 2(86) \quad + 3(-22) \quad + 0$$
$$= 106.$$

Shortcut method:
Place columns 1 and 2 as shown, then multiply along arrows.

Example:

$$\begin{vmatrix} 2 & 1 & 4 \\ -3 & 5 & -8 \\ 0 & 7 & 6 \end{vmatrix} \begin{matrix} 2 \\ -3 \\ 0 \end{matrix} \begin{matrix} 1 \\ 5 \\ 7 \end{matrix} = 2 \cdot 5 \cdot 6 + 1(-8) \cdot 0 +$$

$$4(-3) \cdot 7 - 0 \cdot 5 \cdot 4 - 7(-8) \cdot 2$$
$$- 6(-3) \cdot 1$$
$$= 60 + 0 - 84 - 0 + 112 + 18$$
$$= 106.$$

Cramer's Rule

Example:

To solve the linear system

$$2x + 3y = 4$$
$$-x + 2y = 3$$

form these determinants:

$$D = \begin{vmatrix} 2 & 3 \\ -1 & 2 \end{vmatrix} = 7, \quad D_x = \begin{vmatrix} 4 & 3 \\ 3 & 2 \end{vmatrix} = -1,$$

$$D_y = \begin{vmatrix} 2 & 4 \\ -1 & 3 \end{vmatrix} = 10$$

Then the *solutions* are given by

$$x = \frac{D_x}{D} = \frac{-1}{7}, \quad y = \frac{D_y}{D} = \frac{10}{7}.$$

Cramer's Rule for three unknowns is similar to this method.

CHAPTER 7 · REVIEW EXERCISES   *Answers, page A73–A75*

7.1

Graph these lines:

1. $y = 3x - 1$ **2.** $2x - 4 + y = 0$ **3.** $2x + 3y = 6$ **4.** $\dfrac{x}{4} + \dfrac{y}{5} = 1$ **5.** $2x + 6 = 0$ **6.** $7 - 2y = 0$

7. Find the x- and y-intercepts of the line $2x + 5y = 10$. Then find the area of the shaded triangle in the first quadrant formed by this line and the x- and y-axes.

8. Decide whether $(-2,5)$ is on the line $7x + 2y = -3$.

9. Decide whether $\left(\dfrac{-1}{2}, \dfrac{2}{3}\right)$ is on the line $y = \dfrac{-5x}{3} - \dfrac{1}{6}$.

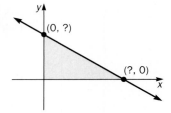

Problem 7

7.2

Find the slope of the line through the given pairs of points:

10. $(2,-1)$ and $(-4,7)$

11. $(3,-2)$ and $(1,2)$

12. $\left(\dfrac{1}{6}, \dfrac{2}{5}\right)$ and $\left(\dfrac{-2}{3}, \dfrac{3}{10}\right)$

13. $\left(\dfrac{1}{2}, \dfrac{2}{3}\right)$ and $\left(\dfrac{-1}{4}, \dfrac{5}{6}\right)$

14. $(-7,4)$ and $(-7,-1)$

15. $(\sqrt{2}, -2\sqrt{3})$ and $(3\sqrt{2}, 2\sqrt{3})$

16. $(-1,1)$ and $(\sqrt{2}, \sqrt{2})$

17. $(4a,4b)$ and $(8b,2a)$

18. $(3,9)$ and (x,x^2)

Put each equation into the slope–intercept form of a line, if it is not already in that form. Give the slope and y-intercept, and use them to graph each line.

19. $y = 2x - 3$ **20.** $y = \dfrac{2}{3}x + 1$ **21.** $2x + 5y = 10$ **22.** $3x + 4y = 0$

Which of the following pairs of lines are parallel, perpendicular, or neither?

23. Line 1 has slope $\dfrac{-3}{4}$;   line 2 has slope $\dfrac{3}{4}$.

24. Line 1 has slope $4\sqrt{2}$;   line 2 has slope $\dfrac{8}{\sqrt{2}}$.

25. Line 1: $2x + 3y - 9 = 0$;   line 2: $6x = 4y + 11$.

26. Line 1 goes through $(2,-1)$ and $(4,7)$;   line 2 goes through $(3,0)$ and $(-1,-16)$.

27. Line 1 goes through $(5,2)$ and $(7,-8)$;   line 2 goes through $(1,-1)$ and $(16,2)$.

28. Line 1 has slope $\dfrac{\sqrt{2} + \sqrt{6}}{2}$;   line 2 has slope $\dfrac{\sqrt{2} - \sqrt{6}}{2}$.

29. Line 1: $5x - 4y + 7 = 0$;   line 2: $10x + 19 = 8y$.

30. Line 1: $2x + 8 = 0$;   line 2: $y = -3$.

Find *k* satisfying the following conditions:

31. The line through (2,5) and (4,*k*) is parallel to the line $y = \frac{2}{5}x + 13$.

32. The line through (4,1) and (*k*,7) is parallel to the line through (−3,2) and (1,*k*).

33. The line through (0,2) and (4,*k*) is perpendicular to the line through (1,*k*) and (9,−2).

34. Show that the accompanying polygon is a *parallelogram*. *Hint:* Use slopes to show that both pairs of opposite sides are parallel.

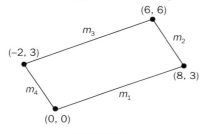

Problem 34

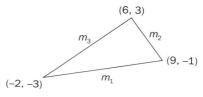

Problem 35

35. Show that the accompanying triangle is a *right triangle*. *Hint:* Use slopes to show that two sides are perpendicular.

7.3

Find the equation of the line satisfying the given conditions. Express your answer in slope–intercept form: $y = mx + b$.

36. $m = 2$, y-intercept (0,5)

37. $m = \dfrac{-3}{4}$, goes through (5,−2)

38. Goes through (2,−1) and (4,3)

39. Goes through (−3,5) and (1,−1)

40. x-intercept (−3,0), y-intercept (0,5)

41. Goes through (3,2) and (−5,2)

42. Goes through (2,1) and is *parallel* to the line $y = -3x + 9$

43. Goes through (2,1) and is *perpendicular* to the line $y = -3x + 9$

44. Goes through (−4,3) and is parallel to the line $5x - 2y = 9$

45. Goes through (−4,3) and is perpendicular to the line $5x - 2y = 9$

46. A line goes through (3,9) and (10,2).
   a. Find the equation of this line.
   b. Use your equation to find its x- and y-intercepts.
   c. Find the area of the shaded triangle.

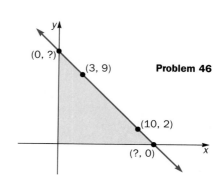

Problem 46

7.4

Solve for x and y by the addition or substitution method:

47. $3x + 4y = 2$
$\quad\ 5x + 2y = 8$

48. $3x + 7y = 9$
$\quad\ 4x - 5y = 12$

49. $5x + 7y = -24$
$\quad\ 2y + 19 = x$

50. $-6x + 15y = 4$
$\quad\ \ 4x - 10y = 3$

51. $\ \ 5x + 2y = -3$
$\quad 15x + 6y = -9$

52. $\dfrac{5}{x} + \dfrac{2}{y} = 4$

$\quad\ \dfrac{7}{x} - \dfrac{3}{y} = 23$

53. $\dfrac{x-1}{3} - \dfrac{2y-2}{5} = -1$

$\quad\ \dfrac{x}{3} - \dfrac{y}{4} = \dfrac{17}{6}$

54. $\dfrac{x-1}{y+2} = \dfrac{1}{2}$

$\quad\ \dfrac{x}{y} = \dfrac{3}{4}$

55. $3x + 4y = 7a$
$\quad\ 2x - 3y = -a + 17b$

56. $cx - dy = c^2 + d^2$
$\quad\ \ x + \ \ y = 2c$

7.5

Solve using two unknowns:

57. A collection of 34 coins in dimes and quarters is worth $5.80. How many of each coin are there?

58. Twenty-two coins worth $2.09 consist of equal numbers of pennies and quarters and equal numbers of nickels and dimes. How many of each coin are there?

59. A cash register has a total of 18 bills in $10's and $20's. If the total value of each denomination is the same, how many of each bill are there?

60. $20,000 is invested, part at 6% for 3 years and the balance at $9\frac{1}{2}$% for 4 years. How much is invested at each rate if the total interest is $5100?

61. $5000 is deposited into an account paying 7%. How much money should be deposited into an account paying 12% so that the amount deposited earns the equivalent of $11\frac{1}{2}$%? Assume the deposits are for 1 year.

62. $1000 is invested, part at 4% and the rest at 6%. How much is invested at each rate for 2 years if the interest at 6% is $40 more than the interest at 4%?

63. How many cubic centimeters of a 20% acid solution should be mixed with 10 cc of a 50% acid solution to obtain a solution that is 40% acid? How many cc of 40% acid will be produced?

64. How many pounds of candy selling for 75¢ a pound, and how many pounds of candy selling for 60¢ a pound, should be mixed to make a 30-pound mix selling for 69¢ a pound?

65. How many liters of a solution that is 80% alcohol, and how many liters of a 25% alcohol solution, should be mixed to form 22 liters of a 30% alcohol solution?

66. In 1960, John F. Kennedy defeated Richard Nixon by only 118,550 votes out of a total of 68,335,642 votes cast for these two candidates. How many votes did each receive? (*Historical note:* This was the closest popular vote in history.)

67. How many cubic centimeters of water should a person add to 40 cc of pure antifreeze in order to dilute it to a 25% antifreeze solution? *Hint:* Water is 0% A.F. and pure antifreeze is 100% A.F.

68. On a 150-question multiple-choice test, a student got 75% correct in part A and 90% correct in part B. If her overall score was 86% correct, how many problems were in each part? How many questions did she get correct in each part?

69. When reduced to lowest terms, a fraction equals $\frac{7}{10}$. If both numerator and denominator of the fraction are increased by 4, the new fraction that is formed equals $\frac{3}{4}$. Find the original fraction.

70. If the width of a certain rectangle were increased by 2 and the length decreased by 2, the new rectangle formed would have the same area as the original. If the width of the original rectangle were decreased by 2 and the length increased by 6, this new rectangle would also have the same area as the original. Find the dimensions of the original rectangle, as well as those of the other two rectangles.

71. A boat can go a certain speed in still water. On a river, it can travel 50 miles downstream in 2 hours and 15 miles upstream in 1 hour. Find the rate of the boat in still water and the rate of the current.

72. A cyclist can average a certain speed when there is no wind. One day, he rode for 2 hours with a wind of 4 miles per hour at his back, then turned around and returned to his starting point in 3 hours, while fighting the same headwind. Find the cyclist's average speed with no wind and the distance traveled in each direction.

73. A cyclist left home and rode for 2 hours. She then turned around and returned home. Because of fatigue, her return trip took 3 hours, and her average speed was reduced by 7 miles per hour. Find her speeds in each direction and the total distance she rode.

74. Find the equation of the line through point (1,7) that is perpendicular to the line $y = \frac{1}{2}x + 4$. Then find the point of intersection of the two lines.

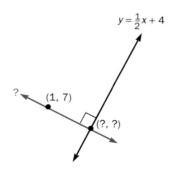

7.6

Solve for x, y, and z:

75. $x + 2y + z = 8$
$3x - 4y - z = -8$
$5x + 3y + z = 14$

76. $2x + 3y + 5z = 13$
$x - y - z = -1$
$5x - 3y + 4z = 1$

77. $2x - 7y - z = 35$
$x + y = 1$
$2y + z = -5$

78. $x - 2y = 4$
$x + 3z = 8$
$y - 4z = -9$

79. $\dfrac{1 - y}{x + z} = \dfrac{-1}{2}$

$\dfrac{x + 3}{y + 5} = \dfrac{6}{7}$

$\dfrac{x}{y - z} = 3$

80. $\dfrac{x + 1}{y - 3} = \dfrac{x}{y - 2}$

$\dfrac{z + 5}{x} = \dfrac{z + 3}{x - 1}$

$(y + 3)(z - 1) = z(y + 1)$

Solve using three unknowns:

81. The sum of three numbers is ten. Twice the second equals the difference between the first and the third. Twice the third equals the difference between the first and the second. Find the three numbers.

82. Eighteen coins in nickels, dimes, and quarters are worth $2.30. If the value of the nickels is the same as the value of the dimes, how many of each coin are there?

83. A $10,000 investment was distributed among stocks A, B, and C. In the first year, A paid 7%, B paid 5%, and C lost 9%. The net earnings were $320. Stock C was then sold, but the other two investments were continued. During the next two years, A paid 8% and B paid 3%; the combined earnings were $980. How much was originally invested in each stock?

7.7

Compute these determinants:

84. $\begin{vmatrix} -3 & 11 \\ 8 & -5 \end{vmatrix}$

85. $\begin{vmatrix} \dfrac{1}{2} & 6 \\ \dfrac{3}{4} & \dfrac{2}{3} \end{vmatrix}$

86. $\begin{vmatrix} 2\sqrt{3} & \sqrt{5} \\ \sqrt{10} & \sqrt{6} \end{vmatrix}$

87. $\begin{vmatrix} 2i & 2 \\ -1 & i \end{vmatrix}$

88. $\begin{vmatrix} 1 + \sqrt{3} & 2 - \sqrt{3} \\ \sqrt{3} & 1 - \sqrt{3} \end{vmatrix}$

89. $\begin{vmatrix} 2 - i & 1 - i \\ 1 + i & 3i \end{vmatrix}$

90. $\begin{vmatrix} 2 & 0 & -1 \\ 3 & -5 & 4 \\ 7 & 0 & -3 \end{vmatrix}$

91. $\begin{vmatrix} 5 & -1 & 7 \\ 8 & 10 & -2 \\ 4 & 1 & 0 \end{vmatrix}$

92. $\begin{vmatrix} 3 & \dfrac{1}{2} & -2 \\ 3 & \dfrac{1}{4} & 3 \\ -1 & 5 & \dfrac{1}{2} \end{vmatrix}$

93. $\begin{vmatrix} 2 & -1 & i \\ i & 2i & 1 \\ -i & 2 & -i \end{vmatrix}$

94. $\begin{vmatrix} 1 & x & x^2 \\ a & b & c \\ x & x^2 & x^3 \end{vmatrix}$

Solve by Cramer's Rule:

95. $3x + 2y = 0$
$5x - 7y = 31$

96. $5x + y = 7$
$4x + 3y = -3$

97. $5\sqrt{2}\,x + 2y = 8$
$3\sqrt{2}\,x - y = 7$

98. $ix + 2y = 2i$
$x + 2iy = -2$

99. $a^2x + y = a$
$b^2x + y = b$

100. $ax + 2y = a^2$
$2x + ay = 4$

101. $2x + y - z = 0$
$x \quad\; - 2z = -4$
$3y + z = 0$

102. $x + y - z = 7$
$x + 2y + z = 7$
$4x - y \quad\;= 0$

103. $2x - y - z = 6$
$3x + 2y + 2z = 2$
$x + y - 3z = 8$

CHAPTER 7 · TEST   *Answers, page A76*

Graph these lines using any method:

1. $y = 2x - 1$

2. $3x + 4y = 12$

3. a. $2y = 6$ **b.** $x + 2 = 0$

4. Use slopes to decide whether the line $3x + 2y - 7 = 0$ and the line through $(7, -4)$ and $(1, 0)$ are parallel, perpendicular, or neither.

Solve by the addition or the substitution method:

5. $2x + 5y = -14$
$4x - 3y = 11$

6. $2x - 3y = 4$
$y = 5x - 10$

7. $\dfrac{5x}{2} - 3y = -1$
$3x - \dfrac{y}{2} = 5$

8. $2x + y - z = 2$
$3x - y + 3z = 5$
$x + y - z = 7$

Solve by Cramer's Rule:

9. $5x + 2y = -2$
$3x + y = 0$

10. $3x + 2y \quad\;= 0$
$2x \quad\; - z = 4$
$y + 2z = -3$

11. Find the equation of the line through $(2, 5)$ that is parallel to the line $y = 3x + 1$.

12. Find the equation of the line with slope 2 that goes through $(0,1)$.

13. Find the equation of the line through $(0,2)$ and $(4,8)$.

14. Find the point of intersection of the lines obtained in Problems 12 and 13.

15. Find the equation of the line through $(3,2)$ that is perpendicular to the line $x - 2y - 4 = 0$.

16. Find the point of intersection of the two lines in Problem 15.

Solve using more than one unknown:

17. How many liters of 8% H_2SO_4 should be added to 18 liters of 20% H_2SO_4 in order to produce 17% H_2SO_4? How many liters of 17% acid will be produced?

18. $10,000 is invested, part at 6% for 4 years, and the balance at 10% for 3 years. How much is invested at each rate if the total interest earned is $2550?

19. The perimeter of a rectangle is 28 inches. When each width is increased by 2 inches and each length is decreased by the same amount, a square is formed. Find the sides of each figure.

20. A collection of 19 coins in nickels, dimes, and quarters is worth $2.10. There are twice as many nickels as quarters. How many of each coin are there?

8 THE CONIC SECTIONS

Parabolas

This chapter is devoted to four types of curves: parabolas, circles, ellipses, and hyperbolas. Together they are called **conic sections,** because they can be formed by passing a plane through cones at different angles:

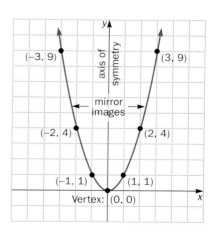

circle

ellipse

parabola

hyperbola

We saw in Section 7.1 that the equations of straight lines are first-degree equations. Here we will see that the equations of the conic sections are all second-degree. Let us begin by graphing the **parabola** $y = x^2$. To do so, we prepare a table and then plot points.

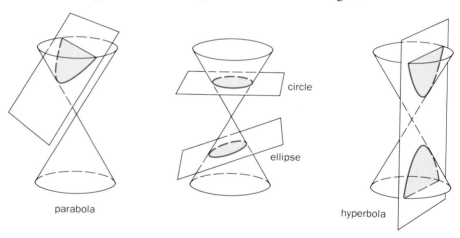

x	$y = x^2$
0	0
1	1
2	4
3	$9 \leftarrow 3^2$
-1	1
-2	$4 \leftarrow (-2)^2$
-3	9

The graph of $y = x^2$ is a *parabola*.

339

The lowest or highest point on a parabola is called the **vertex,** and it is the point $(0,0)$ on the parabola $y = x^2$ on the previous page. The line through the vertex separating the parabola into two "mirror-image" halves is the *axis of symmetry* and is the y-axis in the accompanying graph.

Next we graph the parabola $y = -2x^2 + 4x + 1$ in the same manner.

x	$y = -2x^2 + 4x + 1$
0	$1 \leftarrow -2 \cdot 0^2 + 4 \cdot 0 + 1 = 0 + 0 + 1$
1	$3 \leftarrow -2 \cdot 1^2 + 4 \cdot 1 + 1 = -2 + 4 + 1$
2	1
3	-5
-1	$-5 \leftarrow -2(-1)^2 + 4(-1) + 1 = -2 - 4 + 1$

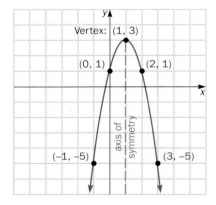

On the basis of these examples, we can state that the more general equation

$$y = ax^2 + bx + c, \quad \text{with } a \neq 0$$

represents the graph of a parabola. If $a > 0$ (as in $y = 1x^2$), the parabola will open upward; if $a < 0$ (as in $y = -2x^2 + 4x + 1$), the parabola will open downward. We now derive a formula for quickly locating the vertex of any parabola, using the fact that it is either the lowest or the highest point on the curve. To do so, we complete the square in a manner similar to the derivation of the Quadratic Formula in Section 6.4:

$$\frac{y}{a} = x^2 + \frac{b}{a}x \qquad + \frac{c}{a} \qquad \qquad \text{Divide } y = ax^2 + bx + c \text{ by } a.$$

We now add *and* subtract $\left(\frac{1}{2} \cdot \frac{b}{a}\right)^2 = \frac{b^2}{4a^2}$ on the right side of this equation:

$$\frac{y}{a} = \left(x^2 + \frac{b}{a}x + \frac{b^2}{4a^2}\right) + \frac{c}{a} - \frac{b^2}{4a^2} \qquad \text{Add } \frac{b^2}{4a^2} - \frac{b^2}{4a^2} = 0.$$

$$\frac{y}{a} = \left(x + \frac{b}{2a}\right)^2 + \frac{4ac - b^2}{4a^2} \qquad \text{Factor; } \frac{c \cdot 4a}{a \cdot 4a} = \frac{4ac}{4a^2}$$

To isolate y, we multiply both sides of this equation by a:

$$a \cdot \frac{y}{a} = a \cdot \left(x + \frac{b}{2a}\right)^2 + a \cdot \frac{4ac - b^2}{4a^2}$$

$$y = a\left(x + \frac{b}{2a}\right)^2 + \frac{4ac - b^2}{4a}.$$

Because the lowest (minimum) or highest (maximum) point on a parabola occurs at the vertex, we must find the value of the variable x in the term $a\left(x + \frac{b}{2a}\right)^2$ at which this happens. There are two possibilities:

If $a > 0$, then $a\left(x + \frac{b}{2a}\right)^2 \geq 0$ because the square $\left(x + \frac{b}{2a}\right)^2$ is ≥ 0; the *minimum value* of this term occurs when

$$a\left(x + \frac{b}{2a}\right)^2 = 0, \quad \text{or} \quad x = \frac{-b}{2a}.$$

If $a < 0$, then $a\left(x + \frac{b}{2a}\right)^2 \leq 0$ because $\left(x + \frac{b}{2a}\right)^2 \geq 0$; the *maximum value* of this term also occurs when

$$a\left(x + \frac{b}{2a}\right)^2 = 0, \quad \text{or} \quad x = \frac{-b}{2a}.$$

We see in either case that the vertex of the parabola

$$y = ax^2 + bx + c,$$

$$\text{or} \quad y = a\left(x + \frac{b}{2a}\right)^2 + \frac{4ac - b^2}{4a}$$

occurs when the term $a\left(x + \frac{b}{2a}\right)^2 = 0$, or $x = \frac{-b}{2a}$. At this point,

$$y = 0 \qquad \qquad + \frac{4ac - b^2}{4a} = \frac{4ac - b^2}{4a}.$$

Let us summarize, on the basis of this discussion and the two previous examples:

Equation of a Parabola

$$y = ax^2 + bx + c, \quad a \neq 0$$

has its **vertex** at the point with coordinates

$$x = \frac{-b}{2a}, \quad y = \frac{4ac - b^2}{4a}.$$

If $a > 0$, the parabola opens *upward*
and the vertex is the minimum point:

If $a < 0$, the parabola opens *downward*
and the vertex is the maximum point:

Alternatively, the y-coordinate of the vertex can be obtained by substituting the x-coordinate into the original equation of the parabola.

EXAMPLE 1

Graph each parabola by first locating the vertex:

a. $y = x^2 - 4$

Here $a = 1$, $b = 0$, and $c = -4$. The *vertex* is at

$$x = \frac{-0}{2 \cdot 1} = 0. \quad x = \frac{-b}{2a}$$

In this case, it is easiest to obtain the y-coordinate by substituting into $y = x^2 - 4$

$$y = 0^2 - 4 = -4$$

We now prepare a table, selecting two points greater than $x = 0$, (say $x = 1, 2$), and two points less than $x = 0$ (say $x = -1, -2$).

x	$y = x^2 - 4$
0	$-4 \leftarrow$ Vertex
1	$-3 \leftarrow 1^2 - 4$
2	$0 \leftarrow 2^2 - 4$
-1	$-3 \leftarrow (-1)^2 - 4$
-2	0

2 points > 0 { 1, 2

2 points < 0 { -1, -2

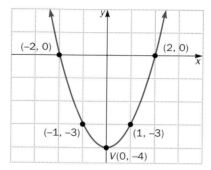

b. $y = -2x^2 - 6x + 1$.

Here $a = -2$, $b = -6$, and $c = 1$. The *vertex* is at

$$x = \frac{-(-6)}{2(-2)} = \frac{6}{-4} = \frac{-3}{2} = -1.5.$$

In this case, it is easiest to obtain the y coordinate by substituting into $y = \dfrac{4ac - b^2}{4a}$.

$$y = \frac{4(-2) \cdot 1 - (-6)^2}{4(-2)} = \frac{-8 - 36}{-8} = \frac{-44}{-8} = \frac{11}{2} = 5.5$$

x	$y = -2x^2 - 6x + 1$
-1.5	$5.5 \leftarrow$ Vertex
-1	$5 \leftarrow -2(-1)^2 - 6(-1) + 1$
0	1
-2	5
-3	$1 \leftarrow -2(-3)^2 - 6(-3) + 1$

2 points > -1.5 { -1, 0

2 points < -1.5 { -2, -3

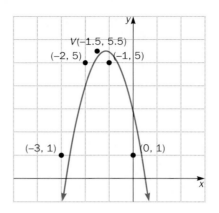

If x and y are interchanged in the equation $y = ax^2 + bx + c$, the result is $x = ay^2 + by + c$. This represents a parabola that opens to the *left or right*, depending on the sign of a. The vertex is obtained by using the same formula by which we found the vertex of a parabola opening upward or downward, with x and y interchanged:

Equation of a Parabola

$$x = ay^2 + by + c, \quad a \neq 0$$

has its **vertex** at the point with coordinates

$$y = \frac{-b}{2a}, \quad x = \frac{4ac - b^2}{4a},$$

If $a > 0$, the parabola opens to the *right*:

If $a < 0$, the parabola opens to the *left*:

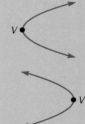

EXAMPLE 2   Graph each parabola by first locating the vertex:

a. $x = y^2 - 2y$.

Here $a = 1$, $b = -2$, and $c = 0$. The vertex is at

$$y = \frac{-(-2)}{2 \cdot 1} = 1. \qquad y = \frac{-b}{2a}$$

It is now easiest to substitute into $x = y^2 - 2y$:

$$x = 1^2 - 2 \cdot 1 = -1$$

As in Example 1, we select two points greater than $y = 1$ (say $y = 2, 3$) and two points less than $y = 1$ (say $x = 0, -1$).

$y^2 - 2y = x$	y	
-1	1	← Vertex
$2^2 - 2 \cdot 2 \rightarrow 0$	2	
$3^2 - 2 \cdot 3 \rightarrow 3$	3	2 points > 1
0	0	
$(-1)^2 - 2(-1) \rightarrow 3$	-1	2 points < 1

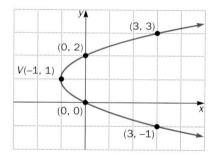

b. $x = -y^2 + y + 2$.

Here $a = -1$, $b = 1$, and $c = 2$. The vertex is at

$$y = \frac{-1}{2(-1)} = \frac{1}{2}.$$

It is now easiest to substitute into $x = \frac{4ac - b^2}{4a}$:

$$x = \frac{4(-1) \cdot 2 - 1^2}{4(-1)} = \frac{-8 - 1}{-4} = \frac{9}{4} = 2\frac{1}{4}$$

We select two points greater than $y = \frac{1}{2}$ (say $y = 1, 2$) and two points less than $y = \frac{1}{2}$ (say $y = 0, -1$).

(continued)

$-y^2 + y + 2 = x$	y	
$2\dfrac{1}{4}$	$\dfrac{1}{2}$	$\leftarrow$ Vertex
$\begin{array}{l} -(1)^2 + 1 + 2 \ \rightarrow\ 2 \\ -(2)^2 + 2 + 2 \ \rightarrow\ 0 \end{array}$	$\left.\begin{array}{l} 1 \\ 2 \end{array}\right\}$	2 points $> \dfrac{1}{2}$
$\begin{array}{l} 2 \\ -(-1)^2 + (-1) + 2 \ \rightarrow\ 0 \end{array}$	$\left.\begin{array}{l} 0 \\ -1 \end{array}\right\}$	2 points $< \dfrac{1}{2}$

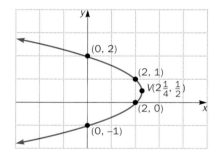

The next two examples use the fact that the vertex of a parabola represents the maximum or the minimum point on the curve. The first example is similar to Exercise 1.3, Problem 37; Exercise 3.5, Problem 73; Exercise 6.3, Problem 37; and Exercise 6.4, Problem 61.

EXAMPLE 3

When a ball is thrown upward, the formula

$$h = -16t^2 + 80t + 96$$

gives its height h (in feet) above the ground after t seconds. When does the ball reach its maximum height? What is the maximum height?

Solution
Here $a = -16$, $b = 80$, and $c = 96$. Because $a < 0$, this parabola on the t–h axis system will open downward, so the vertex will represent the *maximum* value of h. The *vertex* is

$$t = \frac{-80}{2(-16)} = 2.5,$$
$$h = -16(2.5)^2 + 80(2.5) + 96 = 196.$$

The following table and graph illustrate this:

t	h
2.5	196
2	192
3	192
1	160
4	160

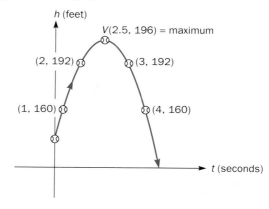

Answer: The maximum height of the ball is 196 feet after 2.5 seconds.

EXAMPLE 4

A rectangular grazing area is to be fenced off along a river, and 400 feet of fencing are to be used. The river acts as one side, so only three sides of fencing are needed. Find the dimensions that give the largest grazing area. What is the largest area?

Solution
Let x = width and y = the length.
Because 400 feet of fencing are available,

$$2x + y = 400, \quad \text{or} \quad y = 400 - 2x.$$

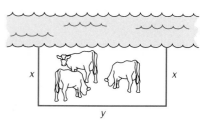

The rectangle has area $A = xy$.
Substituting $y = 400 - 2x$ into $A = xy$ gives

$$A = x(400 - 2x) = 400x - 2x^2,$$

$$\text{or} \quad A = -2x^2 + 400x.$$

This represents a parabola opening downward in the x–A axis system, so the vertex will be a *maximum* point. Using $a = -2$ and $b = 400$, we get

$$x = \frac{-400}{2(-2)} = 100.$$

(continued)

$$y = 400 - 2x$$
$$= 400 - 2(100) = 200.$$
$$A = xy = (100)(200) = 20,000.$$

Answer: The maximum area occurs when $x = 100$ feet and $y = 200$ feet. The maximum area is $A = 20,000$ square feet.

EXERCISE 8.1   *Answers, pages A76–A77*

A

Graph each parabola by first finding the vertex, as in EXAMPLES 1 and 2:

1. $y = x^2 - 1$
2. $y = -x^2 + 4$
3. $y = -x^2 + 4x$
4. $y = x^2 - 6x$
5. $y = x^2 + 2x - 3$
6. $y = -x^2 - 4x + 5$
7. $y = x^2 + 3x - 4$
8. $y = x^2 - 5x + 1$
9. $y = -2x^2 + x + 3$
10. $y = 2x^2 + x - 1$
11. $x = y^2$
12. $x = -y^2 + 1$
13. $x = -y^2 + 2y$
14. $x = y^2 + 4y$
15. $x = 3y^2 - 6y + 2$
16. $x = -2y^2 - 10y + 5$
17. $x = y^2 - y$
18. $x = y^2 + 3y - 4$
19. $x = -2y^2 + 7y - 3$
20. $x = 3y^2 + 5y - 7$

Solve as in EXAMPLE 3:

21. When a ball is thrown upward, the formula

$$h = -16t^2 + 64t + 80$$

gives its height h (in feet) above the ground after t seconds. When does the ball reach its maximum height? What is the maximum height?

22. Answer the same questions as in Problem 21, using the formula

$$h = -4.9t^2 + 34.3t$$

where h is measured in meters and t is measured in seconds.

23. A company manufacturing bicycles finds that its profit is given by

$$P = -x^2 + 500x - 6500$$

where P is the profit (in dollars) if x bicycles are manufactured. How many bikes should the company make in order to maximize profits? Find the maximum profit.

24. A company manufacturing shoes finds that its costs are given by

$$C = 2x^2 - 1400x + 250,000$$

where C is the cost (in dollars) to make x pairs of shoes. How many pairs of shoes should the company make in order to minimize costs? Find the minimum total cost.

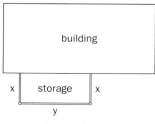

Solve as in EXAMPLE 4:

25. A rectangular storage area is to be fenced off on the side of a building, and 500 feet of fencing are to be used. The building acts as one side, so only three sides of fencing are needed. Find the dimensions that maximize the storage area. What is the maximum storage area?

Problem 25

26. The perimeter of a rectangle is 20 inches. Find the dimensions that give the largest area. What is this area? *Hint:* $2x + 2y = 20$; isolate y and then substitute into $A = xy$.

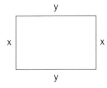

27. One number is twelve more than another. Find the two numbers such that their product is as small as possible. What is this minimum product? *Hint:* The numbers are x and $x + 12$; then form their product P.

Problem 26

28. Find two numbers whose difference is 8 such that the sum of their squares is a minimum. What is this sum? *Hint:* $x - y = 8$; isolate either variable and substitute into $S = x^2 + y^2$.

B

29. An apartment manager finds that the total monthly income I is given by

$$I = \$(600 + 20x)(40 - x),$$

where x is the number of vacant apartments. Find the value of x that maximizes the income, and then find this income.

30. In the retail marketplace, the demand for an item is affected by its selling price. The demand for a particular jacket selling for $\$x$ is $150 - \frac{1}{2}x$ jackets. The revenue R generated from sales is given by the formula

$$R = (\text{selling price of jacket})(\text{demand for jacket}).$$

Find the selling price that maximizes revenue. Then give the demand and the revenue.

C

31. A rectangular area is to be fenced off; 600 feet of fencing are to be used, including a fenced partition inside, as shown in the accompanying drawing. Find the dimensions that maximize the area. What is the maximum area? *Hint:* Write an expression set equal to 600, solve for one of the variables, and then substitute into $A = xy$.

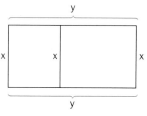

Problem 31

32. Repeat Problem 31, but here 600 feet of fencing are to be used, including the two fenced partitions shown.

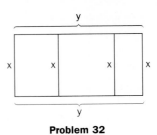

Problem 32

33. In Example 4 in the text, the largest grazing area = 20,000 square feet occurred when $x = 100$ feet and $y = 200$ feet.

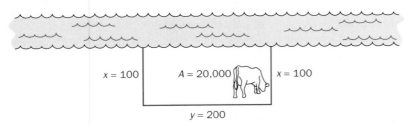

To further illustrate that this *is* the maximum area, find the area when $x = 99$ and when $x = 101$. Both of your answers should be less than 20,000. *Hint:* Recall that $y = 400 - 2x$.

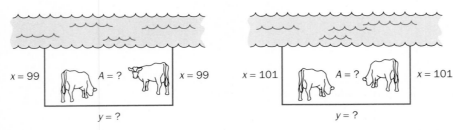

34. The equation $y = ax^2 + bx + c$, with $a \neq 0$, represents the graph of a parabola. Describe the graph if $a = 0$.

The coefficients a, b, and c in the parabola

$$y = ax^2 + bx + c$$

can be determined, given that the points $(-1,0)$, $(1,-6)$, and $(2,-3)$ lie on the graph. Substituting them into the preceding equation produces a system of three equations in three unknowns:

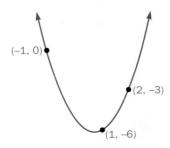

$(-1,0)$: $0 = a(-1)^2 + b(-1) + c$,   or $a - b + c = 0$

$(1,-6)$: $-6 = a \cdot 1^2 + b \cdot 1 + c$,   or $a + b + c = -6$

$(2,-3)$: $-3 = a \cdot 2^2 + b \cdot 2 + c$,   or $4a + 2b + c = -3$

Using the methods of Section 7.6, we get the solution $a = 2$, $b = -3$, and $c = -5$. The parabola containing these three points, therefore, is

$$y = 2x^2 - 3x - 5.$$

In like manner, find the parabola containing the indicated points:

35. $y = ax^2 + bx + c$ contains $(1,4)$, $(-1,0)$, and $(3,16)$.

36. $y = ax^2 + bx + c$ contains $(-1,0)$, $(2,-3)$, and $(3,0)$.

37. $x = ay^2 + by + c$ contains $(-5,1)$, $(0,2)$, and $(-6,0)$

Hint: The first equation is $-5 = a \cdot 1^2 + b \cdot 1 + c$.

38. $x = ay^2 + by + c$ contains $(0,2)$, $(0,-2)$, and $(-8,0)$.

8.2 Distance and Circles

The picture in Section 7.2 used to obtain the slope formula can also be used to find the distance D between any two points P_1 and P_2:

By the Pythagorean Theorem,

$$D^2 = (x_2 - x_1)^2 + (y_2 - y_1)^2.$$

Taking the positive square root of both sides yields our desired result:

$$D = \sqrt{(x_2 - x_1)^2 + (y_2 - y_1)^2}.$$

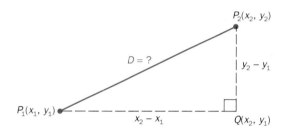

The Distance Formula

The distance between points $P_1(x_1, y_1)$ and $P_2(x_2, y_2)$ is

$$D = \sqrt{(x_2 - x_1)^2 + (y_2 - y_1)^2}.$$

Just as in the case of the slope formula, it does not matter which point is labeled P_1 and which is labeled P_2. See Problem 82 in the exercises.

EXAMPLE 1

Find the distance between the following pairs of points:

a. $(-1,2)$ and $(3,5)$

$$D = \sqrt{[3 - (-1)]^2 + (5 - 2)^2}$$
$$= \sqrt{(4)^2 + (3)^2}$$
$$= \sqrt{16 + 9} = \sqrt{25}$$
$$D = 5.$$

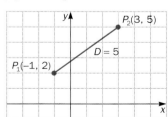

(continued)

b. $(-2, -1)$ and $(4, -5)$

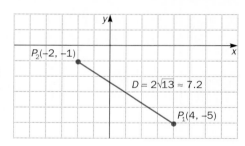
$P_2(-2, -1)$
$D = 2\sqrt{13} \approx 7.2$
$P_1(4, -5)$

$$D = \sqrt{(-2 - 4)^2 + [-1 - (-5)]^2}$$
$$= \sqrt{(-6)^2 + (4)^2} = \sqrt{36 + 16}$$
$$= \sqrt{52} = \sqrt{4 \cdot 13}$$
$$D = 2\sqrt{13}.$$

WARNING! $\sqrt{36 + 16} \neq 6 + 4$, because $\sqrt{52} \neq 10$.
The numbers under the radical must be added first before the square root is taken.

The distance formula enables us to derive the equation of a circle. A **circle** is the set of all points at a fixed distance (called the **radius**) from a fixed point (called the **center**). Suppose the radius is r and the center is the point (h,k). For any point (x,y) on the circle, the distance from (x,y) to (h,k) will always be r. By the distance formula,

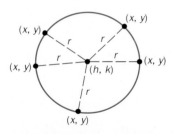
(x, y)
(x, y)
r r
(x, y) r —— r (x, y)
(h, k)
r
(x, y)

$$\sqrt{(x - h)^2 + (y - k)^2} = r.$$

Squaring both sides yields our desired equation:

$$(x - h)^2 + (y - k)^2 = r^2.$$

Equation of a Circle

$$(x - h)^2 + (y - k)^2 = r^2$$

has center (h,k) and radius r.

EXAMPLE 2

Graph each of the following circles, indicating center and radius:

a. $(x - 2)^2 + (y - 1)^2 = 9$

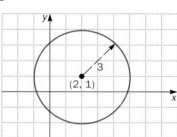
3
$(2, 1)$

Comparing this to the equation of a circle, we see that the center (h,k) is $(2,1)$. Also, the radius satisfies $r^2 = 9$, or $r = \sqrt{9} = 3$.

b. $x^2 + (y + 3)^2 = 8$

Rewriting this as $(x - 0)^2 + [y - (-3)]^2 = 8$, we see that the center is $(0, -3)$. The radius satisfies $r^2 = 8$, or $r = \sqrt{8} = 2\sqrt{2} \approx 2.8$.

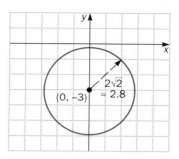

c. $4x^2 + 4y^2 = 25$

Dividing by 4 gives $x^2 + y^2 = \frac{25}{4}$. Rewriting this as $(x - 0)^2 + (y - 0)^2 = \frac{25}{4}$, we see that the center is $(0, 0)$ and the radius satisfies $r^2 = \frac{25}{4}$, or $r = \frac{5}{2} = 2.5$.

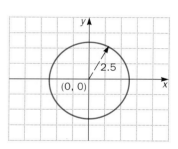

In the next example, you are asked to do the reverse of what you just did. That is, given information about the circle, find its equation.

EXAMPLE 3 Find the equation of each circle satisfying the given conditions:

a. Center $= (3, -5)$, radius $= 4$.

Substituting $(h, k) = (3, -5)$ and $r = 4$ into the equation of a circle gives

$$(x - 3)^2 + [y - (-5)]^2 = 4^2,$$

or $(x - 3)^2 + (y + 5)^2 = 16.$

b. Center is the origin, radius $= 5\sqrt{3}$

The center is $(0, 0)$ and $r = 5\sqrt{3}$, so we get

$$(x - 0)^2 + (y - 0)^2 = (5\sqrt{3})^2,$$

or $x^2 + y^2 = 75.$

c. Center $= (-2, 1)$, circle goes through the point $(5, -4)$.

All we need is the radius, which is the distance   *(continued)* between the center and any point on the circle; thus the radius is the distance between $(-2, 1)$ and $(5, -4)$. By the distance formula,

$$r = \sqrt{[5 - (-2)]^2 + (-4 - 1)^2}$$
$$= \sqrt{(7)^2 + (-5)^2}$$
$$= \sqrt{49 + 25} = \sqrt{74}.$$

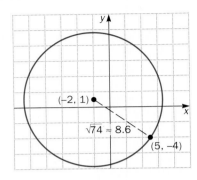

Substituting the center $(-2, 1)$ and the radius $\sqrt{74}$ into the equation for a circle gives

$$[x - (-2)]^2 + (y - 1)^2 = (\sqrt{74})^2,$$
$$\text{or} \qquad (x + 2)^2 + (y - 1)^2 = 74.$$

In the next example, we put the equation into the standard form of a circle by first *completing the square*.

EXAMPLE 4   Graph $x^2 + y^2 - 4x + 9y + 11 = 0$ by first completing the square.

Solution
This is similar to the method in Section 6.3, except that we are completing the square for *both* variables.

$$(x^2 - 4x \quad) + (y^2 + 9y \quad) = -11 \qquad \text{Isolate constant.}$$

$$(x^2 - 4x + 4) + \left(y^2 + 9y + \frac{81}{4}\right) = -11 + 4 + \frac{81}{4} \qquad \text{Add } \left(\frac{-4}{2}\right)^2 = 4$$
$$\text{and } \left(\frac{9}{2}\right)^2 = \frac{81}{4}.$$

$$(x - 2)^2 \quad + \quad \left(y + \frac{9}{2}\right)^2 = \frac{-11 \cdot 4}{1 \cdot 4} + \frac{4 \cdot 4}{1 \cdot 4} + \frac{81}{4} \longleftarrow \text{L.C.D.} = 4$$

$$(x - 2)^2 + \left(y + \frac{9}{2}\right)^2 = \frac{53}{4}.$$

This is the standard form of a circle.

We read off the center $= \left(2, \dfrac{-9}{2}\right) = (2, -4.5)$;

the radius satisfies

$$r^2 = \frac{53}{4}, \quad \text{or} \quad r = \frac{\sqrt{53}}{2} \approx 3.6.$$

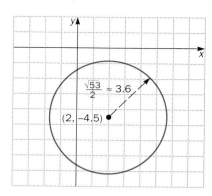

EXERCISE 8.2   *Answers, pages A77–A79*

Find the distance between the following pairs of points, as in EXAMPLE 1:

1. $(1,2)$ and $(4,6)$

2. $(2,1)$ and $(7,13)$

3. $(-3,1)$ and $(3,-7)$

4. $(-6,-2)$ and $(6,7)$

5. $(2,-3)$ and $(-4,-3)$

6. $(-1,5)$ and $(-1,-2)$

7. $(-2,1)$ and $(2,-3)$

8. $(4,-2)$ and $(-2,2)$

9. $(0,-2)$ and $(8,0)$

10. $(8,-12)$ and the origin

Graph each of the following circles, indicating center and radius. See EXAMPLE 2.

11. $(x + 5)^2 + (y - 4)^2 = 25$

12. $(x - 2)^2 + (y + 3)^2 = 9$

13. $(x - 1)^2 + y^2 = 1$

14. $x^2 + (y + 2)^2 = 4$

15. $(x - 3)^2 + (y + 3)^2 = 12$

16. $(x + 4)^2 + (y - 4)^2 = 20$

17. $16x^2 + 16y^2 = 25$ *Hint:* Divide by 16.

18. $9x^2 + 9y^2 = 4$

19. $4x^2 + 4(y - 1)^2 = 9$

20. $9(x + 3)^2 + 9y^2 = 1$

21. $\dfrac{x^2}{25} + \dfrac{y^2}{25} = 1$ *Hint:* Clear fractions.

22. $\dfrac{x^2}{9} + \dfrac{(y - 3)^2}{9} = 1$

Find the equation of each circle satisfying the given conditions, as in EXAMPLE 3:

23. Center = $(2,5)$, radius = 4

24. Center = $(-3,2)$, radius = 6

25. Center = $(-5,0)$, radius = $2\sqrt{5}$

26. Center is the origin, radius = $4\sqrt{7}$

27. Center = $(0,4)$, radius = $\dfrac{2}{3}$

28. Center = $(-3,0)$, radius = $\dfrac{4}{3}$

29. Center = $(3,-1)$, the circle goes through the point $(6,4)$

30. Center = $(-4,4)$, the origin lies on the circle

Graph each of the following circles by first completing the square, as in EXAMPLE 4. Indicate center and radius.

31. $x^2 + y^2 + 6x - 8y + 9 = 0$ **32.** $x^2 + y^2 - 10x + 12y - 20 = 0$ **33.** $x^2 + y^2 - 2x - 4y - 4 = 0$

34. $x^2 + y^2 + 2x + 8y + 8 = 0$ **35.** $x^2 + y^2 - 10y + 17 = 0$ **36.** $x^2 + y^2 + 6y - 3 = 0$

37. $x^2 - 6x + y^2 - 3 = 0$ **38.** $x^2 + y^2 + 12x = 0$

B

39. $x^2 + y^2 - 3x + 5y - 2 = 0$ **40.** $x^2 + y^2 + x - 7y + 8 = 0$

The distance between $\left(-1,\frac{2}{3}\right)$ **and** $\left(\frac{1}{4},\frac{-1}{6}\right)$ **can be found by first expressing all coordinates in terms of the entire L.C.D. = 12:**

$$\left(-1,\frac{2}{3}\right) = \left(\frac{-12}{12},\frac{8}{12}\right) \quad \text{and} \quad \left(\frac{1}{4},\frac{-1}{6}\right) = \left(\frac{3}{12},\frac{-2}{12}\right).$$

Thus

$$D = \sqrt{\left(\frac{3}{12} - \left(\frac{-12}{12}\right)\right)^2 + \left(\frac{-2}{12} - \frac{8}{12}\right)^2} = \sqrt{\left(\frac{15}{12}\right)^2 + \left(\frac{-10}{12}\right)^2} = \sqrt{\frac{225}{144} + \frac{100}{144}}$$

$$= \sqrt{\frac{325}{144}} = \sqrt{\frac{25 \cdot 13}{144}} = \frac{5\sqrt{13}}{12}.$$

In like manner, find the distances between the following pairs of points:

41. $\left(2,\frac{1}{4}\right)$ and $\left(\frac{-3}{4},\frac{1}{8}\right)$ **42.** $\left(\frac{-3}{2},\frac{3}{4}\right)$ and $\left(-2,\frac{1}{2}\right)$ **43.** $\left(\frac{1}{2},-\frac{3}{8}\right)$ and $\left(-\frac{1}{3},\frac{1}{4}\right)$

44. $\left(\frac{1}{3},\frac{-1}{4}\right)$ and $\left(\frac{-1}{6},\frac{5}{12}\right)$ **45.** $\left(\frac{2}{3},2\right)$ and $\left(\frac{-1}{3},\frac{1}{2}\right)$ **46.** $\left(\frac{1}{6},1\right)$ and $\left(\frac{-1}{10},\frac{13}{15}\right)$

The distance between $(1,-1)$ **and** $(\sqrt{2},\sqrt{2})$ **involves methods used in Section 5.4:**

$$D = \sqrt{(\sqrt{2} - 1)^2 + (\sqrt{2} - (-1))^2} = \sqrt{(\sqrt{2} - 1)^2 + (\sqrt{2} + 1)^2}$$

$$= \sqrt{\underbrace{2 - 2\sqrt{2} + 1}_{\text{FOIL.}} + \underbrace{2 + 2\sqrt{2} + 1}_{\text{FOIL.}}} = \sqrt{6}.$$

In like manner, find the distances between the following pairs of points:

47. $(-2,2)$ and $(\sqrt{2},\sqrt{2})$ **48.** $(\sqrt{3},\sqrt{3})$ and $(-1,1)$ **49.** $(0,0)$ and $(2\sqrt{2},1)$

50. $(2\sqrt{3},-2)$ and $(0,2)$ **51.** $(-\sqrt{6},\sqrt{6})$ and $(\sqrt{3},\sqrt{3})$ **52.** $(\sqrt{2},-\sqrt{2})$ and $(\sqrt{7},\sqrt{7})$

In geometry, triangles are often described in terms of equality of their sides, as shown:

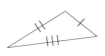
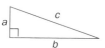

isosceles:
2 or more
sides equal

equilateral:
3 sides equal

scalene:
no sides equal

right triangle:
$a^2 + b^2 = c^2$

In each of the following problems, draw the triangle formed by the given points; then
a. Find the length of each side, using the distance formula.
b. Decide which type of triangle it is. (More than one answer is possible.)
c. Find each *perimeter* (the sum of the lengths of the sides).

53. $(-1,4)$, $(3,0)$, and $(7,8)$

54. $(2,-2)$, $(8,4)$, and $(6,6)$

55. $(1,10)$, $(2,3)$, and $(9,4)$

56. $(1,0)$, $(3,0)$, and $(2,\sqrt{3})$

A *parallelogram* is a four-sided polygon in which both pairs of *opposite sides are equal* in length. (See also Exercise 7.2, Problem 65.) A *rhombus* is a four-sided polygon with *all sides equal* in length, such as a square or a diamond. Note that a rhombus is always a parallelogram, but a parallelogram is not always a rhombus.

parallelogram

rhombus

In each of the following problems, connect the four points in sequence. Then determine whether the figure is a *parallelogram* or a *rhombus* by finding the lengths of the sides.

57. $(1,1)$, $(7,3)$, $(8,6)$, and $(2,4)$

58. $(-2,1)$, $(3,3)$, $(5,8)$, and $(0,6)$

In Problems 59 and 60, you will be given a circle and a parabola.
a. Find the center of the circle.
b. Find the vertex of the parabola. See Section 8.1.
c. Find the distance between the center and the vertex, using the distance formula.
d. Find the equation of the line between these two points. See Section 7.3.

59. Circle: $(x + 4)^2 + (y - 1)^2 = 9$; parabola: $y = x^2 - 4x + 1$

60. Circle: $x^2 + y^2 - 10x + 4y - 3 = 0$; parabola: $x = y^2 - 6y - 4$

For $a > 0$, the distance between $(3a,4a)$ and $(a,-2a)$ is given by

$$D = \sqrt{(a - 3a)^2 + (-2a - 4a)^2} = \sqrt{(-2a)^2 + (-6a)^2} = \sqrt{4a^2 + 36a^2}$$
$$= \sqrt{40a^2} = \sqrt{4a^2 \cdot 10} = 2a\sqrt{10}.$$

In like manner, find the distances between the following pairs of points. Assume all constants are positive.

61. $(a,-a)$ and $(4a,3a)$　　**62.** $(-2b,5b)$ and $(3b,-7b)$　　**63.** $(-2c,3c)$ and $(2c,7c)$　　**64.** $(5d,-d)$ and $(3d,3d)$

65. (m,n) and $(3m,3n)$　　**66.** $(-p,2q)$ and $(8p,-7q)$　　**67.** (m,n) and (n,m)　　**68.** (km,kn) and $(-kn,km)$

Consider this problem:

　　　"Find all values of x such that the distance between the
　　　points $(x,6)$ and $(4,2)$ is 5 units."

This can be solved using the distance formula:

$$\sqrt{(4 - x)^2 + (2 - 6)^2} = 5$$
$$(4 - x)^2 + (-4)^2 = 25 \qquad \text{Square both sides.}$$
$$16 - 8x + x^2 + 16 = 25 \qquad \text{FOIL.}$$
$$x^2 - 8x + 7 = 0$$
$$(x - 7)(x - 1) = 0$$
$$x = 7, \quad x = 1. \qquad \text{Answers.}$$

In like manner, find all values of the variable such that

69. The distance between $(x,4)$ and $(2,7)$ is 5 units.

70. The distance between $(6,-3)$ and $(x,2)$ is 13 units.

71. The distance between $(-3,y)$ and $(1,4)$ is 8 units.

72. The distance between $(1,-2)$ and $(3,y)$ is $6\sqrt{2}$ units.

73. The distance between $(2,-x)$ and $(x,3)$ is 5 units.

74. The distance between $(4,x)$ and $(x,2)$ equals the distance between $(-3,1)$ and $(x,3)$.

The area of a circle is given by $A = \pi r^2$. Find the radius of each of the following circles, and then use it to find the area of the circle.

75. $(x - 2)^2 + (y + 3)^2 = 25$　　　　**76.** $x^2 + (y - 7)^2 = 16$　　　　**77.** $\dfrac{x^2}{10} + \dfrac{y^2}{10} = 2$

78. $4x^2 + 4y^2 = 49$　　　　**79.** $x^2 + y^2 - 2x + 4y - 11 = 0$　　　　**80.** $x^2 + y^2 + 6x - 8y + 7 = 0$

81. In Section 7.2 we proved that if two lines are perpendicular, then the product of their slopes is -1: $m_1 m_2 = -1$. The proof required the slope formula. In this problem, we prove the same result using the distance formula. In the picture, *perpendicular lines* with slopes m_1 and m_2, respectively, intersect at the origin.

Let us assume that these lines go through the points $(1,b)$ and $(1,a)$, as shown, and that $b > a$.

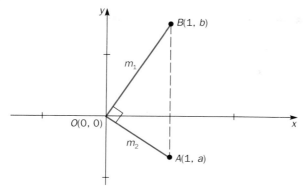

a. Use the distance formula to show that $OA = \sqrt{1 + a^2}$, $OB = \sqrt{1 + b^2}$, and $AB = b - a$.

b. By the Pythagorean Theorem, $(AB)^2 = (OA)^2 + (OB)^2$. Substitute the results of part a into this equation and obtain the result that $ab = -1$.

c. Use the slope formula to show that $m_1 = b$ and $m_2 = a$.

d. Substitute these results into the equation $ab = -1$ obtained in part b. What is the final result?

82. It was mentioned in the text that it does not matter which of two points is labeled $P_1(x_1, y_1)$ and which is labeled $P_2(x_2, y_2)$, because the distance between them will be the same in either case. Verify this fact by explaining why

$$\sqrt{(x_2 - x_1)^2 + (y_2 - y_1)^2} = \sqrt{(x_1 - x_2)^2 + (y_1 - y_2)^2}.$$

8.3 Ellipses and Hyperbolas

The equation $(x - 0)^2 + (y - 0)^2$, or $x^2 + y^2 = r^2$, is a circle with center $(0,0)$ and radius r. Dividing both sides of this equation by r^2 gives

$$\frac{x^2}{r^2} + \frac{y^2}{r^2} = 1.$$

The more general equation $\dfrac{x^2}{a^2} + \dfrac{y^2}{b^2} = 1$,

in which a and b are not necessarily equal, is called an **ellipse**.

Let us illustrate by graphing the equation

$$\frac{x^2}{25} + \frac{y^2}{9} = 1.$$

First we let $x = 0$:

$$\frac{0^2}{25} + \frac{y^2}{9} = 1$$

$$\frac{y^2}{9} = 1$$

$$y^2 = 9, \quad \text{or} \quad y = \pm 3.$$

Thus $(0,3)$ and $(0,-3)$ are the y-intercepts of the graph.
Now we let $y = 0$:

$$\frac{x^2}{25} + \frac{0^2}{9} = 1$$

$$\frac{x^2}{25} = 1$$

$$x^2 = 25, \quad \text{or} \quad x = \pm 5.$$

Thus $(5,0)$ and $(-5,0)$ are the x-intercepts of the graph. It can be shown that our ellipse looks like the accompanying figure.

The graph of

$$\frac{x^2}{25} + \frac{y^2}{9} = 1$$

is an *ellipse* with center $(0,0)$.

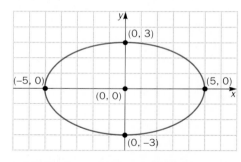

Using this as a prototype, we make the following generalization:

Equation of an Ellipse

$$\frac{x^2}{a^2} + \frac{y^2}{b^2} = 1$$

1. It has center $(0,0)$.
2. It has x-intercepts $(\pm a,0)$.
3. It has y-intercepts $(0,\pm b)$.

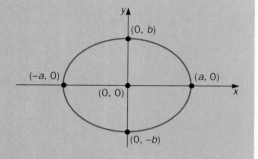

EXAMPLE 1

Graph each of these ellipses:

a. $\dfrac{x^2}{16} + \dfrac{y^2}{4} = 1$

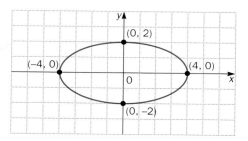

Here $a^2 = 16$, or $a = \pm 4$;
the x-intercepts are $(\pm 4, 0)$.
Also $b^2 = 4$, or $b = \pm 2$;
the y-intercepts are $(0, \pm 2)$.

b. $\dfrac{x^2}{8} + \dfrac{y^2}{25} = 1$

Here $a^2 = 8$, or $a = \pm\sqrt{8} = \pm 2\sqrt{2} \approx \pm 2.8$;
the x-intercepts are $(\pm 2\sqrt{2}, 0)$.
Also $b^2 = 25$, or $b = \pm 5$;
the y-intercepts are $(0, \pm 5)$.

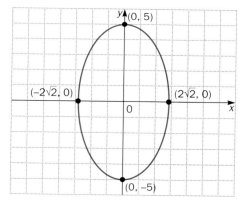

c. $4x^2 + 9y^2 = 9$

To put this into standard form, we first divide by 9:

$$\dfrac{4x^2}{9} + \dfrac{9y^2}{9} = \dfrac{9}{9}, \quad \text{or} \quad \dfrac{4x^2}{9} + \dfrac{y^2}{1} = 1.$$

The first term here is not in standard form, so we multiply its numera-

tor and denominator by $\dfrac{1}{4}$:

$$\dfrac{1/4 \cdot 4x^2}{1/4 \cdot 9} + \dfrac{y^2}{1} = 1,$$

or $\dfrac{x^2}{\dfrac{9}{4}} + \dfrac{y^2}{1} = 1.$

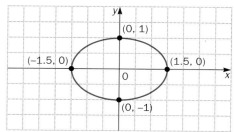

(continued)

This standard form gives $a^2 = \dfrac{9}{4}$, or $a = \pm\dfrac{3}{2} = \pm 1.5$;

and $b^2 = 1$, or $b = \pm 1$.

Our final conic section, the **hyperbola,** is illustrated by the equation

$$\frac{x^2}{9} - \frac{y^2}{4} = 1.$$

Here the terms are subtracted instead of added, as with the ellipse, whose equation and graph are closely akin to those of the circle. Consequently, the hyperbola looks far different from either of these conic sections, and to graph it, we need to plot points carefully. We start by solving the preceding equation for x (see Section 6.2, Example 3b, where we solved for y).

$$\frac{x^2 \cdot 4}{9 \cdot 4} - \frac{y^2 \cdot 9}{4 \cdot 9} = \frac{1 \cdot 36}{1 \cdot 36} \qquad \text{L.C.D.} = 36$$

$$4x^2 - 9y^2 = 36 \qquad \text{Equate numerators.}$$

$$4x^2 = 36 + 9y^2$$

$$x^2 = \frac{9(4 + y^2)}{4} \qquad \begin{array}{l}\text{Factor numerator; then}\\ \text{divide by 4.}\end{array}$$

$$x = \pm\sqrt{\frac{9(4 + y^2)}{4}} \qquad \text{Extract roots.}$$

$$x = \frac{\pm 3\sqrt{4 + y^2}}{2} \qquad \frac{\pm\sqrt{9}\sqrt{4 + y^2}}{\sqrt{4}}$$

$$\textbf{WARNING!} \quad \sqrt{4 + y^2} \neq 2 + y$$

We now prepare a table by substituting the values $y = 0, \pm 1, \pm 2, \pm 3, \ldots$ into this equation and evaluating x:

$\dfrac{\pm 3\sqrt{4 + y^2}}{2} = x$	y
$\dfrac{\pm 3\sqrt{4 + 0^2}}{2} = \dfrac{\pm 3\sqrt{4}}{2} = \pm 3$	0
$\dfrac{\pm 3\sqrt{4 + (\pm 1)^2}}{2} = \dfrac{\pm 3\sqrt{5}}{2} \approx \pm 3.4$	± 1
$\dfrac{\pm 3\sqrt{4 + (\pm 2)^2}}{2} = \dfrac{\pm 3\sqrt{8}}{2} \approx \pm 4.2$	± 2
± 5.4	± 3

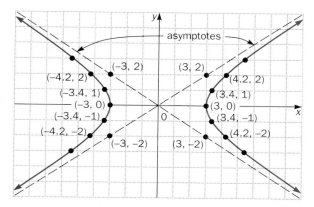

The graph of $\dfrac{x^2}{9} - \dfrac{y^2}{4} = 1$ is a *hyperbola*.

The hyperbola has two "branches" which, in this case, open to the right and to the left. It can be shown that each branch gets closer and closer to the diagonal lines, which are called **asymptotes,** but never touches them. It can also be shown that the asymptotes intersect at the origin and go through the points $(3,2)$, $(3,-2)$, $(-3,2)$, and $(-3,-2)$, as shown on the graph. On the basis of this, and of the picture above, we can make the following generalizations about a hyperbola:

Equation of a Hyperbola

$$\frac{x^2}{a^2} - \frac{y^2}{b^2} = 1$$

1. This hyperbola opens *left and right*.
2. Its asymptotes go through $(\pm a, \pm b)$ and intersect at $(0,0)$.
3. It has x-intercepts $(\pm a, 0)$.
4. It has no y-intercepts.

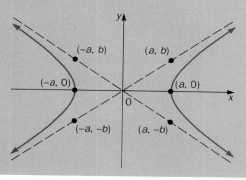

EXAMPLE 2   Graph each of these hyperbolas:

a. $\dfrac{x^2}{16} - \dfrac{y^2}{4} = 1$

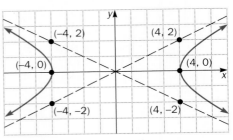

Here $a^2 = 16$,   or $a = \pm 4$;
and $b^2 = 4$,   or $b = \pm 2$.

Thus the asymptotes go through
$(\pm 4, \pm 2)$. The x-intercepts
are $(\pm 4, 0)$.

b. $4x^2 - y^2 = 4$

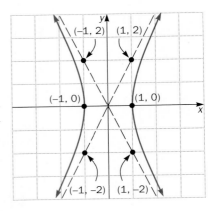

To put this into standard form,
we divide by 4:

$\dfrac{4x^2}{4} - \dfrac{y^2}{4} = \dfrac{4}{4}$,   or $\dfrac{x^2}{1} - \dfrac{y^2}{4} = 1$.

We read off $a^2 = 1$,   or $a = \pm 1$;
and $b^2 = 4$,   or $b = \pm 2$.

The asymptotes go through
$(\pm 1, \pm 2)$. The x-intercepts are
$(\pm 1, 0)$.

Earlier, the equation $\dfrac{x^2}{9} - \dfrac{y^2}{4} = 1$ was used to introduce you to hyperbolas.
Solving for x by extracting roots gave $x = \dfrac{\pm 3\sqrt{4 + y^2}}{2}$, and the resulting table
of points produced a hyperbola opening left and right. Now, by exchanging
terms in the original equation, we get

$$\dfrac{y^2}{4} - \dfrac{x^2}{9} = 1.$$

Solving for y in a similar fashion yields

$$y = \dfrac{\pm 2\sqrt{9 + x^2}}{3}.$$

Substituting the values $x = 0, \pm 1, \pm 2, \pm 3, \ldots$ into this equation and evaluating y produces the following table of points and a hyperbola *opening up and down:*

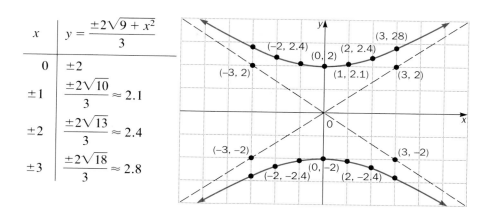

x	$y = \dfrac{\pm 2\sqrt{9 + x^2}}{3}$
0	± 2
± 1	$\dfrac{\pm 2\sqrt{10}}{3} \approx 2.1$
± 2	$\dfrac{\pm 2\sqrt{13}}{3} \approx 2.4$
± 3	$\dfrac{\pm 2\sqrt{18}}{3} \approx 2.8$

The graph of $\dfrac{y^2}{4} - \dfrac{x^2}{9} = 1$ is a *hyperbola.*

We see that the branches of this hyperbola open up and down, instead of left and right. The asymptotes go through the same points as those of the hyperbola $\dfrac{x^2}{9} - \dfrac{y^2}{4} = 1$, namely $(3,2)$, $(3,-2)$, $(-3,2)$, and $(-3,-2)$. The y intercepts are $(0,2)$ and $(0,-2)$, but there are no x-intercepts. Let us generalize:

Equation of a Hyperbola

$$\frac{y^2}{b^2} - \frac{x^2}{a^2} = 1$$

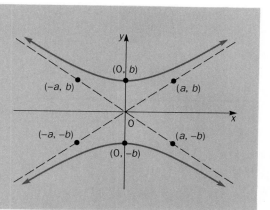

1. This hyperbola opens *up and down.*
2. Its asymptotes go through $(\pm a, \pm b)$ and intersect at $(0,0)$.
3. It has y-intercepts $(0, \pm b)$.
4. It has no x-intercepts.

EXAMPLE 3 | Graph each of these hyperbolas:

a. $\dfrac{y^2}{9} - \dfrac{x^2}{16} = 1$

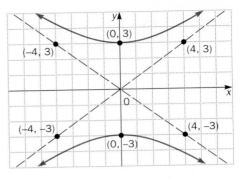

Here $a^2 = 16$, or $a = \pm 4$;
and $b^2 = 9$, or $b = \pm 3$.

Thus the asymptotes go
through $(\pm 4, \pm 3)$.
The y-intercepts are $(0, \pm 3)$.

b. $12x^2 - 6y^2 + 72 = 0$

We put this into standard form
as follows:

$$12x^2 - 6y^2 = -72$$

$$\frac{12x^2}{-72} - \frac{6y^2}{-72} = \frac{-72}{-72}$$

$$\frac{x^2}{-6} + \frac{y^2}{12} = 1, \quad \text{or}$$

$$\frac{y^2}{12} - \frac{x^2}{6} = 1.$$

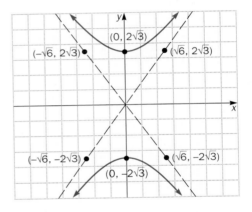

Thus $a^2 = 6$, or $a = \pm\sqrt{6} \approx \pm 2.4$,
and $b^2 = 12$, or $b = \pm 2\sqrt{3} \approx \pm 3.5$.
The asymptotes go through $(\pm\sqrt{6}, \pm 2\sqrt{3})$.
The y-intercepts are $(0, \pm 2\sqrt{3})$.

EXERCISE 8.3   *Answers, pages A79–A81*

A

Graph each of these ellipses, as in EXAMPLE 1. If necessary, put into standard form.

1. $\dfrac{x^2}{16} + \dfrac{y^2}{9} = 1$

2. $\dfrac{x^2}{9} + \dfrac{y^2}{4} = 1$

3. $\dfrac{x^2}{4} + \dfrac{y^2}{25} = 1$

4. $\dfrac{x^2}{16} + \dfrac{y^2}{25} = 1$

5. $\dfrac{y^2}{18} + \dfrac{x^2}{4} = 1$

6. $\dfrac{y^2}{36} + \dfrac{x^2}{12} = 1$

7. $\dfrac{x^2}{9} + \dfrac{y^2}{9} = 1$

8. $\dfrac{x^2}{4} + \dfrac{y^2}{4} = 1$

9. $4x^2 + 9y^2 = 36$

10. $4x^2 = 16 - 16y^2$

11. $100x^2 + 16y^2 - 100 = 0$

12. $4y^2 = 16 - 9x^2$

Graph each of these hyperbolas, as in EXAMPLES 2 and 3. If necessary, put into standard form.

13. $\dfrac{x^2}{16} - \dfrac{y^2}{9} = 1$

14. $\dfrac{x^2}{25} - \dfrac{y^2}{16} = 1$

15. $\dfrac{x^2}{9} - \dfrac{y^2}{25} = 1$

16. $\dfrac{x^2}{4} - \dfrac{y^2}{9} = 1$

17. $\dfrac{x^2}{24} - \dfrac{y^2}{8} = 1$

18. $\dfrac{x^2}{9} - \dfrac{y^2}{20} = 1$

19. $\dfrac{x^2}{36} - \dfrac{y^2}{36} = 1$

20. $x^2 - y^2 = 1$

21. $\dfrac{y^2}{4} - \dfrac{x^2}{16} = 1$

22. $\dfrac{y^2}{9} - \dfrac{x^2}{25} = 1$

23. $\dfrac{y^2}{16} - \dfrac{x^2}{9} = 1$

24. $\dfrac{y^2}{9} - x^2 = 1$

25. $\dfrac{y^2}{10} - \dfrac{x^2}{5} = 1$

26. $y^2 - x^2 = 1$

27. $x^2 - 4y^2 = 4$

28. $4y^2 - 9x^2 = 36$

29. $100y^2 - 9x^2 = 225$

30. $9x^2 - 25y^2 = 100$

31. $4x^2 - 9y^2 = -36$

32. $y^2 - 4x^2 = -4$

33. $36y^2 = 4x^2 - 144$

34. $4x^2 + 25 = 25y^2$

These are the reverse of the problems above. Give the equation, in standard form, of the following ellipses and hyperbolas:

35.
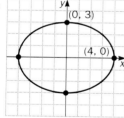

36.
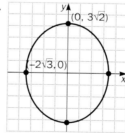

37.
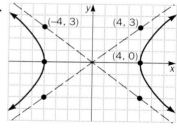

38.
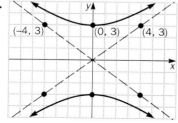

39.
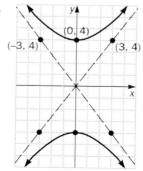

40.
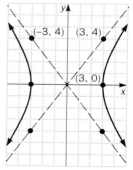

B

In Section 8.2, we saw that the circle $(x - 5)^2 + (y + 1)^2 = 9$ has center $(5, -1)$. Likewise, it can be shown that the ellipse

$$\frac{(x - 5)^2}{9} + \frac{(y + 1)^2}{4} = 1$$

also has center $(5, -1)$. Furthermore, $a^2 = 9$, or $a = \pm 3$; and $b^2 = 4$, or $b = \pm 2$. These values appear in the graph as follows:

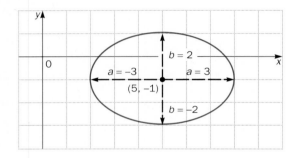

In like manner, graph the following ellipses:

41. $\dfrac{(x - 3)^2}{16} + \dfrac{(y - 2)^2}{9} = 1$

42. $\dfrac{(x + 3)^2}{9} + \dfrac{(y - 1)^2}{4} = 1$

43. $\dfrac{(x - 2)^2}{9} + \dfrac{y^2}{25} = 1$

44. $\dfrac{x^2}{4} + \dfrac{(y + 4)^2}{16} = 1$

45. $\dfrac{(x + 1)^2}{20} + \dfrac{(y + 2)^2}{8} = 1$

46. $\dfrac{(x - 4)^2}{12} + \dfrac{(y + 2)^2}{25} = 1$

In calculus, it is shown that

the area of the ellipse $\dfrac{x^2}{a^2} + \dfrac{y^2}{b^2} = 1$ is $A = \pi|ab|.$

Find the area of these ellipses:

47. $\dfrac{x^2}{16} + \dfrac{y^2}{9} = 1$

48. $\dfrac{x^2}{12} + \dfrac{y^2}{27} = 1$

49. $\dfrac{x^2}{8} + \dfrac{y^2}{8} = 1$

50. $4x^2 + 18y^2 = 72$

51. $4x^2 + 9y^2 = 9$

52. $\dfrac{(x - 1)^2}{36} + \dfrac{(y + 2)^2}{32} = 1$

53. The important facts stated in the text about the hyperbola

$$\frac{x^2}{a^2} - \frac{y^2}{b^2} = 1$$

can be obtained by looking at its intercepts. First, let $x = 0$; then

$$\frac{0^2}{a^2} - \frac{y^2}{b^2} = 1, \quad \text{or} \quad y^2 = -b^2, \quad \text{or} \quad y = \pm bi.$$

Thus the hyperbola has no real y-intercepts and so cannot open up and down. Now let $y = 0$; then

$$\frac{x^2}{a^2} - \frac{0^2}{b^2} = 1, \quad \text{or} \quad x^2 = a^2, \quad \text{or} \quad x = \pm a.$$

Thus the hyperbola has x-intercepts $(a,0)$ and $(-a,0)$ and so must open left and right. In like fashion, show that the hyperbola

$$\frac{y^2}{b^2} - \frac{x^2}{a^2} = 1$$

opens up and down by finding its intercepts.

54. The hyperbolas $\frac{x^2}{16} - \frac{y^2}{9} = 1$ and $\frac{y^2}{9} - \frac{x^2}{16} = 1$ share the same asymptotes shown:

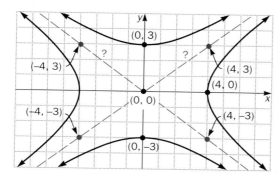

Use the methods of Section 7.3 to find the equation of each line.

55. Find the equation of each asymptote of the more general hyperbolas

$$\frac{x^2}{a^2} - \frac{y^2}{b^2} = 1 \quad \text{and} \quad \frac{y^2}{b^2} - \frac{x^2}{a^2} = 1$$

Hint: See Problem 54.

56. In Problem 54, we saw that two hyperbolas, one opening left and right, the other opening up and down, can share the same asymptotes. It is also possible for two hyperbolas, both opening left and right, to share the same asymptotes. Find the equations of the two asymptotes of

a. $\dfrac{x^2}{9} - \dfrac{y^2}{4} = 1$

b. $\dfrac{x^2}{36} - \dfrac{y^2}{16} = 1$

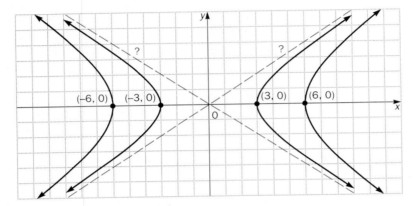

Your equations should be identical in each case.

57. Look at the pictures of the conic sections at the beginning of this chapter. You see that the hyperbola is obtained by passing a vertical plane through the two cones. Describe the graph obtained when this vertical plane passes through the intersection of these cones.

58. In the equation of the ellipse $\dfrac{x^2}{a^2} + \dfrac{y^2}{b^2} = 1$,   what happens if $a = b$?

59. In the text, it was stated that the equation $\dfrac{y^2}{4} - \dfrac{x^2}{9} = 1$, when solved for y, gives $y = \dfrac{\pm 2\sqrt{9 + x^2}}{3}$. You solve the equation $\dfrac{y^2}{25} - \dfrac{x^2}{16} = 1$ for y.

60. Solve the last equation in Problem 59 for x.

61. The "eye" shown consists of two circles and an ellipse. Find the equations of the circles if the ellipse has the equation

$$\frac{x^2}{25} + \frac{y^2}{9} = 1.$$

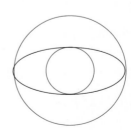

8.4 Non-Linear Systems

We saw in Section 7.4 that the intersection of two lines can be obtained by solving a system of two linear equations. We now extend this to the intersection of lines, parabolas, circles, ellipses, and hyperbolas. In all cases, we will be solving systems of equations containing second-degree terms. These are called *non-linear* systems.

EXAMPLE 1

Solve the non-linear system

$$x^2 + y^2 = 25 \qquad \text{Circle}$$
$$x - y = 1 \qquad \text{Line}$$

Solution

Solve the second equation for x: $x = y + 1$. Then substitute this into the first equation:

$$(y + 1)^2 + y^2 = 25$$
$$y^2 + 2y + 1 + y^2 = 25$$
$$2y^2 + 2y - 24 = 0$$
$$2(y^2 + y - 12) = 0$$
$$2(y + 4)(y - 3) = 0$$
$$y = -4, \quad y = 3.$$

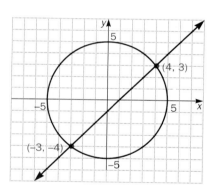

Now substitute these values into $x = y + 1$:

$$\text{If } y = -4, \quad \text{then} \quad x = -4 + 1 = -3.$$
$$\text{If } y = 3, \quad \text{then} \quad x = 3 + 1 = 4.$$

Answers: $(-3, -4)$ and $(4, 3)$.

Check $(-3, -4)$: $(-3)^2 + (-4)^2 = 25$, or $9 + 16 \overset{\checkmark}{=} 25$.
$(-3) - (-4) = 1$, or $-3 + 4 \overset{\checkmark}{=} 1$.

Check $(4, 3)$: You verify this yourself.

As illustrated, these points are the intersection of the circle and the line.

EXAMPLE 2   Solve this system:

$$3x^2 + 2y^2 = 7 \qquad \text{Ellipse}$$
$$5x^2 - 7y^2 = -9 \qquad \text{Hyperbola}$$

Solution

This system is best solved by the addition method:

$$3x^2 + 2y^2 = 7 \quad \xrightarrow{\text{Multiply by 7.}} \quad 21x^2 + 14y^2 = 49$$
$$5x^2 - 7y^2 = -9 \quad \xrightarrow{\text{Multiply by 2.}} \quad \underline{10x^2 - 14y^2 = -18}$$
$$31x^2 \qquad\quad = 31$$
$$x^2 = 1$$
$$x = \pm 1.$$

Now substitute into $3x^2 + 2y^2 = 7$:

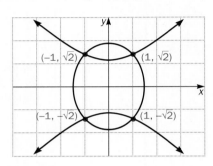

If $x = 1$, then $3 \cdot 1^2 + 2y^2 = 7$
$$2y^2 = 4$$
$$y^2 = 2$$
$$y = \pm\sqrt{2}$$
If $x = -1$, then $3(-1)^2 + 2y^2 = 7$
and likewise $y = \pm\sqrt{2}$.

Answers: $(1, \sqrt{2})$, $(1, -\sqrt{2})$, $(-1, \sqrt{2})$, and $(-1, -\sqrt{2})$
The solutions are the intersection of the ellipse and the hyperbola.

EXAMPLE 3   Solve this system:

$$x^2 + y^2 = 10$$
$$xy = 3$$

Solution

We solve the second equation for y: $y = \dfrac{3}{x}$. Now substitute this into the first equation:

$$x^2 + \left(\frac{3}{x}\right)^2 = 10$$

$$x^2 + \frac{9}{x^2} = 10$$

$$\frac{x^2 \cdot x^2}{1 \cdot x^2} + \frac{9}{x^2} = \frac{10 \cdot x^2}{1 \cdot x^2} \longleftarrow \text{L.C.D.} = x^2$$

$$x^4 + 9 = 10x^2 \qquad \text{Equate numerators.}$$

$$x^4 - 10x^2 + 9 = 0 \qquad \text{Standard form.}$$

$$(x^2 - 9)(x^2 - 1) = 0 \qquad \text{Factor trinomial.}$$

$$x^2 = 9, \quad x^2 = 1$$

$$x = \pm 3, \quad x = \pm 1. \qquad \text{Extract roots.}$$

Now substitute into $y = \dfrac{3}{x}$:

If $x = 3$, then $y = \dfrac{3}{3} = 1$; and if $x = -3$, then $y = \dfrac{3}{-3} = -1$.

If $x = 1$, then $y = \dfrac{3}{1} = 3$; and if $x = -1$, then $y = \dfrac{3}{-1} = -3$.

Answers: $(3,1)$, $(-3,-1)$, $(1,3)$, and $(-1,-3)$

EXERCISE 8.4   *Answers, pages A81–A82*

A

Solve these systems, as in EXAMPLE 1:

1. $x^2 + y^2 = 169$
$\quad x - y = 7$

2. $x^2 + y^2 = 100$
$\quad x + y = 14$

3. $x^2 + 4y^2 = 9$
$\quad x + 2y = 3$

4. $4x^2 + y^2 = 4$
$\quad y - 2x = 2$

5. $x + y = 6$
$\quad xy = 8$

6. $x - y = 5$
$\quad xy + 4 = 0$

7. $2x^2 - y^2 = 1$
$\quad x - 2y = 3$

8. $3x^2 - 2y^2 = 12$
$\quad x + 2y = 2$

9. $y = x^2 - 2x + 2$
$\quad x + y = 8$

10. $y = 2x^2 - 1$
$\quad 4x - y = 3$

11. $y = 3x^2 - 1$
$\quad x^2 = 8 - y$

12. $y^2 = x + 4$
$\quad x = 2y - y^2$

Solve these systems, as in EXAMPLE 2:

13. $x^2 + y^2 = 25$
$\quad x^2 - y^2 = 7$

14. $x^2 - y^2 = -3$
$\quad x^2 + y^2 = 5$

15. $2x^2 + 5y^2 = 13$
$\quad 3x^2 + 2y^2 = 14$

16. $5x^2 + 3y^2 = 47$
$\quad 2x^2 - 7y^2 = -55$

17. $3x^2 - 2y = -7$
$\quad x^2 + 4y = 28$

18. $5x + 3y^2 = 4$
$\quad 2x - 4y^2 = -14$

Solve these systems, as in EXAMPLE 3:

19. $x^2 + y^2 = 13$
$\quad xy = 6$

20. $x^2 + y^2 = 37$
$\quad xy = 6$

21. $x^2 - y^2 = 8$
$\quad xy + 3 = 0$

22. $2x^2 - y^2 = 0$
$\quad xy - 2 = 0$

B

Solve these systems, using any method:

23. $x^2 + xy - y^2 = 1$
$\quad x - y = 3$

24. $2x^2 - xy - 3y^2 = 7$
$\quad 2x - y = 5$

25. $2x + 3xy + 4y = -1$
$\quad xy = -1$

26. $3x^2 - 2xy + y^2 = 9$
$\quad xy = 6$

27. $x^2 + \dfrac{y^2}{9} = 1$ *Hint:*
First
clear
$\quad x + \dfrac{y}{3} = 1$ fractions.

28. $\dfrac{x^2}{25} + \dfrac{y^2}{4} = 1$
$\quad \dfrac{x}{5} + \dfrac{y}{2} = 1$

29. $\dfrac{1}{x} + \dfrac{1}{y} = \dfrac{3}{4}$
$x + y = 6$

30. $\dfrac{1}{x} - \dfrac{1}{y} = \dfrac{1}{12}$
$xy = 24$

31. $x^2 + y^2 = 1$
$x - y = 1$

32. $2x - 3y = 12$
$xy + 6 = 0$

33. $x + y = 2$
$xy = -1$

34. $x^2 + y^2 = 24$
$x + y = 6$

C

Solve for x and y. Assume all constants are positive. See Section 7.4, Problems 47–62.

35. $x^2 + y^2 = 10a^2$
$x + y = 4a$

36. $x^2 - y^2 = 16b^2$
$x - y = 2b$

37. $x^2 - y^2 = m^2 + 2mn$
$x - y = m$

38. $x + y = 2c$
$xy = c^2 - d^2$

39. $3x^2 + 2y^2 = 7p^2$
$y^2 - 5x^2 = -3p^2$

40. $n^2y^2 + x^2 = 20$
$\dfrac{x}{2} = ny$

41. $x^2 + 4y^2 = 8c^2$
$xy = -2c^2$

42. $2x^2 + 3y^2 = 30d^2$
$xy = -6d^2$

| **8.5** | **Word Problems** |

The word problems in this section lead to systems of non-linear equations.

EXAMPLE 1

The perimeter of a rectangle is 26 inches and its area is 40 square inches. Find the width and the length.

Solution

Let l = length and w = width. Then

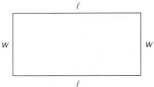

$$2l + 2w = 26 \qquad \text{Perimeter}$$
$$lw = 40 \qquad \text{Area}$$

Multiply the first equation by $\dfrac{1}{2}$ and then isolate l:

$$\overset{x\ 1/2}{2l + 2w = 26} \to l + w = 13, \text{ or } l = 13 - w$$

Now substitute into the second equation:

$$(13 - w)w = 40$$
$$13w - w^2 = 40$$
$$0 = w^2 - 13w + 40$$
$$0 = (w - 5)(w - 8)$$
$$w = 5, \quad w = 8$$

Substitute into $l = 13 - w$:

$$\text{If } w = 5, \quad \text{then} \quad l = 13 - 5 = 8.$$
$$\text{If } w = 8, \quad \text{then} \quad l = 13 - 8 = 5.$$

Answers: 5 in. by 8 in.

EXAMPLE 2

The perimeter of a right triangle is 24 and its hypotenuse is 10. Find the two legs. Assume all measurements are in centimeters.

Solution

Let x = one leg and y = other leg. Then

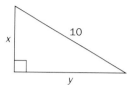

$$x + y + 10 = 24 \qquad \text{Perimeter}$$
$$x^2 + y^2 = 10^2 \qquad \text{Pythagorean Theorem}$$

Solving the first equation for y gives $y = 14 - x$; substituting into the second equation gives

$$x^2 + (14 - x)^2 = 10^2$$
$$x^2 + 196 - 28x + x^2 = 100$$
$$2x^2 - 28x + 96 = 0$$
$$2(x^2 - 14x + 48) = 0$$
$$2(x - 6)(x - 8) = 0$$
$$x = 6, \quad x = 8.$$

Now substitute these values into $y = 14 - x$:

$$\text{If } x = 6, \text{ then } y = 14 - 6 = 8.$$
$$\text{If } x = 8, \text{ then } y = 14 - 8 = 6.$$

Answers: 6 cm and 8 cm

EXAMPLE 3

A man can row a boat 12 miles per hour on a lake with no current. On a river, he can row 7 miles downstream in the same time that he can row 5 miles upstream. Find the rate of the current and the time spent rowing in each direction.

Solution

Let c = rate of current and t = time in each direction. This is similar to Section 7.5, Example 4.

	R	$\cdot T =$	D
downstream (with current)	$12 + c$	t	$(12 + c)t = 7$
upstream (against current)	$12 - c$	t	$(12 - c)t = 5$

When multiplied out, our system of equations becomes

$$12t + ct = 7$$
$$\underline{12t - ct = 5}$$
$$24t \qquad = 12$$

$$t = \frac{1}{2}.$$

(continued)

Substituting $t = \frac{1}{2}$ into the first equation gives

$$12\left(\frac{1}{2}\right) + c\left(\frac{1}{2}\right) = 7$$

$$6 + \frac{c}{2} = 7$$

$$\frac{c}{2} = 1, \quad \text{or} \quad c = 2.$$

Answer: current = 2 mph and time = $\frac{1}{2}$ hour in each direction

EXERCISE 8.5   *Answers, page A82*

Solve using two unknowns, as in **EXAMPLE 1:**

1. Find the dimensions of a rectangle whose perimeter is 18 inches and whose area is 14 square inches.

2. A rectangular grazing area is to be fenced off along a river, and 200 feet of fencing are to be used. The river acts as one side, so only three sides of fencing are needed. If the area enclosed is 5000 square feet, find the dimensions of the rectangle. See Section 8.1, Example 4.

3. Find two numbers whose sum is 9 and whose product is 18.

4. Find two numbers whose difference is 8 and whose product is −12.

Solve using two unknowns, as in **EXAMPLE 2:**

5. Find the legs of a right triangle whose hypotenuse is 5 meters and whose perimeter is 12 meters.

6. The perimeter of a rectangle is 10 and its diagonal is $\sqrt{13}$. Find the width and the length. All measurements are in feet.

7. Find two numbers whose sum is 6 and the sum of whose squares is 20.

8. The difference between two numbers is 2 and the sum of their squares is 2. Find the numbers.

Solve using two unknowns, as in EXAMPLE 3:

9. A powerboat that can travel 60 miles per hour on a lake with no current is taken to a river. On the river it can travel 25 miles downstream in the same time in which it can travel 15 miles upstream. Find the rate of the current and the time in each direction.

10. A plane can fly a certain speed in still air. Flying 300 miles into a 25-mph headwind takes the same time as flying 390 miles with a 35-mph tailwind. Find the plane's speed in still air and the time in each direction.

11. A Ferrari, traveling 30 kilometers per hour faster than a Maserati, goes 400 kilometers in the same time in which the Maserati goes 340 kilometers. Find the rate and the time of each car.

12. You drive 300 miles at a certain speed. On the return trip, you increase your speed by 10 mph and, as a result, you save 1 hour. Find your speed and time going, and your speed and time returning.

B

Solve using two unknowns:

13. The sum of two numbers is 6 and the sum of their reciprocals is $\frac{3}{4}$. What are the numbers? Recall that the *reciprocal* of x is $\frac{1}{x}$.

14. The product of two numbers is 2 and the difference between their reciprocals is $\frac{1}{2}$. Find the numbers.

15. A rectangular area is to be fenced off; 180 feet of fencing are to be used, including two fenced partitions inside, as shown. If the enclosed area is 1000 square feet, find the width and the length of the rectangle. See also Exercise 8.1, Prob. 32.

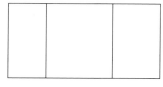

Problem 15

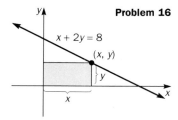

Problem 16

$x + 2y = 8$

(x, y)

16. Find the coordinates of a point (x,y) on the line $x + 2y = 8$ in the first quadrant such that the shaded rectangle has area 6. There may be more than one point.

17. Find x and y in the figure, given that $AB = 8$.

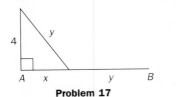

Problem 17

Problem 18

18. Find the sides of two squares, the sum of whose perimeters is 32 inches and the sum of whose areas is 34 square inches. *Formulas: $P = 4$(side), $A = $ (side)^2.*

19. The area of a rectangle is 24 square inches. When the width is decreased by 2 inches and the length is increased by 6 inches, another rectangle is formed whose area is the same as that of the original. Find the width and the length of each rectangle.

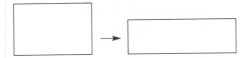

20. The area of one square is 16 square inches greater than the area of a smaller square. If the sides of the smaller square were increased by one inch, the larger square would then have an area only 9 square inches greater than that of the smaller square. Find the sides of all three squares.

21. A wire 12 inches long is cut into two pieces, each of which is bent into a square. Find the length of each piece if the total area of the squares is 5 square inches. *Hint:* Let $x = $ the length of one piece; when this piece is bent into a square, each side of the square will be $\frac{1}{4}x$; $A = \left(\frac{1}{4}x\right)^2$.

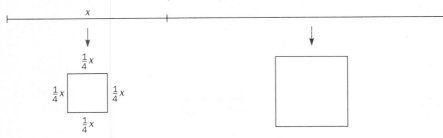

22. The area of a rectangle is 6 square feet and its diagonal is $\sqrt{13}$ feet. Find the width and the length.

23. The area of a right triangle is 2 square inches, and its hypotenuse is $\sqrt{10}$ inches. Find the two legs. *Recall:* $A = \frac{1}{2}bh$.

24. Find x and y in the picture. *Hint:* There are two right triangles present.

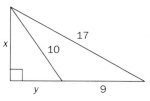

Problem 24

25. It costs $24,000 to charter a plane, the cost to be split equally among the passengers. Originally, a certain number of people were to pay a certain price each. At the last minute, however, 20 people canceled; as a result, the price increased by $200 per person for those who flew. Find the original number of passengers and the original price per passenger. Then find the actual number who flew, and the actual price each one paid. *Hint:* Let x = original number of passengers and y = original price per passenger; then $xy = \$24,000$. A similar equation holds for those who actually flew. *Hint:* complete the chart shown.

	number of passengers ×	price per passenger =	total cost
original	x	y	$24,000
actual			

CHAPTER 8 · SUMMARY

8.1 Parabolas

$y = ax^2 + bx + c$

Has *vertex* at $x = \dfrac{-b}{2a}$.

Opens *up* if $a > 0$.
Opens *down* if $a < 0$.

Example:

$$y = x^2 - 2x + 3$$

Vertex: $x = \dfrac{-(-2)}{2 \cdot 1} = 1,$

$y = 1^2 - 2 \cdot 1 + 3 = 2$

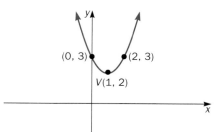

CHAPTER 8 · SUMMARY

$x = ay^2 + by + c$

Has *vertex* at $y = \dfrac{-b}{2a}$.

Opens *right* if $a > 0$.
Opens *left* if $a < 0$.

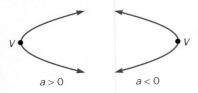

$a > 0$ $a < 0$

Example:

$$x = -y^2 + 4y - 1$$

Vertex: $y = \dfrac{-4}{2(-1)} = 2,$

$$x = -(2)^2 + 4(2) - 1 = 3$$

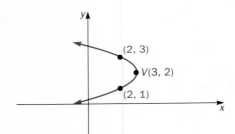

8.2 Distance

$$D = \sqrt{(x_2 - x_1)^2 + (y_2 - y_1)^2}$$

Example:

Distance between $(-2,3)$ and $(4,-1)$:

$$D = \sqrt{(4 - (-2))^2 + (-1 - 3)^2}$$
$$= \sqrt{36 + 16} = \sqrt{52} = 2\sqrt{13}$$

Circles

$(x - h)^2 + (y - k)^2 = r^2$

Has center at (h,k).
Has radius $= r$.

Example:

$(x - 3)^2 + (y + 1)^2 = 4$

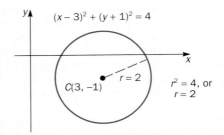

8.3 Ellipses

$$\dfrac{x^2}{a^2} + \dfrac{y^2}{b^2} = 1$$

Has center at $(0,0)$.
Has x-intercepts $(\pm a, 0)$.
Has y-intercepts $(0, \pm b)$.

Example:

$$\dfrac{x^2}{9} + \dfrac{y^2}{4} = 1$$

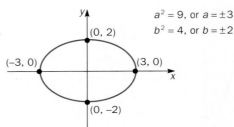

$a^2 = 9$, or $a = \pm 3$
$b^2 = 4$, or $b = \pm 2$

CHAPTER 8 · SUMMARY

Hyperbolas

$$\frac{x^2}{a^2} - \frac{y^2}{b^2} = 1$$

Opens *left and right*,
Asymptotes go through $(\pm a, \pm b)$.
Has x-intercepts $(\pm a, 0)$.
Has no y-intercepts.

Example:

$$\frac{x^2}{9} - \frac{y^2}{4} = 1 \qquad a^2 = 9, \text{ or } a = \pm 3$$
$$b^2 = 4, \text{ or } b = \pm 2$$

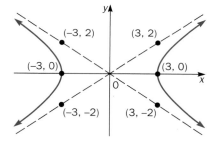

$$\frac{y^2}{b^2} - \frac{x^2}{a^2} = 1$$

Opens *up and down.*
Asymptotes go through $(\pm a, \pm b)$.
Has y-intercepts $(0, \pm b)$.
Has no x-intercepts.

Example:

$$\frac{y^2}{4} - \frac{x^2}{9} = 1 \qquad a^2 = 9, \text{ or } a = \pm 3$$
$$b^2 = 4, \text{ or } b = \pm 2$$

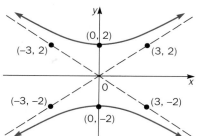

8.4 Non-Linear Systems

Example:

$$x - y = 2 \xrightarrow{\text{Solve for } x} x = y + 2 \qquad (1)$$
$$xy = 8 \xrightarrow{\text{Substitute}} (\quad)y = 8 \qquad (2)$$

Solve (2)
$$y^2 + 2y - 8 = 0$$
$$(y + 4)(y - 2) = 0$$
$$y = -4, \quad y = 2$$

If $y = -4$, $x = -4 + 2 = -2$. $\left.\rule{0pt}{1.5em}\right\}$ Substitute y into
If $y = 2$, $x = 2 + 2 = 4$. $\left\{\rule{0pt}{1.5em}\right.$ (1) to find x.

Answers: $(-2, -4)$ and $(4, 2)$.

8.5 Word Problems

Example:

"Find the dimensions of a rectangle with perimeter 14 and diagonal 5."

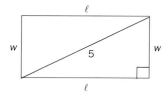

Solution

Write two equations:

$$2\ell + 2w = 14 \xrightarrow{\text{Solve for } \ell.} \ell = 7 - w$$
$$\ell^2 + w^2 = 5^2 \xrightarrow{\text{Substitute.}} (7 - w)^2 + w^2 = 25$$
$$2w^2 - 14w + 24 = 0$$
$$2(w - 3)(w - 4) = 0$$
$$w = 3, \ w = 4.$$
If $w = 3$, $\ell = 4$.
If $w = 4$, $\ell = 3$.

Answer: 3 by 4

CHAPTER 8 ▪ REVIEW EXERCISES   *Answers, pages A82–A86*

8.1

Graph each parabola by first finding the vertex:

1. $y = x^2 - 2$
2. $y = x^2 - 2x$
3. $y = x^2 + 4x + 5$
4. $y = -x^2 + 2x + 3$

5. $y = -x^2 + 3x$
6. $y = 2x^2 + 2x - 3$
7. $x = y^2 - 1$
8. $x = -y^2 + 4$

9. $x = y^2 - 2y$
10. $x = 2y^2 + 6y + 3$

11. When a ball is thrown upward, the formula

$$h = -16t^2 + 96t + 20$$

gives its height h (in feet) above the ground after t seconds. When does the ball reach its maximum height? What is the maximum height?

12. Find two numbers whose sum is 12 and whose product is a maximum. What is the maximum product?

13. Find two numbers whose difference is 6 and the sum of whose squares is a minimum. What is this sum?

14. The perimeter of a rectangle is 28 inches. Find the width and the length that maximize the area. What is this area?

15. A rectanglar storage area is to be fenced off on the side of a building; 600 feet of fencing are to be used, including a fenced partition inside, as shown. Find the dimensions that maximize the area. What is this area? *Hint: A = xy.* Note that fencing is not needed along the side of the building.

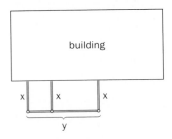

16. The coefficients a, b, and c in the parabola

$$y = ax^2 + bx + c$$

can be determined, given that the points $(-1,0)$, $(0,5)$, and $(1,6)$ all lie on the parabola.

a. Substitute each point into the equation, producing a system of three equations in the unknowns a, b, and c.
b. Solve this system.
c. Substitute your answers into the foregoing equation to find the equation of this parabola.
d. Find the vertex of this parabola.

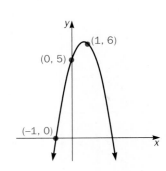

8.2

Find the distance between the following pairs of points:

17. $(-3,2)$ and $(1,-1)$

18. $(-4,0)$ and $(8,5)$

19. $(2,3)$ and $(-2,-1)$

20. $(0,-3)$ and $(8,-1)$

21. $\left(\dfrac{-1}{6},\dfrac{-5}{24}\right)$ and $\left(\dfrac{1}{12},\dfrac{1}{8}\right)$

22. $(3,-3)$ and $(\sqrt{3},\sqrt{3})$

Graph each of the following circles, indicating center and radius:

23. $x^2 + y^2 = 16$

24. $x^2 + (y-3)^2 = 9$

25. $(x-2)^2 + (y+1)^2 = 4$

26. $(x+3)^2 + y^2 = 12$

27. $9(x-1)^2 + 9(y+2)^2 = 16$ *Hint:* Divide by 9.

28. $\dfrac{x^2}{25} + \dfrac{(y-3)^2}{25} = 2$ *Hint:* Clear fractions.

Find the equation of each circle satisfying the given conditions:

29. Center $= (2,-3)$, radius $= 5$

30. Center $= (4,0)$, radius $= 3\sqrt{2}$

31. Center $= (3,4)$, circle goes through the point $(2,7)$

32. Center is in second quadrant, circle touches x-axis only at $(-4,0)$ and touches y-axis only at $(0,4)$. *Hint:* Draw a picture.

Graph each of the following circles by first completing the square. Indicate the center and the radius.

33. $x^2 + y^2 - 4x + 6y - 12 = 0$

34. $x^2 + y^2 - 2x - 3 = 0$

35. $x^2 + y^2 + 4y = 0$

36. $x^2 + y^2 + 3x + y = 0$

37. The coefficients a, b, and c in the circle

$$x^2 + y^2 + ax + by + c = 0$$

can be determined, given that the points $(5,3)$, $(0,-2)$, and $(1,-5)$ all lie on the circle. See also Problem 16.

  a. Substitute each point into the equation, producing a system of three equations in the three unknowns a, b, and c.

  b. Solve this system.

  c. Substitute your answers into the equation above.

  d. Find the center and radius of this circle by completing the square.

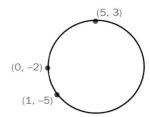

38. Find the center of the circle $(x+1)^2 + (y-4)^2 = 25$ and the vertex of the parabola $y = 2x^2 - 12x + 19$. Then find the distance between these two points.

39. Find the y-intercept of the line $2x - 3y + 9 = 0$ and the center of the circle $x^2 + y^2 - 8x + 2y + 8 = 0$. Then find the distance between these two points.

40. a. Find the length of each side of the triangle shown.
   b. Use the Pythagorean Theorem to show that it is a right triangle.

   c. Find its area using the formula $A = \frac{1}{2}bh$.

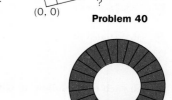

Problem 40

41. Find the area of the circle $x^2 + y^2 + 2x - 6y - 2 = 0$ by first putting it into standard form and then using the formula $A = \pi r^2$.

42. Find x such that the distance from $(1,-2)$ to $(4,x)$ is 5 units.

43. Find x such that the distance from $(3,1)$ to $(x,2x)$ is $\sqrt{85}$ units.

44. Find the area of the "annulus," or washer-like region, between the circles $x^2 + y^2 = 20$ and $x^2 + y^2 = 9$. *Hint:* Area of annulus = area of outer circle *minus* area of inner circle.

Problem 44

45. See the picture below. The perpendicular distance D from the point $(4,8)$ to the line $y = -\frac{1}{2}x + 5$ can be found by following these steps:

   a. Find the equation of the line through $(4,8)$ that is perpendicular to the line $y = -\frac{1}{2}x + 5$. See Section 7.3, Example 2b.

   b. Find the point of intersection of the line obtained in part a and the line $y = -\frac{1}{2}x + 5$. See Section 7.1.

   c. Find the distance D between the point obtained in part b and the point $(4,8)$.

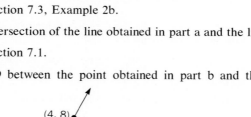

8.3

Graph each of the following ellipses and hyperbolas. If necessary, put the equations into standard form:

46. $\dfrac{x^2}{25} + \dfrac{y^2}{9} = 1$

47. $\dfrac{x^2}{12} + \dfrac{y^2}{24} = 1$

48. $4y^2 + 9x^2 = 36$

49. $9x^2 + 36y^2 = 225$

50. $\dfrac{x^2}{36} - \dfrac{y^2}{16} = 1$

51. $\dfrac{y^2}{16} - \dfrac{x^2}{36} = 1$

52. $4y^2 - 9x^2 = 36$

53. $9y^2 - x^2 = -9$

54. $\dfrac{(x-4)^2}{16} + \dfrac{(y+2)^2}{9} = 1$

Give, in standard form, the equation of each of these ellipses and hyperbolas:

55.

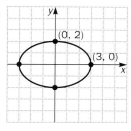

56.

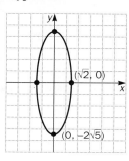

57.

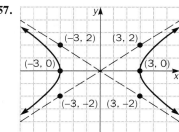

58.

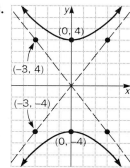

8.4

Solve these systems for x and y:

59. $x^2 + y^2 = 10$
$x - y = -4$

60. $x^2 + 2xy - y^2 = -2$
$x - y = 2$

61. $3x^2 + 2y^2 = 18$
$5x^2 - 3y^2 = 11$

62. $3x - 5y = 14$
$xy = 8$

63. $x^2 + y^2 = 10$
$xy = 3$

64. $x^2 + xy + 3y^2 = 25$
$xy + 6 = 0$

65. $x + y = 5$
$\dfrac{1}{x} + \dfrac{1}{y} = \dfrac{5}{6}$

66. $\dfrac{1}{y} + \dfrac{1}{x} = \dfrac{-1}{2}$
$xy = -2$

67. $x + 3y = -7a$
$xy = -6a^2$

8.5

Solve using two unknowns:

68. Find two numbers whose sum is 12 and whose product is 32.

69. The product of two numbers is -6. When one of the numbers is decreased by twice the other number, the result is 8. Find the numbers.

70. Find two numbers whose difference is 1 and the sum of whose squares is 13.

71. Find two numbers whose sum is 10, and such that the sum of their reciprocals is $\frac{5}{12}$.

72. Find two numbers whose product is 2, and such that the difference between their squares is 3.

73. A rectangular grazing area enclosing 600 square feet uses 100 feet of fencing. Find the width and the length of this pasture.

Problem 73

Problem 74

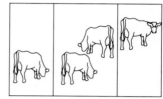

Problem 75

74. A rectangular grazing area along a river enclosing 600 square feet uses 70 feet of fencing. Only three sides are needed, because the river acts as one side. Find the width and the length.

75. A rectangular grazing area enclosing 600 square feet uses 160 feet of fencing, including two inner partitions. Find the width and the length.

76. The area of one square is 7 square inches more than the area of a smaller square. When the sides of the smaller square are each doubled, a third square is formed whose area is 20 square inches more than that of the larger of the first two squares. Find the sides of all three squares. *Hint:* Draw pictures.

77. A plane can fly 150 miles per hour in still air. Flying 240 miles against the wind takes the same time as flying 360 miles with the same wind. Find the speed of the wind and the time in each direction.

78. A boat can travel a certain speed in still water. On a river flowing 5 mph, the boat can go 10 miles upstream in the same time that it goes 15 miles downstream. Find the speed of the boat in still water and the time in each direction.

79. A boat can travel 10 mph on a lake with no current. On a river, it takes 1 hour longer to go 24 miles upstream than to go the same distance downstream. Find the rate of the current and the time in each direction.

80. The current of a river is 1 mph. A boat travels 20 miles upstream and then 24 miles downstream. The total travel time is 9 hours. Find the speed of the boat on a lake with no current and the time in each direction *Hint:* Let t = time upstream; then $9 - t$ = time downstream.

81. A family drove 300 miles to the mountains. On the trip home, it slowed down by 10 mph, so returning took 1 hour longer than going. Find the rate and the time in each direction.

82. Find the dimensions of a rectangle with perimeter 12 and diagonal $2\sqrt{5}$.

83. Find the dimensions of a rectangle with area 4 and diagonal $2\sqrt{2}$.

84. Find a point (x,y) in the first quadrant on the line $2x + y = 8$ such that the shaded rectangle has area 6.

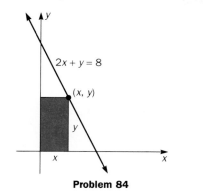

Problem 84

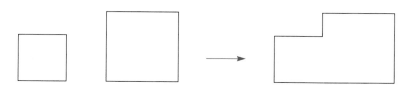

Problem 85

85. Find a point (x,y) in the first quadrant on the line $x + 2y = 20$ such that the shaded right triangle has hypotenuse $\sqrt{85}$.

86. The combined areas of the two squares below is 13 square inches. When they are placed together, the third figure shown is formed; it has an outside perimeter of 16 inches. Find the sides of the two squares.

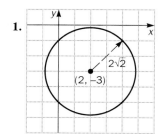

Problem 86

87. Find x and y.

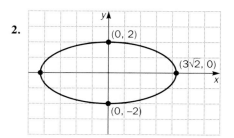

Problem 87

CHAPTER 8 · TEST *Answers, page A86*

Give, in standard form, the equation of each graph:

1.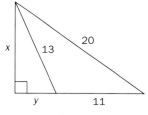

2.

3.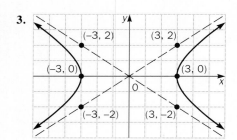

4.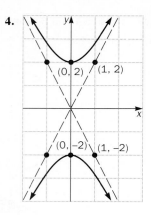

5.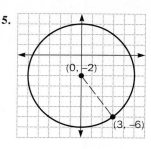

Graph each equation, showing all important points. If necessary, put the equation into standard form.

6. $(x + 3)^2 + (y - 2)^2 = 4$

7. $\dfrac{x^2}{16} + \dfrac{y^2}{9} = 1$

8. $\dfrac{x^2}{9} - \dfrac{y^2}{16} = 1$

9. $16y^2 - 9x^2 = 144$

10. $y = x^2 - 2x + 1$

11. $x^2 + y^2 + 4x - 2y - 4 = 0$

Solve for x and y:

12. $2x - y = 4$
$\quad\ xy = 6$

13. $y = x^2 - 2x + 3$
$\quad\ y - 2x = 8$

14. $x^2 + y^2 = 5$
$\quad\ x + y = 3$

15. Find the vertex of the parabola $x = -2y^2 + 8y - 9$ and the center of the circle $x^2 - 6x + y^2 - 7 = 0$. Then find the distance between these two points.

16. When a ball is thrown upward, the formula

$$h = -16t^2 + 64t$$

gives the height h (in feet) of the ball above the ground after t seconds. When does the ball reach its maximum height? What is the maximum height?

Solve using two unknowns:

17. Find the width and the length of a rectangle whose perimeter is 12 inches and whose area is 8 square inches.

18. Find the legs of a right triangle whose hypotenuse is 5 centimeters and whose perimeter is 12 centimeters.

19. A boat can go 35 miles per hour on a lake with no current. On a river, it can go 20 miles downstream in the same time that it goes 15 miles upstream. Find the rate of the current and the time in each direction.

20. Find the width and the length of a rectangle whose perimeter is 12 inches and whose area is a maximum. What is the maximum area?

9 FUNCTIONS

Relations and Functions

In Section 7.1, Example 1, the equation relating x and y,

$$-2x + y = 1$$

gave rise to the set of ordered pairs satisfying this equation:

$$A = \{(0,1),\ (1,3),\ (2,5),\ (-1,-1),\ (-2,-3),\ \left(\tfrac{1}{2},2\right),\ \ldots\}.$$

These were plotted and the graph was a straight line. Thus, associated with any equation relating x and y is the set of ordered pairs satisfying the equation. This example illustrates a more general concept, according to the following definitions:

> A **relation** is any set of ordered pairs.
> The **domain** of a relation is the set consisting of the first coordinates (the x-values).
> The **range** of a relation is the set consisting of the second coordinates (the y-values).

According to this definition, the set A of ordered pairs defined by $-2x + y = 1$ is a relation. The *domain* of A consists of *all real numbers*, illustrated above by the set consisting of its x-coordinates:

$$\text{domain } A = \{0,\ 1,\ 2,\ -1,\ -2,\ \tfrac{1}{2},\ \ldots\} = \text{all real numbers.}$$

The *range* of A also consists of *all real numbers*, illustrated above by the set consisting of its y-coordinates:

$$\text{range } A = \{1,\ 3,\ 5,\ -1,\ -3,\ 2,\ \ldots\} = \text{all real numbers.}$$

Relations need not arise from equations. For example, the set

$$B = \{(2,-3),\ (5,70),\ (-9,-3),\ (4,12)\}$$

is a relation, in which

domain $B = \{2,\ 5,\ -9,\ 4\}$ and   range $B = \{-3,\ 70,\ 12\}$.

Note that -3 appears twice in the relation but is listed only once in the range.

Relation B, which we describe by listing its elements, is far less important to us than relation A, whose points arise from an equation. Consequently, it will be convenient to refer to "the relation $-2x + y = 1$" rather than "the relation associated with the equation $-2x + y = 1$."

Let us now examine two relations, $y = x^2$ and $y^2 = x$, by means of their tables of values:

$y = x^2$:

x	0	1	2	3	-1	-2	-3
y	0	1	4	9	1	4	9

$y^2 = x$:

x	0	1	4	9	16	25
y	0	± 1	± 2	± 3	± 4	± 5

In the relation $y = x^2$, each x-value produces *exactly one* y-value. For example, if $x = 3$, then $y = 9$; and if $x = -2$, then $y = 4$. However, the relation $y^2 = x$ produces *two* y-values for each positive x-value. For example, if $x = 1$, then $y = \pm 1$; and if $x = 9$, then $y = \pm 3$. These ideas lead to the following definition:

A **function** is a relation such that, for each value of the first coordinate, there is *exactly one* value of the second coordinate.

The key phrase here is *exactly one*. According to this definition, $y = x^2$ *is* a function but $y^2 = x$ *is not* a function.

Because relations defined by equations are our main concern, we can "fine-tune" the foregoing definition to suit our needs, in the following manner:

Suppose x and y are related by an equation. We say that **y is a function of x** if each x value produces *exactly one* y value.

Thus y *is* a function of x in the equation $y = x^2$, because each x-value produces *exactly one* y-value. But y is *not* a function of x in the equation $y^2 = x$, because each positive x-value produces *two* y-values.

The graph of a relation can be used to decide whether or not y is a function of x. We illustrate with the same two equations, both of which yield parabolas:

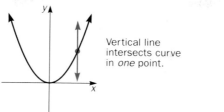
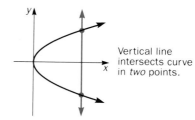

Vertical line intersects curve in *one* point.

Vertical line intersects curve in *two* points.

We see that any vertical line passing through $y = x^2$ (a function) intersects the curve in exactly one point. On the other hand, vertical lines passing through $y^2 = x$ (not a function) intersect the curve in more than one point. This leads to a useful geometrical criterion:

The Vertical Line Test

A relation *is a function* if every vertical line passing through its graph intersects it in *exactly one* point.

This test will be used in the following example.

EXAMPLE 1

For each of the following relations, tell whether or not it is a function, and give its domain and range.

a. $\{(3,-2), (3,4), (5,6), (7,0), (0,8)\}$
It is *not* a function, because 3 appears with both −2 and 4.
Domain = {3, 5, 7, 0}.
Range = {−2, 4, 6, 0, 8}.

b. $y = -x + 4$
It *is* a function, because it passes the vertical line test.
Domain = $\{x \mid x$ is any real number$\}$, and
Range = $\{y \mid y$ is any real number$\}$,
because the graph extends infinitely far in both positive and negative x and y directions.

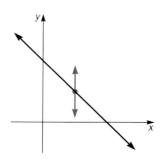

c. $\dfrac{x^2}{25} + \dfrac{y^2}{9} = 1$

It is *not* a function, because it fails the vertical line test.

Domain = $\{x \mid -5 \le x \le 5\}$.
Range = $\{y \mid -3 \le y \le 3\}$.

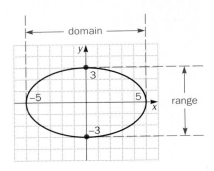

d. $y = x^2 - 2x + 3$

This parabola opens upward and has vertex (1,2). See Section 8.1.

It *is* a function, because it passes the vertical line test.

Domain = $\{x \mid x \text{ is any real number}\}$, because the graph extends infinitely far in the positive and negative x directions.

Range = $\{y \mid y \ge 2\}$, because the lowest point on the graph is at $y = 2$.

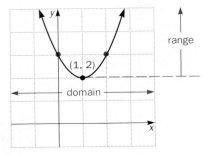

e. $\dfrac{x^2}{9} - \dfrac{y^2}{4} = 1$

This hyperbola opens left and right, with x-intercepts $(\pm 3, 0)$. See Section 8.3.

It is *not* a function, because it fails the vertical line test.

Domain = $\{x \mid x \le -3\} \cup \{x \mid x \ge 3\}$. Note that the domain is the union of two "disconnected" parts.

Range = $\{y \mid y \text{ is any real number}\}$, because the graph extends infinitely far in the positive and negative y directions.

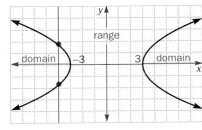

We saw earlier that the relation

$$y = x^2$$

defines y as a function of x. To symbolize this, we write

$$f(x) = x^2$$

which is read "function of x equals x-squared," or simply "f of x equals x-squared." The symbol $f(x)$ does *not* denote the product fx.

By substituting the value $x = 5$, we evaluate

$$f(5) = 5^2 = 25;$$

likewise, other substitutions give

$$f(-3) = (-3)^2 = 9,$$
and $\quad f(x + h) = (x + h)^2 = x^2 + 2xh + h^2.$

Students going on to calculus will encounter the expression

$$\frac{f(x + h) - f(x)}{h}.$$

This is called the **difference quotient** for the function $f(x)$, and it concerns the behavior of the graph of $y = f(x)$. As algebra students, you need only learn how to simplify it in such a way that h is divided out in the denominator. For example, the difference quotient for the function $f(x) = x^2$ becomes

$$\frac{f(x + h) - f(x)}{h} = \frac{(x + h)^2 - x^2}{h}$$

$$= \frac{x^2 + 2xh + h^2 - x^2}{h}$$

$$= \frac{2xh + h^2}{h}$$

$$= \frac{h(2x + h)}{h} \qquad \text{Factor.}$$

$$= 2x + h. \qquad \text{Reduce.}$$

EXAMPLE 2 | Let $f(x) = 3x - 2;$ then we evaluate by substitution:

$$f(0) = 3 \cdot 0 - 2$$
$$= -2.$$

$$f(-7) = 3(-7) - 2$$
$$= -23.$$

$$f\left(\frac{1}{2}\right) = 3 \cdot \frac{1}{2} - 2$$

$$= \frac{3}{2} - \frac{4}{2} \qquad \text{L.C.D.} = 2$$

$$= \frac{-1}{2}.$$

(continued)

$$f(x + h) = 3(x + h) - 2$$
$$= 3x + 3h - 2.$$

$$\frac{f(x + h) - f(x)}{h} = \frac{3x + 3h - 2 - (3x - 2)}{h} \qquad \text{"Difference quotient."}$$

$$= \frac{3x + 3h - 2 - 3x + 2}{h}$$

$$= \frac{3h}{h} \qquad \text{Combine terms.}$$

$$= 3. \qquad \text{Reduce.}$$

Other letters are used to denote functions, such as $g(x)$, $F(x)$, and $G(x)$.

EXAMPLE 3 Let $g(x) = 5x^2 - 9x + 7$;   then we evaluate by substitution:

$$g(0) = 5 \cdot 0^2 - 9 \cdot 0 + 7$$
$$= 7.$$

$$g(2) = 5 \cdot 2^2 - 9 \cdot 2 + 7$$
$$= 20 - 18 + 7$$
$$= 9.$$

$$g\left(-\frac{1}{2}\right) = 5\left(-\frac{1}{2}\right)^2 - 9\left(-\frac{1}{2}\right) + 7$$

$$= 5\left(\frac{1}{4}\right) + \frac{9}{2} + 7$$

$$= \frac{5}{4} + \frac{18}{4} + \frac{28}{4} \qquad \text{L.C.D.} = 4$$

$$= \frac{51}{4}.$$

$$g(1 + \sqrt{2}) = 5(1 + \sqrt{2})^2 - 9(1 + \sqrt{2}) + 7$$
$$= 5(1 + 2\sqrt{2} + 2) - 9 - 9\sqrt{2} + 7 \qquad (A + B)^2 = A^2 + 2AB + B^2$$
$$= 5 + 10\sqrt{2} + 10 - 9 - 9\sqrt{2} + 7$$
$$= 13 + \sqrt{2}.$$

$$g(x + h) = 5(x + h)^2 - 9(x + h) + 7$$
$$= 5(x^2 + 2xh + h^2) - 9x - 9h + 7$$
$$= 5x^2 + 10xh + 5h^2 - 9x - 9h + 7.$$

$$\frac{g(x + h) - g(x)}{h} = \frac{5x^2 + 10xh + 5h^2 - 9x - 9h + 7 - (5x^2 - 9x + 7)}{h}$$

$$= \frac{5x^2 + 10xh + 5h^2 - 9x - 9h + 7 - 5x^2 + 9x - 7}{h}$$

$$= \frac{10xh + 5h^2 - 9h}{h} \qquad \text{Combine terms.}$$

$$= \frac{h(10x + 5h - 9)}{h} \qquad \text{Factor.}$$

$$= 10x + 5h - 9. \qquad \text{Reduce.}$$

EXAMPLE 4   Let $F(x) = \dfrac{6}{x}$;   then we evaluate by substitution:

$$F(3) = \frac{6}{3} = 2.$$

$$F(-4) = \frac{6}{-4} = \frac{-3}{2}.$$

$$F\left(\frac{3}{4}\right) = \frac{6}{\frac{3}{4}} = \frac{6}{1} \cdot \frac{4}{3} = 8.$$

$$F(\sqrt{2}) = \frac{6}{\sqrt{2}} = \frac{6 \cdot \sqrt{2}}{\sqrt{2} \cdot \sqrt{2}} \qquad \text{Rationalize denominator.}$$

$$= \frac{6\sqrt{2}}{2} = 3\sqrt{2}.$$

$$F(x + h) = \frac{6}{x + h}.$$

$$\frac{F(x + h) - F(x)}{h} = \frac{\dfrac{6}{x + h} - \dfrac{6}{x}}{h} \qquad \text{Complex fraction.}$$

$$= \frac{\dfrac{6 \cdot x}{(x + h) \cdot x} - \dfrac{6 \cdot (x + h)}{x \cdot (x + h)}}{h} \qquad \longleftarrow \text{L.C.D.} = x(x + h)$$

$$= \frac{\dfrac{6x - 6x - 6h}{x(x + h)}}{h} \qquad \longleftarrow \text{Note signs.}$$

$$= \frac{\dfrac{-6h}{x(x + h)}}{\dfrac{h}{1}}$$

$$= \frac{-6h}{x(x + h)} \cdot \frac{1}{h} \qquad \text{Invert and multiply.}$$

$$= \frac{-6}{x(x + h)}. \qquad \text{Divide out } h\text{'s.}$$

EXERCISE 9.1   *Answers, page A87*

Answers, page A87

A

For each of the following relations, a) tell whether or not it is a function, and b) give its domain and range. See EXAMPLE 1. Sketch graphs, if necessary.

1. $\{(4,2),\ (5,3),\ (4,7),\ (0,9)\}$

2. $\{(5,3),\ (8,-11),\ (0,3),\ (13,1)\}$

3. $\{(6,6),\ (5,6),\ (-1,3),\ (4,6),\ (0,3)\}$

4. $\left\{(2,7),\ \left(-9,\frac{1}{2}\right),\ (7,7),\ (2,17),\ (0,0)\right\}$

5. $2x - y = 6$

6. $y = 5 - \frac{1}{2}x$

7. $y = x^2 - 2x$

8. $y = -x^2 + 4x - 1$

9. $x = -y^2 + 4$

10. $x = y^2 - 6y$

11. $\dfrac{x^2}{16} + \dfrac{y^2}{4} = 1$

12. $x^2 + y^2 = 9$

13. $\dfrac{y^2}{25} - \dfrac{x^2}{16} = 1$

14. $\dfrac{x^2}{9} - \dfrac{y^2}{4} = 1$

For each of the following functions, evaluate the given functional values, as in EXAMPLES 2, 3, and 4:

15. $f(x) = 5x + 3;\quad f(0),\ f(-2),\ f\left(\frac{1}{4}\right),\ f(\sqrt{2}),\ f(x+h),\ \dfrac{f(x+h) - f(x)}{h}$

16. $g(x) = 7 - 3x;\quad g(0),\ g(-4),\ g\left(-\frac{1}{2}\right),\ g(\sqrt{3}),\ g(x+h),\ \dfrac{g(x+h) - g(x)}{h}$

17. $F(x) = 2x^2 - 3x - 1;\quad F(5),\ F\left(-\frac{1}{2}\right),\ F(-\sqrt{5}),\ F(1+\sqrt{2}),\ F(x+h),\ \dfrac{F(x+h) - F(x)}{h}$

18. $G(x) = -x^2 + 5x - 2;\quad G(-2),\ G\left(\frac{1}{2}\right),\ G(\sqrt{7}),\ G(1-\sqrt{3}),\ G(x+h),\ \dfrac{G(x+h) - G(x)}{h}$

19. $f(x) = \dfrac{8}{x};\quad f(2),\ f\left(\dfrac{-4}{5}\right),\ f(\sqrt{2}),\ f(2+\sqrt{2}),\ f(x+h),\ \dfrac{f(x+h) - f(x)}{h}$

20. $F(x) = \dfrac{-12}{x};\quad F(-8),\ F\left(\dfrac{2}{3}\right),\ F(\sqrt{6}),\ F(3-\sqrt{3}),\ F(x+h),\ \dfrac{F(x+h) - F(x)}{h}$

21. $f(x) = x^3 + 3x^2 - 4x - 1;\quad f(0),\ f(5),\ f(-2),\ f\left(\frac{1}{2}\right),\ f(\sqrt{3})$

22. $g(x) = -2x^3 - x^2 + 3x + 2;\quad g(0),\ g(3),\ g(-2),\ g\left(\frac{2}{3}\right),\ g(-\sqrt{2})$

B

23. $F(x) = x^3$; $\quad F(x + h)$, $\dfrac{F(x + h) - F(x)}{h}$ $\quad$ *Hint:* $(x + h)^3 = (x + h)(x + h)^2$

24. $G(x) = 2x^3$; $\quad G(x + h)$, $\dfrac{G(x + h) - G(x)}{h}$

For each of the following relations, a) tell whether or not it is a function, and b) give its domain and range.

25. The relation whose graph is

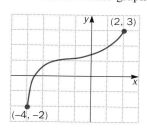

26. The relation whose graph is

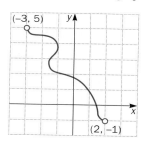

27. The relation whose graph is

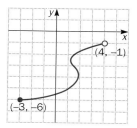

28. The relation whose graph is

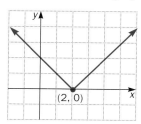

29. The relation whose graph is

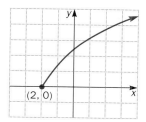

30. The relation whose graph is

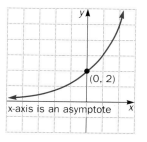

31. The line $x = 2$

32. The line $y = -3$

C

33. $(x - 2)^2 + (y + 3)^2 = 16$

34. $\dfrac{(x + 1)^2}{9} + \dfrac{y^2}{16} = 1$ $\quad$ *Hint*: See Exercise 8.3, Problems 41–46.

9.2   Special Functions

We learned in earlier sections that the

$$\textbf{linear function} \quad y = f(x) = 2x + 1$$

graphs as a straight line, and that the

$$\textbf{quadratic function} \quad y = f(x) = x^2 - 6x + 1$$

graphs as a parabola. We now graph the

$$\textbf{rational function} \quad y = f(x) = \frac{4}{x}, \quad x \neq 0,$$

$$\textbf{square root function} \quad y = f(x) = \sqrt{x + 1},$$

and $\qquad$ **absolute-value** function $\quad y = f(x) = |x - 1|,$

by preparing a table and then plotting points.

EXAMPLE 1

Graph $\quad y = f(x) = \dfrac{4}{x}, \quad x \neq 0.$

Solution

x	$y = \dfrac{4}{x}$
1	4
2	$2 \leftarrow \dfrac{4}{2}$
4	1
-1	-4
-2	-2
-4	$-1 \leftarrow \dfrac{4}{-4}$

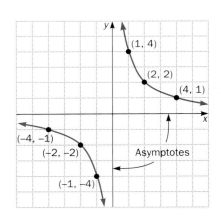

The graph of

$$y = \frac{4}{x}$$

is a *hyperbola*.

This graph is a "non-standard" hyperbola, with the x- and y-axes serving as asymptotes. You should compare this to the "standard" hyperbolas of Section 8.3, illustrated by $\dfrac{x^2}{9} - \dfrac{y^2}{4} = 1$.

Next, we graph the

square root function $y = f(x) = \sqrt{x + 1}$.

Because we can graph only real-valued functions, we must use only values of x for which the radicand is non-negative: $x + 1 \geq 0$, or $x \geq -1$. For if we used $x = -2$, say, then $f(-2) = \sqrt{-2 + 1} = \sqrt{-1} = i$, which is imaginary and cannot be plotted on either axis. Hence we must use only values of x that lie in the domain of this function, which is $\{x | x \geq -1\}$.

EXAMPLE 2   Graph $y = f(x) = \sqrt{x + 1}$.

Solution
Because the domain $= \{x | x \geq -1\}$, we use only these x-values:

x	$y = \sqrt{x + 1}$
-1	0
0	1
1	$1.4 \leftarrow \sqrt{1 + 1} = \sqrt{2}$
2	$1.7 \leftarrow \sqrt{2 + 1} = \sqrt{3}$
3	$2 \quad\leftarrow \sqrt{4}$
4	$2.2 \leftarrow \sqrt{5}$
8	$3 \quad\leftarrow \sqrt{9}$

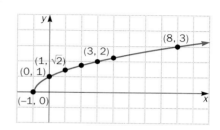

Finally, we graph the

absolute-value function $y = f(x) = |x - 1|$.

Because there is no denominator that can equal zero, and no square root that can be imaginary, we can substitute any real number x; thus the domain $= \{x | x \text{ is any real number}\}$. See Section 2.6 for more on absolute values.

EXAMPLE 3   Graph $y = f(x) = |x - 1|$.

Solution

x	$y = |x - 1|$
0	$1 \leftarrow |0 - 1| = |-1|$
1	0
2	1
3	$2 \leftarrow |3 - 1| = |2|$
4	3
-1	2
-2	$3 \leftarrow -2 - 1| = |-3|$

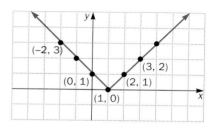

We see that the absolute-value function is a V-shaped graph consisting of two lines intersecting at right angles.

EXERCISE 9.2 *Answers, pages A87–A88*

A

Graph each function by first completing the table and then plotting points. See EXAMPLES 1, 2, and 3.

1. $y = \dfrac{6}{x}$

x	y
1	
2	
3	
6	
−1	
−2	
−3	
−6	

2. $f(x) = \dfrac{-8}{x}$

x	y
1	
2	
4	
8	
−1	
−2	
−4	
−8	

3. $xy = -4$

x	y
1	
2	
4	
−1	
−2	
−4	

Hint: First solve for *y*.

4. $xy = 2$

x	y
1	
2	
4	
$\frac{1}{2}$	
−1	
−2	
−4	
$-\frac{1}{2}$	

5. $y = \sqrt{x}$

x	y
0	
1	
2	
3	
4	
9	

6. $f(x) = \sqrt{x - 2}$

x	y
2	
3	
4	
5	
6	
11	

7. $g(x) = \sqrt{1 - x}$

x	y
1	
0	
−1	
−2	
−3	
−8	

8. $F(x) = -\sqrt{x + 2}$

x	y
−2	
−1	
0	
1	
2	
7	

9. $y = |x|$

x	y
0	
1	
2	
3	
−1	
−2	
−3	

10. $f(x) = |x - 2|$

x	y
0	
1	
2	
3	
4	
5	
−1	

11. $G(x) = -|3 + x|$

x	y
0	
1	
−1	
−2	
−3	
−4	
−5	
−6	

12. $F(x) = |1 - x| - 1$

x	y
0	
1	
2	
3	
4	
−1	
−2	

B

13. $y = f(x) = x^3$

x	y
0	
1	
1.5	
2	
-1	
-1.5	
-2	

14. $y = f(x) = \sqrt[3]{x}$

x	y
0	
1	
4	
8	
-1	
-4	
-8	

15. $y = \sqrt{x^2}$

x	y
0	*Hint:*
1	Perform
2	inside
3	first.
-1	
-2	
-3	

16. $f(x) = \sqrt{|x|}$

x	y
0	*Hint:*
1	Perform
4	inside
9	first.
-1	
-4	
-9	

17. $g(x) = \dfrac{|x|}{x}$

x	y
0	
$\frac{1}{2}$	
1	
2	
3	
$-\frac{1}{2}$	
-1	
-2	
-3	

18. $y = x^{2/3}$

x	y
0	*Hint:*
1	Express
4	$x^{2/3}$
8	in terms of
-1	a radical.
-4	
-8	

19. $f(x) = \sqrt{25 - x^2}$

x	y
0	
± 1	
± 2	
± 3	
± 4	
± 5	

20. $F(x) = -\sqrt{x^2 - 9}$

x	y
± 3	
± 4	
± 5	
± 6	

The *domain* of the function

$$f(x) = \frac{1}{x^2 - 3x - 4} = \frac{1}{(x - 4)(x + 1)}$$

consists of all values of x for which the denominator is not zero:

$$(x - 4)(x + 1) \neq 0, \quad \text{or} \quad x \neq 4, -1.$$

See Section 4.1, Example 1. Thus the domain $= \{x | x \neq 4, -1\}$.
Find the domain of each function:

21. $f(x) = \dfrac{1}{x^2 + x - 6}$

22. $g(x) = \dfrac{3x}{2x^2 - 9x + 4}$

23. $F(x) = \dfrac{x + 7}{x^2 - 4}$

24. $y = \dfrac{x - 3}{x^2 - 4x}$

25. $G(x) = \dfrac{-x}{x^2 - 8}$

26. $f(x) = \dfrac{x + 3}{2x^2 - 9}$

The *domain* of the function

$$f(x) = \sqrt{x^2 - 5x + 4} = \sqrt{(x - 4)(x - 1)}$$

consists of all values of x for which the radicand is non-negative (to avoid imaginary numbers):

$$(x - 4)(x - 1) \geq 0.$$

Using the methods of Section 6.7, we obtain the solution to this inequality, which becomes the domain of the function:

$$\text{domain} = \{x | x \geq 4\} \cup \{x | x \leq 1\}.$$

Find the domain of each function:

27. $f(x) = \sqrt{x - 3}$

28. $y = \sqrt{1 - 2x}$

29. $g(x) = \sqrt{x^2 + 3x - 10}$

30. $F(x) = \sqrt{x^2 - 6x + 8}$

31. $f(x) = \sqrt{x^2 - 4}$

32. $G(x) = \sqrt{9 - x^2}$

33. $y = \sqrt{\dfrac{x - 4}{x + 2}}$ (See Section 6.7, Example 2.)

34. $f(x) = \sqrt{\dfrac{x}{2 - x}}$

35. Find the domain of
  a. $f(x) = 2x - 5$
  b. $f(x) = x^2 - 7x + 2$
  c. $f(x) =$ any polynomial function

36. Find the range of
  a. The function in Example 2 in the text
  b. The function in Example 3

37. The difference quotient for the function $f(x) = \sqrt{x}$ is

$$\frac{f(x + h) - f(x)}{h} = \frac{\sqrt{x + h} - \sqrt{x}}{h}.$$

Simplify this expression so that the h in the denominator divides out.
Hint: Rationalize the *numerator* using the conjugate.

C

38. Do likewise for the function $f(x) = \dfrac{1}{\sqrt{x}}$,   whose difference quotient is

$$\frac{f(x + h) - f(x)}{h} = \frac{\dfrac{1}{\sqrt{x + h}} - \dfrac{1}{\sqrt{x}}}{h}.$$

9.3   Variations

This section covers a class of functions having a wide variety of applications. To illustrate, we know that the circumference C of a circle is a function of its radius r:

$$C = 2\pi r.$$

Likewise, the area A of a circle depends on the 2nd power of its radius r:

$$A = \pi r^2.$$

And the volume V of a sphere is a function of the 3rd power if its radius r:

$$V = \frac{4}{3}\pi r^3.$$

These functions are called *direct variations*, according to the following definition:

> y **varies directly** as the nth power of x
> if there is a constant k such that
> $$y = k \cdot x^n, \quad n > 0.$$

Under such conditions, we also say that y is **directly proportional** to x^n. The number k is called the **constant of variation.** Let us summarize the preceding examples:

Direct variation	Equation	Constant
C varies directly as r.	$C = 2\pi r$	$k = 2\pi$
A is directly proportional to the square of r.	$A = \pi r^2$	$k = \pi$
V varies directly as r^3.	$V = \frac{4}{3}\pi r^3$	$k = \frac{4}{3}\pi$

In many instances, the constant of variation k is not immediately known but must be determined from a given set of data. Once k is found, other data can then be determined.

EXAMPLE 1

Suppose y varies *directly* as x.

a. Given that $y = 6$ when $x = 8$, find k.
b. Then find y when $x = 12$.

(continued)

Solution
Because y varies directly with x, we write $y = kx$. It is helpful to use a chart:

Direct variation	Data a	Data b
$y = kx$	$y = 6, x = 8, k = ?$	$y = ?, x = 12$

a. Substitute **Data a** into $y = kx$ to find k:

$$6 = k \cdot 8, \quad \text{or} \quad \frac{3}{4} = k.$$

b. Now substitute **Data b** to find y:

$$y = \frac{3}{4} \cdot 12, \quad \text{or} \quad y = 9.$$

Another type of variation is illustrated by the function given in Section 9.2, Example 1:

$$y = \frac{4}{x}.$$

It is easy to see that as the denominator x increases, the value of y decreases. This is called an *inverse variation* according to the following definition:

> y **varies inversely** as the nth power of x
> if there is a constant k such that
> $$y = \frac{k}{x^n}, \quad n > 0.$$

We also say that y is **inversely proportional** to x^n. Let us illustrate:

Inverse variation	Equation
P varies inversely as V.	$P = \dfrac{k}{V}$
F is inversely proportional to the 2nd power of d.	$F = \dfrac{k}{d^2}$
v varies inversely as the square root of m.	$v = \dfrac{k}{\sqrt{m}} = \dfrac{k}{m^{1/2}}$

EXAMPLE 2

The illumination I from a light source varies *inversely* as the *square* of the distance d from the source. Illumination is measured in "foot-candles," the amount of light produced by one candle at a distance of one foot.

a. Suppose that, for a particular lamp, $I = 40$ foot-candles at a distance $d = 6$ feet. Find k.

b. Then find the distance at which the illumination is 80 foot-candles.

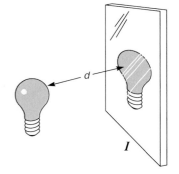

Solution

Because I varies inversely as the square of d, we write $I = \dfrac{k}{d^2}$. We use a chart as in Example 1:

Inverse variation	Data a	Data b
$I = \dfrac{k}{d^2}$	$I = 40,\ d = 6,\ k = ?$	$I = 80,\ d = ?$

a. Substitute **Data a** to find k:

$$40 = \frac{k}{6^2}, \quad \text{or} \quad 40 \cdot 6^2 = k, \quad \text{or} \quad k = 1440.$$

b. Now substitute **Data b** to find d:

$$80 = \frac{1440}{d^2}, \quad \text{or} \quad 80d^2 = 1440, \quad \text{or} \quad d^2 = 18, \quad \text{or} \quad d = 3\sqrt{2} \approx 4.2 \text{ feet.}$$

Quite often, an equation requires several variables to describe a physical situation. For example, the volume V of a cylinder is given by

$$V = \pi r^2 h,$$

where $r = $ its radius
and $h = $ its height.

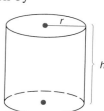

Here, V varies directly as the *product* of r^2 and h and is called a *joint variation*, which is defined as follows:

> z **varies jointly** as x^n and y^m, if there is a constant k such that
> $$z = k \cdot x^n \cdot y^m, \quad m, n > 0.$$

We also say that z is **directly proportional to the product** of x^n and y^m.

A variation can be direct, inverse, and joint all at the same time! Examples of such **combined variations** are given below:

Combined variation	Equation	
z varies jointly as x and y^2.	$z = k \cdot x \cdot y^2$	
w varies directly as the square of u and inversely as the cube of v.	$w = \dfrac{k \cdot u^2}{v^3}$	k is always in the numerator.
F varies jointly as m and M and inversely as d^2.	$F = \dfrac{k \cdot m \cdot M}{d^2}$	

EXAMPLE 3

The lifting force F on a hang-glider varies *jointly* as the wing area A and the square of its airspeed V.

a. On a typical glider, whose wing area is 180 square feet, there are 200 pounds of lifting force at a speed of 20 mph; find k.

b. Then find the lift at 30 mph on the same type of glider having 200 square feet of wing area.

Solution

This joint variation is written $F = kAV^2$.

Joint variation	Data a	Data b
$F = k \cdot A \cdot V^2$	$F = 200,\ A = 180,\ V = 20,\ k = ?$	$F = ?,\ A = 200,\ V = 30$

a. Substitute **Data a** to find k:

$$200 = k \cdot 180 \cdot 20^2, \quad \text{or} \quad 200 = 72{,}000k, \quad \text{or} \quad k = \frac{1}{360}.$$

b. Now substitute **Data b** to find F:

$$F = \frac{1}{360} \cdot 200 \cdot 30^2 = \frac{180,000}{360}, \quad \text{or} \quad F = 500 \text{ pounds of lift.}$$

EXERCISE 9.3   *Answers, page A88*

A

Set up and solve these variation problems, as in EXAMPLE 1:

1. y varies directly as x.
  a. Given that $y = 8$ when $x = 2$, find k.
  b. Find y when $x = -5$.

2. w varies directly as z.
  a. If $w = 4$ when $z = -6$, find k.
  b. Find w when $z = 9$.

3. v is directly proportional to the square of u.
  a. If $v = 36$ when $u = -4$, find k.
  b. Find v when $u = 6$.

4. s is directly proportional to t^2.
  a. If $s = -64$ when $t = 2$, find k.
  b. Find s when $t = 5$.

5. F varies directly as v^2.
  a. If $F = \frac{243}{2}$ when $v = 9$, find k.
  b. Find v when $F = 50$.

6. z varies directly as the 3rd power of x.
  a. If $z = \frac{-9}{2}$ when $x = -3$, find k.
  b. Find x when $z = 36$.

Set up and solve, as in EXAMPLE 2:

7. y varies inversely as x.
  a. If $y = 6$ when $x = 4$, find k.
  b. Find y when $x = -9$.

8. L varies inversely as W.
  a. If $L = 12$ when $W = 4$, find k.
  b. Find W when $L = 16$.

9. F is inversely proportional to the 2nd power of d.
  a. Find k if $F = 120$ when $d = 3$.
  b. Find d when $F = 90$.

10. z is inversely proportional to the cube of w.
  a. If $z = 4$ when $w = -2$, find k.
  b. Find z when $w = 4$.

Set up and solve, as in EXAMPLE 3:

11. y varies jointly with x and t^3.
  a. If $y = 12$ when $x = 3$ and $t = 2$, find k.
  b. Find y when $x = 1$ and $t = -3$.

12. c varies jointly with a^3 and the square of b.
  a. If $c = 75$ when $a = 1$ and $b = 5$, find k.
  b. Find c when $a = \frac{2}{3}$ and $b = -\frac{1}{2}$.

13. z varies directly with x and inversely as y^2.
  a. Find k if $z = 4$ when $x = 36$ and $y = -4$.
  b. Find x when $z = -6$ and $y = \frac{1}{2}$.

14. s varies directly with the second power of t and inversely as r.
   a. Given that $s = 10$ when $t = 2$ and $r = 5$, find k.
   b. Find t when $r = 4$ and $s = 3$.

15. Hooke's Law in physics states that
       "the tension T in a spring varies directly
         with the distance d that it is stretched
         or compressed."

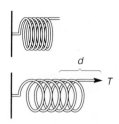

   a. If there are 10 pounds of tension on a spring when it is stretched 4 inches, find
       the "spring constant" k.
   b. Find the tension when this spring is stretched 14 inches.
   c. How many pounds are needed to compress it 6 inches?
   d. How many inches does the spring stretch when a 25-pound weight is attached
       to it?

16. The distance d traveled by a freely falling body is directly proportional to the
   square of the time t that it falls.
   a. If the body falls 64 feet in 2 seconds, find the "gravity constant" k.
   b. How far will it fall in 3 seconds?
   c. How long will it take to fall 256 feet?
   d. How long will it take to fall 512 feet?

17. Ohm's Law states that
       "the current I in an electrical circuit is
         inversely proportional to the square of the
         resistance R."

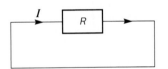

   a. If $I = 12$ amps when $R = 4$ ohms, find k.
   b. Find I when $R = 8$ ohms.
   c. Find the resistance when the current is 8 amps.

18. The Ideal Gas Law states that
       "the volume V of a gas varies directly with
         its temperature T (in degrees Kelvin, K) and
         inversely with its pressure P."

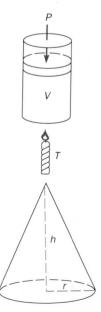

   a. If $V = 30$ cubic inches when $T = 200$ K and $P = 25$ pounds per square inch,
       find k.
   b. Find the volume of the same gas at 240 K under 45 psi of pressure.
   c. Find the pressure exerted by 50 cubic inches of the same gas at 300 K.

B

19. The volume V of a cone varies jointly as its height h and the square of its base
   radius r.
   a. If $V = 12\pi$ when $r = 3$ and $h = 4$, find k.
   b. Now write the formula for the volume of a cone.
   c. Then find the volume of a cone with radius 6 and height 9.

20. The Universal Law of Gravitation states that
       "the force F of attraction between two

bodies varies jointly with their masses m_1 and m_2 and inversely as the square of the distance d between them."

a. If $F = 30$ when $m_1 = 100$, $m_2 = 150$, and $d = 200$, find k.
b. Find F for the same two masses, but now $d = 100$.
c. Find d when $F = 20$, $m_1 = 50$, and $m_2 = 200$.
   (As in Problem 19, the units of measurement are not specified.)

21. The frictional force F needed to keep a car from skidding on a curve varies jointly with its mass m and the square of its velocity v and inversely as the radius r of the curve.

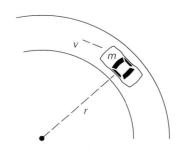

 a. If a 2500-lb car traveling 40 mph on a curve with radius 800 ft requires 1500 lb of friction to prevent it from skidding, find the "coefficient of friction" k, which depends on the road surface and the tires of the car.
 b. How much frictional force would be necessary to keep a 4000-lb car from skidding on a curve with radius 600 ft, if the car is traveling 30 mph, and the road surface and tires are the same as in part a?
 c. A 6400-lb truck approaches the same curve as in part b, equipped with the same tires. If there are 2000 lb of frictional force between tires and road, find the maximum velocity of the truck before it starts to skid.

22. The frequency F of a vibrating musical string is directly proportional to the square root of its tension T and inversely proportional to its length L.

 a. On a 30-in. guitar string with 400 lb of tension, the frequency is 440 cycles per second (the key of A). Find k for this string.
 b. If the same string is retuned to 484 lb of tension, and "fretted" down to 15 in., what is the resulting frequency?

In any variation, it is interesting to see what happens when any of the variables are changed. Suppose w varies jointly as x and y^2 and inversely as z^3:

$$w = \frac{k \cdot x \cdot y^2}{z^3}.$$

When x and z are doubled, and y is tripled,

$$w = \frac{k \cdot (2x) \cdot (3y)^2}{(2z)^3} = \frac{k \cdot 2x \cdot 9y^2}{8z^3} = \frac{9k \cdot x \cdot y^2}{4z^3}.$$

Comparing this result to the original, we see that

$$w \text{ is now } \frac{9}{4}, \text{ or } 2\frac{1}{4}, \text{ times as large as originally.}$$

In like manner, solve the following problems:

23. Suppose y varies directly as the square of x. Find the effect on y when
    a. x is doubled.
    b. x is tripled.
    c. x is "halved."

24. Suppose z varies jointly as x and y^2. Find the effect on z when
    a. x is doubled and y is tripled.
    b. x is doubled and y is "halved."

25. Suppose z varies jointly as u and the square root of v and inversely as the square of w. Find the effect on z when
    a. u is halved, v is quadrupled, and w is doubled.
    b. u is tripled, v is doubled, and w is halved.

26. In Problem 21, the frictional force is given by

$$F = \frac{k \cdot m \cdot v^2}{r}.$$

    Find the effect on F when
    a. m is doubled, v is halved, and r is tripled.
    b. m is halved and all other variables remain unchanged.
    c. v is halved and all other variables remain unchanged.
    Which of these three options is the safest? In other words, which requires the least friction to keep the car from skidding?

In Problems 23–26, we found the effect on one variable when all the other variables were changed. The constant of variation k had no effect on the outcome, because it never changed; in fact, we had no numerical way to find the actual value of k. In problems 1–22, our *"Data a"* enabled us to find k, which we then substituted into *"Data b."* Surprisingly, it is possible to solve all of these problems without ever finding the value of k! Suppose the chart of a typical variation problem looks like this:

Variation	Data a	Data b
$z = \dfrac{k \cdot x}{y^2}$	$z = 3,\ x = 6,\ y = 2,\ k = ?$	$z = ?,\ x = 4,\ y = -3$

Instead of first substituting Data for a to find k, as we did earlier, let us first solve the variation equation for k:

$$z = \frac{k \cdot x}{y^2}.$$
$$y^2 z = kx$$
$$\frac{y^2 z}{x} = k.$$

Now we substitute both *Data a* and *Data b* into this equation:

$$\frac{2^2 \cdot 3}{6} = k \quad \text{and} \quad \frac{(-3)^2 \cdot z}{4} = k.$$

Because both quantities equal k, they must equal each other:

$$\frac{2^2 \cdot 3}{6} = \frac{(-3)^2 \cdot z}{4}$$

Simplifying and then solving for z give

$$2 = \frac{9z}{4}, \quad \text{or} \quad z = \frac{8}{9}.$$

Note that we found z *without* ever finding k.

In like manner, solve the following problems *without* finding k:

27. y varies directly with x. If $y = 8$ when $x = 6$, find y when $x = 9$.

28. y varies inversely with x^2. If $y = 6$ when $x = -2$, find y when $x = 3$.

29. According to the Ideal Gas Law (Problem 18),

$$V = \frac{kT}{P}.$$

If $P = 20$ when $V = 80$ and $T = 150°$, find P when $V = 40$ and $T = 75°$.

30. Suppose that, in applying the Ideal Gas Law, we have two sets of data: P_1, V_1, T_1 and P_2, V_2, T_2. By equating these data as above, show that

$$\frac{P_1 V_1}{T_1} = \frac{P_2 V_2}{T_2}.$$

Then find V_2 when $P_1 = 30$, $V_1 = 25$, $T_1 = 180°$, $P_2 = 50$, and $T_2 = 270°$.

9.4   Composite and Inverse Functions

Let $f(x) = 2x + 3$ and $g(x) = x^2 - 4$. If $x = 5$, then

$$g(5) = 5^2 - 4$$
$$= 21;$$
$$\text{and} \quad f[g(5)] = f(21)$$
$$= 2 \cdot 21 + 3$$
$$= 45.$$

Thus the value of $f[g(5)]$ is found by first evaluating the innermost quantity, $g(5) = 21$, and then substituting this value, 21, into the outer function $f(x)$ and obtaining $f(21) = 45$. In like manner, let $x = -8$; then

$$g(-8) = (-8)^2 - 4$$
$$= 60,$$
$$\text{and} \quad f[g(-8)] = f(60)$$
$$= 2 \cdot 60 + 3$$
$$= 123.$$

More generally, for any value of x,

$$g(x) = x^2 - 4,$$
$$\text{and} \quad f[g(x)] = 2(x^2 - 4) + 3$$
$$= 2x^2 - 5.$$

The function $f[g(x)]$ is called the **composite function of g followed by f.**

We can interchange functions and get $g[f(x)]$, which is the **composite function of f followed by g.** If we again let $x = -8$, then

$$f(-8) = 2(-8) + 3$$
$$= -13,$$
$$\text{and} \quad g[f(-8)] = g(-13)$$
$$= (-13)^2 - 4$$
$$= 165.$$

More generally, for any value of x,

$$f(x) = 2x + 3,$$
$$\text{and} \quad g[f(x)] = (2x + 3)^2 - 4$$
$$= 4x^2 + 12x + 5.$$

Our results tell us that $f[g(-8)] = 123 \neq 165 = g[f(-8)]$ and that $f[g(x)] = 2x^2 - 5 \neq 4x^2 + 12x + 5 = g[f(x)]$. This illustrates that, in general, "composition of functions" is not commutative: $f[g(x)] \neq g[f(x)]$. See Section 1.4 for more on the Commutative Property.

EXAMPLE 1

Let $f(x) = 5 - 3x$ and $g(x) = 2x^2 + 1$. Compute:

a. $f[g(3)]$

$$\begin{aligned} \text{First,} \quad g(3) &= 2 \cdot 3^2 + 1 \\ &= 19, \\ \text{then} \quad f[g(3)] &= f(19) \\ &= 5 - 3 \cdot 19 \\ &= 5 - 57 \\ &= -52. \end{aligned}$$

b. $g[f(-2)]$

$$\begin{aligned} \text{First,} \quad f(-2) &= 5 - 3(-2) \\ &= 11, \\ \text{then} \quad g[f(-2)] &= g(11) \\ &= 2 \cdot 11^2 + 1 \\ &= 2 \cdot 121 + 1 \\ &= 243. \end{aligned}$$

c. $f[f(6)]$

$$\begin{aligned} \text{First,} \quad f(6) &= 5 - 3 \cdot 6 \\ &= -13, \\ \text{then} \quad f[f(6)] &= f(-13) \\ &= 5 - 3(-13) \\ &= 5 + 39 \\ &= 44. \end{aligned}$$

$f(x)$ is used twice.

d. $f[g(x)]$

$$\begin{aligned} \text{First,} \quad g(x) &= 2x^2 + 1, \\ \text{then} \quad f[g(x)] &= f(2x^2 + 1) \\ &= 5 - 3(2x^2 + 1) \\ &= 5 - 6x^2 - 3 \\ &= 2 - 6x^2. \end{aligned}$$

If we substitute $x = 3$ into this last result, $f[g(x)] = 2 - 6x^2$, we get $f[g(3)] = 2 - 6 \cdot 3^2 = 2 - 6 \cdot 9 = -52$. This is the same result we obtained in part a, which means we can always "double-check" our answers.

Now suppose $f(x) = 3x - 4$ and $g(x) = \dfrac{x + 4}{3}$. Their composite functions are

$$f[g(x)] = f\left(\frac{x + 4}{3}\right) \quad \text{and} \quad g[f(x)] = g(3x - 4)$$

$$= 3\left(\frac{x + 4}{3}\right) - 4 \qquad\qquad = \frac{(3x - 4) + 4}{3}$$

$$= x + 4 - 4 \qquad\qquad\qquad = \frac{3x}{3}$$

$$= x. \qquad\qquad\qquad\qquad = x.$$

We see that $f[g(x)] = x$ and $g[f(x)] = x$. These functions are called *inverses* of each other, according to the following definition:

> Let $f(x)$ and $g(x)$ be functions such that
>
> $$f[g(x)] = x \text{ and } g[f(x)] = x$$
>
> for all x for which all quantities are defined, respectively. Then g is called the **inverse function** of f, and we denote g by f^{-1}. Thus
>
> $$f[f^{-1}(x)] = x \text{ and } f^{-1}[f(x)] = x.$$

The symbol f^{-1} is read "f inverse." It does *not* mean $\frac{1}{f}$, as might be suggested by its negative exponent. The last two equations in the box suggest that f and f^{-1} "undo each other," or "cancel out," leaving the original x. This property of inverse functions is analogous to the Inverse Properties of real numbers, which we discussed in Section 1.4. Finally, we can also say that f is the inverse function of g and can equally well denote f by g^{-1}.

According to this definition, the function

$$f(x) = 3x - 4 \text{ has the inverse } f^{-1}(x) = \frac{x + 4}{3}.$$

Let us graph these two functions on the same axis system:

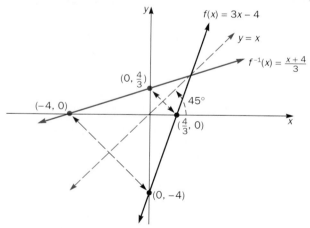

We see that $f(x)$ and $f^{-1}(x)$ are "mirror images" of each other when reflected through the line $y = x$, which is a 45° line through quadrants I and III. The picture shows that the points $(-4,0)$ and $(0,-4)$ are "reflections" of each other, as are the points $\left(0,\frac{4}{3}\right)$ and $\left(\frac{4}{3},0\right)$. This illustrates two important facts about any function $f(x)$ and its inverse $f^{-1}(x)$:

1. The graphs of $f(x)$ and $f^{-1}(x)$ are mirror images when reflected through the 45° line $y = x$.

2. The point (x,y) is on the graph of $f(x)$, provided that the point (y,x) is on the graph of $f^{-1}(x)$.

Now let $y = f(x) = 2 - 7x$. The second fact in the preceding list says that if (x,y) satisfies this equation, then (y,x) satisfies the equation of $f^{-1}(x)$. This means that the equation of $f(x)$,

$$y = 2 - 7x,$$

becomes the equation of $f^{-1}(x)$ when x and y are *interchanged*:

$$x = 2 - 7y.$$

We now solve for y to obtain the inverse function:

$$7y = 2 - x$$

$$y = \frac{2 - x}{7}.$$

$$\text{Thus,} \quad f^{-1}(x) = \frac{2 - x}{7}.$$

Check: $\quad f[f^{-1}(x)] = f\left(\dfrac{2 - x}{7}\right) \qquad$ and $\quad f^{-1}[f(x)] = f^{-1}(2 - 7x)$

$$= 2 - 7\left(\frac{2 - x}{7}\right) \qquad\qquad = \frac{2 - (2 - 7x)}{7}$$

$$= 2 - (2 - x) \qquad\qquad\qquad = \frac{2 - 2 + 7x}{7}$$

$$\overset{\checkmark}{=} x. \qquad\qquad\qquad\qquad\quad \overset{\checkmark}{=} x.$$

We summarize the steps used *to find the inverse* of a function $f(x)$:

Step 1 Write the function $f(x) = \ldots$ as $y = \ldots$
Step 2 Interchange x and y in the equation.
Step 3 Solve the equation for y; the resulting equation represents $f^{-1}(x)$.

EXAMPLE 2 Find the inverse of each function:

a. $f(x) = 5x - 6$

$$y = 5x - 6 \qquad \text{Step 1}$$

$$x = 5y - 6 \qquad \text{Step 2}$$

$$x + 6 = 5y$$

$$\frac{x + 6}{5} = y$$

$$f^{-1}(x) = \frac{x + 6}{5}. \qquad \text{Step 3}$$

(continued)

b. $f(x) = x^3 + 2$

$$y = x^3 + 2 \qquad \text{Step 1}$$
$$x = y^3 + 2 \qquad \text{Step 2}$$
$$\left. \begin{array}{c} x - 2 = y^3 \\ \sqrt[3]{x - 2} = y \\ f^{-1}(x) = \sqrt[3]{x - 2}. \end{array} \right\} \quad \text{Step 3}$$

Check:
$$\begin{aligned} f[f^{-1}(x)] &= f(\sqrt[3]{x - 2}) \\ &= (\sqrt[3]{x - 2})^3 + 2 \\ &\overset{\checkmark}{=} x - 2 + 2 \\ &\overset{\checkmark}{=} x. \end{aligned} \qquad \begin{aligned} f^{-1}[f(x)] &= f^{-1}(x^3 + 2) \\ &= \sqrt[3]{(x^3 + 2) - 2} \\ &= \sqrt[3]{x^3} \\ &\overset{\checkmark}{=} x. \end{aligned}$$

c. $f(x) = \dfrac{x}{x + 1}$

$$y = \frac{x}{x + 1} \qquad \text{Step 1}$$
$$x = \frac{y}{y + 1} \qquad \text{Step 2}$$
$$xy + x = y \qquad \text{Clear fraction.}$$
$$x = y - xy \qquad \text{y-terms to one side.}$$
$$x = y(1 - x) \qquad \text{Factor.}$$
$$\frac{x}{1 - x} = y \qquad \text{Divide.}$$
$$f^{-1}(x) = \frac{x}{1 - x}. \qquad \text{Inverse function.}$$

Recall from Section 9.1 that a relation is any set of ordered pairs and that a function is a special type of relation. Let us therefore extend the ideas with a definition. If R is a relation, its **inverse relation** R^{-1} is the set of points obtained by *interchanging* the coordinates of the points in R:

$$R^{-1} = \{(y,x) \mid (x,y) \in R\}.$$

For example, let $R = \{(1,5), (2,-7), (3,5)\}$; then $R^{-1} = \{(5,1), (-7,2), (5,3)\}$. Note that R is a function but R^{-1} is not a function, because when $x = 5$, $y = 1$ *and* $y = 3$.

EXAMPLE 3

For each relation R, obtain its inverse relation R^{-1}. Identify any relations that are functions. Sketch R and R^{-1} on the same set of axes (except in part a).

a. $R = \{(2,9), (5,0), (2,-3), (0,8)\}$

R is not a function, because when $x = 2$, $y = 9$ *and* $y = -3$.
$R^{-1} = \{(9,2), (0,5), (-3,2), (8,0)\}$
R^{-1} is a function.

b. $R: y = x^2 + 1$

R is a function, because it passes the vertical line test.
$R^{-1}: x = y^2 + 1$
R^{-1} is not a function, because it fails the vertical line test.

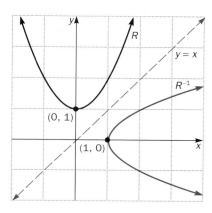

c. $R: \dfrac{x^2}{16} + \dfrac{y^2}{4} = 1$

$R^{-1}: \dfrac{y^2}{16} + \dfrac{x^2}{4} = 1$

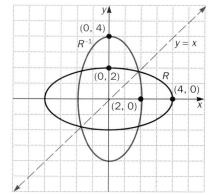

Neither of these ellipses is a function; both fail the vertical line test.

EXERCISE 9.4   *Answers, pages A89–A90*

A

Let $f(x) = 3x - 2$ and $g(x) = x^2 + 5$. Compute, as in EXAMPLE 1:

1. $f[g(4)]$ **2.** $g[f(-2)]$ **3.** $f[f(-5)]$ **4.** $g[g(3)]$

5. $f[g(x)]$ **6.** $g[f(x)]$ **7.** $f[f(x)]$ **8.** $g[g(x)]$

Let $f(x) = x + 1$, $g(x) = 3 - x^2$, and $h(x) = \sqrt{x}$. Compute, as in **EXAMPLE 1:**

9. $f[g(0)]$ **10.** $f[h(4)]$ **11.** $g[h(8)]$ **12.** $h[f(3)]$

13. $f[h(x)]$ **14.** $g[h(x)]$ **15.** $f[g[h(4)]]$ **16.** $h[f[g(-1)]]$

17. $f[g[h(x)]]$ **18.** $h[g[f(x)]]$

Find the inverse of each function, as in **EXAMPLE 2:**

19. $f(x) = 2x + 3$ **20.** $f(x) = 5x - 2$ **21.** $g(x) = 4 - x$ **22.** $g(x) = 6 - 2x$

23. $f(x) = \dfrac{3x - 4}{12}$ **24.** $F(x) = \dfrac{9x}{5} + 32$ **25.** $h(x) = x^3 - 1$ **26.** $G(x) = x^3 + 4$

27. $f(x) = 8x^3$ **28.** $g(x) = \dfrac{x^3}{8}$ **29.** $f(x) = \sqrt[3]{x + 1}$ **30.** $F(x) = \sqrt[3]{x} + 1$

31. $f(x) = \dfrac{2}{x}$ **32.** $g(x) = \dfrac{-3}{x - 1}$ **33.** $F(x) = \dfrac{x}{x - 2}$ **34.** $G(x) = \dfrac{3x}{x + 3}$

35. $f(x) = \dfrac{2x + 1}{x}$ **36.** $h(x) = \dfrac{x - 2}{2x}$

For each relation R, obtain its inverse relation R^{-1}. Identify any relations that are functions. Sketch R and R^{-1} on the same set of axes (except in Exercises 37 and 38). See **EXAMPLE 3.**

37. $\{(3, -1), (-2, 0), (9, -1), (-2, 4)\}$ **38.** $\{(4, 6), (6, 4), (-3, 1), (0, 1)\}$ **39.** $2x + 3y = 6$

40. $5x - 2y = 10$ **41.** $y = x^2$ **42.** $y = x^2 + 2$

43. $x = y^2 + 1$ **44.** $x = -y^2 - 1$ **45.** $\dfrac{x^2}{25} + \dfrac{y^2}{9} = 1$ **46.** $\dfrac{x^2}{4} + \dfrac{y^2}{9} = 1$

47. $\dfrac{x^2}{9} - \dfrac{y^2}{9} = 1$ **48.** $\dfrac{y^2}{16} - \dfrac{x^2}{4} = 1$ **49.** $x^2 + y^2 = 9$ **50.** $(x - 2)^2 + (y + 2)^2 = 4$

B

How good an artist are you? Using the "mirror image" concept, sketch the inverse R^{-1} of each relation R shown:

51.
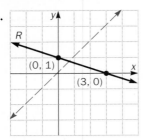

52.
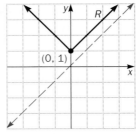

53.
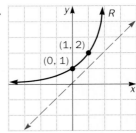

54.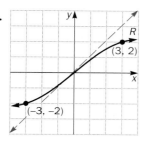

55.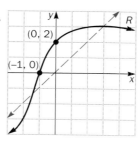

56.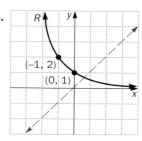

Sketch the graph of a relation R and its inverse R^{-1} such that:

57. Both R and R^{-1} are functions.

58. R is a function but R^{-1} is not.

59. R is not a function but R^{-1} is.

60. Neither R nor R^{-1} is a function.

61. Sketch the graph of a relation that is its own inverse.

62. For the relations

$$R = \{(2,3),\ (-1,4),\ (5,8)\} \quad \text{and} \quad R^{-1} = \{(3,2),\ (4,-1),\ (8,5)\},$$

list the elements of these sets:

Domain $R = ?$ Range $R = ?$ Domain $R^{-1} = ?$ Range $R^{-1} = ?$

On the basis of your answers, what conclusions can you make regarding these four sets *for any relation R and its inverse R^{-1}*?

In the text preceding Example 1, it was shown that if

$$f(x) = 2x + 3 \quad \text{and} \quad g(x) = x^2 - 4,$$

then $f[g(x)] = 2x^2 - 5$ and $g[f(x)] = 4x^2 + 12x + 5$;   thus

$$f[g(x)] \neq g[f(x)].$$

In other words, "composition of functions" is generally not commutative. However, in some instances, it is commutative; that is, there are functions for which

$$f[g(x)] = g[f(x)].$$

For each pair of functions, obtain expressions for both $f[g(x)]$ and $g[f(x)]$. Then tell whether the composition of these functions is commutative.

63. $f(x) = x^2$, $g(x) = x^3$

64. $f(x) = x + 1$, $g(x) = x - 6$

65. $f(x) = 2x$, $g(x) = \dfrac{x}{2}$

66. $f(x) = -x + 1$, $g(x) = x - 1$

67. $f(x) = 2x + 1$, $g(x) = \dfrac{x - 1}{2}$

68. $f(x) = \dfrac{x}{2}$, $g(x) = \dfrac{2}{x}$

Find the inverse of each function:

69. $f(x) = \dfrac{x + 1}{x - 1}$

70. $g(x) = \dfrac{2x - 3}{3x + 2}$

71. $F(x) = 8x^3 - 1$

C

72. $G(x) = \dfrac{x^3}{8} + 1$

73. $f(x) = \dfrac{27x^3}{5} + 4$

74. $g(x) = \dfrac{27x^3 - 4}{5}$

75. $f(x) = \sqrt[3]{2x - 1}$

76. $g(x) = \sqrt[3]{2x} - 1$

77. $f(x) = 2 - \dfrac{8x^3}{27}$

78. $F(x) = 3 - \dfrac{2\sqrt[3]{3x + 2}}{3}$

CHAPTER 9 ▪ SUMMARY

9.1 Functions and Relations

A **relation** is any set of ordered pairs.

y is a **function** of x if each x-value produces **exactly one** y-value.

Vertical Line Test: y is a function of x if every vertical line intersects the graph in *exactly one* point.

Examples:

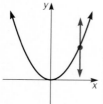

Vertical line intersects graph in *one* point.

$y = x^2$ *is* a function.

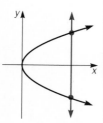

Vertical line intersects graph in *two* points.

$y^2 = x$ is *not* a function.

$y = f(x)$ means y is a function of x.

$f(a)$ = value obtained by substituting $x = a$ into the function

Examples:

$f(x) = x^2$

$f(5) = 5^2 = 25$

$f(-3) = (-3)^2 = 9$

The **difference quotient** is

$$\dfrac{f(x + h) - f(x)}{h}$$

$$\dfrac{f(x + h) - f(x)}{h} = \dfrac{(x + h)^2 - x^2}{h}$$

$$= \dfrac{\cancel{x^2} + 2xh + h^2 - \cancel{x^2}}{h}$$

$$= \dfrac{h(2x + h)}{h} = 2x + h.$$

CHAPTER 9 · SUMMARY

9.2 Special Functions

$f(x) = \frac{6}{x}, x \neq 0$

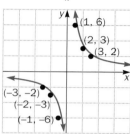

$f(x) = \sqrt{x-1}, x \geq 1$

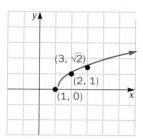

$f(x) = |x - 2|$

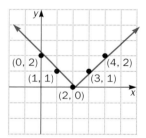

9.3 Variations

y varies **directly** as x^2.

Example:

a. If $y = 16$ when $x = 2$, find k.
b. Find y when $x = 3$.

Solution

Variation	Data for a	Data for b
$y = kx^2$	$y = 16, x = 2,$ $k = ?$	$y = ?$ $x = 3$

a. $16 = k \cdot 2^2$, or $k = 4$
b. $y = 4 \cdot 3^2$, or $y = 36$

Type of Variations	Equation
y varies **directly** as x^2.	$y = kx^2$
y varies **inversely** as x.	$y = \dfrac{k}{x}$
z varies **jointly** as x and y^2.	$z = kxy^2$
F varies directly as m and inversely as d^2.	$F = \dfrac{km}{d^2}$

9.4 Composite Functions

Let $f(x) = 3x - 2$

$$g(x) = x^2 + 4$$

The **composite functions** are

$$f[g(x)] \text{ and } g[f(x)]$$

Examples:

Let $x = 5$; then

$$g(5) = 5^2 + 4 = 29,$$

and $f[g(5)] = f(29)$

$$= 3 \cdot 29 - 2 = 85.$$

Also, $f[g(x)] = f(x^2 + 4)$
$$= 3(x^2 + 4) - 2$$
$$= 3x^2 + 10,$$

and $g[f(x)] = g(3x - 2)$
$$= (3x - 2)^2 + 4$$
$$= 9x^2 - 12x + 8.$$

Inverse Functions

If $f(x)$ and $g(x)$ satisfy

$$f[g(x)] = x = g[f(x)],$$

then f and g are **inverse** functions, and we denote $g(x)$ by $f^{-1}(x)$.

(continued)

CHAPTER 9 · SUMMARY

Example:

Let $f(x) = 2x - 5$. To find $f^{-1}(x)$:
1. Write $y = 2x - 5$.
2. *Interchange* x and y;

$$x = 2y - 5$$

3. Solve for y:

$$x + 5 = 2y$$

$$\frac{x + 5}{2} = y$$

then

$$f^{-1}(x) = \frac{x + 5}{2}.$$

CHAPTER 9 · REVIEW EXERCISES   *Answers, pages A90–A93*

9.1

For each relation,
a. Tell whether it is a function.   **b. Give its domain and range. Sketch the graphs, if necessary.**

1. $\{(5,0), (-2,3), (5,4), (8,7)\}$

2. $3x - 2y = 6$

3. $y = -x^2 + 4$

4. $x = y^2 + 2$

5. $x^2 + y^2 = 25$

6. $\dfrac{x^2}{16} + \dfrac{y^2}{9} = 1$

For each function, evaluate the given functional values:

7. $f(x) = 2x + 1$; $f(0), f(8), f(-3), f\left(\dfrac{2}{3}\right), f(\sqrt{3}), f(x + h), \dfrac{f(x + h) - f(x)}{h}$

8. $f(x) = 3x^2 - 4x - 2$; $f(0), f(5), f(-2), f\left(\dfrac{1}{2}\right), f(1 + \sqrt{2}), f(x + h), \dfrac{f(x + h) - f(x)}{h}$

9. $g(x) = \dfrac{-6}{x}$; $g(3), g(-8), g\left(\dfrac{3}{4}\right), g(\sqrt{2}), g(\sqrt{3} - 1), \dfrac{g(x + h) - g(x)}{h}$

10. $F(x) = x^3 - 2x^2 + x - 4$; $F(0), F(2), F(-3), F\left(-\dfrac{1}{2}\right), F(\sqrt{3})$

11. $G(x) = \sqrt{x - 2}$; $G(2), G(3), G(6), G(10), G\left(\dfrac{9}{4}\right), G\left(\dfrac{7}{2}\right)$

12. $f(x) = \dfrac{1}{x^2}$; $f(-3), f\left(\dfrac{1}{2}\right), f\left(\dfrac{-2}{3}\right), f(\sqrt{2}), f(x + h), \dfrac{f(x + h) - f(x)}{h}$

13. $g(x) = 4x^3$; $g(x + h), \dfrac{g(x + h) - g(x)}{h}$

The following are the graphs of relations. In each case,
a. Tell whether it is a function.   b. Give its domain and range.

14.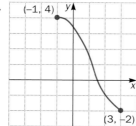

15.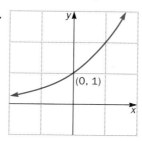

16.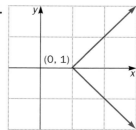

 9.2

Graph each function by first completing the table and then plotting points:

17. $f(x) = \dfrac{8}{x}$

x	y
1	
2	
4	
8	
-1	
-2	
-4	
-8	

18. $xy = -2$

x	y
1	
2	
4	
$\frac{1}{2}$	
-1	
-2	
-4	
$-\frac{1}{2}$	

19. $g(x) = \sqrt{x + 3}$

x	y
-3	
-2	
-1	
0	
1	
6	

20. $y = -\sqrt{2 - x}$

x	y
2	
1	
0	
-1	
-2	
-7	

21. $f(x) = |x - 3|$

x	y
0	
1	
2	
3	
4	
5	
6	
-1	

22. $F(x) = |x + 2| - 2$

x	y
0	
1	
2	
3	
-1	
-2	
-3	
-4	
-5	

23. $y = x^3 + 1$

x	y
0	
1	
1.5	
2	
-1	
-1.5	
-2	

24. $f(x) = \sqrt{16 - x^2}$

x	y
0	
± 1	
± 2	
± 3	
± 4	

Find the domain of each function:

25. $f(x) = \dfrac{3}{2x - 6}$

26. $g(x) = \dfrac{x + 1}{x^2 + 4x - 21}$

27. $y = \dfrac{2x}{x^2 - 9}$

28. $F(x) = \dfrac{4}{x^2 - 9x}$

29. $G(x) = \dfrac{x + 2}{x^2 - 12}$

30. $f(x) = \dfrac{x - 1}{25 - 3x^2}$

31. $f(x) = \sqrt{2x - 6}$

32. $y = -\sqrt{4 - x}$

33. $F(x) = \sqrt{x^2 - 9}$

34. $g(x) = \sqrt{9 - x^2}$

35. $G(x) = \sqrt{x^2 - 7x + 6}$

36. $y = \sqrt{x^2 - 9x}$

37. $f(x) = \dfrac{1}{\sqrt{x - 2}}$

38. $g(x) = \sqrt{\dfrac{x - 1}{x + 3}}$

39. The difference quotient for the function $f(x) = \sqrt{2x}$ is

$$\frac{f(x + h) - f(x)}{h} = \frac{\sqrt{2(x + h)} - \sqrt{2x}}{h}.$$

Rationalize the numerator so that, after simplifying, the h in the denominator divides out.

9.3

Solve these variation problems:

40. y varies directly as x.
   a. If $y = 12$ when $x = 8$, find k.
   b. Find y when $x = 6$.

41. y is inversely proportional to the square of x.
   a. If $y = 4$ when $x = -3$, find k.
   b. Find x when $y = 2$.

42. w varies directly as u and inversely as the cube of v.
   a. Given that $w = -4$ when $u = 4$ and $v = -2$, find k.
   b. Find w when $u = -6$ and $v = 3$.

43. Boyle's Law states that

   "at a fixed temperature, the volume V of a gas varies inversely with its pressure P."

   a. If $V = 1.5$ liters when $P = 1.8$ atmospheres, find k.
   b. Find V when $P = 1.2$ atm.
   c. Find P when $V = 2.7$ liters.

44. The velocity v of a freely falling body is directly proportional to the square root of the distance d that it has fallen.
   a. If $v = 32$ feet per second when $d = 16$ feet, find k.
   b. Find v when $d = 64$.
   c. Find v when $d = 128$.
   d. Find d when $v = 40$.

45. The load L that can be supported by a wooden beam varies jointly as its width w and the square of its height h and inversely as its length l.
   a. If a beam 96 in. long that is 2 in. wide and 4 in. high can support a load of 800 pounds, find k.
   b. What load can the same beam support if it is rotated such that it is now 4 in. wide and 2 in. high?

46. The weight W of an object is inversely proportional to the square of its distance d from the *center* of the earth. The radius of the earth is 4000 miles.
   a. If a person weighs 175 lb on the earth's *surface*, find k.
   b. Find the weight of the same person orbiting the earth 200 mi above the earth's surface.

47. Charles' Law states that

   "at a fixed pressure, the volume V of a gas is directly proportional to its temperature T."

   a. Write this variation.
   b. Solve it for k in terms of V and T.
   c. Suppose V_1, T_1 and V_2, T_2 represent two sets of data for a gas at a fixed pressure. Show that

$$\frac{V_1}{T_1} = \frac{V_2}{T_2}.$$

9.4

Let $f(x) = 3 - 4x$ and $g(x) = 2x^2 + 1$. Compute:

48. $f[g(0)]$ **49.** $f[g(-2)]$ **50.** $g[f(3)]$ **51.** $g[f(-4)]$

52. $f[f(7)]$ **53.** $g[g(-1)]$ **54.** $f[g(x)]$ **55.** $g[f(x)]$

56. $f[f(x)]$ **57.** $g[g(x)]$ **58.** $f[g[f(2)]]$

Let $f(x) = x + 1$, $g(x) = |x|$, $F(x) = \sqrt{x}$, and $G(x) = \dfrac{12}{x}$. Compute:

59. $f[g(-3)]$ **60.** $F[f(8)]$ **61.** $G[f(-7)]$ **62.** $G[F(3)]$

63. $G\left[f\left(\dfrac{1}{2}\right)\right]$ **64.** $g[G(-8)]$ **65.** $F[g[f(-9)]]$ **66.** $G[F[g[f(-3)]]]$

67. $G[f(x)]$ **68.** $F[f(x)]$.   **69.** $F[G(x)]$ **70.** $G[F(x)]$

Find the inverse of each function:

71. $f(x) = 3x - 2$

72. $g(x) = 1 - 2x$

73. $y = \dfrac{2x}{5} + 3$

74. $F(x) = x$

75. $f(x) = x^3 + 3$

76. $G(x) = 8x^3 - 1$

77. $y = \dfrac{4}{x}$

78. $f(x) = \dfrac{x}{x - 3}$

79. $F(x) = \dfrac{2x - 1}{x + 2}$

80. $f(x) = \sqrt[3]{x - 1}$

81. $g(x) = \sqrt[3]{x} - 1$

For each relation R, obtain its inverse relation R^{-1}. Identify any relations that are functions. Sketch R and R^{-1} on the same set of axes.

82. $3x - 4y = 12$

83. $y = x^2 - 4$

84. $y = |x| + 1$

85. $\dfrac{x^2}{16} + \dfrac{y^2}{4} = 1$

86. $\dfrac{x^2}{16} - \dfrac{y^2}{4} = 1$

87. $y = 3$

CHAPTER 9 ▪ TEST   *Answers, pages A93–A94*

For each of the following relations,
a. Tell whether it is a function.   **b. Give its domain and range.**

1. $\{(2, -1), (3, 2), (0, 2), (5, 0)\}$

2. $\dfrac{x^2}{16} + \dfrac{y^2}{25} = 1$

3. $y = x^2 - 1$

Find the domain of each function:

4. $f(x) = \dfrac{1}{x - 4}$

5. $g(x) = \sqrt{x - 4}$

6. $y = x^2 - 4$

7. $F(x) = \dfrac{1}{x^2 - 4}$

8. $f(x) = \sqrt{x^2 - 4}$

For each function, evaluate the given functional values:

9. $f(x) = 3 + 5x;$ $f(0), f(-2), f\left(\dfrac{1}{2}\right), \dfrac{f(x + h) - f(x)}{h}$

10. $g(x) = x^2 - 3x + 2;$ $g(0), g(1), g\left(-\dfrac{1}{2}\right), g(1 + \sqrt{2}), \dfrac{g(x + h) - g(x)}{h}$

Let $f(x) = x + 1$ and $g(x) = x^2 + 2x - 3$. Compute:

11. $f[g(1)],$ $f[g(-3)],$ and $f[g(x)]$

12. $g[f(0)],$ $g[f(-5)],$ and $g[f(x)]$

Find the inverse of each function:

13. $f(x) = 2 - x$

14. $y = x^3 - 8$

15. $g(x) = \dfrac{2x}{x + 1}$

Graph each function by first completing the table and then plotting points:

16. $f(x) = \dfrac{10}{x}$

x	y
1	
2	
5	
10	
-1	
-2	
-5	
-10	

17. $y = \sqrt{1 - x}$

x	y
0	
1	
-1	
-3	
-8	

18. The distance d traveled by a freely falling body varies directly as the 2nd power of time t.
  a. If $d = 144$ ft when $t = 3$ s, find k.
  b. How far will the body have fallen after 6 s?

19. The kinetic energy E of a moving object varies jointly with its mass m and the square of its velocity v.
  a. If $E = 2000$ joules when $m = 10$ grams and $v = 20$ meters/second, find k.
  b. Find v when $E = 1500$ and $m = 30$.

20. The distance d traveled by an object is directly proportional to the square of its velocity v and inversely proportional to its acceleration a.
  a. If $d = 125$ when $v = 50$ and $a = 10$, find k.
  b. Find a when $v = 75$ and $d = 100$.

CHAPTERS 1–9 · CUMULATIVE REVIEW EXERCISES   *Answers, pages A94–A95*

Simplify each expression, solve each equation, and solve and graph each inequality:

1. $x^2 - 22x + 120 = 0$

2. $x^2 + 23x + 120 = 0$

3. $x^2 + 26x - 120 = 0$

4. $x^2 - 29x + 120 = 0$

5. $\dfrac{2}{x - 2} - \dfrac{x}{x + 3} + \dfrac{2x^2 - 3}{x^2 + x - 6}$

6. $\left(\dfrac{a^4 b^{-2} c^0}{a^{-1} b^2 c^3}\right)^{-3}$

7. $\dfrac{a^{-1} - a^{-3}}{a + a^{-2}}$

8. $\dfrac{4}{y} = y - 2$

9. $\dfrac{4z}{2z - 3} = \dfrac{3}{z}$

10. $\dfrac{25}{6} = \dfrac{x^2}{2}$

11. $\sqrt{16 - x^2} - \dfrac{x^2}{\sqrt{16 - x^2}} = 0$

12. $\left|\dfrac{1 - 2x}{3}\right| < 3$

13. $\dfrac{a^2 + 3a - 10}{a^2 + 2a - 15} \div \dfrac{4a - 2a^2}{a^2 + a - 12}$

14. $\sqrt{x + 7} + \sqrt{x} = 7$

15. $(8x^3)^{-2/3}$

16. $\dfrac{\sqrt{6} + \sqrt{3}}{\sqrt{6} - \sqrt{3}}$

Solve these systems:

17. $3x + 2y = 6$
$5x - 3y = 10$

18. $2x + 3y - 4z = -4$
$x + y + z = 6$
$x + y - z = 0$

19. $x^2 + y^2 = 10$
$x - y = 2$

20. $(x + 2)(y - 1) = 24$
$xy = 24$

Solve these systems by Cramer's Rule:

21. $2x + 4y = -5$
$3x - 5y = 9$

22. $x + y + z = 2$
$y - 2z = 1$
$x + y - z = 0$

Graph:

23. $2x - 5y = 10$

24. $y = -x^2 + 2x - 1$

25. $\dfrac{x^2}{9} + \dfrac{y^2}{4} = 1$

26. $\dfrac{x^2}{9} - \dfrac{y^2}{4} = 1$

27. $x^2 + y^2 - 4x + 2y + 1 = 0$
Hint: Complete the square.

28. $f(x) = \sqrt{x - 1}$
Use $x = 1, 2, 3, 5,$ and 10

Let $f(x) = 2x - 3$ and $g(x) = x^2 + 2x$. Compute:

29. $g[f(5)]$

30. $f[g(-3)]$

31. $f[g(x)]$

32. $g[f(x)]$

33. $\dfrac{f(x + h) - f(x)}{h}$

34. $\dfrac{g(x + h) - g(x)}{h}$

Find the inverse of each function:

35. $f(x) = 5 - 10x$

36. $g(x) = \dfrac{x^3}{2} + 4$

37. $F(x) = \dfrac{x - 2}{x + 2}$

38. Find the equation of the line going through the points $(2, -3)$ and $(-2, 3)$.

39. Find the equation of the circle with center $(1, 0)$ and radius $2\sqrt{3}$.

40. Find the equation of the circle with center $(-3, 1)$ such that the point $(1, -3)$ lies on the circumference. Then find the area of this circle.

41. Find the equation of the line going through the point $(2, 1)$ that is perpendicular to the line $4x + 2y - 7 = 0$.

42. F varies directly with x and inversely as the square of y.
   a. If $F = 4$ when $x = 6$ and $y = -3$, find k.
   b. Find F when $x = 8$ and $y = 4$.

43. Old MacDonald's farm has 52 animals, consisting only of chickens and pigs. The total number of feet on these animals is 132. How many chickens and how many pigs are on the farm?

44. Find the dimensions of a rectangle with perimeter 12 and area 8.

45. Find the dimensions of a rectangle with area 12 and diagonal 5.

46. Equal amounts of 7% alcohol and 12% alcohol are mixed with 6.5% alcohol to obtain 15 liters of 8% alcohol. How many liters of each strength are used?

47. When reduced to lowest terms, a fraction equals $\frac{3}{4}$. If both numerator and denominator of the fraction are decreased by 4, the new fraction that is formed reduces to $\frac{1}{2}$. Find the original fraction.

48. A family drives 200 miles to the mountains at a certain speed. On the trip home, it increases its speed by 10 mph. As a result, it takes one hour less time than it did going. Find the speed and the time in each direction.

49. Nine coins in nickels, dimes, and quarters are worth $1.35. If, suddenly, the nickels turned into dimes, the dimes into quarters, and the quarters into nickels, the collection would only be worth $1.00. Find the actual number of each coin.

50. The perimeter of a rectangle is 12 inches. Find the width and the length that maximize the area. What is this maximum area?

51. The perimeter of a rectangle is 20 inches. If the width is doubled and the length is reduced by 3 inches, a new rectangle is formed having the same area as the original rectangle. Find the width and the length of each rectangle.

52. Find the vertex of the parabola $y = x^2 - 4x - 1$ and the center of the circle $x^2 + y^2 - 12x - 6y + 33 = 0$. Then find the distance between these points.

53. Line 1 goes through the points $(0, -3)$ and $(3, 3)$. Line 2 goes through the point $(6, -1)$ and is parallel to the line $x + 2y - 17 = 0$.
  a. Find the equation of each line.
  b. Find their point of intersection.

54. Find x. Then find the three sides of the triangle.

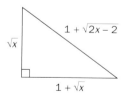

$1 + \sqrt{2x - 2}$

$\sqrt{x}$

$1 + \sqrt{x}$

10 EXPONENTIAL AND LOGARITHMIC FUNCTIONS

Exponential Functions

In Chapter 5, you learned the meaning of integer and rational exponents. For example,

$$5^0 = 1$$

$$4^{-2} = \frac{1}{4^2} = \frac{1}{16}$$

$$75^{1/2} = \sqrt[2]{75} = 5\sqrt{3}$$

$$8^{-5/3} = (\sqrt[3]{8})^{-5} = 2^{-5} = \frac{1}{32}.$$

For decimal exponents, we have the interpretation

$$6^{0.4} = 6^{4/10} = 6^{2/5} = (\sqrt[5]{6})^2.$$

We can obtain its approximate value on a scientific calculator by using the $\boxed{x^y}$ or $\boxed{y^x}$ button:

$$6^{0.4} = 2.0476725 \qquad\qquad 6 \boxed{x^y} .4 \boxed{=}$$

For irrational exponents, we also have the approximation

$$2^{\sqrt{3}} = 3.3219971^* \qquad\qquad 2 \boxed{x^y} 3 \boxed{\sqrt{\ }} \boxed{=}$$

EXAMPLE 1

Find the value of each expression using a calculator. Round to three decimal places.

a. $7^{-1.2} = 0.0968015 = 0.097$ $\qquad\qquad 7 \boxed{x^y} 1.2 \boxed{+/-} \boxed{=}$

b. $(\sqrt{5})^\pi = 12.530$ $\qquad\qquad 5 \boxed{\sqrt{\ }} \boxed{x^y} \boxed{\pi} \boxed{=}$

c. $\left(\dfrac{5}{8}\right)^{28/60} = 0.803$ $\qquad\qquad 5 \boxed{\div} 8 \boxed{=} \boxed{x^y} \boxed{(} 28 \boxed{\div} 60 \boxed{)} \boxed{=}$

---

*Compare the complexity of $2^{\sqrt{3}}$ to the ease of $(\sqrt{3})^2 = 3$.

d. $3500\left(1 + \dfrac{0.065}{12}\right)^{12(5)} = 4839.861$

$1\boxed{+}.065\boxed{\div}12\boxed{=}\boxed{x^y}\boxed{(}12\boxed{\times}5\boxed{)}\boxed{\times}3500\boxed{=}$

Other button sequences are possible, depending on your method and your particular calculator.

We will assume that the foregoing approximations and all other values of b^x represent real numbers for all real-number exponents x and all *positive* bases b. Note that a negative base can lead to an imaginary number; for example, $(-9)^{0.5} = (-9)^{1/2} = \sqrt{-9} = 3i$. This is undesirable for our present purposes. We shall also assume that the properties of exponents introduced in earlier chapters are valid for all real-number powers b^x.

EXAMPLE 2

Simplify each expression:

a. $2^x \cdot 2^y = 2^{x+y}$ Add exponents.

b. $\dfrac{5^{3x}}{5^{x-1}} = 5^{3x-(x-1)}$ Subtract exponents.

      $= 5^{2x+1}$ $3x - x + 1$

c. $(3^s \cdot 7^t)^2 = 3^{2s} \cdot 7^{2t}$ Multiply exponents.

d. $10^x + 10^{-x} = 10^x + \dfrac{1}{10^x}$ Invert.

               $= \dfrac{10^x \cdot 10^x}{1 \cdot 10^x} + \dfrac{1}{10^x}$ L.C.D. $= 10^x$

               $= \dfrac{10^{2x} + 1}{10^x}$ Add exponents: $10^{x+x} = 10^{2x}$

The equation of the parabola $y = x^2$ has a variable base x and a constant exponent 2. By interchanging them, we get

$$y = 2^x$$

which has a constant base 2 and a variable exponent x. We graph it, as usual, by making a table and then plotting points:

x	$y = 2^x$
0	$1 \longleftarrow 2^0$
1	$2 \longleftarrow 2^1$
2	4
3	$8 \longleftarrow 2^3$
-1	$\dfrac{1}{2}$
-2	$\dfrac{1}{4} \longleftarrow 2^{-2} = \dfrac{1}{2^2}$
-3	$\dfrac{1}{8}$
$\dfrac{1}{2}$	$1.4 \longleftarrow 2^{1/2} = \sqrt{2}$
$\dfrac{3}{2}$	$2.8 \longleftarrow 2^{3/2} = (\sqrt{2})^3 = 2\sqrt{2}$
$-\dfrac{1}{2}$	$0.7 \longleftarrow 2^{-1/2} = \dfrac{1}{\sqrt{2}},\ \text{ or } 2^{-0.5}$

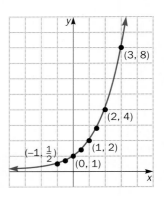

The graph tells us that the relation $y = 2^x$ defines y as a function of its exponent x, because it passes the vertical line test with flying colors. Thus we can write

$$f(x) = 2^x$$

and call this an *exponential function*. The graph also tells us that its

$$\text{domain} = \{x \mid x \text{ is any real number}\},$$

because the graph extends infinitely far along the x-axis in both directions. The

$$\text{range} = \{y \mid y > 0\},$$

because the graph always stays above the x-axis but never touches it; the x-axis is, therefore, an asymptote for this graph. (See Section 8.3 for more on asymptotes.)

More generally, any function of the form

$$f(x) = b^x, \quad b > 0,\ b \neq 1$$

is called an **exponential function**. It was explained before Example 2 why we require that b be positive. Can you tell why we also require that $b \neq 1$?

EXAMPLE 3 Graph these exponential functions:

a. $f(x) = \left(\dfrac{1}{2}\right)^x$

x	$y = \left(\dfrac{1}{2}\right)^x$
0	$1 \leftarrow \left(\dfrac{1}{2}\right)^0$
1	$\dfrac{1}{2}$
2	$\dfrac{1}{4} \leftarrow \left(\dfrac{1}{2}\right)^2$
3	$\dfrac{1}{8}$
-1	$2 \leftarrow \left(\dfrac{1}{2}\right)^{-1} = \dfrac{1}{\dfrac{1}{2}}$
-2	4
-3	$8 \leftarrow \left(\dfrac{1}{2}\right)^{-3} = \dfrac{1}{\left(\dfrac{1}{2}\right)^3} = \dfrac{1}{\dfrac{1}{8}}$

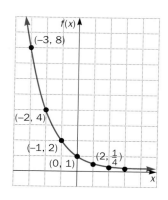

b. $g(x) = 3^{x-1}$

x	$y = 3^{x-1}$
0	$\dfrac{1}{3} \leftarrow 3^{0-1} = 3^{-1}$
1	$1 \leftarrow 3^{1-1} = 3^0$
2	$3 \leftarrow 3^{2-1} = 3^1$
3	9
-1	$\dfrac{1}{9} \leftarrow 3^{-1-1} = 3^{-2}$
-2	$\dfrac{1}{27}$

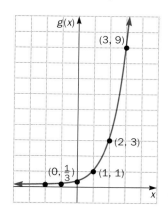

The natural and physical sciences, as well as business, are rich in exponential functions. Examples of phenomena exhibiting exponential characteristics are population growth, radioactive decay of certain chemicals, monetary growth and inflation under compound interest, heat transfer, current in electrical currents, and falling objects subject to air resistance.

EXAMPLE 4 Microorganisms, such as amoebae, paramecia, and bacteria, reproduce by splitting in half. One certain amoeba culture reproduces according to the exponential function

$$P = 100,000(2)^{t/3},$$

where P is the population of the culture after t minutes. Find P when $t = 0, 3$, 4.5, and 6 minutes.

Solution
When $t = 0$, $P = 100,000(2)^{0/3}$ $2^{0/3} = 2^0 = 1$
                       $= 100,000$ (the starting population);
when $t = 3$, $P = 100,000(2)^{3/3}$ $2^{3/3} = 2^1 = 2$
                       $= 200,000;$
when $t = 4.5$, $P = 100,000(2)^{4.5/3}$
                       $= 282,843;$ $2\;\boxed{x^y}\;\boxed{(}\;4.5\;\boxed{\div}\;3\;\boxed{)}\;\boxed{\times}\;100,000\;\boxed{=}$
when $t = 6$, $P = 100,000(2)^{6/3}$ $2^{6/3} = 2^2 = 4$
                       $= 400,000.$
These figures indicate that the population of amoebae in this culture doubles every 3 minutes.

Certain chemicals are radioactive, meaning that they constantly emit particles into the atmosphere during a process called **radioactive decay**. The rate at which this decay occurs is determined by the **half-life** of the element, which is the *time* required for any given amount to *decay to one-half* of that amount. For example, suppose we start with 10 grams of radioactive radium, whose half-life is approximately 1600 years. After 1600 years, there will be 5 grams remaining; and 1600 years later, or after 3200 years, there will be 2.5 grams remaining; and so forth. The following table illustrates this:

Time (years)	Grams of radium remaining
0	10
1600	5
3200	2.5
4800	1.25
6400	0.625
8000	0.3125

EXAMPLE 5 Radioactive carbon-14 has a half-life of 5600 years. The formula

$$A = 100\left(\frac{1}{2}\right)^{t/5600}$$

gives the amount A remaining (in grams) after t years. Find A when $t = 0$, 2800, 5600, 11,200, and 15,000 years.

Solution

When $t = 0$, $A = 100\left(\dfrac{1}{2}\right)^{0/5600}$ $\left(\dfrac{1}{2}\right)^{0} = 1$

$= 100$ grams (the starting amount);

when $t = 2800$, $A = 100\left(\dfrac{1}{2}\right)^{2800/5600}$

$= 100\left(\dfrac{1}{2}\right)^{1/2}$ .5 $\boxed{x^y}$ .5 $\boxed{x}$ 100 $\boxed{=}$

$= 70.7$ grams;

when $t = 5600$, $A = 100\left(\dfrac{1}{2}\right)^{5600/5600}$ $\left(\dfrac{1}{2}\right)^{1} = \dfrac{1}{2}$

$= 50$ grams (amount at half-life);

when $t = 11{,}200$, $A = 100\left(\dfrac{1}{2}\right)^{11{,}200/5600}$

$= 100\left(\dfrac{1}{2}\right)^{2}$ $\left(\dfrac{1}{2}\right)^{2} = \dfrac{1}{4}$

$= 25$ grams (amount at second half-life);

when $t = 15{,}000$, $A = 100\left(\dfrac{1}{2}\right)^{15{,}000/5600}$ .5 $\boxed{x^y}$ $\boxed{(}$ 15,000 $\boxed{\div}$ 5600 $\boxed{)}$ $\boxed{x}$ 100 $\boxed{=}$

$= 15.6$ grams.

The population growth function in Example 4 used base 2, which is ideal for amoebae because they reproduce by splitting in half. Likewise, the radioactive decay function in Example 5 used base $\dfrac{1}{2}$, which is convenient because of the half-life property. These bases are limited in scope, however. For many other complex phenomena in the sciences, such as human population growth, we use base e, an irrational number given by the decimal

$$e = 2.718281828459045\ldots \approx 2.718$$

This curious number comes from calculus, just as $\pi = 3.141592654\ldots$ comes from geometry. We can obtain powers of e on a scientific calculator by using the $\boxed{\text{INV}}$ and $\boxed{\text{ln}}$ keys, or the $\boxed{\text{2nd}}$ and $\boxed{\text{ln}}$ keys, as shown below:

Powers of e	Calculator sequence
$e^{1} = 2.7182818$	1 $\boxed{\text{INV}}$ $\boxed{\text{ln}}$ or 1 $\boxed{\text{2nd}}$ $\boxed{\text{ln}}$
$e^{0.7} = 2.0137527$	.7 $\boxed{\text{INV}}$ $\boxed{\text{ln}}$
$e^{-2.8} = 0.0608101$	2.8 $\boxed{+/-}$ $\boxed{\text{INV}}$ $\boxed{\text{ln}}$

Powers of e can also be found on the Table of Powers of e in the Appendix. The letter e is used in honor of the eighteenth-century Swiss mathematician Leonhard Euler (pronounced "oiler"). Euler is considered one of the most prolific mathematicians in history and is credited with introducing i to represent the imaginary number $\sqrt{-1}$.

EXAMPLE 6

The population of Las Vegas, Nevada, is given by the exponential function

$$P = 63,000e^{0.07t},$$

where P is the population t years after the census year 1960, and $0.07 = 7\% =$ the annual growth rate. This is in accordance with the "Malthusian growth-rate model." Compute the population of the city in the years 1960, 1970, and 1985.

Solution

In 1960, $t = 0$: $P = 63,000e^{0.07(0)}$ $e^0 = 1$
                       $= 63,000$

In 1970, $t = 10$: $P = 63,000e^{0.07(10)}$
                       $= 126,866$ .07 $\boxed{\times}$ 10 $\boxed{=}$ $\boxed{\text{INV}}$ $\boxed{\text{ln}}$ $\overset{e^x}{}$ $\boxed{\times}$ 63,000 $\boxed{=}$

In 1985, $t = 25$: $P = 63,000e^{0.07(25)}$
                       $= 362,540$ .07 $\boxed{\times}$ 25 $\boxed{=}$ $\boxed{\text{INV}}$ $\boxed{\text{ln}}$ $\overset{e^x}{}$ $\boxed{\times}$ 63,000 $\boxed{=}$

In Section 2.3, we used the formula $I = Prt$ to compute simple interest. More often, money deposited into a bank account earns **compound interest,** which is "interest on interest." As a result, the account grows exponentially, according to the

> **Compound Interest Formulas**
>
> $$A = P\left(1 + \frac{r}{n}\right)^{nt} \quad \text{and} \quad I = A - P$$

where A = accumulated amount n = number of times
           in the account   compounded per year
       P = original principal deposited t = number of years
       r = interest rate per year I = interest earned

EXAMPLE 7

Find the accumulated amount A and the interest earned I:

a. $P = \$2500$ is deposited at $r = 6\% = 0.06$ per year compounded $n = 12$ times per year (monthly) for $t = 4$ years.

$$A = \$2500\left(1 + \frac{0.06}{12}\right)^{4(12)}$$

$$= \$3176.22$$

$$I = \$3176.22 - \$2500$$
$$= \$676.22$$

$$\boxed{1}\ \boxed{+}\ .06\ \boxed{\div}\ 12\ \boxed{=}\ \boxed{x^y}\ \boxed{(}\ 4\ \boxed{\times}\ 12\ \boxed{)}\ \boxed{\times}\ 2500\ \boxed{=}$$

b. \$40,000 is deposited at $8\frac{1}{2}\%$ interest compounded daily for $3\frac{1}{4}$ years.

We have $P = \$40,000$, $r = 8\frac{1}{2}\% = 0.085$, $n = 365$ times per year (daily) and $t = 3.25$ years.

$$A = \$40,000\left(1 + \frac{0.085}{365}\right)^{365(3.25)}$$

$$= \$52,725.35$$
$$I = \$52,725.35 - \$40,000$$
$$= \$12,725.35$$

EXERCISE 10.1   *Answers, pages A95–A96*

A

Find the value of each expression, as in **EXAMPLE 1**. For powers of e, use $\boxed{\text{INV}}\ \boxed{\text{ln}}$, as in the discussion before **EXAMPLE 5**. Round to three decimal places.

1. $(1.06)^{4.5}$
2. $(5.7)^{0.9}$
3. $8^{-0.62}$
4. $(2.05)^{-2.3}$

5. $5^{\sqrt{2}}$
6. $2^{-\sqrt{5}}$
7. $(\sqrt{3})^{-\pi}$
8. $\pi^{\sqrt{6}}$

9. $(\sqrt{7})^{\sqrt{8}}$
10. $(\sqrt{2})^{-\sqrt{3}}$
11. $\left(\frac{3}{4}\right)^{0.5(3.5)}$
12. $\left(\frac{1}{2}\right)^{0.03(7.2)}$

13. $\left(\frac{1}{2}\right)^{9800/5600}$
14. $\left(\frac{2}{3}\right)^{-27/54}$
15. $2500\left(1 + \frac{0.0825}{365}\right)^{365(8)}$
16. $12,500\left(1 + \frac{0.07}{4}\right)^{4(12)}$

17. $e^{2.7}$
18. $e^{-0.53}$
19. $24,000e^{0.08(12)}$
20. $915,000e^{-0.015(6)}$

21. $20[1 + e^{-0.12(3)}]$
22. $108[1 - e^{0.02(14)}]$
23. $\dfrac{e^2 - e^{-2}}{e^2 + e^{-2}}$
24. $(e^{3.1} + e^{-2.7})^3$

Simplify each expression, as in **EXAMPLE 2:**

25. $3^x \cdot 3^{x+1}$
26. $2^{y+9} \cdot 2^{2y-1}$
27. $\dfrac{e^{5x}}{e^{2x}}$
28. $\dfrac{10^{2t+7}}{10^{t-1}}$

29. $(5^x)^5$
30. $(7^{3t})^4$
31. $(2^x \cdot 2^{3x})^4$
32. $(e^{-t} \cdot e^{3t+4})^5$

33. $(2^x \cdot 3^x)^4$ **34.** $(3^t \cdot 5^t)^8$ **35.** $2^x + 2^{-x}$ **36.** $1 + e^{-2t}$

37. $\dfrac{10^x - 10^{-x}}{10^x + 10^{-x}}$ *Hint:* This becomes a complex fraction.   **38.** $(3^x + 4^x)^2$ *Hint:* $(A + B)^2 = A^2 + 2AB + B^2$

Graph these exponential functions, as in EXAMPLE 3:

39. $f(x) = 3^x$ **40.** $f(x) = 4^x$ **41.** $g(x) = \left(\dfrac{1}{4}\right)^x$ **42.** $F(x) = \left(\dfrac{1}{3}\right)^x$ **43.** $y = 2^{x-1}$

44. $y = 2^{-x}$ **45.** $f(x) = 3^{x+1}$ **46.** $g(x) = 2^{2-x}$ **47.** $f(x) = e^x$ **48.** $g(x) = e^{-x}$

Solve as in EXAMPLES 4, 5, 6, and 7:

49. The population of a bacteria culture is given by the formula

$$P = 500,000(2)^t,$$

where P is the population of the culture after t minutes. Find P when $t = 0, 1, 3,$ 4.5, and 6.2 minutes.

50. The population of an ant colony is given by the formula

$$P = 1000(1.5)^{2t},$$

where P is the number of ants after t days. Find P when $t = 0, 1, 2, 4.5,$ and 9.3 days.

51. The 1986 Chernobyl nuclear plant explosion in the Soviet Union sent masses of radioactive cesium-137 into the atmosphere. Cesium-137's half-life is about 30 years, meaning that it will take hundreds of years for it to disintegrate to a non-radioactive, harmless state. The formula

$$A = 1000\left(\frac{1}{2}\right)^{t/30}$$

gives the amount A remaining (in kilograms) after t years. Find A when $t = 0, 15,$ 30, 45, 100, and 200 years.

52. Radium-214 is so unstable that its half-life is 0.000164 seconds. The formula

$$A = 200(2)^{-t/0.000164}$$

gives the amount A remaining (in grams) after t seconds. Find A when $t = 0,$ 0.000164, 0.000328, 0.001, and 0.01 seconds.

53. The United States population is given by the exponential function

$$P = 150,000,000e^{0.02t},$$

where P is the population t years after the census year 1950, assuming an average growth rate of $2\% = 0.02$ per year. Find the population in the years 1950, 1965, 1970, and 1973. *Hint:* $150,000,000 = 1.5 \times 10^8$ is 1.5 $\boxed{\text{EXP}}$ 8

54. As a result of economic factors and a high crime rate, Detroit, Michigan, has been slowly declining in population since 1960, according to the formula

$$P = 1,600,000e^{-0.015t}.$$

Find the population in 1960, 1975, 1980, and 1986.

Use the formulas

$$A = P\left(1 + \frac{r}{n}\right)^{nt} \quad \text{and} \quad I = A - P$$

to find the accumulated amount A and the interest I for the following compound-interest accounts:

55. $P = \$5000$, $r = 8\% = 0.08$, $n = 4$ (compounded quarterly), $t = 6$ years

56. $P = \$12,500$, $r = 9\frac{1}{2}\%$, $n = 12$ (compounded monthly), $t = 7.5$ years

B

57. $P = \$3600$, $r = 10\frac{1}{4}\%$, $n = 1$ (compounded annually), $t = 5\frac{1}{4}$ years

58. $P = \$150,000$, $r = 12.4\%$, $n = 365$ (compounded daily), $t = 2\frac{1}{2}$ years

Money deposited into an account can be compounded annually ($n = 1$), quarterly ($n = 4$), monthly ($n = 12$), or daily ($n = 365$). It can even be compounded hourly, by the minute, or by the second! Even more often than this, money is said to be *compounded continuously* **if it is compounded "all the time" or "at every instant." The formula**

$$A = Pe^{rt}$$

gives the accumulated amount A in an account compounded continuously at interest rate r for t years. Find the accumulated amount A in each of the following accounts, compounded continuously. Then find the interest, using the formula $I = A - P$.

59. $P = \$10,000$, $r = 7\% = 0.07$, $t = 5$ years

60. $P = \$3750$, $r = 5\frac{1}{2}\%$, $t = 4.5$ years

In an economy undergoing *inflation* **at $r\%$ per year, the formula**

$$C = P(1 + r)^t$$

gives the inflated cost C of an item t years after it originally was purchased at price P.

61. California real estate prices soared dramatically in the 1970s and 1980s, with an average inflation rate of 15% per year in some areas. If a house there sold for $37,500 in 1972, what would be its market value in 1988?

62. If a loaf of bread cost $0.75 in 1960, find the cost in 1970, assuming an inflation rate of 7% per year.

63. The velocity of a sky diver under free-fall conditions (before opening the parachute) is given by

$$v = 120(1 - e^{-1.6t}),$$

where v is the velocity (in miles per hour) t seconds after the sky diver jumps from the plane. Find the velocity at $t = 0$ (the moment of the jump), 1, 2, and 3.5 seconds. Find the "terminal velocity," which is usually attained in about 5 seconds.

64. The current in an electrical circuit is given by

$$I = 10(1 - e^{-2t}),$$

where I is the current (in amperes) t seconds after the power is turned on. Find the current at $t = 0$ (the initial current), 1, and 2 seconds. The maximum, or "steady-state," current is attained at about 2.5 seconds; find it.

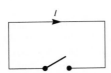

65. In this problem, you will compare the benefits of using different compounding periods to calculate compound interest. Assume $P = \$100$ is deposited at $r = 6\%$ for $t = 1$ year. Using the formulas

$$A = P\left(1 + \frac{r}{n}\right)^{nt}, \quad A = Pe^{rt}, \quad \text{and} \quad I = A - P,$$

find the interest earned when the account is compounded:
a. Annually (This represents simple interest: $n = 1$.)   **b.** Quarterly ($n = 4$)
c. Monthly   **d.** Daily   **e.** Continuously
Your results should show that daily compounding and continuous compounding yield the most interest, followed in order by monthly, quarterly, and annual compounding.

66. When $100 is deposited into an account paying 6% interest per year, compounded monthly for 1 year, the amount and interest are

$$A = \$100\left(1 + \frac{0.06}{12}\right)^{12 \cdot 1} = \$106.17$$

$$I = \$106.17 - \$100 = \$6.17$$

Thus a $100 deposit yields $6.17 interest in 1 year. This represents an **effective interest rate** of 6.17%. Give the effective interest rates for the five accounts in Problem 65.

C

67. $1500 is deposited at 7% interest compounded daily, and $2500 is deposited at $9\frac{1}{2}\%$ interest compounded monthly. After 2 years, both accounts are "rolled over" into an account paying $11\frac{1}{4}\%$ interest compounded continuously for 3 years. Find the final accumulated amount and the interest earned.

68. The **present value** of $20,000 at 8% interest compounded monthly for 4 years is the *principal P* that must be *deposited now*, so that the accumulated amount A becomes $20,000 after 4 years. Find this present value P. Use $A = P\left(1 + \dfrac{r}{n}\right)^{nt}$.

69. The "polynomial/exponential" equation

$$x^2 = 2^x$$

has a solution $x = -0.76667$ approximately, because $(-0.76667)^2 = 2^{-0.76667} = 0.58778$. (Verify this on your calculator.) Its value was obtained by observing that the graphs of $y = x^2$ and $y = 2^x$ intersect somewhere between $x = 0$ and $x = -1$. After repeatedly substituting values of x between 0 and -1 into the equation above, the value $x = -0.76667$ was obtained.

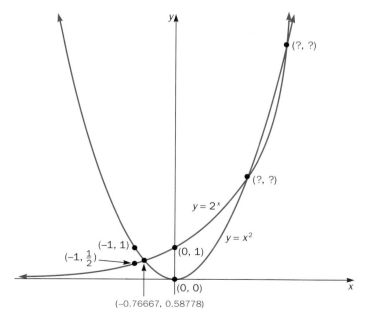

(continued)

Surprisingly, there are two positive-integer solutions of the equation $x^2 = 2^x$ that can be found without the "number crunching" that was needed to find $x = -0.76667$. What are they?

70. In finance, the expression

$$\left(1 + \frac{1}{n}\right)^n$$

is used to compute the accumulated amounts of money, when compounded n times per year, for 1 year. In calculus, it is shown that as n gets very large, the "limiting value" of this expression is $e = 2.71828 \dots$. Calculus students will see this written as

$$\lim_{n \to \infty} \left(1 + \frac{1}{n}\right)^n = 2.71828 \dots = e$$

As an algebra student, you fill in the rest of this table:

n	$\left(1 + \dfrac{1}{n}\right)^n$
1	$2 \longleftarrow \left(1 + \dfrac{1}{1}\right)^1 = (2)^1$
2	$2.25 \longleftarrow \left(1 + \dfrac{1}{2}\right)^2 = (1.5)^2$
3	$2.3703704 \leftarrow$ $\boxed{(}\,\boxed{1}\,\boxed{+}\,\boxed{1}\,\boxed{\div}\,\boxed{3}\,\boxed{)}\,\boxed{x^y}\,\boxed{3}\,\boxed{=}$
5	?
10	?
50	?
100	?
1000	?
100,000	?
1,000,000	?

71. In the text, it was stated that b^x is a real number for all real-number exponents x and all *positive* bases b. Then it was shown that if $b < 0$, then b^x can be imaginary: $(-9)^{1/2} = \sqrt{-9} = 3i$.

a. If $b = 0$, there is one value of x for which b^x is undefined. What is it?

b. For $b > 0$ and x positive, negative, or zero, can b^x ever be zero or negative?

c. In the text, an exponential function was described as

$$f(x) = b^x, \quad b > 0, b \neq 1.$$

Tell why we required that $b \neq 1$.

10.2   Logarithms

The population of a bacteria culture after t seconds is given by

$$P = 10,000(2)^t.$$

When $t = 2$,

$$P = 10,000(2)^2 = 40,000.$$

when $t = 3$,

$$P = 10,000(2)^3 = 80,000.$$

It is natural to ask how we proceed when the population P is 60,000 bacteria. That is,

$$60,000 = 10,000(2)^t, \quad t = ?$$

The preceding equations indicate that $2 < t < 3$, although at this point we have no method of solving this equation for t. Dividing both sides by 10,000 gives $6 = 2^t$, or

$$2^t = 6.$$

The following results were obtained on a calculator:

$$2^{2.5} = 5.6568542 \qquad \text{Too small.}$$
$$2^{2.6} = 6.0628663 \qquad \text{Too big.}$$
$$2^{2.55} = 5.8563428 \qquad \text{Closer.}$$

$$\vdots$$

$$2^{2.585} = 6.0001560 \qquad \text{Close enough.}$$

Finding the approximate solution $t = 2.585$ by trial-and-error "number crunching" certainly illustrates the *need for a more direct method* of solving the "exponential equation" $2^t = 6$.* This method uses **logarithms**.

Recall from Section 9.4 that the inverse of a relation is obtained by interchanging x and y. Thus

$$y = 2^x \quad \text{has its inverse} \quad x = 2^y.$$

Unfortunately, we have no direct method for isolating y in the equation $x = 2^y$, so we simply give y a name. We say that

$$y = \log_2 x \quad \text{is the solution of} \quad x = 2^y,$$

---

*Compare the difficulty of solving $2^t = 6$ by calculator to the ease of solving $t^2 = 6$ by extracting roots: $t = \pm \sqrt{6} = 2.4494897$.

and call the solution y "the logarithm, in base 2, of x." Let us graph these inverse relations on the same set of axes, using the "mirror image" concept:

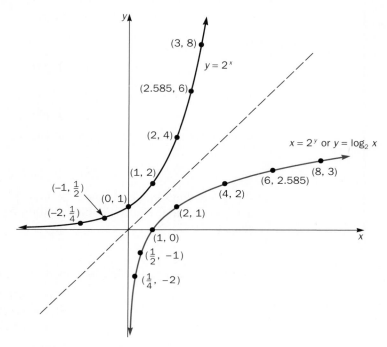

This inverse relation

$$x = 2^y \quad \text{or} \quad y = \log_2 x$$

is a function, because it passes the vertical line test. We see that its

$$\text{domain} = \{x \mid x > 0\},$$

because the graph extends horizontally to the right of the y-axis. Its

$$\text{range} = \{y \mid y \text{ is any real number}\},$$

because the graph extends vertically upward and downward. The point (8,3) on the graph of $x = 2^y$, which is equivalent to $y = \log_2 x$, means that

$$8 = 2^3 \quad \text{is equivalent to} \quad 3 = \log_2 8.$$

Likewise, the points (6, 2.585) and $\left(\frac{1}{2}, -1\right)$ on this graph mean, respectively, that

$$6 = 2^{2.585} \quad \text{is equivalent to} \quad 2.585 = \log_2 6$$

and $\qquad \dfrac{1}{2} = 2^{-1}$ is equivalent to $-1 = \log_2 \dfrac{1}{2}$

The second statement suggests that the solution to the exponential equation $2^t = 6$, which is $t = 2.585 = \log_2 6$, can be obtained directly by using logarithms. We shall pursue this idea in Section 10.4.

Instead of base 2, we can use any positive number $b \neq 1$ as the base of a logarithm, which we now define:

Definition of a Logarithm

$$\log_b x = y \quad \text{provided} \quad b^y = x, \quad b, x > 0, b \neq 1.$$

The definition means that

"the *logarithm* of a positive number *x is the exponent y*
to which the base *b* must be raised in order to equal *x*."

Thus "logarithm" is a fancy name for "exponent." For example,

"the logarithm of 25, in base 5, is 2" because $5^2 = 25$.

This is written

$$\log_5 25 = 2 \quad \text{because} \quad 5^2 = 25.$$

Likewise,

$$\log_4 64 = 3 \quad \text{because} \quad 4^3 = 64.$$

The definition requires that the base b be positive. As stated in Section 10.1, a negative base can lead to an imaginary number, such as $(-9)^{1/2} = 3i$, which is undesirable for our purposes. Thus the exponential function $y = b^x$ must have a positive base b, and so, therefore, must its inverse function $y = \log_b x$. The definition also requires that x be positive. This is because the domain of the function $y = \log_b x$ consists only of $x > 0$. Algebraically, a positive base b raised to any power y (positive, negative, or zero) is always a positive number x. Thus $9^2 = 81$, $9^{-1} = \dfrac{1}{9}$, and $9^0 = 1$ are all positive.

In addition to solving exponential equations such as $2^t = 6$, we will use logarithms to estimate the age of fossils, to find the pH of orange juice, to compute the magnitudes of earthquakes, and to find out how loud a rock concert is.

Let us now give examples that relate the logarithmic and exponential forms of a power.

EXAMPLE 1 The table gives equivalent logarithmic and exponential forms of a power:

Logarithmic	Exponential
$\log_b x = y$	$b^y = x$
$\log_2 8 = 3$	$2^3 = 8$
$\log_3 81 = 4$	$3^4 = 81$
$\log_7 7 = 1$	$7^1 = 7$
$\log_6 1 = 0$	$6^0 = 1$
$\log_5 \dfrac{1}{25} = -2$	$5^{-2} = \dfrac{1}{25}$
$\log_9 3 = \dfrac{1}{2}$	$9^{1/2} = 3$
$\log_6 \sqrt{6} = \dfrac{1}{2}$	$6^{1/2} = \sqrt{6}$
$\log_8 4 = \dfrac{2}{3}$	$8^{2/3} = (\sqrt[3]{8})^2 = 4$
$\log_{16} \dfrac{1}{8} = \dfrac{-3}{4}$	$16^{-3/4} = (\sqrt[4]{16})^{-3} = \dfrac{1}{8}$
$\log_{1/2} \dfrac{1}{4} = 2$	$\left(\dfrac{1}{2}\right)^2 = \dfrac{1}{4}$
$\log_9 (-81)$ is undefined	$9^y \neq -81;$ must have $x > 0$
$\log_7 0$ is undefined	$7^y \neq 0;$ must have $x > 0$
$\log_{-2} 5$ is undefined	$(-2)^y \neq 5;$ must have $b > 0$

When we know any two variables in $\log_b x = y$, we can find the third by first converting to exponential mode. The following examples illustrate this and provide an excellent opportuntiy for you to review the properties of exponents.

EXAMPLE 2 Solve for x:

a. $\log_2 x = 5$ This is equivalent to
$$2^5 = x, \quad \text{or} \quad x = 32.$$

b. $\log_{10} x = -3$ This is equivalent to
$$10^{-3} = x, \quad \text{or} \quad x = \frac{1}{1000}.$$

c. $\log_{25} x = \dfrac{3}{2}$ This is equivalent to
$$25^{3/2} = x, \quad \text{or} \quad x = (\sqrt{25})^3 = 5^3 = 125.$$

EXAMPLE 3 | Solve for b:

a. $\log_b 49 = 2$ This is equivalent to $b^2 = 49$. Raising both sides to the $\frac{1}{2}$ power gives

$$(b^2)^{1/2} = (49)^{1/2}$$
$$b^{2/2} = \sqrt{49}, \quad \text{or} \quad b = 7.$$

Note that $b = -7$ also satisfies $b^2 = 49$, but a negative base of a logarithm is not allowed.

b. $\log_b 8 = \frac{3}{4}$ This is equivalent to $b^{3/4} = 8$. Raising both sides to the $\frac{4}{3}$ power gives

$$(b^{3/4})^{4/3} = (8)^{4/3}$$
$$b^{12/12} = (\sqrt[3]{8})^4$$
$$b^1 = (2)^4, \quad \text{or} \quad b = 16.$$

c. $\log_b 25 = \frac{-2}{3}$ In exponential form, this is $b^{-2/3} = 25$. Raising both sides to the $-\frac{3}{2}$ power gives

$$(b^{-2/3})^{-3/2} = (25)^{-3/2}$$
$$b^{6/6} = (\sqrt{25})^{-3}$$
$$b^1 = (5)^{-3}, \quad \text{or} \quad b = \frac{1}{125}.$$

In the next example, we use the fact that if $b^y = b^z$, then $y = z$; in other words, if two powers with the same base are equal, then the exponents must also be equal.

EXAMPLE 4 | Solve for y:

a. $\log_5 125 = y$ This is equivalent to

$$5^y = 125 = 5^3; \quad \text{thus} \quad y = 3.$$

We expressed 125 as the power 5^3.

b. $\log_{10} \dfrac{1}{10,000} = y$ This is equivalent to

$$10^y = \frac{1}{10,000} = 10^{-4}; \quad \text{thus} \quad y = -4.$$

(continued)

c. $\log_{27} 81 = y$ This is equivalent to

$$27^y = 81.$$

We now express both 27 and 81 as powers of their least common base 3:

$$(3^3)^y = 3^4, \quad \text{or} \quad 3^{3y} = 3^4.$$

Equating exponents gives

$$3y = 4, \quad \text{or} \quad y = \frac{4}{3}.$$

Check: $27^{4/3} = (\sqrt[3]{27})^4 = 3^4 \overset{\checkmark}{=} 81.$

Logarithms were invented in the seventeenth century by John Napier and Henry Briggs as a computational tool. Multiplying numbers was done by adding their logarithms, dividing was done by subtracting logs, and powers and roots were obtained by multiplying and dividing logs, respectively. Each of these operations was reduced to a "simpler" operation by the use of base-10 logarithms. Today calculators have replaced logarithms, but those still interested should consult Selected Topic G: "Calculating with Logarithms" in the Appendix. The word *logarithm* comes from the Greek: *logos + arithmos*. *Logos* means "the study of" or "reckoning," and *arithmos* means "numbers." Hence logarithms were used for number-reckoning, or calculating.

Logarithms in base 10 are called **common logarithms** and are written without the base:

$$\log x \quad \text{means} \quad \log_{10} x$$

The following common logs are direct consequences of the definition of a logarithm:

Common log	Exponential form
$\log_{10} 1000 = \log 1000 = 3$	$10^3 = 1000$
$\log 100 = 2$	$10^2 = 100$
$\log 10 = 1$	$10^1 = 10$
$\log 1 = 0$	$10^0 = 1$
$\log 0.1 = -1$	$10^{-1} = \dfrac{1}{10} = 0.1$
$\log 0.01 = -2$	$10^{-2} = 0.01$
$\log 0.001 = -3$	$10^{-3} = 0.001$

The common log of any positive number can be found on the Table of Common Logarithms in the Appendix. To interpret this table, you should first consult Selected Topic F: "Common Logarithms." It is far more convenient to obtain the common log of a number on a scientific calculator: simply enter the number and then press the log button. For example,

$$\log 7.64 = 0.8830934 \qquad 7.64 \boxed{\text{log}}$$

$$\log 43.2 = 1.6354837 \qquad 43.2 \boxed{\text{log}}$$

$$\log 0.23 = -0.6382722 \qquad .23 \boxed{\text{log}}$$

$$\log 0.0000000057 = \log 5.7 \times 10^{-9} = -8.2441251 \qquad 5.7 \boxed{\text{EXP}} \; 9 \boxed{+/-} \; \boxed{\text{log}}$$

These are approximations; the exact values of these logs are unending decimals. In exponential form, the first two are

$$7.64 = 10^{0.8830934} \qquad .8830934 \boxed{\text{INV}} \boxed{\text{log}}$$

$$43.2 = 10^{1.6354837} \qquad 1.6354837 \boxed{\text{INV}} \boxed{\text{log}}$$

To see that these logs make sense, we show that they have the correct "order of magnitude" compared to the exact values displayed above:

$$\log 1000 = 3$$
$$\log 100 = 2$$
$$\log 43.2 = 1.6355$$
$$\log 10 = 1$$
$$\log 7.64 = 0.8831$$
$$\log 1 = 0$$
$$\log 0.23 = -0.6383$$
$$\log 0.1 = -1$$
$$\log 0.0447 = -1.3497$$

Increasing logs

Decreasing logs

Chemists use common logarithms to measure the activity of a solution, which is determined by its concentration of the hydrogen ion H^+. This concentration is written $[H^+]$ and is measured in "moles per liter." (You need not know what a "mole" is to understand the discussion.) The **pH** (hydrogen potential) of a chemical solution is defined logarithmically by

$$pH = -\log [H^+]$$

EXAMPLE 5

Find the pH of the following substances:

a. Pure water: $[H^+] = 0.0000001$ moles per liter

$$pH = -\log 0.0000001 = 7 \qquad 0.0000001 \boxed{\text{log}} \boxed{+/-}$$

(continued)

b. Orange juice: $[H^+] = 0.0000052$

$$pH = -\log 0.0000052 = 5.28 \text{ (rounded off)}$$

Solutions with pH < 7 are called **acidic;**
thus orange juice is acidic.

c. Hand soap: $[H^+] = 0.00000000871$

$$pH = -\log 8.71 \times 10^{-9} = 8.06 \qquad 8.71 \boxed{\text{EXP}}\; 9 \boxed{+/-}\; \boxed{\log}\; \boxed{+/-}$$

Solutions with pH > 7 are called **basic** or
alkaline; thus hand soap is basic.

The number $e = 2.71828 \ldots$ was used in Section 10.1 as the base of exponential functions describing population growth. Logarithms in base e are called **natural logs** and are used extensively in the natural sciences and in calculus. Like common logs, natural logs are written without the base:

$$\ln x \quad \text{means} \quad \log_e x$$

To find the natural log of a number on a scientific calculator, enter the number and then press the $\boxed{\text{ln}}$ button. For example,

$$\ln 9.86 = 2.2884862 \qquad 9.86 \boxed{\text{ln}}$$

In exponential form, this means

$$9.86 = e^{2.2884862} \qquad 2.2884862 \boxed{\text{INV}}\boxed{\text{ln}}$$

EXAMPLE 6

Suppose the population of a city was P_0 in its census year. If the population, growing at an annual rate of r, becomes P after t years, then

$$t = \frac{\ln P - \ln P_0}{r}.$$

A city had a population of $P_0 = 25{,}000$ in 1965. Assuming an annual growth rate of $r = 8\%$, find the time t it will take for the population to become $P = 60{,}000$. In what year will this occur?

Solution
By direct substitution, we get

$$t = \frac{\ln 60{,}000 - \ln 25{,}000}{0.08}$$

$$= 10.94 \text{ years.} \qquad 60{,}000 \boxed{\text{ln}}\; \boxed{-}\; 25{,}000 \boxed{\text{ln}}\; \boxed{=}\; \boxed{\div}\; .08 \boxed{=}$$

The population will become 60,000 in the year

$$1965 + 10.94 = 1976 \text{ (approximately)}.$$

EXERCISE 10.2 *Answers, pages A96–A97*

A

Convert to exponential form, as in the table of EXAMPLE 1:

1. $\log_4 64 = 3$

2. $\log_8 8 = 1$

3. $\log_6 1 = 0$

4. $\log_3 \dfrac{1}{9} = -2$

5. $\log_{1/2} 16 = -4$

6. $\log_{125} 25 = \dfrac{2}{3}$

7. $\log 100{,}000 = 5$

8. $\ln 7 = 1.9459101$

Convert to logarithmic form, as in the table of EXAMPLE 1:

9. $3^2 = 9$

10. $13^0 = 1$

11. $2^{-5} = \dfrac{1}{32}$

12. $36^{1/2} = 6$

13. $\left(\dfrac{1}{2}\right)^6 = \dfrac{1}{64}$

14. $12^{1/2} = 2\sqrt{3}$

15. $64^{2/3} = 16$

16. $\left(\dfrac{8}{27}\right)^{-1/3} = \dfrac{3}{2}$

17. $e^{3.02} = 20.491$

18. $10^{-4.6} = 0.0000251$

Solve for x as in EXAMPLE 2:

19. $\log_3 x = 4$

20. $\log_4 x = 2$

21. $\log_6 x = 0$

22. $\log_6 x = 1$

23. $\log_2 x = -3$

24. $\log_9 x = \dfrac{1}{2}$

25. $\log_{20} x = \dfrac{1}{2}$

26. $\log_{27} x = \dfrac{2}{3}$

27. $\log_{64} x = \dfrac{-1}{3}$

28. $\log_{125} x = -\dfrac{2}{3}$

29. $\log x = 4$

30. $\ln x = 6.31$

Solve for b as in EXAMPLE 3:

31. $\log_b 25 = 2$

32. $\log_b 216 = 3$

33. $\log_b 4 = \dfrac{2}{3}$

34. $\log_b 3 = \dfrac{1}{2}$

35. $\log_b \dfrac{1}{8} = 3$

36. $\log_b 49 = -2$

37. $\log_b 16 = \dfrac{2}{3}$

38. $\log_b \dfrac{1}{27} = \dfrac{-3}{4}$

Solve for y as in EXAMPLE 4:

39. $\log_2 16 = y$

40. $\log_4 64 = y$

41. $\log_3 3 = y$

42. $\log_5 1 = y$

43. $\log_5 \dfrac{1}{125} = y$

44. $\log 10{,}000 = y$

45. $\log_9 3 = y$

46. $\log_{64} 4 = y$

47. $\log_8 16 = y$

48. $\log_9 27 = y$

49. $\log_{125} 25 = y$

50. $\log_{32} 8 = y$

51. $\log_4 \dfrac{1}{2} = y$

52. $\log_{25} \dfrac{1}{125} = y$

Find the pH of the following substances, using the formula

$$pH = -\log\ [H^+],$$

where $[H^+]$ is given in moles per liter. Classify the substances as acidic or basic. See **EXAMPLE 5.**

53. Lemonade: $[H^+] = 0.0000072$

54. Vinegar: $[H^+] = 0.0000195$

55. Human blood: $[H^+] = 0.000000039$

56. NaOH: $[H^+] = 0.00000000056$

57. Baking soda: $[H^+] = 0.0000000092$

58. H_2SO_4: $[H^+] = 0.0000561$

When a population increases from P_0 to P at the annual rate r, the time t that it takes for this increase to occur is given by the formula

$$t = \frac{\ln P - \ln P_0}{r}.$$

In Problems 59 and 60, find the time t, and then find the year in which P occurs. See **EXAMPLE 6.**

59. $P_0 = 30,000$ in 1950, $P = 75,000$, $r = 6\% = 0.06$

60. $P_0 = 150,000,000$ in 1960, $P = 225,000,000$, $r = 3\frac{1}{2}\%$

B

Solve for x, b, or y:

61. $\log_5 x = \frac{3}{2}$

62. $\log_{12} x = \frac{3}{2}$

63. $\log_{1/2} x = 3$

64. $\log_{3/4} x = -2$

65. $\log_{4/9} x = \frac{3}{2}$

66. $\log_{16/25} x = -\frac{1}{2}$

67. $\log_b \frac{4}{9} = 2$

68. $\log_b \frac{4}{3} = -\frac{1}{2}$

69. $\log_b \frac{9}{4} = -2$

70. $\log_b 125 = -3$

71. $\log_4 \frac{1}{8} = y$

72. $\log_5 \sqrt{5} = y$

73. $\log_{4/9} \frac{8}{27} = y$

74. $\log_8 2\sqrt{2} = y$

75. $\log 0.0000001 = y$

76. $\log_{1/64} 32 = y$

The *magnitude M* of an earthquake is measured logarithmically on the *Richter Scale*, according to the formula

$$M = \log \frac{E}{E_0},$$

where E is the energy (in ergs) released by the earthquake being measured, and E_0 is the energy of the smallest measurable quake used for comparison. Find the magnitude M of the earthquake with the given ratio $\frac{E}{E_0}$:

77. 1971 Los Angeles earthquake: $\frac{E}{E_0} = 3,160,000$

78. 1906 San Francisco quake: $\frac{E}{E_0} = 178{,}000{,}000$

79. 1964 Alaska quake: $\frac{E}{E_0} = 795{,}000{,}000$

80. 1985 Mexico City quakes: $\frac{E}{E_0} = 1.26 \times 10^8$ and 3.16×10^7

81. 1988 Armenia earthquake: $\frac{E}{E_0} = 7{,}950{,}000$

82. How much more powerful is an $M = 5$ earthquake than an $M = 4$ quake?

$$Answer:\ 10^{5-4} = 10^1 = 10 \text{ times as powerful}$$

In like manner, compare each of these pairs of quakes:
a. $M = 7$ vs $M = 5$
b. $M = 6$ vs $M = 2$
c. $M = 8.5$ vs $M = 6.1$

The *age of a fossil* is given by the formula

$$t = \frac{5600 \log R}{\log 0.5},$$

where $R = $ the ratio of carbon-14 to carbon-12 in the fossil. **Find the age t of each of the following fossils, whose values of R are given:**

83. $R = 25\% = 0.25$ **84.** $R = 15\%$

When a principal of P is deposited into an account compounded yearly at an interest rate r, the *time t* for the accumulated amount to be A is given by the formula

$$t = \frac{\ln \dfrac{A}{P}}{\ln (1 + r)}.$$

Find the time t when

85. $P = \$10{,}000$, $A = \$25{,}000$, $r = 7\% = 0.07$

86. $P = \$27{,}500$, $A = \$55{,}000$, $r = 11\frac{1}{2}\%$

Loudness of sound is measured in *decibels* and is given by the formula

$$db = 10 \log \frac{P}{P_0},$$

where P is the power (in watts) of the sound being measured, and P_0 is the power of sound at the lowest threshold of hearing. As in the Richter Scale, loudness is a relative measurement, not an absolute one. Find the number of decibels db for the following sounds:

87. A classical music concert: $\frac{P}{P_0} = 81{,}000{,}000$

88. A heavy-metal concert: $\frac{P}{P_0} = 263{,}000{,}000$

89. Sound of a jet plane: $\dfrac{P}{P_0} = 1 \times 10^{11}$

90. Conversation at a party: $\dfrac{P}{P_0} = 25,100,000$

91. Conversation at a fraternity party: $\dfrac{P}{P_0} = 360,000,000$

92. A barking dog: $\dfrac{P}{P_0} = 4.56 \times 10^8$

93. A barking tree spider: $\dfrac{P}{P_0} = 2.7 \times 10^5$

94. The lowest threshold of hearing: $\dfrac{P}{P_0} = 1$

We can simplify the "composition of logs" expression $\log_3(\log_2 8)$ **by first evaluating the inside:**

$$\begin{aligned} \log_3(\log_2 8) &= \log_3(3) & \log_2 8 = 3 \quad \text{Why?}\\ &= 1. & \log_3 3 = 1 \quad \text{Why?} \end{aligned}$$

Simplify these expressions by starting with the innermost log:

95. $\log_4(\log_3 81)$ **96.** $\log_5(\log_2 32)$ **97.** $\log_7(\log_6 6)$

98. $\log_2(\log_3 81)$ **99.** $\log_2(\log_2 256)$ **100.** $\log_5[\log_3(\log_5 125)]$

10.3 Properties of Logarithms

In the previous section, we defined the inverse of the exponential function $f(x) = b^x$ to be the logarithmic function $f^{-1}(x) = \log_b x$. According to the definition of inverse function given in Section 9.4,

$$f[f^{-1}(x)] = x \quad \text{and} \quad f^{-1}[f(x)] = x,$$

which means that

$$b^{[\log_b x]} = x \quad \text{and} \quad \log_b[b^x] = x.$$

In other words, obtaining the logarithm of a number x and then "exponentiating" the result give the original number x. Likewise, exponentiating a number x and then obtaining the logarithm of the result give the original number x. For example,

$$2^{\log_2 8} = 8 \quad \text{and} \quad \log_2 2^3 = 3$$

Check:

$$\begin{aligned} 2^{\log_2 8} &\overset{\checkmark}{=} 2^3 & \log_2 2^3 &\overset{\checkmark}{=} \log_2 8\\ &\overset{\checkmark}{=} 8. & &\overset{\checkmark}{=} 3. \end{aligned}$$

These are our first two properties of this section:

> **Inverse Properties of Logarithms***
> $$b^{\log_b x} = x \quad \text{and} \quad \log_b b^x = x$$

EXAMPLE 1

a. $3^{\log_3 9} = 9$ $3^{\log_3 9} = 3^2 \overset{\checkmark}{=} 9$

b. $10^{\log 4.72} = 10^{\log_{10} 4.72}$
$$= 4.72$$
        4.72 $\boxed{\log}$ $\boxed{\text{INV}}$ $\boxed{\log}$ $\boxed{=}$ 4.72

c. $e^{\ln 0.3} = e^{\log_e 0.3}$
$$= 0.3$$
        0.3 $\boxed{\ln}$ $\boxed{\text{INV}}$ $\boxed{\ln}$ $\boxed{=}$ 0.3

d. $\log_5 125 = \log_5 5^3$
$$= 3$$
        $\log_5 5^3 = 3$ because $5^3 \overset{\checkmark}{=} 125!$

e. $\log 10{,}000 = \log_{10} 10^4$
$$= 4$$
        4 $\boxed{\text{INV}}$ $\boxed{\log}$ $\boxed{\log}$ $\boxed{=}$ 4

f. $\log 0.001 = \log_{10} 10^{-3}$
$$= -3$$

g. $\ln e^2 = \log_e e^2$
$$= 2$$
        2 $\boxed{\text{INV}}$ $\boxed{\ln}$ $\boxed{\ln}$ $\boxed{=}$ 2

h. $\log_6 \sqrt{6} = \log_6 6^{1/2}$
$$= \frac{1}{2}$$

i. $\log_b b = \log_b b^1$
$$= 1$$

j. $\log_b 1 = \log_b b^0$
$$= 0$$
        $\log_b 1 = 0$ because $b^0 \overset{\checkmark}{=} 1$

The first of these properties, $b^{\log_b x} = x,$ is valid for positive numbers x and y, as well as for their product: $b^{\log_b xy} = xy.$ But we can write this as

$$b^{\log_b xy} = xy = (b^{\log_b x})(b^{\log_b y}) = b^{\log_b x + \log_b y} \qquad \text{Add exponents.}$$

Equating the exponent of the first power with the exponent of the last power gives

$$\log_b xy = \log_b x + \log_b y.$$

---

*See Problems 75 and 76 in the exercises for other proofs of these properties.

This result means that "the logarithm of a product of two *positive* numbers is the sum of their logarithms."

Product Rule for Logarithms

$$\log_b xy = \log_b x + \log_b y.$$

For example,

$$\log_2 8 \cdot 16 = \log_2 8 + \log_2 16$$

Check:
$$\log_2 128 = \quad 3 \quad + \quad 4 \qquad 2^3 = 8, \quad 2^4 = 16$$
$$7 \overset{\checkmark}{=} 7 \qquad\qquad 2^7 = 128$$

The Product Rule can be extended to the product of three or more factors:

$$\log_b xyz \ldots = \log_b x + \log_b y + \log_b z + \cdots.$$

The Inverse Property is also valid for the quotient of positive numbers x and y: $b^{\log_b x/y} = \dfrac{x}{y}$. But this can be written as

$$b^{\log_b x/y} = \frac{x}{y} = \frac{b^{\log_b x}}{b^{\log_b y}} = b^{\log_b x - \log_b y} \qquad \text{Subtract exponents.}$$

Equating the first exponent with the last exponent gives

$$\log_b \frac{x}{y} = \log_b x - \log_b y.$$

This says that "the logarithm of a quotient of two positive numbers is the difference between their logarithms."

Quotient Rule for Logarithms

$$\log_b \frac{x}{y} = \log_b x - \log_b y$$

For example,

$$\log_3 \frac{81}{27} = \log_3 81 - \log_3 27$$

Check:
$$\log_3 3 = \quad 4 \quad - \quad 3 \qquad 3^4 = 81, \quad 3^3 = 27$$
$$1 \overset{\checkmark}{=} 1 \qquad\qquad 3^1 = 3$$

The Inverse Property is also valid for any power x^p of a positive number x: $b^{\log_b x^p} = x^p$. But this can be written as

$$b^{\log_b x^p} = x^p = (b^{\log_b x})^p = b^{p \log_b x} \qquad \text{Multiply exponents.}$$

Again, equating the first exponent with the last gives

$$\log_b x^p = p \log_b x$$

In words, "the logarithm of a positive number raised to *any power* equals the power times the logarithm of the number."

Power Rule for Logarithms

$$\log_b x^p = p \log_b x$$

For example,

$$\log_2 4^3 = 3 \log_2 4 \qquad\qquad \log_3 81^{1/2} = \frac{1}{2} \log_3 81$$

Check: $\qquad \log_2 64 = 3 \cdot 2 \qquad\qquad \log_3 \sqrt{81} = \frac{1}{2} \cdot 4$

$$6 \overset{\checkmark}{=} 6 \qquad\qquad\qquad \log_3 9 \overset{\checkmark}{=} 2$$

We now show how these properties can be used to express the logarithm of an expression as a combination of simpler logs.

EXAMPLE 2

Write each log as an algebraic combination of logs:

a. $\log_5 xy^2z^3 = \log_5 x + \log_5 y^2 + \log_5 z^3$ Product Rule
$$= \log_5 x + 2 \log_5 y + 3 \log_5 z. \qquad \text{Power Rule}$$

b. $\log \dfrac{4mv^2}{r^3} = \log 4mv^2 - \log r^3$ Quotient Rule

$$= \log 4 + \log m + \log v^2 - \log r^3 \qquad \text{Product Rule}$$
$$= \log 4 + \log m + 2 \log v - 3 \log r. \qquad \text{Power Rule}$$

c. $\ln \dfrac{2\pi\sqrt{x+1}}{3\sqrt[3]{x^2}} = \ln \dfrac{2\pi(x+1)^{1/2}}{3x^{2/3}}$ Express radicals. using exponents

$$= \ln 2\pi(x+1)^{1/2} - \ln 3x^{2/3} \qquad \text{Quotient Rule}$$
$$= \ln 2 + \ln \pi + \ln (x+1)^{1/2}$$
$$\qquad\qquad - (\ln 3 + \ln x^{2/3}) \qquad \text{Product Rule}$$

$$= \ln 2 + \ln \pi + \frac{1}{2} \ln (x+1)$$

$$\qquad\qquad\qquad - \ln 3 - \frac{2}{3} \ln x. \qquad \text{Power Rule}$$

WARNING! $\ln (x + 1) \neq \ln x + \ln 1$. In general,

$$\log_b (x + y) \neq \log_b x + \log_b y = \log_b xy. \qquad \textit{(continued)}$$

d. $\log_b \sqrt{\dfrac{3x^5}{y^3}} = \log_b \left(\dfrac{3x^5}{y^3}\right)^{1/2}$ Express radical as a power.

$\qquad\qquad = \dfrac{1}{2} \log_b \dfrac{3x^5}{y^3}$ Power Rule

$\qquad\qquad = \dfrac{1}{2} (\log_b 3x^5 - \log_b y^3)$ Quotient Rule

$\qquad\qquad = \dfrac{1}{2}(\log_b 3 + \log_b x^5 - \log_b y^3)$ Product Rule

$\qquad\qquad = \dfrac{1}{2} (\log_b 3 + 5 \log_b x - 3 \log_b y).$ Power Rule

The foregoing steps are reversible, as we now show in illustrating how to convert a combination of logs into a single logarithm or as few logarithms as possible:

EXAMPLE 3   Write with as few logs as possible:

a. $2 \log_3 x + \log_3 y - 5 \log_3 z = \log_3 x^2 + \log_3 y - \log_3 z^5$ Power Rule

$\qquad\qquad\qquad = \log_3 x^2y \quad - \log_3 z^5$ Product Rule

$\qquad\qquad\qquad = \log_3 \dfrac{x^2y}{z^5}.$ Quotient Rule

b. $4 \log x + 3 \log (x + 1) - \dfrac{1}{2} \log (x - 1)$

$\qquad\qquad = \log x^4 + \log (x + 1)^3 - \log (x - 1)^{1/2}$ Power Rule

$\qquad\qquad = \log x^4(x + 1)^3 \quad - \log (x - 1)^{1/2}$ Product Rule

$\qquad\qquad = \log \dfrac{x^4(x + 1)^3}{(x - 1)^{1/2}} \quad \text{or} \quad \log \dfrac{x^4(x + 1)^3}{\sqrt{x - 1}}.$ Quotient Rule

c. $2^{3\log_2 5} = 2^{\log_2 5^3}$ Power Rule

$\qquad = 5^3 \quad \text{or} \quad 125.$ Inverse Property

d. $\dfrac{\ln 6 + \ln 3}{2 \ln 6 - \ln 3} = \dfrac{\ln 6 + \ln 3}{\ln 6^2 - \ln 3}$ Power Rule

$\qquad\qquad = \dfrac{\ln 6 \cdot 3}{\ln \dfrac{6^2}{3}}$ Product Rule / Quotient Rule

$\qquad\qquad = \dfrac{\ln 18}{\ln 12}.$ WARNING! $\dfrac{\ln 18}{\ln 12} \neq \ln \dfrac{18}{12} = \ln 18 - \ln 12$

A **logarithmic equation** contains a variable in at least one logarithm. In the next example, we first express one side of the equation as a single logarithm, as above. We then use the fact that if $\log_b Y = \log_b Z$, then $Y = Z$.

EXAMPLE 4

Solve these logarithmic equations:

a. $2 \log_5 3 + \log_5 (x - 2) = \log_5 18$.

$\qquad \log_5 3^2(x - 2) = \log_5 18 \qquad$ Write left side as single log.

$\qquad\qquad 3^2(x - 2) = \qquad 18 \qquad$ If $\log_5 Y = \log_5 Z$, then $Y = Z$.

$\qquad\qquad\quad 9x - 18 = \qquad 18$

$\qquad\qquad\qquad\quad 9x = 36$

$\qquad\qquad\qquad\quad\ x = 4. \qquad\qquad$ Solution

b. $2 \log x - \log (x + 4) = \log 2$.

$\qquad\qquad \log \dfrac{x^2}{x + 4} = \log 2 \qquad$ Write left side as single log.

$\qquad\qquad\quad \dfrac{x^2}{x + 4} = 2 \qquad$ If $\log Y = \log Z$, then $Y = Z$.

$\qquad\qquad\qquad x^2 = 2(x + 4)$

$\qquad\quad x^2 - 2x - 8 = 0$

$\qquad (x + 2)(x - 4) = 0 \qquad$ Solve algebraically.

$\qquad\ x = -2, \quad x = 4.$

Check $x = -2$: $\qquad 2 \underbrace{\log (-2)}_{\text{Undefined}} - \log (-2 + 4) = \log 2$

Because this log is undefined, we must reject $x = -2$.

Check $x = 4$: $\qquad 2 \log 4 - \log(4 + 4) = \log 2$

Both of these logs are defined.

Answer: $x = 4$ ($x = -2$ is extraneous).

In the next example, we first express one side of the equation as a single logarithm and then convert from logarithmic to exponential form.

EXAMPLE 5

Solve these logarithmic equations:

a. $\log_6 x + \log_6 (x + 5) = 2$.

$\qquad\qquad \log_6 x(x + 5) = 2 \qquad$ Write left side as single log.

$\qquad\qquad\quad x(x + 5) = 6^2 \qquad$ Write in exponential form.

$\qquad\quad x^2 + 5x - 36 = 0$

$\qquad (x + 9)(x - 4) = 0 \qquad$ Solve algebraically.

$\qquad\ x = -9, \quad x = 4.$

We reject $x = -9$ because $\log_6(-9) + \log_6(-9 + 5)$ is *undefined*.

Answer: $x = 4$ ($x = -9$ is extraneous). $\qquad\qquad$ *(continued)*

b. $2 \log 3 + \log x = 1 + \log (x - 1)$.

$$2 \log 3 + \log x - \log (x - 1) = 1 \qquad \text{All logs to left side.}$$

$$\log \frac{3^2 x}{x - 1} = 1 \qquad \text{Write as single log.}$$

$$\frac{3^2 x}{x - 1} = 10^1 \qquad \begin{array}{l} \log Z = 1 \text{ means } \log_{10} Z = 1, \\ \qquad \text{or} \qquad Z = 10^1. \end{array}$$

$$9x = 10(x - 1)$$
$$9x = 10x - 10$$
$$-x = \qquad - 10$$
$$x = 10.$$

Check:
All terms in $2 \log 3 + \log 10 = 1 - \log (10 - 1)$ *are defined.*
Answer: $x = 10$.

EXERCISE 10.3   *Answers, page A97*

Simplify as in EXAMPLE 1, using the Inverse Properties of Logs:

1. $7^{\log_7 49}$ **2.** $5^{\log_5 25}$ **3.** $10^{\log 0.51}$ **4.** $10^{\log 100}$ **5.** $e^{\ln 7}$

6. $e^{\ln 8.30}$ **7.** $b^{\log_b x^2}$ **8.** $b^{\log_b 1}$ **9.** $\log_6 36$ **10.** $\log_3 81$

11. $\log 1000$ **12.** $\log 0.00001$ **13.** $\ln e^{2.3}$ **14.** $\ln e^{-0.16}$ **15.** $\log_5 \sqrt[5]{5}$

16. $\log_7 \sqrt{7}$ **17.** $\log_b b^3$ **18.** $\log_b \sqrt[4]{b}$

Write each log as an algebraic combination of logs, as in EXAMPLE 2:

19. $\log_6 5x^2 y$ **20.** $\log_7 3xy^2$ **21.** $\log \dfrac{mv^2}{2}$ **22.** $\ln \dfrac{4\pi r^3}{3}$ **23.** $\log_b \dfrac{x^2(x + 1)^3}{(x - 1)^4}$

24. $\log_a \dfrac{2x(x - a)^2}{(x + a)^3}$ **25.** $\log_5 \dfrac{3x^3 \sqrt{t}}{2\sqrt[3]{s}}$ **26.** $\log \dfrac{8\sqrt[4]{x^3}}{3\sqrt[3]{y^2}}$ **27.** $\ln \sqrt{\dfrac{7x^3}{y^5}}$ **28.** $\log \sqrt[3]{\dfrac{2b^4}{c^2}}$

Write with as few logs as possible, as in EXAMPLE 3:

29. $\log_3 7 + \log_3 x - \log_3 y$ **30.** $\log_4 6 + \log_4 u - \log_4 v$ **31.** $2 \log 4 + 5 \log x - \dfrac{1}{2} \log y$

32. $3 \ln t + \dfrac{1}{3} \ln x + \dfrac{1}{2} \ln c$ **33.** $\log_b (x^2 + 1) - 2 \log_b (x - 1)$ **34.** $2 \log_3 (x + y) - \dfrac{1}{4} \log_3 (x^2 + y^2)$

35. $\log (x^2 - 1) - \log (x + 1)$ **36.** $2 \ln (x - y) - \ln (x^2 - y^2)$ **37.** $3^{2 \log_3 6}$

38. $5^{4 \log_5 2}$

39. $10^{-2 \log 4}$

40. $e^{1/2 \ln 8}$

41. $\dfrac{\log 8 + \log 4}{\log 8 - \log 4}$

42. $\dfrac{\ln 7 - \ln 2}{\ln 7 + \ln 2}$

43. $\dfrac{\ln 15 - \ln 3}{2 \ln 5}$

44. $\dfrac{3 \log 2}{\log 6 + \log 3}$

45. $\dfrac{\log 4 + 2 \log 3}{3 \log 2 - \log 3}$

46. $\dfrac{2 \ln 5 + 3 \ln 2}{\ln 3 - 2 \ln 5}$

Solve these logarithmic equations, as in EXAMPLE 4. Check for extraneous solutions.

47. $\log_2 3 + \log_2 (x - 2) = \log_2 12$

48. $\log_5 (2x + 1) - \log_5 (x - 1) = \log_5 3$
49. $2 \log x - \log (x - 1) = \log 4$

50. $3 \ln 2 + \ln (x + 1) = \ln 16$

51. $\log_7 x + \log_7 (x + 8) = 2 \log_7 3$
52. $2 \log x = \log 4 + \log (x - 1)$

Solve these logarithmic equations, as in EXAMPLE 5. Check for extraneous solutions.

53. $\log_3 (5x - 1) = 2$

54. $\log_5 (2x + 3) = 3$

55. $\log_2 4 + \log_2 (x - 1) = 5$

56. $\log_3 9 + \log_3 (2x + 1) = 4$

57. $2 \log 5 + \log x = 2$

58. $3 \log_4 2 + \log_4 x = 3$

59. $\log_2 x + \log_2 (x + 2) = 3$

60. $\log x + \log (x + 21) = 2$

61. $\log_5 (x - 3) - \log_5 2 = 3$

62. $\log_8 (2x + 1) - \log_8 3 = 1$

63. $\log_3 x + \log_3 (x + 3) = 2 + \log_3 2$

64. $\log_2 x + \log_2 (x - 5) = 3 + \log_2 3$

B

65. $\log_2 (x^2 + 1) - \log_2 x = 1$

66. $\log (x^2 - 11) - \log x = 1$

67. $\log (x + 6) - \log (x - 3) = 1$

68. $\log_2 (x + 12) - \log_2 (x - 2) = 3$

69. $2 \log_6 x + \log_6 (x^2 - 5) = 2$

70. $1 = \dfrac{1}{2} \log_2 (x + 6) - \log_2 (x - 8)$

Express as an algebraic combination of logs:

71. $\log \dfrac{2\pi}{k} \sqrt{\dfrac{L}{g}}$

72. $\ln \dfrac{3t}{7} \sqrt[3]{\dfrac{x^2}{y}}$

Express as a single log:

73. $2 \ln x + \dfrac{1}{2} \ln y - 3 \ln z - \dfrac{1}{4} \ln w$

74. $\log x + \dfrac{2}{3} \log t + 2 \log c - 2 \log y - \log z$

75. The Inverse Property

$$b^{\log_b x} = x$$

can be proved by letting $\log_b x = y$. You write this in exponential form, and then use substitution, to prove this property.

76. The other Inverse Property

$$\log_b b^x = x$$

can be proved simply by verifying it in exponential form. Do so.

Use the properties of logarithms to show that:

77. $-\log_b \dfrac{x}{y} = \log_b \dfrac{y}{x}$

78. $-\log_b \dfrac{1}{x} = \log_b x$

10.4   Exponential Equations

As mentioned at the beginning of Section 10.2, one way we use logarithms is as a method for solving the **exponential equation** $2^t = 6$, in which the unknown t appears as the exponent. In order to solve this and other such equations, we will use the fact that if

$$Y = Z$$

represents any equation, then we get an equivalent equation by taking the logarithm of both sides:*

$$\log_b Y = \log_b Z.$$

The base b depends on the equation we are solving, as the following examples illustrate.

EXAMPLE 1 Solve these exponential equations:

a. $2^{4-5x} = 8^x$

Both 2 and 8 have the least common base 2, so one way is to take the $\log_2$ of both sides:

$$\log_2 2^{4-5x} = \log_2 8^x$$

$(4 - 5x) \log_2 2 = x \log_2 8 \qquad$ Power Rule

$(4 - 5x) \cdot 1 = x \cdot 3 \qquad\qquad 2^1 = 2, \quad 2^3 = 8$

$$4 - 5x = 3x$$

$$4 = 8x$$

$$\frac{1}{2} = x. \qquad\qquad\qquad \text{Solution}$$

---

*Assuming, of course, that both Y and Z are positive.

b. $81 \cdot 27^{x-1} = \left(\dfrac{1}{9}\right)^{2x}$

The common base here is 3, so we take the $\log_3$ of both sides.

$$\log_3 81 \cdot 27^{x-1} = \log_3 \left(\dfrac{1}{9}\right)^{2x}$$

$$\log_3 81 + (x-1) \log_3 27 = 2x \log_3 \left(\dfrac{1}{9}\right) \qquad \text{Product and Power Rules}$$

$$4 + (x-1) \cdot 3 = 2x \cdot (-2) \qquad\qquad 3^4 = 81, \quad 3^3 = 27, \quad 3^{-2} = \dfrac{1}{9}$$

$$4 + 3x - 3 = -4x$$

$$1 = -7x$$

$$-\dfrac{1}{7} = x. \qquad\qquad\qquad \text{Solution}$$

The next example contains the base e, so we will take the $\log_e$, or ln, of both sides.

EXAMPLE 2 Solve these exponential equations:

a. $e^{0.08t} = 1.5$

We take the ln of both sides:

$$\ln e^{0.08t} = \ln 1.5$$

$$0.08t(\ln e) = \ln 1.5 \qquad\qquad \text{Power Rule.}$$

$$0.08t(1) = \ln 1.5 \qquad\qquad \ln e = \log_e e = 1$$

$$t = \dfrac{\ln 1.5}{0.08} \qquad\qquad \text{Divide by 0.08}$$

$$t = 5.068 \qquad\qquad 1.5 \;\boxed{\text{ln}}\; \boxed{\div}\; .08 \;\boxed{=}$$

b. $15 = 20e^{-1.3x}$

We first divide both sides by 20 and then take the ln of both sides:

$$\dfrac{15}{20} = \dfrac{20e^{-1.3x}}{20} \qquad\qquad \text{Divide by 20.}$$

$$0.75 = e^{-1.3x}$$

$$\ln 0.75 = \ln e^{-1.3x} \qquad\qquad \text{Take ln of both sides.}$$

$$\ln 0.75 = -1.3x(\ln e) \qquad\quad \text{Power Rule}$$

$$\ln 0.75 = -1.3x \qquad\qquad \ln e = 1$$

$$\dfrac{\ln 0.75}{-1.3} = x \qquad\qquad \text{Divide by } -1.3$$

$$0.221 = x \qquad\qquad .75 \;\boxed{\text{ln}}\; \boxed{\div}\; 1.3 \;\boxed{+/-}\; \boxed{=}$$

Finally, the equation $2^t = 6$ from Section 10.2 has neither a common base nor base e. We can solve it by taking either the common log or the natural log of both sides.

EXAMPLE 3

Solve these exponential equations:
a. $2^t = 6$

We take the common log of both sides:

$$\log 2^t = \log 6$$
$$t \log 2 = \log 6 \qquad \text{Power Rule}$$
$$t = \frac{\log 6}{\log 2} \qquad \text{Divide by } \log 2.$$
$$t = 2.585 \qquad 6 \boxed{\log} \boxed{\div} 2 \boxed{\log} \boxed{=}$$

WARNING! $\dfrac{\log 6}{\log 2} \neq \log \dfrac{6}{2} = \log 6 - \log 2.$

The solution $t = 2.585$ obtained by using logs is the same as that which we obtained by trial and error in Section 10.2.

b. $3^{2x-1} = 14$

Let us take the natural log of both sides this time:

$$\ln 3^{2x-1} = \ln 14$$
$$(2x - 1) \ln 3 = \ln 14 \qquad \text{Power Rule}$$
$$2x - 1 = \frac{\ln 14}{\ln 3} \qquad \text{Divide by } \ln 3.$$
$$2x = \frac{\ln 14}{\ln 3} + 1 \qquad \text{Add 1.}$$
$$x = \frac{\dfrac{\ln 14}{\ln 3} + 1}{2} \qquad \text{Divide by 2.}$$
$$x = 1.701 \qquad 14 \boxed{\ln} \boxed{\div} 3 \boxed{\ln} \boxed{=} \boxed{+} 1 \boxed{=} \boxed{\div} 2 \boxed{=}$$

This solution can also be obtained by taking the common log of both sides.

The exponential functions of Section 10.1 become exponential equations when the time is unknown.

EXAMPLE 4

The population of a city is given by

$$P = 35{,}000\ e^{0.065t},$$

where P is the population t years after the census year 1970. In how many years will the population grow to 80,000? During what year will this occur?

Solution

We substitute $P = 80{,}000$ into the equation, and solve for t:

$$80{,}000 = 35{,}000\ e^{0.065t}$$
$$2.2857 = e^{0.065t} \qquad \text{Divide by 35,000.}$$
$$\ln 2.2857 = \ln e^{0.065t} \qquad \text{Take ln of both sides.}$$
$$\ln 2.2857 = 0.065t(\underbrace{\ln e}_{1}) \qquad \text{Power Rule}$$
$$\frac{\ln 2.2857}{0.065} = t \qquad \text{Divide by 0.065; } \ln e = 1.$$
$$12.7 = t \qquad 2.2857\ \boxed{\ln}\ \boxed{\div}\ .065\ \boxed{=}$$

The population will reach 80,000 during the year

$$1970 + 12.7 = 1982.7, \quad \text{or} \quad \text{during 1982.}$$

Our next application comes from Section 10.1, Example 5, in which we saw that the half-life of radioactive carbon-14 is 5600 years. This fact is the basis of **carbon dating**, a method used to calculate the age of fossils. In a living organism, the ratio of carbon-14 to carbon-12 is a constant, because the lost carbon-14 is continually replenished. Upon death, the carbon-14 is no longer replenished, and the C-14/C-12 ratio decreases according to the formula $R = \left(\frac{1}{2}\right)^{t/5600}$, where R is the observed C-14/C-12 ratio found in the fossil t years after the organism died. A fossil that is unearthed is chemically analyzed, and a value of R is determined. This leads to an exponential equation, and the solution t of that equation is the age of the fossil. Developed in the late 1950s, carbon dating earned Professor Willard Libby of UCLA the 1960 Nobel Prize in chemistry.

EXAMPLE 5

The carbon-14/carbon-12 ratio R in a decaying fossil is given by

$$R = \left(\frac{1}{2}\right)^{t/5600}.$$

Find the age of a fossil for which $R = 30\% = 0.30$ of the normal ratio found in the tissues of the living organism.

(continued)

Solution

We substitute $R = 0.30$ into the equation and get

$$0.30 = \left(\frac{1}{2}\right)^{t/5600}.$$

We can solve this exponential equation by taking the natural log of both sides:

$$\ln 0.30 = \ln \left(\frac{1}{2}\right)^{t/5600}$$

$$\ln 0.30 = \frac{t}{5600} \ln \left(\frac{1}{2}\right)$$

$$5600(\ln 0.30) = t \ln (0.5)$$

$$\frac{5600(\ln 0.30)}{\ln 0.5} = t$$

$$9727 = t \qquad .30 \boxed{\text{ln}} \boxed{\times} 5600 \boxed{\div} .5 \boxed{\text{ln}} \boxed{=}$$

The fossil is 9727 years old.

EXERCISE 10.4   *Answers, page A97–A98*

Solve these exponential equations using $\log_b$, where b is the common base of the powers.
See **EXAMPLE 1.**

1. $3^{2x-1} = 27$

2. $5^{3x+2} = 25$

3. $16^{x-3} = 8^x$

4. $9^{3x-2} = 27^x$

5. $25^{x+3} = \left(\frac{1}{125}\right)^{2x-1}$

6. $4^{x+1} = \left(\frac{1}{8}\right)^{2-x}$

7. $8^x \cdot 16^{x-1} = 32^{3-x}$

8. $9^x \cdot 27^{x-2} = 81^{2-x}$

9. $3^{x^2} = 9^{x+4}$

10. $4^{x+3} = 2^{x^2+x}$

Solve these exponential equations using ln, as in **EXAMPLE 2:**

11. $e^{0.06x} = 4.7$

12. $e^{1.2t} = 0.855$

13. $e^{-0.075t} = 4$

14. $6.2 = e^{t/2}$

15. $25e^{0.3x} = 80$

16. $5.6 = 1.9e^{-3.1x}$

Solve these exponential equations using log or ln, as in **EXAMPLE 3:**

17. $2^x = 3$

18. $7^x = 4$

19. $5^{x-1} = 17$

20. $6^{2x+1} = 2.8$

21. $(0.5)^{t/28} = 0.3$

22. $(1.2)^{-3t} = 4.68$

Solve as in EXAMPLES 4 and 5:

23. The population of San Diego, California, is given by

$$P = 350,000 \, e^{0.05t},$$

where P is the population t years after the census year 1950. In how many years will the population reach 525,000? During what year will this occur?

24. As a result of pesticides, a certain insect population is decreasing according to the formula

$$P = 100,000 \, e^{-0.02t},$$

where P is the population after t days. In how many days will there be 50,000 insects?

25. The carbon-14/carbon-12 ratio in a decaying fossil is given by

$$R = \left(\frac{1}{2}\right)^{t/5600}.$$

What is the age of a fossil whose ratio is $R = 18\% = 0.18$?

26. *Art fraud* can be detected by measuring the radioactive decay of one of the ingredients of oil paint. The formula used is

$$R = \left(\frac{1}{2}\right)^{t/125},$$

where R is the percentage of this component remaining in the paint. An art dealer claims to have a 300-year-old original for sale, but chemical analysis reveals that $R = 96\% = 0.96$ for this painting. By finding its true age, show that the painting is a fake.

B

27. The population of a termite colony is given by

$$P = 5000(2.5)^{0.8t},$$

where t is measured in days. In how many days will there be 200,000 termites?

28. The population of an amoeba culture is given by

$$P = 500,000(2)^{3t/4},$$

where t is measured in minutes. When will there be 800,000 amoebae?

29. $15,000 is deposited into an account paying 6% compounded monthly. Use the formula

$$A = P\left(1 + \frac{r}{n}\right)^{nt}$$

(see Section 10.1) to find the number of years t it will take for the accumulated amount to reach $30,000.

30. A real estate investment originally worth $1,200,000 is depreciated annually as a tax write-off. Its value V for tax purposes is given by

$$V = \$1,200,000(1.136)^{-t},$$

where t is the number of years after the purchase. In how many years will this investment lose half of its original value as a tax write-off?

The exponential equation $3^{x-1} = 7^{2x}$

can be solved by taking either the common or the natural log of both sides, as in EXAMPLE 3:

$$\log 3^{x-1} = \log 7^{2x}$$
$$(x - 1) \log 3 = 2x \log 7 \qquad \text{Power Rule.}$$
$$x \log 3 - \log 3 \quad = 2x \log 7 \qquad \text{Distributive Property.}$$

We now isolate all terms containing x on the left side, and then factor out x:

$$x \log 3 - 2x \log 7 \quad = \log 3$$
$$x(\log 3 - 2 \log 7) \quad = \log 3 \qquad \text{Factor out } x.$$

$$x = \frac{\log 3}{\log 3 - 2 \log 7}$$

$$x = \frac{\log 3}{\log \dfrac{3}{7^2}} \quad \longleftarrow \quad \text{Power and Quotient Rules.}$$

$$x = -0.393 \qquad 3 \;\boxed{\log}\; \boxed{\div}\; \boxed{(}\; 3 \;\boxed{\div}\; 49 \;\boxed{)}\; \boxed{\log}\; \boxed{=}$$

In like manner, solve these exponential equations, using log or ln:

31. $2^{x+1} = 5^{3x}$ **32.** $17^{2x} = 3^{x-1}$ **33.** $3^{2x-1} = 7^{x+1}$

34. $e^{1-x} = (4.6)^{x+2}$

In Section 10.2, Example 5, we defined the pH of a substance by

$$pH = -\log_{10}[H^+]$$

Dividing both sides by −1 and then writing the exponential form give

$$-pH = \log_{10}[H^+]$$
$$10^{-pH} = [H^+]$$

Find the hydrogen ion concentration, [H$^+$], (in moles per liter) for each of the following substances with the given pH:

35. Vinegar: **36.** Cola: **37.** Dish soap:
    pH = 4.71   pH = 5.2   pH = 8.2

38. Ammonia: **39.** Human hair: **40.** Brand X shampoo:
    pH = 9.34   pH = 5.5*   pH = 8*

---

*So-called "pH-balanced" shampoos have a pH = 5.5, like that of human hair. Some consider normal shampoos, with pH = 8, too alkaline, or harsh, for the hair.

Common and natural logarithms of numbers can be obtained by using the log and ln buttons on a scientific calculator. Although 10 and e are the most popular bases for logarithms, nothing prevents us from finding the log of a number in a different base. For example, we can find the value of $\log_3 7$ as follows:

$$\text{Let} \quad \log_3 7 = x$$
$$\text{then} \quad 3^x = 7$$

$$x \log 3 = \log 7 \qquad \text{Take the log of both sides.}$$

$$x = \frac{\log 7}{\log 3} \qquad \text{Divide by log 3.}$$

$$\log_3 7 = \frac{\log 7}{\log 3} \qquad \text{Substitute } \log_3 7 \text{ for } x.$$

$$= 1.771 \qquad 7 \boxed{\text{log}} \boxed{\div} 3 \boxed{\text{log}} \boxed{=}$$

We have expressed $\log_3 7$ as the quotient of two common logs. We could have obtained the same result by taking the natural log of both sides,

$$\log_3 7 = \frac{\log 7}{\log 3} = \frac{\ln 7}{\ln 3}.$$

This "change of base" of logarithms can be generalized:

Change-of-Base Formula

$$\log_b x = \frac{\log x}{\log b} = \frac{\ln x}{\ln b}$$

Find these logs via the Change-of-Base formula, using either common or natural logs:

41. $\log_2 6$ **42.** $\log_3 4$ **43.** $\log_{1.2} 60$

44. $\log_6 0.223$ **45.** $\log_\pi e$ **46.** $\log_8 \sqrt{13}$

By "cross-canceling" of fractions, we see that

$$\frac{\log x}{\log b} \frac{\log a}{\log x} = \frac{\log a}{\log b}.$$

By the Change-of-Base formula, this translates into

$$\log_b x \cdot \log_x a = \log_b a$$

Use this "product of logs" formula to obtain, *without using a calculator,* the exact value of:

47. $\log_2 3 \cdot \log_3 4$ **48.** $\log_3 7 \cdot \log_7 9$ **49.** $\log_4 7 \cdot \log_7 4$

C

50. $\ln 10 \cdot \log e$ (*Hint*: Write these logs with their bases).

51. $\log_5 9 \cdot \log_9 18 \cdot \log_{18} 25$

52. $\log_2 3 \cdot \log_3 4 \cdot \log_4 5 \cdot \log_5 6 \cdot \log_6 7 \cdot \log_7 8$

CHAPTER 10 · SUMMARY

10.1 Exponential Functions

$$y = f(x) = b^x, \quad b > 0, \, b \neq 1$$

Example:

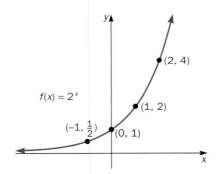

$f(x) = 2^x$

$(2, 4)$
$(1, 2)$
$(-1, \frac{1}{2})$
$(0, 1)$

Calculating exponents

$2^{3.6} = 12.125 \qquad 2 \boxed{x^y} 3.6 \boxed{=}$

$\left(\frac{1}{2}\right)^{4.5} = 0.044 \qquad 0.5 \boxed{x^y} 4.5 \boxed{=}$

$e^{2.1} = 8.166 \qquad 2.1 \boxed{INV} \boxed{ln}$

Applications of exponential functions

Amoeba population: $P = 100{,}000(2)^{t/3}$

Radioactive decay: $A = 100\left(\frac{1}{2}\right)^{t/5600}$

Las Vegas population: $P = 63{,}000e^{0.07t}$,
    where $e = 2.71828 \ldots$.

10.2 Logarithms

$$\log_b x = y \quad \text{provided} \quad b^y = x, \quad b > 0, \, b \neq 1$$

Logarithmic	Exponential
$\log_2 8 = 3$	$2^3 = 8$
$\log_5 \dfrac{1}{25} = -2$	$5^{-2} = \dfrac{1}{25}$
$\log_3 3 = 1$	$3^1 = 3$
$\log_7 1 = 0$	$7^0 = 1$
$\log_6 \sqrt{6} = \dfrac{1}{2}$	$6^{1/2} = \sqrt{6}$

Examples:
Solve for x, b, or y:

$\log_4 x = -3 \quad \text{means} \quad 4^{-3} = x, \quad \text{or} \quad x = \dfrac{1}{64}$.

$\log_b 9 = 2 \quad \text{means} \quad b^2 = 9, \quad \text{or} \quad b = 3$.

$\log_8 16 = y \quad \text{means} \quad 8^y = 16$

$$(2^3)^y = 2^4, \quad \text{or} \quad y = \frac{4}{3}.$$

Common logs

$\log x \quad \text{means} \quad \log_{10} x$

$\log 1000 = 3 \quad \text{because} \quad 10^3 = 1000$

$\text{pH} = -\log [\text{H}^+]$

CHAPTER 10 · SUMMARY

Common logs *(continued)*

Example:

If $[H^+] = 0.000092$, then

$$pH = -\log 0.000092$$
$$= 4.04 \qquad .000092 \boxed{\log} \boxed{+/-}$$

Natural logs

$\ln x$ means $\log_e x$

$\ln 6.32 = 1.8437 \qquad 6.32 \boxed{\ln}$

10.3 Properties of Logarithms

Inverse Properties: $b^{\log_b x} = x$ and $\log_b b^x = x$

Product Rule: $\log_b xy = \log_b x + \log_b y$

Quotient Rule: $\log_b \dfrac{x}{y} = \log_b x - \log_b y$

Power Rule: $\log_b x^p = p \log_b x$

Examples:

$$5^{\log_5 25} = 25 \quad \text{and} \quad \log_3 3^4 = 4$$
$$\log \frac{3x^2}{y^4} = \log 3 + 2 \log x - 4 \log y$$

Logarithmic equation:

$$3 \log_3 2 - \log_3 (x - 1) = 2$$
$$\text{single log} \rightarrow \log_3 \frac{2^3}{x-1} = 2$$
$$\text{exponential form} \rightarrow \frac{8}{x-1} = 3^2$$
$$8 = 9x - 9$$
$$\frac{17}{9} = x.$$

10.4 Exponential Equations

Take the appropriate log of both sides of the equation, and "bring down the exponent" using the Power Rule.

Example:
$$8^{x-1} = 32$$
$$(x-1)\log_2 8 = \log_2 32$$
$$(x-1)\,3 = 5$$
$$3x - 3 = 5$$
$$x = \frac{8}{3}.$$

Example:
$$e^{-1.2t} = 4.7$$
$$-1.2t(\ln e) = \ln 4.7$$
$$-1.2t(1) = \ln 4.7$$
$$t = \frac{\ln 4.7}{-1.2}$$
$$t = -1.29$$

Example:
$$5^x = 3$$
$$x \log 5 = \log 3$$
$$x = \frac{\log 3}{\log 5}$$
$$x = 0.683$$

CHAPTER 10 ▪ REVIEW EXERCISES   *Answers, pages A98–A99*

10.1

Find the value of each expression, using your calculator. Round to three decimal places.

1. $(2.3)^{1.7}$

2. $5^{-\sqrt{2}}$

3. $\left(\dfrac{5}{7}\right)^{3/8}$

4. $4500\left(1 + \dfrac{0.07}{12}\right)^{12(7)}$

5. $e^{3.58}$

6. $75{,}000e^{0.09(12)}$

Simplify each expression:

7. $e^{x+1}e^{2x-3}$

8. $\dfrac{3^{2t+5}}{3^{t+2}}$

9. $(2^x \cdot 5^{2y})^3$

10. $1 + 2^{-t}$

11. $10^x + 10^{-x}$

12. $\dfrac{e^x - e^{-x}}{e^x + e^{-x}}$

Graph these exponential functions:

13. $f(x) = 3^{x-1}$

14. $y = \left(\dfrac{1}{2}\right)^{2x}$

15. $g(x) = e^{1-x}$

16. The population of an ant colony is given by

$$P = 2000(2)^{2t},$$

where P is the number of ants after t days. Find P when $t = 0, 1, 2, 5.5,$ and 7.3 days.

17. Smog City, U.S.A., is losing people. Its population is given by

$$P = 850{,}000e^{-0.07t},$$

using 1960 as the census year. Find its population in 1960, 1970, and 1985.

18. Radioactive strontium-90 decays according to the formula

$$A = 100\left(\dfrac{1}{2}\right)^{t/28},$$

where A is the amount (in grams) remaining after t years. Find the amount remaining when $t = 0, 14, 28, 56,$ and 98 years. What is the half-life of strontium-90?

Use the formulas

$$A = P\left(1 + \dfrac{r}{n}\right)^{nt}, \quad A = Pe^{rt}, \quad \text{and} \quad I = A - P$$

to find the accumulated amount A and the interest I for the following compound-interest accounts:

19. $P = \$15,000,\quad r = 7\% = 0.07,\quad n = 1$ (compounded annually), $t = 6$ years

20. $P = \$3000,\quad r = 8.5\%,\quad n = 4$ (compounded quarterly), $t = 7.5$ years

21. $P = \$7500,\quad r = 6\frac{1}{4}\%,\quad$ compounded monthly for 3 years

22. $P = \$5000,\quad r = 11\%,\quad$ compounded daily for 5 years

23. $P = \$5000,\quad r = 11\%,\quad$ compounded continuously for 5 years

24. *Newton's Law of Cooling* states that the temperature of an object being cooled under particular conditions is given by

$$T = 25 + 12e^{-1.2t},$$

where T is the temperature (in °C) of the object after t seconds. Find its temperature when $t = 0, 1, 2, 4.5,$ and 10 seconds.

10.2

Convert to exponential form:

25. $\log_7 49 = 2$

26. $\log_{16} 8 = \dfrac{3}{4}$

27. $\log 0.0001 = -4$

Convert to logarithmic form:

28. $5^3 = 125$

29. $4^{-2} = \dfrac{1}{16}$

30. $e^{5.2} = 181.27$

Solve for the variable:

31. $\log_7 x = 2$

32. $\log_5 x = -4$

33. $\log_8 x = \dfrac{2}{3}$

34. $\log_9 x = -\dfrac{1}{2}$

35. $\log_{25} x = \dfrac{-3}{2}$

36. $\log_{12} x = \dfrac{1}{2}$

37. $\log_2 x = -\dfrac{1}{2}$

38. $\log_b 125 = 3$

39. $\log_b \dfrac{4}{9} = 2$

40. $\log_b 49 = -2$

41. $\log_b 4 = \dfrac{1}{2}$

42. $\log_b 64 = \dfrac{3}{2}$

43. $\log_b 27 = \dfrac{-3}{2}$

44. $\log_2 32 = y$

45. $\log_8 8 = y$

46. $\log_{10} 1000 = y$

47. $\log_{10} 0.01 = y$

48. $\log_3 1 = y$

49. $\log_5 \dfrac{1}{25} = y$ **50.** $\log_{27} 9 = y$ **51.** $\log_8 16 = y$

52. $\log_4 \dfrac{1}{8} = y$

Find the pH of the following substances, using the formula

$$pH = -\log [H^+],$$

where [H$^+$] is given in moles per liter. Then classify each substance as acidic or basic.

53. Tomato juice: $[H^+] = 0.00000061$ **54.** Bathroom cleanser: $[H^+] = 0.0000000078$

55. The magnitude M of an earthquake is given by

$$M = \log \frac{E}{E_0},$$

where $\dfrac{E}{E_0}$ is the ratio of the energy of the earthquake to the energy of the smallest measurable quake. Find the magnitude of the 1556 earthquake in China, with $\dfrac{E}{E_0} = 316{,}000{,}000$. This quake resulted in the largest human loss in history, an estimated 830,000 dead.

56. A free-falling sky diver, before pulling his parachute, reaches velocity v (in miles per hour) after T seconds, according to the formula

$$T = \frac{\ln 120 - \ln (120 - v)}{0.8}.$$

How many seconds will it take to reach $v = 0$, $v = 100$, and $v = 119$ mph?

57. Sound loudness, measured in decibels, is given by

$$db = 10 \log \frac{P}{P_0}.$$

Find the decibels during a football game at the Seattle Kingdome, where $\dfrac{P}{P_0} = 31{,}600{,}000{,}000$.

10.3

Simplify using the Inverse Properties of Logs:

58. $2^{\log_2 5}$ **59.** $10^{\log 7}$ **60.** $e^{\ln 4.3}$

Write each log as an algebraic combination of logs:

61. $\log_5 \dfrac{4x^3 y}{z^2}$

62. $\log \dfrac{2s^2 \sqrt{t}}{u \sqrt[4]{v^3}}$

63. $\ln \sqrt{\dfrac{x(x^2 + 1)}{x - 1}}$

Write with as few logs as possible:

64. $\log_b 5 + \log_b x - \log_b (x - 5)$

65. $2 \log_5 x + 3 \log_5 y - \log_5 z$

66. $3 \log 2 + \log \pi + \dfrac{1}{2} \log C - 2 \log r$

67. $\ln (x^2 - 1) - 2 \ln (x + 1)$

68. $5^{2 \log_5 4}$

69. $e^{-2 \ln 3}$

70. $\dfrac{2 \log 6 - \log 3}{3 \log 2}$

Solve these logarithmic equations:

71. $\log_2 5 + \log_2 (x + 1) = \log_2 15$

72. $\log x + \log (x - 3) = \log 18$

73. $\log_4 (3x + 1) = 3$

74. $\log_6 3 + \log_6 (x + 4) = 2$

75. $\log_7 x + \log_7 (x - 6) = 1$

76. $\log_2 6 + \log_2 x = 4 + \log_2 (x - 5)$

77. $2 \log_2 x - \log_2 (x + 4) = 1$

78. $\log_2 x - \dfrac{1}{2} \log_2 (x - 1) = 1$

79. Prove the Inverse Property

$$\log_b b^x = x$$

by first using the Power Rule for logarithms.

80. Prove the other Inverse Property,

$$b^{\log_b x} = x$$

by verifying it in its equivalent *logarithmic* form.

10.4

Solve these exponential equations by taking the appropriate logarithm of both sides:

81. $5^{x-1} = 125$

82. $9^{x+1} = 3^{2-x}$

83. $8^{x-3} = 16^{2x}$

84. $25^{3x+1} = \dfrac{1}{125}$

85. $3^{x+1} \cdot 27^{1-2x} = 9^{x-3}$

86. $25 \cdot 125^{x-1} = 625^x$

87. $7^{x-2} = 1$

88. $e^{0.04t} = 1.7$

89. $e^{-1.4x} = 5.82$

90. $15 = 10 e^{0.065t}$

91. $3^x = 7$

92. $5^{x-1} = 37$

93. $(1.06)^{2x} = 8.75$ **94.** $\left(\dfrac{1}{2}\right)^{-t/90} = 0.35$ **95.** $2^{x+1} = 5^x$

96. The population of a city is given by

$$P = 20,000\, e^{0.03t},$$

using 1960 as the census year. In how many years will the population reach 50,000? During what year will this occur?

97. The carbon-14/carbon-12 ratio R is given by

$$R = \left(\frac{1}{2}\right)^{t/5600}.$$

What is the age of a fossil whose ratio is $R = 23\% = 0.23$?

98. \$25,000 is deposited into an account paying 6% annual interest compounded quarterly. Use the formula

$$A = P\left(1 + \frac{r}{n}\right)^{nt}$$

to find the number of years it will take for the amount to double.

Use the Change-of-Base formula

$$\log_b x = \frac{\log x}{\log b} \quad \text{or} \quad \frac{\ln x}{\ln b}$$

to find the values of the following:

99. $\log_3 5$ **100.** $\log_5 \sqrt{3}$ **101.** $\log_{\sqrt{2}} e$

CHAPTER 10 ▪ TEST   *Answers, page A99*

1. Simplify: $5^{\log_5 125}$ **2.** Simplify: $\ln e^{1.32}$

3. Find the pH of H_2SO_4, whose $[H^+] = 7.2 \times 10^{-5}$ moles per liter.
   Formula: $pH = -\log [H^+]$

4. Find the decibel level of an obnoxious leaf-blower used by gardeners,
   if its $\dfrac{P}{P_0} = 3,000,000,000.$ *Formula:* $db = 10 \log \dfrac{P}{P_0}$

5. Graph the exponential function $f(x) = 2^{-x}$.

6. Write as a combination of logs: $\log_3 \dfrac{2xy^3}{5\sqrt{z}}$

7. Write as a single log: $2 \log x + \log y - 3 \log (x - y)$

Solve:

8. $\log_8 x = 2$

9. $\log_x 8 = 2$

10. $\log_8 2 = x$

11. $2 \log x - \log (x + 4) = \log 2$

12. $\log_2 x + \log_2 (x - 6) = 4$

13. $4^{x-1} = 8^x$

14. $e^{0.05t} = 1.8$

15. $2^{x+1} = 6$

16. Find the age of a fossil whose carbon-14/carbon-12 ratio is $R = 8\%$.

   Formula: $R = \left(\dfrac{1}{2}\right)^{t/5600}$

The population of a city is given by $P = 35{,}000e^{0.06t}$, **using 1965 as the census year.**

17. Find the population in 1980.

18. During what year will the population reach 85,000?

The *compound-interest formulas are* $A = P\left(1 + \dfrac{r}{n}\right)^{nt}$ **and** $I = A - P.$

19. Find the accumulated amount and the interest when $25,000 is deposited at 8% annual interest, compounded quarterly, for 3 years.

20. $15,000 is deposited at 6% interest compounded monthly. In how many years will the account contain $45,000?

11 SEQUENCES AND SUMS

11.1   Sequences and the Sigma Notation

Would you rather have a million dollars, *or* a penny today, 2¢ tomorrow, 4¢ on day 3, 8¢ on day 4, 16¢ on day 5, 32¢ on day 6, and so forth, for 30 days? Better to take the money and run, you say? Lest there be any doubts, let's sweeten the deal by making it *ten* million dollars ($10,000,000), plus 10% interest for a month. And you can haul it off in a brand new Ferrari Testarossa worth $150,000.

$10,000,000
+
$150,000

This should clinch it, right? Possibly, but why not wait until Section 11.3, when all the pennies are tallied up.

In the meantime, the sequence of payments 1¢, 2¢, 4¢, 8¢, 16¢, 32¢, . . . and, more important, their total 1¢ + 2¢ + 4¢ + 8¢ + 16¢ + 32¢ + · · · have served to introduce the two major topics of this chapter: sequences and their sums. We start with a definition: a **sequence** is a function whose domain is the set of natural numbers. It is customary to write the function using a variable subscript a_n rather than the functional notation $a(n)$.

EXAMPLE 1

Write the first four terms of the sequences defined as follows:

a. $a_n = \dfrac{n}{n + 1}$ By direct substitution, we get

$$a_1 = \frac{1}{1 + 1} = \frac{1}{2} \qquad a_2 = \frac{2}{2 + 1} = \frac{2}{3}$$

$$a_3 = \frac{3}{3+1} = \frac{3}{4} \qquad a_4 = \frac{4}{4+1} = \frac{4}{5}$$

The sequence is $\dfrac{1}{2}, \dfrac{2}{3}, \dfrac{3}{4}, \dfrac{4}{5}, \cdots$.

b. $a_n = n^2 + 5n$

$$a_1 = 1^2 + 5 \cdot 1 = 6 \qquad a_2 = 2^2 + 5 \cdot 2 = 14$$
$$a_3 = 3^2 + 5 \cdot 3 = 24 \qquad a_4 = 4^2 + 5 \cdot 4 = 36$$

The sequence is   6, 14, 24, 36, ...

c. $a_n = \dfrac{(-1)^{n-1}}{2^n}$

$$a_1 = \frac{(-1)^{1-1}}{2^1} = \frac{(-1)^0}{2} = \frac{1}{2} \qquad a_2 = \frac{(-1)^{2-1}}{2^2} = \frac{(-1)^1}{4} = \frac{-1}{4}$$

$$a_3 = \frac{(-1)^{3-1}}{2^3} = \frac{(-1)^2}{8} = \frac{1}{8} \qquad a_4 = \frac{(-1)^{4-1}}{2^4} = \frac{(-1)^3}{16} = \frac{-1}{16}$$

The sequence is $\dfrac{1}{2}, \dfrac{-1}{4}, \dfrac{1}{8}, \dfrac{-1}{16}, \cdots$.

Obtaining the terms of a sequence is relatively simple if you are given its formula. A more challenging problem is to *discover a formula* for a_n, given the first several terms.

EXAMPLE 2 Discover a formula for a_n for each of the following sequences:

a. 5, 10, 15, 20, . . .   These are multiples of 5, and can be written

$$5 \cdot 1, 5 \cdot 2, 5 \cdot 3, 5 \cdot 4, \cdots, 5 \cdot n, \cdots \quad .$$

Thus a formula is $a_n = 5n$.

b. 1, 4, 9, 16, 25, . . .   These are the perfect squares, written as

$$1^2, 2^2, 3^2, 4^2, 5^2, \ldots, n^2, \ldots \quad .$$

A formula for this sequence is $a_n = n^2$.

c. $\dfrac{1}{2}, \dfrac{8}{3}, \dfrac{27}{4}, \dfrac{64}{5}, \cdots$ The numerators are perfect cubes, and we can write this sequence as

$$\frac{1^3}{1+1}, \frac{2^3}{2+1}, \frac{3^3}{3+1}, \frac{4^3}{4+1}, \cdots, \frac{n^3}{n+1}, \cdots \quad .$$

A formula is $a_n = \dfrac{n^3}{n+1}$.

(continued)

d. $-2, 4, -8, 16, \ldots$ These are powers of 2, but the alternating signs can be obtained by using powers of -2:

$$(-2)^1, (-2)^2, (-2)^3, (-2)^4, \ldots, (-2)^n, \ldots$$

A formula is $a_n = (-2)^n$.

The sum of the first four terms in a sequence $a_1 + a_2 + a_2 + a_4$ can be written in a compact form using **sigma notation**:

$$\sum_{n=1}^{4} a_n = a_1 + a_2 + a_3 + a_4.$$

Σ is the Greek letter *sigma*, and it denotes summation. The subscript n is the *index of summation*; here it takes the values $n = 1, 2, 3$, and 4. Likewise,

$$\sum_{k=2}^{6} b_k = b_2 + b_3 + b_4 + b_5 + b_6,$$

in which k is the index of summation and takes the values $k = 2, 3, 4, 5$, and 6. In the next example, we substitute consecutive values of the index into the expression under the summation symbol sigma.

EXAMPLE 3   Compute the following sums:

$$
\begin{array}{cccc}
& n=1 & n=2 & n=3 & n=4 \\
& \downarrow & \downarrow & \downarrow & \downarrow
\end{array}
$$

a. $\displaystyle\sum_{n=1}^{4} (5n-2)$ $= (5 \cdot 1 - 2) + (5 \cdot 2 - 2) + (5 \cdot 3 - 2) + (5 \cdot 4 - 2)$

$\qquad\qquad\qquad = \quad 3 \quad + \quad 8 \quad + \quad 13 \quad + \quad 18$

$\qquad\qquad\qquad = 42.$

$$
\begin{array}{ccccc}
& k=2 & k=3 & k=4 & k=5 & k=6 \\
& \downarrow & \downarrow & \downarrow & \downarrow & \downarrow
\end{array}
$$

b. $\displaystyle\sum_{k=2}^{6} (-1)^k \cdot k^2$ $= (-1)^2 \cdot 2^2 + (-1)^3 \cdot 3^2 + (-1)^4 \cdot 4^2 + (-1)^5 \cdot 5^2 + (-1)^6 \cdot 6^2$

$\qquad\qquad\qquad\quad = \quad 4 \quad + \quad (-9) \quad + \quad 16 \quad + \quad (-25) \quad + \quad 36$

$\qquad\qquad\qquad\quad = 22.$

c. $\displaystyle\sum_{m=1}^{5} \frac{m}{m+1}$ $= \dfrac{1}{1+1} + \dfrac{2}{2+1} + \dfrac{3}{3+1} + \dfrac{4}{4+1} + \dfrac{5}{5+1}$

$\qquad\qquad\qquad = \dfrac{1}{2} + \dfrac{2}{3} + \dfrac{3}{4} + \dfrac{4}{5} + \dfrac{5}{6}$

$$= \frac{30}{60} + \frac{40}{60} + \frac{45}{60} + \frac{48}{60} + \frac{50}{60}$$

$$= \frac{213}{60}, \quad \text{or} \quad \frac{71}{20}.$$

EXERCISE 11.1   *Answers, page A100*

A

Write the first four terms of the sequences defined below, as in EXAMPLE 1:

1. $a_n = 2n - 3$ **2.** $a_n = 4 - 5n$ **3.** $a_n = n - n^2$

4. $a_n = 3n^2 - 2$ **5.** $b_n = \frac{(n+1)^2}{n}$ **6.** $b_n = \frac{n^3}{n+1}$

7. $c_n = (-2)^{n+1}$ **8.** $c_n = 3^{n-1}$ **9.** $a_n = \left(-\frac{1}{2}\right)^n$

10. $b_n = (-1)^n \cdot 4^{2-n}$ **11.** $c_n = -4$ **12.** $b_n = 5$

13. $a_n = 1 - (-1)^n$ **14.** $b_n = \left(\frac{1}{2}\right)^n + 1$

Discover a formula a_n for the following sequences, as in EXAMPLE 2:

15. 4, 8, 12, 16, . . .   **16.** $-3, -6, -9, -12, \ldots$ **17.** $\frac{1}{2}, \frac{4}{3}, \frac{9}{4}, \frac{16}{5}, \ldots$

18. 1, 8, 27, 64, 125, . . .   **19.** $\frac{-1}{2}, \frac{1}{4}, \frac{-1}{8}, \frac{1}{16}, \frac{-1}{32}, \ldots$ **20.** $-3, 9, -27, 81, \ldots$

21. $\frac{2}{1}, \frac{3}{2}, \frac{4}{3}, \frac{5}{4}, \frac{6}{5}, \ldots$ **22.** $\frac{1}{3}, \frac{2}{4}, \frac{3}{5}, \frac{4}{6}, \frac{5}{7}, \ldots$

Compute the following sums, as in EXAMPLE 3:

23. $\sum_{n=1}^{5} (2n + 3)$ **24.** $\sum_{k=1}^{6} (5 - 2k)$ **25.** $\sum_{m=1}^{6} (m^2 - m + 1)$ **26.** $\sum_{j=2}^{7} (j^2 - 3j)$

27. $\sum_{n=2}^{6} n^3$ **28.** $\sum_{k=3}^{8} (k-1)^3$ **29.** $\sum_{m=1}^{6} \frac{1}{m}$ **30.** $\sum_{k=1}^{5} \frac{1}{2^k}$

31. $\sum_{n=2}^{6} (-1)^{n+1} \cdot 2^n$ **32.** $\sum_{j=1}^{5} (-1)^{j-1} \cdot 3^j$ **33.** $\sum_{m=1}^{5} 6(3)^{m-2}$ **34.** $\sum_{n=1}^{5} 2^{5-n} \cdot 3^n$

B

35. $\displaystyle\sum_{n=1}^{6} 4\left(\frac{1}{2}\right)^{n-1}$

36. $\displaystyle\sum_{k=1}^{8} (\sqrt{2})^{k}$

37. $\displaystyle\sum_{m=1}^{5} \frac{1}{m(m+1)}$

38. $\displaystyle\sum_{n=1}^{10} |n-4|$

39. $\displaystyle\sum_{n=1}^{8} i^{n}$ (Use $i^2 = -1$.)

40. $\displaystyle\sum_{k=1}^{6} \log k$ (List all terms, then use Product Rule to write as a single log.)

Discover a formula a_n for the following sequences:

41. $\dfrac{2}{3}, \dfrac{4}{9}, \dfrac{8}{27}, \dfrac{16}{81}, \ldots$

42. $2, -4, 8, -16, \ldots$

43. $\sqrt{3}, 3, 3\sqrt{3}, 9, \ldots$

44. $3, 4, 5, 6, 7, \ldots$

45. $\dfrac{4}{1}, \dfrac{9}{2}, \dfrac{16}{3}, \dfrac{25}{4}, \ldots$

46. $2, 5, 10, 17, 26, 37, \ldots$

47. $8, 27, 64, 125, \ldots$

48. $0, 7, 26, 63, 124, \ldots$

C

Show that each of the following statements is true by first listing all terms in each sum, and then using the properties of logs to write them as a single log:

49. $\displaystyle\sum_{n=1}^{9} \log\frac{n+1}{n} = 1$

50. $\displaystyle\sum_{n=1}^{9} \log\frac{n}{n+1} = -1$

51. The **Fibonacci Sequence** is

$$1, 1, 2, 3, 5, 8, 13, 21, \ldots \quad .$$

a. Discover the pattern, and write the next eight terms in the sequence.
b. Is 2584 in the sequence? What about 4183?
c. This pattern has been observed in new stem growths on plants. After any stem is formed, it takes *two months* to sprout a new stem but only one month thereafter to sprout another. Starting with a single stem Jan in January, a new stem Mar appears in March. In April, only the Jan stem sprouts a new Apr stem. In May, both the Jan and the Mar stems sprout a new May stem. Note that the number of stems on the plant each month forms the Fibonacci Sequence. Can you draw the plant as it will appear in June and July?
d. In what month will the plant have 233 stems?

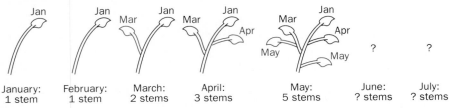

January: February: March: April: May: June: July:
1 stem   1 stem   2 stems   3 stems   5 stems ? stems ? stems

52. Discover a pattern for the sequence

$$1, 1, 2, 4, 7, 13, 24, 44, \ldots$$

and then write the next five terms.

53. Although it is not apparent, the sequence

$$2, 3, 5, 7, 11, 13, 17, \ldots$$

does have a pattern. Discover it, and then write the next ten terms.

54. For the sequence defined by

$$a_n = \left(1 + \frac{1}{n}\right)^n,$$

use your calculator to obtain the terms $a_1, a_{10}, a_{100}, a_{1000}, a_{10,000}, a_{100,000},$ and $a_{1,000,000}.$ Round your answers to the nearest 6 decimal places.

11.2 Arithmetic Sequences

A special type of sequence is illustrated by

$$2, 5, 8, 11, 14, 17, \ldots$$

wherein each term after the first can be obtained by adding 3 to its predecessor. It has a special name, according to the following definition:

> An **arithmetic sequence*** is a sequence in which each term after the first can be obtained by **adding** to the previous term a fixed number $d,$ called the **common difference** between consecutive terms.

Thus the sequence above is arithmetic, with common difference $d = 3.$ Likewise

$$3, -1, -5, -9, -13, \ldots \text{ is an arithmetic sequence,}$$
$$\text{with } \quad d = -4 \quad = -1 - 3 = -5 - (-1) = \cdots \quad .$$

But

$$1, 4, 9, 16, 25, \ldots \text{ is } not \text{ arithmetic,}$$
$$\text{because } 4 - 1 \neq 9 - 4 \neq 16 - 9 \neq \cdots \quad .$$

In the previous section you were asked to discover a formula for a sequence, given its first few terms. This can sometimes be done by quick inspection, but more often than not the formula is difficult to discover. For arithmetic sequences, however, the task becomes simple when we use a for-

---

*Also called an **arithmetic progression.**

mula that we now derive. Let a_1 = the first term and d = the common difference; then

$$a_2 = a_1 + d$$
$$a_3 = a_1 + 2d$$
$$a_4 = a_1 + 3d$$

$+ d$
$+ d$

$$a_n = a_1 + (n - 1)d$$

This is our desired formula:

> The **nth term** of an **arithmetic sequence** is given by
> $$a_n = a_1 + (n - 1)d$$
> where a_1 = the first term   and d = the common difference.

Using this formula, we can find any desired term in the sequence.

EXAMPLE 1

Obtain the indicated term for the arithmetic sequences *only* below:

a. 2, 5, 8, 11, 14, . . . ; a_{100}

The first term is $a_1 = 2$ and the common difference is $d = 3$; by the formula, the 100th term is

$$a_{100} = 2 + (100 - 1) \cdot 3$$
$$= 2 + 99 \cdot 3$$
$$= 299.$$

b. $1, \frac{1}{4}, -\frac{1}{2}, -1\frac{1}{4}, \ldots ; a_{50}$

It is helpful to write this sequence as

$$\frac{4}{4}, \frac{1}{4}, \frac{-2}{4}, \frac{-5}{4}, \ldots .$$

Here $a_1 = \frac{4}{4}$ and $d = \frac{-3}{4}$; by the formula, the 50th term is

$$a_{50} = \frac{4}{4} + (50 - 1)\left(\frac{-3}{4}\right)$$

$$= \frac{4}{4} + \left(\frac{-147}{4}\right) \qquad\qquad \frac{49}{1} \cdot \frac{-3}{4} = \frac{-147}{4}$$

$$= \frac{-143}{4}, \quad \text{or} \quad -35\frac{3}{4}.$$

c. 4.6, 4.9, 5.2, 5.6, 5.9, . . . ; a_{35}

This is *not* an arithemetic sequence, because

$$4.9 - 4.6 = 5.2 - 4.9 \neq 5.6 - 5.2$$

We cannot predict the 35th term.

Karl Friedrich Gauss is considered one of the greatest mathematicians ever. One day two centuries ago, when he was a mere lad of six, his grammar school class was unruly and the German schulemeister's stern discipline wasn't working. So he made the pupils add up all the numbers from 1 to 100. Before he could say "*Danke schoen,*" young Gauss had the correct answer of 5050. Calling this sum S, he astounded everyone by writing the sum forwards and backwards, and then adding:

$$
\begin{array}{l}
S = \quad\; 1 + \quad 2 + \quad 3 + \cdots + \quad 99 + 100 \\
S = 100 + \quad 99 + \quad 98 + \cdots + \quad 2 + \quad 1 \\
\hline
2S = 101 + 101 + 101 + \cdots + 101 + 101
\end{array}
$$

$$\underbrace{}_{100 \text{ terms}}$$

$$2S = 100(101)$$

$$S = \frac{100(101)}{2} = 5050.\text{*}$$

Not only did Gauss obtain the correct answer, but in so doing, this *wunderkind* had discovered a formula for the sum of the first n terms of an arithmetic sequence. Writing the sum, denoted by S_n, forwards and backwards, Gauss proceeded as above:

Forwards $S_n = \quad a_1 \quad + (a_1 + d) \; + (a_1 + 2d) + \cdots + (a_n - d) \; + \quad a_n$
Backwards $S_n = \quad a_n \quad + (a_n - d) \; + (a_n - 2d) + \cdots + (a_1 + d) \; + \quad a_1$

$$2S_n = (a_1 + a_n) + (a_1 + a_n) + (a_1 + a_n) + \cdots + (a_1 + a_n) + (a_1 + a_n)$$

$$\underbrace{}_{n \text{ terms}}$$

$$2S_n = n(a_1 + a_n), \quad \text{or} \quad S_n = \frac{n(a_1 + a_n)}{2}.$$

---

*While his schoolmates were struggling with sums like this, Gauss was already thinking about infinity! See Section 11.4 for more on infinity.

This is our desired formula:

> **Sum of an Arithmetic Sequence**
>
> The sum of the first n terms of an arithmetic sequence is
>
> $$S_n = \frac{n(a_1 + a_n)}{2}$$
>
> where a_1 = first term, a_n = last term = $a_1 + (n-1)d$, and n = number of terms.

EXAMPLE 2

Compute the sums of these arithmetic sequences:

a. $\displaystyle\sum_{k=1}^{100} (5k - 2) = 3 + 8 + 13 + 18 + \cdots + 498.$

The first term is $a_1 = 3$, the last term is $a_{100} = 498$, and there are $n = 100$ terms. By the formula, the sum of these 100 terms is

$$S_{100} = \frac{100(3 + 498)}{2}$$
$$= 25{,}050. \qquad 50(501)$$

b. $-1 + 3 + 7 + 11 + \cdots$ (30 terms)

Here $a_1 = -1$, $d = 4$, and $n = 30$, so the last term is

$$a_{30} = -1 + (30 - 1) \cdot 4$$
$$= -1 + 29 \cdot 4$$
$$= 115.$$

The sum of the first 30 terms is

$$S_{30} = \frac{30(-1 + 115)}{2}$$
$$= 1710. \qquad 15(114)$$

c. $4 + 6 + 8 + 10 + \cdots + 180.$

We have $a_1 = 4$, $d = 2$, and $a_n = 180$, but $n = ?$ To find n = the number of terms, we use the formula

$$a_n = a_1 + (n-1)d, \quad \text{which becomes}$$
$$180 = 4 + (n-1) \cdot 2$$
$$180 = 4 + 2n - 2$$
$$178 = 2n, \quad \text{or} \quad n = 89 \text{ terms.}$$

The summation formula now gives

$$S_{89} = \frac{89(4 + 180)}{2}$$

$$= 8188. \qquad \frac{89(184)}{2} = 89(92)$$

EXERCISE 11.2   *Answers, pages A100–A101*

A

Obtain the indicated term *only* for the arithmetic sequences below, as in EXAMPLE 1:

1. 2, 7, 12, 17, . . .; a_{50}

2. 2, 9, 16, 23, . . .; a_{80}

3. −1, 3, 7, 11, . . .; a_{105}

4. −4, −2, 0, 2, . . .; a_{215}

5. 5, 1, −3, −7, . . .; a_{86}

6. −4, −7, −10, −13, . . .; a_{62}

7. 3, 8, 13, 19, . . .; a_{60}

8. 1, −3, −9, −13, . . .; a_{26}

9. $1\frac{3}{4}, 2\frac{1}{4}, 2\frac{3}{4}, 3\frac{1}{4}, \ldots; a_{40}$

10. $7\frac{2}{3}, 8\frac{1}{3}, 9, 9\frac{2}{3}, \ldots; a_{32}$

11. 5, 5, 5, 5, . . .; a_{206}

12. −2, −2, −2, −2, . . .; a_{22}

13. 2, 4, 8, 16, . . .; a_{20}

14. 1, 4, 9, 16, . . .; a_{85}

15. 5, 4.27, 3.54, 2.81, . . .; a_{70}

16. 4.62, 5.15, 5.68, 6.19, . . .; a_{65}

17. $2\frac{1}{2}, 3\frac{3}{8}, 4\frac{1}{4}, 5\frac{3}{8}, \ldots; a_{15}$

18. $3\frac{1}{6}, 3\frac{3}{4}, 4\frac{1}{3}, 4\frac{11}{12}, \ldots; a_{75}$

Compute the sums of these arithmetic sequences, as in EXAMPLE 2:

19. $\displaystyle\sum_{k=1}^{60} (3k + 2)$

20. $\displaystyle\sum_{m=1}^{75} (4m - 3)$

21. $\displaystyle\sum_{j=1}^{105} (5 - 2j)$

22. $\displaystyle\sum_{n=1}^{120} (1 - 3n)$

23. $\displaystyle\sum_{k=1}^{200} \frac{2k}{3}$

24. $\displaystyle\sum_{m=1}^{55} \left(\frac{3m + 4}{6}\right)$

25. 2 + 5 + 8 + 11 + · · ·   (50 terms)

26. 4 + 1 − 2 − 5 − · · ·   (80 terms)

27. $1 + 1\frac{1}{2} + 2 + 2\frac{1}{2} + \cdots$ (60 terms)

28. $7\frac{3}{4} + 8\frac{1}{4} + 8\frac{3}{4} + 9\frac{1}{4} + \cdots$ (100 terms)

29. 5 + 5.1 + 5.2 + 5.3 + · · ·   (200 terms)

30. 6.3 + 6.6 + 6.9 + 7.2 + · · ·   (300 terms)

31. 3 + 6 + 9 + · · · + 117

32. −2 + 5 + 12 + · · · + 271

33. 5 + 1 − 3 − 7 − · · · − 255

34. 7 + 2 − 3 − 8 − · · · − 93

35. 2 + 4 + 6 + 8 + · · · + 500

36. −3 − 6 − 9 − 12 − · · · − 300

B

37. $\sqrt{2} + \sqrt{8} + \sqrt{18} + \sqrt{32} + \cdots$ (30 terms)   *Hint:* Simplify radicals.

38. $\log 2 + \log 4 + \log 8 + \log 16 + \ldots$ (20 terms)
Hint: This is $\log 2 + \log 2^2 + \log 2^3 + \cdots$;   now use the Power Rule.

39. $c + 3c + 5c + 7c + \cdots$ (80 terms)

40. $(a + 2b) + (2a + 4b) + (3a + 6b) + (4a + 8b) + \cdots$ (100 terms)

41. A freely falling body falls 16 feet during the 1st second, 48 feet during the 2nd second, 80 feet during the 3rd second, and so forth.
a. How far does it fall during the 10th second?
b. Find the total distance it falls in 10 seconds.

42. A job pays $25,000 the first year, $26,500 the second year, $28,000 the third year, and so forth.
a. Find the salary in the twelfth year.
b. Find the total income earned on this job in twelve years.

43. A pile of logs (the wooden kind, not the logarithmic kind) is stacked with 30 logs on the bottom, 29 in the second row, 28 in the third row, and so forth.
a. If there are 20 rows in all, how many logs are in the pile?
b. How many additional logs would be needed for the stack to form a perfect triangle?

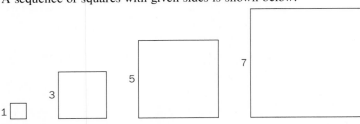

44. A pile of logs (the logarithmic kind, not the wooden kind) is stacked in a triangle, with log 2 on top, log 4 and log 8 in the second row, log 16, log 32, and log 64 in the third row, and so forth.
a. If there are 10 rows in all, how many logs are in the pile?
b. This is a toughie! Find the sum of the logs in the pile. *Hint:* See Problem 38.

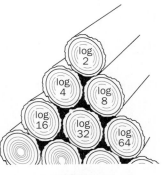

PROBLEM 44

45. In consecutive years, Smith earns $80,000, $83,000, $86,000, . . ., and at the same time, Jones earns $46,000, $51,000, $56,000, ...
a. In what year will Jones earn the same as Smith?
b. Find their common salary the year they earn the same amount.
c. Find the total amount that each person earns during this entire period.

46. A sequence of squares with given sides is shown below:

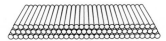

7

5

3

1

· · ·

a. If there are 25 squares in all, find the side of the last square.
b. Use the formula $P = 4s$ to write down the perimeters of the first few squares. If these form an arithmetic sequence, find the sum of the perimeters of the 25 squares.
c. Use the formula $A = s^2$ to write down the areas of the first few squares. Find the sum of the 25 areas only if these form an arithmetic sequence.

47. Under *straight-line depreciation,* a certain apartment building originally purchased for $1,500,000 is worth $1,425,000 for tax purposes after 1 year. It is "worth" $1,350,000 after 2 years, $1,275,000 after 3 years, and so forth.
a. Find its worth, for tax purposes, after 12 years.
b. After how many years will its value be completely written off?

48. Find the product $a^1 a^2 a^3 a^4 \ldots a^{100}$. *Hint:* Product Law of Exponents.

49. Show that the formula for the sum of an arithmetic sequence

$$S_n = \frac{n(a_1 + a_n)}{2} \quad \text{can also be written as} \quad S_n = \frac{n[2a_1 + (n-1)d]}{2}.$$

Hint: Substitute the formula for a_n.

50. Solve the formula $\quad S_n = \frac{n(a_1 + a_n)}{2} \quad$ for $\quad a_n$.

The missing terms in the arithmetic sequence

$$\underline{\quad}, 7, \underline{\quad}, \underline{\quad}, \underline{\quad}, \underline{\quad}, 22, \underline{\quad}, \underline{\quad}, \ldots$$

can be found by adding the common difference d as follows:

$$\underline{\quad}, 7, \underline{7+d}, \underline{7+2d}, \underline{7+3d}, \underline{7+4d}, 7+5d, \underline{\quad}, \underline{\quad}, \ldots .$$

Equating the common terms gives $\quad 7 + 5d = 22, \quad$ or $\quad d = 3$.
Filling in the missing terms gives

$$\underline{\ 4\ }, 7, \underline{\ 10\ }, \underline{\ 13\ }, \underline{\ 16\ }, \underline{\ 19\ }, 22, \underline{\ 25\ }, \underline{\ 28\ }, \ldots .$$

In like manner, fill in the missing terms in these arithmetic sequences:

51. $\underline{\quad}, 4, \underline{\quad}, \underline{\quad}, \underline{\quad}, 28, \underline{\quad}, \underline{\quad}, \ldots$ **52.** $\underline{\quad}, 2, \underline{\quad}, \underline{\quad}, \underline{\quad}, \underline{\quad}, \underline{\quad}, 44, \underline{\quad}, \ldots$

53. $\underline{\quad}, \underline{\quad}, 3, \underline{\quad}, \underline{\quad}, \underline{\quad}, \underline{\quad}, \underline{\quad}, 6, \underline{\quad}, \ldots$ **54.** $5, \underline{\quad}, \underline{\quad}, \underline{\quad}, \underline{\quad}, \underline{\quad}, \underline{\quad}, \underline{\quad}, 3, \underline{\quad}, \ldots$

Use the formula $\quad S_n = \frac{n(a_1 + a_n)}{2} \quad$ to find the value of n such that the given sums equal
the indicated amounts. You will need to solve quadratic equations here.

55. $1 + 2 + 3 + \cdots + n = 36$ **56.** $2 + 4 + 6 + \cdots + 2n = 240$

C

57. $2 + 5 + 8 + \cdots + (3n - 1) = 155$ **58.** $1 + 3 + 5 + \cdots + (2n - 1) = 100$

A sequence is called *harmonic* if it can be written in the form

$$\frac{1}{a_1}, \frac{1}{a_2}, \frac{1}{a_3}, \frac{1}{a_4}, \cdots$$

in which $a_1, a_2, a_3, a_4, \ldots$ is an arithmetic sequence. **Which of the following are harmonic sequences?**

59. $\dfrac{1}{1}, \dfrac{1}{2}, \dfrac{1}{3}, \dfrac{1}{4}, \dfrac{1}{5}, \cdots$

60. $1, \dfrac{1}{3}, \dfrac{1}{5}, \dfrac{1}{7}, \dfrac{1}{9}, \cdots$

61. $\dfrac{1}{3}, \dfrac{1}{7}, \dfrac{1}{11}, \dfrac{1}{17}, \cdots$

62. $2, 1, \dfrac{2}{3}, \dfrac{1}{2}, \dfrac{2}{5}, \dfrac{1}{3}, \dfrac{2}{7}, \cdots$ *Hint:* This is $\dfrac{1}{\frac{1}{2}}, \dfrac{1}{1}, \dfrac{1}{?}, \dfrac{1}{?}, \cdots$

63. An accelerated method of depreciating real estate for tax purposes is called *sum-of-digits depreciation*. Suppose property originally valued at \$2,200,000 is to be written off in 10 years. According to this method, in the first year,

$$\frac{10}{1 + 2 + 3 + \cdots + 10}(\$2{,}200{,}000) = \frac{10}{55}(\$2{,}200{,}000) = \$400{,}000$$

can be written off. In the second year,

$$\frac{9}{1 + 2 + 3 + \cdots + 10}(\$2{,}200{,}000) = \frac{9}{55}(\$2{,}200{,}000) = \$360{,}000$$

can be written off; and so forth, for the remaining 8 years.
a. Find the tax write-off in the 10th year.
b. Find the *total* amount written off in 10 years.

64. Repeat Problem 63 for a \$900,000 apartment building that is to be written off over 8 years.

11.3   Geometric Sequences

The money introduced at the beginning of this chapter,

$$1\cent, \ 2\cent, \ 4\cent, \ 8\cent, \ 16\cent, \ 32\cent, \ \ldots$$

forms a sequence in which each payment after the first can be obtained by multiplying its predecessor by 2. It illustrates a special type of sequence, which we now define:

> A **geometric sequence*** is a sequence in which each term after the first can be obtained by **multiplying** the

---

*Also called a **geometric progression**.

previous term by a fixed number r, called the **common ratio** of consecutive terms.

Thus the sequence above is geometric,

with common ratio $r = 2 = \dfrac{2}{1} = \dfrac{4}{2} = \dfrac{8}{4} = \ldots$.

Likewise,

2, 6, 18, 54, . . . is a geometric sequence,

with $r = 3 = \dfrac{6}{2} = \dfrac{18}{6} = \dfrac{54}{18} = \ldots$,

and

$2, -1, \dfrac{1}{2}, -\dfrac{1}{4}, \ldots$ is a geometric sequence,

with $r = -\dfrac{1}{2} = \dfrac{-1}{2} = \dfrac{\frac{1}{2}}{-1} = \ldots$.

But   2, 6, 10, 14, . . . is *not* geometric, because $\dfrac{6}{2} \neq \dfrac{10}{6} \neq \dfrac{14}{10} \neq \ldots$.

(What kind of a sequence is it?)

As in the case of arithmetic sequences, there is a formula for a_n, the nth term of a geometric sequence. Let $a_1 =$ the first term and $r =$ the common ratio; then

$$
\begin{aligned}
a_2 &= a_1 r \\
a_3 &= a_1 r^2 \\
a_4 &= a_1 r^3 \\
&\ \ . \\
&\ \ . \\
&\ \ . \\
a_n &= a_1 r^{n-1}
\end{aligned}
$$

$\times r$

$\times r$

This is our desired formula:

The ***n*th term** of a **geometric sequence** is given by

$$a_n = a_1 r^{n-1}$$

where $a_1 =$ the first term   and $r =$ the common ratio.

Using this formula, we can find any desired term in the sequence.

EXAMPLE 1

Obtain the indicated term for the geometric sequences *only* below:

a. 2, 6, 18, 54, . . .; a_8

The first term is $a_1 = 2$ and the common ratio is $r = \dfrac{6}{2} = 3$; then, by the formula, the 8th term is

$$\begin{aligned} a_8 &= 2(3)^{8-1} \\ &= 2(3)^7 \\ &= 2(2187) \\ &= 4374. \end{aligned}$$

b. 2, -1, $\dfrac{1}{2}$, $-\dfrac{1}{4}$, . . .; a_{10}

We have $a_1 = 2$ and $r = \dfrac{-1}{2} = -\dfrac{1}{2}$; the 10th term is

$$\begin{aligned} a_{10} &= 2\left(-\dfrac{1}{2}\right)^{10-1} \\ &= 2\left(-\dfrac{1}{2}\right)^9 \\ &= 2\left(\dfrac{-1}{512}\right) \\ &= \dfrac{-1}{256}. \end{aligned}$$

c. 2, -4, 6, -8, . . .; a_{14}

This is *not* a geometric sequence, because

$$\dfrac{-4}{2} \neq \dfrac{6}{-4} \neq \dfrac{-8}{6} \neq \cdots \quad .$$

d. 12, 8, $\dfrac{16}{3}$, $\dfrac{32}{9}$, . . .; a_9

This is a geometric sequence, with common ratio

$$r = \dfrac{8}{12} = \dfrac{16/3}{8} = \dfrac{32/9}{16/3} = \cdots = \dfrac{2}{3}.$$

The first term is $a_1 = 12$, so by the formula,

$$\begin{aligned} a_9 &= 12\left(\dfrac{2}{3}\right)^{9-1} \\ &= 12\left(\dfrac{2}{3}\right)^8 \end{aligned}$$

$$= 12\left(\frac{256}{6561}\right) \qquad \frac{\overset{1}{\cancel{3} \cdot 4 \cdot 256}}{\underset{2187}{\cancel{6561}}}$$

$$= \frac{1024}{2187}.$$

e. 5000, 5000(1.07), 5000(1.07)^2, 5000(1.07)^3, . . .; a_{12}

This is a geometric sequence, with common ratio

$$r = \frac{5000(1.07)}{5000} = \frac{5000(1.07)^2}{5000(1.07)} = \cdots 1.07.$$

The first term is $a_1 = 5000$, so by the formula,

$$a_{12} = 5000(1.07)^{12-1}$$
$$= 5000(1.07)^{11} \qquad 5000 \boxed{\text{x}} \, 1.07 \, \boxed{\text{x}^{\text{y}}} \, 11 \, \boxed{=}$$
$$= 10{,}524.26$$

In order to compute the huge sum of money promised at the beginning of this chapter, we first need a formula for the sum of the first n terms of a geometric sequence:

$$S_n = a_1 + a_1 r + a_1 r^2 + a_1 r^3 + \cdots + a_1 r^{n-1}.$$

In the previous section, we recounted how young Gauss used a trick to find the sum of an arithmetic sequence. Here we must be just as clever. We first multiply the preceding sum by r:

$$rS_n = a_1 r + a_1 rr + a_1 r^2 r + a_1 r^3 r + \cdots + a_1 r^{n-1} r$$
$$= a_1 r + a_1 r^2 + a_1 r^3 + a_1 r^4 + \cdots + a_1 r^n$$

When we subtract this equation from the first one, all the terms conveniently "cancel out" except the first and the last:

$$\begin{aligned} S_n &= a_1 + a_1 r + a_1 r^2 + a_1 r^3 + \cdots + a_1 r^{n-1} \\ -rS_n &= \quad\;\; - a_1 r - a_1 r^2 - a_1 r^3 - \quad\;\; - a_1 r^{n-1} - a_1 r^n \\ \hline S_n - rS_n &= a_1 \qquad\qquad\qquad\qquad\qquad\qquad\qquad\; - a_1 r^n \end{aligned}$$

$$S_n(1 - r) = a_1 - a_1 r^n, \quad \text{or} \quad S_n = \frac{a_1(1 - r^n)}{1 - r} \qquad \text{Factor and divide.}$$

This is our desired formula:

Sum of a Geometric Sequence

The sum of the first n terms of a geometric sequence is

$$S_n = \frac{a_1(1 - r^n)}{1 - r}, \quad r \neq 1$$

where a_1 = first term, r = common ratio, and n = number of terms.

EXAMPLE 2 | Compute the sums of these geometric sequences:

a. $\displaystyle\sum_{k=1}^{8} 3(2)^k = 6 + 12 + 24 + \cdots + 768$

Here $a_1 = 6$, $r = \dfrac{12}{6} = 2$, and $n = 8$ terms; thus the sum is

$$S_8 = \frac{6(1 - 2^8)}{1 - 2}$$

$$= \frac{6(1 - 256)}{-1}$$

$$= \frac{6(-255)}{-1}$$

$$= 1530.$$

Check: $6 + 12 + 24 + 48 + 96 + 192 + 384 + 768 \overset{\checkmark}{=} 1530$

b. $1 - \dfrac{1}{2} + \dfrac{1}{4} - \dfrac{1}{8} + \cdots$ (10 terms)

We have $a_1 = 1$, $r = \dfrac{\dfrac{-1}{2}}{1} = -\dfrac{1}{2}$, and $n = 10$; thus

$$S_{10} = \frac{1\left[1 - \left(-\dfrac{1}{2}\right)^{10}\right]}{1 - \left(-\dfrac{1}{2}\right)} \leftarrow 1 - \left(-\dfrac{1}{2}\right)^{10} \neq 1 + \left(+\dfrac{1}{2}\right)^{10}$$

$$= \frac{1 - \dfrac{1}{1024}}{1 + \dfrac{1}{2}}$$

$$= \frac{\dfrac{1023}{1024}}{\dfrac{3}{2}} \longleftarrow \frac{1024}{1024} - \frac{1}{1024}$$

$$= \frac{341}{512}.$$

$$\begin{array}{c} 341 \quad 1 \\ \dfrac{\cancel{1023}}{\cancel{1024}} \cdot \dfrac{\cancel{2}}{\cancel{3}} \\ 512 \quad 1 \end{array}$$

Check: $1 - \dfrac{1}{2} + \dfrac{1}{4} - \dfrac{1}{8} + \dfrac{1}{16} - \dfrac{1}{32} + \dfrac{1}{64} - \dfrac{1}{128} + \dfrac{1}{256} - \dfrac{1}{512}$

$$= \dfrac{512 - 256 + 128 - 64 + 32 - 16 + 8 - 4 + 2 - 1}{512}$$

$$\overset{\checkmark}{=} \dfrac{341}{512}.$$

c. $20{,}000 + 20{,}000(1.06) + 20{,}000(1.06)^2 + \cdots$ (8 terms)

We have $a_1 = 20{,}000$ and $r = 1.06$ (see Example 1e); thus

$$S_8 = \dfrac{20{,}000[1 - (1.06)^8]}{1 - 1.06}$$

$$= 197{,}949.36$$

Calculator: $20{,}000 \;\boxed{(}\; 1 \;\boxed{-}\; 1.06 \;\boxed{x^y}\; 8 \;\boxed{)}\; \boxed{\div}\; \boxed{(}\; 1 \;\boxed{-}\; 1.06 \;\boxed{)}\; \boxed{=}$

Finally, the sum of the 30 payments promised at the beginning of this chapter,

$$1¢ + 2¢ + 4¢ + 8¢ + 16¢ + 32¢ + \cdots$$

has $a_1 = 1¢$, $r = 2$, and $n = 30$. Thus

$$S_{30} = \dfrac{1¢(1 - 2^{30})}{1 - 2}$$

$$= 1{,}073{,}740{,}000¢ \qquad 1 \;\boxed{-}\; 2 \;\boxed{x^y}\; 30 \;\boxed{=}\; \boxed{\div}\; 1 \;\boxed{+/-}\; \boxed{=}$$

$$= \$10{,}737{,}400. \qquad\quad ¢ \text{ to } \$: \text{ divide by } 100.$$

The other offer, you recall, was

$$\$10{,}000{,}000 + 10\% \text{ interest for 30 days} + \$150{,}000 \text{ Ferrari}$$

The interest for 30 days $= \dfrac{30}{365}$ years is

$$I = Prt = \$10{,}000{,}000(0.10)\left(\dfrac{30}{365}\right) = \$82{,}192.$$

The total package is worth "only"

$$\$10{,}000{,}000 + \$82{,}192 + \$150{,}000 = \$10{,}232{,}192.$$

Thus waiting 30 days would make you

$$\$10{,}737{,}400 - \$10{,}232{,}192 = \$505{,}208 \text{ richer.}$$

Patience is profitable. Let the early bird drive off in his Ferrari; for the difference you can buy three of them and gasoline for a lifetime!

EXERCISE 11.3   *Answers, page A101*

A

Obtain the indicated term *only* for the geometric sequences below, as in EXAMPLE 1:

1. 3, 6, 12, 24, . . .; a_8

2. $-4, -12, -36, -108, . . .; a_9$

3. $\frac{1}{4}, -\frac{1}{2}, 1, -2, . . .; a_{12}$

4. $-\frac{1}{2}, 2, -8, 32, . . .; a_{10}$

5. $2, \frac{4}{3}, \frac{8}{9}, \frac{16}{27}, . . .; a_8$

6. $2, \frac{-3}{2}, \frac{9}{8}, \frac{-27}{32}, . . .; a_8$

7. $4, -2, \frac{1}{2}, -\frac{1}{4}, . . .; a_{12}$

8. $1, \frac{1}{4}, \frac{1}{9}, \frac{1}{16}, . . .; a_{10}$

9. 4, 8, 12, 16, . . .; a_{10}

10. 3, 3, 3, 3, . . .; a_{15}

11. $24, -18, \frac{27}{2}, \frac{-81}{8}, . . .; a_7$

12. 72, 60, 50, . . .; a_6

13. $\frac{-3}{4}, \frac{-1}{2}, \frac{-1}{3}, \frac{-2}{9}, . . .; a_8$

14. $\frac{1}{3}, \frac{-1}{2}, \frac{3}{4}, \frac{-9}{8}, . . .; a_8$

15. $2\frac{2}{9}, 3\frac{1}{3}, 5, 7\frac{1}{2}, . . .; a_7$

Hint: Write as improper fractions.

16. $6\frac{3}{4}, 4\frac{1}{2}, 3, 2, . . .; a_9$

17. $2000, 2000(1.05), 2000(1.05)^2, 2000(1.05)^3, . . .; a_8$

18. $15,000, 15,000(0.92), 15,000(0.92)^2, 15,000(0.92)^3, . . .; a_{10}$

19. $1, \sqrt{2}, 2, \sqrt{8}, . . .; a_{10}$

20. $1, i, -1, -i, . . .; a_{26}$ *Hint*: See Exercise 6.1, Problems 53–62.

Compute the sums of these geometric sequences, as in EXAMPLE 2:

21. $\sum_{k=1}^{8} 2(3)^k$

22. $\sum_{n=1}^{9} 3(5)^n$

23. $\sum_{m=1}^{7} \left(\frac{-1}{3}\right)^{m-1}$

24. $\sum_{j=1}^{8} \left(\frac{2}{3}\right)^j$

25. $\sum_{n=1}^{20} (-2)^n$

26. $\sum_{k=1}^{18} 3^{k-1}$

27. $3 + \frac{9}{2} + \frac{27}{4} + \frac{81}{8} + \cdots$ (6 terms)

28. $3 - \frac{3}{2} + \frac{3}{4} - \frac{3}{8} + \cdots$ (8 terms)

29. $8000 + 8000(1.09) + 8000(1.09)^2 + 8000(1.09)^3 + \cdots$ (12 terms)

30. $25,000 + 25,000(0.88) + 25,000(0.88)^2 + 25,000(0.88)^3 + \cdots$ (10 terms)

B

Find the indicated term *only* for the geometric sequences below:

31. $x, x^4, x^7, x^{10}, . . .; a_{12}$

32. $x^{-1}, x^{-2}, x^{-4}, x^{-8}, . . .; a_9$

33. $xy^3, x^2y, \frac{x^3}{y}, \frac{x^4}{y^3}, . . .; a_{10}$

34. $\dfrac{c}{d}, d, \dfrac{d^3}{c}, \dfrac{d^5}{c^2}, \ldots; a_9$ **35.** $\dfrac{1}{\sqrt{2}}, \dfrac{-1}{2}, \dfrac{\sqrt{2}}{4}, \ldots; a_8$ **36.** $\sqrt{6}, 2\sqrt{3}, \sqrt{24}, 4\sqrt{3}, \ldots; a_{10}$

37. $1, 3, 5, 7, \ldots$ is an arithmetic sequence. Show that

$$2^1, 2^3, 2^5, 2^7, \ldots$$

is a *geometric* sequence by exhibiting the common ratio r.

38. $2, 4, 8, 16, \ldots$ is a geometric sequence. Show that

$$\log 2, \log 4, \log 8, \log 16, \ldots$$

is an *arithmetic* sequence by exhibiting the common difference d.
Hint: $\log 2, \log 2^2, \log 2^3, \log 2^4, \ldots$

39. Rank these payment plans, from best to worst:
  a. $1¢ + 2¢ + 4¢ + 8¢ + 16¢ + \cdots$ (for 30 days)
  b. $1¢ + 3¢ + 9¢ + 27¢ + 81¢ + \cdots$ (for 20 days)
  C. $\$100{,}000 + \$200{,}000 + \$300{,}000 + \$400{,}000 + \cdots$ (for 15 days)

40. In the sequence of payments $1¢, 2¢, 4¢, 8¢, \ldots$ over 30 days, which was described in the text, how much was actually paid on the *last day*?

41. The first swing of a pendulum is 12 feet. Each swing thereafter is $\dfrac{3}{4}$ of the preceding swing.
  a. Find the distance traveled by the pendulum on the 8th swing.
  b. Find the total distance the pendulum has traveled after 8 swings.

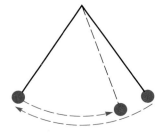

Problem 41

42. A ball is thrown upward to a height of 50 feet and returns to the ground. After each bounce, it returns to a height that is 90% of its previous height.
  a. To what height does it rise after the 6th bounce?
  b. Find the total vertical distance it travels at the moment of the 7th bounce.

43. A job pays $25,000 the first year, with 8% annual raises thereafter. The salary in the second year will be

$$\$25{,}000 + \$25{,}000(0.08) = \$25{,}000(1 + 0.08) = \$25{,}000(1.08).$$

Continuing in this manner, it can be shown that the annual salaries will be

$$\$25{,}000, \$25{,}000(1.08), \$25{,}000(1.08)^2, \$25{,}000(1.08)^3, \ldots$$

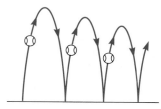

Problem 42

  a. Find the salary in the 10th year.
  b. Find the total income earned in the first 10 years.

44. In order to provide for their child's college education, a family deposits $1000 each year into a fund paying 9% interest compounded annually. Such a sequence of equal payments made on a regular basis is called an **annuity**. Suppose these payments start at birth and continue on each birthday, up to and including the 18th birthday, at which time the annuity ends and the total amount is withdrawn. Because the first deposit draws interest for 18 years, it will grow to $\$1000(1.09)^{18}$. (See Section 10.1, Example 7.) Likewise the second deposit, which runs for 17 years, will accumulate to $\$1000(1.09)^{17}$, and so forth. The annuity schedule looks like this:

Child's age	Deposit	Accumulated amount on 18th birthday
at birth	$1000	$\$1000(1.09)^{18}$
1	$1000	$\$1000(1.09)^{17}$
2	$1000	$\$1000(1.09)^{16}$
.	.	.
.	.	.
.	.	.
.	.	.
17	$1000	$\$1000(1.09)^{1}$
18	$1000	$1000

Find the total value of this annuity.

The missing terms in the geometric sequence

$$\_\_, 2, \_\_, 8, \_\_, \_\_, \cdots$$

can be found by multiplying by the common ratio *r* as follows:

$$\_\_, 2, \underline{2r}, 2r^2, \_\_, \_\_, \cdots$$

Equating the common terms, as in Exercise 11.2, Problems 51–54, gives

$$2r^2 = 8, \quad \text{or } r^2 = 4, \quad \text{or } r = \pm 2$$

If $r = 2$, the sequence becomes $\underline{1}$, 2, $\underline{4}$, 8, $\underline{16}$, 32, ...
If $r = -2$, the sequence becomes $\underline{-1}$, 2, $\underline{-4}$, 8, $\underline{-16}$, $\underline{32}$, ...

In like manner, fill in the missing terms in these geometric sequences:

45. __, 5, __, 45, __, __, · · ·

46. __, 12, __, 3, __, __, · · ·

47. __, __, 6, __, 12, __, · · ·

48. __, __, 6, __, 2, __, __, · · ·

49. __, 6, __, __, 48, __, · · ·

50. __, __, 4, __, __, __, $\dfrac{1}{4}$, · · ·

51. Fill in the missing terms in such a way that the sequence

$$\_\_, 3, \_\_, \_\_, -24, \_\_, \_\_, \cdots$$

     a. is geometric;   **b.** is arithmetic.

52. Is it possible for a sequence to be both arithmetic and geometric? If so, give an example of such a sequence. If not, tell why.

Find these sums, in simplest form:

53. $\sqrt{2} + 2 + \sqrt{8} + \cdots$ (12 terms)

54. $\sqrt{3} + \sqrt{12} + \sqrt{48} + \cdots$ (9 terms)

C

55. $1 - \dfrac{1}{\sqrt{2}} + \dfrac{1}{2} - \dfrac{\sqrt{2}}{4} + \cdots$ (8 terms)

56. $i + i^2 + i^3 + \cdots$ (15 terms)

11.4   Infinite Geometric Sequences

By adding "all of the terms" of a geometric sequence, we obtain the sum of an *infinite* number of terms, as illustrated by

$$\frac{1}{2} + \frac{1}{4} + \frac{1}{8} + \frac{1}{16} + \frac{1}{32} + \frac{1}{64} + \cdots = ?$$

To interpret the meaning of this "sum," let us compute the "partial sums":

$$S_1 = \frac{1}{2}$$

$$S_2 = \frac{1}{2} + \frac{1}{4} = \frac{3}{4}$$

$$S_3 = \frac{1}{2} + \frac{1}{4} + \frac{1}{8} = \frac{7}{8}$$

$$S_4 = \frac{1}{2} + \frac{1}{4} + \frac{1}{8} + \frac{1}{16} = \frac{15}{16}$$

$$S_5 = \frac{1}{2} + \frac{1}{4} + \frac{1}{8} + \frac{1}{16} + \frac{1}{32} = \frac{31}{32}$$

$$S_\infty = \frac{1}{2} + \frac{1}{4} + \frac{1}{8} + \frac{1}{16} + \frac{1}{32} + \frac{1}{64} + \cdots \overset{?}{=} 1$$

The pattern suggests that the sequence of partial sums "approaches" 1:

$$\frac{1}{2}, \frac{3}{4}, \frac{7}{8}, \frac{15}{16}, \frac{31}{32}, \cdots \longrightarrow 1$$

Certainly these partial sums never "reach" 1, but they can get "as close to 1 as we please." Calculus students study this concept, which is called "limits," and would conclude that

$$S_\infty = \frac{1}{2} + \frac{1}{4} + \frac{1}{8} + \frac{1}{16} + \frac{1}{32} + \frac{1}{64} + \cdots = 1$$

To derive a formula for the sum of infinitely many terms of *any* geometric sequence, let us compute the fifteenth partial sum of this sequence, S_{15}, using the formula given in the previous section:

$$S_{15} = \frac{1}{2} + \frac{1}{4} + \frac{1}{8} + \cdots + \frac{1}{32768}$$

$$= \frac{\frac{1}{2}\left[1 - \left(\frac{1}{2}\right)^{15}\right]}{1 - \frac{1}{2}} \qquad a_1 = \frac{1}{2}, \ r = \frac{1}{2}, \ n = 15$$

$$= \frac{\frac{1}{2}\left[1 - \frac{1}{32768}\right]}{\frac{1}{2}} \qquad \text{Approximately 0}$$

$$= \frac{32767}{32768} \qquad \frac{32768}{32768} - \frac{1}{32768}$$

$$= 0.9999694$$

Note how close to 1 this partial sum is. In the above expression, the term $\left(\frac{1}{2}\right)^{15} = \frac{1}{32768} = 0.0000305175$ is approximately zero. This term will be crucial in finding the sum of an infinite number of terms in any geometric sequence:

$$S_\infty = a_1 + a_1 r + a_1 r^2 + a_1 r^3 + \cdots = ?$$

As before, the nth partial sum S_n is given by

$$S_n = \frac{a_1(1 - r^n)}{1 - r} \qquad \begin{array}{l}\text{Approximately 0 if } |r| < 1 \\ \text{and } n \text{ is large}\end{array}$$

Calculus students also learn that "if $|r| < 1$, and n approaches infinity, then the term r^n approaches zero." This fact leads to a surprisingly simple result:

$$S_\infty = \frac{a_1(1 - 0)}{1 - r} = \frac{a_1}{1 - r}. *$$

_____

*See Problem 32 in the exercises for *another proof* of this formula.

> **Sum of an Infinite Geometric Sequence**
>
> The sum of an infinite geometric sequence is
>
> $$S_\infty = \frac{a_1}{1-r}$$
>
> where a_1 = first term   and r = common ratio, $|r| < 1$.

It is crucial that $|r| < 1$. For example, the sum

$$1 + 2 + 4 + 8 + 16 + \cdots$$

increases without bound and, therefore, has *no finite sum*. This is because its common ratio $r = 2 \not< 1$. See also Problems 30 and 31 in the exercises. The sum of an infinite sequence is also called an *infinite series*.

EXAMPLE 1   Compute the sums of these infinite geometric sequences:

a. $2 - 1 + \dfrac{1}{2} - \dfrac{1}{4} + \cdots$

We have $a_1 = 2$ and $r = -\dfrac{1}{2}$.

Because $\left|-\dfrac{1}{2}\right| = \dfrac{1}{2} < 1$, we can use the preceding formula:

$$S_\infty = \frac{2}{1 - \left(-\dfrac{1}{2}\right)}$$

$$= \frac{2}{1 + \dfrac{1}{2}}$$

$$= \frac{2}{\dfrac{3}{2}}$$

$$= \frac{4}{3}. \qquad\qquad \frac{2}{1} \cdot \frac{2}{3} = \frac{4}{3}$$

Thus $2 - 1 + \dfrac{1}{2} - \dfrac{1}{4} + \cdots = \dfrac{4}{3}$.

b. $\displaystyle\sum_{n=1}^{\infty} \left(\frac{1}{3}\right)^{n-1} = 1 + \frac{1}{3} + \frac{1}{9} + \frac{1}{27} + \cdots$

(continued)

Here $a_1 = 1$ and $r = \dfrac{1}{3}$.

Because $\left|\dfrac{1}{3}\right| = \dfrac{1}{3} < 1$, we can use the formula:

$$S_\infty = \dfrac{1}{1 - \dfrac{1}{3}}$$

$$= \dfrac{1}{\dfrac{2}{3}}$$

$$= \dfrac{3}{2}. \qquad\qquad \dfrac{1}{1} \cdot \dfrac{3}{2}$$

Thus $1 + \dfrac{1}{3} + \dfrac{1}{9} + \dfrac{1}{27} + \cdots = \dfrac{3}{2}$.

In Section 1.4, we saw that a terminating or repeating decimal represents a rational number. For example, these terminating decimals represent fractions:

$$0.6 = \dfrac{6}{10} = \dfrac{3}{5}$$

$$2.45 = 2\dfrac{45}{100} = \dfrac{245}{100} = \dfrac{49}{20}.$$

But what rational number does the repeating decimal

$$0.21212121 \ldots$$

represent? In the next example, we will express this decimal as the sum of an infinite geometric sequence, resulting in the correct rational number.

EXAMPLE 2   Convert these repeating decimals into rational numbers:
a. 0.21212121 . . .

This is the sum of the infinite geometric sequence

$$.21 + .0021 + .000021 + .00000021 + \cdots$$

Here $a_1 = .21$ and $r = \dfrac{.0021}{.21} = .01$; thus

$$S_\infty = \dfrac{.21}{1 - .01}$$

$$= \dfrac{.21}{.99}$$

$$= \dfrac{21}{99} \qquad\qquad \dfrac{.21}{.99} \cdot \dfrac{100}{100}$$

$$= \dfrac{7}{33}.$$

Thus $0.21212121\ldots = \frac{7}{33}$.

You can verify this by dividing 7 by 33.

b. $1.4\overline{312} = 1.4312312312312\ldots$

This is the sum

$$1.4 + .0312 + .0000312 + .0000000312 + \cdots$$

Temporarily omitting 1.4, we find that the sum of the infinite geometric sequence

$$.0312 + .0000312 + .0000000312 + \cdots$$

has $a_1 = .0312$ and $r = \frac{.0000312}{.0312} = .001$. We get

$$S_\infty = \frac{.0312}{1 - .001}$$

$$= \frac{.0312}{.999}$$

$$= \frac{312}{9990} \qquad \frac{.0312}{.999} \cdot \frac{10000}{10000}$$

$$= \frac{52}{1665}. \qquad \frac{52 \cdot \cancel{6}}{1665 \cdot \cancel{6}}$$

Thus $.0312312312\ldots = \frac{52}{1665}$.

Adding back the first term,

$$1.4 = 1\frac{4}{10} = 1\frac{2}{5} = \frac{7}{5}$$

gives

$$\frac{7}{5} + \frac{52}{1665} = \frac{7 \cdot 333}{5 \cdot 333} + \frac{52}{1665}$$

$$= \frac{2331}{1665} + \frac{52}{1665}$$

$$= \frac{2383}{1665}$$

and so, $1.4312312312\ldots = \frac{2383}{1665}$.

You should check this by dividing out the fraction.

EXERCISE 11.4   *Answers, pages A101–A102*

A

Compute the sums of these infinite geometric sequences, as in EXAMPLE 1:

1. $1 + \dfrac{1}{2} + \dfrac{1}{4} + \dfrac{1}{8} + \cdots$ **2.** $1 - \dfrac{1}{3} + \dfrac{1}{9} - \dfrac{1}{27} + \cdots$ **3.** $\dfrac{2}{3} - \dfrac{4}{9} + \dfrac{8}{27} - \dfrac{16}{81} + \cdots$ **4.** $-\dfrac{3}{4} - \dfrac{1}{2} - \dfrac{1}{3} - \dfrac{2}{9} - \cdots$

5. $\displaystyle\sum_{n=1}^{\infty} \left(\dfrac{1}{4}\right)^{n-1}$ **6.** $\displaystyle\sum_{k=1}^{\infty} \left(\dfrac{-3}{4}\right)^{k}$ **7.** $\displaystyle\sum_{m=0}^{\infty} 6\left(\dfrac{-1}{3}\right)^{m}$ **8.** $\displaystyle\sum_{j=2}^{\infty} 4\left(\dfrac{1}{2}\right)^{j-2}$

Convert these repeating decimals into rational numbers in lowest terms, as in EXAMPLE 2:

9. $0.333333\ldots$ **10.** $0.66666\ldots$ **11.** $0.12121212\ldots$ **12.** $0.545454\ldots = 0.\overline{54}$

13. $0.\overline{123}$ **14.** $0.\overline{117}$ **15.** $0.3555555\ldots$ **16.** $1.3696969\ldots$

B

Compute the sums of these infinite geometric sequences, and simplify your answers:

17. $\sqrt{2} + 1 + \dfrac{1}{\sqrt{2}} + \dfrac{1}{2} + \cdots$ **18.** $3 + \sqrt{3} + 1 + \dfrac{\sqrt{3}}{3} + \cdots$ **19.** $1 - \dfrac{\sqrt{3}}{3} + \dfrac{1}{3} - \dfrac{\sqrt{3}}{9} + \cdots$

20. $2\sqrt{2} - 2 + \sqrt{2} - 1 + \cdots$ **21.** $1 - \dfrac{1}{e} + \dfrac{1}{e^2} - \dfrac{1}{e^3} + \cdots$ **22.** $1 + \dfrac{1}{1+e} + \dfrac{1}{(1+e)^2} + \dfrac{1}{(1+e)^3} + \cdots$

23. Using the method in the text, show that

$$0.999999\ldots = 1.000000\ldots$$

Does this paradox violate any rules of mathematics, or is it actually true?

24. The first swing of a pendulum is 12 feet. Each swing thereafter is $\dfrac{3}{4}$ of the preceding swing. (See Exercise 11.3, Problem 41.) Theoretically, the pendulum comes to rest after an infinite number of swings. How far will it have traveled before it comes to rest?

25. A ball is thrown upward to a height of 50 feet and returns to the ground. After each bounce, it returns to a height that is 90% of its previous height. (See Exercise 11.3, Problem 42.) Theoretically, the ball comes to rest after infinitely many bounces. Find the total vertical distance (up and back down) it will have traveled before it comes to rest.

26. Solve the equation $\quad S_\infty = \dfrac{a_1}{1-r} \quad$ for r.

27. If the sum of the infinite geometric sequence

$$3 + 3r + 3r^2 + 3r^3 + \cdots = \frac{9}{5},$$

find the common ratio r.

An infinite sequence of squares is shown below, and the sides of the first four squares are given. Use this sequence to do Problems 28 and 29.

   ...

28. Write down the *perimeters* of the squares shown. If these form a geometric sequence, find the sum of the perimeters of this infinite sequence of squares. *Formula: P = 4(side)*

29. Write down the *areas* of the squares shown. If these form a geometric sequence, find the sum of the areas of this infinite sequence of squares. *Formula: A = (side)²*

30. The sum $1 - 2 + 4 - 8 + 16 - 32 + \cdots$ has $a_1 = 1$ and $r = 2$; thus

$$S_\infty = \frac{2}{1 - (-2)} = \frac{1}{3}.$$

This means $1 - 2 + 4 - 8 + 16 - 32 + \cdots = \frac{1}{3}$, which is absurd! Tell where the mistake is.

31. The summation formulas for n terms and for infinitely many terms of a geometric sequence are, respectively,

$$S_n = \frac{a_1(1 - r^n)}{1 - r} \quad \text{and} \quad S_\infty = \frac{a_1}{1 - r}.$$

These are *invalid* if $r = 1$, because the denominators would be zero.
a. Give an example of a geometric sequence of non-zero terms that has common ratio $r = 1$.

b. Find the sum of the first n terms of the sequence in part a.

c. What happens to the sum of infinitely many of these terms?

32. The formula for the sum of an infinite geometric sequence

$$S_\infty = a_1 + a_1 r + a_1 r^2 + a_1 r^3 + \cdots = \frac{a_1}{1-r}, \quad |r| < 1$$

can be proved by *factoring* out r as follows:

$$S_\infty = a_1 + a_1 r + a_1 rr + a_1 r^2 r + \cdots$$
$$= a_1 + r(\underbrace{a_1 + a_1 r + a_1 r^2 + \cdots}_{?})$$

You should recognize the expression in parentheses. Make the correct substitution for it, and then to proceed to prove that

$$S_\infty = \frac{a_1}{1-r}, \quad |r| < 1.$$

Note: Calculus students learn that if $|r| < 1$, this sum "converges" to a finite limit and that simple algebraic operations involving it are valid.

C

33. The middle third of a line of length 1 is shaded in color and then removed. The middle thirds of the remaining two segments are shaded in color and then removed. The middle thirds of the remaining four segments are shaded in color and then removed. This process is repeated an infinite number of times. Find the total length of the colored segments that are removed.

0	$\frac{1}{3}$	$\frac{2}{3}$	1

| 0 | $\frac{1}{9}$ $\frac{2}{9}$ $\frac{3}{9}$ | $\frac{6}{9}$ $\frac{7}{9}$ $\frac{8}{9}$ | 1 |

| 0 $\frac{1}{27}$ $\frac{2}{27}$ $\frac{3}{27}$ | $\frac{6}{27}$ $\frac{7}{27}$ $\frac{8}{27}$ $\frac{9}{27}$ | $\frac{18}{27}$ $\frac{19}{27}$ $\frac{20}{27}$ $\frac{21}{27}$ | $\frac{24}{27}$ $\frac{25}{27}$ $\frac{26}{27}$ 1 |

$\cdot$ $\cdot$ $\cdot$ $\cdot$
$\cdot$ $\cdot$ $\cdot$ $\cdot$
$\cdot$ $\cdot$ $\cdot$ $\cdot$

34. The large square has side 2. If its midpoints are connected, a smaller square is formed. If the midpoints of that smaller square are connected, yet a smaller square is formed. This is repeated indefinitely, forming an infinite sequence of squares:

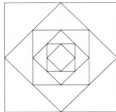

Write the sequence consisting of the sides of these squares. *Hint:* Pythagorean Theorem.

35. Find the *perimeters* of the first four squares, using your results from Problem 34. If these form a geometric sequence, find the sum of this infinite sequence of perimeters. *Formula: P = 4*(side)

36. Find the *areas* of the first four squares, using your results from Problem 34. If these form a geometric sequence, find the sum of this infinite sequence of areas. *Formula: A = *(side)^2

11.5   The Binomial Expansion

We conclude this chapter by expanding the powers of a binomial. The first few are done by repeated use of the Distributive Property:

$$(a + b)^2 = a^2 + 2ab + b^2.$$

$$(a + b)^3 = (a + b)(a^2 + 2ab + b^2) \qquad (a + b)(a + b)^2$$
$$= a^3 + 2a^2b + ab^2 + a^2b + 2ab^2 + b^3$$
$$= a^3 + 3a^2b + 3ab^2 + b^3.$$

$$(a + b)^4 = (a + b)(a^3 + 3a^2b + 3ab^2 + b^3) \qquad (a + b)(a + b)^3$$
$$= a^4 + 3a^3b + 3a^2b^2 + ab^3 + a^3b + 3a^2b^2 + 3ab^3 + b^4$$
$$= a^4 + 4a^3b + 6a^2b^2 + 4ab^3 + b^4.$$

Let us compile the following list, with the purpose of "predicting" $(a + b)^5$:

$$(a + b)^0 = \qquad\qquad 1$$
$$(a + b)^1 = \qquad\qquad a + b$$
$$(a + b)^2 = \qquad\qquad a^2 + 2ab + b^2$$
$$(a + b)^3 = \qquad\qquad a^3 + 3a^2b + 3ab^2 + b^3$$
$$(a + b)^4 = \qquad\qquad a^4 + 4a^3b + 6a^2b^2 + 4ab^3 + b^4$$
$$(a + b)^5 = a^5 + \text{?}a^4b + \text{?}a^3b^2 + \text{?}a^2b^3 + \text{?}ab^4 + b^5$$

A few observations will enable us to write the next few powers of $(a + b)^n$:

1. The first and last terms are a^n and b^n. The second and second-to-last coefficients are both n.

$$(a + b)^n = a^n + n\, a^{n-1}b^1 + \cdots + n\, a^1b^{n-1} + b^n$$

2. The exponents of a decrease from n to 1, whereas the exponents of b increase from 1 to n:
3. The sum of the exponents in any term is always n.
4. $(a + b)^n$ contains $n + 1$ terms.
5. The coefficient of any term is the *sum of the coefficients diagonally above it* in the triangular array.

Armed with these facts, we get

$$(a + b)^4 = \qquad\qquad a^4 + 4a^3b + 6a^2b^2 + 4ab^3 + b^4$$

$$(a + b)^5 = \qquad\quad a^5 + 5a^4b + 10a^3b^2 + 10a^2b^3 + 5ab^4 + b^5$$

$$(a + b)^6 = \quad a^6 + 6a^5b + 15a^4b^2 + 20a^3b^3 + 15a^2b^4 + 6ab^5 + b^6$$

$$(a + b)^7 = a^7 + 7a^6b + 21a^5b^2 + 35a^4b^3 + 35a^3b^4 + 21a^2b^5 + 7ab^6 + b^7$$

When we exclude variables, the coefficients in the expansion of $(a + b)^n$ form a triangular array:

$$
\begin{array}{rccccccccc}
(a + b)^0 = & & & & & 1 & & & & \\
(a + b)^1 = & & & & 1 & & 1 & & & \\
(a + b)^2 = & & & 1 & & 2 & & 1 & & \\
(a + b)^3 = & & 1 & & 3 & & 3 & & 1 & \\
(a + b)^4 = & 1 & & 4 & & 6 & & 4 & & 1 \\
(a + b)^5 = & 1 & 5 & & 10 & & 10 & & 5 & 1 \\
(a + b)^6 = & 1 & 6 & 15 & & 20 & & 15 & 6 & 1 \\
(a + b)^7 = & 1 & 7 & 21 & 35 & & 35 & 21 & 7 & 1 \\
(a + b)^8 = & 1\ 8 & 28 & 56 & 70 & 56 & 28 & 8 & 1 \\
\end{array}
$$

. . . .

This is called **Pascal's Triangle** of binomial coefficients; it is named after Blaise Pascal, a French mathematician who discovered it in the seventeenth century. Using it, we see that

$$(a + b)^8 = 1a^8 + 8a^7b + 28a^6b^2 + 56a^5b^3 + 70a^4b^4 + 56a^3b^5 + 28a^2b^6 + 8ab^7 + 1b^8.$$

EXAMPLE 1

Expand these binomial powers and simplify each term:

a. $(x + 2)^6$

The expansion of $(a + b)^6$ has coefficients

$$1 \qquad 6 \qquad 15 \qquad 20 \qquad 15 \qquad 6 \qquad 1$$

Substituting $a = x$ and $b = 2$ into its expansion gives

$$
\begin{aligned}
(x + 2)^6 &= 1x^6 + 6x^5 \cdot 2 + 15x^4 \cdot 2^2 + 20x^3 \cdot 2^3 + 15x^2 \cdot 2^4 + 6x \cdot 2^5 + 1 \cdot 2^6 \\
&= x^6 + 12x^5 \quad + 15x^4 \cdot 4 \quad + 20x^3 \cdot 8 \quad + 15x^2 \cdot 16 + 6x \cdot 32 + 64 \\
&= x^6 + 12x^5 \quad + 60x^4 \qquad + 160x^3 \qquad + 240x^2 \qquad + 192x \quad + 64
\end{aligned}
$$

b. $\left(a^2 - \dfrac{1}{a}\right)^4$, or $\left[a^2 + \left(\dfrac{-1}{a}\right)\right]^4$

The expansion of $(a + b)^4$ has coefficients

$$1 \quad 4 \quad 6 \quad 4 \quad 1$$

Substituting $a = a^2$ and $b = \dfrac{-1}{a}$ into its expansion gives

$$\left(a^2 - \frac{1}{a}\right)^4 = 1(a^2)^4 + 4(a^2)^3\left(\frac{-1}{a}\right) + 6(a^2)^2\left(\frac{-1}{a}\right)^2 + 4(a^2)\left(\frac{-1}{a}\right)^3 + 1\left(\frac{-1}{a}\right)^4$$

$$= a^8 + 4a^6 \cdot \frac{-1}{a} + 6a^4 \cdot \frac{1}{a^2} + 4a^2 \cdot \frac{-1}{a^3} + \frac{1}{a^4}$$

$$= a^8 - 4a^5 + 6a^2 - \frac{4}{a} + \frac{1}{a^4}.$$

The expansion of $(a + b)^{15}$ has sixteen terms, the first two and the last two of which we are certain:

$$a^{15} + 15a^{14}b + ?a^{13}b^2 + ?a^{12}b^3 + \cdots + ?a^3b^{12} + ?a^2b^{13} + 15ab^{14} + b^{15}.$$

The missing coefficients depend on those of $(a + b)^{14}$, which depend on those of $(a + b)^{13}$, and so forth, until we get to $(a + b)^8$, whose coefficients are listed in Pascal's Triangle. Fortunately, there is a direct method for finding the binomial coefficients without obtaining those of smaller powers. It is based on the observation that the seven coefficients in the expansion of $(a + b)^6$ can be expressed as the following quotients:

1	6	15	20	15	6	1
↑	↑	↑	↑	↑	↑	↑
1	$\dfrac{6}{1}$	$\dfrac{6 \cdot 5}{1 \cdot 2}$	$\dfrac{6 \cdot 5 \cdot 4}{1 \cdot 2 \cdot 3}$	$\dfrac{6 \cdot 5 \cdot 4 \cdot 3}{1 \cdot 2 \cdot 3 \cdot 4}$	$\dfrac{6 \cdot 5 \cdot 4 \cdot 3 \cdot 2}{1 \cdot 2 \cdot 3 \cdot 4 \cdot 5}$	$\dfrac{6 \cdot 5 \cdot 4 \cdot 3 \cdot 2 \cdot 1}{1 \cdot 2 \cdot 3 \cdot 4 \cdot 5 \cdot 6}$
6 $\boxed{\text{nCr}}$ 0	6 $\boxed{\text{nCr}}$ 1	6 $\boxed{\text{nCr}}$ 2	6 $\boxed{\text{nCr}}$ 3	6 $\boxed{\text{nCr}}$ 4	6 $\boxed{\text{nCr}}$ 5	6 $\boxed{\text{nCr}}$ 6

Note how the quotients reduce to the binomial coefficients listed directly above them. For example,

$$\frac{\cancel{6} \cdot 5 \cdot 4}{1 \cdot \cancel{2} \cdot \cancel{3}} = 20, \quad \frac{6 \cdot 5 \cdot \cancel{4} \cdot \cancel{3}}{1 \cdot 2 \cdot \cancel{3} \cdot \cancel{4}} = 15, \quad \text{and} \quad \frac{6 \cdot \cancel{5} \cdot \cancel{4} \cdot \cancel{3} \cdot \cancel{2}}{1 \cdot \cancel{2} \cdot \cancel{3} \cdot \cancel{4} \cdot \cancel{5}} = 6.$$

These coefficients can be obtained on a calculator equipped with an $\boxed{\text{nCr}}$ button, as shown.

An inspection of Pascal's Triangle reveals that the binomial coefficients are "symmetric" in any given row: the coefficients to the left of the middle are identical to those to the right of the middle. In expanding any power of $(a + b)^n$, once the coefficients start to repeat, you can save time by using the ones already listed. We now state the general binomial expansion, which is proved in more advanced algebra books:

The Binomial Expansion

$$(a + b)^n = a^n + na^{n-1}b + \frac{n(n-1)}{1 \cdot 2}a^{n-2}b^2 + \frac{n(n-1)(n-2)}{1 \cdot 2 \cdot 3} a^{n-3}b^3 + \cdots + b^n$$

Returning to our expansion of $(a + b)^{15}$, we find that the first four coefficients are

$$1, \quad \frac{15}{1} = 15, \quad \frac{15 \cdot \overset{7}{\cancel{14}}}{1 \cdot \cancel{2}} = 105, \quad \text{and} \quad \frac{\overset{5}{\cancel{15}} \cdot \overset{7}{\cancel{14}} \cdot 13}{1 \cdot \cancel{2} \cdot \cancel{3}} = 455.$$

$$15 \boxed{\text{nCr}} \, 2 \qquad\qquad 15 \boxed{\text{nCr}} \, 3$$

By the symmetry of Pascal's Triangle, these are also the last four coefficients. Without listing the middle terms, we have the expansion

$$(a + b)^{15} = a^{15} + 15a^{14}b + 105a^{13}b^2 + 455a^{12}b^3 + \cdots + 455a^3b^{12} + 105a^2b^{13} + 15ab^{14} + b^{15}$$

EXAMPLE 2 Obtain the indicated terms in the following expansions:

a. $(x + h)^{12}$, first four terms.

The Binomial Expansion formula gives

$$(x + h)^2 = x^{12} + 12x^{11}h + \frac{\overset{6}{\cancel{12}} \cdot 11}{1 \cdot \cancel{2}}x^{10}h^2 + \frac{\overset{2}{\cancel{12}} \cdot 11 \cdot 10}{1 \cdot \cancel{2} \cdot \cancel{3}}x^9h^3 + \cdots$$

$$= x^{12} + 12x^{11}h + 66x^{10}h^2 + 220x^9h^3 + \cdots .$$

$$\uparrow \qquad\qquad \uparrow$$
$$12 \boxed{\text{nCr}} \, 2 \quad 12 \boxed{\text{nCr}} \, 3$$

b. $\left(3a - \dfrac{1}{a^2}\right)^{17}$, last three terms

By the symmetry of Pascal's Triangle, and by the Commutative Property, we can rewrite this as

Last term

$$\left(\frac{-1}{a^2} + 3a\right)^{17} = \left(\frac{-1}{a^2}\right)^{17} + 17\left(\frac{-1}{a^2}\right)^{16}(3a) + \frac{17 \cdot \overset{8}{\cancel{16}}}{1 \cdot \cancel{2}}\left(\frac{-1}{a^2}\right)^{15}(3a)^2 + \cdots$$

$$= \frac{-1}{a^{34}} \quad + \quad 17 \cdot \frac{1}{a^{32}} \cdot 3a \quad + \quad 136 \cdot \frac{-1}{a^{30}} \cdot 9a^2 \quad + \cdots$$

$$= \frac{-1}{a^{34}} \quad + \quad \frac{51}{a^{31}} \quad + \quad \frac{-1224}{a^{28}} \quad + \cdots$$

Turning things back around as they originally were, we get

$$\left(3a - \frac{1}{a^2}\right)^{17} = \cdots - \frac{1224}{a^{28}} + \frac{51}{a^{31}} - \frac{1}{a^{34}} \qquad \text{Last 3 terms.}$$

EXERCISE 11.5   *Answers, page A102*

Expand these binomial powers and simplify each term, as in **EXAMPLE 1:**

1. $(a + 1)^3$

2. $(2x + 1)^4$

3. $(2x - y)^5$

4. $(a - 2b)^6$

5. $\left(a - \dfrac{1}{a}\right)^7$

6. $\left(x + \dfrac{3}{x}\right)^5$

7. $\left(2x + \dfrac{1}{x^2}\right)^4$

8. $\left(a^2 - \dfrac{1}{a}\right)^8$

Obtain and simplify the indicated terms in these expansions, as in **EXAMPLE 2:**

9. $(x + 1)^{15}$, first four terms

10. $(a - b)^{11}$, first four terms

11. $(a^2 - 2)^{18}$, first four terms

12. $(x + 3)^{10}$, first four terms

13. $\left(x + \dfrac{1}{x}\right)^{13}$, last three terms

14. $\left(2x - \dfrac{1}{x^2}\right)^{19}$, last three terms

15. $(a^2 - a)^{14}$, 5th term

16. $(x^3 + x)^{20}$, 6th term

B

17. $(x + h)^{18}$, 15th term   **18.** $(x + h)^{21}$, 18th term   **19.** $(a + b)^{10}$, middle term   **20.** $(x + 1)^{12}$, middle term

21. Find the approximate value of $(1.01)^8$ by obtaining, and then adding, the first four terms in the expansion of $(1 + 0.01)^8$. Compare your answer to that obtained directly by calculator.

22. Repeat Problem 21 for the value of $(0.99)^6 = (1 - 0.01)^6$.

23. The sums of the coefficients in the expansions of $(a + b)^4$ and $(a + b)^5$ are both powers of 2:

$$(a + b)^4: \qquad 1 + 4 + 6 + 4 + 1 \quad = 16, \quad \text{or} \quad 2^4$$
$$(a + b)^5: \quad 1 + 5 + 10 + 10 + 5 + 1 = 32, \quad \text{or} \quad 2^5$$

a. Verify that the sum of the coefficients in $(a + b)^6$ is 2^6.

b. Explain this phenomenon by writing 2^6 as $(1 + 1)^6$ and then expanding:

$$2^6 = (1 + 1)^6 = ? + ? + ? + ? + ? + ? + ?$$

24. The sums of the coefficients in the expansions of $(a - b)^3$ and $(a - b)^4$ are both zero:

$$(a - b)^3: \qquad 1 - 3 + 3 - 1 \quad = 0$$
$$(a - b)^4: \quad 1 - 4 + 6 - 4 + 1 = 0$$

    a. Verify that the sum of the coefficients in $(a - b)^7$ is 0.
    b. Show why this is true by writing 0 as $(1 - 1)^7$ and then expanding:

$$0 = (1 - 1)^7 = ? + ? + ? + ? + ? + ? + ? + ?$$

Expand, simplify, and then combine like terms:

25. $(1 + \sqrt{2})^8$ **26.** $(1 + \sqrt{3})^7$ **27.** $(1 - i)^6$ **28.** $(1 + i)^9$

In Section 9.1, Examples 2, 3, and 4, we found the difference quotient

$$\frac{f(x + h) - f(x)}{h}$$

for several functions. Using Pascal's Triangle of coefficients, we can obtain the difference quotient for the function $f(x) = x^4$ **as follows:**

$$\frac{f(x + h) - f(x)}{h} = \frac{(x + h)^4 - x^4}{h}$$

$$= \frac{x^4 + 4x^3h + 6x^2h^2 + 4xh^3 + h^4 - x^4}{h}$$

$$= \frac{4x^3h + 6x^2h^2 + 4xh^3 + h^4}{h}$$

$$= \frac{h(4x^3 + 6x^2h + 4xh^2 + h^3)}{h} \qquad \text{Factor.}$$

$$= 4x^3 + 6x^2h + 4xh^2 + h^3. \qquad \text{Reduce.}$$

In like manner, obtain the difference quotient for the following functions:

29. $f(x) = x^5$ **30.** $f(x) = x^6$ **31.** $f(x) = 4x^3$ *Hint:* $\dfrac{4(x + h)^3 - 4x^3}{h}$

C

32. $f(x) = 2x^7$ **33.** $f(x) = x^4 + x^3$ **34.** $f(x) = x^5 - x^3$

             Hint: $\dfrac{(x + h)^4 + (x + h)^3 - (x^4 + x^3)}{h}$

CHAPTER 11 · SUMMARY

11.1 Sequences and the Sigma Notation

A **sequence** is a function whose domain is the set of natural numbers. It is written a_n, rather than $a(n)$.

Example:

$$a_n = \frac{n^2}{n+1}$$

$$a_1 = \frac{1^2}{1+1} = \frac{1}{2}, \quad a_2 = \frac{2^2}{2+1} = \frac{4}{3}, \ldots$$

The sequence is $\frac{1}{2}, \frac{4}{3}, \frac{9}{4}, \frac{16}{5}, \ldots$.

The Greek letter **sigma** indicates the **sum** of the terms in a sequence.

Example:

$$\sum_{n=1}^{4} n^2 = 1^2 + 2^2 + 3^2 + 4^2$$
$$= 1 + 4 + 9 + 16 = 30.$$

11.2 Arithmetic Sequences

This is a sequence in which each term after the *first term* a_1 can be obtained by *adding* to the previous term a fixed number d called the **common difference** between consecutive terms. The **nth term** of an A.S. is given by

$$a_n = a_1 + (n-1)d.$$

Example:

2, 5, 8, 11, 14, . . . is an A.S. with $a_1 = 2$, $d = 3$.

The 80th term is

$$a_{80} = 2 + (80 - 1) \cdot 3$$
$$= 2 + 79 \cdot 3$$
$$= 239.$$

The **sum** of the first n terms of an A.S. is given by

$$S_n = \frac{n(a_1 + a_n)}{2},$$

where a_n = the last term.

Example:

The sum of the first 80 terms
$2 + 5 + 8 + 11 + \cdots + 239$ is

$$S_{80} = \frac{80(2 + 239)}{2}$$
$$= 9640.$$

11.3 Geometric Sequences

This is a sequence in which each term after the *first term* a_1 can be obtained by *multiplying* the previous term by a fixed number r called the **common ratio** between consecutive terms. The **nth term** of a G.S. is given by

$$a_n = a_1 r^{n-1}.$$

Example:

2, -4, 8, -16, 32, . . . is a G.S. with $a_1 = 2$, $r = -2$.

The 10th term is

$$a_{10} = 2(-2)^{10-1}$$
$$= 2(-2)^9$$
$$= -1024.$$

CHAPTER 11 · SUMMARY

The **sum** of the first n terms of a G.S. is given by

$$S_n = \frac{a_1(1 - r^n)}{1 - r}, \quad r \neq 1.$$

Example:

The sum of the first 10 terms
$2 - 4 + 8 - 16 + 32 - \cdots$ is

$$S_{10} = \frac{2[1 - (-2)^{10}]}{1 - (-2)}$$

$$= \frac{2[1 - 1024]}{3}$$

$$= -682.$$

11.4 Infinite Geometric Sequences

The sum of an **infinite number** of terms of a G.S. is given by

$$S_\infty = \frac{a_1}{1 - r}, \quad |r| < 1.$$

Example:

The sum of infinitely many terms
$1 + \dfrac{1}{2} + \dfrac{1}{4} + \dfrac{1}{8} + \dfrac{1}{16} + \cdots$ with $a_1 = 1$, $r = \dfrac{1}{2}$, is

$$S_\infty = \frac{1}{1 - \dfrac{1}{2}} = \frac{1}{\dfrac{1}{2}} = 2.$$

11.5 The Binomial Expansion

$$(a + b)^n = a^n + na^{n-1}b + \frac{n(n-1)}{1 \cdot 2}a^{n-2}b^2$$

$$+ \frac{n(n-1)(n-2)}{1 \cdot 2 \cdot 3}a^{n-3}b^3 + \cdots + b^n$$

Examples:

$$(a + b)^5 = a^5 + 5a^4b + 10a^3b^2 + 10a^2b^3 + 5ab^4$$
$$+ b^5 \text{ (6 terms)}$$

$$(x - 1)^{12} = x^{12} + 12x^{11}(-1) + \frac{12 \cdot 11}{1 \cdot 2}x^{10}(-1)^2$$

$$+ \frac{12 \cdot 11 \cdot 10}{1 \cdot 2 \cdot 3}x^9(-1)^3 + \cdots$$

$$= x^{12} - 12x^{11} + 66x^{10} - 220x^9 + \cdots$$

$$\text{(13 terms)}$$

CHAPTER 11 · REVIEW EXERCISES   *Answers, pages A102–A104*

11.1

Write the first four terms of the sequences defined below:

1. $a_n = (n - 1)^2$

2. $a_n = \dfrac{n^2 - 1}{n + 1}$

3. $a_n = (\sqrt{2})^{n+1}$

Discover a formula a_n for the following sequences:

4. $\dfrac{1}{4}, \dfrac{2}{9}, \dfrac{3}{16}, \dfrac{4}{25}, \dots$

5. $1, \dfrac{2}{3}, \dfrac{3}{5}, \dfrac{4}{7}, \dots$

6. $2, \dfrac{5}{4}, \dfrac{10}{9}, \dfrac{17}{16}, \dots$

Compute these sums:

7. $\displaystyle\sum_{n=1}^{5} \dfrac{n}{n + 1}$

8. $\displaystyle\sum_{k=1}^{7} (2k^2 - k)$

9. $\displaystyle\sum_{m=1}^{4} (-1)^{m+1} 2^m \cdot 3^{4-m}$

10. $\displaystyle\sum_{j=0}^{9} (\sqrt{3})^j$

11. $\displaystyle\sum_{n=2}^{6} \log n$
(Express answer as single log.)

12. $\displaystyle\sum_{m=1}^{10} i^m$ (Recall: $i^2 = -1$.)

11.2

Obtain the indicated term for the arithmetic sequences *only* below:

13. $-2, 1, 4, 7, \dots;\ a_{75}$

14. $5, 3, 1, -1, -3, \dots;\ a_{90}$

15. $2, -4, 6, -8, \dots;\ a_{60}$

16. $1, 3\dfrac{1}{2}, 6, 8\dfrac{1}{2}, \dots;\ a_{100}$

17. $\dfrac{1}{4}, \dfrac{2}{3}, 1\dfrac{1}{12}, 1\dfrac{1}{2}, \dots;\ a_{25}$

18. $4.2, 4.5, 4.8, 5.1, \dots;\ a_{50}$

19. $\sqrt{3}, \sqrt{12}, \sqrt{27}, \dots;\ a_{10}$
(Simplify radicals.)

20. $\log 2, \log 6, \log 18, \log 54, \dots;\ a_{10}$
(Factor each number, then use log properties.)

21. $2x, 5x, 8x, 11x, \dots;\ a_{80}$

22. $x + 1, 2x + 3, 3x + 5, 5x + 7, \dots;\ a_{40}$

Compute the sums of these arithmetic sequences:

23. $\displaystyle\sum_{k=1}^{100} (3k - 1)$

24. $\displaystyle\sum_{m=1}^{205} (2 - 4m)$

25. $\displaystyle\sum_{n=1}^{75} \dfrac{2n + 1}{3}$

26. $1 + 4 + 7 + 10 + \cdots$ (80 terms)

27. $2 + 1\dfrac{1}{2} + 1 + \dfrac{1}{2} + \cdots$ (120 terms)

28. $5.4 + 5.6 + 5.8 + \cdots$ (300 terms)

29. $\dfrac{2}{3} + \dfrac{7}{6} + \dfrac{5}{3} + \cdots$ (175 terms)

30. $\sqrt{5} + \sqrt{20} + \sqrt{45} + \cdots$ (15 terms)

31. $\log 3 + \log 9 + \log 27 + \cdots$ (12 terms)
(Give exact answer, in terms of a log.)

32. $(x + 1) + (3x + 2) + (5x + 3) + \cdots$ (40 terms)

33. $4 + 8 + 12 + \cdots + 124$

34. $5 + 11 + 17 + \cdots + 485$

35. $5 + 2 - 1 - \cdots - 142$

36. $1 + 1\dfrac{1}{2} + 2 + \cdots + 85\dfrac{1}{2}$

37. A job pays $28,000 the first year, with $3500 annual raises thereafter.
  a. Find the salary in the 14th year.
  b. Find the total income earned on this job in 14 years.

38. In successive years, A earns $36,000, $38,500, $41,000, . . ., while B earns $20,000, $23,500, $27,000, ...
  a. In what year will B earn the same as A?
  b. Find their common salary from part a.
  c. Find the total amount earned by each person during this entire period.

Fill in the missing terms in these arithmetic sequences:

39. __, 5, __, __, __, 21, __, . . .

40. __, __, 2, __, __, __, __, 0, __, . . .

Use the formula $S_n = \dfrac{n(a_1 + a_n)}{2}$ to find the value of n such that:

41. $1 + 2 + 3 + \cdots + n = 55$

42. $3 + 6 + 9 + \cdots + 3n = 84$

43. In an arithmetic sequence, $a_4 = 10$ and $a_{15} = 43$.
  a. Use the formula $a_n = a_1 + (n - 1)d$ to write two equations in unknowns a_1 and d.
  b. Solve these equations for a_1 and d.
  c. List the first 15 terms in this sequence.

44. In an arithmetic sequence, $a_1 = 2$ and $S_{10} = 155$. Find a_{10} and d.

11.3

Obtain the indicated term for the geometric sequences *only* below:

45. $1, -2, 4, -8, 16, \ldots;$ a_{15}

46. $8, 4, 2, 1, \dfrac{1}{2}, \ldots;$ a_{12}

47. $18, 6, 2, \dfrac{2}{3}, \ldots;$ a_{10}

48. $64, -48, 36, -27, \ldots;$ a_{12}

49. $12, 8, 5\dfrac{1}{3}, 3\dfrac{5}{6}, \ldots;$ a_8

50. $\sqrt{2}, \sqrt{8}, 4\sqrt{2}, \ldots;$ a_6

51. $e, e^2, e^4, e^8, \ldots;$ a_{10}

52. $i, -1, -i, 1, \ldots;$ a_{27}

53. $2000, 2000(1.12), 2000(1.12)^2, 2000(1.12)^3, \ldots;$ a_{15}

54. $\dfrac{x^3}{y^2}, \dfrac{x}{y}, \dfrac{1}{x}, \dfrac{y}{x^3}, \ldots;$ a_{10}

55. $\log 3, \log 9, \log 81, \ldots;$ a_6

Compute the sums of these geometric sequences:

56. $\dfrac{1}{4} + \dfrac{1}{2} + 1 + \cdots$ (9 terms)

57. $12 - 8 + \dfrac{16}{3} - \cdots$ (7 terms)

58. $\displaystyle\sum_{k=1}^{8} 3^k$

59. $\displaystyle\sum_{n=1}^{9} 3(-2)^n$

60. $\displaystyle\sum_{m=1}^{12} 5000(1.04)^{m-1}$

61. $\displaystyle\sum_{m=1}^{8} (\sqrt{2})^m$

62. Plan A pays 1¢, 4¢, 16¢, 64¢, . . . for 15 days.
Plan B pays 1¢, 2¢, 4¢, 8¢, . . . for 30 days.
a. Which plan pays more on the last day, or are they the same?
b. Which plan results in the most total amount paid, or are they the same?

63. A job pays $40,000 the first year, with 9% annual raises thereafter. It can be shown that the annual salaries will be
$$\$40,000, \$40,000(1.09), \$40,000(1.09)^2, \$40,000(1.09)^3, \ldots$$
a. Find the salary in the 8th year.
b. Find the total income earned in the first 8 years.

64. You are 8 feet from a wall, and each step you take toward the wall measures half of your distance from it.
a. How far do you travel on the 10th step?
b. Find the total distance you have traveled after taking 10 steps.
c. Find your distance from the wall after 10 steps.

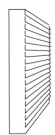

Problem 64

65. A ball is thrown upward to a height of 64 feet and returns to the ground. After each bounce, it returns to a height that is $\dfrac{3}{4}$ of its previous height.
a. To what height does it return after the 6th bounce?
b. How many vertical feet will it have traveled when it strikes the ground for the 7th time?

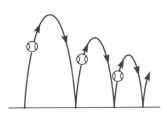

Problem 65

Fill in the missing terms in these geometric sequences:

66. __, 2, __, 18, __, __, . . .

67. __, -6, __, __, 48, __, __, . . .

68. __, __, 4, __, 8, __, . . .

69. __, 18, __, 8, __, __, . . .

70. Fill in the missing terms so that the sequence

$$\text{__, 3, __, __, 24, __, __, . . .}$$

   a. is geometric;   **b.** is arithmetic.

71. The sum of the first 8 terms of a geometric sequence is $\frac{85}{32}$. If $r = -\frac{1}{2}$, find a_1.

11.4

Compute the sums of these infinite geometric sequences:

72. $4 + 2 + 1 + \frac{1}{2} + \cdots$

73. $6 - 2 + \frac{2}{3} - \frac{2}{9} + \cdots$

74. $4 - 3 + \frac{9}{4} - \frac{27}{16} + \cdots$

75. $2 + \sqrt{2} + 1 + \frac{\sqrt{2}}{2} + \cdots$

76. $\displaystyle\sum_{n=1}^{\infty} 1000(0.92)^{n-1}$

77. $\displaystyle\sum_{k=1}^{\infty} e^{-k}$

Convert these repeating decimals into rational numbers in lowest terms:

78. $0.444444\ldots$

79. $0.\overline{36}$

80. $0.396396396\ldots$

81. $0.5666666\ldots$

82. $1.3212121\ldots$

83. $0.0\overline{3}$

84. You are 8 feet from a wall, and each step you take toward the wall measures half of your distance from it. (See Problem 64.) Theoretically, how far will you have traveled after infinitely many steps?

85. A ball is thrown upward to a height of 64 feet and returns to the ground. After each bounce, it returns to a height that is $\frac{3}{4}$ of its previous height. (See Problem 65.)

Theoretically, the ball will come to a rest after infinitely many bounces. How many vertical feet will it have traveled before it comes to rest?

The sum of an infinite geometric sequence is given. In each case, find the common ratio r :

86. $4 + 4r + 4r^2 + 4r^3 + \cdots = \frac{8}{3}$

87. $2 + 2r + 2r^2 + 2r^3 + \cdots = 6$

88. $2 + 2r + 2r^2 + 2r^3 + \cdots = 4 + 2\sqrt{2}$

An infinite sequence of circles is shown below, and the radii of the first four circles are given. Use this sequence to do Problems 89 and 90.

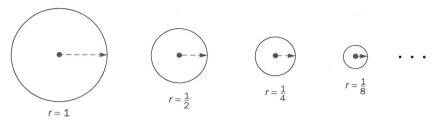

$r = 1$ $r = \frac{1}{2}$ $r = \frac{1}{4}$ $r = \frac{1}{8}$ $\cdots$

89. Write down the *circumferences* of the circles shown. If these form a geometric sequence, find the sum of the circumferences of this infinite sequence of circles. *Formula:* $C = 2\pi r$

90. Write down the *areas* of the circles shown. If these form a geometric sequence, find the sum of the areas of this infinite sequence of circles. *Formula:* $A = \pi r^2$.

11.5

Expand these binomial powers and simplify each term:

91. $(x + h)^3$ **92.** $(a + 2b)^4$ **93.** $(x^2 + x)^6$ **94.** $\left(x - \dfrac{1}{x}\right)^5$ **95.** $(2x - 1)^7$ **96.** $(1 - \sqrt{2})^8$

Obtain and simplify the indicated terms in these expansions:

97. $(a^2 - a)^{16}$, first four terms   **98.** $(x + 2y)^{14}$, first four terms

99. $\left(2x + \dfrac{1}{x}\right)^{15}$, last four terms   **100.** $\left(a^2 - \dfrac{1}{a}\right)^{20}$, last three terms

101. $(x^2 - 1)^{13}$, 5th term   **102.** $(x + 1)^{18}$, 15th term

103. $(x - 1)^8$, middle term

Obtain, in simplest form, the difference quotient

$$\frac{f(x + h) - f(x)}{h}$$

for each of the following functions:

104. $f(x) = x^3$ **105.** $f(x) = 2x^5$ *Hint:* $\dfrac{2(x + h)^5 - 2x^5}{h}$

106. $f(x) = x^6 + x^2$ *Hint:* $\dfrac{(x + h)^6 + (x + h)^2 - (x^6 + x^2)}{h}$

107. Find the approximate value of $(1.1)^9$ by finding, and then adding, the first five terms in the expansion of $(1 + 0.1)^9$. Compare your answer to that obtained directly by calculator.

108. Expand $(x^2 + x + 1)^3$ as follows:

$$(x^2 + x + 1)^3 = [x^2 + (x + 1)]^3$$
$$= 1(x^2)^3 + 3(x^2)^2(x + 1) + 3(????) + 1(????)$$

Finish the expansion, and then simplify your result.

109. A sequence of lines is shown passing through Pascal's Triangle of binomial coefficients. The *sum* of the numbers along each line forms a sequence that was introduced in one of the exercises in this chapter. Write down this sequence, and then list the next six terms in it.

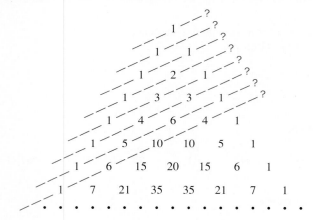

CHAPTER 11 ▪ TEST   *Answers, page A104*

Give the indicated term in each sequence:

1. 2, 6, 10, 14, . . .; a_{100}

2. 3, 6, 12, 24, . . .; a_{10}

3. $\dfrac{1}{2}, \dfrac{8}{3}, \dfrac{27}{4}, \dfrac{64}{5}, \ldots$; a_{10}

4. $\dfrac{2}{3}, -1, \dfrac{3}{2}, \dfrac{-9}{4}, \ldots$; a_{10}

5. 1, $\sqrt{2}$, 2, $\sqrt{8}$, . . .; a_{10}

6. 5, 2, -1, -4, . . .; a_{150}

Fill in the missing terms so that

$$\underline{\quad}, 8, \underline{\quad}, \underline{\quad}, 27, \underline{\quad}, \ldots$$

7. is a geometric sequence

8. is an arithmetic sequence

Compute these sums:

9. $\sum_{k=1}^{100} (9 + 2k)$

10. $\sum_{n=0}^{\infty} \left(\frac{1}{4}\right)^n$

11. $\sum_{m=2}^{7} \frac{m}{m-1}$

12. $1 + 4 + 7 + 10 + \cdots$ (50 terms)

13. $8 + 14 + 20 + \cdots + 482$

14. $2 - 1 + \frac{1}{2} - \frac{1}{4} + \cdots$

15. $\sqrt{2} + 1 + \frac{\sqrt{2}}{2} + \frac{1}{2} + \cdots$

Expand and simplify:

16. $(a^2 - a)^7$

17. $\left(x + \frac{2}{x}\right)^{12}$, first four terms

18. Convert 0.51515151. . . into a rational number in lowest terms.

19. A job pays \$25,000 in the first year, with annual raises of \$4500 thereafter.
    a. Find the salary in the 12th year.
    b. Find the total income earned during the first 12 years on the job.

20. Another job pays \$35,000 the first year, with 9% annual raises thereafter. It can be shown that the sequence of salaries is

    $$\$35,000, \ \$35,000(1.09), \ \$35,000(1.09)^2, \ \ldots$$

    a. Find the salary in the 8th year.
    b. Find the total income earned during the first 8 years on the job.

SAMPLE FINAL EXAMINATIONS

Your most challenging hurdle, the final exam, still remains. Even countless hours of studying sometimes bring unhappy results. "We never even covered that topic!" and "What chapter was Problem 7 in?" can often be heard in the hallways after the final. Legitimate gripes or sour grapes? Well, you studied 20 hours for the final and went over all your old tests as well as the chapter review exercises. The one mistake you probably made was to study each topic in the order in which it appeared in the book. And final exam problems have a way of being "out of sequence." We now present problems taken from the author's previous finals. They will appear in *random order,* much like they might on your final. In Final Exam 1, references are given to sections in the text to help you review.

Here are some tips for achieving "peak performance" on your final exam:

Tip 1 Make each sample final a "dress rehearsal" for the real thing. Limit yourself to 2 hours, say, and if feasible, take the sample final in the *actual seat* you will be using during the final.

Tip 2 If time permits, write down on a slip of paper each problem given by your instructor on each test during the entire course. Draw them *randomly* out of a hat and solve them.

Tip 3 *Do not cram* the night before the final. Trying to strengthen your weaknesses will actually weaken your strengths.

Tip 4 *Do not rush* through the final exam. By "going slower," you will actually go faster!

FINAL EXAM 1   *Answers, pages A104–A105*

Simplify or solve. For the inequalities, be sure to graph your solutions.

1. $\dfrac{(a^2b^3)^2}{(ab^{10})^4}$

2. $\dfrac{x^2 - 4}{x^2 + 3x + 2} \cdot \dfrac{x^2 - 3x - 4}{4x - 8}$

3. $x^4 - 7x^2 - 18 = 0$ [6.6]

4. $\dfrac{1 + a^{-3}}{a^{-1} - a^{-3}}$ [5.1]

5. $2\sqrt{12} - 5\sqrt{27} + \sqrt{300}$

6. $\log_{25} 125 = y$ [10.1]

7. $\dfrac{5}{x + 2} - \dfrac{3}{2 - x} - \dfrac{3x - 14}{x^2 - 4}$

8. $\dfrac{1}{x} + \dfrac{1}{x + 2} = \dfrac{3}{4}$

9. $\dfrac{\sqrt{6} + \sqrt{2}}{\sqrt{6} - \sqrt{2}}$ [5.5]

10. $1 \le \dfrac{3 - 2x}{5} \le 3$

11. $x^2 - y^2 = 5$
    $x + 2y = 7$

12. $\dfrac{b}{ax} + \dfrac{1}{2a} = \dfrac{a}{4bx} - \dfrac{1}{4b}$; for x [4.7]

13. $\sqrt{x + 5} - \sqrt{x} = 1$ [6.5]

14. $2x + 2y + \ z = -1$
    $\ x \qquad\ \ - 3z = -1$ (Use Cramer's Rule.)
    $\ x + \ y + \ z = 0$

15. $3^{x+1} = 7$ [10.4]

16. $(2 + \sqrt{-9})(4 - \sqrt{-4})$ [6.1]

17. $|2 - 3x| < 1$ [2.6]

18. $2x^3 + 3x^2 - 8x - 12 = 0$ [6.6]

Graph:

19. $\dfrac{x^2}{16} + \dfrac{y^2}{9} = 1$ [8.3]

20. $x^2 + y^2 - 4x + 6y - 3 = 0$ [8.2]

21. Find the equation of the line through $(-2,3)$ and $(2,-3)$.   [7.3]

22. Find the sum $2 + 5 + 8 + \ldots + 434$.   [11.2]

23. The perimeter of a rectangle is 24. Find its dimensions, given that
    a. the area is 35.
    b. the diagonal is $4\sqrt{5}$.

24. $20,000 is invested, part at 9% for 3 years and the balance at 12% for 4 years. How much is invested at each rate if the total interest is $6450?

25. The population of a city is given by

$$P = 50,000e^{0.065t},$$

using 1960 as the census year.
a. Find the population in 1975.   [10.1]
b. In what year will the population be 200,000?   [10.4]

FINAL EXAM 2   *Answers, pages A105–A106*

Simplify or solve. For the inequalities, be sure to graph your solutions.

1. $\left(\dfrac{x^{-3}y^{-5}}{xy^{-7}}\right)^{-6}$

2. $\sqrt{8x^3} - 2x\sqrt{50x}, \quad x > 0$

3. $\dfrac{3a^2 + a - 2}{4a^2 - 12a^3} \div \dfrac{6a^2 - a - 2}{6a^2 - 2a}$

4. $\dfrac{1}{x} = \dfrac{x}{x + 1}$

5. $(25x^4)^{-3/2}$

6. $x^2 - 5x - 6 \ge 0$

7. $\dfrac{2}{x+3} - \dfrac{54}{x^3+27} - \dfrac{1}{x^2-3x+9}$

8. $\log_b 125 = \dfrac{3}{2}$

9. $\dfrac{a-x}{b} = \dfrac{b-x}{a}$; for x

10. $\log_5 x + \log_5 (x-24) = 2$

11. $\dfrac{\dfrac{2}{x+4} - \dfrac{1}{x-4}}{1 - \dfrac{128}{x^2-16}}$

12. $\begin{aligned}5x - 3y &= 7\\ 3x + 7y &= 9\end{aligned}$ (Use Cramer's Rule.)

13. $|x^2 + 2x - 4| = 4$

14. $x^6 + 7x^3 - 8 = 0$

Graph:

15. $f(x) = 2^{x-1}$

16. $y = -x^2 + 4x + 2$

17. Obtain and simplify the first four terms in the expansion of $\left(x + \dfrac{1}{x}\right)^{14}$.

18. Express $0.216216216\ldots$ as a rational number in lowest terms.

19. \$15,000 is deposited into an account paying $7\frac{1}{2}\%$ annual interest compounded quarterly.
    a. Find the accumulated amount, and the interest earned, after 5 years.
    b. In how many years will the original deposit double?
    Formulas:

    $$A = P\left(1 + \dfrac{r}{n}\right)^{nt} \quad \text{and} \quad I = A - P$$

20. Let $f(x) = 2x^2 + 3x - 7$. Obtain and simplify: $f(0)$, $f(-4)$, $f\left(\dfrac{1}{2}\right)$, $f(1 + \sqrt{2})$, and

    the difference quotient $\dfrac{f(x+h) - f(x)}{h}$.

21. W varies jointly with x and the square of y and inversely with the cube of z.
    a. If $W = 3$ when $x = 4$, $y = -3$, and $z = 6$, find k.
    b. Find W when $x = 6$, $y = 8$, and $z = -4$.

22. Find the equation of the line through the point $(-1,4)$ that is
    a. parallel to the line $3x + 5y = 1$.
    b. perpendicular to the line $3x + 5y = 1$.

23. If the width of a rectangle is decreased by 1 and the length is increased by 3, a second rectangle is formed with the same area as the first. If the width of the first rectangle is increased by 2 and the length is decreased by 3, a third rectangle is formed whose area is also the same as the first. Find the width and length of each rectangle.

First

Second

Third

24. Find x and then y:

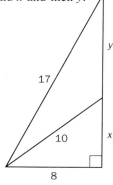

PROBLEM 24

25. A square of side 2 is inscribed in a circle.
   a. Find the radius of the circle.
   b. Find the shaded area. *Hint:* Circle minus square.

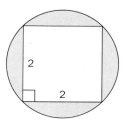

PROBLEM 25

FINAL EXAM 3   *Answers, page A106*

Simplify or solve. For the inequalities, be sure to graph your solutions.

1. $\dfrac{e^x + e^{-x}}{e^x - e^{-x}}$

2. $\dfrac{a^{1/2}a^{2/3}}{a^{5/6}}$

3. $\dfrac{18 - 2x}{x^2 - 9} + \dfrac{4x}{x + 3} - \dfrac{2}{x - 3}$

4. $|5 + 2x| \geq 3$

5. $-\left|\dfrac{7}{24} - \dfrac{5}{16}\right|$

6. $2^{x+1} \cdot 4^x = 8^{3-2x}$

7. $\dfrac{x + 3}{x - 1} > 2$

8. $\dfrac{(x + h)^5 - x^5}{h}$

9. $\dfrac{a - \dfrac{8}{a^2}}{a - \dfrac{4}{a}}$

10. $\dfrac{3x^2 + x - 13}{x^2 - 9} + \dfrac{3}{x + 3} = \dfrac{x + 1}{x - 3}$

11. $\sqrt{2x + 3} + x = 6$

12. $x + y = 6$
   $\dfrac{1}{x} + \dfrac{1}{y} = \dfrac{3}{4}$

13. $\log_{27} x = \dfrac{-2}{3}$

14. $P = 2\pi\sqrt{\dfrac{L}{g}}$; for g

Graph:

15. $\dfrac{y^2}{9} - \dfrac{x^2}{4} = 1$

16. $f(x) = \sqrt{x + 1}$

17. Let $f(x) = 3x - 1$ and $g(x) = x^2 + 2$.
   Find $f[g(3)]$, $g[f(-4)]$, $f[g(x)]$, $g[f(x)]$, and $f^{-1}(x)$.

18. Find the pH of NH_4OH, whose $[H^+] = 0.00000000023$ moles per liter.
   Formula: $pH = -\log [H^+]$

Find these sums:

19. $\displaystyle\sum_{k=1}^{\infty} (-1)^{k+1} \left(\frac{1}{2}\right)^k$

20. $6 + 9 + 12 + 15 + \ \dots$ (150 terms)

21. On a training ride, a cyclist rides for $1\frac{1}{2}$ hours. She then turns around and returns to her starting point, which she reaches 2 hours after she turned around. Because of head winds and fatigue, her speed on the return trip is slower by 5 miles per hour. Find her speed in each direction. Find the distance in each direction.

22. A job pays $25,000 the first year, with $3500 yearly raises thereafter.
   a. Find the salary in the 14th year.
   b. Find the total income earned during the first 14 years.
   c. In what year will the salary be $84,500?

23. How many liters of water should be added to 4 liters of 35% alcohol in order to weaken it to 15% alcohol?

24. The first swing of a pendulum is 18 feet. Each swing thereafter is $\frac{2}{3}$ of the preceding swing.
   a. Find the distance the pendulum travels on the 6th swing.
   b. Find the total distance the pendulum has traveled after 6 swings.
   c. Theoretically, the pendulum will come to rest after an infinite number of swings. Find the total distance it travels before coming to rest.

25. Find the unknown sides of this right triangle:

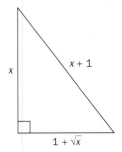

SELECTED TOPICS

A. Sets

Groups of objects are familiar to everyone. We speak of a flock of geese, a batch of letters, a stamp collection, a set of dishes, a pack of wolves, a bunch of grapes, a tool kit, and a squad of players. In mathematics, we use the word **set** to mean any collection of objects. Sets are denoted by capital letters; their **elements**, or **members**, are listed inside braces. Simple examples of sets are

$$A = \{2, 3, 5, 7, 11, 13\}$$
$$\text{and} \quad B = \{\text{red, yellow, blue}\}.$$

If an object x is an element of a set S, we write $x \in S$; if x is not a member of S, we write $x \notin S$. Thus, for the foregoing sets,

$$3 \in A \quad \text{but} \quad 6 \notin A;$$
$$\text{and} \quad \text{red} \in B \quad \text{but} \quad \text{green} \notin B.$$

These two sets can also be described using **set-builder notation**. For example, set A can be written

$$A = \{x \mid x \text{ is a prime number less than } 14\}.$$

This is read

$$A = \text{"the set of all objects } x \text{ such that } x \text{ is a}$$
$$\text{prime number less than } 14.\text{"}$$

The vertical bar | means "such that." Likewise, the set B given above can be written as

$$B = \{x \mid x \text{ is a primary color}\}.$$

"the set of all x"   "such that"

If you tried to list the elements of the set

$$\{x \mid x \text{ is a 3-wheeled bicycle}\},$$

you would come up empty-handed. This set, which contains no elements, is called the **empty set** (or **null set** or **void set**) and is denoted by the symbol $\varnothing$. Thus

$$\{x \mid x \text{ is a 3-wheeled bicycle}\} = \varnothing.$$

EXAMPLE 1 | List the elements of these sets:

a. $\{x \mid x \text{ satisfies } x^2 = 25\} = \{5, -5\}$.

b. $\{x \mid x \text{ is a secondary color}\} = \{\text{orange, green, purple}\}$.

c. $\{x \mid x \text{ is a prime number, } 25 < x < 50\} = \{29, 31, 37, 41, 43, 47\}$.

d. $\{x \mid x \text{ satisfies } x + 3 = x + 9\} = \varnothing$.

    Just as arithmetic operations between numbers produce new numbers, so there are operations between sets that produce new sets. If A and B are sets, we define their **intersection**, written $A \cap B$, to be the set consisting of all elements **common** to both sets:

$$A \cap B = \{x \mid x \in A \quad \text{and} \quad x \in B\}.$$

For example,

$$\{2, 3, 4\} \cap \{0, 2, 3, 5\} = \{2, 3\};$$
$$\text{and} \quad \{a, b, c\} \cap \{d, e, f, g\} = \varnothing. \qquad \text{No elements in common.}$$

    The **union** of sets, written $A \cup B$, is the set formed by **combining** all the elements of A and B:

$$A \cup B = \{x \mid x \in A \quad \text{or} \quad x \in B \quad \text{or} \quad \text{both}\}.$$

For example,

$$\{2, 3, 4\} \cup \{0, 2, 3, 5\} = \{0, 2, 3, 4, 5\};$$
$$\text{and} \quad \{a, b, c\} \cup \{d, e, f, g\} = \{a, b, c, d, e, f, g\}.$$

Note in the first example that 2 and 3 appear in both sets but that each is listed only once in the union of the sets.

EXAMPLE 2 | Let $A = \{1, 2, 3\}$, $B = \{3, 4, 5\}$, and $C = \{4, 5, 6, 7\}$; then

a. $A \cap B = \{3\}$

b. $A \cap C = \varnothing$

c. $A \cup B = \{1, 2, 3, 4, 5\}$

d. $A \cap (B \cap C) = \{1, 2, 3\} \cap \{4, 5\}$ $B \cap C = \{4, 5\}$
                  $= \varnothing$.

e. $(B \cup C) \cap A = \{3, 4, 5, 6, 7\} \cap \{1, 2, 3\}$
                $= \{3\}$.

    Instead of considering all the elements of a set, we may look at only part of it. For example, the offensive players form a part of an entire football squad.

The United States stamps form part of an entire stamp collection. We say that

A is a **subset** of B, written $A \subseteq B$,

provided that every element of A is also in B. Thus the offensive players form a subset of the entire squad, and the U.S. stamps form a subset of the entire collection:

offensive players $\subseteq$ entire squad;
U.S. stamps $\subseteq$ entire collection.

Other examples of subsets are

$$\{1, 3\} \subseteq \{1, 2, 3\};$$
$$\{a, b, d\} \subseteq \{a, b, c, d, e\}.$$

For our purposes, the most important sets are the many number systems discussed in Sections 1.4 and 6.1:

Complex numbers	$\mathbf{C} = \{2 + 3i, 7 + 0i, 0 - 8i, \frac{1}{2} + i\sqrt{2}, \ldots\}$
Real numbers	$\mathbf{Re} = \{2, \frac{1}{2}, -\sqrt{3}, 0, \pi, \sqrt[3]{5}, \ldots\}$
Imaginary numbers	$\mathbf{Im} = \{3i, -8i, i\sqrt{2}, i, \ldots\}$
Rational numbers	$\mathbf{Ra} = \{2, \frac{1}{2}, -8, 0, \frac{-3}{5}, 1.63, 0.393939\ldots, \ldots\}$
Irrational numbers	$\mathbf{Ir} = \{-\sqrt{3}, \pi, \sqrt[3]{5}, 0.1010010001\ldots, \ldots\}$
Integers	$\mathbf{In} = \{\ldots, -2, -1, 0, 1, 2, \ldots\}$
Whole numbers	$\mathbf{W} = \{0, 1, 2, 3, 4, 5, \ldots\}$
Natural numbers	$\mathbf{N} = \{1, 2, 3, 4, 5, \ldots\}$
Even numbers	$\mathbf{E} = \{\ldots, -4, -2, 0, 2, 4, \ldots\}$
Odd numbers	$\mathbf{Od} = \{\ldots, -5, -3, -1, 1, 3, 5, \ldots\}$
Prime numbers	$\mathbf{P} = \{2, 3, 5, 7, 11, 13, 17, \ldots\}$

EXAMPLE 3

a. $\mathbf{E} \cup \mathbf{Od} = \mathbf{In}$ Evens combined with odds produce all integers.

b. $\mathbf{E} \cap \mathbf{Od} = \varnothing$ No number is both even and odd.

c. $\mathbf{Ir} \subseteq \mathbf{Re}$ Every irrational number is a real number.

d. $\mathbf{Ir} \cap \mathbf{Ra} = \varnothing$ No number is both irrational and rational.

e. $\mathbf{Re} \subseteq \mathbf{C}$ For example, the real number $7 = 7 + 0i$ is complex.

f. $\sqrt{5} \in \mathbf{Ir}$ $\sqrt{5}$ is irrational.

■ **EXERCISE A**   *Answers, page A107*

A

List the elements of these sets, as in EXAMPLE 1:

1. $\{x \mid x \text{ satisfies } x - 2 = 7\}$
2. $\{x \mid x \text{ satisfies } 1 + 3x = 3x - 1\}$
3. $\{x \mid x \text{ satisfies } x^2 - 64 = 0\}$
4. $\{x \mid x \text{ satisfies } 2x^2 - x = 3\}$
5. $\{x \mid x \text{ satisfies } 2x^2 = 25\}$
6. $\{x \mid x \text{ satisfies } x^2 - 2x + 2 = 0\}$
7. $\{x \mid x \text{ is a prime number, } 50 < x < 100\}$
8. $\{x \mid x \text{ is a perfect square}\}$
9. $\{x \mid x \text{ is a positive, perfect cube}\}$
10. $\{x \mid x \in \textbf{In}, -4 \le 2x + 1 \le 11\}$
11. $\{x \mid x \text{ is a state in the U.S.A. beginning with the letter A}\}$
12. $\{x \mid x \text{ is a state in the U.S.A. beginning with the letter B}\}$
13. $\{x \mid x \text{ was assassinated while president of the United States}\}$
14. $\{x \mid x \text{ walked on the moon in 1969}\}$

Let $A = \{1, 2, 4\}$, $B = \{2, 3, 4\}$, $C = \{3, 5, 7, 8\}$, and $D = \{4, 5\}$.
List the elements of these sets, as in EXAMPLE 2:

15. $A \cup B$	16. $A \cup C$	17. $C \cup D$	18. $A \cap B$
19. $B \cap C$	20. $C \cap D$	21. $(A \cup B) \cap C$	22. $(A \cup B) \cup C$
23. $(B \cup C) \cap D$	24. $(A \cap B) \cap C$	25. $(B \cup C) \cup D$	26. $(A \cap B) \cap (C \cap D)$
27. $A \cup \varnothing$	28. $A \cap \varnothing$	29. $A \cap (B \cup C)$	30. $(A \cap B) \cup (A \cap C)$

Indicate whether each of the following statements is true or false. See Example 3.

31. $\textbf{In} \subseteq \textbf{Re}$	32. $\textbf{E} \subseteq \textbf{In}$	33. $\textbf{Ra} \cup \textbf{Ir} = \textbf{Re}$	34. $\sqrt{2} \in \textbf{Ra}$
35. $-3 \in \textbf{Ra}$	36. $\textbf{In} \cap \textbf{Ra} = \varnothing$	37. $\textbf{P} \subseteq \textbf{Od}$	38. $\textbf{P} \cap \textbf{E} = \{2\}$
39. $\dfrac{3}{4} \in \textbf{Re}$	40. $\pi \in \textbf{Ra}$	41. $\textbf{Ra} \cap \textbf{Ir} = \{0\}$	42. $\textbf{E} \cap \textbf{Od} = \{0\}$
43. $\textbf{Im} \subseteq \textbf{C}$	44. $5 \in \textbf{C}$	45. $127 \in \textbf{P}$	46. $221 \in \textbf{P}$

B

47. Is the statement $\{1, 2\} \cap \{3, 4\} = \{\varnothing\}$ true or false? If false, how can you make it true?

B.   Synthetic Division

It is a curious fact that in the division of a polynomial by a binomial (Section 4.4), the essential quantities involved are the numerical coefficients. Not that the powers x, x^2, x^3, and so forth are useless, but they serve only as place holders. In applying a quick method called **synthetic division**, we utilize this fact by throwing away all the letters! This method is *restricted* to binomial divisors of the form $x \pm a$, such as $x + 5$ and $x - 3$. It does not work when the coefficient of x is different from 1; thus $2x + 3$, $5x - 1$, and $-x + 3$ do *not* work as divisors in synthetic division.

EXAMPLE 1   Divide by synthetic division: $\dfrac{3x^2 + 5x - 9}{x - 4}$.

Solution

Step 1   Change the sign of the constant in the divisor: -4 to 4. Then display the coefficients of the numerator and leave a blank line:

$$\text{Change } -4 \text{ to } 4 \longrightarrow 4 \rfloor \quad 3\text{-} \quad 5\text{-} \quad\quad -9 \leftarrow \text{Coefficients of numerator}$$
$$\leftarrow \text{Blank}$$

Step 2   Bring down the first coefficient 3, multiply it by 4, and enter 12 under the 5:

$$
\begin{array}{r|rrr}
4 & 3 & 5 & -9 \\
  &   & {}^{\nearrow}12 \\
\hline
  & 3
\end{array}
$$

Step 3   Add: $5 + 12 = 17$; multiply: $4 \cdot 17 = 68$; enter 68 under the -9; then add: $-9 + 68 = 59$.

$$
\begin{array}{r|rrr}
4 & 3 & 5 & -9 \\
  &   & 12 & {}^{\nearrow}68 \\
\hline
  & 3 & 17 & 59
\end{array}
$$

Coefficients of quotient $\longrightarrow$ Remainder $\nwarrow$

Dividing a polynomial by a first-degree polynomial gives a quotient whose degree is *one less* than that of the original polynomial. In our case, the second-degree polynomial $3x^2 + 5x - 9$ has the first-degree quotient $3x - 17$, plus the remainder term.

Answer: $\dfrac{3x^2 + 5x - 9}{x - 4} = 3x + 17 + \dfrac{59}{x - 4}$.

Check: $3x + 17 + \dfrac{59}{x - 4} = \dfrac{(3x + 17) \cdot (x - 4)}{1 \cdot (x - 4)} + \dfrac{59}{x - 4}$

$$= \frac{3x + 5x - 68 + 59}{x - 4} \overset{\checkmark}{=} \frac{3x^2 + 5x - 9}{x - 4}.$$

EXAMPLE 2 | Divide by synthetic division: $\dfrac{5x^3 - x - 12}{x + 2}$

Solution

As in Section 4.4, Example 3, we must add the term $0x^2$ to the numerator as a place holder. The full synthetic division treatment gives:

Change 2 to -2 $\longrightarrow$ $\begin{array}{r|rrrr} & 5 & 0 & -1 & -12 \\ & & -10 & 20 & -38 \\ \hline & 5 & -10 & 19 & -50 \end{array}$

Coefficients of quotient   Remainder

The degree of the quotient must be 2 (one less than that of the dividend, whose degree is 3.)

Answer: $\dfrac{5x^3 - x - 12}{x + 2} = 5x^2 - 10x + 19 - \dfrac{50}{x - 2}.$

Besides being quick and clean, synthetic division avoids the need always to change the sign of each term in the bottom row before adding, as we did in Section 4.4. The single sign change of the constant in the divisor accomplishes this at every step throughout the synthetic division process.

EXERCISE B   *Answers, page A107*

A

Divide by synthetic division, as in EXAMPLE 1:

1. $\dfrac{2x^2 + x - 15}{x + 3}$

2. $\dfrac{3x^2 + 2x - 16}{x - 2}$

3. $\dfrac{3x^2 - x - 7}{x - 2}$

4. $\dfrac{4x^2 + 3x - 6}{x + 4}$

5. $\dfrac{5x + 7}{x + 1}$

6. $\dfrac{2x - 3}{x + 2}$

7. $\dfrac{4x^3 - 7x^2 + x + 2}{x - 1}$

8. $\dfrac{2x^3 + 4x^2 - 5x - 10}{x + 2}$

9. $\dfrac{3x^3 + 11x^2 - 5x - 9}{x + 3}$

10. $\dfrac{2x^3 + 8x^2 + 7x - 19}{x - 5}$

11. $\dfrac{4x^4 + 8x^3 + 5x^2 + 10x - 3}{x + 2}$

12. $\dfrac{3x^4 - 6x^3 + 4x^2 - 5x - 6}{x - 2}$

Divide by synthetic division, after first adding place holders $(0, 0x, 0x^2, 0x^3, ...)$, as in EXAMPLE 2:

13. $\dfrac{2x^3 + 7x - 9}{x - 1}$

14. $\dfrac{5x^3 + 9x - 9}{x + 3}$

15. $\dfrac{x^3 + 2x^2 - 3}{x + 2}$

16. $\dfrac{7x^3 + 4x^2 - 2}{x - 4}$

17. $\dfrac{2x^4 + 5x^2 - 9x + 1}{x - 2}$

18. $\dfrac{13x^4 + x^2 + 2x - 6}{x + 3}$

19. $\dfrac{x^4 + x^2 + 3x}{x + 2}$

20. $\dfrac{2x^4 + 7x^3 - 3}{x - 3}$

21. $\dfrac{-2x^3 + 5x}{x - 4}$ **22.** $\dfrac{-x^4 + 3x^2 + x}{x + 6}$ **23.** $\dfrac{x^3 + 8}{x + 2}$ **24.** $\dfrac{x^3 - 27}{x - 3}$

B

25. $\dfrac{x^5 - 1}{x - 1}$ **26.** $\dfrac{x^5 + 32}{x + 2}$ **27.** $\dfrac{x^6 - 64}{x - 2}$ **28.** $\dfrac{x^7 - 1}{x + 1}$

C.   Changing the Index of a Radical

Using the definition of a rational exponent (Section 5.6), we can write

$$\sqrt[nk]{a^{mk}} = a^{mk/nk} = a^{m/n} = \sqrt[n]{a^m} \qquad \text{Reduce exponent.}$$

In other words, we can alter the index of a radical according to the following formula (all variables are assumed to be positive):

Changing the Index of a Radical
$$\sqrt[nk]{a^{mk}} = \sqrt[n]{a^m}$$

EXAMPLE 1   Reduce the index of each radical:

a. $\sqrt[8]{a^6} = \sqrt[4\cdot2]{a^{3\cdot2}}$
    $= \sqrt[4]{a^3}.$

b. $\sqrt[6]{125} = \sqrt[2\cdot3]{5^{1\cdot3}}$ $125 = 5^3 = 5^{1\cdot3}$
    $= \sqrt[2]{5}.$

c. $\sqrt[15]{x^5y^{10}} = \sqrt[3\cdot5]{x^{1\cdot5}y^{2\cdot5}}$
    $= \sqrt[3]{xy^2}.$

In the next example, we multiply and divide radicals with different indices by first increasing each radical to the "least common index."

EXAMPLE 2   Increase to the least common index, and then perform the indicated operation:

a. $\sqrt[3]{a}\sqrt[2]{a} = \sqrt[3\cdot2]{a^{1\cdot2}}\sqrt[2\cdot3]{a^{1\cdot3}}$ Least Common Index $= 6$
    $= \sqrt[6]{a^2}\sqrt[6]{a^3}$
    $= \sqrt[6]{a^5}.$

(continued)

b. $\sqrt[3]{x^2}\,\sqrt[4]{y} = \sqrt[3\cdot4]{x^{2\cdot4}}\,\sqrt[4\cdot3]{y^{1\cdot3}}$ L.C.I. = 12

$\qquad\qquad = \sqrt[12]{x^8}\,\sqrt[12]{y^3}$

$\qquad\qquad = \sqrt[12]{x^8 y^3}.$

c. $\dfrac{\sqrt[3]{4}}{\sqrt[5]{8}} = \dfrac{\sqrt[3\cdot5]{4^{1\cdot5}}}{\sqrt[5\cdot3]{8^{1\cdot3}}} = \dfrac{\sqrt[15]{4^5}}{\sqrt[15]{8^3}}$ L.C.I. = 15

$\qquad = \sqrt[15]{\dfrac{4^5}{8^3}}$

$\qquad = \sqrt[15]{2}.$ $\dfrac{4^5}{8^3} = \dfrac{4\cdot4\cdot4\cdot4\cdot4}{8\cdot8\cdot8} = \dfrac{64\cdot16}{64\cdot8} = 2$

By multiplying exponents, we can express "a root of a root" as a single root:

$$\sqrt[n]{\sqrt[m]{a}} = (a^{1/m})^{1/n} = a^{1/mn} = \sqrt[mn]{a}\qquad \frac{1}{m}\cdot\frac{1}{n} = \frac{1}{mn}$$

This is our desired result:

A Root of a Root

$$\sqrt[n]{\sqrt[m]{a}} = \sqrt[mn]{a}$$

EXAMPLE 3   Express as a single root:

a. $\sqrt[4]{\sqrt[2]{x}} = \sqrt[8]{x}$ Multiply indices.

b. $\sqrt{\sqrt[3]{16}} = \sqrt[6]{16}$ Multiply indices: $2\cdot3 = 6$

$\qquad\quad = \sqrt[6]{2^4}$

$\qquad\quad = \sqrt[3]{2^2}$ Reduce the index.

$\qquad\quad = \sqrt[3]{4}.$

EXERCISE C   *Answers, page A107*

A

Reduce the index of each radical, as in EXAMPLE 1:

1. $\sqrt[6]{x^3}$ **2.** $\sqrt[8]{y^2}$ **3.** $\sqrt[15]{a^{10}}$ **4.** $\sqrt[12]{b^8}$ **5.** $\sqrt[4]{25}$ **6.** $\sqrt[4]{49}$

7. $\sqrt[6]{27}$ **8.** $\sqrt[6]{625}$ **9.** $\sqrt[12]{x^4 y^8}$ **10.** $\sqrt[9]{a^6 b^3}$ **11.** $\sqrt[8]{64}$ **12.** $\sqrt[6]{16}$

13. $\sqrt[4]{144}$ **14.** $\sqrt[4]{324}$ **15.** $\sqrt[6]{729}$ **16.** $\sqrt[6]{512}$ **17.** $\sqrt[8]{256x^{10}}$ **18.** $\sqrt[4]{1024a^6 y^2}$

Increase to the least common index, and then perform the indicated operation, as in EXAMPLE 2:

19. $\sqrt[3]{x}\ \sqrt[5]{x^2}$ **20.** $\sqrt[4]{y^3}\ \sqrt[3]{y^2}$ **21.** $\sqrt[4]{a}\ \sqrt[5]{a}$ **22.** $\sqrt{b}\ \sqrt[3]{b}$

23. $\sqrt[3]{x^2}\ \sqrt[4]{y^3}$ **24.** $\sqrt[3]{a^2}\ \sqrt{b}$ **25.** $\sqrt[3]{4}\ \sqrt{2}$ **26.** $\sqrt[4]{8}\ \sqrt[3]{4}$

27. $\sqrt{a}\ \sqrt[3]{a^2}\ \sqrt[4]{b^3}$ **28.** $\sqrt[3]{x}\ \sqrt{y}\ \sqrt[5]{y^4}$ **29.** $\sqrt{6}\ \sqrt[3]{4}\ \sqrt{3}$ **30.** $\sqrt[3]{5}\ \sqrt{10}\ \sqrt[3]{6}$

31. $\dfrac{\sqrt{a}}{\sqrt[3]{a}}$ **32.** $\dfrac{\sqrt[3]{b}}{\sqrt[4]{b}}$ **33.** $\dfrac{\sqrt[4]{x^3}}{\sqrt[3]{x^2}}$ **34.** $\dfrac{\sqrt[3]{y}}{\sqrt[5]{y^2}}$

35. $\dfrac{\sqrt[3]{4}}{\sqrt{2}}$ **36.** $\dfrac{\sqrt[4]{27}}{\sqrt[3]{9}}$ **37.** $\dfrac{\sqrt{x}\ \sqrt[3]{xy^2}}{\sqrt[4]{x^2y}}$ **38.** $\dfrac{\sqrt[3]{a^2}\ \sqrt{b}}{\sqrt[5]{ab^3}}$

Express as a single root, as in EXAMPLE 3:

39. $\sqrt[4]{\sqrt[3]{a}}$ **40.** $\sqrt{\sqrt[5]{y}}$ **41.** $\sqrt[4]{\sqrt{x^3}}$ **42.** $\sqrt[5]{\sqrt[3]{b^2}}$ **43.** $\sqrt[3]{\sqrt[4]{8}}$ **44.** $\sqrt{\sqrt[3]{81}}$

B

45. $\sqrt[4]{\sqrt[3]{625}}$ **46.** $\sqrt[3]{\sqrt[4]{64}}$ **47.** $\sqrt{\sqrt[4]{100}}$ **48.** $\sqrt{\sqrt{8}}$ **49.** $\sqrt[3]{\sqrt[5]{\sqrt{b^3}}}$ **50.** $\sqrt[3]{\sqrt{\sqrt[4]{x^6}}}$

Decide which root is larger *without using their decimal values*. Hint: Use a common index.

51. $\sqrt[3]{4}$ or $\sqrt{2}$ **52.** $\sqrt{3}$ or $\sqrt[3]{9}$ **53.** $\sqrt[3]{25}$ or $\sqrt[4]{125}$

54. $\sqrt[3]{4}$ or $\sqrt[4]{8}$ **55.** $\sqrt[4]{2}$ or $\sqrt[6]{3}$ **56.** $\sqrt[6]{64}$ or $\sqrt[8]{81}$

Reduce the index of each radical, and then combine:

57. $\sqrt{2}+\sqrt[4]{4}$ **58.** $\sqrt{27}-\sqrt[4]{9}$ **59.** $\sqrt[6]{25}-\sqrt[3]{5}$ **60.** $\sqrt[4]{64}+\sqrt[6]{8}$

D.   Inequalities in Two Variables

We saw in Chapters 7 and 8 that equations in the variables x and y have graphs that are lines, parabolas, circles, ellipses, and hyperbolas. We will now see that their companion *inequalities* have graphs that are entire *regions* in the plane.

The graph of the inequality

$$y < 2x + 1$$

will serve as an illustration. We start by graphing the "boundary line" (Section 7.1)

$$y = 2x + 1,$$

which divides the plane into two regions:

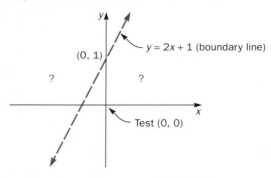

In order to decide which of the two regions is the solution, we pick a *test point* from one of the regions, say (0,0), and substitute it into the original inequality $y < 2x + 1$. We get

$$0 < 2 \cdot 0 + 1, \quad \text{or} \quad 0 < 1, \quad \text{which is TRUE.}$$

The answer consists of all points in the shaded region containing (0,0), but *excluding* the line itself:

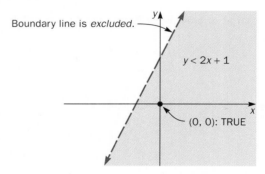

Note that the test point (0,0) *does not lie* on the boundary line. You must make sure that any test point you use lies inside one of the two regions, not on the boundary itself. (Why?) Note also that the boundary line must be excluded from the shaded "solution region," because the original inequality was "less than," or $<$. You should recognize this method as the two-dimensional analogue of the one-dimensional solution of quadratic inequalities on the x-axis using test points (Section 6.7).

EXAMPLE 1   Graph the region $2x + 3y \geq 6$.

Solution
We first graph the boundary line

$$2x + 3y = 6$$

using the intercept method. Because (0,0) does not lie on this line, we can use it as a test point:

$$2 \cdot 0 + 3 \cdot 0 \ge 6, \quad \text{or} \quad 0 \ge 6, \quad \text{which is FALSE.}$$

The solution consists of all points in the shaded region not containing (0,0), and *including* the boundary line. (Why?)

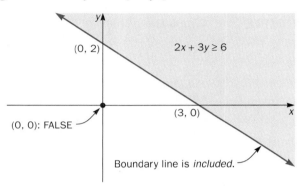

EXAMPLE 2

Graph the region $y > x^2 - 2$.

Solution
The boundary is the parabola

$$y = x^2 - 2,$$

which opens upward and has vertex at $(0,-2)$. See Section 8.1. Because (0,0) does not lie on this parabola, we can use it as a test point:

$$0 > 0^2 - 2, \quad \text{or} \quad 0 > -2, \quad \text{which is TRUE.}$$

The graph of the inequality is the shaded region containing (0,0), but *excluding* the parabola itself. (Why?)

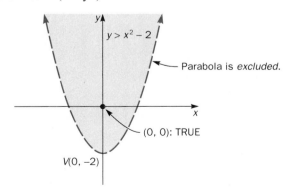

To graph a system of two inequalities in two variables, we first obtain the region defined by each inequality, using test points as before. The graph of the system, therefore, is the intersection of these regions; that is, it is the region common to both inequalities.

EXAMPLE 3

Graph the system

$$x - y \leq -4$$
$$2x + y \leq 0$$

Solution

We first graph each inequality, using test points. Note that $(0,0)$ *cannot* be used in the *second* inequality, because it lies *on* the boundary line $2x + y = 0$. Instead we use the test point $(-1,0)$ in that inequality.

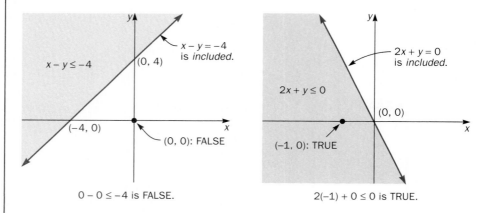

$0 - 0 \leq -4$ is FALSE. $2(-1) + 0 \leq 0$ is TRUE.

The graph of the original system of inequalities is the intersection, or overlapping region common to both inequalities.

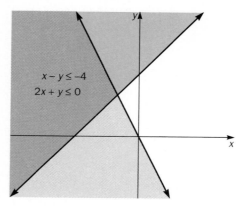

EXAMPLE 4 | Graph the system

$$x^2 + y^2 < 16$$
$$\frac{x^2}{9} - \frac{y^2}{4} > 1$$

Solution
The circle $x^2 + y^2 = 16$ has center $(0,0)$ and radius 4. (Section 8.2). The hyperbola $\frac{x^2}{9} - \frac{y^2}{4} = 1$ opens left and right, with x-intercepts $(\pm 3,0)$ and no y-intercepts. (Section 8.3). In each case, we can use $(0,0)$ as a test point.

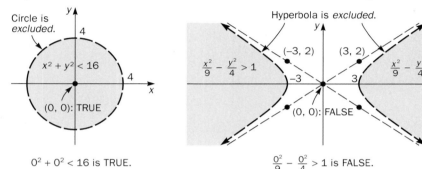

$0^2 + 0^2 < 16$ is TRUE. $\frac{0^2}{9} - \frac{0^2}{4} > 1$ is FALSE.

The graph of the original system is the overlapping region common to both inequalities.

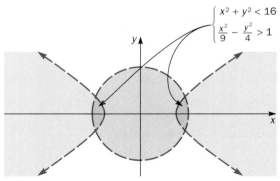

EXERCISE D   *Answers, pages A108–A109*

A

Graph these regions, as in EXAMPLES 1 and 2:

1. $y \geq 3x - 1$ **2.** $y < 1 - 2x$ **3.** $3x - 5y > 15$ **4.** $3x + 7y + 21 \leq 0$

5. $y \leq x$ **6.** $y > 2x$ **7.** $y > x^2 - 1$ **8.** $y \leq x^2 + 1$

9. $x \geq y^2$

10. $x < 1 - y^2$

11. $x^2 + y^2 \leq 4$

12. $x^2 + y^2 > 9$

13. $\dfrac{x^2}{16} + \dfrac{y^2}{9} > 1$

14. $\dfrac{x^2}{4} + \dfrac{y^2}{9} < 1$

15. $\dfrac{x^2}{25} - \dfrac{y^2}{9} \leq 1$

16. $\dfrac{y^2}{9} - \dfrac{x^2}{25} > 1$

Graph these systems, as in EXAMPLES 3 and 4:

17. $x + 2y \geq 4$
$\quad\ 2x - y \leq 6$

18. $2x - 5y < 10$
$\quad\ 3x + 2y > 6$

19. $y > 2x - 3$
$\quad\ y < x$

20. $y \geq x - 3$
$\quad\ y \geq 3x$

21. $x + y < 4$
$\quad\ x - y < 4$

22. $x - 2y \leq 6$
$\quad\quad\ y < x - 3$

23. $y \leq 4 - x^2$
$\quad\ y \geq x - 2$

24. $x^2 + y^2 \leq 9$
$\quad\ x + y \geq 3$

25. $x^2 + y^2 < 4$
$\quad\quad\ y > x^2$

26. $x^2 + y^2 \leq 16$
$\quad\ x^2 + y^2 \geq 4$

27. $x^2 + y^2 \leq 25$
$\quad\ \dfrac{x^2}{9} - \dfrac{y^2}{9} \geq 1$

28. $\dfrac{x^2}{16} + \dfrac{y^2}{4} < 1$
$\quad\ \dfrac{x^2}{4} + \dfrac{y^2}{16} > 1$

B

Graph these regions:

29. $y > 2^x$ *Hint:* Section 10.1

30. $y \leq \left(\dfrac{1}{2}\right)^x$

31. $y \geq |x|$ *Hint:* Section 9.2

32. $y < |x - 1|$

33. $x + y > 6$
$\quad\ y < 3$

34. $x - 2y \leq 8$
$\quad\quad\ x \geq 2$

Describe these shaded regions using inequalities:

35.

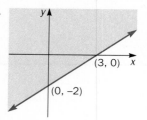

36.

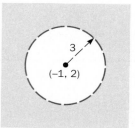

37.

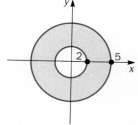

38. Find the area of the shaded annulus in Problem 37.

E.   Matrix Methods

Linear systems can be solved by a streamlined version of the addition method you learned in Section 7.4. Just as in synthetic division, this method treats the variables only as place holders. It puts their coefficients into the form of a **matrix,** which is a rectangular array of numbers. The **dimension** of a matrix is determined by the number of its rows and columns. The first matrix shown below has dimension 2×3, denoting 2 rows and 3 columns:

$$\begin{bmatrix} 5 & 1 & 3 \\ 2 & 4 & 7 \end{bmatrix} \qquad \begin{bmatrix} -4 & 0 \\ 13 & 5 \end{bmatrix} \qquad \begin{bmatrix} 9 \\ 7 \\ 0 \\ -2 \end{bmatrix} \qquad \begin{bmatrix} 4 & -1 & 8 \\ 7 & 3 & 5 \\ 2 & 0 & -6 \end{bmatrix}$$

$$2 \times 3 \qquad\qquad 2 \times 2 \qquad\quad 4 \times 1 \qquad\qquad 3 \times 3$$

(2 rows, 3 columns)   (4 rows, 1 column)

Associated with the standard linear system

$$\begin{matrix} a_1x + b_1y = c_1 \\ a_2x + b_2y = c_2 \end{matrix} \quad \text{is the \textbf{augmented matrix}} \quad \begin{bmatrix} a_1 & b_1 & \vdots & c_1 \\ a_2 & b_2 & \vdots & c_2 \end{bmatrix}$$

consisting of the coefficients of x and y and the constants. For example, associated with the linear system

$$\begin{matrix} 3x + 5y = -9 \\ 6x - 7y = 33 \end{matrix} \quad \text{is the augmented matrix} \quad \begin{bmatrix} 3 & 5 & \vdots & -9 \\ 6 & -7 & \vdots & 33 \end{bmatrix}.$$

In order to solve this system by the addition method, we would multiply the first equation by -2 and then add it to the second equation. This would eliminate x and we could then solve for y. Analogous to these steps are the **elementary row operations** on matrices:

1. Any row may be multiplied by a non-zero constant.

2. A non-zero constant multiple of one row may be added to a non-zero constant multiple of another row.

3. Any two rows may be interchanged.

When an augmented matrix is subjected to any of these row operations, the solutions obtained from the resulting matrix are the same as those of the original system. In the following examples, we show how matrices are used to solve linear systems.

EXAMPLE 1 | Solve by matrix methods: $\quad 3x + 5y = -9$
$\qquad\qquad\qquad\qquad\qquad\qquad\quad\; 6x - 7y = 33$

Solution
The augmented matrix of this system is shown above. Let us multiply row 1 by -2 and add it to row 2, procedures denoted by $-2R_1 + R_2$:

$$\begin{bmatrix} 3 & 5 & | & -9 \\ 6 & -7 & | & 33 \end{bmatrix} \xrightarrow{-2R_1 + R_2} \begin{bmatrix} 3 & 5 & | & -9 \\ -2\cdot 3 + 6 & -2\cdot 5 - 7 & | & -2(-9) + 33 \end{bmatrix}$$

$$= \begin{bmatrix} 3 & 5 & | & -9 \\ 0 & -17 & | & 51 \end{bmatrix}$$

Row 2 now means that $\qquad\qquad\qquad\qquad -17y = 51$
$$y = -3.$$

Substitute $y = -3$ into row 1: $\quad 3x + 5(-3) = -9$
$$3x \qquad\quad = 6$$
$$x = 2.$$

Answer: $x = 2$, $y = -3$ or $(2, -3)$.

Based on this example, the goal is to transform the augmented matrix to a matrix with a 0 in the lower left corner, using elementary row operations:

$$\begin{bmatrix} a & b & | & c \\ d & e & | & f \end{bmatrix} \longrightarrow \begin{bmatrix} a & b & | & c \\ 0 & g & | & h \end{bmatrix}$$

$\qquad\qquad\qquad$ Augmented $\qquad\qquad\quad$ 0 in lower
$\qquad\qquad\qquad\quad$ matrix $\qquad\qquad\qquad$ left corner

We then solve for y using row 2, and then solve for x by substituting the value of y into row 1.

EXAMPLE 2 | Solve by matrices: $\qquad\qquad 5x - 3y = \quad 2$
$\qquad\qquad\qquad\qquad\qquad\qquad\quad\; 4x + 7y = -36$

Solution
In the augmented matrix, we multiply row 1 by -4 and add it to 5 times row 2. This will create a 0 in the lower left corner:

$$\begin{bmatrix} 5 & -3 & | & 2 \\ 4 & 7 & | & -36 \end{bmatrix} \xrightarrow{-4R_1 + 5R_2} \begin{bmatrix} 5 & -3 & | & 2 \\ -4\cdot 5 + 5\cdot 4 & -4(-3) + 5\cdot 7 & | & -4\cdot 2 + 5(-36) \end{bmatrix}$$

$$= \begin{bmatrix} 5 & -3 & | & 2 \\ 0 & 47 & | & -188 \end{bmatrix}$$

Row 2 now means that $\qquad\qquad\qquad 47y = -188$
$$y = -4.$$

Substitute $y = -4$ into row 1:

$$5x - 3(-4) = 2$$
$$5x \qquad\;\; = -10$$
$$x = -2.$$

Answer: $x = -2$, $y = -4$ or $(-2, -4)$.

Although matrices might seem like more trouble than they are worth to solve for x and y, they definitely show their value in the case of three unknowns. Recall from Section 7.6 that we first eliminated one of the unknowns, say x, twice, which resulted in a **reduced system** in y and z. We then solved the reduced system as in Section 7.4. In terms of matrices, we will transform the augmented matrix first into the matrix representing the reduced system, and then into the matrix with a triangle of 0's in the lower left corner:

$$\begin{bmatrix} a & b & c & | & d \\ e & f & g & | & h \\ i & j & k & | & l \end{bmatrix} \longrightarrow \begin{bmatrix} a & b & c & | & d \\ 0 & m & n & | & p \\ 0 & q & r & | & s \end{bmatrix} \longrightarrow \begin{bmatrix} a & b & c & | & d \\ 0 & m & n & | & p \\ 0 & 0 & t & | & u \end{bmatrix}$$

augmented   reduced system   triangle of 0's
matrix   matrix   matrix

EXAMPLE 3

Solve by matrix methods:

$$x + y + z = 2$$
$$-x + 2y - 3z = -9$$
$$3x + y + 2z = 6$$

Solution
We transform the augmented matrix as follows:

$$\begin{bmatrix} 1 & 1 & 1 & | & 2 \\ -1 & 2 & -3 & | & -9 \\ 3 & 1 & 2 & | & 6 \end{bmatrix} \xrightarrow[3R_1 - R_3]{R_1 + R_2} \begin{bmatrix} 1 & 1 & 1 & | & 2 \\ 0 & 3 & -2 & | & -7 \\ 0 & 2 & 1 & | & 0 \end{bmatrix} \xrightarrow{2R_2 - 3R_3} \begin{bmatrix} 1 & 1 & 1 & | & 2 \\ 0 & 3 & -2 & | & -7 \\ 0 & 0 & -7 & | & -14 \end{bmatrix}$$

Row 3 of the last matrix means that $-7z = -14$
$$z = 2.$$

Substitute $z = 2$ into row 2: $3y - 2 \cdot 2 = -7$
$$3y \qquad\;\; = -3$$
$$y = -1.$$

Substitute $y = -1$, $z = 2$ into row 1: $x + (-1) + 2 = 2$
$$x = 1.$$

Answer: $x = 1$, $y = -1$, $z = 2$ or $(1, -1, 2)$.

EXAMPLE 4 | Solve by matrices:

$$3x + 2y - z = 12$$
$$2x + 3z = 4$$
$$-5x + 3y + 2z = -1$$

Solution

We transform the augmented matrix of the system as follows:

$$
\begin{bmatrix} 3 & 2 & -1 & | & 12 \\ 2 & 0 & 3 & | & 4 \\ -5 & 3 & 2 & | & -1 \end{bmatrix}
\xrightarrow[\underset{5 \cdot 2 + 3 \cdot 3}{5R_1 + 3R_3}]{2R_1 - 3R_2}
\begin{bmatrix} 3 & 2 & -1 & | & 12 \\ 0 & 4 & -11 & | & 12 \\ 0 & 19 & 1 & | & 57 \end{bmatrix}
\xrightarrow[]{19R_2 - 4R_3}
\begin{bmatrix} 3 & 2 & -1 & | & 12 \\ 0 & 4 & -11 & | & 12 \\ 0 & 0 & -213 & | & 0 \end{bmatrix}
$$

$$19(-11) - 4 \cdot 1$$
$$19 \cdot 12 - 4 \cdot 57$$

Row 3 of the last matrix means that $-213z = 0$
$$z = 0.$$

Substitute $z = 0$ into row 2: $4y - 11 \cdot 0 = 12$
$$y = 3.$$

Substitute $y = 3, z = 0$ into row 1: $3x + 2 \cdot 3 - 1 \cdot 0 = 12$
$$3x = 6$$
$$x = 2.$$

Answer: $x = 2, y = 3, z = 0$ or $(2,3,0)$.

EXERCISE E   *Answers, page A109*

Solve by matrices, as in EXAMPLES 1 and 2:

1. $x + 4y = -5$
  $2x - 3y = 12$

2. $x - 7y = -17$
  $5x + 2y = 26$

3. $3x - 7y = 1$
  $2x + 5y = 20$

4. $5x + 4y = -2$
  $3x + 7y = 8$

5. $6x - 7y = 18$
  $-x + 8y = -3$

6. $5x + 3y = -10$
  $x - 2y = -2$

7. $2x + 3y = 3$
  $6x - 5y = 2$

8. $9x - 8y = 4$
  $3x + 4y = 3$

Solve by matrices, as in EXAMPLES 3 and 4:

9. $x + y + z = 3$
  $-x + 2y + 7z = 8$
  $x + 4y + 8z = 13$

10. $x + y + z = 7$
  $x + 4y - 3z = -6$
  $-x + 2y + 2z = 8$

11. $x + 2y + z = 8$
  $3x - 4y - z = -8$
  $5x + 3y + z = 14$

12. $x - y - z = -1$
$2x + 3y + 5z = 13$
$5x - 3y + 4z = 1$

13. $2x + 3y + 5z = -7$
$4x - y + 2z = -3$
$3x + 4y - 3z = 30$

14. $2x + 3y - 4z = -1$
$3x + 6y + 2z = 0$
$4x + 7y + 5z = -1$

15. $4x - y + 3z = -4$
$3y + z = -1$
$3x + 6y + 2z = -4$

16. $3x - 5y + 2z = -2$
$x - y = 2$
$5x + 7y - 2z = 3$

17. $y + z = 1$
$x - z = 0$
$x + y + z = 3$
Hint: Interchange rows.

B

18. $x - 3z = 0$
$x + y = 5$
$y + z = 3$

19. $x + y + z + w = 0$
$-x + y + 2z = 0$
$2x - y + z + w = -1$
$x - y - z + w = -2$
Hint: Get a triangle
of 0's in the lower left
corner.

20. $x + y + z + w = 4$
$2x - 3y + 2z - 5w = 0$
$-x + 2y + w = 3$
$x - y + z = -1$

F.   ## Common Logarithms

In Section 10.2, you obtained the values of base-10 logarithms on a scientific calculator. Common logs of numbers between 1.00 and 9.99 can also be found in the Table of Common Logarithms, pages A28–A29, where they are approximated to four decimal places.

EXAMPLE 1

a. $\log 2.00 = 0.3010$

b. $\log 4.72 = 0.6739$

c. $\log 7.65 = 0.8837$

To find the common log of positive numbers less than 1.00 or greater than 9.99, we first write them in scientific notation and then use two properties of logarithms. For example,

$$\log 8540 = \log (8.54 \times 10^3) \qquad \text{Scientific notation.}$$
$$= \log 8.54 + \log 10^3 \qquad \text{Product Rule.}$$
$$= 0.9135 + 3 \qquad \text{Inverse Property: } \log_{10} 10^n = n$$
$$= 3.9135$$

The four-place decimal 0.9135 obtained from the Table is called the **mantissa** of the logarithm; the integer exponent 3 is the **characteristic.** In like manner,

$$\log 0.0000468 = \log (4.68 \times 10^{-5})$$
$$= \log 4.68 + \log 10^{-5}$$
$$= \underbrace{0.6702}_{\text{Mantissa}} - \overset{\text{Characteristic}}{5}$$

Note that we did not complete the subtraction. Had we done so, we would have obtained −4.3298. This is a negative logarithm in which neither the mantissa 0.6702 nor the characteristic −5 is visible. Therefore, *in the case of negative characteristics, do not complete the subtraction.*

EXAMPLE 2

a. $\log 651 = \log (6.51 \times 10^2)$
   $= 0.8136 + 2$ $\log 6.51 + \log 10^2$
   $= 2.8136$

b. $\log 40{,}200 = \log (4.02 \times 10^4)$
   $= 0.6042 + 4$
   $= 4.6042$

c. $\log 0.00856 = \log (8.56 \times 10^{-3})$
   $= 0.9325 - 3$ Do not subtract.

d. $\log 0.927 = \log (9.27 \times 10^{-1})$
   $= 0.9671 - 1$ Do not subtract.

The Table tells us that $\log 4.72 = 0.6739$. In exponential form, this means that $10^{0.6739} = 4.72$. Another name for a power of 10 is **antilogarithm.** Thus

$$\text{antilog } 0.6739 = 10^{0.6739} = 4.72$$

Finding the antilog of a number is the inverse of finding its log. First locate the four-place mantissa on the *body* of the Table; this gives a three-place number between 1.00 and 9.99. Then multiply the three-place number by the power of 10 indicated by the characteristic. For example,

$$\text{antilog } 2.8525 = 10^{2.8525} = ?$$

is obtained as follows:

1. Locate the mantissa 0.8525 in the *body* of the Table; it is under the heading 7.12.

2. Multiply 7.12 by the power of 10 indicated by the characteristic 2: 7.12×10^2. Thus

$$\text{antilog } 2.8525 = 10^{2.8525} = 7.12 \times 10^2$$
$$= 712$$

EXAMPLE 3

a. antilog $5.6096 = 4.07 \times 10^5$

Mantissa

Characteristic

   $= 407,000$ or $10^{5.6096} = 407,000$

b. antilog $3.8136 = 6.51 \times 10^3$

   $= 6510$ or $10^{3.8136} = 6510$

c. antilog $0.4771 = 3.00 \times 10^0$

   $= 3.00$ or $10^{0.4771} = 3.00$

d. antilog $(0.9325 - 4) = 8.56 \times 10^{-4}$

   $= 0.000856$

e. antilog $(0.5105 - 1) = 3.24 \times 10^{-1}$

   $= 0.324$

f. antilog $(2.5478 - 7) = $ antilog $(0.5478 + 2 - 7)$

   $= $ antilog $(0.5478 - 5)$ Characteristic $= 2 - 7 = -5$

   $= 3.53 \times 10^{-5}$

   $= 0.0000353$

g. antilog $(1.8768 - 3) = $ antilog $(0.8768 - 2)$ Characteristic $= 1 - 3 = -2$

   $= 7.53 \times 10^{-2}$

   $= 0.0753$

Note that the characteristic of $2.5478 - 7$ is -5,
and the characteristic of $1.8768 - 3$ is -2.

It was mentioned above that negative logarithms, such as -5.3161, display neither the mantissa nor the characteristic. Here 0.3161 is *not* the mantissa and -5 is *not* the characteristic, because

$$-5.3161 \neq 0.3161 - 5$$

In order to obtain

$$\text{antilog } (-5.3161) = 10^{-5.3161} = ?$$

we *add and subtract* 6, which converts this to a positive mantissa and a negative characteristic:

antilog $(-5.3161) = $ antilog $(-5.3161 + 6.0000 - 6)$ Add and subtract 6.

   $= $ antilog $(\quad 0.6839 \quad - 6)$

   $= 4.83 \times 10^{-6}$

   $= 0.00000483$

Note that $+6.0000 - 6 = 0$, hence the value inside the parentheses is unchanged.

EXAMPLE 4

a. antilog (-2.7218) = antilog $(-\underbrace{2.4134 + 3.0000}_{} - 3)$ Add and subtract 3.

                = antilog $(\quad 0.5866 \quad - 3)$ or $\quad 10^{-2.7218} = $

                = 3.86×10^{-3} 0.00386

                = 0.00386

b. antilog (-4.1785) = antilog $(-4.1785 + 5.0000 - 5)$

                = antilog $(0.8215 - 5)$

                = 6.63×10^{-5}

                = 0.0000663

c. antilog (-0.6861) = antilog $(-0.6861 + 1.0000 - 1)$

                = antilog $(0.3139 - 1)$

                = 2.06×10^{-1}

                = 0.206

If a mantissa is not in the body of the Table, it always lies between two mantissas that are. For example, antilog 0.9407 lies between antilog 0.9405 = 8.72 and antilog 0.9410 = 8.73; thus antilog 0.9407 = 8.72?. We write these values as shown:

$$\begin{array}{l} \text{antilog } 0.9410 = 8.73 \\ .0005 \quad\quad \text{antilog } 0.9407 = 8.72\underline{?} \\ \quad\quad .0002 \quad \text{antilog } 0.9405 = 8.72 \end{array}$$

By subtraction, note $0.9407 - 0.9405 = .0002$, and $0.9410 - 0.9405 = .0005$. Thus

$$0.9407 \text{ is } \frac{.0002}{.0005} = .4 \text{ of the way from the bottom number.}$$

Therefore,

$$\text{antilog } 0.9407 = 8.72\underline{4}$$

Approximately an unknown value lying between two known values in this manner is called **linear interpolation.**

EXAMPLE 5

Find these antilogs by interpolation:

a. antilog 3.4694 = antilog $(0.4694 + 3)$

The mantissa 0.4694 lies between the two mantissas shown:

$$\begin{array}{l} \text{antilog } 0.4698 = 2.95 \\ .0015 \quad\quad \text{antilog } 0.4694 = 2.94? \\ \quad\quad .0011 \quad \text{antilog } 0.4683 = 2.94 \end{array}$$

From the discussion above, the required fraction is

$$\frac{.0011}{.0015} = .733 = .7 \text{ (rounded off)}$$

Thus

$$\begin{aligned} \text{antilog } 3.4694 &= 2.94\underline{7} \times 10^3 \\ &= 294\overline{7} \end{aligned}$$

b. antilog $(0.6120 - 5)$

The mantissa 0.6120 lies between the two mantissas shown:

antilog 0.6128 = 4.10

.0011   .0003 antilog 0.6120 = 4.09?

antilog 0.6117 = 4.09

The required fraction is

$$\frac{.0003}{.0011} = .272 = .3 \text{ (rounded off)}$$

Thus

$$\begin{aligned} \text{antilog } (0.6120 - 5) &= 4.09\underline{3} \times 10^{-5} \\ &= 0.0000\overline{4093} \end{aligned}$$

EXERCISE F   *Answers, page A110*

Use the Table of Common Logarithms to find each log, as in EXAMPLES 1 and 2:

1. log 3.24
2. log 6.92
3. log 58.9

4. log 21.6
5. log 8270
6. log 10,200

7. log 1000
8. log 100,000
9. log 0.00529

10. log 0.0605
11. log 0.000042
12. log 0.73

13. log 0.000005
14. log 0.00001

Find these antilogs, as in EXAMPLE 3:

15. antilog 0.6542
16. antilog 0.9624
17. antilog 1.7796

18. antilog 2.5079
19. antilog 4.9031
20. antilog 5.6532

21. antilog $(0.4814 - 3)$
22. antilog $(0.9576 - 2)$
23. antilog $(0.3054 - 1)$

24. antilog $(0.7782 - 6)$ **25.** antilog $(1.3010 - 5)$ **26.** antilog $(2.9542 - 6)$

27. antilog $(4.7143 - 5)$ **28.** antilog $(7.6990 - 10)$

Find these antilogs by first converting to a positive mantissa and a negative characteristic, as in EXAMPLE 4:

29. antilog (-2.1858) **30.** antilog (-1.3224) **31.** antilog (-5.0424)

32. antilog (-3.6003) **33.** antilog (-0.2190) **34.** antilog (-6.2993)

Find these antilogs by interpolation, as in EXAMPLE 5:

35. antilog 0.6558 **36.** antilog 0.6077 **37.** antilog 3.9551

38. antilog 2.9357 **39.** antilog 6.2397 **40.** antilog 1.3123

41. antilog $(0.5557 - 4)$ **42.** antilog $(0.4142 - 2)$ **43.** antilog $(2.4022 - 7)$

44. antilog $(1.8533 - 2)$ **45.** antilog (-2.5146) **46.** antilog (-4.7120)

47. $10^{0.4171}$ **48.** $10^{4.9135}$

G.   Calculating with Logarithms

Logarithms were originally invented in the seventeenth century as a computational tool. They have since given way to the calculator. In this section, we revive a method first used over 350 years ago.

EXAMPLE 1   Compute $(523)(47.1)$ by logarithms.

Solution

$$\text{Let}\quad N = (523)(47.1); \text{ then}$$

$$\log N = \log 523 + \log 47.1 \qquad \text{Product Rule.}$$
$$= 2.7185 + 1.6730 \qquad \text{From previous section.}$$
$$= 4.3915$$

Thus

$$N = \text{antilog } 4.3915 \qquad \begin{array}{l}\log N = x \text{ provided} \\ N = 10^{x} = \text{antilog } x.\end{array}$$

$$= 2.46\underline{3} \times 10^{4} \qquad \text{By interpolation.}$$

Answer: $(523)(47.1) = 24{,}630$ $24{,}633.3$ by calculator!

EXAMPLE 2 | Compute $\dfrac{0.000385}{0.0719}$ by logarithms.

Solution

Let $\quad N = \dfrac{0.000385}{0.0719}$; then

$$\begin{aligned}
\log N &= \log 0.000385 - \log 0.0719 \qquad &\text{Quotient Rule.}\\
&= (0.5855 - 4) - (0.8567 - 2) \qquad &\text{From previous section.}\\
&= 0.5855 - 4 - 0.8567 + 2\\
&= 0.5855 - 0.8567 - 4 + 2\\
&= -0.2712 - 2\\
&= -2.2712
\end{aligned}$$

Thus

$$\begin{aligned}
N &= \text{antilog } (-2.2712)\\
&= \text{antilog } (-2.2712 + 3.0000 - 3) \qquad &\text{See Example 4}\\
&= \text{antilog } (0.7288 - 3) \qquad &\text{of previous section.}\\
&= 5.35\underline{5} \times 10^{-3} \qquad &\text{By interpolation.}
\end{aligned}$$

Answer: $\dfrac{0.000385}{0.0719} = 0.005355$ $\qquad\qquad$ 0.0053547 by calculator.

EXAMPLE 3 | Compute $(5270)^4$ by logarithms.

Solution

Let $\quad N = (5270)^4$; then

$$\begin{aligned}
\log N &= 4(\log 5270) \qquad &\text{Power Rule.}\\
&= 4(3.7218)\\
&= 14.8872
\end{aligned}$$

Thus

$$\begin{aligned}
N &= \text{antilog } 14.8872\\
&= 7.71\underline{2} \times 10^{14} \qquad &\text{By interpolation.}
\end{aligned}$$

Answer: $\qquad (5270)^4 = 7.712 \times 10^{14}$ $\qquad$ 7.7133×10^{14} by calculator.

EXAMPLE 4 | Compute $\sqrt{8370}$ by logarithms.

Solution

Let $\quad N = (8370)^{1/2}$; then $\qquad$ Write $\sqrt{8370}$ as $(8370)^{1/2}$.

$$\log N = \frac{1}{2}(\log 8370) \qquad \text{Power Rule.}$$

$$= \frac{1}{2}(3.9227)$$

$$= 1.9614 \qquad \text{Rounded off.}$$

Thus

$$N = \text{antilog } 1.9614$$
$$= 9.15 \times 10^1 \qquad \text{No interpolation needed.}$$

Answer: $\sqrt{8370} = 91.5$ 91.487704 by calculator.

EXAMPLE 5 Compute $\sqrt[3]{0.000413}$ by logarithms.

Solution

Let $N = (0.000413)^{1/3}$; then Write $\sqrt[3]{0.000413}$ as $(0.000413)^{1/3}$.

$$\log N = \frac{1}{3}(\log 0.000413) \qquad \text{Product Rule.}$$

$$= \frac{1}{3}(0.6160 - 4)$$

Now $\frac{1}{3}$ of -4 would give a fractional characteristic. Therefore, it is desirable to convert -4 to -6, which is divisible by 3. To do this, we *add* 2.0000 to the mantissa and *subtract* 2 from the characteristic.

$$\log N = \frac{1}{3}\underbrace{(0.6160 + 2.0000}\ \underbrace{- 2 - 4)}$$

$$= \frac{1}{3}(2.6160 - 6)$$

$$= 0.8720 - 2 \qquad \text{Distributive Property.}$$

Thus

$$N = \text{antilog } (0.8720 - 2)$$
$$= 7.44\underline{7} \times 10^{-2} \qquad \text{By interpolation.}$$

Answer: $\sqrt[3]{0.000413} = 0.07447$ 0.0744703 by calculator.

EXERCISE G *Answers, page A110*

Compute by logarithms, as in EXAMPLES 1 through 5:

1. $(2.00)(4.00)$ **2.** $(3.60)(2.50)$ **3.** $(53.9)(4200)$

4. $(68.0)(97,700)$ **5.** $(0.000542)(0.0216)$ **6.** $(0.00345)(0.322)$

7. $\dfrac{8.80}{4.00}$ **8.** $\dfrac{8.00}{2.00}$ **9.** $\dfrac{772,000}{0.00028}$

10. $\dfrac{9,090,000}{0.0000035}$

11. $\dfrac{0.00672}{2130}$

12. $\dfrac{0.0000089}{673}$

13. $(43.7)^2$

14. $(162)^3$

15. $(0.000515)^4$

16. $(0.068)^5$

17. $\sqrt{4.00}$

18. $\sqrt[3]{8.00}$

19. $\sqrt[3]{27.0}$

20. $\sqrt[5]{32.0}$

21. $\sqrt{5280}$

22. $\sqrt[3]{78,700}$

23. $\sqrt[3]{0.00413}$

24. $\sqrt{0.00085}$

25. $\sqrt{0.00592}$

26. $\sqrt[3]{0.000978}$

27. $\dfrac{(87.2)(4670)}{9.43}$

B

28. $\dfrac{(827)(0.00556)}{0.0913}$

29. $(427)^3(78.7)^2$

30. $(0.00092)^2(32,700)$

31. $\dfrac{(2300)\sqrt[3]{617}}{(0.45)^2}$

32. $\dfrac{\sqrt{0.0794}\,(672)^3}{21.3}$

Table of Common Logarithms

N	0	1	2	3	4	5	6	7	8	9
1.0	.0000	.0043	.0086	.0128	.0170	.0212	.0253	.0294	.0334	.0374
1.1	.0414	.0453	.0492	.0531	.0569	.0607	.0645	.0682	.0719	.0755
1.2	.0792	.0828	.0864	.0899	.0934	.0969	.1004	.1038	.1072	.1106
1.3	.1139	.1173	.1206	.1239	.1271	.1303	.1335	.1367	.1399	.1430
1.4	.1461	.1492	.1523	.1553	.1584	.1614	.1644	.1673	.1703	.1732
1.5	.1761	.1790	.1818	.1847	.1875	.1903	.1931	.1959	.1987	.2014
1.6	.2041	.2068	.2095	.2122	.2148	.2175	.2201	.2227	.2253	.2279
1.7	.2304	.2330	.2355	.2380	.2405	.2430	.2455	.2480	.2504	.2529
1.8	.2553	.2577	.2601	.2625	.2648	.2672	.2695	.2718	.2742	.2765
1.9	.2788	.2810	.2833	.2856	.2878	.2900	.2923	.2945	.2967	.2989
2.0	.3010	.3032	.3054	.3075	.3096	.3118	.3139	.3160	.3181	.3201
2.1	.3222	.3243	.3263	.3284	.3304	.3324	.3345	.3365	.3385	.3404
2.2	.3424	.3444	.3464	.3483	.3502	.3522	.3541	.3560	.3579	.3598
2.3	.3617	.3636	.3655	.3674	.3692	.3711	.3729	.3747	.3766	.3784
2.4	.3802	.3820	.3838	.3856	.3874	.3892	.3909	.3927	.3945	.3962
2.5	.3979	.3997	.4014	.4031	.4048	.4065	.4082	.4099	.4116	.4133
2.6	.4150	.4166	.4183	.4200	.4216	.4232	.4249	.4265	.4281	.4298
2.7	.4314	.4330	.4346	.4362	.4378	.4393	.4409	.4425	.4440	.4456
2.8	.4472	.4487	.4502	.4518	.4533	.4548	.4564	.4579	.4594	.4609
2.9	.4624	.4639	.4654	.4649	.4683	.4698	.4713	.4728	.4742	.4757
3.0	.4771	.4786	.4800	.4814	.4829	.4843	.4857	.4871	.4886	.4900
3.1	.4914	.4928	.4942	.4955	.4969	.4983	.4997	.5011	.5024	.5038
3.2	.5051	.5065	.5079	.5092	.5105	.5119	.5132	.5145	.5159	.5172
3.3	.5185	.5198	.5211	.5224	.5237	.5250	.5263	.5276	.5289	.5302
3.4	.5315	.5328	.5340	.5353	.5366	.5378	.5391	.5403	.5416	.5428
3.5	.5441	.5453	.5464	.5478	.5490	.5502	.5514	.5527	.5539	.5551
3.6	.5563	.5575	.5587	.5599	.5611	.5623	.5635	.5647	.5658	.5670
3.7	.5682	.5694	.5705	.5717	.5729	.5740	.5752	.5763	.5775	.5786
3.8	.5798	.5809	.5821	.5832	.5843	.5855	.5866	.5877	.5888	.5899
3.9	.5911	.5922	.5933	.5944	.5955	.5966	.5977	.5988	.5999	.6010
4.0	.6021	.6031	.6042	.6053	.6064	.6075	.6085	.6096	.6107	.6117
4.1	.6128	.6138	.6149	.6160	.6170	.6180	.6191	.6201	.6212	.6222
4.2	.6232	.6243	.6253	.6263	.6274	.6284	.6294	.6304	.6314	.6325
4.3	.6335	.6345	.6355	.6365	.6375	.6385	.6395	.6405	.6415	.6425
4.4	.6435	.6444	.6454	.6464	.6474	.6484	.6493	.6503	.6513	.6522
4.5	.6532	.6542	.6551	.6561	.6571	.6580	.6590	.6599	.6609	.6618
4.6	.6628	.6637	.6646	.6656	.6665	.6675	.6684	.6693	.6702	.6712
4.7	.6721	.6730	.6739	.6749	.6758	.6767	.6776	.6785	.6794	.6803
4.8	.6812	.6821	.6830	.6839	.6848	.6857	.6866	.6875	.6884	.6893
4.9	.6902	.6911	.6920	.6928	.6937	.6946	.6955	.6964	.6972	.6981
5.0	.6990	.6998	.7007	.7016	.7024	.7033	.7042	.7050	.7059	.7067
5.1	.7076	.7084	.7093	.7101	.7110	.7118	.7126	.7135	.7143	.7152
5.2	.7160	.7168	.7177	.7185	.7193	.7202	.7210	.7218	.7226	.7235
5.3	.7243	.7251	.7259	.7267	.7275	.7284	.7292	.7300	.7308	.7316
5.4	.7324	.7332	.7340	.7348	.7356	.7364	.7372	.7380	.7388	.7396

Examples: log 4.72 = 6739   antilog .6739 = $10^{.6739}$

$$= 4.72$$

Table of Common Logarithms (continued)

N	0	1	2	3	4	5	6	7	8	9
5.5	.7404	.7412	.7419	.7427	.7435	.7443	.7451	.7459	.7466	.7474
5.6	.7482	.7490	.7497	.7505	.7513	.7520	.7528	.7536	.7543	.7551
5.7	.7559	.7566	.7574	.7582	.7589	.7597	.7604	.7612	.7619	.7627
5.8	.7634	.7642	.7649	.7657	.7664	.7672	.7679	.7686	.7694	.7701
5.9	.7709	.7716	.7723	.7731	.7738	.7745	.7752	.7760	.7767	.7774
6.0	.7782	.7789	.7796	.7803	.7810	.7818	.7825	.7832	.7839	.7846
6.1	.7853	.7860	.7868	.7875	.7882	.7889	.7896	.7903	.7910	.7917
6.2	.7924	.7931	.7938	.7945	.7952	.7959	.7966	.7973	.7980	.7987
6.3	.7993	.8000	.8007	.8014	.8021	.8028	.8035	.8041	.8048	.8055
6.4	.8062	.8069	.8075	.8082	.8089	.8096	.8102	.8109	.8116	.8122
6.5	.8129	.8136	.8142	.8149	.8156	.8162	.8169	.8176	.8182	.8189
6.6	.8195	.8202	.8209	.8215	.8222	.8228	.8235	.8241	.8248	.8254
6.7	.8261	.8267	.8274	.8280	.8287	.8293	.8299	.8306	.8312	.8319
6.8	.8325	.8331	.8338	.8344	.8351	.8357	.8363	.8370	.8376	.8382
6.9	.8388	.8395	.8401	.8407	.8414	.8420	.8426	.8432	.8439	.8445
7.0	.8451	.8457	.8463	.8470	.8476	.8482	.8488	.8494	.8500	.8506
7.1	.8513	.8519	.8525	.8531	.8537	.8543	.8549	.8555	.8561	.8567
7.2	.8573	.8579	.8585	.8591	.8597	.8603	.8609	.8615	.8621	.8627
7.3	.8633	.8639	.8645	.8651	.8657	.8663	.8669	.8675	.8681	.8686
7.4	.8692	.8698	.8704	.8710	.8716	.8722	.8727	.8733	.8739	.8745
7.5	.8751	.8756	.8762	.8768	.8774	.8779	.8785	.8791	.8797	.8802
7.6	.8808	.8814	.8820	.8825	.8831	.8837	.8842	.8848	.8854	.8859
7.7	.8865	.8871	.8876	.8882	.8887	.8893	.8899	.8904	.8910	.8915
7.8	.8921	.8927	.8932	.8938	.8943	.8949	.8954	.8960	.8965	.8971
7.9	.8976	.8982	.8987	.8993	.8998	.9004	.9009	.9015	.9020	.9025
8.0	.9031	.9036	.9042	.9047	.9053	.9058	.9063	.9069	.9074	.9079
8.1	.9085	.9090	.9096	.9101	.9106	.9112	.9117	.9122	.9128	.9133
8.2	.9138	.9143	.9149	.9154	.9159	.9165	.9170	.9175	.9180	.9186
8.3	.9191	.9196	.9201	.9206	.9212	.9217	.9222	.9227	.9232	.9238
8.4	.9243	.9248	.9253	.9258	.9263	.9269	.9274	.9279	.9284	.9289
8.5	.9294	.9299	.9304	.9309	.9315	.9320	.9325	.9330	.9335	.9340
8.6	.9345	.9350	.9355	.9360	.9365	.9370	.9375	.9380	.9385	.9390
8.7	.9395	.9400	.9405	.9410	.9415	.9420	.9425	.9430	.9435	.9440
8.8	.9445	.9450	.9455	.9460	.9465	.9469	.9474	.9479	.9484	.9489
8.9	.9494	.9499	.9504	.9509	.9513	.9518	.9523	.9528	.9533	.9538
9.0	.9542	.9547	.9552	.9557	.9562	.9566	.9571	.9576	.9581	.9586
9.1	.9590	.9595	.9600	.9605	.9609	.9614	.9619	.9624	.9628	.9633
9.2	.9638	.9643	.9647	.9652	.9657	.9661	.9666	.9671	.9675	.9680
9.3	.9685	.9689	.9694	.9699	.9703	.9708	.9713	.9717	.9722	.9727
9.4	.9731	.9736	.9741	.9745	.9750	.9754	.9759	.9763	.9768	.9773
9.5	.9777	.9782	.9786	.9791	.9795	.9800	.9805	.9809	.9814	.9818
9.6	.9823	.9827	.9832	.9836	.9841	.9845	.9850	.9854	.9859	.9863
9.7	.9868	.9872	.9877	.9881	.9886	.9890	.9894	.9899	.9903	.9908
9.8	.9912	.9917	.9921	.9926	.9930	.9934	.9939	.9943	.9948	.9952
9.9	.9954	.9961	.9965	.9969	.9974	.9978	.9983	.9987	.9991	.9996

Powers of e

x	e^x	e^{-x}
0.0	1.000	1.000
0.1	1.105	0.905
0.2	1.221	.819
0.3	1.350	.741
0.4	1.492	.670
0.5	1.649	.607
0.6	1.822	.549
0.7	2.014	.497
0.8	2.226	.449
0.9	2.460	.407
1.0	2.718	.368
1.1	3.004	.333
1.2	3.320	.301
1.3	3.669	.273
1.4	4.055	.247
1.5	4.482	.223
1.6	4.953	.202
1.7	5.474	.183
1.8	6.050	.165
1.9	6.686	.150
2.0	7.389	.135
2.1	8.166	.122
2.2	9.025	.111
2.3	9.974	.100
2.4	11.023	.091
2.5	12.182	.082
2.6	13.464	.074
2.7	14.880	.067
2.8	16.445	.061
2.9	18.174	.055
3.0	20.086	.050
3.1	22.198	.045
3.2	24.533	.041
3.3	27.113	.037
3.4	29.964	.033
3.5	33.116	.030
3.6	36.598	.027
3.7	40.447	.025
3.8	44.701	.022
3.9	49.402	.020
4.0	54.598	.018
4.1	60.340	.017
4.2	66.686	.015
4.3	73.700	.014
4.4	81.451	.012
4.5	90.017	.011
4.6	99.484	.010
4.7	109.947	.009
4.8	121.510	.008
4.9	134.290	.007
5.0	148.413	.007

Roots and Powers

N	N^2	$\sqrt{N}$	N^3	$\sqrt[3]{N}$
0	0	0	0	0
1	1	1	1	1
2	4	1.414	8	1.260
3	9	1.732	27	1.442
4	16	2	64	1.537
5	25	2.236	125	1.710
6	36	2.449	216	1.817
7	49	2.646	343	1.913
8	64	2.828	512	2
9	81	3	729	2.080
10	100	3.162	1000	2.154
11	121	3.317	1331	2.224
12	144	3.464	1728	2.289
13	169	3.606	2197	2.351
14	196	3.742	2744	2.410
15	225	3.873	3375	2.466
16	256	4	4096	2.520
17	289	4.124	4913	2.571
18	324	4.243	5832	2.621
19	361	4.359	6859	2.668
20	400	4.472	8000	2.714
21	441	4.583	9261	2.759
22	484	4.690	10648	2.802
23	529	4.796	12167	2.844
24	576	4.899	13824	2.884
25	625	5	15625	2.924
26	676	5.099	17576	2.962
27	729	5.196	19683	3
28	784	5.292	21952	3.037
29	841	5.385	24389	3.072
30	900	5.477	27000	3.107
31	961	5.568	29791	3.141
32	1024	5.657	32768	3.175
33	1089	5.745	35937	3.208
34	1156	5.831	39304	3.240
35	1225	5.916	42875	3.271
36	1296	6	46656	3.302
37	1369	6.083	50653	3.332
38	1444	6.164	54872	3.362
39	1521	6.245	59319	3.391
40	1600	6.325	64000	3.420
41	1681	6.403	68921	3.448
42	1764	6.481	74088	3.476
43	1849	6.557	79507	3.503
44	1936	6.633	85184	3.530
45	2025	6.708	91125	3.557
46	2116	6.782	97336	3.583
47	2209	6.856	103823	3.609
48	2304	6.928	110592	3.634
49	2401	7	117649	3.659
50	2500	7.071	125000	3.684

Table of Algebraic Phrases

Phrases	*Algebraic Expressions*

Addition Phrases

the sum of a number and 6 six more than a number a number increased by six	$x + 6$

Subtraction Phrases

the difference between x and 4 four less than a number a number decreased by four 4 subtracted from a number	$x - 4$
the difference between 4 and x four decreased by a number	$4 - x$

Multiplication and Mixed Phrases

the product of 3 and x	$3x$
the sum of twice a number and 7 seven more than twice a number	$2x + 7$
twice the sum of a number and 7	$2(x + 7)$
5 less than four times a number 5 subtracted from four times a number	$4x - 5$
five decreased by four times a number	$5 - 4x$
twice the square of a number	$2x^2$
the square of twice a number	$(2x)^2$
the sum of the squares of x and y	$x^2 + y^2$
the square of the sum of x and y	$(x + y)^2$
eight more than twice the difference between a number and five	$2(x - 5) + 8$
nine less than three times the sum of one and a number	$3(1 + x) - 9$
two numbers whose sum is 32	x and $32 - x$

Consecutive Integer Phrases

sum of three consecutive integers	$n + (n + 1) + (n + 2)$
sum of three consecutive *even* or *odd* integers	$n + (n + 2) + (n + 4)$

Phrases	*Algebraic Expressions*

Consecutive Integer Phrases (continued)

sum of squares of three consecutive integers	$n^2 + (n + 1)^2 + (n + 2)^2$
sum of squares of three consecutive *even* or *odd* integers	$n^2 + (n + 2)^2 + (n + 4)^2$
square of the sum of two consecutive integers	$(n + n + 1)^2$, or $(2n + 1)^2$
product of two consecutive *even* or *odd* integers	$n(n + 2)$

Fractional Phrases

two-thirds of a number	$\frac{2}{3}x$, or $\frac{2}{3} \cdot \frac{x}{1}$, or $\frac{2x}{3}$
one-half of the sum of a number and three	$\frac{1}{2}(x + 3)$, or $\frac{x + 3}{2}$
four-fifths of the difference between two and a number	$\frac{4}{5}(2 - x)$, or $\frac{4(2 - x)}{5}$
half the result of decreasing a number by one	$\frac{1}{2}(x - 1)$, or $\frac{x - 1}{2}$
six less than one-third of the result of increasing eight by a number	$\frac{1}{3}(8 + x) - 6$, or $\frac{8 + x}{3} - 6$
a fraction whose denominator is four more than its numerator	$\frac{x}{x + 4}$
a fraction such that the sum of its numerator and denominator is 20	$\frac{x}{20 - x}$
the *reciprocal* of x; of $2x$; of $\frac{x}{y}$	$\frac{1}{x}$; $\frac{1}{2x}$; $\frac{y}{x}$
twice the reciprocal of x	$2\left(\frac{1}{x}\right)$, or $\frac{2}{x}$
the sum of the reciprocals of two consecutive integers	$\frac{1}{n} + \frac{1}{n + 1}$
the *ratio* of x to y; of y to x	$\frac{x}{y}$; $\frac{y}{x}$

Square Root Phrases

the square root of the sum of a number and 7	$\sqrt{x + 7}$
one more than the square root of a number	$\sqrt{x} + 1$

ANSWERS

Chapter 1

Exercise 1.1
pp. 7–9

1. -28 **3.** 9 **5.** 7 **7.** 4 **9.** $\dfrac{5}{4}$ **11.** $\dfrac{23}{24}$

13. $\dfrac{-55}{16}$ **15.** -288 **17.** 16 **19.** -125 **21.** -64 **23.** -25

25. $\dfrac{1}{10}$ **27.** $\dfrac{-20}{9}$ **29.** $\dfrac{-1}{42}$ **31.** $\dfrac{9}{4}$ **33.** $\dfrac{63}{16}$ **35.** -72

37. $\dfrac{16}{625}$ **39.** $\dfrac{27}{64}$ **41.** $\dfrac{-1}{18}$ **43.** $2^2 \cdot 3^2$ **45.** $2^3 \cdot 3 \cdot 5$ **47.** $2^4 \cdot 3^2$

49. $2^3 \cdot 3^2 \cdot 5^2$ **51.** $2^3 \cdot 3^2 \cdot 5 \cdot 7$ **53.** $53,59,61,67,71,73,79,83,89,97$ **55.** $\dfrac{1}{8}$ inch   **57.** -50

59. 5050 **61.** $\dfrac{161}{240}$ **63.** $\dfrac{95}{180} = \dfrac{19}{36}$ **65.** $\dfrac{157}{450}$

67. If $\dfrac{6}{0} = x$, then $6 = 0 \cdot x = 0$, which is impossible.

Exercise 1.2
pp. 12–13

1. 23 **3.** 63 **5.** 37 **7.** -45 **9.** -32

11. 256 **13.** -12 **15.** 0 **17.** $\dfrac{1}{16}$ **19.** $\dfrac{125}{64}$

21. $\dfrac{-7}{10}$ **23.** -70 **25.** -1 **27.** 83 **29.** 4

31. -1 **33.** 8 **35.** -80 **37.** -327 **39.** 19

41. 2 **43.** -7 **45.** $(2 + 5) \cdot 7 = 49$ **47.** $8 - 2 \cdot (3 - 4) = 10$

49. $(8 - 2) \cdot 3 - 4 = 14$ **51.** $8 - (2 \cdot 3 - 4) \cdot 8 = -8$ **53.** $6 \cdot 8 \div (4 - 2) = 24$ **55.** $(5 - 7)^2 = 4$

57. $(2 - 3) \cdot (4 + 1)^2 = -25$ **59.** $(4 - 4)(4 + 4) = 0$; $\dfrac{4}{4} + 4 - 4 = 1$; $4 - \dfrac{4 + 4}{4} = 2$; $\dfrac{4 + 4 + 4}{4} = 3$;

$4 + 4(4 - 4) = 4$; $\dfrac{4}{4} + \sqrt{4} \cdot \sqrt{4} = 5$; $4 + \dfrac{4 + 4}{4} = 6$;   etc.

Exercise 1.3   *pp. 18–21*

1. $-6, 0, \dfrac{-33}{4}, 18, 0$ **3.** $0, 0, 0, -168, \dfrac{75}{8}$ **5.** 78

7. 4 **9.** $\dfrac{-71}{30}$ **11.** $P = 22$ ft, $A = 24$ ft^2 **13.** $P = \dfrac{77}{6}$ ft, $A = 10$ ft^2

15. $P = 25$ ft, $A = 39.0625$ ft^2 **17.** $C = 10\pi = 31.42$ cm, $A = 25\pi = 78.54$ cm^2

19. $C = 8\pi = 25.13$ cm, $A = 16\pi = 50.27$ cm^2 **21.** $C = \dfrac{3\pi}{2} = 4.71$ cm, $A = \dfrac{9\pi}{16} = 1.77$ cm^2

23. $C = \dfrac{4\pi}{5} = 2.51$ cm, $A = \dfrac{4\pi}{25} = 0.50$ cm^2 **25.** Yes   **27.** Yes   **29.** No   **31.** Yes

33. No   **35.** Yes   **37. a.** 0 ft, 48 ft, 64 ft, 48 ft, 0 ft   **b.** 64 ft at 2 sec   **c.** 4 sec

39. $32°, 77°, 143.6°, 212°, -40°$ **41. a.** $78{,}825$ **b.** 264

43. 5,12,13;   7,24,25;   9,40,41;   11,60,61;   13,84,85;   15,112,113;   17,144,145

45. 20 **47.** 40 **49.** $16\pi = 50.27$ **51.** $100 - 25\pi = 21.46$ **53.** 13 in.^{2}

Exercise 1.4   *pp. 28–31*

1. real, rational, integer   **3.** real, irrational   **5.** real, rational

7. real, irrational   **9.** real, rational, integer, whole, natural, prime

11. real, rational, integer, whole, natural, prime   **13.** real, rational, integer, whole, natural

15. real, rational   **17.** real, irrational   **19.** real, rational, integer, whole, natural, prime

21. $8a - 5$ **23.** $4x^2 - 7x - 4$ **25.** $-5x^2 + 5xy - 40y^2$

27. $-2a - 4b$ **29.** $-2x + 35y$ **31.** $-178x + 25$

33. $(-7x + 2) + 2(8 - 3x) = -13x + 18$

35. $[(4a + b) - (a - 7b)] + [(2a + b) + (b - 2a)] = 3a + 10b$

37. $2x^2 + 3x - (x - 3x^2) = 5x^2 + 2x$

39. $2(3x - 5) + 7 = 6x - 3$ **41.** $-5a + 20$ or $20 - 5a$ **43.** $5x + 2y$ *(continued)*

Exercise 1.4, continued

45. $2^3 + 2^4 \neq 2^7$, or $8 + 16 \neq 128$ **47.** 3417   **49.** 1584

51. 41,916   **53.** 521,040   **55.** $a(b + c) = ab + ac$

57. $8 \div (4 \div 2) \neq (8 \div 4) \div 2$, or $8 \div 2 \neq 2 \div 2$

59. $2^3 \neq 3^2$, or $8 \neq 9$; $2^4 = 4^2$ is true, and $a^b = b^a$ is always true if $a = b$

61. $8 \div (4 + 2) \neq 8 \div 4 + 8 \div 2$, or $\dfrac{8}{6} \neq 2 + 4$ **63.** $2 + 3 = 5$ is prime

65. $1 + 3 = 4$ is even   **67.** True: $\dfrac{1}{2} + \dfrac{2}{3} = \dfrac{7}{6}, \dfrac{1}{2} \cdot \dfrac{2}{3} = \dfrac{1}{3}$

69. False: $2^{(2^3)} \neq (2^2)^3$, or $2^8 \neq 4^3$, or $256 \neq 64$

71. True, because $a \div b = b \div a$ is false: $8 \div 2 \neq 2 \div 8$

73. True: both 1 and -1 are their own reciprocals, because $1 = \dfrac{1}{1}$ and $-1 = \dfrac{1}{-1}$

75. False: $\dfrac{2}{0}$ is not a real number

Chapter 1 Review Exercises   *pp. 33–38*

1. -6 **2.** $\dfrac{9}{8}$ **3.** -112 **4.** $\dfrac{-9}{4}$ **5.** $\dfrac{113}{144}$

6. $\dfrac{-1}{4}$ **7.** $\dfrac{7}{3}$ **8.** $-\dfrac{8}{45}$ **9.** -216 **10.** $\dfrac{81}{256}$

11. $-\dfrac{1}{50}$ **12.** 31   **13.** 0   **14.** -8 **15.** 20

16. 16   **17.** 1125   **18.** -41 **19.** -65 **20.** $\dfrac{143}{4}$

21. $\dfrac{-5}{34}$ **22.** $-\dfrac{3}{4}$ **23.** 14   **24.** 389   **25.** $2^3 \cdot 3^2$

26. $2^2 \cdot 3^3 \cdot 5$ **27.** $2^2 \cdot 5 \cdot 7 \cdot 11$ **28.** $2^3 \cdot 3^2 \cdot 7^2$ **29.** $3^4 \cdot 5^2 \cdot 7$

30. $(5 - 2) \cdot 3 + 7 = 16$ **31.** $5 - 2 \cdot 3 + 7 = 6$ **32.** $5 - 2 \cdot (3 + 7) = -15$ **33.** $(5 - 2) \cdot (3 + 7) = 30$

34. $101, 103, 107, 109, 113$ **35.** $\dfrac{115}{1440} = \dfrac{23}{288}$ **36.** $P = \dfrac{1}{6}, S = \dfrac{273}{60} = \dfrac{91}{20}$ **37.** $-18, 0, 12, 0, \dfrac{-85}{4}$

38. $4, 49, 0, 22, 0$ **39.** $0, 8, 0, 0, \dfrac{525}{64}$ **40.** 65   **41.** 26

42. $P = 26$ in., $A = 40$ in.^{2} **43.** $P = 54$ in., $A = 162$ in.^{2} **44.** $P = 14.4$ cm, $A = 12.96$ cm^2

45. $P = \dfrac{23}{3} = 7\dfrac{2}{3}$ ft, $A = \dfrac{65}{18} = 3\dfrac{11}{18}$ ft^2

46. $P = \dfrac{19}{4} = 4\dfrac{3}{4}$ in., $A = \dfrac{39}{32} = 1\dfrac{7}{32}$ in.^{2} **47.** $C = 14\pi = 43.98$ m, $A = 49\pi = 153.94$ m^2

48. $C = 11.2\pi = 35.19$ m, $A = 31.36\pi = 98.52$ m^2 **49.** $C = 12\pi = 37.70$ m, $A = 36\pi = 113.10$ m^2

50. $C = 7\pi = 21.99$ m, $A = \dfrac{49\pi}{4} = 38.48$ m^2 **51.** $C = \dfrac{4\pi}{3} = 4.19$ m, $A = \dfrac{4\pi}{9} = 1.40$ m^2

52. $C = \dfrac{19\pi}{8} = 7.46$ m, $A = \dfrac{361\pi}{256} = 4.43$ m^2

53. a. 80, 128, 144, 128, 80, 0   **b.** 144 ft in 2 sec   **c.** 5 sec   **d.** 80 ft $(t = 0)$

54. 0°, 100°, 37°, −40°, 20°   **55.** 16, $20\dfrac{1}{2}$, 23, $28\dfrac{1}{2}$, 36, 14; they are each 14

56. a. $\dfrac{17}{6}$ **b.** $\dfrac{14}{3}$ **c.** $\dfrac{235}{6}$ **d.** $\dfrac{40}{3}$ **57. a.** 165 mi   **b.** 290 mi   **c.** 30 mi   **d.** 562.5 mi

58. a. \$10,050   **b.** \$10,400   **59.** \$4725   **60.** Yes   **61.** Yes
62. No   **63.** Yes   **64.** Yes   **65.** Yes   **66.** Yes

67. Yes   **68.** 46   **69.** 64   **70.** $\dfrac{25\pi}{4} = 19.63$ **71.** $7\pi = 21.99$

72. $144 - 36\pi = 30.90$ **73.** $80 + 8\pi = 105.13$ **74. a.** \$720   **b.** Yes   **75.** real, rational

76. real, rational, integer, whole   **77.** real, rational, integer, whole, natural, prime

78. real, irrational   **79.** real, rational   **80.** real, rational   **81.** real, rational

82. real, rational, integer, whole, natural   **83.** real, rational, integer, whole, natural, prime

84. real, rational, integer, whole, natural   **85.** real, rational

86. real, rational, integer, whole, natural   **87.** none of these   **88.** real, irrational

89. real, rational, integer, whole, natural, prime   **90.** real, rational

91. real, irrational   **92.** $7x - 30$ **93.** $33a - 53b$ **94.** $-13x^2 - 24x + 26$

95. $2a^2 - 37b^2$ **96.** $-2y^2 - 11y + 10$ **97.** $-7t^2 + 12t$ **98.** $14x - 39$

99. $23a - 16b$ **100.** $256 - 51a$ **101.** $2120a - 711$

102. $2(2x - 5) - [7 + (-4x)] = 8x - 17$ **103.** $3[x + (-3y)] + 4[-2y - (-5x)] = 23x - 17y$

104. $x + 2x^2 - 2(x^2 - x) = 3x$ **105.** $2(x + 7) - (2x + 7) = 7$

106. $(3x - 5) - 3(x - 5) = 10$ **107.** $2(6 + 5x) - 9 = 10x + 3$

108. $2[2a - (-3b)] - (-3a + 2b) = 7a + 4b$ **109.** 1414   **110.** 4646

111. 2574   **112.** 46,953   **113.** 849,830   **114.** 254,127

115. $5 + 2 = 7$ and $5 - 2 = 3$ are each prime   **116.** $11 + 2 = 13$ is prime, but $11 - 2 = 9$ is *not* prime

117. $7 - 2 = 5$ is prime, but $7 + 2 = 9$ is *not* prime

118. $7 + 3 = 10$, $7 - 3 = 4$, and $7 \cdot 3 = 21$ are all composite

119. a. irrational   **b.** $2 \cdot \sqrt{3}$ is irrational, but $0 \cdot \sqrt{3} = 0$ is rational

120. False: $(4 \cdot 8) \div 2 \neq (4 \div 2)(8 \div 2)$, or $16 \neq 8$ **121.** True: $2^{(1 + 2)} = (2^1)(2^2)$, or $2^3 = (2)(4)$, or $8 = 8$

122. Identity Property of 0; Distributive Property; Identity Property of 0; Uniqueness of 0 in Identity Property

Chapter 1 Test   *p. 39*

1. 1   **2.** -72 **3.** $\dfrac{5}{8}$ **4.** $\dfrac{5}{8}$ **5.** -20 **6.** 8

7. $\dfrac{25}{24}$ **8.** -23.8 **9.** $-\dfrac{1}{216}$ **10.** 15   **11.** $-2xy$ **12.** $13x - 14$

13. $96 - 59a$ **14.** $3(2x - 1) + 5 = 6x + 2$ **15.** $2(x + 4) - (2x + 4) = 4$

16. a. 0 ft, 64 ft, 96 ft, 100 ft, 96 ft, 64 ft, 0 ft   **b.** 100 ft in 2.5 sec   **c.** 5 sec

17. $13,826.34   **18. a.** No   **b.** Yes   **c.** Yes   **19.** 88   **20.** $240 - 18\pi = 183.45$

Chapter 2

Exercise 2.1   *pp. 44–46*

1. $x = 5$ **3.** $z = -5$ **5.** $x = 2$ **7.** $m = \dfrac{5}{2}$ **9.** $x = 0$ **11.** $x = 0$

13. $x = -6$ **15.** All real numbers   **17.** No solution   **19.** $x = \dfrac{2}{5}$ **21.** $y = \dfrac{1}{9}$

23. $n = 7$ **25.** 14   **27.** 7   **29.** -20 **31.** 6   **33.** 3, 8

35. 6, 15   **37.** 2   **39.** $\dfrac{1}{2}$ **41.** 5, 10, 14, 27   **43.** 1   **45.** $x = 7.5$

47. $x = 2$

Exercise 2.2   *pp. 48–49*

1. 16 dimes, 7 quarters
5. 28 $5 bills, 7 $20 bills
9. 20,000 at $25, 40,000 at $35, 30,000 at $50
13. 7 nickels, 7 dimes, 18 quarters

3. 38,000 at $8, 24,000 at $12
7. 8 dimes, 15 nickels, 30 quarters
11. *fries:* $0.75, *hamburger:* $2.00
15. *box A:* 14 nickels, 8 dimes; *box B:* 8 nickels, 14 dimes

Exercise 2.3   *pp. 52–54*

1. $20,000 at 8%, $30,000 at 11%

5. $20,000 at 9%, $15,000 at 6%; $7200 interest

9. *Smith:* $20,000 at 6%; *Jones:* $30,000 at 8%

3. $12,000 in bonds, $8000 in stocks

7. $5000 at 8%, $8000 at $7\frac{1}{2}$%; $2400 interest

11. $2000 at $5\frac{1}{2}$%, $2000 at 7%, $3000 at 8%

13. $7000 at 7%, $8000 at 9%

15. $4000

17. $7\frac{1}{2}$% for 2 years, 6% for 4 years

19. $15,000 in bonds, $25,000 in partnership

21. $4500

Exercise 2.4   *pp. 57–59*

1. *Toyota:* 60 mph, *Nissan:* 65 mph

3. 1.5 hours

5. 4 hrs, 8:00 PM, 600 km from Rome

7. *Mazda:* 70 mph, *Audi:* 85 mph; 1190 miles

9. *jog:* $\frac{1}{2}$ hr, 3 miles: *run:* 2 hrs, 20 miles

11. *city:* 40 mph, 40 miles; *highway:* 60 mph, 180 miles

13. 30 miles

15. *swim:* 1 hr, 2 miles; *bike:* 3 hrs, 60 miles; *run:* 2 hrs, 20 miles

17. $3\frac{1}{2}$ hours

19. city: $\frac{1}{2}$ hr, 25 miles; highway: $\frac{1}{4}$ hr, 15 miles

21. 48 mph

Exercise 2.5   *pp. 64–67*

1. $\{x|x < -2\}$:

3. $\{x|2 \geq x\}$:

5. $\{x|x \geq 3\}$:

7. $\{x|x > 3\}$:

9. $\left\{x\Big|\frac{3}{4} < x\right\}$:

11. $\{x|x > 0\}$:

13. $\{x|0 \geq x\}$:

15. $\{x|0 > x\}$: A

17. $\{x|x < 1\}$:

19. $\left\{x\Big|\frac{1}{5} \geq x\right\}$:

21. $\left\{x\Big|x > \frac{1}{10}\right\}$:

23. $\{x|x > -2\}$:

25. $\{x|x \leq -3\}$:

27. $\{x|3 < x < 5\}$:

29. $\{x|-8 \leq x \leq 2\}$:

31. $\{x|-1 > x \geq -3\}$:

33. $\{x|-2 < x \leq 0\}$:

35. 3 nickels, 6 dimes; 2n,7d; 1n,8d; 0n,9d

37. between $3000 and $6000

39. between 1:30 PM and 3:00 PM

41. between 2 and $5\frac{1}{2}$, inclusive

43. $167 \leq$ score ≤ 200 (actually < 247)

45. $\{x|2 \geq x \geq -1\}$

47. $\{x|1 > x > -2\}$

(continued)

Exercise 2.5, continued

49. $\dfrac{5}{12} > \dfrac{7}{18}$ **51.** $\dfrac{7}{16} < \dfrac{11}{24}$ **53.** $\dfrac{-5}{8} > -\dfrac{3}{4}$ **55.** $\dfrac{27}{47} > \dfrac{4}{7}$ **57.** $\dfrac{26}{39} = \dfrac{34}{51}$

59. $0.12 < 0.\overline{12}$ **61.** Yes   **63.** No   **65.** Yes   **67.** Yes

69. Yes   **71.** No   **73.** $3 < x < 13$ **75.** $x > 4$ **77.** $1 < x < 6$

Exercise 2.6   *pp. 72–74*

1. 6   **3.** 3   **5.** 17   **7.** $\dfrac{1}{4}$ **9.** $\dfrac{41}{36}$

11. $\dfrac{43}{48}$ **13.** $\dfrac{319}{180}$ **15.** -5 **17.** -4 **19.** 4

21. -9 **23.** -3 **25.** $x = 5, -5$ **27.** $y = 6, -6$ **29.** $x = 7, -5$

31. $t = 5, \dfrac{-19}{3}$ **33.** $x = -4, 6$ **35.** $x = 4$ **37.** $x = \dfrac{5}{2}, -\dfrac{3}{2}$ **39.** $x = 7.1, 6.9$

41. $x = -1.495, -1.505$ **43.** $x = 3, -2$ **45.** $x = -1, 4$ **47.** $x = \dfrac{1}{2}, -\dfrac{3}{2}$

49. $\{x \mid -4 < x < 4\}$:   **51.** $\{x \mid 4 \geq x \geq -4\}$:

53. $\{x \mid -2 < x < 4\}$:   **55.** $\{x \mid -1 \leq x \leq 4\}$:

57. $\{x \mid 4 > x > 2\}$:   **59.** $\left\{x \mid \dfrac{7}{2} \geq x \geq \dfrac{3}{2}\right\}$:

61. $\left\{x \mid 1 > x > \dfrac{-7}{3}\right\}$:   **63.** $\{x \mid 4.9 < x < 5.1\}$:

65. $\{x \mid -2.51 \leq x \leq -2.49\}$:   **67.** No solution

69. $\{x \mid x \geq 2\} \cup \{x \mid x \leq -2\}$:   **71.** $\{x \mid x > 4\} \cup \{x \mid x < 0\}$:

73. $\{x \mid x > 3\} \cup \{x \mid x < -3\}$:   **75.** $\{x \mid x < 2\} \cup \{x \mid x > 3\}$:

77. $\{x \mid x > 2.1\} \cup \{x \mid x < 1.9\}$:   **79.** $\{x \mid x \leq -3.505\} \cup \{x \mid x \geq -3.495\}$:

81. $\{x \mid x < 0\} \cup \{x \mid x > 3\}$:   **85.** $\{x \mid x > 2\} \cup \{x \mid x < 2\}$:

83. $\{x \mid x \geq 10\} \cup \{x \mid x \leq 0\}$:

87. $x = -2$ ($x = -6$ is extraneous)   **89.** No solution ($x = -6, 3$ are extraneous)

91. No solution ($|\ |$ is never < 0)   **93.** $z = 4$ ($z = 8$ is extraneous)

95. No solution ($y = -4$ is extraneous)   **97.** $x = -3$

99. $x = 6$ **101.** $x = -4, 8$

103. $x = 4, -10, 2, -8$

105. $\dfrac{9 + 3 + |9 - 3|}{2} = \dfrac{12 + 6}{2} = 9$ (the larger),

$\dfrac{9 + 3 - |9 - 3|}{2} = \dfrac{12 - 6}{2} = 3$ (the smaller); etc.

107. $5; 7; 5; \dfrac{1}{8}; 0$

Chapter 2 Review Exercises   *pp. 77–81*

1. $x = 4$

2. $y = \dfrac{9}{2}$

3. $z = 0$

4. No solution

5. All real numbers

6. $w = 0$

7. $x = 0.5$

8. $z = \dfrac{1}{2}$

9. 5

10. -1

11. 6

12. 5 and 7

13. -1

14. 5, 8, 12

15. 4 and 6

16. -4

17. 5 dimes, 16 quarters

18. 9500 at \$20, 4000 at \$35

19. 18 nickels, 9 dimes

20. \$10 general admission, \$20 reserved

21. 4 quarters, 6 nickels, 13 dimes

22. 7 nickels, 7 quarters, 11 dimes

23. 9 \$10 bills, 11 \$5 bills, 22 \$20 bills

24. 3000 at \$15, 5000 at \$40, 15,000 at \$25

25. 25 women's, 15 men's

26. \$7500 at 11%, \$12,500 at 9%

27. \$5000 at 6%, \$7000 at 8%

28. bond = \$4000, stock = \$5000, real estate = \$10,000

29. \$20,000 at 8%, \$16,000 at 5%

30. \$3000 at 5%, \$6000 at 7%, \$6000 at 4%

31. 6% = 3 yrs, 9% = 4 yrs, 7% = 6 yrs

32. Chevy = 65 mph, Ford = 55 mph

33. $8\dfrac{1}{2}$ hrs, 9:30 PM, 892.5 miles

34. city = 1.5 hrs, 135 km; highway = 2 hrs, 270 km

35. 75 miles

36. cops = 100 mph, robbers = 80 mph

37. drive = $\dfrac{1}{2}$ hr, 30 miles; bus = $\dfrac{1}{2}$ hr, 15 miles; plane = 5 hrs, 2375 miles

38. $\{x|x < 3\}$:

39. $\{x|x \geq -3\}$:

40. $\{x|x < 3\}$:

41. $\{x|2 \leq x\}$:

42. $\{x|x \leq 0\}$:

43. $\{x|-2 < x\}$:

44. $\{x|-1 < x < 3\}$:

45. $\left\{x \left| \dfrac{3}{2} \geq x \geq -2\right.\right\}$:

46. $\{x|0 \geq x > -2\}$:

47. $\{x|1 < x < 3\}$:

48. $\{x|2 \geq x \geq -3\}$:

49. $\{x|2 < x < 3\}$:

50. 9 nickels, 5 dimes; 10n,4d; 11n,3d; 12n,2d; 13n,1d; 14n,0d

51. 10 dimes, 2 quarters; 9d,3q; 8d,4q; 7d,5q; 6d,6q; 5d,7q

(continued)

Chapter 2 Review Exercises, continued

52. more than $12,000

53. between $1\frac{1}{2}$ and 2 hours

54. $121 \le$ score < 191

55. 8751 rods (to make profit); 8750 rods (break even point)

56. $\dfrac{11}{32} < \dfrac{17}{48}$

57. $\dfrac{7}{24} < \dfrac{5}{16}$

58. $-\dfrac{7}{9} > \dfrac{-5}{6}$

59. $0.2\overline{1} < 0.\overline{21}$

60. $0.1\overline{21} > 0.\overline{121}$

61. $0.\overline{12} = 0.\overline{1212}$

62. Yes

63. No

64. No

65. No

66. Yes

67. No

68. Yes

69. No

70. No

71. $x > 4$

72. $4 < x < 8$

73. $x > 1$

74. 6

75. 2

76. $\dfrac{29}{24}$

77. -3

78. -1

79. 1

80. $x = 9, -5$

81. $x = -4, 7$

82. $x = 3, \dfrac{-11}{3}$

83. $x = 5, -6$

84. $x = \dfrac{8}{3}, -\dfrac{4}{3}$

85. $x = 1, -2$

86. $x = 6, -3$

87. No solution $\left(x = -6, \dfrac{4}{3} \text{ are extraneous} \right)$

88. $x = 2, -4$

89. $\{x | -5 < x < 3\}$:

90. $\{x | 2 \ge x \ge 1\}$:

91. $\{x | 4.9 < x < 5.1\}$:

92. $\{x | 6 \ge x \ge -4\}$:

93. $\{x | x > 4\} \cup \{x | x < -3\}$:

94. $\{x | x \ge 1\} \cup \left\{ x | x \le -\dfrac{7}{3} \right\}$:

95. $\{x | x > 3\} \cup \{x | x < -3\}$:

96. $\{x | x \ge -1.99\} \cup \{x | x \le -2.01\}$:

97. $\{x | 0 > x\} \cup \{x | x > 4\}$:

98. True

Chapter 2 Test   *p. 81*

1. -8

2. $\dfrac{15}{8}$

3. 2

4. -6

5. $x = 3$

6. $x = 0$

7. No solution

8. All real numbers

9. $x = 1$

10. $x = 7, -6$

11. $x = 5.9, 6.1$

12. $x = 9$ ($x = -3$ is extraneous)

13. $x = 1$ ($x = -5$ is extraneous)

14. $\{x|x > -2\}$:

15. $\left\{x\left|-\dfrac{5}{3} \le x \le 1\right.\right\}$:

16. $\{x|x < -2\} \cup \{x|x > 3\}$:

17. 8 dimes, 11 nickels, 22 quarters

18. \$3000 at 6%, \$3000 at $9\dfrac{1}{2}$%, \$4000 at 7%

19. 3:00 PM

20. $2 <$ number < 6

Chapter 3

Exercise 3.1   *pp. 87–89*

1. a^{11}

3. b^{17}

5. $(a - b)^{16}$

7. 3^{12}

9. $14a^4b^{10}$

11. $-15m^6n^7$

13. $(x + 1)^9(y - 2)^7$

15. $6x^{10}(x - 1)^5(x + 2)^{10}$

17. $a^{10n}b^{6n}$

19. $-24x^{6n - 3}y^{4m + 7}$

21. $8a^3 - 14a^2 - 22a$

23. $-3x^3y - x^2y^2 + 3xy^3$

25. $2x^5y + 7x^4y^2 - x^3y^3 - 2x^2y^4$

27. $10a^5(a + b)^{10} - 14a^6(a + b)^9$

29. $x^3 - 61x^2 + 6x$

31. $x^2 + 10x + 16$

33. $2a^2 - 11ab + 5b^2$

35. $10y^2 + 51yz - 22z^2$

37. $8x^4y^2 - 2x^2yz^3 - 45z^6$

39. $-10a^2b^2 - 51abc + 91c^2$

41. $9x^2 - 24xz + 21xy - 56yz$

43. $6x^{2n} - 37x^ny^m + 56y^{2m}$

45. $x^2 + 18x + 81$

47. $4a^2 - 20ab + 25b^2$

49. $16x^4 - 56x^2y^3 + 49y^6$

51. $4x^{2n} + 12x^ny^n + 9y^{2n}$

53. $4x^2 - 81$

55. $36y^2 - 169b^2$

57. $1 - 361x^2y^2$

59. $25x^6 - 289y^4$

61. $4x^{2n} - 9y^{2m}$

63. $25x^{4n} - 81y^{6m}$

65. $18x^3 + 57x^2 - 21x$

67. $-18a^4b + 96a^3b^2 - 128a^2b^3$

69. $3x^3y - 75xy^3$

71. $12a^3 - 27a^2b - 4ab^2 - 9b^3$

73. $6x^3y - 19x^2y^2 + 24xy^3 - 20xy + 20$

75. $a^4 - b^4$

77. $x^8 - 1$

79. $4x^4 - 61x^2 + 225$

81. $a^2 - 2ab + b^2 + 14a - 14b + 49$

83. $x^2 + 6xy + 9y^2 - 4xz - 12yz + 4z^2$

85. $x^2 - 2xy + y^2 - 121$

87. $a^2 + 4ab + 4b^2 - 49c^2$

89. $3x^3 - x^2 - 27x - 26$

91. $4a^3 + a^2b - 25ab^2 - 14b^3$

93. $8x^3 - 27y^3$

95. $x^3 + 3x^2h + 3xh^2 + h^3$

97. $x^4 - 1$

99. $3m^4 - 5m^3 + 17m^2 + 11m - 14$

101. $x^4 + x^2 + 1$

103. $2x^3 + 9x^2 - 26x - 105$

105. $x^6 - 1$

107. $a^4 - 8a^2 + 16$

109. $x^5 + 1$

111. $2xh + h^2 - 2h$

113. $3hx^2 + 3h^2x + h^3 + 4hx + 2h^2 - 9h$

115. 6399

117. 39,999

119. 3596

121. 9991

123. 2601

125. 10,201

127. 7921

129. 9801

131. 3844

133. 3364

135. $(a + b)(c + d) = ac + ad + bc + bd$

Exercise 3.2

pp. 92–93

1. $7a^2(a - 2)$

3. $2ab^2(a^2 + b - 2a)$

5. $2y(2x^2 + 3xy^2 - 5y)$

7. $P(1 + rt)$

9. $13x^2yz^2(2xyz + 4y^2 - 3z^2)$

11. $11a^2b^2(9ac^2 - 8bc + 12)$

13. $5x^5(1 + 2x^5 - 3x^{10} + 4x^{15})$

15. $4x^m(x^2 - 2x + 3)$

17. $a^{n+1}(a^2 - a + 1)$

19. $x^{3n}y^{m+2}(x^{2n} + x^ny + y^2)$

21. $-7a(2x + 3y - 5z)$

23. $-6a(a^2 - 2a + 3)$

25. $-4ab(2a^2 - 5b^2)$

27. $-1(2x^2 + x - 7)$

29. $-x^n(x^2 + x - 1)$

31. $(x + y)(2a - 3)$

33. $(m - n)(4x + 3y - z)$

35. $5(b - c)[(b - c)^2 - 3(b - c) + 2]$

37. $(x + 2)(x + y)$

39. $(x + y)(z + 3)$

41. $(4x - 3)(x + a)$

43. $(ax - 7)(x - y)$

45. $(5ab - 3)(c + 7)$

47. $(2uv + 3)(5u - 1)$

49. $(2a - b)(x + 1)$

51. $(x + c)(ax + b)$

53. $(b + 2)(bc - 3)$

55. $(b - 10y)(5by + 3c)$

57. $(ax + y)(x - 2z)$

59. $2(x + 2)^7(x - 3)^5(7x - 6)$

61. $2(2x - 1)^4(x + 4)^7(13x + 16)$

63. $10x^5(2x + 5)^3(2x + 3)$

65. $2x^9(1 - 2x)^3(5 - 14x)$

Exercise 3.3

pp. 98–100

1. $(x + 5)(x + 1)$

3. $(a - 7)(a - 1)$

5. $(x + 19y)(x - y)$

7. $(ax - 13)(ax + 1)$

9. $(x + 4)^2$

11. $(x - 5y)^2$

13. $(u + 9v)(u - 2v)$

15. $(x^2 - 5)(x^2 + 4)$

17. $(a^2 + 13b^2)^2$

19. $(m^3 + 14)(m^3 + 2)$

21. $(x^2y - 17)(x^2y + 4)$

23. $(x^n + 21)(x^n + 5)$

25. $(x - 12)(x - 6)$

27. $(x + 9)(x + 15)$

29. $(a - 17b)^2$

31. $(2x + 1)(x + 7)$

33. $(5a - 17b)(a + b)$

35. $(11x - 2y)(x - 2y)$

37. $(4x + 3)(x + 2)$

39. $(3s - 5t)(2s - t)$

41. $(3a^2 + 5)(2a^2 - 7)$

43. $(2x + 3)^2$

45. $(4ax - 1)^2$

47. $(3x - 2)(2x + 5)$

49. $(6a - 5b)(a + 3b)$

51. $2(x - 13)(x - 3)$

53. $5ab(a - 6b)(a - 8b)$

55. $-3(m - 7n)(m + 2n)$

57. $5x(x - 3)(x + 20)$

59. $3a(a + 12b)(a - 10b)$

61. $-x(x + 14)(x - 9)$

63. $-(x - 11y)(x + 2y)$

65. $2(x - 65y)(x - 5y)$

67. $2(3x - 1)(2x - 5)$

69. $5xy(4x - 3y)(x + 2y)$

71. $4xy(2x + 3y)(x - 5y)$

73. $2(5x - 7y)^2$

75. $x^n(x - 9)(x + 2)$

77. $(8x + 13y)(x + 2y)$

79. $(4x^2 - 5)(3x^2 - 2)$

81. $(5x^2 + 3)(3x^2 - 4)$

83. $(8x^n - 3)(x^n - 4)$

85. $(6u + 7v)(4u + 5v)$

87. $2x^2(9x^2 - 2)(2x^2 - 3)$

89. $4x^6(3x + 8)(x + 7)$

91. $x^n(5x - 4)(2x + 3)$

93. $-2x^ny^n(3x^n - y^n)(2x^n + 3y^n)$

95. $(24x + 5)(2x - 3)$

97. $(9xy + 4)(4xy - 9)$

99. $x^2 + bx + c = x^2 + mx + nx + mn = x(x + m) + n(x + m) = (x + m)(x + n)$.

Exercise 3.4   *pp. 102–104*

1. $(x + 10)(x - 10)$

3. $(ab - 6)(ab + 6)$

5. $(3u + 11vw)(3u - 11vw)$

7. $(17a^2 - 20b)(17a^2 + 20b)$

9. $(25x + 19y)(25x - 19y)$

11. $(37b^3 + 43c^2)(37b^3 - 43c^2)$

13. $(a^n + b^m)(a^n - b^m)$

15. $(x + y + 2z)(x + y - 2z)$

17. $(a + b + 2c)(a - b - 2c)$

19. $(a^2 + b^2)(a + b)(a - b)$

21. $(x^4 + 16)(x^2 + 4)(x + 2)(x - 2)$

23. $(x^8 + 1)(x^4 + 1)(x^2 + 1)(x + 1)(x - 1)$

25. $(x + 5)(x^2 - 5x + 25)$

27. $(2a - 3b)(4a^2 + 6ab + 9b^2)$

29. $(xy^2 - 6z)(x^2y^4 + 6xy^2z + 36z^2)$

31. $(9a + 11b^2c^3)(81a^2 - 99ab^2c^3 + 121b^4c^6)$

33. $(15 - 4x)(225 + 60x + 16x^2)$

35. $(x^n + y^n)(x^{2n} - x^ny^n + y^{2n})$

37. $(x + 2)(x^2 - 11x + 49)$

39. $(y - 1)(y^2 + 4y + 7)$

41. $h(3x^2 + 3xh + h^2)$

43. $2(x + 5y)(x - 5y)$

45. $3ab(a - 7b)(a + 7b)$

47. $3x^3y^2(2x + 15y)(2x - 15y)$

49. $3xy(x^2 + 4y^2)(x + 2y)(x - 2y)$

51. $x^ny^n(x^n + y^n)(x^n - y^n)$

53. $2(x + 2)(x^2 - 2x + 4)$

55. $3ab(2a - 3b)(4a^2 + 6ab + 9b^2)$

57. $3x(4x - 1)(16x^2 + 4x + 1)$

59. $-2ab(a + 5)(a^2 - 5a + 25)$

61. $(x + 5)(x - 5)(x + 1)(x - 1)$

63. $(2x + y)(2x - y)(x + 3y)(x - 3y)$

65. $(x - 2y)(x^2 + 2xy + 4y^2)(x + y)(x^2 - xy + y^2)$

67. $(x + y)(x^2 - xy + y^2)(x - y)(x^2 + xy + y^2)$

69. $3x(5x + 2)(5x - 2)(x + 1)(x - 1)$

71. $-2x(x - 2)^2(x^2 + 2x + 4)^2$

73. $(x + 3)(x - 3)(x + 2y)(x - 2y)$

75. $(x + 1)(x - 1)(x + y)(x^2 - xy + y^2)$

77. $(x - 2)(x^2 + 2x + 4)(x + y)(x^2 - xy + y^2)$

79. $(x + 3)(x - 3)(x + 2)(x^2 - 2x + 4)$

81. $(2x - y)(4x^2 + 2xy + y^2)(2a + 3b)(2a - 3b)$

83. $(x + 5 + 2y)(x + 5 - 2y)$

85. $(x - 3y + 5)(x - 3y - 5)$

87. $(2a + 3b + 4c)(2a + 3b - 4c)$

89. $(x^2 + 2x + 4)(x^2 - 2x + 4)$

91. $(x^2 + 4xy + 5y^2)(x^2 - 4xy + 5y^2)$

93. $(A - B)(A^2 + AB + B^2) = A^3 + A^2B + AB^2 - A^2B - AB^2 - B^3 \overset{\checkmark}{=} A^3 - B^3.$

Exercise 3.5   *pp. 109–112*

1. $x = 5, 3$

3. $z = 9, -4$

5. $x = \dfrac{3}{2}, -5$

7. $m = \dfrac{3}{4}, 3$

9. $x = -\dfrac{5}{8}, 6$

11. $x = 3, -3$

13. $z = \dfrac{-7}{4}, \dfrac{7}{4}$

15. $x = 0, 9$

17. $x = -\dfrac{3}{2}$

19. $n = \dfrac{6}{7}$

21. $x = -3, 1$

23. $x = \dfrac{1}{3}, \dfrac{3}{2}$

25. $y = \dfrac{-2}{3}$

27. $x = 0, 16$

29. $x = 4, -4$

31. $t = 5, -1$

(continued)

Exercise 3.5, continued

33. $z = \dfrac{11}{9}, \dfrac{-11}{9}$

35. $z = 0, \dfrac{3}{2}$

37. $t = 0, 1$

39. $n = \dfrac{3}{5}, -\dfrac{9}{4}$

41. $x = \dfrac{5}{2}, -2$

43. $y = 7, -5$

45. $z = -12, 5$

47. $x = 5, -1$

49. $x = \dfrac{5}{3}, -2$

51. $1, \dfrac{1}{3}$

53. $0, 1$

55. $5, 1$

57. 2 and 7; -4 and 1

59. 5 and 13; $\dfrac{-13}{2}$ and -10

61. 28, 29, 30

63. 15, 17, 19, 21

65. 3, 4, 5; $-5, -4, -3$

67. 7, 9; $-9, -7$

69. $-2, 0, 2$

71. $-3, -2; 2, 3$

73. 4 seconds

75. $x = 0, 8, 4$

77. $x = -7, 5, -5, 3$

79. $x = 2, 1$ ($x = -3, -6$ are extraneous)

81. $x = 0, 11, -2$

83. $x = 0, \dfrac{7}{2}, \dfrac{5}{2}$

85. $x = 3, -3, 1, -1$

87. $x = 0, 4, -4, 1, -1$

89. $x = -3, 2, -2$

91. $x = -5, \dfrac{3}{2}, -\dfrac{3}{2}$

93. 1, 2, 3; $-1, 0, 1; -3, -2, -1$

95. Division by $a - b$ is impossible, because $a - b = 0$.

Exercise 3.6   *pp. 116–119*

1. a. 7 ft by 17 ft   **b.** 3 ft by 9 ft

3. a. 7 m by 15 m   **b.** 4 m by 12 m

5. a. 4 in.   **b.** 6 in.

7. 5 ft by 5 ft, 9 ft by 9 ft

9. 2 in.

11. 5 m by 13 m; 7 m by 15 m

13. 12 in., 16 in.

15. 8 in., 15 in., 17 in.

17. 10 in., 24 in., 26 in. ($x = 5$)

19. 15 in., 20 in.

21. 6, 8, 10

23. 1 in.

25. a. 6 ft by 15 ft; 12 ft by 30 ft   **b.** 2 ft by 7 ft; 4 ft by 14 ft

27. 3 by 6

29. 12 ft

31. 10 ft

33. $BC = 3$, $AC = CD = 5$

35. 3 in. by 3 in.; 7 in. by 7 in.; 13 in. by 13 in.

Chapter 3 Review Exercises   *pp. 121–124*

1. $15x^4y^3$

2. 4^{12}

3. $(a + b)^{11}$

4. $x^{7m}y^{4n+2}$

5. $-12m^4n^5$

6. $x^3 + 10x^2 + 16x$

7. $2x^2 - x - 21$

8. $6a^2 - 29ab + 35b^2$

9. $2a^4b^2 + 9a^2bc + 10c^2$

10. $-10x^2 - 11xy + 6y^2$

11. $8a^2 + 6ab - 28a - 21b$

12. $4x^2 + 12xy + 9y^2$

13. $25a^2 - 20ab + 4b^2$

14. $9a^4 - 42a^2b^3 + 49b^6$

15. $9x^2 - 64$

16. $169 - 289c^8$

17. $15x^3 - 36x^2 + 21x$

18. $-18a^3b + 60a^2b^2 - 50ab^3$

19. $-32a^2 + 130a - 106$

20. $a^8 - b^8$

21. $a^2 + 2ab + b^2 - 25$

22. $x^2 - 2xy + y^2 - z^2$

23. $3x^3 - 16x^2 + 25x - 12$

24. $27a^3 + 54a^2 + 36a + 8$

25. $8a^3 - 27b^3$

26. $x^4 + 4x^2 + 16$

27. $x^5 - 1$

28. $4x^4 - 61x^2 + 225$

29. $x^4 - 2x^2 + 1$

30. $4a^4 - 65a^2 + 196$

31. $2xh + h^2 + 3h$

32. $3x^2h + 3xh^2 + h^3 + 2xh + h^2 - 4h$

33. $2xy(6x^2 - 2xy^2 - 3y)$

34. $-4a(2x + 3y - 5z)$

35. $(x + 2y)(5x + 3)$

36. $(a - 3)(7ab - 2c)$

37. $(x + 1)(5x + 2y)$

38. $(a - 7b)(x - 2)$

39. $(x + 7y)(x + 6y)$

40. $(a - 12)(a - 4)$

41. $(2x - 7)(x + 3)$

42. $(3a + 8b)(a - 2b)$

43. $(4m - 5n)(m + 2n)$

44. $(2x - 15)(2x - 3)$

45. $(4x + 5y)(x - 9y)$

46. $(6u + 5v)(u + 2v)$

47. $(5x + 7)^2$

48. $(3a - 10b)^2$

49. $2x(x - 12)(x + 6)$

50. $-3x(x + 9)(x - 6)$

51. $2mn(5m - 2n)(m + 4n)$

52. $-5ab(3a + 4b)(2a - 5b)$

53. $(8t - 3)(2t - 5)$

54. $-2x^n(4x + 7)(3x - 2)$

55. $(8x - 3)(3x + 8)$

56. $(6x - 7y)(4x + 3y)$

57. $(2x + 3y)(2x - 3y)$

58. $9(4x + 5yz)(4x - 5yz)$

59. $(x^2y^2 + 4z^2)(xy + 2z)(xy - 2z)$

60. $3x^7(x + 11y)(x - 11y)$

61. $(x + y + 4z)(x + y - 4z)$

62. $a^3b^4c(a^2c + b^2)(a^2c - b^2)$

63. $(x - 2y)(x^2 + 2xy + 4y^2)$

64. $(ab + 3c)(a^2b^2 - 3abc + 9c^2)$

65. $(4x + 11)(16x^2 - 44x + 121)$

66. $(5ab^2 - 9c^3)(25a^2b^4 + 45ab^2c^3 + 81c^6)$

67. $3ab(a - 10b)(a^2 + 10ab + 100b^2)$

68. $(x - 3y)(x^2 + 3y^2)$

69. $(3x + 1)(3x - 1)(x + 2)(x - 2)$

70. $(x + 2y)^2(x^2 - 2xy + 4y^2)^2$

71. $2x(5x + 3)(5x - 3)(x^2 + 4)$

72. $(2x + y)(4x^2 - 2xy + y^2)(2x - y)(4x^2 + 2xy + y^2)$

73. $2xy(x - 3y)(x^2 + 3xy + 9y^2)(x + y)(x^2 - xy + y^2)$

74. $(x + 2y)(x - 2y)(x^2 + 4)(x + 2)(x - 2)$

75. $(x + 1)(x^2 - x + 1)(2x + 5)(2x - 5)$

76. $(x + 3y + 2z)(x + 3y - 2z)$

77. $x = 11, -2$

78. $y = \dfrac{5}{2}, -1$

79. $z = 13, 7$

80. $t = \dfrac{-2}{3}, \dfrac{7}{2}$

81. $v = 0, \dfrac{9}{4}$

82. $v = \dfrac{3}{2}, -\dfrac{3}{2}$

83. $x = 3, -3$

84. $x = 0, 9$

85. $t = 5, -1$

86. $t = 0, 5$

87. $w = \dfrac{3}{5}, 3$

88. $y = -18, 14$

89. $x = \dfrac{5}{3}, \dfrac{1}{2}$

90. $n = \dfrac{2}{9}, -3$

91. $x = \dfrac{7}{6}, -3$

92. $y = \dfrac{3}{4}, -2$

93. $x = 29, -3$

94. $x = -19, -3$

95. $x = -14, 9$

96. $x = -21, 6$

97. $x = 33, -4$

(continued)

Chapter 3 Review Exercises, continued

98. $x = -44, 3$

99. $x = -\dfrac{100}{3}, 50$

100. $x = 10, \dfrac{-66}{5}$

101. $x = \dfrac{1}{12}, 2$

102. $x = \dfrac{3}{4}, \dfrac{4}{3}$

103. $t = 0, \dfrac{6}{7}$

104. $x = 16, \dfrac{1}{2}$

105. $y = 1, -1$

106. $z = 2, \dfrac{1}{11}$

107. $m = -\dfrac{5}{8}, 7$

108. $x = \dfrac{-3}{8}, \dfrac{5}{6}$

109. $x = 0, -4, -2$

110. $x = 6, -6, 0$

111. $x = 5, -3, 3, -1$

112. $x = 6, 2$ ($x = -4, -12$ are extraneous)

113. $x = 0, -3, 2$

114. $y = 0, \dfrac{5}{2}, \dfrac{-3}{2}$

115. $x = 3, -3, 2, -2$

116. $y = 0, 2, -2, \dfrac{1}{3}, -\dfrac{1}{3}$

117. $x = -3, 5, -5$

118. $z = \dfrac{5}{2}, \dfrac{3}{2}, -\dfrac{3}{2}$

119. 3 or $-\dfrac{1}{2}$

120. 1

121. $2, 5; \dfrac{-2}{5}, \dfrac{1}{5}$

122. $4, 6; -6, -4$

123. $-2, -1, 0, 1$

124. **a.** 5 in. by 8 in.   **b.** 9 in. by 12 in.

125. 7 m; 10 m

126. 3 ft by 8 ft; 2 ft by 12 ft

127. 1 foot

128. 5 in. by 9 in.

129. 5, 12

130. 15, 25 ($x = 5$)

131. 12, 16, 20 ($x = 6$)

132. 5 in., 12 in., 13 in.

133. 25 feet

134. $x = 12$

135. $x = 8$ ($x = 32$ is too big)

Chapter 3 Test   *pp. 124–125*

1. $24a^6 b^9$

2. $x^2 + x - 16$

3. $x^4 - 16$

4. $100x^4 - 229x^2 y^2 + 9y^4$

5. $(x + 9)(x - 4)$

6. $(3x + 13y)(3x - 13y)$

7. $(3a - 5b)(a - 2b)$

8. $(6a - 5)(a + 3)$

9. $(3x + 10y)(9x^2 - 30xy + 100y^2)$

10. $(2ax - b)(x + 3b)$

11. $-2xy(3x - 2y)(2x + 3y)$

12. $(x - 2)(x + 2)^2(x^2 - 2x + 4)$

13. $x = -\dfrac{1}{3}, 2$

14. $z = \dfrac{5}{2}, \dfrac{-5}{2}$

15. $z = 0, \dfrac{25}{4}$

16. $x = \dfrac{10}{3}, -\dfrac{1}{2}$

17. $x = 5, -5, \dfrac{1}{2}, -\dfrac{1}{2}$

18. 4 or -2

19. 8 in. by 18 in.

20. 3, 4, 5

Chapters 1–3 Cumulative Review Exercises   *pp. 125–126*

1. $\dfrac{3}{2}$ **2.** 4   **3.** $20a^6b^4$ **4.** $\dfrac{25}{24}$ **5.** $10x^2 + 13xy - 3y^2$

6. $136a - 56$ **7.** $25y^2 - 81z^2$ **8.** $49x^2 + 154x + 121$ **9.** -17

10. $27a^3 - 8$ **11.** $x^4 + 9x^2 + 81$ **12.** $4x^4 - 13x^2 + 9$ **13.** $7x + 22$

14. $5x^3y - 8x^2y^2 + 4xy^3$ **15.** No   **16.** Yes   **17.** Yes   **18.** Yes

19. 56 sq. meters   **20.** $60 + 9\pi = 88.27$ sq. meters   **21.** $(2a + b)(a - 3b)$

22. $(2a - 3b)(a - b)$ **23.** $(3x - 5)(2x - 3)$ **24.** $(6x - 5)(x + 3)$

25. $(x + 8)(x - 8)$ **26.** $(x - 4)(x^2 + 4x + 16)$ **27.** $(a + 3bc)(a^2 - 3abc + 9b^2c^2)$

28. $2x(x + 13)(x - 7)$ **29.** $3xy(x + 7y)(x - 7y)$ **30.** $(3x + 5a)(2x - 3)$

31. $(x^2 + 4)(x + 2)(x - 2)$ **32.** $(x + 3)(x - 3)(x + 2)(x^2 - 2x + 4)$ **33.** $x = 34$

34. $x = 0$ **35.** $x = \dfrac{2}{3}, -1$ **36.** $x = 4, -3$ **37.** $x = 3, -3$

38. $x = 12$ **39.** $\{x|x < 2\}$:

40. $\{x|4 \geq x \geq -2\}$:

41. $x = 4, -2$

42. $y = \dfrac{-1}{3}, 2$ **43.** $z = \dfrac{5}{2}, \dfrac{-5}{2}$ **44.** $z = 0, \dfrac{25}{4}$ **45.** $x = 1$

46. $x = 6, -2$ **47.** $\{x|x > -3\} \cup \{x|x < -5\}$:

48. $\{x|3 \geq x \geq 0\}$:   **49.** $t = 1, -1, 3, -3$
$\qquad\qquad\qquad\qquad\qquad\qquad \dfrac{}{2} \quad \dfrac{-1}{2}$

49. $t = 1, \dfrac{-1}{2}, 3, -3$

50. $w = \dfrac{2}{3}, -\dfrac{2}{3}$ **51.** -1 **52.** $-1, \dfrac{3}{2}$ **53.** 1, 2, 3

54. 22 nickels, 15 dimes   **55.** 4 quarters, 5 dimes, 10 nickels

56. \$15,000 at 7%, \$15,000 at 12%, \$20,000 at $8\frac{1}{2}\%$ **57.** 8:30 PM, 450 mi from town

58. a. 7 in. by 17 in.   **b.** 6 in. by 15 in.   **59.** 5, 12, 13 ($x = 1$ is not valid here)

60. 8 and 17 ($x = 4$)   **61.** 6, 8, 10

Chapter 4

Exercise 4.1   *pp. 132–135*

1. $x \neq 0$ **3.** No restrictions   **5.** $x \neq 3$

7. $x \neq 5, -2$ **9.** $x \neq -3, 2$ **11.** $z \neq 0, 4$

(continued)

Exercise 4.1, continued

13. a^6

15. $\dfrac{1}{x^6}$

17. $\dfrac{-2a^7}{3b^2}$

19. $\dfrac{y^3}{x^{5n}}$

21. $\dfrac{a}{2a^2 + 3}$

23. $\dfrac{3x^2 - 2y}{2xy}$

25. $\dfrac{1}{2}$

27. $\dfrac{b - 3}{3}$

29. $\dfrac{x + 3}{6(2x + 3)}$

31. $\dfrac{x + 3}{2x - 1}$

33. $\dfrac{x^2 + 2xy + 4y^2}{2x + 3y}$

35. $\dfrac{x(x + 4)}{x^2 + 4x + 16}$

37. $\dfrac{5b - a}{2}$

39. $\dfrac{2x^3}{3y^2}$

41. $-10x^2y^3$

43. $\dfrac{-16a^2c^2}{b}$

45. $-2x^3$

47. $\dfrac{1}{2}$

49. $\dfrac{1}{3}$

51. $\dfrac{3b}{2}$

53. 1

55. $\dfrac{x + 3}{x + 2}$

57. $\dfrac{4x^2 + 6x + 9}{3x + 2}$

59. $-(x + 3)$ or $-x - 3$

61. $-\dfrac{2}{3}$

63. $\dfrac{-(x + 3)}{x^2 + 3x + 9}$

65. $\dfrac{3}{2}$

67. $\dfrac{-2}{x - 1}$ or $\dfrac{2}{1 - x}$

69. $\dfrac{x + 2y}{2}$

71. $\dfrac{x^2 - 3x + 9}{2}$

73. $\dfrac{8(a + 1)}{5a(a + 3)}$

75. $\dfrac{ax + 5}{x - c}$

77. $\dfrac{x^2 + 2xy + 4y^2}{x + 3y}$

79. $\dfrac{3a - 2}{a + 6}$

81. $\dfrac{2}{3x}$

83. $\dfrac{ade}{bcf}$

85. $\dfrac{adeh}{bcfg}$

87. $\dfrac{2 + 5}{2 + 5} = \dfrac{2}{2}$; $\dfrac{A + C}{B + C} = \dfrac{A}{B}$ if $A = B$ and $C =$ any number

89. $4\dfrac{2}{3}$ eggs

Exercise 4.2   *pp. 139–142*

1. $\dfrac{3a}{x}$

3. $\dfrac{-4a + 3b - 7c}{5xy}$

5. 5

7. -3

9. $\dfrac{33b^2 + 16ab - 30a^2}{24a^2b^2}$

11. $\dfrac{12y^2 - 30x^2 + 35x^3}{20x^3y^2}$

13. $\dfrac{2a^2 + a - 1}{a^2}$ or $\dfrac{(2a - 1)(a + 1)}{a^2}$

15. $\dfrac{x^2 - 2}{x - 1}$

17. $\dfrac{-h}{x(x + h)}$

19. $\dfrac{2x}{(x + 1)(x - 1)}$

21. $\dfrac{2(a^2 + b^2)}{(a + b)(a - b)}$

23. $\dfrac{8}{3(x + 3)}$

25. $\dfrac{-8}{x + 2}$

27. $\dfrac{12}{x + 3}$

29. $\dfrac{2(x + 2y)}{x - y}$

31. $\dfrac{x}{x^2 + 2x + 4}$

33. $\dfrac{x - 1}{(x + 1)^2}$

35. $\dfrac{x - 3}{2(x + 1)}$

37. $\dfrac{15a^2 + 46a - 10}{12(a + 2)(a - 1)}$

39. $\dfrac{-2}{x - y}$ or $\dfrac{2}{y - x}$

41. -2

43. $\dfrac{1}{x + y}$

45. $\dfrac{3x - a}{2(x - 2a)}$ or $\dfrac{a - 3x}{2(2a - x)}$

47. $\dfrac{2}{x - 1}$

49. $\dfrac{5x + 3}{(x + 3)(x - 3)}$

51. $\dfrac{2x - 1}{x + 1}$

53. $\dfrac{x + 3}{x^2 + 3x + 9}$

55. $\dfrac{-x^2 - 15xy - 6y^2}{10(x - 2y)(x + 2y)}$

57. $\dfrac{-16}{(3a - 4)(3a + 4)(9a^2 - 12a + 16)}$

59. $\dfrac{-y^2 + 3y + 7}{(y + 3)(y - 2)(y + 5)}$

61. $\dfrac{8x^2 - 5xy}{(3x - 2y)(2x + y)(x - y)}$

63. $\dfrac{2x(x^2 + 6x - 9)}{(x + 3)^2(x - 3)^2}$

65. $\dfrac{24x^3 - 39x^2 + 14x + 7}{6x(x - 1)^2}$

67. $\dfrac{-2x^2 + ax - 9a^2}{(x + a)(x + 2)(x - 3a)}$

69. $\dfrac{1}{2} + \dfrac{3}{4} \neq \dfrac{1 + 3}{2 + 4}$, or $\dfrac{5}{4} \neq \dfrac{4}{6}$; $\dfrac{A}{B} + \dfrac{C}{D} = \dfrac{AD + BC}{BD}$

71. $1\dfrac{1}{12}$ in.

Exercise 4.3   *pp. 145–150*

1. $\dfrac{3b^2}{2a}$

3. $\dfrac{-5x^2y^2}{2}$

5. $\dfrac{5}{2z^3}$

7. $-\dfrac{1}{6rq}$

9. $\dfrac{2xy}{3}$

11. $\dfrac{ab(b + a)}{b^2 + ab + a^2}$

13. $\dfrac{x + 3}{x + 1}$

15. $\dfrac{2(x - 1)}{3}$

17. $\dfrac{(x + 2)(x + 5)}{(x + 3)^2}$

19. $\dfrac{50}{29}$

21. $\dfrac{-7}{10}$

23. $\dfrac{70}{43}$

25. $\dfrac{1}{2(x + 1)}$

27. $a - b$

29. $\dfrac{a + 5b}{a - b}$

31. $\dfrac{-(x + y)}{xy}$

33. $\dfrac{4y^2 + 2xy + x^2}{2xy(2y + x)}$

35. $\dfrac{5x(5x - 1)}{25x^2 - 5x + 1}$

37. $\dfrac{x}{x + y}$

39. $\dfrac{-1}{x(x + h)}$

41. $\dfrac{-2}{3ax}$

43. $\dfrac{8}{3}$

45. $\dfrac{1}{3}$

47. $\dfrac{1}{x - 15}$ *(continued)*

Exercise 4.3, continued

49. $\dfrac{23}{7}$

51. $\dfrac{77}{23}$

53. $\dfrac{8x-3}{2x-1}$

55. $\dfrac{x^3+2x}{x^2+1}$

57. $\dfrac{-(x^2+xy+y^2)}{x+y}$

59. $\dfrac{2rR}{r+R}$

61. $\dfrac{x(x-2)}{(x-1)(x+1)}$

63. $\dfrac{2xy}{(r-x)(r+x)}$

65. $\dfrac{1}{x+3a}$

67. $\dfrac{34}{13}$

69. $\dfrac{(x^2+1)(x-3)}{(x^2-3)(x+1)}$

71. $\dfrac{54}{41}$

73. $1-x$

75. False

77. False

79. $(1\div x)\div(1+1\div x)$

81. $(1\div a+1\div b)\div(1\div a^2-1\div b^2)$

83. $x+1\div[x-1\div(x+1\div(x+1))]$

85. $\dfrac{\dfrac{a}{b}}{c}$

87. $\dfrac{1+\dfrac{a}{b}}{1-\dfrac{a^2}{b^2}}$

89. $1+\dfrac{x}{1-\dfrac{x}{1+x}}$

Exercise 4.4   *pp. 152–154*

1. $2x^2-x+4$

3. $1+rt$

5. $3x^2+3xh+h^2$

7. $2a^2+3a-1+\dfrac{1}{2a}$

9. $2xy-3+\dfrac{5}{y}$

11. $b^4-\dfrac{2b}{a}+\dfrac{3b}{2}-\dfrac{9}{4a}$

13. $2x-5$

15. $3x+1-\dfrac{2}{2x-3}$

17. $2-\dfrac{13}{x+5}$

19. $3x-\dfrac{2}{2x+1}$

21. $10x+1.5-\dfrac{2.5}{2x+1}$

23. $2.5+\dfrac{19.5}{2x-5}$

25. $-8x+43-\dfrac{217}{x+5}$

27. $2x^2+x-7$

29. $3x^2+5x-6-\dfrac{3}{3x-2}$

31. $5x^2+6-\dfrac{2}{x+2}$

33. $4x^2+x+2.5+\dfrac{3.5}{2x-1}$

35. $5x^3-22x^2+74x-221+\dfrac{661}{x+3}$

37. $2x^2+2x+7$

39. $7x^2-10x+20-\dfrac{42}{x+2}$

41. x^2+2x+4

43. $1-\dfrac{1}{x+1}$

45. $x^3+3x^2+11x+33+\dfrac{90}{x-3}$

47. $2x^2-4x+1-\dfrac{1}{2x+1}$

49. x^3+2x^2+4x+8

51. $8x^4-9x^2+16-\dfrac{23}{x^2+2}$

53. $2x^2-x-7$

55. $a^3+4a^2+4a+13+\dfrac{31a+12}{a^2-2a-1}$

Exercise 4.5   *pp. 157–159*

1. $x = -\dfrac{5}{2}$ **3.** $z = 19$ **5.** $x = -4,\ 2$ **7.** $z = \dfrac{3}{2}, \dfrac{-3}{2}$ **9.** $x = -60$

11. $t = -1$ **13.** No solution ($x = 4$ is extraneous)   **15.** $x = 1$ **17.** $x = \dfrac{1}{3},\ 1$

19. $x = -2$ **21.** $x = 5,\ -1$ **23.** $t = 5$ ($t = -2$ is extraneous)   **25.** $z = 0,\ 5$

27. $r = \dfrac{119}{23}$ **29.** $x = 4$ ($x = -2$ is extraneous)   **31.** $z = 13$ ($z = -2$ is extraneous)

33. $x = 5,\ -1$ **35.** $x = 1$ **37.** $z = 5$ **39.** $x = -2$

41. $\{x | x < 5\}$:

43. $\{x | -7 \geq x\}$:

45. $\{x | x < 3\}$:

47. $\{x | x > 0\}$:

49. $\{x | -8 < x < 2\}$:

51. $\left\{ x \left| \dfrac{-8}{3} \geq x > \dfrac{-20}{3} \right. \right\}$:

53. $\left\{ x \left| \dfrac{11}{5} > x > 1 \right. \right\}$:

55. $z = -1$ **57.** $x = \dfrac{160}{47}$

59. $y = -\dfrac{27}{35}$ **61.** $x = \dfrac{25}{3}$ **63.** $x = -10,\ 1$ **65.** $x = 9,\ -2$ **67.** $x = -2,\ 1$

69. $x = \dfrac{5}{8}, \dfrac{1}{8}$ **71.** $x = 13,\ \dfrac{-19}{5}$ **73.** $\{x | 8 > x > -7\}$ **75.** $\left\{ x | x \geq \dfrac{7}{2} \right\} \cup \left\{ x | x \leq \dfrac{1}{2} \right\}$

77. $x = -3$ **79.** $\dfrac{3x + 25}{20}$ **81.** $x = 5$ **83.** $\dfrac{15x - 4}{12}$

Exercise 4.6   *pp. 162–165*

1. 43   **3.** 22   **5.** $\dfrac{5}{11}; \dfrac{9}{15}$ ($x = 5$)   **7.** $\dfrac{3}{5}; \dfrac{2}{8}$ ($x = 5$)

9. 3   **11.** $\dfrac{16}{24}$ ($x = 16$)   **13.** 14, 15, 16   **15.** 6, 8

17. $\dfrac{3}{4}$ or $\dfrac{4}{3}$ **19.** 6   **21.** $\dfrac{1}{2}$ or 1   **23.** 2; $\dfrac{5}{10}$ and $\dfrac{7}{14}$

25. 6 in. by 9 in.   **27.** 3, 4, 5   **29.** 13 in. by 6 in.   **31.** $h = 3$ m, $b = 8$ m

33. $b = 3$ in., $h = 2\dfrac{1}{2}$ in.   **35.** estate = \$1,000,000; eldest = \$500,000, next eldest = \$125,000, youngest = \$250,000

Exercise 4.7   *pp. 167–168*

1. $L = \dfrac{V}{WH}$

3. $h = \dfrac{2A}{b}$

5. $W = \dfrac{LS}{200H^2}$

7. $x = 2a$

9. $P = \dfrac{A}{1 + rt}$

11. $y = \dfrac{2x - 6}{3}$

13. $n = \dfrac{L - a + d}{d}$ or $\dfrac{L - a}{d} + 1$

15. $y = \dfrac{4a}{3}$

17. $y = -3x - 13$

19. $x = \dfrac{a + 2b}{2}$

21. $x = \dfrac{ab}{2(a - 3b)}$

23. $y = \dfrac{x}{1 - x}$

25. $C = \dfrac{5(F - 32)}{9}$

27. $c_2 = \dfrac{cc_1}{c_1 - c}$

29. $x = a + b$

31. $a = \dfrac{2A - hb}{h}$ or $\dfrac{2A}{h} - b$

33. $y = a + b$

35. $x = a^2 + ab + b^2$

37. $x = \dfrac{-4b}{9}$

39. $r = \dfrac{S - a}{S - an}$ or $\dfrac{a - S}{an - S}$

41. $y = a^2 - ab + b^2$

43. $R_3 = \dfrac{RR_1R_2}{R_1R_2 - RR_2 - RR_1}$

45. $a = \dfrac{2w}{w - 2}$

47. $m = \dfrac{EM}{M - E}$

Chapter 4 Review Exercises   *pp. 171–174*

1. $x \neq 0$

2. $x \neq 1$

3. $y \neq -3$

4. $x \neq -7, 2$

5. $z \neq -4, 2$

6. $t \neq 3, -\dfrac{7}{6}$

7. $\dfrac{3x^2y^3}{4z}$

8. $\dfrac{5}{8}$

9. $\dfrac{x}{2y}$

10. $\dfrac{-(x + 1)}{x + 8}$

11. $\dfrac{4a^2 + 6ab + 9b^2}{a - 3b}$

12. $\dfrac{x - 2}{2x - 3b}$

13. $\dfrac{9a^3b^4}{4c^2}$

14. $\dfrac{-m}{4np^4}$

15. $\dfrac{x - 9}{4}$

16. $\dfrac{x - y}{6}$

17. $\dfrac{2y(2x - 3)}{x(2x + 3)}$

18. $\dfrac{4(3a + 1)(3a - 1)}{(2a - 5)(a - 7)}$

19. 4

20. -4

21. $\dfrac{30y^2 - 27xy + 14x^2}{72x^2y^2}$

22. $\dfrac{6}{x - 4y}$

23. $\dfrac{-2}{x + 2}$

24. $\dfrac{6}{2a - 5}$

25. $\dfrac{3(x^2 + 2x + 25)}{4(x - 3)^2(x + 3)}$

26. $\dfrac{2(x^2 + h^2)}{x(x + h)(x - h)}$

27. $\dfrac{8b}{a^2 + 2ab + 4b^2}$

28. $\dfrac{x^2 + 8x + 4}{(x + 1)(x - 1)(x + 2)}$

29. $\dfrac{-2ab^2c}{3}$

30. $\dfrac{-1}{10xy}$

31. $\dfrac{b}{b-a}$

32. $\dfrac{-(x^2+xy+y^2)}{xy(x+y)}$

33. $x-3$

34. $\dfrac{x}{2x-3}$

35. $x+2$

36. $4x^2-2x+3$

37. $-2a^2+3-\dfrac{1}{2b}$

38. $2-\dfrac{13}{2x+3}$

39. $1-\dfrac{1}{x+1}$

40. $3x-2-\dfrac{9}{x-2}$

41. $4x-3+\dfrac{8}{2x+1}$

42. $5x^2-2x+6$

43. $2x^3-4x^2+15x-39+\dfrac{81}{x+2}$

44. $2x^3+3x^2+3x-7+\dfrac{-9x+2}{x^2-x+1}$

45. $x=5$

46. $x=-11$

47. No solution ($x=5$ is extraneous)

48. $y=5$

49. $x=10$

50. $z=7$

51. $n=8,\ \dfrac{-10}{9}$

52. $x=\dfrac{1}{3},\ 2$

53. $x=-4,\ 3$

54. $x=-12,\ 2$

55. $x=-\dfrac{3}{2},\ \dfrac{2}{3}$

56. $u=\dfrac{62}{3}$

57. $x=\dfrac{4}{5},\ -\dfrac{4}{5}$

58. $t=50,\ -\dfrac{100}{3}$

59. $x=2$

60. $x=4$

61. $x=5,\ -3$

62. $x=3,\ -1$

63. $x=\dfrac{1}{2},\ \dfrac{1}{18}$

64. $z=\dfrac{25}{2},\ \dfrac{-35}{2}$

65. $x=23,\ \dfrac{17}{7}$

66. $x=3,\ -3$

67. $\{x\,|\,x<3\}:$ 3

68. $\{x\,|\,x\le -33\}:$ -33

69. $\{x\,|\,1\ge x\ge -5\}:$ -5   1

70. $\{x\,|\,0>x\}:$ 0

71. $\left\{x\,\left|\,-\dfrac{5}{2}<x<\dfrac{7}{2}\right.\right\}:$ $-\dfrac{5}{2}$ $\dfrac{7}{2}$

72. $\{x\,|\,x\ge 1\}\cup\{x\,|\,x\le -2\}:$ -2   1

73. 19

74. $\dfrac{5}{11};\ \dfrac{2}{8}\ (x=5)$

75. 4

76. 12 and 28

77. $1;\ \dfrac{6}{8}$ and $\dfrac{9}{12}$

78. 4 or $\dfrac{1}{4}$

79. 5, 6

80. 4

81. $-48.9°\le C\le 6.7°$

82. $-126.9°\le F\le 136.0°$

83. $6\dfrac{1}{2}$ in. by 6 in.

84. $3\dfrac{1}{2}$ in. by 2 in.

85. 8 in.

86. 8, 6, 10

87. $h=4$ in., $b=13$ in.

88. $m=\dfrac{E}{c^2}$

89. $L=\dfrac{P-2W}{2}$

90. $T_2=\dfrac{T_1P_2V_2}{P_1V_1}$

91. $d_i=\dfrac{fd_0}{d_0-f}$

92. $y=\dfrac{2x}{x-1}$

93. $y=\dfrac{-2x-4}{3}$

94. $x=5$

95. $x=a+3$

96. $R_2=\dfrac{RR_1}{R-R_1}$

97. $x=a+b$

98. $y=\dfrac{m^2+mn+n^2}{m+n}$

99. $r=\dfrac{IR}{E-I}$

Chapter 4 Test   *p. 175*

1. $\dfrac{-a^3b}{2}$

2. $\dfrac{x^2}{y}$

3. $\dfrac{x^2 - 2x + 4}{x - 2}$

4. $\dfrac{-(2a + b)}{4b^2 + 6ab + 9a^2}$

5. $\dfrac{x - 2}{2x - 3}$

6. $\dfrac{-a}{a - 1}$ or $\dfrac{a}{1 - a}$

7. $\dfrac{x^2 - x + 1}{x(x - 1)}$

8. $\dfrac{-4}{x(x + h)}$

9. 3

10. $x = 6$

11. $x = -6$ ($x = 3$ is extraneous)

12. $\dfrac{4}{x - 2}$

13. $\dfrac{6y^2 + 20xy - 9x^2}{48x^2y^2}$

14. $\left\{ x \middle| x > \dfrac{1}{6} \right\}$: ⟵———○———————⟶
$\qquad\qquad\qquad\qquad\quad \frac{1}{6}$

15. $5a^2 + 3b - \dfrac{3a}{5b}$

16. $5x^2 - 10x + 18 - \dfrac{33}{x + 2}$

17. $y = \dfrac{cd}{c + d}$

18. $x = a + b$

19. $\dfrac{5}{8}$ and $\dfrac{3}{9}$ $(x = 8)$

20. 6, 8; $-8, -6$

Chapter 5

Exercise 5.1   *pp. 183–185*

1. x^{36}

3. a^7b^{14}

5. $81c^{16}d^{28}$

7. $-x^{12}$

9. $10^6 = 1,000,000$

11. $\dfrac{27x^3y^6}{64z^9}$

13. $\dfrac{-x^{72}}{y^{18}}$

15. $(a - b)^{72}$

17. $16x^{12m}y^{4n+4}$

19. $x^{40}y^{60}$

21. 1

23. 1

25. -5

27. $0^0 =$ undefined

29. $\dfrac{1}{x^8}$

31. $\dfrac{1}{64}$

33. $-\dfrac{1}{243}$

35. 1

37. a^4

39. $\dfrac{32}{x^3}$

41. $\dfrac{z^7}{x^5y^2w^4}$

43. $\dfrac{z^3}{x^3w}$

45. a^{35}

47. b^{48}

49. $\dfrac{x^{30}}{y^{55}}$

51. x^{24}

53. $\dfrac{b^{11}}{a^2}$

55. $-72x^{11}y^{13}$

57. x^{14m}

59. $\dfrac{1}{x^7}$

61. a^2

63. $\dfrac{1}{x^3y^9}$

65. $\dfrac{1}{x^{20}}$

67. $\dfrac{16}{9}$

69. $x^{15}y^{20}$

71. $\dfrac{x^6}{64}$

73. $a^{16}b^{36}$

75. $\dfrac{a + b}{ab}$

77. $\dfrac{1 - x^3}{x^2y^2}$

79. $\dfrac{xy}{y - x}$

81. $\dfrac{1}{a-1}$ **83.** 6×10^5 **85.** 4.9×10^{10} **87.** 3×10^{-5}

89. 4.3×10^{-1} **91.** 9.3×10^7 **93.** 7.5×10^{-6} **95.** 1×10^{100}

97. 200,000 **99.** 730,000,000 **101.** 0.0003 **103.** 0.000009025

105. 480,000,000 **107.** 0.00000153 **109.** x^{120} **111.** $x^{14}y^{42}$

113. $\dfrac{c^{2m+6}}{d^{2m+2}}$ **115.** $\dfrac{x^{11}}{6y^7z^8}$ **117.** $b-a$ **119.** $\dfrac{88}{39}$

121. $x+y$ **123.** 80 **125.** 4×10^7 **127.** 6×10^{10}

129. 9×10^{-12} **131.** 5.87×10^{12} miles

Exercise 5.2 *pp. 189–190*

1. 8 **3.** $-\dfrac{5}{7}$ **5.** $\dfrac{18}{19}$ **7.** 5 **9.** -9

11. $\dfrac{-7}{6}$ **13.** $\dfrac{3}{2}$ **15.** $\dfrac{-1}{2}$ **17.** $\dfrac{2}{3}$ **19.** 4

21. 3 **23.** 2 **25.** 2 **27.** $3, -5$ **29.** $5, -\dfrac{2}{3}$

31. $\dfrac{9}{4}, 0$ **33.** 5 **35.** 15 **37.** $2|x|$ **39.** $4a^8$

41. $-3xy^9$ **43.** $\dfrac{2|xy^2|}{3|z^3|}$ **45.** $2x^2y^3$ **47.** $2|xy^2|$ **49.** $|x-1|$

51. $x+y$ **53.** $|x+1|$ **55.** $2a^2+1$ **57.** $3x^2+4y^2$ **59.** $x^2|x+y|$

61. $|xy^2(2x^2-3y^2)|$ **63.** 2 **65.** $\dfrac{5}{6}$ **67.** 150 **69.** 165

71. 315 **73.** 42 **75.** 66

77. $|x^2| = x^2$, because $x^2 \geq 0$ for all real x, so you should be marked correct; however, $|x^3| = x^3$ only when $x \geq 0$, so you are incorrect here.

Exercise 5.3 *pp. 196–198*

1. $2\sqrt{6}$ **3.** $-3\sqrt{3}$ **5.** $4\sqrt{5}$ **7.** $5\sqrt{6}$

9. $2\sqrt[3]{3}$ **11.** $-3\sqrt[3]{4}$ **13.** $2\sqrt[4]{5}$ **15.** $2ab^4\sqrt{2ab}$

17. $-6x^2y^4z^4\sqrt{2z}$ **19.** $-2xy^2z^3\sqrt[3]{9xy^2}$ **21.** $-cd^3e^2\sqrt[3]{5de}$ **23.** $2xyz^2w^5\sqrt[4]{2yz^2w^3}$

25. $2x\sqrt{2x+1}$ **27.** $-3\sqrt{x^2+1}$ **29.** $1+\sqrt{2}$ **31.** $\dfrac{-1-\sqrt{3}}{2}$

(continued)

Exercise 5.3, continued

33. $\dfrac{\sqrt{2} + \sqrt{3}}{2}$

35. $1 \pm \sqrt{5}$

37. $\dfrac{-3 \pm 3\sqrt{5}}{2}$

39. $4\sqrt{5}$

41. $5\sqrt{5}$

43. $7\sqrt{2}$

45. $2\sqrt{3}$

47. $12\sqrt{2} + 3\sqrt{3}$

49. $5\sqrt[3]{2}$

51. $5\sqrt[3]{5} + \sqrt[3]{3}$

53. $21x\sqrt{2}$

55. $-5a\sqrt{3a}$

57. $x\sqrt[3]{2x}$

59. $\dfrac{82\sqrt{2}}{105}$

61. $\dfrac{51\sqrt{3}}{10}$

63. $\dfrac{13\sqrt{2} - 4\sqrt{3}}{12}$

65. $18\sqrt{2}$

67. $2a^2\sqrt[3]{3a^2}$

69. $\dfrac{2\sqrt{x^2 + 1}}{3}$

71. $\dfrac{-\sqrt{9 + x^2}}{3}$

73. 5 ft; $2\sqrt{5} = 4.5$ ft; $5\sqrt{2} = 7.1$ ft; $10\sqrt{3} = 17.3$ ft

75. $18\sqrt{5}$

77. $28\sqrt{3}$

79. $42\sqrt{3}$

81. $6\sqrt[3]{12}$

83. False: $\sqrt[3]{1^3 + 2^3} \neq 1 + 2$, or $\sqrt[3]{9} \neq 3$

85. False: $\sqrt{36a^{36}} = 6a^{18} \neq 6a^6$

87. True: $\sqrt[3]{2^9} = 2^3 = 8$

89. False: $1.260 + 1.414 \neq 1.149$

91. $\sqrt[n]{\dfrac{a}{b}} \overset{?}{=} \dfrac{\sqrt[n]{a}}{\sqrt[n]{b}} \rightarrow \left(\sqrt[n]{\dfrac{a}{b}}\right)^n \overset{?}{=} \left(\dfrac{\sqrt[n]{a}}{\sqrt[n]{b}}\right)^n \rightarrow \left(\sqrt[n]{\dfrac{a}{b}}\right)^n \overset{?}{=} \dfrac{(\sqrt[n]{a})^n}{(\sqrt[n]{b})^n} \rightarrow \dfrac{a}{b} \overset{\checkmark}{=} \dfrac{a}{b}$

Exercise 5.4   *pp. 200–202*

1. 12

3. -3

5. $9x^5$

7. $2a$

9. $2x$

11. 180

13. 8

15. 25

17. $25x - 25$

19. $x^2 + 2x$

21. $10\sqrt{2}$

23. $2x^4\sqrt{10x}$

25. $-5x\sqrt[3]{2x^2}$

27. $27\sqrt{3}$

29. $128\sqrt{2}$

31. $10\sqrt{6} - 40 + 10\sqrt{3}$

33. $2x\sqrt{10} - 15x^2 + 5x\sqrt{3x} + 2\sqrt{5x}$

35. $-17 + 11\sqrt{3}$

37. $72 + 38\sqrt{3}$

39. $20\sqrt{3} - 60\sqrt{2} - 5\sqrt{6} + 30$

41. $8 + 2\sqrt{15}$

43. $57 - 18\sqrt{2}$

45. $16 + 8\sqrt{x} + x$

47. $x + 5 - 4\sqrt{x + 1}$

49. $2x - 1 + 2\sqrt{x(x - 1)}$

51. 2

53. 12

55. $x - a$

57. h

59. $A = 24\sqrt{3} + 8$; $P = 10\sqrt{2} + 6\sqrt{6}$

61. $A = 18\sqrt{3}$

63. $A = \dfrac{1}{2}$

65. $x - 8$

67. $x - 1$

69. $\sqrt{13} - \sqrt{10}$

71. $3\sqrt{5} - 2\sqrt{10}$

73. $4\sqrt{5} - 5\sqrt{3}$

75. $6\sqrt{5} - 5\sqrt{6}$

77. $\dfrac{9}{5} - \sqrt{3}$

79. 14

81. 4

83. $2\sqrt{6}$

85. $9 + 2\sqrt{2} - 2\sqrt{6} - 4\sqrt{3}$

Exercise 5.5   *pp. 207–210*

1. 4

3. $3x$

5. $-2a^2$

7. $\dfrac{5x}{2y^2}$

9. $2\sqrt{3}$

11. $2x^4z^4\sqrt{2xyz}$

13. $3a^2$

15. $3a^4$

17. $2\sqrt{2} + 5$

19. $2\sqrt{b} - 2\sqrt{3} + 2\sqrt{5c}$

21. $\dfrac{\sqrt{14}}{7}$

23. $2\sqrt{2}$

25. $\dfrac{2\sqrt{10}}{5}$

27. $\dfrac{3\sqrt{2}}{2}$

29. $3x\sqrt{3x}$

31. $\dfrac{\sqrt{3x}}{x}$

33. $\dfrac{\sqrt{xy}}{x^4 y^3}$

35. $\dfrac{\sqrt{2x+y}}{2x+y}$

37. $\dfrac{\sqrt{6} + 2\sqrt{3} - 4}{2}$

39. $\dfrac{x\sqrt{y} + \sqrt{xy}}{x}$

41. $\dfrac{5\sqrt{2} - 8 + \sqrt{21}}{2}$

43. $\dfrac{6\sqrt{10} - 15\sqrt{2} + 10\sqrt{6}}{30}$

45. $\dfrac{3\sqrt{2}}{4}$

47. $\dfrac{\sqrt[3]{4}}{2}$

49. $2\sqrt[3]{5}$

51. $\dfrac{\sqrt[3]{6xy^2}}{3y}$

53. $6\sqrt[3]{2x}$

55. $5\sqrt[3]{25ab^2}$

57. $2a\sqrt[4]{2a^2}$

59. $\dfrac{\sqrt[5]{9x^3 y^2}}{3xy}$

61. $2(\sqrt{6} - \sqrt{2})$

63. $2(\sqrt{6} + 3)$

65. $2(3\sqrt{3} + \sqrt{2})$

67. $-\dfrac{3\sqrt{7} + 7\sqrt{3}}{4}$

69. $\dfrac{x\sqrt{y} - y\sqrt{x}}{x - y}$

71. $\sqrt{25 + h} + 5$

73. $2 - \sqrt{3}$

75. $4 + \sqrt{15}$

77. $\dfrac{-(6 + 5\sqrt{2})}{2}$

79. $\sqrt{a} + \sqrt{b}$

81. $-(x + \sqrt{y})$

83. $\dfrac{\sqrt{30} + 3\sqrt{2} + \sqrt{10} + \sqrt{6}}{2}$

85. $2 + \sqrt{3}$

87. $-2 + \sqrt{3}$

89. $-x\sqrt{x+h} - (x+h)\sqrt{x}$

91. $\dfrac{1}{\sqrt{x+h} + \sqrt{x}}$

93. $\dfrac{2}{\sqrt{2x+2h} + \sqrt{2x}}$

95. $\dfrac{1}{\sqrt{x} + \sqrt{a}}$

97. $\sqrt[3]{x^2} + \sqrt[3]{xy} + \sqrt[3]{y^2}$

99. $\sqrt[4]{x^3} + \sqrt[4]{x^2} + \sqrt[4]{x} + 1$

101. $\dfrac{2\sqrt{3} + 3\sqrt{2} - \sqrt{30}}{12}$

Exercise 5.6   *pp. 213–215*

1. 5

3. $2x^2$

5. 216

7. $16x^2 y^6$

9. $\dfrac{1}{12}$

11. $\dfrac{1}{27x^3}$

13. $\dfrac{-1}{3}$

15. $\dfrac{1}{4y^4}$

17. $\dfrac{3}{4}$

19. $\dfrac{27x^3}{64y^6}$

21. $\dfrac{5}{4}$

23. $\dfrac{9y^6}{4x^8}$

25. $2\sqrt{3}$

27. $2x\sqrt{2x}$

29. $3ab^2\sqrt[3]{2b}$

31. $40\sqrt{5}$

33. $54x^4\sqrt{2x}$

35. $\dfrac{\sqrt{2}}{4}$

37. $\dfrac{\sqrt[3]{4x^2}}{4x^2}$

39. $x^{17/12}$

41. $a^{17/24} b^{7/5}$

43. $x^{1/2}$

45. $\dfrac{a^{1/4}}{b^{7/6}}$

47. $\dfrac{x^8}{y^{10}}$

49. $a^6 b^9$

51. $2x^{4/3} - 5x + 4$

53. $12a^{8/5} + 6 - 3a$

55. $a - b$

(continued)

Exercise 5.6, continued

57. $3x^{2/3} + 5x^{1/3}y^{2/3} - 2y^{4/3}$ **59.** $\dfrac{(x-1)^2}{x}$ **61.** $x + y$ **63.** $\dfrac{x+1}{x^{1/2}}$

65. $\dfrac{x+2}{x^{5/3}}$ **67.** $\dfrac{x}{(x+1)^{3/2}}$ **69.** $5x^{1/3} - x^{3/2} + 3x^{-2/5}$

71. $(2x)^{1/2} + 2x^{1/2} - (x+1)^{2/3} + 3x^{-3}$ **73.** True   **75.** False

77. False   **79.** True   **81.** $\dfrac{\sqrt[n]{a}}{\sqrt[n]{b}} = \dfrac{a^{1/n}}{b^{1/n}} = \left(\dfrac{a}{b}\right)^{1/n} \stackrel{\checkmark}{=} \sqrt[n]{\dfrac{a}{b}}$

Chapter 5 Review Exercises   *pp. 218–221*

1. a^{27} **2.** $\dfrac{-8x^9y^3}{27z^6}$ **3.** $\dfrac{b^{16}}{c^{20}}$ **4.** $\dfrac{y^{13}}{x^2}$ **5.** $-4c^8d^{13}$ **6.** x^{15n-5}

7. $-x^{24}$ **8.** -6 **9.** $\dfrac{1}{625}$ **10.** $\dfrac{-1}{216x^6}$ **11.** $8a^6$ **12.** $\dfrac{1}{8}$

13. $\dfrac{c^2}{a^6b^9}$ **14.** $\dfrac{x^8}{y^{28}}$ **15.** $\dfrac{d^{27}}{c^{21}}$ **16.** $\dfrac{a^{45}}{b^{60}}$ **17.** $\dfrac{25y^6}{9x^4}$ **18.** $\dfrac{x^{16}}{16y^{12}}$

19. $\dfrac{ab}{b-a}$ **20.** $\dfrac{x(2x+1)}{4x^2-2x+1}$ **21.** 4×10^6 **22.** 5.35×10^{11} **23.** 2×10^{-5} **24.** 6.2×10^{-9}

25. 5.23×10^9 **26.** 7×10^{-9} **27.** $700{,}000$ **28.** $8{,}200{,}000{,}000$

29. 0.00102 **30.** 0.000003270 **31.** $6{,}600{,}000{,}000{,}000{,}000{,}000{,}000$

32. 0.0786 **33.** 6×10^{10} **34.** 3×10^8 **35.** $7{,}500{,}000$

36. 8×10^{15} **37.** 1 is the unique number satisfying $1 \cdot a^n = a^n$; since $a^0 \cdot a^n = a^{0+n} = a^n$, it must be true that $a^0 = 1$.

38. $\dfrac{a^{-n}a^n}{a^{-n}} = \dfrac{1}{a^{-n}}$, or $a^n = \dfrac{1}{a^{-n}}$ **39.** True   **40.** False   **41.** True   **42.** False

43. True   **44.** True   **45.** False   **46.** False   **47.** False   **48.** False

49. $11a^2b^3$ **50.** $\dfrac{-5x^2y^4}{6z^3}$ **51.** 3 **52.** $\dfrac{5}{3}, -1$

53. 15 **54.** $x + 4$ **55.** $7\sqrt{2}$ **56.** $-3\sqrt[3]{3}$

57. $2xy^3z^4\sqrt{3xz}$ **58.** $2abc^2\sqrt[3]{2bc^2}$ **59.** $3x\sqrt{x^2+4}$ **60.** $\dfrac{3\sqrt{x^2+1}}{4}$

61. $1 - \sqrt{3}$ **62.** $\dfrac{-2+3\sqrt{2}}{4}$ **63.** $2 - \sqrt{2x}$ **64.** $2\sqrt{2a} + \sqrt[3]{4a^2}$

65. $2\sqrt{10}$ **66.** $-2 \pm 2\sqrt{5}$ **67.** $27\sqrt{2}$ **68.** $9\sqrt[3]{3}$

69. $25\sqrt{5} + 3\sqrt{3}$ **70.** $-8x\sqrt{2x}$ **71.** $\dfrac{156\sqrt{2} + 35\sqrt{5}}{60}$ **72.** $42\sqrt{5}$

73. 9 **74.** $-3x^3$ **75.** 180 **76.** 45

77. $16(x + 1)$ **78.** $3\sqrt{10}$ **79.** $5x^5\sqrt{2x}$ **80.** $-2x^2\sqrt[3]{6x}$

81. $25\sqrt{5}$ **82.** $20 - 30\sqrt{3} + 20\sqrt{6}$ **83.** $-25 - 9\sqrt{5}$ **84.** $-38 + 34\sqrt{3}$

85. 5 **86.** $9 + 6\sqrt{2}$ **87.** $5 - 4\sqrt{x + 1} + x$ **88.** $1 + 2\sqrt{x(1 - x)}$

89. h **90.** 16 **91.** $x - a$

92. $P = 6\sqrt{2} + 2\sqrt{6}, A = 4\sqrt{3} + 4$ **93.** $P = 10\sqrt{3} - 2\sqrt{2}, A = 14 - 4\sqrt{6}$

94. $A = 6\sqrt{3}$ **95.** $A = 1$ **96.** $3x$ **97.** $-3x$

98. $\dfrac{a}{b^2}$ **99.** $2ab^2$ **100.** $2x\sqrt{10x}$ **101.** $2ab\sqrt[3]{3ab^2}$

102. $2\sqrt{3} - 3\sqrt{2}$ **103.** $\sqrt[3]{2}$ **104.** $2\sqrt{y} - 2\sqrt{3} + \sqrt{5}$

105. $2\sqrt{13}$ **106.** $\dfrac{\sqrt{10}}{4}$ **107.** $\dfrac{\sqrt{21}}{7}$ **108.** $\dfrac{\sqrt{6xy}}{3y}$

109. $\dfrac{x\sqrt[3]{y^2}}{y}$ **110.** $\dfrac{\sqrt{x + 2}}{x + 2}$ **111.** $2\sqrt[3]{3}$ **112.** $\dfrac{\sqrt{5x}}{5xy}$

113. $\dfrac{\sqrt{2} + 2\sqrt{3} - \sqrt{6}}{2}$ **114.** $\dfrac{36\sqrt{5} - 5\sqrt{6}}{30}$ **115.** $2(\sqrt{7} + \sqrt{2})$ **116.** $\sqrt{2 + h} + \sqrt{2}$

117. $\sqrt{x} + 4$ **118.** $-3 + 2\sqrt{2}$ **119.** $\dfrac{7 + 2\sqrt{6}}{5}$ **120.** $\dfrac{11 - 7\sqrt{3}}{26}$ **121.** $\sqrt{3} + 1$

122. $3 - 2\sqrt{2}$ **123.** -5 **124.** $3x^2$ **125.** 8 **126.** $125x^3y^9$

127. $\dfrac{1}{8}$ **128.** $\dfrac{1}{343x^9}$ **129.** $\dfrac{1}{9x^4}$ **130.** $\dfrac{8}{27}$ **131.** $\dfrac{5}{4}$

132. $\dfrac{64b^9}{125a^3}$ **133.** $5\sqrt{2}$ **134.** $-5\sqrt[3]{2}$ **135.** $24\sqrt{3}$ **136.** $4\sqrt[3]{4}$

137. $\dfrac{\sqrt{2}}{4}$ **138.** $\dfrac{\sqrt[3]{2x^2}}{2x}$ **139.** $x^{13/4}$ **140.** $\dfrac{a^{1/2}}{b^{1/10}}$ **141.** $x^{11/3}y^{1/4}$

142. $a^{4/5}b^{3/5}c$ **143.** $6x - 10x^{3/4} + 14$ **144.** $a^{8/3} + a - 1$ **145.** $x - y$ **146.** $\dfrac{(1 - a)^2}{a}$

147. $a - b$ **148.** $\dfrac{x - 1}{x^{2/3}}$ **149.** $\dfrac{x + 1}{x^{3/2}}$

Chapter 5 Test   *p. 221*

1. 6.02×10^8 **2.** 0.0000052 **3.** $\dfrac{4a^6}{9c^8}$ **4.** $\dfrac{a^7c^4}{b^6}$ **5.** $\dfrac{1}{9x^8}$

6. $\dfrac{27y^3}{125x^3}$ **7.** $5x^2y^8$ **8.** $9a^4b^6$ **9.** $x + 2$ **10.** $\dfrac{4y^2}{3x}$

(continued)

Chapter 5 Test, continued

11. $3x\sqrt{3x}$ **12.** 0 **13.** $a^{7/4}$ **14.** $2\sqrt{6}$ **15.** $2x\sqrt[3]{5x^2}$

16. $\dfrac{ab}{b-a}$ **17.** $\dfrac{(a+1)^2}{a}$ **18.** $\dfrac{3+\sqrt{5}}{2}$ **19.** $2\sqrt{2}-2$ **20.** $\sqrt{x+h}+\sqrt{x}$

Chapter 6

Exercise 6.1
pp. 226–228

1. $7i$ **3.** $-12i$ **5.** $\dfrac{4i}{5}$ **7.** $5i\sqrt{2}$ **9.** $-5i\sqrt{3}$

11. $\dfrac{i\sqrt{2}}{4}$ **13.** $\dfrac{5i\sqrt{3}}{3}$ **15.** $\pm\dfrac{i\sqrt{15}}{12}$ **17.** $2+12i$ **19.** $-22+15i$

21. $23+2i$ **23.** $16+11i$ **25.** $-18+29i\sqrt{2}$ **27.** $48+12i\sqrt{5}$ **29.** $2\pm2i$

31. $\dfrac{-3\pm3i\sqrt{3}}{2}$ **33.** $-21-20i$ **35.** $-9-46i$ **37.** $-5i$ **39.** $\dfrac{3-i}{2}$

41. $-2i\sqrt{5}$ **43.** $\dfrac{-8i\sqrt{3}}{3}$ **45.** $4-6i$ **47.** $\dfrac{-12+9i}{5}$ **49.** $\dfrac{4-7i}{5}$

51. $-i\sqrt{2}$ **53.** 1 **55.** -1 **57.** $-i$ **59.** i

61. $-i$ **63.** -1 **65.** 1 **67.** $-i$ **69.** i

71. True   **73.** True   **75.** False   **77.** False   **79.** True

81. True   **83.** True

85. $\sqrt{(-1)(-1)}=\sqrt{-1}\sqrt{-1}$ is incorrect; the Product Rule $\sqrt{ab}=\sqrt{a}\sqrt{b}$ is not valid if both radicals represent imaginary numbers.

87. a. $(\sqrt[2]{-1})^3=i^3=i^2i=-i$ but $\sqrt[2]{(-1)^3}=\sqrt[2]{-1}=i$ **b.** The definition of $a^{m/n}$ is valid only for real numbers.

Exercise 6.2
pp. 232–235

1. $x=\pm3\sqrt{6}$ **3.** $y=\pm5$ **5.** $w=\pm2i\sqrt{3}$ **7.** $z=\dfrac{\pm11}{2}$

9. $x=\pm\dfrac{3}{2}$ **11.** $y=\pm4\sqrt{2}$ **13.** $x=\pm\dfrac{3\sqrt{2}}{2}$ **15.** $z=\pm4\sqrt{3}$

17. $x=\dfrac{\pm\sqrt{105}}{10}$ **19.** $y=\dfrac{\pm1}{3}$ **21.** $x=3,-9$ **23.** $y=2\pm2\sqrt{3}$

25. $x = 3, -2$

27. $x = \dfrac{-1 \pm 3\sqrt{2}}{2}$

29. $x = \pm 2\sqrt{3}$

31. $c = 4\sqrt{5}$

33. $b = 12$

35. $c = 2\sqrt{2}$

37. $c = \dfrac{5}{12}$

39. $c = 5\sqrt{2}$

41. $a = b = 5\sqrt{2}$

43. $a = b = \dfrac{3\sqrt{2}}{2}$

45. $a = \sqrt{3}, c = 2\sqrt{3}$

47. $a = \dfrac{4\sqrt{3}}{3}, c = \dfrac{8\sqrt{3}}{3}$

49. $c = \sqrt{6}$

51. $x = \pm\dfrac{b}{a}$

53. $c = \dfrac{\sqrt{Em}}{m}$

55. $x = \dfrac{\pm n\sqrt{m}}{m}$

57. $T = \dfrac{2\pi\sqrt{Lg}}{g}$

59. $x = \pm\sqrt{1 - y^2}$

61. $y = \dfrac{\pm 4\sqrt{25 - x^2}}{5}$

63. $y = \pm\dfrac{b\sqrt{x^2 - a^2}}{a}$

65. $x = a\sqrt{3}$

67. $x = \dfrac{b\sqrt{2}}{2}$

69. $x = \sqrt{a^2 + b^2}$

71. $x = a + 1$

73. $w = \sqrt{2}, x = \sqrt{3}, y = 2, z = \sqrt{5}$

75. $y = 5, z = 35$, so $x = 30$

77. side $= -2 + 2\sqrt{3}$; $A = 16 - 8\sqrt{3}, P = -8 + 8\sqrt{3}$

79. a. $x = \dfrac{a\sqrt{2}}{2}$ **b.** $A = \dfrac{a^2}{2}$

81. a. $r = \sqrt{13}$ **b.** $A = 13\pi - 24 = 16.84$

83. $t = 3$ sec.

85. $x = \pm 2\sqrt{3},\ \pm 2\sqrt{2}$

87. $x = \pm 2i,\ \pm 2i\sqrt{2}$

Exercise 6.3
pp. 240–242

1. $x = 2 \pm 2\sqrt{2}$

3. $y = -3 \pm 2i\sqrt{3}$

5. $z = -12, 2$

7. $y = 14, 12$

9. $t = 1 \pm \sqrt{5}$

11. $x = 2 \pm 2i$

13. $x = 1 \pm \sqrt{2}$

15. $x = \dfrac{3 \pm 3\sqrt{5}}{2}$

17. $z = \dfrac{1 \pm i\sqrt{7}}{2}$

19. $x = -3, -4$

21. $u = \dfrac{9 \pm \sqrt{105}}{2}$

23. $x = \dfrac{-1 \pm \sqrt{5}}{2}$

25. $x = \dfrac{3}{2}, -5$

27. $y = \dfrac{1 \pm \sqrt{29}}{4}$

29. $z = \dfrac{-1 \pm \sqrt{73}}{6}$

31. $W = -3 + 2\sqrt{5} = 1.47$ m, $L = 3 + 2\sqrt{5} = 7.47$ m

33. $1 + 3\sqrt{3}$ or $1 - 3\sqrt{3}$

35. $x = -2 + \sqrt{46} = 4.8, x + 4 = 2 + \sqrt{46} = 8.8$

37. $t = 2 + 2\sqrt{2} = 4.8$ sec

39. a. $\left(x + \dfrac{b}{2}\right)^2 = x^2 + 2 \cdot x \cdot \dfrac{b}{2} + \left(\dfrac{b}{2}\right)^2 \overset{\checkmark}{=} x^2 + bx + \left(\dfrac{b}{2}\right)^2$ **b.** $\dfrac{b}{2}$ and $\dfrac{b}{2}$ **c.** $\left(\dfrac{b}{2}\right)^2$ **d.** $\left(\dfrac{b}{2}\right)^2$

Exercise 6.4
pp. 246–249

1. $x = 6, -3$

3. $y = \dfrac{1}{2}, -3$

5. $w = \dfrac{4}{3}, -\dfrac{3}{2}$

7. $u = \dfrac{1}{5}$

9. $x = 0, \dfrac{-2}{3}$

11. $z = \dfrac{\pm 3}{2}$

13. $x = \dfrac{\pm 4i\sqrt{3}}{3}$

15. $x = -4, 3$

17. $z = 2 \pm \sqrt{3}$

19. $x = \dfrac{-1 \pm i\sqrt{3}}{6}$

(continued)

Exercise 6.4, continued

21. $z = 3 \pm 2\sqrt{3}$ **23.** $y = \dfrac{1 \pm \sqrt{3}}{2}$ **25.** $x = \dfrac{-19}{7}, 1$ **27.** $x = \dfrac{13 \pm \sqrt{211}}{14}$ **29.** $x = 2, \dfrac{1}{8}$

31. $t = \dfrac{13 \pm 3\sqrt{129}}{4}$ **33.** $r = 50, -5$ **35.** 72; *not* factorable; 2 irrational

37. 0; factorable; 1 rational   **39.** -4; *not* factorable; 2 complex   **41.** 81; factorable; 2 rational

43. 961; factorable; 2 rational   **45.** $x = -\sqrt{2} \pm 3$ **47.** $x = -\sqrt{2} \pm 2$

49. $x = 4i, -i$ **51.** $x = \dfrac{7}{8}, \dfrac{11}{5}$ **53.** $x = \dfrac{20}{9}, \dfrac{-11}{4}$ **55.** $x = -\dfrac{9}{4}, \dfrac{-14}{3}$

57. $x = \dfrac{3 \pm 3\sqrt{5}}{4}$ **59.** $\dfrac{1 \pm \sqrt{5}}{2}$ **61.** $\dfrac{3 + \sqrt{57}}{2} = 5.3$ sec   **67. a.** $D = 4$ **b.** $x = -i \pm 1$

Exercise 6.5   *pp. 251–254*

1. $x = -64$ **3.** $x = 41$ **5.** $y = 2, -1$

7. $u = 12$ ($u = 0$ is extraneous)   **9.** $x = 3, 1$ **11.** $w = 0, 3$

13. $x = 9$ ($x = -4$ is extraneous)   **15.** $x = -2$ ($x = 8$ is extraneous)   **17.** $x = 2$ ($x = -1$ is extraneous)

19. $x = 3$ ($x = 16$ is extraneous)   **21.** $x = 3$ ($x = -3$ is extraneous)   **23.** $y = -2$

25. $x = 5$ **27.** $z = 9$ **29.** $y = 13, 5$

31. $x = 5$ ($x = 65$ is extraneous)   **33.** $m = 49$ ($m = 4$ is extraneous)   **35.** $v = 5$ ($v = \dfrac{5}{2}$ is extraneous)

37. $x = 2$ **39.** $x = 10, 9$ **41.** $y = 2$

43. $x = -1$ **45.** $w = 5, \dfrac{-13}{3}$ **47.** $A = \pi r^2$

49. $F = \dfrac{Gm_1 m_2}{d^2}$ **51.** $x = \pm \sqrt{r^2 - y^2}$ **53.** 9 (4 is extraneous)

55. 9 (1 is extraneous)   **57.** 4 (-5 is extraneous)   **59.** 9

61. $x = 16$; sides are 4, 3, 5 ($x = 4$ is extraneous)

63. a. $y = \sqrt{x^2 + 64}$ **b.** $z = 10 - x$ **c.** $\sqrt{x^2 + 64} + 10 - x = 14$ **d.** $x = 6$ **e.** $y = 10, z = 4$

65. Squaring both sides of the equation $x = 1$ creates the *extraneous* solution $x = -1$.

Exercise 6.6   *pp. 257–259*

1. $x = \pm 3, \pm 2i$ **3.** $y = \dfrac{\pm \sqrt{3}}{2}, \pm i\sqrt{2}$ **5.** $x = 0, \pm 2\sqrt{3}, \pm i\sqrt{3}$

7. $t = 0, -1 \pm 2\sqrt{3}$ **9.** $x = -2, 1, 1 \pm i\sqrt{3}, \dfrac{-1 \pm i\sqrt{3}}{2}$ **11.** $x = 1, -1, \dfrac{-1 \pm i\sqrt{3}}{2}, \dfrac{1 \pm i\sqrt{3}}{2}$

13. $x = -\dfrac{5}{2}, \ \pm 2\sqrt{2}$

15. $z = 5, \ -2, \ 1 \pm i\sqrt{3}$

17. $t = \dfrac{-4}{3}, \ \pm 1, \ \pm i$

19. $x = -64, \ 8$

21. $z = 343, \ 125$

23. $x = 125, \ -125$

25. $x = 4, \ 9$

27. $x = \dfrac{-1}{9}, \dfrac{1}{3}$

29. $y = \pm 1, \ \dfrac{\pm\sqrt{3}}{3}$

31. $x = \pm\dfrac{3i}{2}, \ \pm\dfrac{3}{2}$

33. $z = 0, \ \dfrac{\pm 2\sqrt{3}}{3}, \ \pm 3i$

35. $y = 0, \ \dfrac{-3}{2}, \ -1, \ \dfrac{3 \pm 3i\sqrt{3}}{4}, \ \dfrac{1 \pm i\sqrt{3}}{2}$

37. $z = 0, \ \pm 2, \ \pm 2i, \ \pm 1, \ \pm i$

39. $x = 1, \ \dfrac{-1 \pm i\sqrt{3}}{2}, \ \pm 2i$

41. $z = \pm 1, \ \pm i$

43. $x = \dfrac{64}{27}, \ -1$

45. $x = -\dfrac{1}{343}, \dfrac{1}{8}$

47. $x = \pm 1, \ \pm 2i\sqrt{2}$

49. $x = 3, \ -i\sqrt{2} \ (x = -3, \ i\sqrt{2} \text{ are extraneous})$

51. $x = \pm 3$

53. $x = 4, \ -3, \ 2, \ -1$

55. $y = 3, \ -1, \ 1 \pm i\sqrt{3}$

57. $t = \pm 1, \ \pm 3i$

59. $x = -1, \ \dfrac{-2 \pm \sqrt{14}}{2}$

61. $x = -8, \ 1, \ 4, \ -2$

63. $r = -10, \ 1, \ 5, \ -2$

65. $x = -\dfrac{5}{6}, \ -2$

67. $x = \pm\sqrt{3}$; sides are 3 and 4 $(x = \pm 2i \text{ are extraneous})$

69. $x = 3$; leg $= 1$, hypotenuse $= 3$ $(x = -3, \ \pm i \text{ are extraneous})$

Exercise 6.7   *pp. 263–266*

1. $\{x|x > 6\} \cup \{x|x < -1\}$:

3. $\{x|-2 \le x \le 5\}$:

5. $\left\{x\middle|-4 < x < \dfrac{3}{2}\right\}$:

7. $\{x|-2 \le x \le 2\}$:

9. $\{x|x > 3\} \cup \{x|x < -3\}$:

11. $\{x|x > 2\} \cup \{x|x < 2\}$:

13. $\left\{x\middle|x = \dfrac{5}{2}\right\}$:

15. $\{x|x > 1\} \cup \{x|x < 0\}$:

17. $\left\{x\middle|-1 \le x \le \dfrac{1}{3}\right\}$:

19. $\{x|x > 7\} \cup \{x|x < -5\}$:

21. $\{x|x > 5\} \cup \{x|x < -2\}$:

23. $\left\{x\middle|-4 < x < \dfrac{1}{2}\right\}$:

25. $\{x|x \ge 3\} \cup \{x|x < 0\}$:

27. $\{x|x < -3\}$:

29. $\{x|3 < x < 9\}$:

31. $\left\{x\middle|\dfrac{-2}{3} \le x < 1\right\}$:

(continued)

Exercise 6.7, continued

33. $\{x|x > 2\} \cup \{x|x < 0\}$: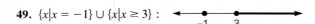

35. 13 hooks (least number for profit);
12 hooks (break even point)

37. side > 4

39. width < 3

41. $\dfrac{1}{4}, \dfrac{2}{5}, \dfrac{3}{6}, \dfrac{4}{7}, \dfrac{5}{8}, \dfrac{6}{9}, \dfrac{7}{10}, \dfrac{8}{11}$

43. $n > 8$: 9, 10, 11, 12, ...

45. $\{x|-3 < x < 2\} \cup \{x|x > 5\}$:

47. $\{x|x \le -4\} \cup \{x|0 \le x \le 3\}$:

49. $\{x|x = -1\} \cup \{x|x \ge 3\}$:

51. $\{x|x < -3\} \cup \{x|0 < x < 2\}$:

53. $\{x|-3 < x < -1\} \cup \{x|x > 2\}$:

55. $\{x|x < -3\} \cup \{x|0 < x < 3\}$:

57. $\{x|x \le 0\} \cup \{x|1 < x \le 3\}$:

59. $\{x|-3 < x < -1\} \cup \{x|1 < x < 3\}$:

61. $\{x|-4 \le x \le -2\} \cup \{x|x \ge 2\}$:

63. $\{x|x < -1 - \sqrt{3}\} \cup \{x|x > -1 + \sqrt{3}\}$:

65. $\{x|3 - 2\sqrt{2} \le x \le 3 + 2\sqrt{2}\}$:

67. $\{x|x \ge \sqrt{2}\} \cup \{x|x \le -\sqrt{2}\}$:

69. $\left\{x\left|\dfrac{-5\sqrt{3}}{3} < x < \dfrac{5\sqrt{3}}{3}\right.\right\}$:

Chapter 6 Review Exercises   *pp. 268–273*

1. $8i$

2. $2i\sqrt{10}$

3. $\dfrac{7i\sqrt{3}}{3}$

4. $\dfrac{-3i\sqrt{14}}{10}$

5. $12 + 3i$

6. $-25 + i$

7. $38 + 27i$

8. $5 - 12i$

9. $-2 - 2i$

10. $20 - 8i\sqrt{2}$

11. $\dfrac{i}{2}$

12. $-2i\sqrt{2}$

13. $\dfrac{-5 - 3i}{2}$

14. $-3i\sqrt{3}$

15. $6 - 8i$

16. $-2\sqrt{3} + 4i$

17. $5i$

18. $\dfrac{5 - 11i\sqrt{5}}{45}$

19. $x = \pm 5$

20. $x = \pm 2\sqrt{3}$

21. $y = \pm 2i\sqrt{6}$

22. $z = \pm\dfrac{7}{2}$

23. $x = \dfrac{\pm 2\sqrt{10}}{5}$

24. $x = \pm\dfrac{3\sqrt{2}}{2}$

25. $x = 2 \pm 3\sqrt{2}$

26. $y = 2, \dfrac{-8}{3}$

27. $x = \pm\dfrac{a\sqrt{5}}{2b}$

28. $v = \pm\sqrt{2as}$

29. $y = \dfrac{\pm 3\sqrt{4 - x^2}}{2}$

30. $h = \dfrac{\sqrt{FLkw}}{kw}$

31. $x = 2\sqrt{10}$ **32.** $x = 4\sqrt{2}$ **33.** $x = 2\sqrt{3}$ **34.** $x = \dfrac{a\sqrt{2}}{2}$ **35.** $x = 2\sqrt{5}$

36. $x = \sqrt{a^2 + 1}$ **37.** $x = 6$, $y = 17$ **38. a.** $r = \sqrt{5}$ **b.** $A = 5\pi - 8 = 7.7$

39. side $= 1 + 2\sqrt{2}$; $A = 9 + 4\sqrt{2}$, $P = 4 + 8\sqrt{2}$ **40.** $\dfrac{\pm 3\sqrt{2}}{2}$ **41.** $x = 8, -2$

42. $y = -2 \pm 2\sqrt{3}$ **43.** $z = -5 \pm 2i\sqrt{2}$ **44.** $w = \dfrac{-3 \pm \sqrt{13}}{2}$ **45.** $x = \dfrac{1}{2}, -3$

46. $y = \dfrac{3 \pm \sqrt{7}}{2}$ **47.** $W = -2 + 2\sqrt{5} = 2.47$ in., $L = 2 + 2\sqrt{5} = 6.47$ in.

48. $t = 3 \pm \sqrt{6} = 5.4$ sec, 0.6 sec   **49.** $-1 + \sqrt{31} = 4.6$, $1 + \sqrt{31} = 6.6$

50. $2 + 2\sqrt{7} = 7.3$, $-2 + 2\sqrt{7} = 3.3$ **51.** $x = \dfrac{4}{3}, -1$ **52.** $x = -1 \pm \sqrt{3}$

53. $z = \pm \dfrac{3i}{2}$ **54.** $z = 0, -\dfrac{9}{4}$ **55.** $t = 3 \pm 2i\sqrt{2}$ **56.** $x = 7, -37$

57. $x = \dfrac{1 \pm \sqrt{5}}{2}$ **58.** $y = \dfrac{3 \pm i\sqrt{23}}{8}$ **59.** $x = \dfrac{-2 \pm \sqrt{6}}{2}$ **60.** $x = \sqrt{2}, -3\sqrt{2}$

61. $x = 3i, -\dfrac{1}{2}i$ **62.** $x = 2\sqrt{3}, -\sqrt{3}$ **63.** 44; *not* factorable; 2 irrational

64. 0; factorable; 1 rational   **65.** -39; *not* factorable; 2 complex

66. -900; *not* factorable; 2 complex   **67.** 625; factorable; 2 rational

68. 1156; factorable; 2 rational   **69.** 1849; factorable; 2 rational

70. -1607; *not* factorable; 2 complex   **71.** $\dfrac{5 \pm \sqrt{65}}{2}$ **72.** $\dfrac{3 \pm \sqrt{13}}{2}$

73. $x = 1 \pm \sqrt{5} = 3.236, -1.236$ **74.** $x = \dfrac{-1 \pm \sqrt{5}}{2} = 0.618, -1.618$

75. $y = \dfrac{1 \pm \sqrt{17}}{4} = 1.281, -0.781$ **76.** $x = 9, -1$

77. $y = 4$ ($y = 0$ is extraneous)   **78.** $x = \dfrac{65}{2}$ **79.** $z = 7$

80. $w = 6$ **81.** $x = 7$ **82.** $v = 8, -\dfrac{4}{3}$

83. $x = 2\sqrt{3}$ ($x = -2\sqrt{3}$ is extraneous)   **84.** 5 (0 is extraneous)

85. 9   **86.** $x = 4$ ($x = \dfrac{4}{9}$ is extraneous); sides are 3, 4, 5

87. a. $y = 8 - x$ **b.** $y = \sqrt{x^2 + 16}$ **c.** $8 - x = \sqrt{x^2 + 16}$ **d.** $x = 3$ **e.** $y = 5$ *(continued)*

Chapter 6 Review Exercises, continued

88. a. $y = \sqrt{x^2 + 1}$, $z = \sqrt{x^2 + 6}$ **b.** $\sqrt{x^2 + 1} + \sqrt{x^2 + 6} = 5$ **c.** $x = \sqrt{3}$ **d.** $y = 2$, $z = 3$

89. $x = 0$, $\pm 3i$, $\pm 2\sqrt{2}$

90. $y = \pm \dfrac{2\sqrt{3}}{3}$, ± 3

91. $z = 2$, $-1 \pm i\sqrt{3}$

92. $w = -2$, $1 \pm i\sqrt{3}$, $\pm 2\sqrt{2}$

93. $t = -\dfrac{3}{2}$, ± 2, $\pm 2i$

94. $x = \dfrac{-27}{8}$, 27

95. $x = 2$, $-\dfrac{1}{3}$

96. $x = \dfrac{-1}{729}$, $\dfrac{1}{8}$

97. $x = 4$, -1, 1, 2

98. $z = \pm \sqrt{2}$, ± 2, $\pm 2i$

99. $x = \pm 2$, $\pm i\sqrt{6}$

100. $x = 2$ ($x = -2$, $\pm i\sqrt{3}$ are extraneous); $\sqrt{3}$

101. $x \pm \sqrt{6}$; 6 and 8

102. $\{x \mid -2 < x < 3\}$:

103. $\{x \mid x \geq 1\} \cup \left\{x \mid x \leq \dfrac{-5}{2}\right\}$:

104. $\{x \mid -2 \leq x \leq 2\}$:

105. $\{x \mid x > 9\} \cup \{x \mid x < 0\}$:

106. $\{x \mid x \geq 3\} \cup \{x \mid x \leq -3\}$:

107. $\{x \mid x > 2\} \cup \{x \mid x < -3\}$:

108. $\left\{x \mid -\dfrac{1}{2} < x < 4\right\}$:

109. $\{x \mid 0 < x \leq 1\}$:

110. $\{x \mid x > 4\}$:

111. $\{x \mid x < -1\} \cup \left\{x \mid x > \dfrac{3}{2}\right\}$:

112. $\{x \mid x < -2\} \cup \{x \mid 0 < x < 2\}$:

113. $\{x \mid -1 < x < 2\} \cup \{x \mid x > 4\}$:

114. $\{x \mid -1 < x < 3\} \cup \{x \mid x < -3\}$

115. $\{x \mid x \geq \sqrt{5}\} \cup \{x \mid x \leq -\sqrt{5}\}$:

116. $\{x \mid 1 - \sqrt{3} < x < 1 + \sqrt{3}\}$:

117. $1 < t < 3$ sec

118. 10 frammises $\left(x = \dfrac{19}{2} = 9\dfrac{1}{2}$ is a critical point$\right)$

119. $0 < x < 4$

120. $\dfrac{1}{5}$, $\dfrac{2}{6}$, $\dfrac{3}{7}$, $\dfrac{4}{8}$, $\dfrac{5}{9}$, $\dfrac{6}{10}$, $\dfrac{7}{11}$

Chapter 6 Test   *pp. 273*

1. $12 - 5i$

2. $x = \pm 2\sqrt{3}$

3. $y = \dfrac{\pm 5i\sqrt{6}}{3}$

4. $z = 4 \pm 2\sqrt{5}$

5. $w = \dfrac{1 \pm i\sqrt{3}}{6}$

6. $r = 5$ ($r = 1$ is extraneous)

7. $t = 6$

8. $x = 0, \pm 2\sqrt{2}, \pm 2i$ **9.** $x = -216, 27$ **10.** $x = \dfrac{\pm 14\sqrt{3}}{3}$ **11.** $x = \dfrac{2 \pm \sqrt{2}}{2}$

12. $x = 2$ ($x = -2$ is extraneous)   **13.** $r = \dfrac{\sqrt{\pi A}}{2\pi}$

14. $\{x | x \geq 1\} \cup \{x | x \leq -4\}$:

15. $\{x | -3 < x < 0\}$:

16. $W = -2 + 2\sqrt{3} = 1.5, L = 2 + 2\sqrt{3} = 5.5$ **17.** 6 (1 is extraneous)

18. $x = 4\sqrt{2}$ **19.** $x = \dfrac{a\sqrt{3}}{3}$ **20.** $x = 4, y = -3 + 4\sqrt{3}$

Chapters 1-6 Cumulative Review Exercises   *pp. 274–277*

1. 269   **2.** -23 **3.** $x = 5$ **4.** $5x - 17$

5. $-\dfrac{9x}{y^4}$ **6.** $t = 0$ **7.** $\dfrac{16a^{16}}{81b^8}$ **8.** $\dfrac{a^2 - a + 1}{a - 1}$

9. $z = \dfrac{-107}{9}$ **10.** $\dfrac{1}{16x^6}$ **11.** $\dfrac{a^{24}b^{24}}{c^{20}}$ **12.** $y = \dfrac{2}{3}, -\dfrac{3}{2}$

13. $\dfrac{x^2 + x + 1}{x(x + 1)}$ **14.** -5 **15.** $\{x | x < 2\}$:

16. $\dfrac{4}{x - 2}$ **17.** $-6\sqrt{2}$ **18.** $x = 6, -8$ **19.** $29 - 2i$

20. $x = \pm \dfrac{5}{4}$ **21.** $\dfrac{81x^4}{256y^8}$ **22.** $\{x | x \leq -33\}$:

23. $x = 0, \dfrac{25}{16}$ **24.** $\dfrac{1}{3x^3}$ **25.** $x = -1$ ($x = 2$ is extraneous)

26. $\dfrac{-(x + 1)}{2x(2x + 1)}$ **27.** $\{x | -2 < x < 5\}$:

    28. $x^{23/12}$

29. $x + 3$ **30.** $z = 3 \pm 2\sqrt{3}$ **31.** $\{x | 1 \geq x \geq -5\}$:

32. $w = \pm 2\sqrt{2}$ **33.** $\dfrac{(x + 1)^2}{x}$ **34.** $\dfrac{2a(3a - 2)(a + 5)}{2a + 3}$

35. $v = \dfrac{\pm 5i\sqrt{3}}{3}$ **36.** $\{x | x \leq -2\} \cup \{x | x \geq 3\}$:

37. $a(a - 1)$ **38.** $\dfrac{x + 2}{x^2 + 2x + 4}$ **39.** $5i\sqrt{2}$ **40.** $w = \dfrac{-3 \pm 3i\sqrt{3}}{2}$

(continued)

Chapters 1–6 Cumulative Review Exercises, continued

41. $\{x|x \geq 2\} \cup \{x|x \leq -4\}$:

42. $2x + h - 2$

43. $x = 0, \pm 2\sqrt{3}, \pm 2i$

44. $\{x|-3 < x < 1\}$:

45. $x = 1$ ($x = -3$ is extraneous)

46. $x = \dfrac{1 \pm \sqrt{5}}{2}$

47. $x = 0, 2, -1, -1 \pm i\sqrt{3}, \dfrac{1 \pm i\sqrt{3}}{2}$

48. $\dfrac{-2(x^2 + 2x - 6)}{(x-3)(x-1)^2}$

49. $x = \pm 2, 3, \dfrac{-3 \pm 3i\sqrt{3}}{2}$

50. $y = 4$

51. $\{x|x \geq 2\} \cup \{x|x < -2\}$:

52. $z = -343, 8$

53. $x = \pm 3, \pm i\sqrt{3}$

54. $x = 5$ ($x = 0$ is extraneous)

55. $x = 6, -4, 4, -2$

56. $x = 7, -5, 1$

57. a. 5.13×10^8 **b.** 3.02×10^{-5}

58. a. 0.00067 **b.** $790,100,000$

59. $x - 3 + \dfrac{4}{x}$

60. $4x^2 - 2x + 7$

61. $2x^3 + 7x^2 + 14x + 21 + \dfrac{51}{x-2}$

62. $1 - \dfrac{1}{x+1}$

63. $4x\sqrt{2x}$

64. $2 - \sqrt{3}$

65. $\dfrac{4 + 3i}{5}$

66. $3\sqrt[3]{9x}$

67. $\sqrt{4 + h} + 2$

68. $\dfrac{3i\sqrt{5}}{2}$

69. $\dfrac{-3 - 4i}{2}$

70. $\sqrt{x} + \sqrt{a}$

71. $B = \dfrac{2AC}{C - A}$

72. $x = a + b$

73. $x = \dfrac{-3a}{2}, 2a$

74. $y = \pm\sqrt{r^2 - x^2}$

75. $x = \dfrac{\pm 3\sqrt{y^2 + 16}}{2}$

76. $g = \dfrac{4\pi^2 L}{P^2}$

77. $c = \dfrac{\sqrt{Em}}{m}$

78. $c_2 = \dfrac{cc_1c_3}{c_1c_3 - cc_3 - cc_1}$

79. $x = \pm a\sqrt{3}$

80. $x = a + b$

81. -35

82. $3, 5, 7$

83. $-3, -2, -1, 0$

84. $\dfrac{77}{2}$

85. $\dfrac{3}{4}, \dfrac{4}{3}$

86. 7 nickels, 7 dimes, 9 quarters

87. 12 nickels, 6 dimes

88. \$5000 in bond, \$5000 in stock, \$12,000 in commodity

89. \$5000

90. 300 miles ($t = 6$)

91. 2 hrs *going,* 3 hrs *returning;* 120 miles

92. 25 miles *city,* 20 miles *highway*

93. a. 7 in. by 10 in. **b.** 8 in. by 11 in.
c. 9 in. by 12 in.

94. 12 in. by 12 in.; 18 in. by 8 in.

95. 5 m by 5 m; 9 m by 9 m

96. 25 ft

97. 6, 8

98. $2\sqrt{2}$ by $4\sqrt{2}$

99. $x = 6\sqrt{5}$

100. $x = 2\sqrt{3}$

101. $x = \sqrt{2}$

102. $x = \sqrt{5}$

103. $x = 9$ ($x = 25$ is extraneous)

104. $x = 2$ ($x = -2, \pm i$ are extraneous)

Chapter 7

Exercise 7.1   *pp. 282–283*

1.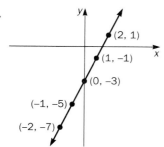

3.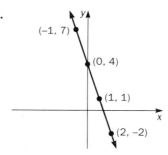

5.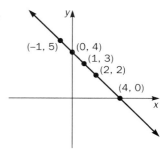

7.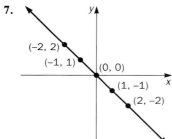

9.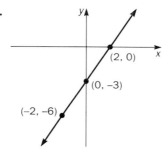

11.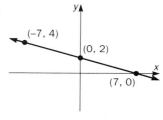

13.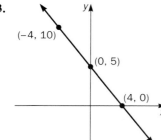

15.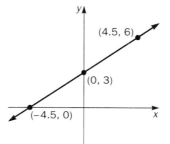

17.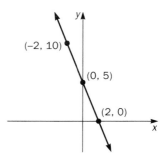

19.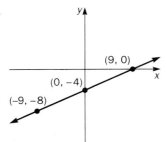

21.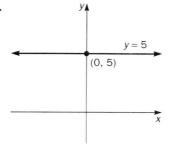

23.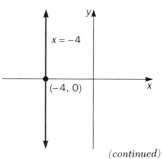

(continued)

Exercise 7.1, continued

25.

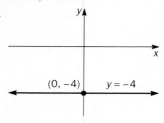

27.

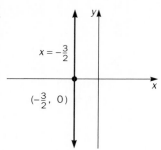

29.

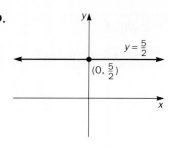

31. (2, 0), (0, 4); A = 4   **33.** (5, 0), (0, −3): $A = \dfrac{15}{2}$ **35.** (−1, 0), (0, −2); A = 1

37. Yes   **39.** No   **41.** Yes   **43.** Yes   **45.** No

47.

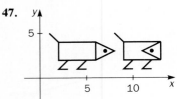

Exercise 7.2   *pp. 290–293*

1. $m = 2$ **3.** $m = \dfrac{-3}{2}$ **5.** $m = \dfrac{-1}{3}$ **7.** $m = \dfrac{3}{2}$ **9.** $m = 0$ **11.** m is undefined

13. $m = -2 = \dfrac{-2}{1}$, $b = 3$

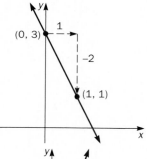

15. $m = \dfrac{1}{2}$, $b = -2$

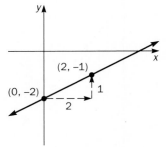

17. $y = 3x - 1$:

     $m = 3 = \dfrac{3}{1}$, $b = -1$

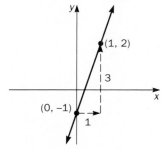

19. $y = \dfrac{-3x}{4} + 3$:

     $m = \dfrac{-3}{4}$, $b = 3$

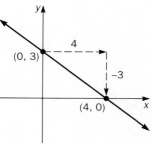

21. $y = \dfrac{2}{3}x + \dfrac{7}{3}$:

$m = \dfrac{2}{3}, b = \dfrac{7}{3}$

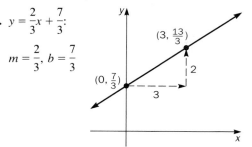

23. $y = \dfrac{-2x}{5}$:

$m = \dfrac{-2}{5}, b = 0$

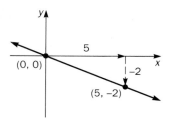

25. $y = -3x + 6$:

$m = -3 = \dfrac{-3}{1}, b = 6$

27. Perpendicular

29. Parallel

31. Perpendicular

33. Parallel

35. Perpendicular

37. Neither

39. Parallel

41. Perpendicular

43. Perpendicular

45. Perpendicular

47. Parallel

49. $m = \dfrac{-13}{6}$

51. $m = 7$

53. $m = \dfrac{-9}{44}$

55. $m = -3 + 2\sqrt{2}$

57. $m = 6$

59. $m = -2\sqrt{3}$

61. $m = \sqrt{7} + \sqrt{2}$

63. a. P_1: $-5 \overset{\checkmark}{=} 2(-1) - 3$; similarly for P_2, P_3, P_4

  b. slope through P_1 and P_2 = slope through P_3 and P_4 = 2

65. $m_1 = m_3 = \dfrac{3}{7}$ (parallel); $m_2 = m_4 = 1$ (parallel)

67. $m_1 = \dfrac{3}{5}, m_2 = \dfrac{-7}{11}, m_3 = \dfrac{-5}{3}$; $m_1 m_3 = -1$ (perpendicular)

69. $m = x + 3$

71. $m = 2$

73. $m = \dfrac{2b}{3a}$

75. $m = x^2 + 2x + 4$

77. $m = \dfrac{-1}{ax}$

79. $m = \dfrac{-(2 + x)}{4x^2}$

81. $k = \dfrac{8}{3}$

83. $k = 5, -1$

Exercise 7.3   *pp. 297–298*

1. $y = -3x - 13$

3. $y = \dfrac{2x}{5} - 16$

5. $y = -1$

7. $y = -2x$

9. $y = 4x + 5$

11. $y = -3x + 5$

13. $y = -\dfrac{2}{3}x + \dfrac{7}{3}$

15. $y = 3$

(continued)

Exercise 7.3, continued

17. $y = \dfrac{-5x}{2} + 5$

19. $y = -3x + \dfrac{19}{6}$

21. a. $y = -2x + 2$ **b.** $y = \dfrac{1}{2}x - 8$

23. a. $y = \dfrac{-2x}{3} + \dfrac{14}{3}$ **b.** $y = \dfrac{3x}{2} - 4$

25. a. $y = \dfrac{1}{4}x - \dfrac{49}{4}$ **b.** $y = -4x - 8$

27. $F = \dfrac{9C}{5} + 32; F = 77°$

29. a. $y = -3x + 20$ **b.** $\left(\dfrac{20}{3}, 0\right), (0, 20)$ **c.** $A = \dfrac{200}{3}$

Exercise 7.4   *pp. 304–307*

1. $x = 7, y = 4$

3. $x = 1, y = 1$

5. $u = 4, v = 0$

7. $s = 0, t = 0$

9. $x = \dfrac{1}{2}, y = \dfrac{-3}{2}$

11. $r = 2, s = 5$

13. $m = \dfrac{1}{7}, n = -3$

15. $x = 7, y = -2$

17. $x = 1, y = 8$

19. $z = \dfrac{1}{2}, w = \dfrac{5}{2}$

21. $u = 7, v = 11$

23. $x = 4, y = 2$

25. $x = 0, y = 0$

27. No solution

29. Infinitely many solutions

31. Infinitely many solutions

33. $x = 2, y = 3$

35. $x = \dfrac{-14}{3}, y = -4$

37. $x = 8, y = -2$

39. $x = -3, y = -13$

41. $x = \dfrac{1}{2}, y = \dfrac{-1}{4}$

43. $x = \dfrac{1}{2}, y = \dfrac{1}{3}$

45. $x = \dfrac{-3}{2}, y = \dfrac{4}{3}$

47. $x = a + b, y = 2a - b$

49. $x = \dfrac{2}{m}, y = \dfrac{2}{n}$

51. $x = a + b, y = -ab$

53. $x = -c^2d - cd^2, y = c^2 + cd + d^2$

55. $x = 3a, y = 2b$

57. $x = r + 2s, y = 2r + s$

59. $x = 2a - b, y = a - 2b$

61. $x = a + 3b, y = 3a + b$

63. $x = \sqrt{2}, y = 2\sqrt{3}$

65. $x = i, y = -1$

67. $x = 1 + i, y = -1 + i$

Exercise 7.5   *pp. 311–315*

1. 27 dimes, 16 quarters

3. 8 nickels, 21 dimes, 8 quarters

5. $8000 at 12%, $16,000 at 6%

7. $6000

9. 45 liters of 20%, 150 liters of 55%

11. 75 cc water, 125 cc total

13. 30 liters of 25%, 80 liters of 80%

15. 6.2 oz beef, 1.8 oz cheese

17. boat = 35 mph, current = 5 mph

19. wind = 5 mph, distance = 75 miles

21. hamburger = $2.00, fries = $0.75

23. $A = B = 10$ liters, $D = 120$ liters

25. 3 pounds at $1.50, 7 pounds at $2.00

27. 3 years at 9%, 6 years at 11%

29. $A = 35, B = 15$

31. $\dfrac{15}{21}$ $(x = 15, y = 21)$

33. $W = 4$ in., $L = 9$ in.

35. 8 by 12; 4 by 4

37. a. Line 1: $y = x - 3$, Line 2: $y = -x + 7$ **b.** $(5, 2)$

Exercise 7.6   *pp. 319–321*

1. $(1, 1, 1)$

3. $(1, 0, -1)$

5. $(-2, -1, 3)$

7. $(5, 2, -1)$

9. $(-2, -4, -8)$

11. $(0, 0, 0)$

13. 2, 3, and 5

15. 8 pennies, 12 nickels, 13 dimes

17. 1 liter of 10%, 2 liters of 40%, 6 liters of 60%

19. $(1, 2, -1)$

21. $(1, 2, 3)$

23. $(4, 3, 2)$

25. $(a, 2a, a)$

27. $(c + d, c - d, 2d)$

Exercise 7.7   *pp. 326–329*

1. 14

3. 0

5. 136

7. 39

9. -46

11. -42

13. $\left(\dfrac{-3}{19}, \dfrac{2}{19}\right)$

15. $(1, 1)$

17. $(-2, 0)$

19. $(5, -4, 3)$

21. $(1, 1, 4)$

23. $(1, 1, 1)$

25. $\left(\dfrac{1}{2}, \dfrac{1}{4}, -\dfrac{3}{4}\right)$

27. $\dfrac{-3}{16}$

29. $6\sqrt{3} - 4\sqrt{10}$

31. 2

33. $2\sqrt{2}$

35. $4ab$

37. $-\dfrac{81}{16}$

39. 0

41. $y = -4x + 13$

43. $y = mx + b$

45. Parallel

47. $(\sqrt{2}, 2\sqrt{3})$

49. $(i, -1)$

51. $(a + b, -ab)$

53. $(cd, d - c)$

55. No solution

57. Infinitely many solutions

59. a. $-a_2b_1c_3 + a_2b_3c_1 + a_1b_2c_3 - a_3b_2c_1 - a_1b_3c_2 + a_3b_1c_2$

b. $a_1b_2c_3 + a_3b_1c_2 + a_2b_3c_1 - a_3b_2c_1 - a_1b_3c_2 - a_2b_1c_3$

Chapter 7 Review Exercises   *pp. 332–337*

1.

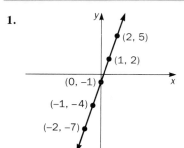

2.

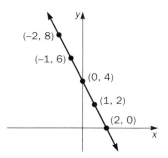

3.

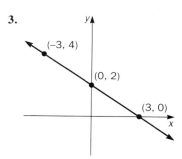

(continued)

Chapter 7 Review Exercises, continued

4.

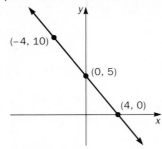

5.

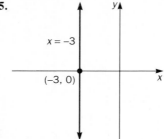

6.

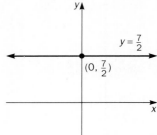

7. $(5, 0), (0, 2); A = 5$

8. No

9. Yes

10. $m = -\dfrac{4}{3}$

11. $m = -2$

12. $m = \dfrac{3}{25}$

13. $m = \dfrac{-2}{9}$

14. m is undefined

15. $m = \sqrt{6}$

16. $m = 3 - 2\sqrt{2}$

17. $m = -\dfrac{1}{2}$

18. $m = x + 3$

19. $m = 2 = \dfrac{2}{1}, b = -3$

20. $m = \dfrac{2}{3}, b = 1$

21. $y = \dfrac{-2x}{5} + 2$:

$m = \dfrac{-2}{5}, b = 2$

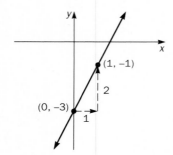

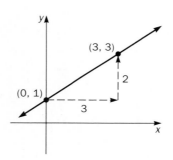

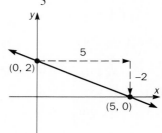

22. $y = \dfrac{-3x}{4}$:

$m = \dfrac{-3}{4}, b = 0$

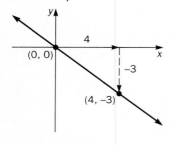

23. Neither

24. Parallel

25. Perpendicular

26. Parallel

27. Perpendicular

28. Perpendicular

29. Parallel

30. Perpendicular

31. $k = \dfrac{29}{5}$

32. $k = 8, -2$

33. $k = \pm 6$

34. $m_1 = m_3 = \dfrac{3}{8}$ (parallel); $m_2 = m_4 = \dfrac{-3}{2}$ (parallel)

35. $m_2 m_3 = \dfrac{-4}{3} \cdot \dfrac{3}{4} = -1$ (perpendicular)

36. $y = 2x + 5$

37. $y = \dfrac{-3x}{4} + \dfrac{7}{4}$

38. $y = 2x - 5$

39. $y = -\dfrac{3}{2}x + \dfrac{1}{2}$

40. $y = \dfrac{5}{3}x + 5$

41. $y = 2$

42. $y = -3x + 7$

43. $y = \dfrac{x}{3} + \dfrac{1}{3}$

44. $y = \dfrac{5x}{2} + 13$

45. $y = \dfrac{-2x}{5} + \dfrac{7}{5}$

46. a. $y = -x + 12$ **b.** $(12, 0), (0, 12)$ **c.** $A = 72$

47. $x = 2, y = -1$

48. $x = 3, y = 0$

49. $x = 5, y = -7$

50. No solution

51. Infinitely many solutions   **52.** $x = \dfrac{1}{2}, y = -\dfrac{1}{3}$

53. $x = 28, y = 26$

54. $x = 6, y = 8$

55. $x = a + 4b, y = a - 3b$ **56.** $x = c + d, y = c - d$

57. 18 dimes, 16 quarters

58. 4 pennies, 4 quarters, 7 nickels, 7 dimes

59. 12 $10 bills, 6 $20 bills

60. $12,500 at 6%, $7500 at $9\dfrac{1}{2}$%

61. $45,000 at 12%

62. $400 at 4%, $600 at 6%

63. 5 cc of 20%, 15 cc of 40%

64. 18 lbs of 75¢, 12 lbs of 60¢

65. 2 liters of 80%, 20 liters of 25%

66. Kennedy 34,227,096; Nixon 34,108,546

67. 120 cc water

68. A: 40 problems, 30 correct; B: 110 problems, 99 correct

69. $\dfrac{14}{20}$ $(x = 14, y = 20)$

70. 4 by 6; 6 by 4; 2 by 12

71. boat = 20 mph, current = 5 mph

72. 20 mph; 48 miles

73. 21 mph going, 14 mph returning; total 84 miles

74. $y = -2x + 9; (2,5)$

75. $(1, 2, 3)$

76. $(2, 3, 0)$

77. $(5, -4, 3)$

78. $(2, -1, 2)$

79. $\left(7, \dfrac{20}{3}, \dfrac{13}{3}\right)$

80. $(3, -1, 1)$

81. 6, 2, 2

82. 8 nickels, 4 dimes, 6 quarters

83. $A = \$5000, B = \$3000, C = \$2000$

84. -73

85. $\dfrac{-25}{6}$

86. $\sqrt{2}$

87. 0

88. $1 - 2\sqrt{3}$

89. $1 + 6i$

90. -5

91. -206

92. $\dfrac{-619}{8}$

93. $-1 - i$

94. 0

95. $x = 2, y = -3$

96. $x = \dfrac{24}{11}, y = \dfrac{-43}{11}$

97. $x = \sqrt{2}, y = -1$

98. $x = 0, y = i$

99. $x = \dfrac{1}{a + b}, y = \dfrac{ab}{a + b}$

100. $x = \dfrac{a^2 + 2a + 4}{a + 2}, y = \dfrac{-2a}{a + 2}$

101. $(2, -1, 3)$

102. $(1, 4, -2)$

103. $(2, 0, -2)$

Chapter 7 Test   *pp. 337–338*

1.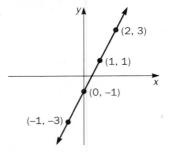

2.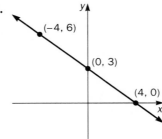

3. a.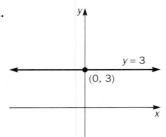

b.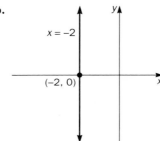

4. Neither

5. $x = \dfrac{1}{2}, y = -3$

6. $x = 2, y = 0$

7. $x = 2, y = 2$

8. $(-5, 28, 16)$

9. $x = 2, y = -6$

10. $(2, -3, 0)$

11. $y = 3x - 1$

12. $y = 2x + 1$

13. $y = \dfrac{3}{2}x + 2$

14. $(2, 5)$

15. $y = -2x + 8$

16. $(4, 0)$

17. 6 liters of 8%, 24 liters of 17%

18. $7500 at 6%, $2500 at 10%

19. 5 in. by 9 in.; 7 in. by 7 in.

20. 8 nickels, 7 dimes, 4 quarters

Chapter 8

Exercise 8.1   *pp. 348–351*

1.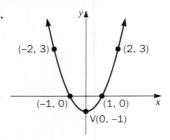

3.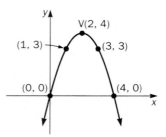

5.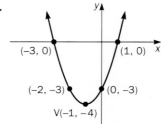

7.

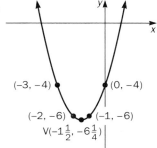

(−3, −4)   (0, −4)
(−2, −6)   (−1, −6)
$V(-1\frac{1}{2}, -6\frac{1}{4})$

9.

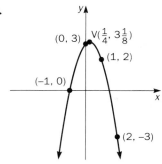

(0, 3) $V(\frac{1}{4}, 3\frac{1}{8})$
(1, 2)
(−1, 0)
(2, −3)

11.

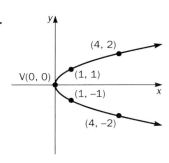

(4, 2)
V(0, 0)   (1, 1)
(1, −1)
(4, −2)

13.

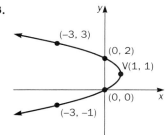

(−3, 3)
(0, 2)
V(1, 1)
(0, 0)
(−3, −1)

15.

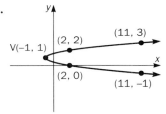

(11, 3)
V(−1, 1)   (2, 2)
(2, 0)
(11, −1)

17.

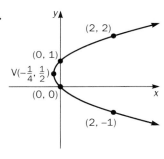

(2, 2)
(0, 1)
$V(-\frac{1}{4}, \frac{1}{2})$
(0, 0)
(2, −1)

19.

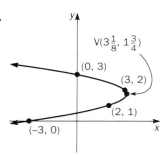

$V(3\frac{1}{8}, 1\frac{3}{4})$
(0, 3)
(3, 2)
(2, 1)
(−3, 0)

21. 2 sec; 144 ft

23. 250; $56,000

25. 125 ft by 250 ft; 31,250 ft^2

27. −6 and 6; $P = -36$

29. $x = 5$; $I = \$24,500$

31. 100 ft by 150 ft; 15,000 ft^2

33. 99 ft by 202 ft: $A = 19{,}998$ ft^2; 101 ft by 198 ft:
$A = 19{,}998$ ft^2

35. $y = 1x^2 + 2x + 1$

37. $x = 2y^2 - 1y - 6$

Exercise 8.2   *pp. 355–359*

1. $D = 5$ **3.** $D = 10$ **5.** $D = 6$ **7.** $D = 4\sqrt{2}$ **9.** $D = 2\sqrt{17}$

11.

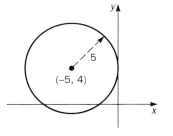

5
(−5, 4)

13.

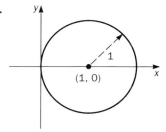

1
(1, 0)

15.

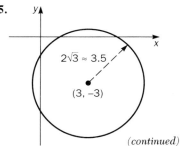

$2\sqrt{3} \approx 3.5$
(3, −3)

(continued)

Exercise 8.2, continued

17.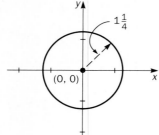

19.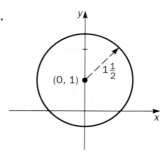

21.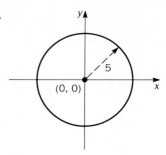

23. $(x - 2)^2 + (y - 5)^2 = 16$ **25.** $(x + 5)^2 + y^2 = 20$ **27.** $x^2 + (y - 4)^2 = \dfrac{4}{9}$ **29.** $(x - 3)^2 + (y + 1)^2 = 34$

31. $(x + 3)^2 + (y - 4)^2 = 16$ **33.** $(x - 1)^2 + (y - 2)^2 = 9$ **35.** $x^2 + (y - 5)^2 = 8$

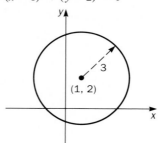

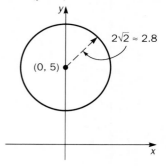

37. $(x - 3)^2 + y^2 = 12$

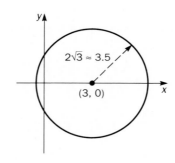

39. $\left(x - \dfrac{3}{2}\right)^2 + \left(y + \dfrac{5}{2}\right)^2 = \dfrac{21}{2}$

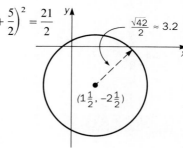

41. $D = \dfrac{\sqrt{485}}{8}$ **43.** $D = \dfrac{25}{24}$ **45.** $D = \dfrac{\sqrt{13}}{2}$

47. $D = 2\sqrt{3}$ **49.** $D = 3$ **51.** $D = 3\sqrt{2}$

53. Isosceles; $P = 4\sqrt{2} + 4\sqrt{5} + 4\sqrt{5} = 4\sqrt{2} + 8\sqrt{5}$

55. Isosceles, right triangle; $P = 5\sqrt{2} + 5\sqrt{2} + 10 = 10\sqrt{2} + 10$

57. Parallelogram, but not a rhombus: sides are $2\sqrt{10}$, $\sqrt{10}$, $2\sqrt{10}$, $\sqrt{10}$

59. $C(-4,1)$, $V(2,-3)$, $D = 2\sqrt{13}$, $y = \dfrac{-2x}{3} - \dfrac{5}{3}$ **61.** $D = 5a$

63. $D = 4c\sqrt{2}$ **65.** $D = 2\sqrt{m^2 + n^2}$ **67.** $D = |m - n|\sqrt{2}$

69. $x = 6, -2$ **71.** $y = 4 \pm 4\sqrt{3}$ **73.** $x = -3, 2$

75. $A = 25\pi$ **77.** $A = 20\pi$ **79.** $A = 16\pi$

81. a. $OA = \sqrt{(1 - 0)^2 + (a - 0)^2} = \sqrt{1 + a^2}$; similarly $OB = \sqrt{1 + b^2}$ and $AB = b - a$

b. $b^2 - 2ab + a^2 = 1 + a^2 + 1 + b^2$, or $-2ab = 2$, or $ab = -1$

c. $m_1 = \dfrac{b - 0}{1 - 0} = \dfrac{b}{1} = b$; similarly $m_2 = a$

d. Since $ab = -1$, substitution gives $m_1 m_2 = -1$

Exercise 8.3   *pp. 366–370*

1.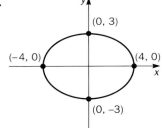

3.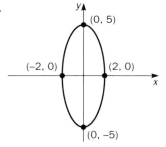

5.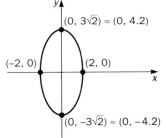

7.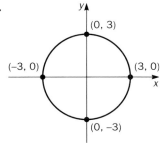

9.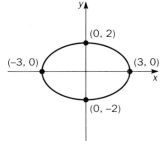

11.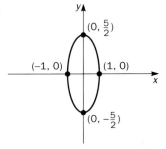

13.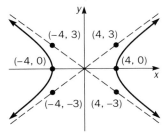

15.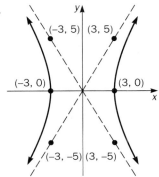

17.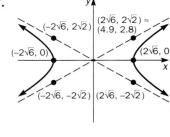

(continued)

Exercise 8.3, continued

19.
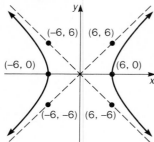

21.
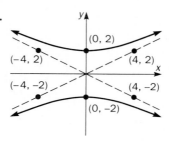

23.
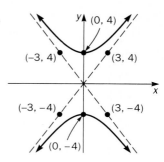

25.
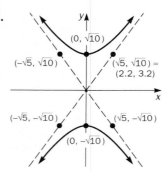

27. $\dfrac{x^2}{4} - \dfrac{y^2}{1} = 1$
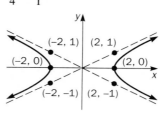

29. $\dfrac{y^2}{9/4} - \dfrac{x^2}{25} = 1$
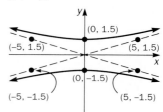

31. $\dfrac{y^2}{4} - \dfrac{x^2}{9} = 1$
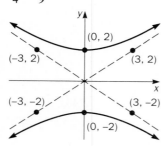

33. $\dfrac{x^2}{36} - \dfrac{y^2}{4} = 1$
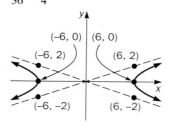

35. $\dfrac{x^2}{16} + \dfrac{y^2}{9} = 1$

37. $\dfrac{x^2}{16} - \dfrac{y^2}{9} = 1$

39. $\dfrac{y^2}{16} - \dfrac{x^2}{9} = 1$

41.
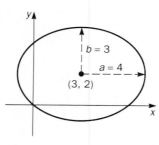

43.
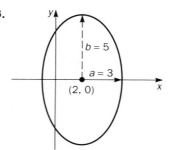

45.
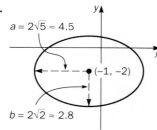

47. $A = 12\pi$ **49.** $A = 8\pi$ **51.** $A = \dfrac{3\pi}{2}$

53. Let $x = 0$: $\dfrac{y^2}{b^2} - \dfrac{0^2}{a^2} = 1$, or $y^2 = b^2$, or $y = \pm b$; y-intercepts are $(0,b)$ and $(0,-b)$.

Let $y = 0$: $\dfrac{0^2}{b^2} - \dfrac{x^2}{a^2} = 1$, or $x^2 = -a^2$, or $x = \pm ai$; no real x-intercepts.

55. $y = \dfrac{b}{a} x$ and $y = -\dfrac{b}{a} x$ for both hyperbolas

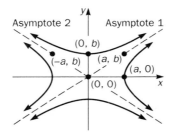

57. Two intersecting lines that are the asymptotes of all the hyperbolas formed

59. $y = \dfrac{\pm 5\sqrt{16 + x^2}}{4}$ **61.** $x^2 + y^2 = 25$ and $x^2 + y^2 = 9$

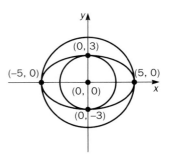

Exercise 8.4   *pp. 373–374*

1. $(12, 5)$, $(-5, -12)$ **3.** $(3, 0)$, $\left(0, \dfrac{3}{2}\right)$ **5.** $(2, 4)$, $(4, 2)$

7. $(1, -1)$, $\left(\dfrac{-13}{7}, \dfrac{-17}{7}\right)$ **9.** $(3, 5)$, $(-2, 10)$ **11.** $\left(\dfrac{3}{2}, \dfrac{23}{4}\right)$, $\left(\dfrac{-3}{2}, \dfrac{23}{4}\right)$

13. $(4, 3)$, $(4, -3)$, $(-4, 3)$, $(-4, -3)$ **15.** $(2, 1)$, $(2, -1)$, $(-2, 1)$, $(-2, -1)$

17. $\left(\sqrt{2}, \dfrac{13}{2}\right)$, $\left(-\sqrt{2}, \dfrac{13}{2}\right)$ **19.** $(3, 2)$, $(-3, -2)$, $(2, 3)$, $(-2, -3)$

21. $(3, -1)$, $(-3, 1)$, $(i, 3i)$, $(-i, -3i)$ **23.** $(-5, -8)$, $(2, -1)$

25. $\left(2, -\dfrac{1}{2}\right)$, $(-1, 1)$ **27.** $(0, 3)$, $(1, 0)$ **29.** $(4, 2)$, $(2, 4)$

(continued)

Exercise 8.4, continued

31. $(1, 0)$, $(0, -1)$

33. $(1 + \sqrt{2}, 1 - \sqrt{2})$, $(1 - \sqrt{2}, 1 + \sqrt{2})$

35. $(a, 3a)$, $(3a, a)$

37. $x = m + n$, $y = n$

39. $(p, p\sqrt{2})$, $(p, -p\sqrt{2})$, $(-p, p\sqrt{2})$, $(-p, -p\sqrt{2})$

41. $(2c, -c)$, $(-2c, c)$

Exercise 8.5   *pp. 376–379*

1. 2 in. by 7 in.

3. 3 and 6

5. 3 in. and 4 in.

7. 2 and 4

9. 15 mph, $\dfrac{1}{3}$ hr

11. Maserati: 170 kph, Ferrari: 200 kph; 2 hrs each

13. 2 and 4

15. $W = 20$ ft, $L = 50$ ft; $W = 25$ ft, $L = 40$ ft

17. $x = 3$, $y = 5$

19. 4 in. by 6 in.; 2 in. by 12 in.

21. 4 in., 8 in.

23. $\sqrt{2}$ in., $2\sqrt{2}$ in.

25. original: 60 passengers, \$400 each; actual: 40 passengers, \$600 each

Chapter 8 Review Exercises   *pp. 382–387*

1.
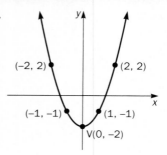

2.
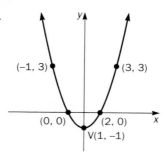

3.
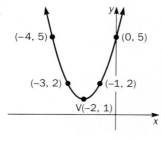

4.
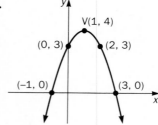

5.
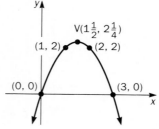

6.
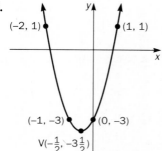

7.

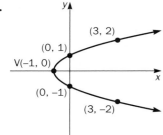

8.

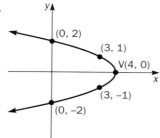

9.

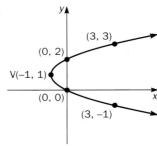

10.

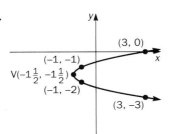

11. 3 sec, 164 ft

13. 3, −3; 18

15. $x = 100$ ft, $y = 300$ ft; $A = 30{,}000$ ft^2

16. c. $y = -2x^2 + 3x + 5$ **d.** $V\left(\dfrac{3}{4}, \dfrac{49}{8}\right)$

12. 6 and 6; 36

14. 7 in. by 7 in.; 49 sq. in.

17. $D = 5$

20. $D = 2\sqrt{17}$

18. $D = 13$

21. $D = \dfrac{5}{12}$

19. $D = 4\sqrt{2}$

22. $D = 2\sqrt{6}$

23.

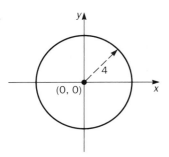

24.

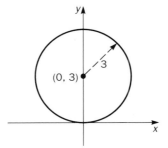

25.

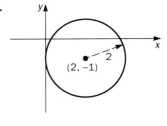

26.

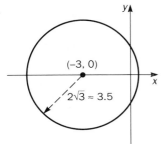

$2\sqrt{3} \approx 3.5$

27.

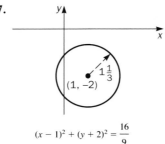

$(x - 1)^2 + (y + 2)^2 = \dfrac{16}{9}$

28.

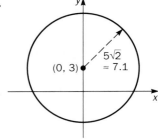

$x^2 + (y - 3)^2 = 50$

(continued)

Chapter 8 Review Exercises, continued

29. $(x - 2)^2 + (y + 3)^2 = 25$

30. $(x - 4)^2 + y^2 = 18$

31. $(x - 3)^2 + (y - 4)^2 = 10$

32. $(x + 4)^2 + (y - 4)^2 = 16$

33.

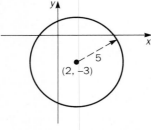

$(x - 2)^2 + (y + 3)^2 = 25$

34.

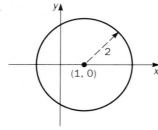

$(x - 1)^2 + y^2 = 4$

35.

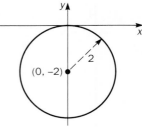

$x^2 + (y + 2)^2 = 4$

36.

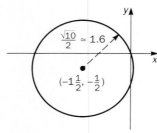

$\left(x + \dfrac{3}{2}\right)^2 + \left(y + \dfrac{1}{2}\right)^2 = \dfrac{5}{2}$

37. a. $5a^2 + 3b + c = -34$
  $-2b + c = -4$
  $a - 5b + c = -2b$
  b. $a = -10$, $b = 4$, $c = 4$
  c. $x^2 + y^2 - 10x + 4y + 4 = 0$
  d. $(x - 5)^2 + (y + 2)^2 = 25$, so $C = (5, -2)$, $r = 5$

38. $C(-1, 4)$, $V(3, 1)$; $D = 5$

39. $(0, 3)$, $C(4, -1)$; $D = 4\sqrt{2}$

40. a. $2\sqrt{17}$, $\sqrt{17}$, $\sqrt{85}$
  b. $(2\sqrt{17})^2 + (\sqrt{17})^2 = (\sqrt{85})^2$, or $68 + 17 \overset{\checkmark}{=} 85$
  c. $A = \dfrac{1}{2}(2\sqrt{17})(\sqrt{17}) = 17$

41. $(x + 1)^2 + (y - 3)^2 = 12$; $A = \pi r^2 = \pi(2\sqrt{3})^2 = 12\pi$

42. $x = 2, -6$

43. $x = 5, -3$

44. $A = \pi(2\sqrt{5})^2 - \pi(3)^2 = 11\pi$

45. a. $y = 2x$ **b.** $(2, 4)$ **c.** $D = 2\sqrt{5}$ art 8-67

46.

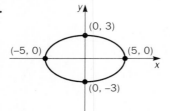

47.

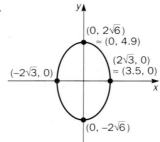

48.

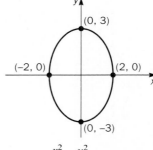

$\dfrac{x^2}{4} + \dfrac{y^2}{9} = 1$

49.

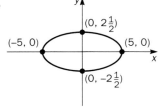

$$\frac{x^2}{25} + \frac{y^2}{25/4} = 1$$

50.

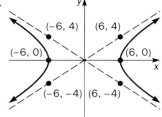

51.

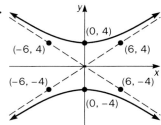

52.

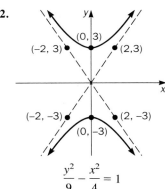

$$\frac{y^2}{9} - \frac{x^2}{4} = 1$$

53.

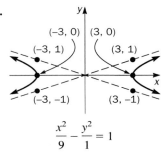

$$\frac{x^2}{9} - \frac{y^2}{1} = 1$$

54.

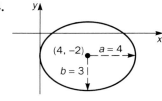

55. $\dfrac{x^2}{9} + \dfrac{y^2}{4} = 1$

56. $\dfrac{x^2}{2} + \dfrac{y^2}{20} = 1$

57. $\dfrac{x^2}{9} - \dfrac{y^2}{4} = 1$

58. $\dfrac{y^2}{16} - \dfrac{x^2}{9} = 1$

59. $(-1, 3), (-3, 1)$

60. $(-1, -3), (1, -1)$

61. $(2, \sqrt{3}), (2, -\sqrt{3}), (-2, \sqrt{3}), (-2, -\sqrt{3})$

62. $(-2, -4), \left(\dfrac{20}{3}, \dfrac{6}{5}\right)$

63. $(1, 3), (-1, -3), (3, 1), (-3, -1)$

64. $(2, -3), (-2, 3), \left(3\sqrt{3}, \dfrac{-2\sqrt{3}}{3}\right), \left(-3\sqrt{3}, \dfrac{2\sqrt{3}}{3}\right)$

65. $(2, 3), (3, 2)$

66. $(2, -1), (-1, 2)$

67. $\left(-9a, \dfrac{2a}{3}\right), (2a, -3a)$

68. 4 and 8

69. 6 and -1; 2 and -3

70. 3 and 2; -2 and -3

71. 4 and 6

72. 2 and 1; -2 and -1; i and $-2i$; $-i$ and $2i$

73. 20 ft by 30 ft

74. 40 ft by 15 ft; 30 ft by 20 ft

75. 10 ft by 60 ft; 30 ft by 20 ft

76. 3 in. by 3 in.; 4 in. by 4 in.; 6 in. by 6 in.

77. 30 mph, 2 hrs

(continued)

Chapter 8 Review Exercises, continued

78. 25 mph, $\frac{1}{2}$ hr

79. 2 mph; 2 hrs downstream, 3 hrs upstream

80. 5 mph; 5 hrs upstream, 4 hrs downstream

81. going: 60 mph, 5 hrs; returning: 50 mph, 6 hrs

82. 2 by 4

83. 2 by 2

84. (3, 2), (1, 6)

85. (2, 9), (6, 7)

86. 2 by 2, 3 by 3

87. $x = 12$, $y = 5$

Chapter 8 Test   *pp. 387–388*

1. $(x - 2)^2 + (y + 3)^2 = 8$

2. $\frac{x^2}{18} + \frac{y^2}{4} = 1$

3. $\frac{x^2}{9} - \frac{y^2}{4} = 1$

4. $\frac{y^2}{4} - \frac{x^2}{1} = 1$

5. $x^2 + (y + 2)^2 = 25$

6.

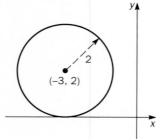

7.

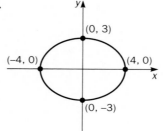

8.

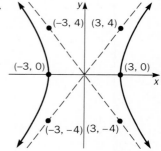

9.

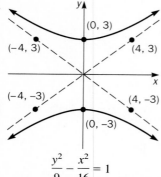

$$\frac{y^2}{9} - \frac{x^2}{16} = 1$$

10.

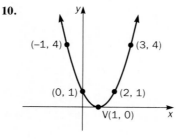

11.

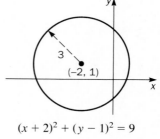

$$(x + 2)^2 + (y - 1)^2 = 9$$

12. (3,2), (−1,−6)

13. (5,18), (−1,6)

14. (1,2), (2,1)

15. $V(-1,2)$, $C(3,0)$; $D = 2\sqrt{5}$

16. 2 sec; 64 ft

17. 2 in. by 4 in.

18. 3 cm and 4 cm

19. 5 mph, $\frac{1}{2}$ hr

20. 3 in. by 3 in.; $A = 9$ sq. in.

Chapter 9

Exercise 9.1   *pp. 396–397*

1. not a function; $D = \{4, 5, 0\}$, $R = \{2, 3, 7, 9\}$ 3. function; $D = \{6, 5, -1, 4, 0\}$, $R = \{6, 3\}$

5. function; $D = \{x | x$ is any real number$\}$, $R = \{y | y$ is any real number$\}$

7. function; $D = \{x | x$ is any real number$\}$, $R = \{y | y \geq -1\}$

9. not a function; $D = \{x | x \leq 4\}$, $R = \{y | y$ is any real number$\}$

11. not a function; $D = \{x | -4 \leq x \leq 4\}$, $R = \{y | -2 \leq y \leq 2\}$

13. not a function; $D = \{x | x$ is any real number$\}$, $R = \{y | y \geq 5\} \cup \{y | y \leq -5\}$

15. $3, -7, \dfrac{17}{4}, 5\sqrt{2} + 3, 5x + 5h + 3, 5$

17. $34, 1, 9 + 3\sqrt{5}, 2 + \sqrt{2}, 2x^2 + 4xh + 2h^2 - 3x - 3h - 1, 4x + 2h - 3$

19. $4, -10, 4\sqrt{2}, 8 - 4\sqrt{2}, \dfrac{8}{x + h}, \dfrac{-8}{x(x + h)}$ 21. $-1, 179, 11, \dfrac{-17}{8}, 8 - \sqrt{3}$

23. $x^3 + 3x^2h + 3xh^2 + h^3, 3x^2 + 3xh + h^2$

25. function; $D = \{x | -4 \leq x \leq 2\}$, $R = \{y | -2 \leq y \leq 3\}$

27. not a function; $D = \{x | -3 \leq x < 4\}$, $R = \{y | -6 \leq y < -1\}$

29. function; $D = \{x | x \geq -2\}$, $R = \{y | y \geq 0\}$

31. not a function; $D = \{2\}$, $R = \{y | y$ is any real number$\}$

33. not a function; $D = \{x | -2 \leq x \leq 6\}$, $R = \{y | -7 \leq y \leq 1\}$

Exercise 9.2   *pp. 400–402*

1.
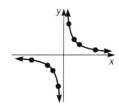

x	1	2	3	6	–1	–2	–3	–6
y	6	3	2	1	–6	–3	–2	–1

3.
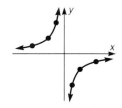

x	1	2	4	–1	–2	–4
y	–4	–2	–1	4	2	1

5.
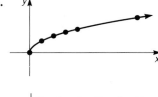

x	0	1	2	3	4	9
y	0	1	1.4	1.7	2	3

(continued)

Exercise 9.2, continued

7.

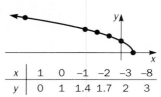

x	1	0	−1	−2	−3	−8
y	0	1	1.4	1.7	2	3

9.

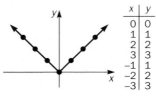

x	y
0	0
1	1
2	2
3	3
−1	1
−2	2
−3	3

11.

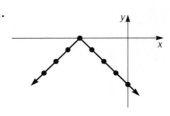

13.

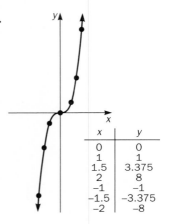

x	y
0	0
1	1
1.5	3.375
2	8
−1	−1
−1.5	−3.375
−2	−8

15.

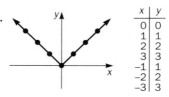

x	y
0	0
1	1
2	2
3	3
−1	1
−2	2
−3	3

x	0	1	−1	−2	−3	−4	−5	−6
y	−3	−4	−2	−1	0	−1	−2	−3

17.

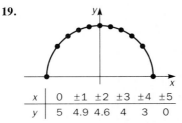

x	0	$\frac{1}{2}$	2	3	$-\frac{1}{2}$	−2	−3
y	undef.	1	1	1	−1	−1	−1

19.

x	0	±1	±2	±3	±4	±5
y	5	4.9	4.6	4	3	0

21. $\{x \mid x \neq -3, 2\}$

23. $\{x \mid x \neq \pm 2\}$

25. $\{x \mid x \neq \pm 2\sqrt{2}\}$

27. $\{x \mid x \geq 3\}$

29. $\{x \mid x \geq 2\} \cup \{x \mid x \leq -5\}$

31. $\{x \mid x \geq 2\} \cup \{x \mid x \leq -2\}$

33. $\{x \mid x \geq 4\} \cup \{x \mid x < -2\}$

35. a., b., c. $\{x \mid x \text{ is any real number}\}$ **37.** $\dfrac{1}{\sqrt{x+h} + \sqrt{x}}$

Exercise 9.3   *pp. 407–411*

1. $y = kx$; $k = 4$; $y = -20$

3. $v = ku^2$; $k = \dfrac{9}{4}$; $v = 81$

5. $F = kv^2$; $k = \dfrac{3}{2}$; $v = \dfrac{\pm 10\sqrt{3}}{3}$

7. $y = \dfrac{k}{x}$; $k = 24$; $y = \dfrac{-8}{3}$

9. $F = \dfrac{k}{d^2}$; $k = 1080$; $d = \pm 2\sqrt{3}$

11. $y = kxt^3$; $k = \dfrac{1}{2}$; $y = \dfrac{-27}{2}$

13. $z = \dfrac{kx}{y^2}$; $k = \dfrac{16}{9}$; $x = \dfrac{-27}{32}$

15. $T = kd$; $k = 2.5$; $T = 35$ lb; $T = 15$ lb; $d = 10$ in.

17. $I = \dfrac{k}{R^2}$; $k = 192$; $I = 3$ amps; $R = 2\sqrt{6} = 4.9$ ohms

19. $V = khr^2$; $k = \dfrac{\pi}{3}$; $V = \dfrac{\pi r^2 h}{3}$; $V = 108\pi$

21. $F = \dfrac{kmv^2}{r}$; $k = 0.3$; $F = 1800$ lb; $v = 25$ mph

23. 4 times as large; 9 times; $\dfrac{1}{4}$ times

25. $\frac{1}{4}$ times as large; $12\sqrt{2}$ times   **27.** $y = 12$ **29.** $P = 20$

Exercise 9.4   *pp. 417–420*

1. 61   **3.** -53 **5.** $3x^2 + 13$ **7.** $9x - 8$

9. 4   **11.** -5 **13.** $\sqrt{x} + 1$ **15.** 0

17. $4 - x$ **19.** $\dfrac{x - 3}{2}$ **21.** $4 - x$ **23.** $\dfrac{12x + 4}{3}$

25. $\sqrt[3]{x + 1}$ **27.** $\dfrac{1}{2}\sqrt[3]{x}$ **29.** $x^3 - 1$ **31.** $\dfrac{2}{x}$

33. $\dfrac{2x}{x - 1}$ **35.** $\dfrac{1}{x - 2}$

37. $R^{-1} = \{(-1,3), (0,-2), (-1,9), (4,-2)\}$; neither R nor R^{-1} is a function.

39.
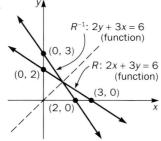

41.
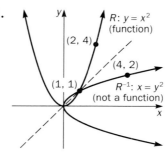

43.
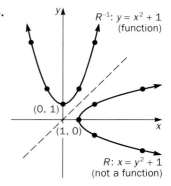

45.
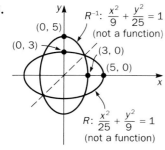

47.
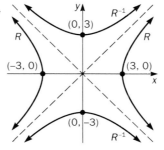

49.
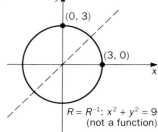

(continued)

Exercise 9.4, continued

51.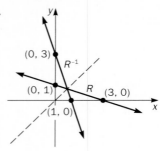

53.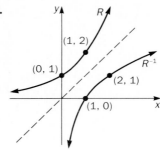

55.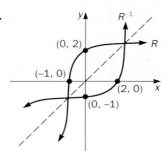

57.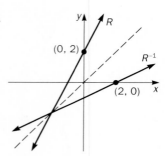

59.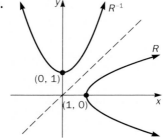

61.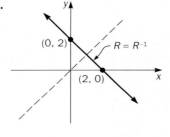

63. $f[g(x)] = g[f(x)] = x^6$; commutative

65. $f[g(x)] = g[f(x)] = x$; commutative

67. $f[g(x)] = g[f(x)] = x$; commutative

69. $\dfrac{x + 1}{x - 1}$

71. $\dfrac{1}{2}\sqrt[3]{x + 1}$

73. $\dfrac{\sqrt[3]{5x - 20}}{3}$

75. $\dfrac{x^3 + 1}{2}$

77. $\dfrac{3\sqrt[3]{2 - x}}{2}$

Chapter 9 Review Exercises   *pp. 424–426*

1. not a function; $D = \{5, -2, 8\}$, $R = \{0, 3, 4, 7\}$

2. function; $D = \{x|x \text{ is any real number}\}$, $R = \{y|y \text{ is any real number}\}$

3. function; $D = \{x|x \text{ is any real number}\}$, $R = \{y|y \leq 4\}$

4. not a function; $D = \{x|x \geq 2\}$, $R = \{y|y \text{ is any real number}\}$

5. not a function; $D = \{x|-5 \leq x \leq 5\}$, $R = \{y|-5 \leq y \leq 5\}$

6. not a function; $D = \{x|-4 \le x \le 4\}$, $R = \{y|-3 \le y \le 3\}$

7. $1, 17, -5, \dfrac{7}{3}, 2\sqrt{3} + 1, 2x + 2h + 1; 2$

8. $-2, 53, 18, \dfrac{-13}{4}, 3 + 2\sqrt{2}, 3x^2 + 6xh + 3h^2 - 4x - 4h - 2, 6x + 3h - 4$

9. $-2, \dfrac{3}{4}, -8, -3\sqrt{2}, -3\sqrt{3} - 3, \dfrac{6}{x(x + h)}$

10. $-4, -2, -52, \dfrac{-41}{8}, 4\sqrt{3} - 10$

11. $0, 1, 2, 2\sqrt{2}, \dfrac{1}{2}, \dfrac{\sqrt{6}}{2}$

12. $\dfrac{1}{9}, 4, \dfrac{9}{4}, \dfrac{1}{2}, \dfrac{1}{(x + h)^2}, \dfrac{-2x - h}{x^2(x + h)^2}$

13. $4x^3 + 12x^2h + 12xh^2 + 4h^3, 12x^2 + 12xh + 4h^2$

14. function; $D = \{x|-1 \le x \le 3\}$, $R = \{y|-2 \le y \le 4\}$

15. function; $D = \{x|x$ is any real number$\}$, $R = \{y|y > 0\}$

16. not a function; $D = \{x|x \ge 1\}$, $R = \{y|y$ is any real number$\}$

17.

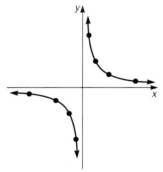

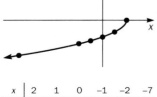

x	1	2	4	8	–1	–2	–4	–8
y	8	4	2	1	–8	–4	–2	–1

18.

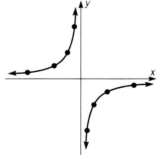

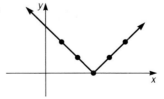

x	1	2	4	$\frac{1}{2}$	–1	–2	–4	$-\frac{1}{2}$
y	–2	–1	$-\frac{1}{2}$	–4	2	1	$\frac{1}{2}$	4

19.

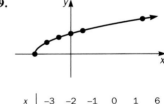

x	–3	–2	–1	0	1	6
y	0	1	1.4	1.7	2	3

20.

x	2	1	0	–1	–2	–7
y	0	–1	–1.4	–1.7	–2	–3

21.

x	0	1	2	3	4	5	6	–1
y	3	2	1	0	1	2	3	4

22.

x	0	1	2	3	–1	–2	–3	–4	–5
y	0	1	2	3	–1	–2	–1	0	1

(continued)

Chapter 9 Review Exercises, continued

23.

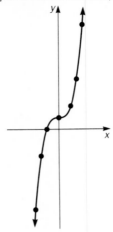

x	0	1	1.5	2	−1	−1.5	−2
y	1	2	4.375	9	0	−2.375	−7

24.

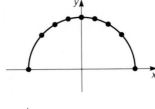

x	0	±1	±2	±3	±4
y	4	3.8	3.5	2.6	0

32. $\{x \mid x \leq 4\}$

34. $\{x \mid -3 \leq x \leq 3\}$

36. $\{x \mid x \geq 9\} \cup \{x \mid x \leq 0\}$

38. $\{x \mid x \geq 1\} \cup \{x \mid x < -3\}$

25. $\{x \mid x \neq 3\}$

26. $\{x \mid x \neq -7, 3\}$

27. $\{x \mid x \neq \pm 3\}$

28. $\{x \mid x \neq 0, 9\}$

29. $\{x \mid x \neq \pm 2\sqrt{3}\}$

30. $\left\{ x \mid x \neq \dfrac{\pm 5\sqrt{3}}{3} \right\}$

31. $\{x \mid x \geq 3\}$

33. $\{x \mid x \geq 3\} \cup \{x \mid x \leq -3\}$

35. $\{x \mid x \geq 6\} \cup \{x \mid x \leq 1\}$

37. $\{x \mid x > 2\}$

39. $\dfrac{2}{\sqrt{2x + 2h} + \sqrt{2x}}$

40. $y = kx;\ k = \dfrac{3}{2};\ y = 9$

41. $y = \dfrac{k}{x^2};\ k = 36;\ x = \pm 3\sqrt{2}$

42. $w = \dfrac{ku}{v^3};\ k = 8;\ w = \dfrac{-16}{9}$

43. $V = \dfrac{k}{P};\ k = 2.7;\ V = 2.25$ liter; $P = 1.0$ atm

44. $v = k\sqrt{d};\ k = 8;\ v = 64$ ft/sec; $v = 64\sqrt{2}$ ft/sec; $d = 25$ ft

45. $L = \dfrac{kwh^2}{\ell};\ k = 2400;\ L = 400$ lb

46. $W = \dfrac{k}{d^2};\ k = 2.8 \times 10^9;\ W = 158.7$ lb

47. $V = kT;\ k = \dfrac{V}{T};\ k = \dfrac{V_1}{T_1}$ and $k = \dfrac{V_2}{T_2}$, so $\dfrac{V_1}{T_1} = \dfrac{V_2}{T_2}$

48. -1

49. -33

50. 163

51. 723

52. 103

53. 19

54. $-1 - 8x^2$

55. $19 - 48x + 32x^2$

56. $16x - 9$

57. $8x^4 + 8x^2 + 3$

58. -201

59. 4

60. 3

61. -2

62. $4\sqrt{3}$

63. 8

64. $\dfrac{3}{2}$

65. $2\sqrt{2}$

66. $6\sqrt{2}$

67. $\dfrac{12}{x + 1}$

68. $\sqrt{x + 1}$

69. $\dfrac{2\sqrt{3x}}{x}$

70. $\dfrac{12\sqrt{x}}{x}$

71. $\dfrac{x + 2}{3}$

72. $\dfrac{1 - x}{2}$

73. $\dfrac{5x - 15}{2}$

74. x

75. $\sqrt[3]{x - 3}$

76. $\dfrac{1}{2}\sqrt[3]{x + 1}$

77. $\dfrac{4}{x}$

78. $\dfrac{3x}{x - 1}$

79. $\dfrac{2x + 1}{2 - x}$

80. $x^3 + 1$

81. $(x + 1)^3$

82.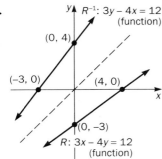
R^{-1}: $3y - 4x = 12$
(function)
(0, 4)
(-3, 0)
(4, 0)
(0, -3)
R: $3x - 4y = 12$
(function)

83.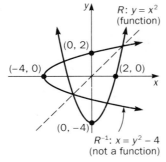
R: $y = x^2$
(function)
(0, 2)
(-4, 0)
(2, 0)
(0, -4)
R^{-1}: $x = y^2 - 4$
(not a function)

84.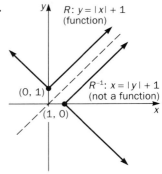
R: $y = |x| + 1$
(function)
(0, 1)
R^{-1}: $x = |y| + 1$
(not a function)
(1, 0)

85.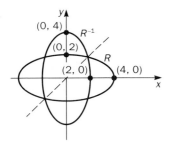
(0, 4)
R^{-1}
(0, 2)
R
(2, 0)
(4, 0)

R: $\dfrac{x^2}{16} + \dfrac{y^2}{4} = 1$ (not a function)

R^{-1}: $\dfrac{y^2}{16} + \dfrac{x^2}{4} = 1$ (not a function)

86.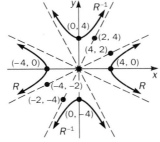
R^{-1}
(0, 4)
(2, 4)
(4, 2)
(-4, 0)
(4, 0)
R
(-4, -2)
R
(-2, -4)
(0, -4)
R^{-1}

R: $\dfrac{x^2}{16} - \dfrac{y^2}{4} = 1$ (not a function)

R^{-1}: $\dfrac{y^2}{16} - \dfrac{x^2}{4} = 1$ (not a function)

87.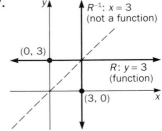
R^{-1}: $x = 3$
(not a function)
(0, 3)
R: $y = 3$
(function)
(3, 0)

Chapter 9 Test   *pp. 426–427*

1. function; $D = \{2, 3, 0, 5\}$, $R = \{-1, 2, 0\}$

2. not a function; $D = \{x | -4 \le x \le 4\}$, $R = \{y | -5 \le y \le 5\}$

3. function; $D = \{x | x$ is any real number$\}$, $R = \{y | y \ge -1\}$

4. $\{x | x \ne 4\}$

5. $\{x | x \ge 4\}$

6. $\{x | x$ is any real number$\}$

7. $\{x | x \ne \pm 2\}$

8. $\{x | x \ge 2\} \cup \{x | x \le -2\}$

9. $3, -7, \dfrac{11}{2}, 5$

10. $2, 0, \dfrac{15}{4}, 2 - \sqrt{2}, 2x + h - 3$

11. $1, 1, x^2 + 2x - 2$

12. $0, 5, x^2 + 4x$

13. $2 - x$

14. $\sqrt[3]{x + 8}$

15. $\dfrac{x}{2 - x}$

(continued)

Chapter 9 Test, continued

16.

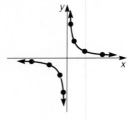

x	1	2	5	10	−1	−2	−5	−10
y	10	5	2	1	−10	−5	−2	−1

17.

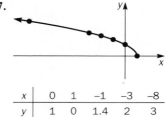

x	0	1	−1	−3	−8
y	1	0	1.4	2	3

18. $d = kt^2$; $k = 16$; $d = 576$ ft

19. $E = kmv^2$; $k = \dfrac{1}{2}$; $v = \pm 10$ m/sec

20. $d = \dfrac{kv^2}{a}$; $k = \dfrac{1}{2}$; $a = 28.125$

Chapters 1–9 Cumulative Review Exercises   *pp. 427–429*

1. $x = 12, 10$

2. $x = -15, -8$

3. $x = -30, 4$

4. $x = 24, 5$

5. $\dfrac{x+1}{x-2}$

6. $\dfrac{b^{12}c^9}{a^{15}}$

7. $\dfrac{a-1}{a(a^2-a+1)}$

8. $y = 1 \pm \sqrt{5}$

9. $z = \dfrac{3 + 3i\sqrt{3}}{4}$

10. $x = \pm \dfrac{5\sqrt{3}}{3}$

11. $x = \pm 2\sqrt{2}$

12. $\{x \mid 5 > x > -4\}$:

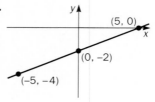

13. $\dfrac{-a-4}{2a}$

14. $x = 9$

15. $\dfrac{1}{4x^2}$

16. $3 + 2\sqrt{2}$

17. $(2,0)$

18. $(1,2,3)$

19. $(-1,-3), (3,1)$

20. $(6,4), (-8,-3)$

21. $\left(\dfrac{1}{2}, -\dfrac{3}{2}\right)$

22. $(-2,3,1)$

23.

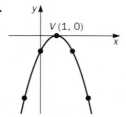

24.

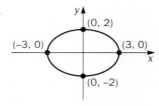

25.

26.

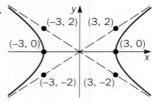

27.

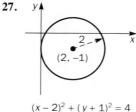

$(x-2)^2 + (y+1)^2 = 4$

28.

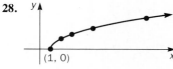

x	1	2	3	5	10
y	0	1	$\sqrt{2}$	2	3

29. 63   **30.** 3   **31.** $2x^2 + 4x - 3$ **32.** $4x^2 - 8x + 3$ **33.** 2

34. $2x + h + 2$ **35.** $f^{-1}(x) = \dfrac{5 - x}{10}$ **36.** $g^{-1}(x) = \sqrt[3]{2x - 8}$

37. $F^{-1}(x) = \dfrac{2x + 2}{1 - x}$ **38.** $y = -\dfrac{3}{2}x$ **39.** $(x - 1)^2 + y^2 = 12$

40. $(x + 3)^2 + (y - 1)^2 = 32; A = 32\pi$ **41.** $y = \dfrac{1}{2}x$ **42.** $k = 6; F = 3$

43. 38 chickens, 14 pigs   **44.** 2 by 4   **45.** 3 by 4

46. $7\% = 12\% = 3.75$ liters, $6.5\% = 7.5$ liters   **47.** $\dfrac{6}{8}$ $(N = 6, D = 8)$

48. going: 40 mph, 5 hrs; returning: 50 mph, 4 hrs   **49.** 3 nickels, 2 dimes, 4 quarters

50. 3 in. by 3 in.; maximum area = 9 sq. in.   **51.** original: 4 in. by 6 in.;   new: 8 in. by 3 in.

52. $V(2, -5), C(6,3); D = 4\sqrt{5}$ **53.** Line 1: $y = 2x - 3$; line 2: $y = -\dfrac{1}{2}x + 2$; $(2,1)$

54. $x = 9$; sides are 3, 4, 5 $(x = 1$ is extraneous$)$

Chapter 10

Exercise 10.1   *pp. 437–442*

1. 1.300   **3.** 0.275   **5.** 9.739   **7.** 0.178

9. 15.673   **11.** 0.604   **13.** 0.297   **15.** 4836.614

17. 14.880   **19.** 62,680.715   **21.** 33.954   **23.** 0.964

25. $3^{2x + 1}$ **27.** e^{3x} **29.** 5^{5x} **31.** 2^{16x}

33. $2^{4x} \cdot 3^{4x}$ or 6^{4x} **35.** $\dfrac{2^{2x} + 1}{2^x}$ **37.** $\dfrac{10^{2x} - 1}{10^{2x} + 1}$

39.
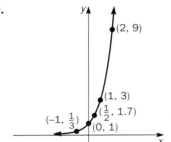

41.
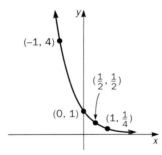

43.
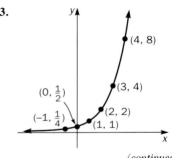

(continued)

Exercise 10.1, continued

45.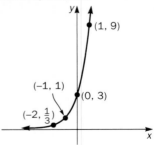

(1, 9)

(−1, 1)
(0, 3)
(−2, $\frac{1}{3}$)

47.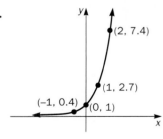

(2, 7.4)

(1, 2.7)

(−1, 0.4)
(0, 1)

49. 500,000; 1,000,000; 4,000,000; 11,313,709; 36,758,347   **51.** 1000; 707; 500; 353.6; 99.2; 9.8

53. 150,000,000; 202,478,000; 223,773,000; 237,611,000

55. $8042.19; $3042.19   **57.** $6008.83; $2408.83   **59.** $14,190.68; $4190.68

61. $350,910   **63.** 0 mph; 95.8 mph; 115.1 mph; 119.6 mph; *t.v.* = 120 mph

65. $6.00; $6.14; $6.17; $6.18; $6.18   **67.** $6651.58; $2651.58

69. $x = 4$, because $4^2 = 2^4$; and $x = 2$, because $2^2 = 2^2$

71. a. $x = 0$, because 0^0 is undefined   **b.** No, because $b^x > 0$ for $b > 0$, x any number

   c. If $b = 1$, then $f(x) = 1^x = 1$ for all x; this constant function represents a horizontal line, not a true exponential curve.

Exercise 10.2   *pp. 451–454*

1. $4^3 = 64$ **3.** $6^0 = 1$ **5.** $\left(\frac{1}{2}\right)^{-4} = 16$ **7.** $10^5 = 100,000$ **9.** $\log_3 9 = 2$

11. $\log_2 \frac{1}{32} = -5$ **13.** $\log_{1/2} \frac{1}{64} = 6$ **15.** $\log_{64} 16 = \frac{2}{3}$ **17.** $\ln 20.491 = 3.02$ **19.** $x = 81$

21. $x = 1$ **23.** $x = \frac{1}{8}$ **25.** $x = 2\sqrt{5}$ **27.** $x = \frac{1}{4}$ **29.** $x = 10,000$

31. $b = 5$ **33.** $b = 8$ **35.** $b = \frac{1}{2}$ **37.** $b = 64$ **39.** $y = 4$

41. $y = 1$ **43.** $y = -3$ **45.** $y = \frac{1}{2}$ **47.** $y = \frac{4}{3}$ **49.** $y = \frac{2}{3}$

51. $y = -\frac{1}{2}$ **53.** 5.14 (acidic)   **55.** 7.41 (basic)   **57.** 8.04 (basic)   **59.** $t = 15.27$; 1965

61. $x = 5\sqrt{5}$ **63.** $x = \frac{1}{8}$ **65.** $x = \frac{8}{27}$ **67.** $b = \frac{2}{3}$ **69.** $b = \frac{2}{3}$

71. $y = \frac{-3}{2}$ **73.** $y = \frac{3}{2}$ **75.** $y = -7$ **77.** 6.5   **79.** 8.9

81. 6.9   **83.** 11,200 yrs   **85.** 13.54 yrs   **87.** 79 db   **89.** 110 db

91. 85.6 db   **93.** 54.3 db   **95.** 1   **97.** 0   **99.** 3

Exercise 10.3   *pp. 460–462*

1. 49   **3.** 0.51   **5.** 7   **7.** x^2 **9.** 2

11. 3   **13.** 2.3   **15.** $\dfrac{1}{3}$

17. 3   **19.** $\log_6 5 + 2\log_6 x + \log_6 y$ **21.** $\log m + 2\log v - \log 2$

23. $2 \cdot \log_b x + 3 \cdot \log_b (x+1) - 4 \cdot \log_b (x-1)$ **25.** $\log_5 3 + 3\log_5 x + \dfrac{1}{2} \cdot \log_5 t - \log_5 2 - \dfrac{1}{3}\log_5 s$

27. $\dfrac{1}{2}(\ln 7 + 3 \cdot \ln x - 5 \cdot \ln y)$ **29.** $\log_3 \dfrac{7x}{y}$ **31.** $\log \dfrac{16x^5}{\sqrt{y}}$

33. $\log_b \dfrac{x^2 + 1}{(x-1)^2}$ **35.** $\log (x-1)$ **37.** 6^2 or 36   **39.** 4^{-2} or $\dfrac{1}{16}$

41. $\dfrac{\log 32}{\log 2}$ or 5   **43.** $\dfrac{\ln 5}{\ln 25}$ or $\dfrac{1}{2}$ **45.** $\dfrac{\log 36}{\log \dfrac{8}{3}}$ **47.** $x = 6$

49. $x = 2$ **51.** $x = 1$ ($x = -9$ is extraneous)   **53.** $x = 2$

55. $x = 9$ **57.** $x = 4$ **59.** $x = 2$ ($x = -4$ is extraneous)

61. $x = 253$ **63.** $x = 3$ ($x = -6$ is extraneous)   **65.** $x = 1$

67. $x = 4$ **69.** $x = 3$ ($x = -3, \pm 2i$ are extraneous)

71. $\log 2 + \log \pi - \log k + \dfrac{1}{2} \log L - \dfrac{1}{2} \log g$ **73.** $\ln \dfrac{x^2 \sqrt{y}}{z^3 \sqrt[4]{w}}$

75. In exponential form, $b^y = x$; substituting $y = \log_b x$ gives $b^{\log_b x} \overset{\checkmark}{=} x$.

77. $-\log_b \dfrac{x}{y} = -(\log_b x - \log_b y) \overset{\checkmark}{=} \log_b y - \log_b x$

Exercise 10.4   *pp. 466–470*

1. $x = 2$ **3.** $x = 12$ **5.** $x = -\dfrac{3}{8}$ **7.** $x = \dfrac{19}{12}$

9. $x = 4, -2$ **11.** $x = 25.793$ **13.** $t = -18.484$ **15.** $x = 3.877$

17. $x = 1.585$ **19.** $x = 2.760$ **21.** $t = 48.635$ **23.** 8.1 yrs; 1958

25. 13,854 yrs   **27.** 5 days   **29.** 11.6 yrs   **31.** $x = 0.168$

(continued)

Exercise 10.4, continued

33. $x = 12.114$

35. 0.0000195

37. 0.00000000631

39. 0.00000316

41. 2.585

43. 22.457

45. 0.874

47. $\log_2 4$ or   2

49. $\log_4 4$ or   1

51. $\log_5 25$ or   2

Chapter 10 Review Exercises   *pp. 472–476*

1. 4.120

2. 0.103

3. 0.881

4. 7334.973

5. 35.874

6. 220,850.97

7. e^{3x-2}

8. 3^{t+3}

9. $2^{3x} \cdot 5^{6y}$

10. $\dfrac{2^t + 1}{2^t}$

11. $\dfrac{10^{2x} + 1}{10^x}$

12. $\dfrac{e^{2x} - 1}{e^{2x} + 1}$

13.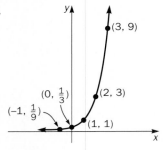

14.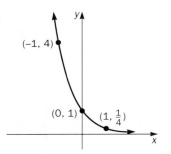

15.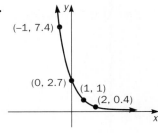

16. 2000; 8000; 32,000; 4,096,000; 49,667,000

17. 850,000; 422,098; 147,708

18. 100 g; 70.7 g; 50 g; 25 g; 8.84 g; 28 yrs

19. $22,510.96; $7510.96

20. $5637.46; $2637.46

21. $9042.33; $1542.33

22. $8665.53; $3665.53

23. $8666.27; $3666.27

24. 37°, 28.6°, 26.1°, 25.1°, 25°

25. $7^2 = 49$

26. $16^{3/4} = 8$

27. $10^{-4} = 0.0001$

28. $\log_5 125 = 3$

29. $\log_4 \dfrac{1}{16} = -2$

30. $\ln 181.27 = 5.2$

31. $x = 49$

32. $x = \dfrac{1}{625}$

33. $x = 4$

34. $x = \dfrac{1}{3}$

35. $x = \dfrac{1}{125}$

36. $x = 2\sqrt{3}$

37. $x = \dfrac{\sqrt{2}}{2}$

38. $b = 5$

39. $b = \dfrac{2}{3}$

40. $b = \dfrac{1}{7}$

41. $b = 16$

42. $b = 16$

43. $b = \dfrac{1}{9}$

44. $y = 5$

45. $y = 1$

46. $y = 3$

47. $y = -2$

48. $y = 0$

49. $y = -2$

50. $y = \dfrac{2}{3}$

51. $y = \dfrac{4}{3}$

52. $y = \dfrac{-3}{2}$

53. 6.21 (acidic)

54. 8.11 (basic)

55. 8.5

56. 0 sec, 2.2, 6.0

57. 105 db

58. 5

59. 7

60. 4.3

61. $\log_5 4 + 3 \log_5 x + \log_5 y - 2 \log_5 z$

62. $\log 2 + 2 \log s + \dfrac{1}{2} \log t - \log u - \dfrac{3}{4} \log v$

63. $\dfrac{1}{2}[\ln x + \ln (x^2 + 1) - \ln (x - 1)]$

64. $\log_b \dfrac{5x}{x - 5}$

65. $\log_5 \dfrac{x^2 y^3}{z}$

66. $\log \dfrac{8\pi \sqrt{C}}{r^2}$

67. $\ln \dfrac{x - 1}{x + 1}$

68. 4^2 or 16

69. 3^{-2} or $\dfrac{1}{9}$

70. $\dfrac{\log 12}{\log 8}$

71. $x = 2$

72. $x = 6$ ($x = -3$ is extraneous)

73. $x = 21$

74. $x = 8$

75. $x = 7$ ($x = -1$ is extraneous)

76. $x = 8$

77. $x = 4$ ($x = -2$ is extraneous)

78. $x = 2$

79. $\log_b b^x = x(\log_b b) = x(1) \stackrel{\checkmark}{=} x$

80. $b^{\log_b x} \stackrel{\checkmark}{=} x$ provided $\log_b x \stackrel{\checkmark}{=} \log_b x$

81. $x = 4$

82. $x = 0$

83. $x = -\dfrac{9}{5}$

84. $x = -\dfrac{5}{6}$

85. $x = \dfrac{10}{7}$

86. $x = -1$

87. $x = 2$

88. $t = 13.266$

89. $x = -1.258$

90. $t = 6.238$

91. $x = 1.771$

92. $x = 3.244$

93. $x = 18.612$

94. $t = -136.316$

95. $x = 0.756$

96. 30.5 yrs; 1990

97. 11,874 yrs

98. 11.6 yrs

99. 1.465

100. 0.341

101. 2.885

Chapter 10 Test   *pp. 476–477*

1. 125

2. 1.32

3. 4.14

4. 94.8 db

5.

6. $\log_3 2 + \log_3 x + 3 \log_3 y - \log_3 5 - \dfrac{1}{2} \log_3 z$

7. $\log \dfrac{x^2 y}{(x - y)^3}$

8. $x = 64$

9. $x = 2\sqrt{2}$

10. $x = \dfrac{1}{3}$

11. $x = 4$ ($x = -2$ is extraneous)

12. $x = 8$ ($x = -2$ is extraneous)

13. $x = -2$

14. $t = 11.76$

15. $x = 1.585$

16. 20,406 yrs

17. 86,086

18. 1979 ($t = 14.79$)

19. $31,706.04; $6706.04

20. 18.4 yrs

Chapter 11

Exercise 11.1

pp. 481–483

1. $-1, 1, 3, 5, \cdots$ **3.** $0, -2, -6, -12, \cdots$ **5.** $4, \dfrac{9}{2}, \dfrac{16}{3}, \dfrac{25}{4}, \cdots$ **7.** $4, -8, 16, -32, \cdots$

9. $-\dfrac{1}{2}, \dfrac{1}{4}, -\dfrac{1}{8}, \dfrac{1}{16}, \cdots$ **11.** $-4, -4, -4, -4, \cdots$ **13.** $2, 0, 2, 0, \cdots$ **15.** $a_n = 4n$

17. $a_n = \dfrac{n^2}{n + 1}$ **19.** $a_n = \left(-\dfrac{1}{2}\right)^n$ **21.** $a_n = \dfrac{n + 1}{n}$ **23.** 45

25. 76   **27.** 440   **29.** $\dfrac{49}{20}$ **31.** -44

33. 242   **35.** $\dfrac{63}{8}$ **37.** $\dfrac{5}{6}$ **39.** 0

41. $a_n = \dfrac{2^n}{3^n}$ or $\left(\dfrac{2}{3}\right)^n$ **43.** $a_n = (\sqrt{3})^n$ **45.** $a_n = \dfrac{(n + 1)^2}{n}$ **47.** $a_n = (n + 1)^3$

49. $\log\dfrac{2}{1} + \log\dfrac{3}{2} + \cdots + \log\dfrac{10}{9} = \log\left(\dfrac{2}{1} \cdot \dfrac{3}{2} \cdot \dfrac{4}{3} \cdot \dfrac{5}{4} \cdot \dfrac{6}{5} \cdot \dfrac{7}{6} \cdot \dfrac{8}{7} \cdot \dfrac{9}{8} \cdot \dfrac{10}{9}\right) = \log\dfrac{10}{1} = \log 10 \overset{\checkmark}{=} 1$

51. a. Each term after the second is the sum of the previous two terms; the next eight terms are 34, 55, 89, 144, 233, 377, 610, 987, $\cdots$

b. 2854 is; 4183 is not   **c.**

June:
8 stems

July:
13 stems

d. The plant will have 233 stems in January of the second year.

53. These are the prime numbers; the next ten primes are 19, 23, 29, 31, 37, 41, 43, 47, 53, 59, $\cdots$

Exercise 11.2

pp. 487–490

1. 247   **3.** 415   **5.** -335 **7.** not an A.S.

9. $\dfrac{85}{4} = 21\dfrac{1}{4}$ **11.** 5   **13.** not an A.S.   **15.** -45.37

17. not an A.S.   **19.** 5610   **21.** $-10,605$ **23.** 13,400

25. 3775   **27.** 945   **29.** 2990   **31.** 2340

33. -8250 **35.** 62,750   **37.** $465\sqrt{2}$ **39.** $6400c$

41. 304 ft; 1600 ft   **43.** 410 logs; 55 logs

45. a. 18th yr   **b.** $131,000   **c.** Smith: $1,899,000; Jones: $1,593,000   **47.** $600,000; 20 yrs

49. $\dfrac{n(a_1 + a_n)}{2} = \dfrac{n[a_1 + a_1 + (n-1)d]}{2} \overset{\checkmark}{=} \dfrac{n[2a_1 + (n-1)d]}{2}$ **51.** $-2, 4, 10, 16, 22, 28, 34, 40, \cdots$

53. $2, 2\frac{1}{2}, 3, 3\frac{1}{2}, 4, 4\frac{1}{2}, 5, 5\frac{1}{2}, 6, 6\frac{1}{2}, \cdots$ **55.** $n = 8$ **57.** $n = 10$

59. Yes   **61.** No   **63.** $40,000; $2,200,000

Exercise 11.3   *pp. 496–499*

1. 384   **3.** -512 **5.** $\dfrac{256}{2187}$ **7.** not a G.S.

9. not a G.S.   **11.** $\dfrac{2187}{512}$ **13.** $\dfrac{-32}{729}$ **15.** $\dfrac{405}{16}$ or $25\dfrac{5}{16}$

17. 2814.20   **19.** $16\sqrt{2}$ **21.** 19,680   **23.** $\dfrac{547}{729}$

25. 699,050   **27.** $\dfrac{1995}{32}$ **29.** 161,125.76   **31.** x^{34}

33. $\dfrac{x^{10}}{y^{15}}$ **35.** $\dfrac{-1}{16}$ **37.** $r = 2^2$ or 4

39. a. $10,737,418   **b.** $17,433,922   **c.** $12,000,000;   so **b.** > **c.** > **a.**

41. a. 1.6 ft   **b.** 43.2 ft   **43. a.** $49,975.12   **b.** $362,164.06

45. $\dfrac{5}{3}, 5, 15, 45, 135, 405, \cdots$ *or* $-\dfrac{5}{3}, 5, -15, 45, -135, 405, \cdots$

47. $3, 3\sqrt{2}, 6, 6\sqrt{2}, 12, 12\sqrt{2}, \cdots$ *or* $3, -3\sqrt{2}, 6, -6\sqrt{2}, 12, -12\sqrt{2}, \cdots$ **49.** $3, 6, 12, 24, 48, 96, \cdots$

51. a. $\dfrac{-3}{2}, 3, -6, 12, -24, 48, -96, \cdots$ **b.** $12, 3, -6, -15, -24, -33, -42, \cdots$ **53.** $126 + 63\sqrt{2}$

55. $\dfrac{30 - 15\sqrt{2}}{16}$

Exercise 11.4   *pp. 504–507*

1. 2   **3.** $\dfrac{2}{5}$ **5.** $\dfrac{4}{3}$ **7.** $\dfrac{9}{2}$

9. $\dfrac{1}{3}$ **11.** $\dfrac{4}{33}$ **13.** $\dfrac{41}{333}$ **15.** $\dfrac{16}{45}$

17. $2\sqrt{2} + 2$ **19.** $\dfrac{3 - \sqrt{3}}{2}$ **21.** $\dfrac{e}{e + 1} = 0.731$

(continued)

Exercise 11.4, continued

23. $0.999999\cdots = .9 + .09 + .009 + \cdots = \dfrac{.9}{1 - .1} = \dfrac{.9}{.9} = 1 = 1.000000\cdots$; it *is* true   **25.** 1000 feet

27. $r = \dfrac{-2}{3}$ **29.** $1 + \dfrac{1}{4} + \dfrac{1}{16} + \dfrac{1}{64} + \cdots = \dfrac{4}{3}$

31. a. $2, 2, 2, 2, \cdots$ **b.** $2n$ **c.** sum is infinite   **33.** 1

35. $8 + 4\sqrt{2} + 4 + 2\sqrt{2} + \cdots = 16 + 8\sqrt{2}$

Exercise 11.5   *pp. 511–512*

1. $a^3 + 3a^2 + 3a + 1$ **3.** $32x^5 - 80x^4y + 80x^3y^2 - 40x^2y^3 + 10xy^4 - y^5$

5. $a^7 - 7a^5 + 21a^3 - 35a + \dfrac{35}{a} - \dfrac{21}{a^3} + \dfrac{7}{a^5} - \dfrac{1}{a^7}$ **7.** $16x^4 + 32x + \dfrac{24}{x^2} + \dfrac{8}{x^5} + \dfrac{1}{x^8}$

9. $x^{15} + 15x^{14} + 105x^{13} + 455x^{12} + \cdots$ **11.** $a^{36} - 36a^{34} + 612a^{32} - 6528a^{30} + \cdots$

13. $\cdots + \dfrac{78}{x^9} + \dfrac{13}{x^{11}} + \dfrac{1}{x^{13}}$ **15.** $1001a^{24}$ **17.** $3060x^4h^{14}$ **19.** $252a^5b^5$

21. $1 + 8(.01) + 28(.01)^2 + 56(.01)^3 = 1.082856$; *by calculator*, $(1.01)^8 = 1.0828567$

23. a. $1 + 6 + 15 + 20 + 15 + 6 + 1 = 64 \overset{\checkmark}{=} 2^6$

    b. $2^6 = (1 + 1)^6 = 1^6 + 6 \cdot 1^5 \cdot 1 + 15 \cdot 1^4 \cdot 1^2 + 20 \cdot 1^3 \cdot 1^3 + 15 \cdot 1^2 \cdot 1^4 + 6 \cdot 1 \cdot 1^5 + 1^6$

    $\overset{\checkmark}{=} 1 + 6 + 15 + 20 + 15 + 6 + 1$

25. $577 + 408\sqrt{2}$ **27.** $8i$ **29.** $5x^4 + 10x^3h + 10x^2h^2 + 5xh^3 + h^4$

31. $12x^2 + 12xh + 4h^2$ **33.** $4x^3 + 6x^2h + 4xh^2 + h^3 + 3x^2 + 3xh + h^2$

Chapter 11 Review Exercises   *pp. 515–520*

1. $0, 1, 4, 9, \cdots$ **2.** $0, 1, 2, 3, \cdots$ **3.** $2, 2\sqrt{2}, 4, 4\sqrt{2}, \cdots$

4. $a_n = \dfrac{n}{(n + 1)^2}$ **5.** $a_n = \dfrac{n}{2n - 1}$ **6.** $a_n = \dfrac{n^2 + 1}{n}$ **7.** $\dfrac{71}{20}$

8. 252   **9.** 26   **10.** $121 + 121\sqrt{3}$ **11.** $\log 720$

12. $-1 + i$ **13.** 220   **14.** -173 **15.** not an A.S.

16. $\dfrac{497}{2}$ or $248\dfrac{1}{2}$ **17.** $\dfrac{123}{12}$ or $10\dfrac{1}{4}$ **18.** 18.9   **19.** $10\sqrt{3}$ or $\sqrt{300}$

20. $\log 39,366$ **21.** $239x$ **22.** $40x + 79$ **23.** 15,050

24. $-84,050$ **25.** 1925   **26.** 9560   **27.** -3450

28. 10,590   **29.** $\dfrac{46375}{6}$ **30.** $120\sqrt{5}$ or $\sqrt{72000}$ **31.** $78 \log 3$

32. $1600x + 820$ **33.** 1984   **34.** 19,845   **35.** -3425

36. $\dfrac{14705}{2}$ or $7352\dfrac{1}{2}$ **37.** \$73,500; \$710,500

38. a. 17 yrs   **b.** \$76,000   **c.** *A:* \$952,000; *B:* \$816,000

39. 1, 5, 9, 13, 17, 21, 25, $\cdots$ **40.** 2.8, 2.4, 2, 1.6, 1.2, 0.8, 0.4, 0, -0.4, $\cdots$

41. $n = 10$ **42.** $n = 7$

43. a. $10 = a_1 + 3d$, $43 = a_1 + 14d$ **b.** $a_1 = 1$, $d = 3$ **c.** 1, 4, 7, 10, 13, 16, 19, 22, 25, 28, 31, 34, 37, 40, 43, $\cdots$

44. $a_{10} = 29$, $d = 3$ **45.** 16,384   **46.** $\dfrac{1}{256}$ **47.** $\dfrac{2}{2187}$

48. $\dfrac{-177147}{65536}$ **49.** not a G.S.   **50.** $32\sqrt{2}$ **51.** not a G.S.

52. $-i$ **53.** 9774.22   **54.** $\dfrac{y^7}{x^{15}}$ **55.** 32 log 3

56. $\dfrac{511}{4}$ **57.** $\dfrac{1852}{343}$ **58.** 9840   **59.** -1026

60. 75,129.03   **61.** $30 + 15\sqrt{2}$ **62.** Plan B; Plan B   **63.** \$73,121.56; \$441,138.95

64. a. $\dfrac{1}{128}$ ft   **b.** $7\dfrac{127}{128}$ ft   **c.** $\dfrac{1}{128}$ ft   **65. a.** $\dfrac{729}{64} = 11.39$ ft   **b.** 443.66 ft

66. $\dfrac{2}{3}$, 2, 6, 18, 54, 162, $\cdots$ *or* $\dfrac{-2}{3}$, 2, -6, 18, -54, 162, $\cdots$

67. 3, -6, 12, -24, 48, -96, 192, $\cdots$

68. $2, 2\sqrt{2}, 4, 4\sqrt{2}, 8, 8\sqrt{2}, \cdots$ *or* $2, -2\sqrt{2}, 4, -4\sqrt{2}, 8, -8\sqrt{2}, \cdots$

69. 27, 18, 12, 8, $\dfrac{16}{3}$, $\dfrac{32}{9}$, $\cdots$ *or* -27, 18, -12, 8, $\dfrac{-16}{3}$, $\dfrac{32}{9}$, $\cdots$

70. a. $\dfrac{3}{2}$, 3, 6, 12, 24, 48, 96, $\cdots$ **b.** -4, 3, 10, 17, 24, 31, 38, $\cdots$

71. $a_1 = 4$ **72.** 8   **73.** $\dfrac{9}{2}$ **74.** $\dfrac{16}{7}$

75. $4 + 2\sqrt{2}$ **76.** 12,500   **77.** $\dfrac{1}{e - 1} = 0.582$ **78.** $\dfrac{4}{9}$

79. $\dfrac{4}{11}$ **80.** $\dfrac{44}{111}$ **81.** $\dfrac{17}{30}$ **82.** $\dfrac{218}{165}$

83. $\dfrac{1}{30}$ **84.** 8 feet   **85.** 512 feet   **86.** $r = -\dfrac{1}{2}$

87. $r = \dfrac{2}{3}$ **88.** $r = \dfrac{\sqrt{2}}{2}$ **89.** $2\pi + \pi + \dfrac{\pi}{2} + \dfrac{\pi}{4} + \cdots = 4\pi$

90. $\pi + \dfrac{\pi}{4} + \dfrac{\pi}{16} + \dfrac{\pi}{64} + \cdots = \dfrac{4\pi}{3}$ **91.** $x^3 + 3x^2h + 3xh^2 + h^3$

(continued)

Chapter 11 Review Exercises, continued

92. $a^4 + 8a^3b + 24a^2b^2 + 32ab^3 + 16b^4$

93. $x^{12} + 6x^{11} + 15x^{10} + 20x^9 + 15x^8 + 6x^7 + x^6$

94. $x^5 - 5x^3 + 10x - \dfrac{10}{x} + \dfrac{5}{x^3} - \dfrac{1}{x^5}$

95. $128x^7 - 448x^6 + 672x^5 - 560x^4 + 280x^3 - 84x^2 + 14x - 1$

96. $577 - 408\sqrt{2}$

97. $a^{32} - 16a^{31} + 120a^{30} - 560a^{29} + \cdots$

98. $x^{14} + 28x^{13}y + 364x^{12}y^2 + 2912x^{11}y^3 + \cdots$

99. $\cdots + \dfrac{3640}{x^9} + \dfrac{420}{x^{11}} + \dfrac{30}{x^{13}} + \dfrac{1}{x^{15}}$

100. $\cdots + \dfrac{190}{a^{14}} - \dfrac{20}{a^{17}} + \dfrac{1}{a^{20}}$

101. $715x^{18}$ **102.** $3060x^4$ **103.** $70x^4$ **104.** $3x^2 + 3xh + h^2$

105. $10x^4 + 20x^3h + 20x^2h^2 + 10xh^3 + 2h^4$

106. $6x^5 + 15x^4h + 20x^3h^2 + 15x^2h^3 + 6xh^4 + h^5 + 2x + h$

107. $1 + 9(.1) + 36(.1)^2 + 84(.1)^3 + 126(.1)^4 = 2.3566$; *by calculator,* $(1.1)^9 = 2.3579477$

108. $(x^2)^3 + 3(x^2)^2(x + 1) + 3(x^2)^1(x + 1)^2 + (x + 1)^3 = x^6 + 3x^5 + 6x^4 + 7x^3 + 6x^2 + 3x + 1$

109. $1, 1, 2, 3, 5, 8, 13, 21$; then $34, 55, 89, 144, 233, 377, \cdots$ (Exercise 11-1, problem 51)

Chapter 11 Test   *pp. 520–521*

1. 398   **2.** 1536   **3.** $\dfrac{1000}{11}$ **4.** $\dfrac{-6561}{256}$

5. $16\sqrt{2}$ or $\sqrt{512}$ **6.** -442 **7.** $\dfrac{16}{3}, 8, 12, 18, 27, \dfrac{81}{2}, \cdots$

8. $\dfrac{5}{3}, 8, \dfrac{43}{3}, \dfrac{62}{3}, 27, \dfrac{100}{3}, \cdots$ **9.** 11,000   **10.** $\dfrac{4}{3}$ **11.** $\dfrac{169}{20}$

12. 3725   **13.** 19,600   **14.** $\dfrac{4}{3}$ **15.** $2 + 2\sqrt{2}$

16. $a^{14} - 7a^{13} + 21a^{12} - 35a^{11} + 35a^{10} - 21a^9 + 7a^8 - a^7$ **17.** $x^{12} + 24x^{10} + 264x^8 + 1760x^6 + \cdots$

18. $\dfrac{17}{33}$ **19.** $74,500; $597,000   **20.** $63,981.37; $385,996.58

Final Examination

Final Exam 1   *pp. 522–523*

1. $\dfrac{1}{b^{34}}$ **2.** $\dfrac{x - 4}{4}$ **3.** $x = \pm 3, \pm i\sqrt{2}$ **4.** $\dfrac{a^2 - a + 1}{a - 1}$

5. $-\sqrt{3}$ **6.** $y = \dfrac{3}{2}$ **7.** $\dfrac{5}{x - 2}$ **8.** $x = -\dfrac{4}{3}, 2$

9. $2 + \sqrt{3}$ **10.** $\{x \mid -1 \geq x \geq -6\}$:   **11.** $(3, 2), \left(\dfrac{-23}{3}, \dfrac{22}{3}\right)$

12. $x = a - 2b$

13. $x = 4$

14. $(2, -3, 1)$

15. $x = 0.771$

16. $14 + 8i$

17. $\left\{ x \mid 1 > x > \dfrac{1}{3} \right\}$: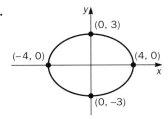

18. $x = -\dfrac{3}{2}, \ \pm 2$

19.

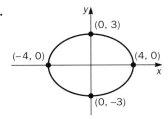

20.

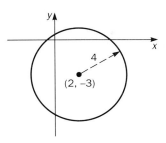

21. $y = \dfrac{-3x}{2}$

22. 31,610

23. a. 5 by 7
b. 4 by 8

24. \$15,000 at 9%, \$5000 at 12%

25. a. 132,558
b. 1981 ($t = 21.3$)

Final Exam 2   *pp. 523–524*

1. $\dfrac{x^{24}}{y^{12}}$

2. $-8x\sqrt{2x}$

3. $\dfrac{-(a + 1)}{2a(2a + 1)}$

4. $x = \dfrac{1 \pm \sqrt{5}}{2}$

5. $\dfrac{1}{125x^6}$

6. $\{x \mid x \geq 6\} \cup \{x \mid x \leq -1\}$:

7. $\dfrac{2x - 1}{x^2 - 3x + 9}$

8. $b = 25$

9. $x = a + b$

10. $x = 25$ ($x = -1$ is extraneous)

11. $\dfrac{1}{x + 12}$

12. $x = \dfrac{19}{11}, \ y = \dfrac{6}{11}$

13. $x = -4, 2, 0, -2$

14. $x = -2, 1, 1 \pm i\sqrt{3}, \dfrac{-1 \pm i\sqrt{3}}{2}$

15.

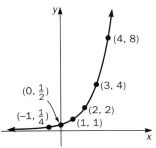

16.

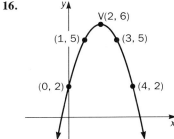

17. $x^{14} + 14x^{12} + 91x^{10} + 364x^8 + \dots$

18. $\dfrac{8}{37}$

(continued)

Final Exam 2, continued

19. a. $21,749.22; $6749.22
  b. 9.3 yrs

20. $-7, 13, -5, 2 + 7\sqrt{2}, 4x + 2h + 3$

21. $W = \dfrac{kxy^2}{z^3}; k = 18; W = -108$

22. $y = \dfrac{-3x}{5} + \dfrac{17}{5}; y = \dfrac{5x}{3} + \dfrac{17}{3}$

23. 4 by 9; 3 by 12; 6 by 6   FE-10

24. $x = 6; y = 9$

25. $r = \sqrt{2}; A = 2\pi - 4 = 2.283$ FE-11

Final Exam 3   *pp. 525–526*

1. $\dfrac{e^{2x} + 1}{e^{2x} - 1}$

2. $a^{1/3}$

3. $\dfrac{4(x - 1)}{x + 3}$

4. $\{x|x \le -4\} \cup \{x|x \ge -1\}$: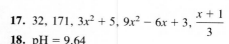

5. $-\dfrac{1}{48}$

6. $x = \dfrac{8}{9}$

7. $\{x|1 < x < 5\}$: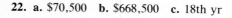

8. $5x^4 + 10x^3h + 10x^2h^2 + 5xh^3 + h^4$

9. $\dfrac{a^2 + 2a + 4}{a(a + 2)}$

10. $x = \dfrac{\pm 5\sqrt{2}}{2}$

11. $x = 3$ ($x = 11$ is extraneous)

12. (2,4), (4,2)

13. $x = \dfrac{1}{9}$

14. $g = \dfrac{4\pi^2 L}{P^2}$

15.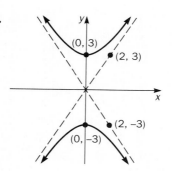

16.

17. $32, 171, 3x^2 + 5, 9x^2 - 6x + 3, \dfrac{x + 1}{3}$

18. pH = 9.64

19. $\dfrac{1}{3}$

20. 34,425

21. 20 mph going, 15 mph returning; 30 miles

22. a. $70,500   **b.** $668,500   **c.** 18th yr

23. $5\dfrac{1}{3}$ liters

24. a. $\dfrac{64}{27}$ ft   **b.** $\dfrac{1330}{27}$ ft   **c.** 54 ft

25. $x = 4$; sides are 4, 3, 5

Selected Topics

Exercise A   *p. A4*

1. {9}

3. {8, −8}

5. $\left\{\dfrac{5\sqrt{2}}{2}, \dfrac{-5\sqrt{2}}{2}\right\}$

7. {53, 59, 61, 67, 71, 73, 79, 83, 89, 97}

9. {1, 8, 27, 64, 125, 216, 343, ...}

11. {Alabama, Alaska, Arizona, Arkansas}

13. {Lincoln, Garfield, McKinley, Kennedy}

15. {1, 2, 3, 4}

17. {3, 4, 5, 7, 8}

19. {3}

21. {3}

23. {4, 5}

25. {2, 3, 4, 5, 7, 8}

27. {1, 2, 4}

29. {2, 4}

31. True

33. True

35. True

37. False

39. True

41. False

43. True

45. True

47. {1, 2} ∩ {3, 4} = {∅} is False; {1, 2} ∩ {3, 4} = ∅ is True. Note: the set {∅} consists of the single element ∅; hence, {∅} is *not* the null set ∅.

Exercise B   *pp. A6–A7*

1. $2x - 5$

3. $3x + 5 + \dfrac{3}{x - 2}$

5. $5 + \dfrac{2}{x + 1}$

7. $4x^2 - 3x - 2$

9. $3x^2 + 2x - 11 + \dfrac{24}{x + 3}$

11. $4x^3 + 5x - \dfrac{3}{x + 2}$

13. $2x^2 + 2x + 9$

15. $x^2 - \dfrac{3}{x + 2}$

17. $2x^3 + 4x^2 + 13x + 17 + \dfrac{35}{x - 2}$

19. $x^3 - 2x^2 + 5x - 7 + \dfrac{14}{x + 2}$

21. $-2x^2 - 8x - 27 - \dfrac{108}{x - 4}$

23. $x^2 - 2x + 4$

25. $x^4 + x^3 + x^2 + x + 1$

27. $x^5 + 2x^4 + 4x^3 + 8x^2 + 16x + 32$

Exercise C   *pp. A8–A9*

1. $\sqrt{x}$

3. $\sqrt[3]{a^2}$

5. $\sqrt{5}$

7. $\sqrt{3}$

9. $\sqrt[3]{xy^2}$

11. $\sqrt[4]{8}$

13. $\sqrt{12}$ or $2\sqrt{3}$

15. $\sqrt{27}$ or $3\sqrt{3}$

17. $\sqrt[3]{2^4 x^5}$ or $2x\sqrt[3]{2x^2}$

19. $\sqrt[15]{x^{11}}$

21. $\sqrt[20]{a^9}$

23. $\sqrt[12]{x^8 y^9}$

25. $\sqrt[6]{4^2 \cdot 2^3} = \sqrt[6]{128} = 2\sqrt[6]{2}$

27. $\sqrt[12]{a^{14} b^9} = a\sqrt[12]{a^2 b^9}$

29. $\sqrt[6]{6^3 \cdot 4^2 \cdot 3^3} = \sqrt[6]{2^7 \cdot 3^6} = 6\sqrt[6]{2}$

31. $\sqrt[6]{a}$

33. $\sqrt[12]{x}$

35. $\sqrt[6]{2}$

37. $\sqrt[12]{x^4 y^5}$

39. $\sqrt[12]{a}$

41. $\sqrt[8]{x^3}$

43. $\sqrt[4]{2}$

45. $\sqrt[3]{5}$

47. $\sqrt[4]{10}$

49. $\sqrt[10]{b}$

51. $\sqrt[3]{4} > \sqrt{2}$

53. $\sqrt[3]{25} < \sqrt[4]{125}$

55. $\sqrt[4]{2} < \sqrt[6]{3}$

57. $2\sqrt{2}$

59. 0

Exercise D   *pp. A13–A14*

1.

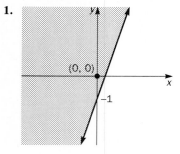

3.

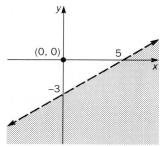

5.

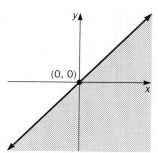

7.

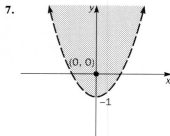

9.

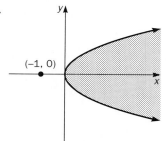

11.

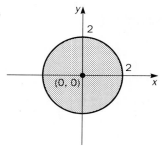

13.

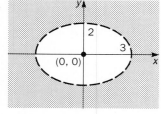

15.

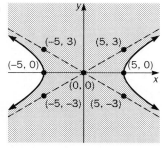

17.

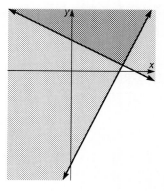

19.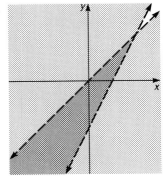

21.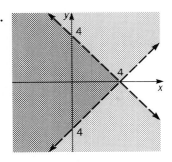

23.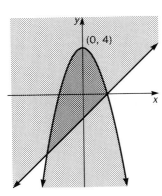

25.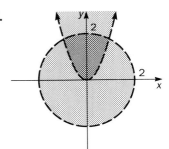

27.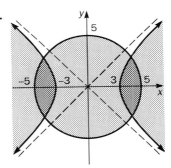

29.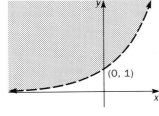

31.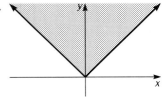

33.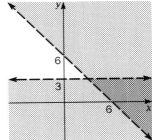

35. $y \geq \dfrac{2}{3}x - 2$

37. $x^2 + y^2 \leq 25$
$x^2 + y^2 \geq 4$

Exercise E   *pp. A18–A19*

1. $(3, -2)$

3. $(5, 2)$

5. $(3, 0)$

7. $x = \dfrac{3}{4},\ y = \dfrac{1}{2}$

9. $(1, 1, 1)$

11. $(1, 2, 3)$

13. $(2, 3, -4)$

15. $x = -\dfrac{2}{3},\ y = -\dfrac{1}{6},\ z = \dfrac{-1}{2}$

17. $(2, -1, 2)$

19. $x = 1,\ y = 1,\ z = 0,\ w = -2$

Exercise F

pp. A23–A24

1. 0.5105
3. 1.7701
5. 3.9175
7. 3.0000
9. 0.7235 − 3
11. 0.6232 − 5
13. 0.6990 − 6
15. 4.51
17. 60.2
19. 80,000
21. 0.00303
23. 0.202
25. 0.0002
27. 0.518
29. 0.00652
31. 0.00000907
33. 0.604
35. 4.527
37. 9018
39. 1,737,000
41. 0.0003595
43. 0.00002525
45. 0.003058
47. 2.613

Exercise G

pp. A26–A27

1. 8.00
3. 226,400
5. 0.00001171
7. 2.20
9. 2,757,000,000
11. 0.000003155
13. 1910
15. 7.034×10^{-14}
17. 2.00
19. 3.00
21. 72.66
23. 0.1604
25. 0.07694
27. 43,180
29. 4.822×10^{11}
31. 96,690

INDEX

Absolute value, 68
  difference of terms, 201
  equations, 69
  as a special function, 398
Absolute-value equation, 69, 76
Absolute-value inequality, 69, 70, 76
Addition
  of complex numbers, 223
  property of an equation, 299
  of real numbers, 3
Addition method, of solving systems, 299
Addition property
  of equations, 299
Additive inverse, *see* Inverse properties
Algebra, 14
Algebra phrases
  table of, 27–28, 160, A31
Algebraic expression, 14, 27–28, 160, A31
Annuity, 497
Antilogarithm, A20
Areas of geometric figures 15, 32, 121
Arithmetic progression, 483n
Arithmetic sequence, 483, 513
  first term, 483, 484, 513
  *n*th term, 484, 513
  sum of an, 486, 513
Associative Properties, 25
Asymptotes, 363, 381
Augmented matrix, A15
Avogadro's number, 185
Axis of symmetry, 340
Art fraud, 467

Bacteria growth, 434, 438
Base
  of a logarithm, 444
  of a power, 5
Baseball diamond, 234
Biathlon, 59
Binomial, 24
  coefficient, 508

expansion formula, 510
  square of a, 84
Boyle's Law, 424
Briggs, Henry, 448
Building factor, 9, 137

Carbon-14 dating, 465
Cartesian coordinate system, 278
Celsius-Fahrenheit formula, 19
Change of logarithmic base, 469
Characteristic, A20
Charles' Law, 425
Chemical mixture problems, 309
Chernobyl, 438
Circle, 352
  area, 16
  circumference, 16
  equation of, 352
Closure Properties, 25
Coefficient, 24
Coin problems 46, 307, 318
Combining like terms, 26
Combined variation, 406
Common difference, 483
Common factor, 89
Common logarithm, 448
Common ratio, 491
Commutative properties, 25
Completing the square, 237
Complex fraction, 142
Complex number, 223
  addition, 223–224, 266
  conjugate of, 224
  form, 223, 266
  multiplication, 223–224, 266
  subtraction, 223, 224
Composite
  function, 412
  number, 6
Compound inequality, 63
  see also Double inequality
Compound interest, 436
Conic sections, 339
  circle, 352
  ellipse, 359

hyperbola, 362
  parabola, 339
Conjugate
  of a binomial, 205
  of a complex number, 224
  involving cube roots, 209
Consecutive integers, 108
Consistent system of equations, 302
Constant, 15
Constant of variation, 403
Coordinate, 278
Cramer's Rule, 324–325
Critical point of inequality, 259
Cube root, 186
  see also Roots and powers, table of

Decibel, 453
Degree of polynomial, 24
Denominator
  least common (L.C.D.), 3, 9, 136
  rationalizing, 203
Dependent system of equations, 302
Depreciation, 489, 490
Descartes, René, 278
Determinant, 321
  expansion by minors, 322
  expansion by shortcut method, 323
Difference
  between two numbers, 27
  of two cubes, 101
  of two squares, 100
Difference quotient, 393
Dimension, of a matrix, A15
Direct variation, 403
Discriminant, 244
Distance formula, 351
Distance-rate-time, 35, 55, 311
Distributive property, 25, 83
Division, 5
  of complex numbers, 224
  of polynomials, 150
  of rational expressions, 130

Division *(continued)*
  synthetic, A5
Domain, 389
Double inequality, 63
  see also Compound inequality

e, 435, 442
  see also Powers of *e,* table of,
    A30
Earthquakes, 452
Effective interest rate, 54, 441
Electrical circuit, 166, 408
Element of a set, A1
Elementary row operations, A15
Elimination method, *see* Addition
    method
Ellipse, 359
  area, 368
  center, 360, 368
  equation, 360
Empty set, A1
Equation
  absolute-value, 69
  exponential, 443, 460
  versus expression, 159
  first-degree, 40
  fractional, 154
  higher-degree, 111, 254
  identity, 42
  linear, 40
  literal, 165
  logarithmic, 459
  multiplication property, 40
  quadratic, 105
  with radicals, 249
  second-degree, 105
  solving by substitution, 258
Equation of a circle, 352
Equations of a line
  point-slope form, 294
  slope-intercept, 286
  standard form, 279
Euler, Leonhard, 436
Even integer, 108
Expansion by minors, 322
Exponent, 5
  fractional, 211
  Generalized power rule, 176
  integer, 176
  negative, 178
  Product rule, 82
  Quotient rule of, 128
  rational, 211

rules for, 176
zero, 178
Exponential
  equations, 443, 462
  functions, 432
Expression, 14
  algebraic, 14, 27-28, 160, A31
  versus equation, 159
  radical, 199
  rational, 127
Extracting roots method, 229
Extraneous solution, 73, 155, 250

Factor
  building 9, 137
  common, 89
  greatest common (G.C.F.), 90
Factoring
  alternative method, 97
  combined methods of, 102
  the difference of two cubes, 101
  the difference of two squares,
    100
  the greatest common factor, 90
  flow-chart 102, 120
  by grouping, 91
  by method of substitution, 258
  polynomials, 90–112
  the sum of two cubes, 101
  second-degree equations, 105
  tips, 96
  trinomials, 93–98
Fahrenheit-Celsius formula, 19, 54,
    168, 297
Fibonacci sequence, 482
First-degree equation, 40
FOIL method, 83
Formula, definition of, 15
Formulas of geometric figures, 15,
    32, 121, 168
Fractions, 2
  complex, 142
  equations containing, 154
  Fundamental Principle of, 127
  inequalities containing, 154, 261
Frictional force, 409
Function, 390
  absolute-value, 398
  composite, 412
  domain of, 389, 401, 402
  exponential, 432

inverse, 414
linear, 398
logarithmic, 444
notation, 392
quadratic, 398
rational, 398
range of, 389
special, 398
square root, 399
vertical line test, 391
Fundamental Principle of Fractions,
  128

Gauss, Karl Friedrich, 485
Generalized Power Rule of Exponents, 176
General term, *see* *n*th term
Geometric progression, 490n
Geometric sequence, 490
  *n*th term of, 491
  sum of, 493
  sum of infinite, 501
Geometry problems, 112
Geometric formulas, 15, 32, 121
Goldbach conjecture, 8
Golden rectangle, 248
Googol, 184
Graph(s)
  of absolute-value equations, 69
  of absolute-value inequalities, 69
  of circles, 352–355
  of ellipses, 359-361
  of fractional inequalities, 157
  of hyperbolas, 363–366
  of linear equations in two vari-
    ables, 298
  of linear functions, 398
  of linear inequalities, 60
  of lines, 278–282
  of logarithmic functions, 432–433
  of numbers on a number line, 3
  of ordered pairs, 278
  of parabolas, 339–346
  of quadratic inequalities, 259
  of rational inequalities, 261–262
Greatest common factor (G.C.F.),
  90
Grouping, factoring by, 91
Growth and decay, 434

Half-life, 434
Hang-gliders, 406

Hero's formula, 197
Hooke's Law, 408
Horizontal line, 282
Hyperbola, 362
Hypotenuse, 17

i, 22
  powers of, 227
Ideal Gas Law, 408
Identity Properties, 25
Illumination of light, 405
Imaginary number, 223
Imaginary unit, 222
Inconsistent system of equations, 302
Index, 186
Inequality, 60
  absolute-value, 69–70
  compound (or double), 63
  fractional, 156
  graph of, 61
  higher-degree, 265
  in two variables, A9
  linear, 60
  Properties of, 60
  quadratic, 259
  rational, 261
  symbols of, 60
  systems of, A12
Infinite geometric sequence, 499
Inflation, 439
Integer, 2
  consecutive, 108
  exponent, 176
  even, 108
  odd, 31, 108
Intercept method, 280
Interest
  compounded, 436
  simple, 35, 49
  compounded, 439
Interpolation, A22
Intersection of sets, A2
Inverse
  function, 414
  relation, 416
  Properties, 25
  variation, 404
Investment problems, 49, 308

Irrational number, 22

Joint variation, 406

Least common denominator (L.C.D.), 3
  methods for obtaining, 9, 136
Leg of a right triangle, 17
Less than, 60
Light year, 185
Like terms, 26
Linear
  equation, 40
  inequality, 60
  interpolation, A22
  systems, 298, 316
Lines
  graph of, 279
  horizontal, 282
  parallel, 288
  perpendicular, 288
  point-slope form of, 294
  slope-intercept form of, 286
  standard form of, 279
  vertical, 282
Logarithm, 443
  base of, 444
  calculating with, A24
  change of base of, 469
  characteristic of, A20
  common, 448
  mantissa of, A20
  natural, 450
  Properties of, 455
  Table of Common, A28–A29
Logarithmic
  equation, 459
  function, 444

Malthusian growth-rate model, 436
Mantissa, A20
Marriage and divorce, 35
Matrix, A15
  augmented, A15
  dimension, A15
  elementary row operations, A15
  reduced system, A17
  methods, A15–A18
  triangle of zeroes, A17
Maximum value, 346
Member of a set, A1

Mile pace, 56
Minimum value, 346
Minor, 321
Mixture problems, 309
Motion problems, 35, 55, 313
Monomial, 23
Multiplication, 5
  of complex problems, 223
  of polynomials, 83
  of radicals, 198
  of rational expressions, 130
  of real numbers, 5
Musical string, 409

*n*th term
  of an arithmetic sequence, 484
  of a geometric sequence, 491
*n*th root, 186
Napier, John, 448
Natural logarithm, 448
Natural number, 23
Negative exponent, 178
Negative number, 3
Non-linear systems of equations, 371
Null set, A1
Number line, 3
Numbers
  complex, 223
  composite, 6
  even, 108
  imaginary, 223
  integers, 2
  irrational, 22
  mixed, 2
  natural, 23
  negative, 3
  odd, 31, 108
  positive, 3
  prime, 6, 23
  rational, 2, 22
  real, 22
  whole, 23
Numerical coefficient, 24

Odd integer, 31, 108
Ohm's Law, 408
Opposites, 25
Order of operations, 10
Ordered pair, 278
Origin, 278

Parabola, 339
  equation, 342
  graph, 339
  vertex, 340
Parallel Axiom, 295
Parallel lines, 288
  and determinants, 328
Parallelogram, 292
Pascal's triangle, 508
Perpendicular Axiom, 295
Perpendicular lines, 288
pH (hygrogen potential), 449
Point-slope form, 294
Polynomial, 24
  binomial, 24
  degree, 24
  division, 150
  factoring, 89-111
  monomial, 23
  multiplication, 83
  synthetic division, A5
  terminology, 24
  trinomial, 24
Polynomial function, 398
Population growth, 434
Power, 5
  Rule of exponents, 176
  Rule for logarithms, 462, 471
Powers of i, 227
Powers and Roots, Table of, A30
Present value, 441
Prime number, 6
Principal n^{th} root, 186
Principle of Random Walk, 197
Product, 27
  of complex numbers, 223
  of polynomials, 83
  of rational expressions, 130
  of real numbers, 5
  Rule of exponents, 82
  Rule for logarithms, 456
  Rule for roots, 191
Progressions, *see* Sequences
Properties
  of an equation, 40
  Associative, 25
  Closure, 25
  Commutative, 25
  of exponents 82, 128, 176, 178, 211
  Identity, 25
  of an inequality, 60
  Inverse, 25
  of logarithms, 454

of roots, 186, 191, 195
of real numbers, 25
Property, Distributive, 25
Proportions, 403
Pure imaginary number, 223
Pythagorean Theorem, 17, 115, 121
Pythagorean triples, 20, 119

Quadrant, 278
Quadratic equation 105, 222
  discriminant of, 244
  solving by completing the square, 237
  solving by extracting roots, 229
  solving by factoring, 105
  solving by the Quadratic Formula, 243
  types of solutions, 245
Quadratic Formula, 243
Quadratic inequality, 259
Quick products, 30, 38, 88
Quotient, 5
  of complex numbers, 244
  of polynomials, 150
  of radicals, 202
  of rational expressions, 130
  Rule of exponents, 128
  Rule for logarithms, 456
  Rule for roots, 195

Radicals, 186
  combining, 194
  division of, 202
  equations, 249
  index of, 186
  multiplying, 198
  simplifying, 191
Radical sign
Radicand, 186
Radioactive decay, 434
Range, 389
Rational exponents, 211
Rational expressions, 127
  combining, 136
  Fundamental Principle of, 128
  restrictions, 127
Rational inequality, 261
Rational numbers, 2, 22
Rationalizing the denominator, 203, 209, 224
Real estate, 439, 489, 490
Real numbers, 22

Properties of, 25
Reciprocal, 25, 160, 162
Rectangle, 112, 292
Reduced system, 316
Reduced system matrix, A15
Richter scale, 452
Right triangle, 17, 292
Root, 186
  cube, 186
  nth, 186
  square, 186
  product rule, 191
  quotient rule, 195
Roots and Powers, Table of, A30
Rule, Product
  of exponents, 82
  for logarithms, 456
  for roots, 191
Rule, Quotient
  of exponents, 128
  for logarithms, 456
  for roots, 195

Scientific notation, 182
Second-degree equation, 105, 222
Second-degree inequality, 259
Sequence, 478
  arithmetic, 483
  Fibonacci, 482
  geometric, 490
  infinite, 501
Set, A1
  element of, A1
  intersection, A2
  subset, A3
  union, A2
  void, A2
Sigma notation, 480
Sign array of determinants, 322
Sign for factoring trinomials, 96
Sky-diving, 440
Slope
  formula, 285
  of a horizontal line, 286
  of parallel lines, 288
  of perpendicular lines, 288
  undefined, 286
  of a vertical line, 286
  zero, 286
Slope-intercept form, 287
Standard form
  of a line, 279
  of a linear system, 303

Standard form (continued)
  of a quadratic equation, 105
  of a rational number, 2
Subset, A3
Substitution method
  for solving equations, 258
  for solving systems, 301
Subtraction
  of complex numbers, 223–225
  of imaginary numbers, 223–225
  of real numbers, 4
Sum of terms, 3, 27
  of an arithmetic sequence, 486
  of an geometric sequence, 493
  of an infinite geometric sequence, 501
  of real numbers, 3
Summation, 480
Symbols of inequality, 61
Symmetry, axis of, 340
Synthetic division, A5
System of equations, 298
  consistent, 302
  dependent, 302
  inconsistent, 302
  linear in three variables, 316
  linear in two variables, 298
  with literal coefficients, 306
  non-linear, 371
  reduced, 316
  solution by addition, 299
  solution by matrices, A16
  solution by substitution, 301
System of inequalities, A12

Tax, 489, 490
Tables
  Algebraic Phrases, 27, 28, 108, 160, A31
  Common Logarithms, A28, A29
  Powers of e, A30

Roots and Powers, A30
Temperature, 19, 297
Terms, 23
  like or similar, 26
  unlike, 26
Terms of a sequence, 483
Test point in inequality, 260
Ticket problems, 46
Tips
  for factoring difficult trinomials, 96
  for setting up word problems, 44
  for solving for a variable, 166
  for taking final exams, 522
Transitive Law of Inequality, 71
Translating word expressions, 27, 28, 108, 160, A31
Triangle
  Inequality, 67
  Pascal's, 508
  Pythagorean Theorem, 17, 115, 120
  right, 17
Triathlon, 59
Trichotomy Law, 66
Trinomial, 24
  factoring, 93
Trapezoid, 168

Union of sets, A2
Universal Law of Gravitation, 408
Unknown, 15
Unlike terms, 26

Variable, 15
  solving for a specific, 165, 231
Variation, 403
  combined, 406
  constant of, 403
  direct, 403
  inverse, 404
  joint, 406
Vertex of parabola, 340

Vertical line test, 391
Void set, A1

Whole number, 23
Word problems
  chemical mixtures, 309
  coin and ticket, 46, 307, 318
  consecutive integer, 108
  distance-rate-time, 35, 55, 311
  with fractional phrases, 160
  geometry, 112
  investment, 49, 308
  with number phrases, 42, 107
  population growth, 438, 467
  maximum/minimum, 346
  mixture, 309
  motion, 35, 55, 313
  setting up, 44
  temperature, 19, 297
  in two variables, 307
  upstream/downstream, 311

x-axis, 278
x-coordinate of vertex, 342, 344
x-intercept
  of a line, 280
  of a hyperbola, 363, 365
  of an ellipse, 360

y-axis, 278
y-coordinate of vertex, 342, 344
y-intercept
  of a line, 280
  of an ellipse, 360
  of a hyperbola, 363, 365

Zero
  division by 2, 9
  n^{th} root of, 186
Zero exponent, 178
Zero product rule, 105

USEFUL RULES AND FORMULAS

Cramer's Rule   *(see section 7.7)*

To solve the system

$$a_1x + b_1y = c_1$$

$$a_2x + b_2y = c_2$$

form the determinants

$$D = \begin{vmatrix} a_1 & b_1 \\ a_2 & b_2 \end{vmatrix}, \quad D_x = \begin{vmatrix} c_1 & b_1 \\ c_2 & b_2 \end{vmatrix}, \quad D_y = \begin{vmatrix} a_1 & c_1 \\ a_2 & c_2 \end{vmatrix}$$

The solutions are $x = \dfrac{D_x}{D}, \ y = \dfrac{D_y}{D}$.

To solve the system

$$a_1x + b_1y + c_1z = d_1$$

$$a_2x + b_2y + c_2z = d_2$$

$$a_3x + b_3y + c_3z = d_3$$

form the determinants $D, D_x, D_y,$ and D_z as above.

The solutions are

$$x = \frac{D_x}{D}, \quad y = \frac{D_y}{D}, \quad z = \frac{D_z}{D}.$$

Parabolas   *(see section 8.1)*

$y = ax^2 + bx + c$

    Vertex: $x = \dfrac{-b}{2a}$

    Opens *up* if $a > 0$

    Opens *down* if $a < 0$

$x = ay^2 + by + c$

    Vertex: $y = \dfrac{-b}{2a}$

    Opens *right* if $a > 0$

    Opens *left* if $a < 0$

Distance   *(see section 8.2)*

$$D = \sqrt{(x_2 - x_1)^2 + (y_2 - y_1)^2}$$

Circle   *(see section 8.2)*

$$(x - h)^2 + (y - k)^2 = r^2$$

$$\text{Center} = (h,k), \quad \text{radius} = r$$

Ellipse   *(see section 8.3)*

$\dfrac{x^2}{a^2} + \dfrac{y^2}{b^2} = 1$

Hyperbolas   *(see section 8.3)*

$\dfrac{x^2}{a^2} - \dfrac{y^2}{b^2} = 1$

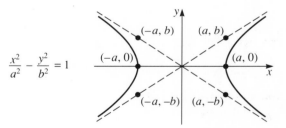

$\dfrac{y^2}{b^2} - \dfrac{x^2}{a^2} = 1$

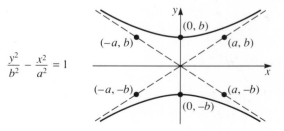